Chrono-biology

By
L. E. Scheving, F. Halberg,
J. E. Pauly

3 Plates in Color
352 Illustrations

1974
Georg Thieme Publishers
Stuttgart
Igaku Shoin Ltd. Tokyo

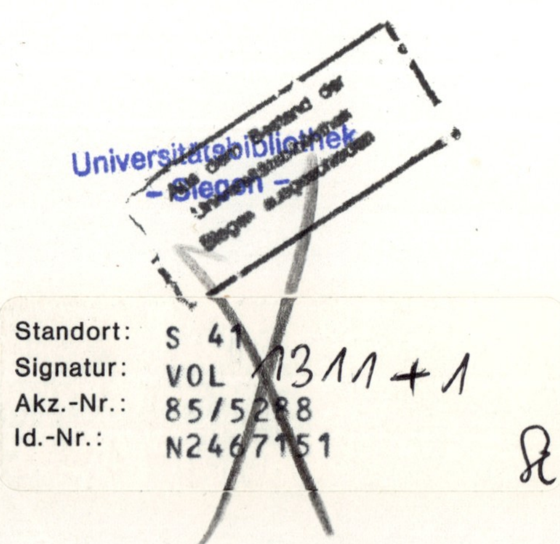

PUBLISHERS

Sole distribution rights for Europe (including the United Kingdom) granted to Georg Thieme Verlag, Stuttgart.

ISBN 3 13 515001 1

Printed and bound in Japan

DEDICATION

This symposium emphasizing a quantitative approach to the study of chronobiology is dedicated to the centennial year of the University of Arkansas.

ACKNOWLEDGEMENTS

The organizers of the Little Rock Meeting of the International Society for the Study of Biological Rhythms gratefully acknowledge help from the following:

Dr. Winston K. Shorey, Dean, University of Arkansas School of Medicine
National Institutes of Health
National Science Foundation
National Aeronautics and Space Administration
Food and Drug Administration
U.S. Department of State
Mr. Winthrop Rockefeller
Icelandic Airlines
KLM Royal Dutch Airlines
The Upjohn Company
Miles Laboratories
Schering Corporation

CONTENTS

CHRONOBIOLOGY, ENDOCRINES, NEUROHUMORS AND REPRODUCTION

PREFACE

The first meeting in the United States of the former International Society for the Study of Biologic Rhythms, now the International Society for Chronobiology, was held in a most hospitable atmosphere in Little Rock, Arkansas, from November 8 to 10, 1971. Several hundred registered participants from many fields, traditional and novel, including chronobiologic pioneers in space and clinical medicine–Hubertus STRUGHOLD of San Antonio, Texas, U.S.A., Werner MENZEL of Hamburg, Germany and Frits GERRITZEN of Wassenaar, Holland–were welcomed by the governor and other State and University officials and private citizens, on several festive occasions. Most contributors documented and exploited the use of quantitative procedures for the study of rhythms. Naturalistic observations and comments also were welcomed. The viewpoints represented include those of laboratories devoted to rhythm research in Paris and Montpellier, France; Hannover Marburg, Munich and Erling-Andechs, Germany; Florence, Italy; Glasgow and Manchester, UK; Leningrad, USSR; New Britain General Hospital, Connecticut, Northwestern University in Illinois, the National Institutes of Health in Bethesda, Maryland, and elsewhere.

The multitude of speakers and discussants from all over the world indicates that, even if its methods, concepts and facts as yet are *in statu nascendi,* chronobiology holds interest if not promise of relevance for many fields. Indeed, this volume reflects diverse views and approaches, inferential and descriptive, all recognizing that the biology of single samples—useful though it has been in the past–must be complemented by the study of time series. Contributions herein document this point for many forms of life—unicellular or multicellular—plants, insects, rodents and primates, to single out but a few examples. Environments studied vary from urban, suburban or rural settings–hospital, school, senior citizen's home to a cave or the moon. The spectrum of rhythms covered in this book extends from the frequencies of the electroencephalogram to circannual rhythms; it ranges from the cellular level, over the endocrines, to the neural functions; it embraces temporal parameters of young and old.

Chronobiologic research reported in this volume focuses upon monitoring and quantifying rhythm characteristics in certain critical variables such as blood pressure, peak expiratory flow and physical as well as mental performance. Self-measure and self-evaluation of records produced by computer have been taught in school and allow an inexpensive assessment of individual health–for eventual disease prevention. We read also about diagnosis and treatment according to rhythms, in illustrative cases. Novel chronobiologic facts, revealed by studies employing partly automated instrumentation await exploitation. Planned population size requires a better knowledge of all rhythmic aspects of the female and also of probable male reproductive cycles. Endpoints of rhythms in peak expiratory flow rate of certain patients with extrinsic asthma serve as effective and pertinent gauges of undesired biologic effects from high levels of pollutants.

Chronobiologic nutritional research may reveal that we may be not only "what we eat" but "when we eat". Those who wish to lose weight and, much more

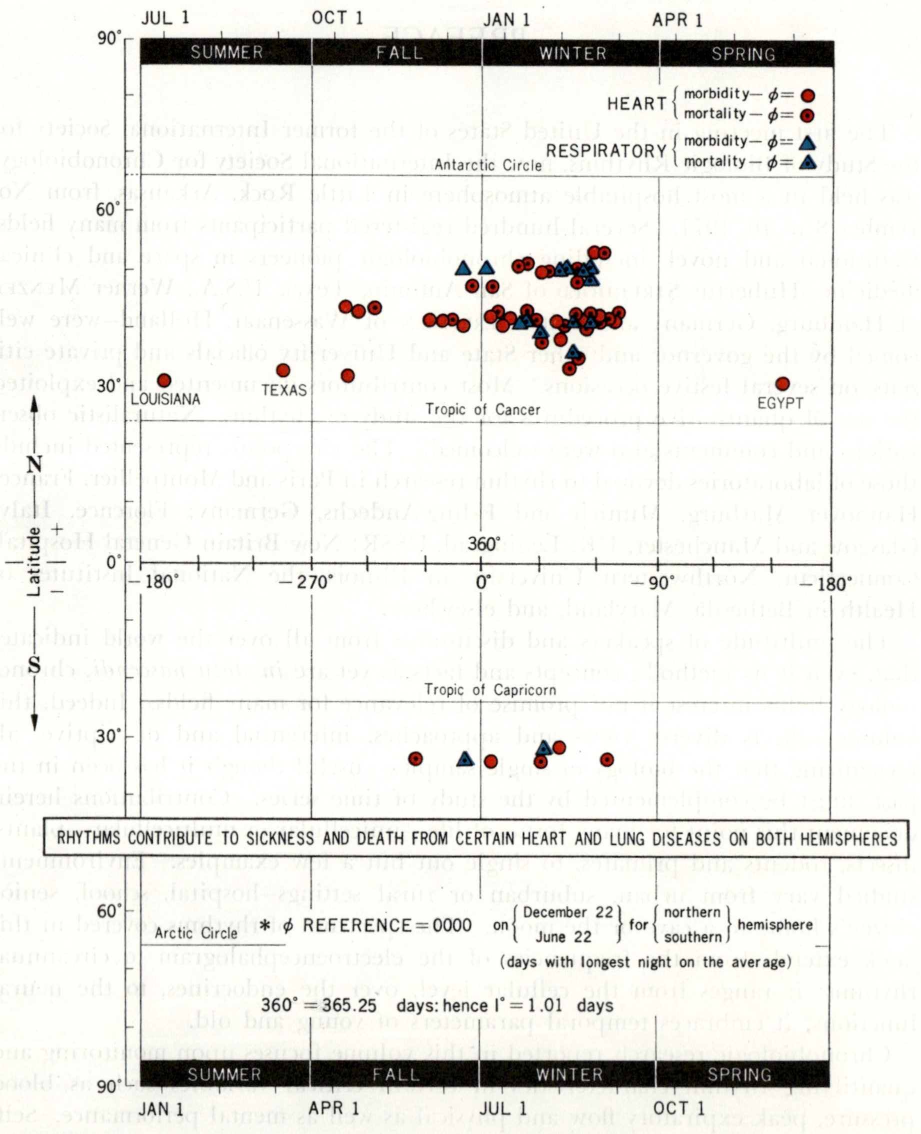

Fig. 1 Changes in morbidity and mortality document the importance of rhythms. Circannual differences in death or sickness from heart and lung disease are shown in this figure. The evidence and related changes along the 24-hour scale are summarized in Chronobiology of the Life Sequence by M. Smolensky, F. Halberg, and F. Sargent II.: *In*: Advances in Climatic Physiology, S. Itoh, K. Ogata, H. Yoshimura, ed.), Igaku Shoin Ltd., Tokyo 1972, pp. 281–318.

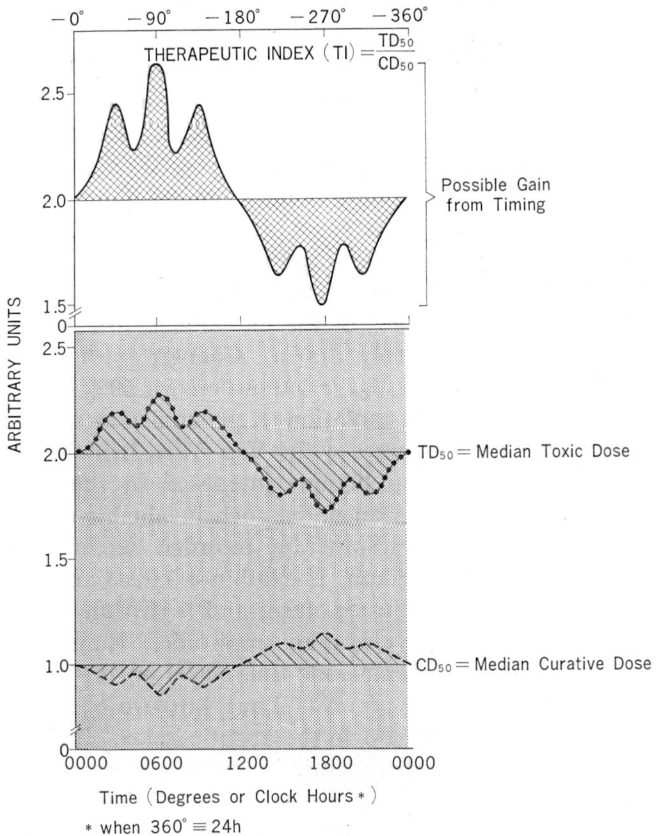

Fig. 2 What we may call "iatrogenic toxicosis" constitutes the major clinical problem in the treatment of several forms of cancer. In view of theoretical considerations documented by facts presented in this volume and elsewhere, the studies on experimental animals concerning the merits of timing treatment and incipient clinical work toward this goal urgently await extension at the bedside.

critically, those who starve would benefit greatly from information on a circadian time-dependent utilization of food, exploited by meal timing.

Agriculture needs more data on factors determining when, along the one-year scale, plants flower and bear fruit or when, along the 24-hour scale, pest control is most effective. Whether drug administration to patients with high blood pressure, cancer, asthma, other allergies or Addison's disease, is more effective and/or less harmful depends at least in part upon the potential development of methods for timing therapy according to the individual's spectrum of rhythms at several levels of organization including that of the cell. Rhythms indeed account for the difference between death and survival; they determine predictable changes in susceptibility to a number of chemical, physical and other agents, including poisons, pollutants and drugs. Much of the original evidence was derived from studies on rodents: the local hosts have so well demonstrated that a given fixed dose of a drug used in treating cancer will kill or be compatible with survival depending upon the stage of the experimental animal's rhythms. The same applies to drugs used in treating heart disease, to the effects of a poison derived from bacteria and even to the harmfulness of noise in susceptible animals.

That rhythms are significant for man is apparent from a tabulation of deaths from certain diseases as a function of clock hour or season. When such data are viewed in conjunction with results of controlled sensitivity tests on man and against the background of experimental animal studies, the critical importance of rhythms in determining susceptibility and resistance is readily apparent (Figure 1).

The abstract Figure 2 shows rhythms with two frequencies that are out of phase with one another and exhibit proportionately equal amplitudes for the undesired effect (toxic dose for 50% of the individuals, briefly TD_{50}) and the desired effect (curative dose for 50% of the individuals, briefly CD_{50}). Actually the abstract graph is conservatively drawn. Changes with larger amplitude already are demonstrated for the LD_{50} (a lethal dose for 50% of the individuals, i.e., a TD_{50}) to whole body x-ray radiation—a phenomenon studied in the late '50's, when shortly after the discovery of the Van Allen belt the National Aeronautics and Space Administration became interested in the optimal time to withstand radiation. What we began under such invaluable sponsorship led to a beginning with human chronoradiotherapy recorded herein.

To turn back to the abstract graph, it exhibits a circadian amplitude of no more than 10% of the rhythm-adjusted mean and a rhythm with a higher frequency involving but half of the circadian amplitude. Nonetheless, the result is a 35% gain in therapeutic index at one time (0600 on the graph and a 26% loss at another time (1800 on the graph). Thus, but two of the many rhythms can account for a total change of 51% in therapeutic index. This change, ignored in the classical time-unqualified therapeutic index, should be exploited at a time when toxicity constitutes the major problem in the clinical treatment of many cancers. Such conservative considerations relate to the *cui bono;* they may elicit an "ah ha" from those who rightly raise the question "so what"—even if this volume is no more than a humble beginning in what may develop into a critical field of basic and applied science.

For the editors,

FRANZ HALBERG
President
International Society for Chronobiology

PHARMACOLOGY, TOXICOLOGY AND TIMED TREATMENT

CIRCADIAN AND CIRCANNUAL MAPS

CHRONOPATHOLOGY, IMMUNOLOGY AND CANCER

CHRONOBIOLOGY, PEDIATRICS AND AGING

CHRONOBIOLOGIC EDUCATION AND HEALTH CARE DELIVERY

SPACE BIOLOGY AND MEDICINE

ULTRADIAN RHYTHMS AND SLEEP

CHRONOBIOLOGY AND RHYTHMS IN PSYCHIATRY AND PSYCHOLOGY

RHYTHMS AND CHANGES IN ENVIRONMENTAL SCHEDULES

CHRONOBIOLOGY AND INSECT RHYTHMS

BIOCHRONOMETRY AND BIRD MIGRATION

CHROBIOLOGY AND PLANTS

ELECTRIC AND MAGNETIC FIELDS AND COMPARATIVE PHYSIOLOGIC TOPICS

CHRONOBIOLOGICAL TECHNIQUES

STATISTICAL CONSIDERATIONS

CONTRIBUTORS

Charles ABULKER	Laboratoire de Physiologie, Fondation A. de Rothschild, Paris, France
Andrew AHLGREN	Center for Educational Development, University of Minnesota, Minneapolis, Minnesota, U.S.A.
Richard V. ANDREWS	Department of Biology, Creighton University, Omaha, Nebraska, U.S.A.
Anthony T. ANGELLAR	Chronobiology Laboratories, Department of Pathology, University of Minnesota, Minneapolis, Minnesota, U.S.A.
Deanna J. ANTHONEY	School of Medicine, University of California at Irvine, Irvine, California, U.S.A.
Sharon L. ANTHONEY	Department of Zoology and School of Medicine, Southern Illinois University, Carbondale, Illinois, U.S.A.
Terence R. ANTHONEY	Department of Zoology and School of Medicine, Southern Illinois University, Carbondale, Illinois, U.S.A.
Marian APFELBAUM	Laboratoire de Physiologie, Fondation A. de Rothschild, Paris, France
Roger ASSAN	Hôpital Bichat, Paris, France
James H. BARNES	Division of Biology and Medical Research, Argonne National Laboratory, Argonne, Illinois, U.S.A.
Frederic C. BARTTER	Endocrinology Branch, National Heart and Lung Institute, National Institutes of Health, Bethesda, Maryland, U.S.A.
Miriam F. BENNETT	Department of Biology, Sweet Briar College, Sweet Briar, Virginia, U.S.A.
Ralph J. BERGER	Department of Psychology, University of California, Santa Cruz, California, U.S.A.
Sue A. BINKLEY	Department of Biology, University of Notre Dame, South Bend, Indiana, U.S.A.
Pierrette BOUDON	Laboratoire de Physiologie, Fondation A. de Rothschild, Paris, France
J. Y. BOURGOIN	Clinique Endocrinologique, Faculté de Medicine, Marseille, France
Frank A. BROWN, Jr.	Department of Biological Sciences, Northwestern University, Evanston, Illinois, U.S.A.
Frederick M. BROWN	Psychology and Psychobiology Programs, Division of Science and Mathematics, Centre College of Kentucky, Danville, Kentucky, U.S.A.

STEVEN D. BROWN — Department of Psychology, University of Virginia, Charlottesville, Virginia, U.S.A.

HENRY BUCHWALD — Department of Surgery, University of Minnesota Hospital, Minneapolis, Minnesota, U.S.A.

NATHANIEL A. BUCHWALD — Departments of Anatomy and Psychiatry, University of California, Los Angeles, California, U.S.A.

E. ROBERT BURNS — Department of Anatomy, University of Arkansas Medical Center, Little Rock, Arkansas, U.S.A.

LINDA CADOTTE — Department of Laboratory Medicine, University of Minnesota, Minneapolis, Minnesota, U.S.A.

MARIO CAGNONI — Clinica Medica Generale, University of Florence, Florence, Italy

BONNALIE O. CAMPBELL — Department of Physiology, Baylor College of Medicine, Texas Medical Center, Houston, Texas, U.S.A.

RALPH A. CARABASI — Department of Radiation Therapy and Nuclear Medicine, Thomas Jefferson University Hospital, Philadelphia, Pennsylvania, U.S.A.

SERGIO S. CARDOSO — Department of Pharmacology, University of Tennessee Medical Units, Memphis, Tennessee, U.S.A.

A. CAVALLERI — Instituto di Medicina del Lavolo, Università di Pavia, Pavia, Italy

K. K. N. CHARYULU — Division of Radiation Therapy, University of Miami School of Medicine, Miami, Florida, U.S.A.

YOSHIHIKO CHIBA — Department of Entomology, Fisheries and Wildlife, University of Minnesota, St. Paul, Minnesota, U.S.A.

L. CHOSY — Department of Preventive Medicine, University of Wisconsin Medical School, Madison, Wisconsin, U.S.A.

HELMUT COPER — Institute for Neuro-Psycho-Pharmacology, Free University of Berlin, Berlin, Germany

ROBERT F. CURLEY — Department of Radiation Therapy and Nuclear Medicine, Thomas Jefferson University Hospital, Philadelphia, Pennsylvania, U.S.A.

GEORGE C. CURTIS — Eastern Pennsylvania Psychiatric Institute, Philadelphia, Pennsylvania, U.S.A.

LAURENCE K. CUTKOMP — Department of Entomology, University of Minnesota, St. Paul, Minnesota, U.S.A.

ROSARIO D'AGATA — Department of Medicine, Division of Endocrinology, Catania University School of Medicine, Catania, Italy

KENNETH B. DAVIS, JR. — Department of Biology, Memphis State University, Memphis, Tennessee, U.S.A.

Colin DAWES	Department of Oral Biology, Faculty of Dentistry, University of Manitoba, Winnipeg, Manitoba, Canada
Catherine S. DELEA	Endocrinology Branch, National Heart and Lung Institute, National Institutes of Health, Bethesda, Maryland, U.S.A.
Abner DELMAN	Department of Medicine, New York Medical College, Center for Chronic Disease, Welfare Island, New York, New York, U.S.A.
J. A. DEMPSEY	Pulmonary Physiology Laboratory, School of Medicine, University of Wisconsin, Madison, Wisconsin, U.S.A.
Alice DENNEY	Plant Science Department, Utah State University, Logan, Utah, U.S.A.
B. DIECKHUES	Augenklinik, Westfalen-Wilhelms Universität, Münster, Germany
G. A. doPICO	Department of Preventive Medicine, University of Wisconsin Medical School, Madison, Wisconsin, U.S.A.
Jean DUPONT	Laboratoire de Physiologie, Fondation A. de Rothschild, Paris, France
Leland N. EDMUNDS, Jr.	Division of Biological Sciences, State University of New York, Stony Brook, New York, U.S.A.
Charles F. EHRET	Division of Biology and Medical Research, Argonne National Laboratory, Argonne, Illinois, U.S.A.
Carl ENNA	United States Public Health Service Hospital, Carville, Louisiana, U.S.A.
Donald S. FARNER	Department of Zoology, University of Washington, Seattle, Washington, U.S.A.
Jerry F. FELDMAN	Department of Biological Sciences, State University of New York, Albany, New York, U.S.A.
Gabriel FERNANDES	Chronobiology Laboratory, Department of Pathology, University of Minnesota, Minneapolis, Minnesota, U.S.A.
Frank W. FINGER	Department of Psychology, University of Virginia, Charlottesville, Virginia, U.S.A.
A. FLECK	University Department of Pathology, Royal Infirmary, Glasgow, Scotland
Paul FRAISSE	Laboratoire de Psychologie Expérimentale et Comparée, Université René Descartes, Paris, France
Alexander H. FRIEDMAN	Department of Pharmacology and Therapeutics, Loyola University, Stritch School of Medicine, Maywood, Illinois, U.S.A.

D. FUKUSHIMA — Montefiore Hospital and Medical Center, Bronx, New York, U.S.A.

T. F. GALLAGHER — Montefiore Hospital and Medical Center, Bronx, New York, U.S.A.

EUGENE GEDGAUDAS — Chronobiology Laboratory, Department of Pathology, University of Minnesota, Minneapolis, Minnesota, U.S.A.

FRITS GERRITZEN — Wassenaar, The Netherlands

PIERRE GERVAIS — Laboratoire de Physiologie, Fondation A. de Rothschild, Paris, France

JEAN GHATA — Laboratoire de Physiologie, Fondation A. de Rothschild, Paris, France

ARTHUR GIESE — Department of Life Sciences, Stanford University, Stanford, California, U.S.A.

WOLFGANG GIRKE — Psychiatric and Neurologic Clinic, Free University of Berlin, Berlin, Germany

LEIV GJESSINGS — Dikemark Hospital, Asker, Norway

ERNEST M. GOLD — St. Paul-Ramsey Hospital, St. Paul, Minnesota, U.S.A.

ROBERT M. GOODMAN — Biodynamics Laboratory, Franklin Institute Research Laboratories, Philadelphia, Pennsylvania, U.S.A.

PEGGY J. GOODRUM — Department of Pharmacology, University of Tennessee Medical Units, Memphis, Tennessee, U.S.A.

MAURICE GRAHAM — Department of Pharmacology, University of Arkansas Medical Center, Little Rock, Arkansas, U.S.A.

S. B. GRAY — Department of Anatomy, Louisiana State University Medical Center, New Orleans, Louisiana, U.S.A.

WILLIAM GRUEN — Calibrated Instruments, Ardsley, New York, U.S.A.

ROBERT GÜNTHER — Department of Medicine, University of Innsbruck, Innsbruck, Austria

RUTH HALABAN — Department of Biology, Brookhaven National Laboratory, Upton, New York, U.S.A.

ERNA HALBERG — Chronobiology Laboratory, Department of Pathology, University of Minnesota, Minneapolis, Minnesota, U.S.A.

FRANCINE HALBERG — Department of Life Sciences, Stanford University, Stanford, California, U.S.A.

FRANZ HALBERG — Chronobiology Laboratory, Department of Pathology, University of Minnesota, Minneapolis, Minnesota, U.S.A.

Julia HALBERG — Chronobiology Laboratory, Department of Pathology, University of Minnesota, Minneapolis, Minnesota, U.S.A.

Karl C. HAMNER — Department of Biological Sciences, University of California, Los Angeles, California, U.S.A.

Edward A. HAND — Coushatta Senior Citizens Home, Coushatta, Louisiana, U.S.A.

Kenneth HANSON — Arctic Health Research Center, College, Alaska, U.S.A.

Richard N. HARNER — Department of Neurology, The Graduate Hospital, University of Pennsylvania, Philadelphia, Pennsylvania, U.S.A.

Eugene K. HARRIS — Laboratory of Applied Studies, Division of Computer Research and Technology, National Institutes of Health, Bethesda, Maryland, U.S.A.

René HARS — Laboratoire de Cytologie et Embryologie Moléculaires, Départment de Biologie Moléculaires, Université Libre de Bruxelles, Bruxelles, Belgium

Ansari HASSEN — Department of Medicine, New York Medical College, Center for Chronic Disease, Welfare Island, New York, New York, U.S.A.

J. W. HASTINGS — Department of Biology, Harvard University, Cambridge, Massachusetts, U.S.A.

Erhard HAUS — Department of Clinical Laboratories, St. Paul-Ramsey Hospital, and Department of Pathology, University of Minnesota, Minneapolis, Minnesota, U.S.A.

Mary P. HAYDEN — Department of Psychiatry, Renard Hospital, St. Louis, Missouri, U.S.A.

Dora K. HAYES — Entomology Research Division, Agricultural Research Service, U.S. Department of Agriculture, Beltsville, Maryland, U.S.A.

Theodor HELLBRÜGGE — Forschungsstelle für Soziale Pädiatrie und Jugendmedizin, Universität München, München, Germany

L. HELLMAN — Montefiore Hospital and Medical Center, Bronx, New York, U.S.A.

A. N. HEWING — Entomology Research Division, Agricultural Research Service, U.S. Department of Agriculture, Beltsville, Maryland, U.S.A.

Günther HILDEBRANDT — Physiologisches Institut der Universität, Marburg an der Lahn, Germany

Dewayne C. HILLMAN — Chronobiology Laboratory, Department of Pathology, University of Minnesota, Minneapolis, Minnesota, U.S.A.

William S. HILLMAN	Department of Biology, Brookhaven National Laboratory, Upton, New York, U.S.A.
Fritz HOLLWICH	Augenklinik, Westfäische-Wilhelms-Universität, Münster, Germany
Takashi HOSHIZAKI	Space Biology Laboratory, Brain Research Institute, University of California, Los Angeles, California, U.S.A.
Marian N. HOYLE	Department of Biological Sciences, State University of New York, Albany, New York, U.S.A.
Bartholomew P. HSI	University of Texas School of Public Health, Houston, Texas, U.S.A.
C. D. HULL	Departments of Anatomy and Psychiatry, University of California, Los Angeles, California, U.S.A.
Jon W. JACKLET	Department of Biological Sciences, State University of New York, Albany, New York, U.S.A.
Robert R. JACOBSON	U.S. Public Health Service Hospital, Carville, Louisiana, U.S.A.
Derrick A. JONES	Electromechanical Research, Central Research Laboratory, 3M Company, St. Paul, Minnesota, U.S.A.
M. M. JOSEPH	Departments of Zoology and Physiology, Louisiana State University, Baton Rouge, Louisiana, U.S.A.
George S. KATINAS	Institute for Experimental Medicine, Leningrad, U.S.S.R.
Lida KECHAVARZ-OLAI	Department of Pediatrics, Albert Einstein Medical Center, Philadelphia, Pennsylvania, U.S.A.
Nancy KELLER	Biodynamics Laboratory, ALZA Corporation, Palo Alto, California, U.S.A.
James R. KING	Department of Zoology, Washington State University, Pullman, Washington, U.S.A.
Rodney W. KING	Department of Plant Science, University of Western Ontario, London, Ontario, Canada
Karl E. KLEIN	Institut für Flugmedizin (DFVLR), Bonn-Bad Godesberg, Germany
Ernst KNAPP	Department of Medicine, University of Innsbruck, Innsbruck, Austria
S. G. KOLAEVA	Laboratory of Endocrinology, Institute of Cytology and Genetics, Siberian Branch, Academy of Sciences of U.S.S.R., Novosibirsk, U.S.S.R.
M. G. KOLPAKOV	Laboratory of Endocrinology, Institute of Cytology and Genetics, Siberian Branch, Academy of Sciences of U.S.S.R., Novosibirsk, U.S.S.R.
Willard L. KOUKKARI	Department of Botany, University of Minnesota, Minneapolis, Minnesota, U.S.A.
Jürgen KRIEBEL	Max-Planck-Institut für Verhaltensphysiologie, Erling-Andechs, Germany

Daniel F. KRIPKE	Department of Psychiatry, University of California, San Diego, California, U.S.A.
Jürgen F. W. KÜHL	Chronobiology Laboratory, Department of Pathology, University of Minnesota, Minneapolis, Minnesota, U.S.A.
Helmut KÜNKEL	Department of Psychiatry, Medizinische Hochschule Hannover, Hannover-Kleefeld, Germany
Dimitri LACATIS	Hôtel-Dieu, Paris, France
David J. LAKATUA	Department of Clinical Laboratories, St. Paul-Ramsey Hospital, St. Paul, Minnesota, U.S.A.
Franklin C. LARIMORE	Electromechanical Research, Central Research Laboratory, 3M Company, St. Paul, Minnesota, U.S.A.
Caroline S. LEACH	Clinical Laboratories, Preventive Medicine Division, NASA Manned Spacecraft Center, Houston, Texas, U.S.A.
Joseph L. LEONG	Department of Pathology, University of Minnesota, Minneapolis, Minnesota, U.S.A.
Howard LEVINE	New Britain General Hospital, New Britain, Connecticut, U.S.A.
Stuart A. LEWIS	Department of Psychiatry, University of Edinburgh, Edinburgh, Scotland
Harry A. LIPSCOMB	Xerox Center for Health Care Research, Baylor College of Medicine, Texas Medical Center, Houston, Texas, U.S.A.
Merle K. LOKEN	Division of Nuclear Medicine, University of Minnesota, Minneapolis, Minnesota, U.S.A.
Russell V. LUCAS	Department of Pediatrics, University of Minnesota, Minneapolis, Minnesota, U.S.A.
Donald McEVOY	Eastern Pennsylvania Psychiatric Institute, Philadelphia, Pennsylvania, U.S.A.
Robert MacGREGOR III	Department of Zoology and Physiology, Louisiana State University, Baton Rouge, Louisiana, U.S.A.
J. P. McMILLAN	Department of Zoology, University of Texas, Austin, Texas, U.S.A.
Laura McMURRY	Department of Biology, Harvard University, Cambridge, Massachusetts, U.S.A.
Wasyl MALYJ, Jr.	Department of Human Physiology, School of Medicine, University of California, Davis, California, U.S.A.
Carl M. MANSFIELD	Department of Radiation Therapy and Nuclear Medicine, Thomas Jefferson University Hospital, Philadelphia, Pennsylvania, U.S.A.

Donn D. MARTIN — Department of Zoology and Physiology, Louisiana State University, Baton Rouge, Louisiana, U.S.A.

R. MARTIN DU PAN — Clinique des Nourrissons, Geneva, Switzerland

Alan MATHER — Center for Disease Control, Atlanta, Georgia, U.S.A.

Heinz von MAYERSBACH — Institut für Anatomie, Medizinische Hochschule Hannover, Hannover-Kleefeld, Germany

Albert H. MEIER — Department of Zoology and Physiology, Louisiana State University, Baton Rouge, Louisiana, U.S.A.

Werner MENZEL — Amalie Sieveking Krankenhaus, Hamburg, Germany

Walter J. MEYER, III — Endocrinology Branch, National Heart and Lung Institute, National Institutes of Health, Bethesda, Maryland, U.S.A.

John N. MILLS — Department of Physiology, The University of Manchester, Manchester, England

Norberto MONTALBETTI — Centro Mallatie Endocrine e Metaboliche Laboratoria, Ospedale-Civile, Magenta, Milan, Italy

Peter J. MORGANE — Worcester Foundation for Experimental Biology, Shrewsbury, Massachusetts, U.S.A.

Martine MORIN — Laboratoire de Physiologie, Fondation A. de Rothschild, Paris, France

Ottfried MÜLLER — Institut für Anatomie, Medizinische Hochschule Hannover, Hannover-Kleefeld, Germany

John A. MUNKBERG — Electromechanical Research, Central Research Laboratory, 3M Company, St. Paul, Minnesota, U.S.A.

Velayudhan NAIR — Department of Pharmacology, The Chicago Medical School, Chicago, Illinois, U.S.A.

J. K. NAYAR — Entomological Research Center, Department of Health and Rehabilitation Services, Vero Beach, Florida, U.S.A.

Walter NELSON — Chronobiology Laboratory, Department of Pathology, University of Minnesota, Minneapolis, Minnesota, U.S.A.

C. NOGEIRE — Montefiore Hospital and Medical Center, Bronx, New York, U.S.A.

D. B. ODESSER — Entomology Research Division, Agricultural Research Service, U.S. Department of Agriculture, Beltsville, Maryland, U.S.A.

Ch. OLIVER — Clinique Endocrinologique, Faculté de Médicine, Marseille, France

R. ORZALESI — Medical School, University of Florence, Florence, Italy

EKKEHARD OTHMER — Department of Psychiatry, Renard Hospital, St. Louis, Missouri, U.S.A.

VIRGINIA PARSONS — Department of Psychology, University of Iowa, Iowa City, Iowa, U.S.A.

PIERRE PASSOUANT — Faculté de Médicine, Université de Montpellier, Montpellier, France

JOHN E. PAULY — Department of Anatomy, University of Arkansas Medical Center, Little Rock, Arkansas, U.S.A.

M. PERLOW — Montefiore Hospital and Medical Center, Bronx, New York, U.S.A.

BETTY ANNE PHILIP — Arctic Health Research Center, College, Alaska, U.S.A.

KAREL M. H. PHILIPPENS — Institut für Anatomie, Medizinische Hochschule Hannover, Hannover-Kleefeld, Germany

CHRISTIAN POIREL — Laboratoire de Psychophysiologie, Université de Toulouse, Toulouse, France

P. POLOSA — Department of Medicine, Division of Endocrinology, Catania University School of Medicine, Catania, Italy

M. G. POLYAK — Laboratory of Endocrinology, Institute of Cytology and Genetics, Siberian Branch, Academy of Science of U.S.S.R., Novosibirsk, U.S.S.R.

VAN R. POTTER — McArdle Laboratory, The Medical School, University of Wisconsin, Madison, Wisconsin, U.S.A.

WILBUR B. QUAY — Department of Zoology, University of California, Berkeley, California, U.S.A.

PAUL C. RAMBAUT — Preventive Medicine Division, NASA Manned Spacecraft Center, Houston, Texas, U.S.A.

WALTER RANDALL — Department of Psychology, University of Iowa, Iowa City, Iowa, U.S.A.

J. RANKIN — Department of Preventive Medicine, University of Wisconsin Medical Center, Madison, Wisconsin, U.S.A.

WILLIAM REDDAN — Department of Preventive Medicine, University of Wisconsin Medical Center, Madison, Wisconsin, U.S.A.

FRITZ REEKER — Chronobiology Laboratory, Department of Pathology, University of Minnesota, Minneapolis, Minnesota, U.S.A.

PHILIP REGAL — Museum of Natural History and Department of Zoology, University of Minnesota, Minneapolis, Minnesota, U.S.A.

HOBART A. REIMANN — Hahnemann Medical College and Hospital, Philadelphia, Pennsylvania, U.S.A.

Alain REINBERG — Laboratoire de Physiologie, Fondation A. de Rothschild, Paris, France

Russel J. REITER — Department of Anatomy, University of Texas Medical School, San Antonio, Texas, U.S.A.

Thomas D. REYNOLDS — Laboratory of Human Behavior, National Institute of Mental Health, St. Elizabeth's Hospital, Washington, D.C., U.S.A.

Donald R. ROBERTS — Department of Medical Entomology, University of Texas School of Public Health, Houston, Texas, U.S.A.

Chester ROIG III — Louisiana State University School of Medicine, New Orleans, Louisiana, U.S.A.

S. ROMANO — Medical School, University of Florence, Florence, Italy

Allen W. ROOT — Department of Pediatrics, Neonatal Research Laboratory, Temple University School of Medicine, Philadelphia, Pennsylvania, U.S.A.

Gordon L. ROSENE — Department of Physiology and Health Science, Ball State University, Muncie, Indiana, U.S.A.

John R. RUBY — Department of Anatomy, Louisiana State University Medical Center, New Orleans, Louisiana, U.S.A.

John A. RUMMEL — Environmental Physiology Division, NASA Manned Spacecraft Center, Houston, Texas, U.S.A.

Walter RUNGE — Chronobiology Laboratory, Department of Pathology, University of Minnesota, Minneapolis, Minnesota, U.S.A.

Frank B. SALISBURY — Plant Science Department, Utah State University, Logan, Utah, U.S.A.

J. SASSIN — Montefiore Hospital and Medical Center, Bronx, New York, U.S.A.

D. M. SAUERMAN, Jr. — Entomological Research Center, Department of Health and Rehabilitation Services, Vero Beach, Florida, U.S.A.

John E. SCANLON — Entomological Research Center, Department of Health and Rehabilitation Services, Vero Beach, Florida, U.S.A.

M. S. SCHECHTER — Entomology Research Division, Agricultural Research Service, U.S. Department of Agriculture, Beltsville, Maryland, U.S.A.

Lawrence A. SCHEVING — Brown University, Providence, Rhode Island, U.S.A.

Lawrence E. SCHEVING — Department of Anatomy, University of Arkansas Medical Center, Little Rock, Arkansas, U.S.A.

MARY SCHILD	Department of Psychology, Lafayette College, Easton, Pennsylvania, U.S.A.
ENID SHAW	Department of Pediatrics, Albert Einstein Medical Center, Philadelphia, Pennsylvania, U.S.A.
RONALD SHIOTSUKA	Chronobiology Laboratory, Department of Pathology, University of Minnesota, Minneapolis, Minnesota, U.S.A.
HUGH W. SIMPSON	University Department of Pathology, Royal Infirmary, Glasgow, Scotland
THOMAS R. C. SISSON	Neonatal Research Laboratory, Department of Pediatrics, Temple University School of Medicine, Philadelphia, Pennsylvania, U.S.A.
ROBERT EL. SMITH	Department of Human Physiology, School of Medicine, University of California, Davis, California, U.S.A.
MICHAEL SMOLENSKY	School of Public Health, University of Texas, Houston, Texas, U.S.A.
ARNE SOLLBERGER	Department of Psychiatry, Yale University School of Medicine, New Haven, Connecticut, U.S.A.
ROBERT B. SOTHERN	Chronobiology Laboratory, Department of Pathology, University of Minnesota, Minneapolis, Minnesota, U.S.A.
J. G. SOWELL	Department of Pharmacology, University of Tennessee Medical Units, Memphis, Tennessee, U.S.A.
M. B. STERMAN	Department of Anatomy, University of California, Davis, California, U.S.A.
W. C. STERN	Worcester Foundation for Experimental Biology, Shrewsbury, Massachusetts, U.S.A.
MILTON H. STETSON	Department of Zoology, University of Texas, Austin, Texas, U.S.A.
JOSEPH E. STONE	Department of Pharmacology, University of Arkansas Medical Center, Little Rock, Arkansas, U.S.A.
F. W. STRATMANN	Diabetikerheim "Haus Berg", Stuttgart, Germany
TH. STRENGERS	Wassenaar, Netherlands
HAROLD C. STROBEL	Minneapolis Southwest High School, Minneapolis, Minnesota, U.S.A.
CHARLES F. STROEBEL	Laboratory for Experimental Psychophysiology, Institute for Living Hospital, Hartford, Connecticut, U.S.A.
HUBERTUS STRUGHOLD	Aerospace Medical Division, Brooks Air Force Base, San Antonio, Texas, U.S.A.
OSIAS STUTMAN	Sloan-Kettering Institute for Cancer Research, New York, New York, U.S.A.

W. N. SULLIVAN — Entomology Research Division, Agricultural Research Service, U.S. Department of Agriculture, Beltsville, Maryland, U.S.A.

F. M. SULZMAN — Division of Biological Sciences, State University of New York, Stony Brook, New York, U.S.A.

BRUNETTO TARQUINI — Department of Medicine, Medical School, University of Florence, Florence, Italy

JOHN M. TAUB — Department of Psychology, University of California, Santa Cruz, California, U.S.A.

JOSEPH W. TERNES — Department of Psychology, University of Puerto Rico, Rio Pidras, Puerto Rico

TERRY NED TROBEC — Department of Zoology and Physiology, Louisiana State University, Baton Rouge, Louisiana, U.S.A.

ROLF ULLNER — Forschungsstelle für Soziale Paediatrie und Poliklinik, Universität München, München, Germany

V. R. R. UPPULURI — Department of Statistics, Division of Mathematics, Oak Ridge National Laboratory, Oak Ridge, Tennessee, U.S.A.

JOHN URQUHART — Biodynamics Laboratory, ALZA Corporation, Palo Alto, California, U.S.A.

PHILIPPE VAGUE — Clinique Endocrinologique, Faculté de Médicine, Marseille, France

THERESE VANDEN DRIESSCHE — Laboratoire de Cytologie et Embryologie Moléculaires, Département de Biologie Moléculaire, Université Libre de Bruxelles, Bruxelles, Belgium

JOAN VERNIKOS-DANELLIS — Environmental Biology Division, Ames Research Center, NASA, Moffett Field, California, U.S.A.

RICCARDO VIGNERI — Department of Medicine, Division of Endocrinology, Catania University School of Medicine, Catania, Italy

CHARLES A. WALKER — Department of Physiology and Pharmacology, School of Veterinary Medicine, Tuskegee Institute, Tuskegee, Alabama, U.S.A.

P. ROY WALKER — McArdle Laboratory, The Medical School, University of Wisconsin, Madison, Wisconsin, U.S.A.

JOHN D. WALLACE — Department of Radiation Therapy and Nuclear Medicine, Thomas Jefferson University Hospital, Philadelphia, Pennsylvania, U.S.A.

W. G. WALTHER — Division of Biological Sciences, State University of New York, Stony Brook, New York, U.S.A.

WILSE B. WEBB — Department of Psychology, University of Florida, Gainesville, Florida, U.S.A.

HANS-MARTIN WEGMANN — Institut für Flugmedizin (DFVLR), Bonn-Bad Godesberg, Germany

ELLIOT D. WEITZMAN — Department of Neurology, Montefiore Hospital and Medical Center, Bronx, New York, U.S.A.

STANLEY G. WELLSO — Entomology Research Division, Agricultural Research Service, U.S. Department of Agriculture, East Lansing, Michigan, U.S.A.

LOTHAR WERTHEIMER — Department of Medicine, New York Medical College, Center for Chronic Disease, Welfare Island, New York, New York, U.S.A.

RÜTGER WEVER — Max-Planck-Institut für Verhaltensphysiologie, Erling-Andechs, Germany

K. WHITE — Department of Anatomy, Louisiana State University Medical Center, New Orleans, Louisiana, U.S.A.

JAMES WICKS — Chronobiology Laboratory, Department of Pathology, University of Minnesota, Minneapolis, Minnesota, U.S.A.

JOHN J. WILLE, JR. — Department of Biological Sciences, University of Cincinnati, Cincinnati, Ohio, U.S.A.

C. W. M. WILSON — Department of Pharmacology, Trinity College, University of Dublin, Dublin, Ireland

CHARLES M. WINGET — Experimental Biology Division, NASA, Ames Research Center, Moffett Field, California, U.S.A.

CHARLES C. WUNDER — Department of Physiology and Biophysics, University of Iowa College of Medicine, Iowa City, Iowa, U.S.A.

AKHTAR YASEEN — Department of Medicine, New York Medical College, Center for Chronic Disease, Welfare Island, New York, New York, U.S.A.

EDMUND J. YUNIS — Department of Laboratory Medicine, University of Minnesota, Minneapolis, Minnesota, U.S.A.

KENNETH E. ZICHAL — Division of Biological and Medical Research, Argonne National Laboratory, Argonne, Illinois, U.S.A.

GLOSSARY OF SELECTED CHRONOBIOLOGIC TERMS*

FRANZ HALBERG and JUNG-KEUN LEE

Chronobiology Laboratories, Department of Laboratory Medicine and Pathology, University of Minnesota, Minneapolis, Minnesota, U.S.A.

1. ACROPHASE, ϕ, φ, Φ: a measure of timing—the lag from a defined acrophase reference of the crest time in the function used to approximate a rhythm: there are three types of acrophases, "computative, external and internal" (Fig. 1).

Units: Angular: degree, radian.
Time: second, minute, hour, day, month, year.
Episodal: integer (such as number of heart beats).

Note: "BATHYPHASE" and "ORTHOPHASE" are the lags from a defined acrophase reference of the "trough time" or the "specified right time", respectively, in the function used to approximate a rhythm.

2. AMPLITUDE, A: a measure of extent of rhythmic change the difference between the maximum and the mesor (q.v.) of a (sinusoidal or other) function used to approximate the rhythm (Fig. 1).

Units: Original physiologic units, e.g., number of heart beats, mm Hg in blood pressure, etc.

3. ANGULAR FREQUENCY, ω: special case of frequency (q.v.), corresponding to the number of repetitions of a periodic process in a unit of time, e.g., ω in $y_i = M + A \cos(\omega t_i + \phi) + e_i$, where M=mesor (q.v.) A=amplinude (q.v.), ϕ= acrophase (q.v.) and t_i=time. Relation: $\omega = 2\pi/\tau = 2\pi f$, since frequency (q. v.), $f = 1/\tau$ (period).

Units: Degrees (or radians) per unit time.

4. AUTORHYTHMOMETRY, AR: special case of rhythmometry with data collection by self-measurement (and/or automatic recording) of physiologic variable(s) as a function of time with ensuing fit of mathematical function(s) for inferential statistical rhythm and/or other temporal parameter description and point-and-interval estimation of characteristics such as mesor, amplitude, acrophase, period (or frequency) and/or waveform.

5. BIOLOGIC NOISE: random or other (useless) components (of a signal), interfering with the (useful) part of the signal (e.g., rhythm) to be evaluated.

Note: Biologic noise—from unidentified and identified sources, separately evaluated or unevaluated—can be computed as the variability remaining after the (e.g. least squares) fit of a model used for approximating a rhythm. In this sense, noise equals the error term in a statistical model and the relative contribution of noise as interference with the predictable (and thus useful) part of the signal is described as Percent Error (q.v.).

* Consisting of abbreviated definitions from glossary of the International Society for Chronobiology (prepared by F. HALBERG and G. S. KATINAS in cooperation with others and published in the International Journal of Chronobiology, Vol. 1, 31–63) and some additional terms. This work was supported by USPHS (5-K6-GM-13,891 and CA 1R01-CA-14445-01S1) and NASA.

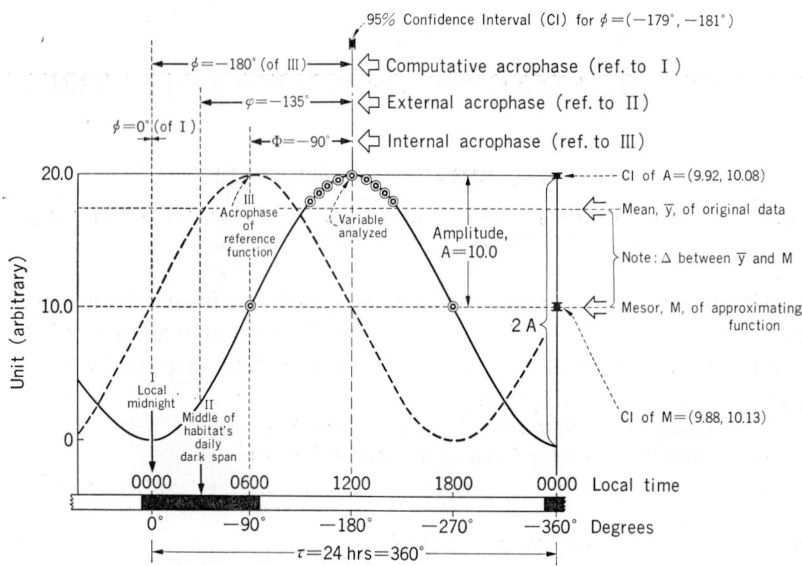

Fig. 1 Rhythm characteristics obtained by the least squares fit of a single cosine curve: the mesor, which can be different from the sample mean in the case of unequidistant data; the amplitude and the extent of total change predictable by the fit of a single cosine curve, namely the double amplitude. Different types of acrophase estimates depend upon the reference point chosen as zero time in estimating the lag of the crest time in the cosine curve approximating all data: (a) the computative acrophase referred to some computationally convenient reference such as midnight local time—on day 1 of studies on circadian rhythms (or on December 22 of previous year, for the case of circannual rhythms, etc.), (b) the external acrophase referred to a point on some environmental cycle such as the midpoint of the dark-span (in one's bedroom or habitat niche) when living on a 24-h cycle of light and darkness, or (c) the internal acrophase referred to the peak of another cosine curve approximating another physiologic series used as the reference series.

6. BIORHYTHM: rhythm (q.v.) persisting as a fundamental property of biologic entities under various conditions of presumed constancy in environmental factors (including those possibly known to synchronize the rhythm).

Note: The term "biorhythm" has been misused in many contexts. It is here reserved for phenomena meeting testable criteria, several of them biologic in nature. First, a biorhythm, as any rhythm, should constitute a statistically significant entity—validated, for instance, by a test showing that its amplitude is not zero. Second, a biorhythm should persist for two or more cycles in an ecosystem, a society, an organism, organ, or other system isolated as far as possible from environmental cycles—e.g., in a cave or another medium with constant temperature, continuous darkness or light and unchanging availability of nutrients. Third, under such conditions, biorhythms usually exhibit a frequency that is statistically significantly different from that of an environmental cycle (the synchronizer) known (or assumed) to synchronize the biorhythm. Fourth, the institution of an abrupt (single-cycle) shift by 90° or more in the synchronizer should be followed by a 'structured' biorhythm (e.g., acrophase) adjustment, its characteristics (such as rate, if not direction) depending on whether the synchronizer was advanced or delayed.

7. CHRONOBIOLOGIC WINDOW: Print-out or graphic display of results obtained by least squares fitting of cosines, usually display of amplitudes (q.v.) and percent errors (q.v.) at chosen trial periods.

8. CHRONOBIOLOGY: science objectively quantifying and investigating mechanisms of biologic time structure, including rhythmic manifestations of life.

Note: Rhythms with different frequencies are found at all levels of biologic integration—ecosystem, population, group, individual, organ-system, organ, tissue, cell and subcellular structure. Their ubiquity and their critical importance to the survival of both the individual and the species have prompted the development of a special methodology to study these temporal characteristics in the context of development, growth and aging, yet in a novel branch of biology separate from embryology, pediatrics and geriatrics. In physiologic terms, chronobiology provides generally applicable concepts and techniques for resolving predictable cycles (q.v.) in organisms and for isolating environmental effects from the underlying endogenous mechanisms. The basic properties of rhythms are important to education, ecology and medicine. Among other subspecialties, chronobiology includes: (a) chronophysiology; (b) chronopathology; (c) chronopharmacology, including chronotoxicology and chronotherapy.

Chronophysiology investigates temporal manifestations of physiologic processes. It evaluates cyclic nervous, endocrine, metabolic and other interactions within the organism which underlie biologic temporal characteristics and their interaction with the environment.

Chronopathology investigates alterations in biologic temporal characteristics as determinants or as resultants of disease states (psychoses, cancers, endocrinopathies etc.).

Chronopharmacology investigates drug effects (1) upon biologic temporal characteristics and (2) as a function of biologic timing. Thus, chronotoxicology investigates undesired or harmful effects from chemical, physical or other agents including poisons, pollutants and overdoses of drugs (1) upon biologic temporal characteristics and (2) as a function of biologic timing. Chronotherapy endeavors to cure (or prevent) disease, with proper regard to temporal characteristics (e.g., corticosteroid therapy timed to simulate the adrenocortical cycle in Addison's disease).

Chronobiology also includes chronology, the review of past events which have to do with the development of environments and the parallel course which animals (and plants) have followed in their evolution.

9. CHRONOBIOLOGIC SERIAL SECTION, SS: analytical results obtained by fitting a fixed-period cosine curve to consecutive overlapping or non-overlapping data sections, called intervals, displaced in increments throughout a time series—displayed with the original data as 'moving' P values (for rhythm description) as moving point and interval estimates for mesor, amplitude and acrophase of a rhythm (Fig. 2).

10. CHRONOGRAM: individual or averaged display of data as a function of time for macroscopic (q.v.) viewing (top of Fig. 2).

11. CIRCADIAN: relating to biologic variations or rhythms with a frequency of 1 cycle in 24±4 h; circa (about, approximately) and dies (24 h) (Table 1).

12. CIRCADISEPTAN: relating to biologic variations or rhythms with a frequency of 1 cycle in 14±3 days.

13. CIRCANNUAL: relating to biologic variations or rhythms with a frequency of 1 cycle in 1 year ±2 months.

14. CIRCASEPTAN: relating to biologic variations or rhythms with a frequency of 1 cycle in 7±3 days.

15. CIRCATRIGINTAN: relating to biologic variations or rhythms with a frequency of 1 cycle in 30±5 days.

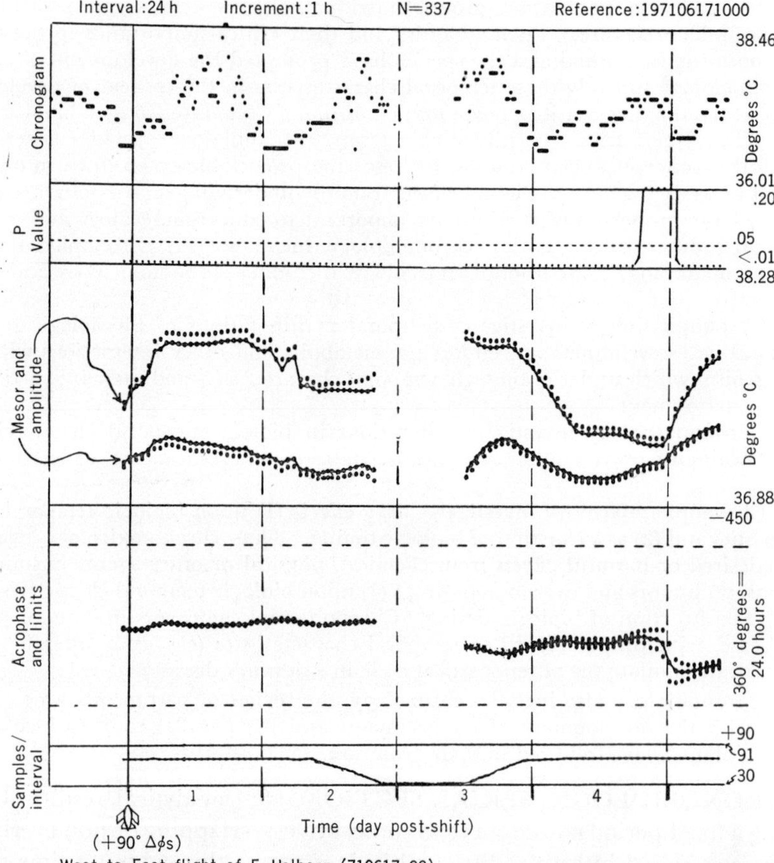

Fig. 2 Chronobiologic serial section: Original temperatures collected from 52-year-old man before, during and after intercontinental flight involving advance in schedule (after crossing six time zones) and followed by a lecture immediately on arrival are displayed in the top row (chronogram) showing the range and pattern of variation (in °C). Some of the data are missing between the end of day 2 and the middle of day 3. Data in chronogram are analyzed by least squares fitting of a 24-h cosin function to consecutive overlapping intervals of 24 hours, displaced in 1-h increments. P-values for the rhythm is description are provided on a logarithmic scale in the second row and indicate that with few exceptions significance is found at (or below) the 0.01 level. In the third row the estimates of amplitudes and mesors by least squares analyses are shown as connected dotted lines with corresponding standard errors as dots below the mesors and above the amplitudes. The acrophases are presented in the forth row with their 95% confidence limits. Immediately following the flight the rhythm's acrophase remains stable, despite a sudden 90° change in routine $(+90° \Delta\phi_S)$. The first jump in acrophase is seen at the middle of day 3 and the second at the beginning of day 5. A major change in chronogram shape occurs on day 4 before the advancing circadian acrophase-jump on day 5. Ultradian frequencies predominate in the temperature record immediately preceding the circadian acrophase jump, concurrent with a transient circadian rhythm obliteration—a drastic drop in circadian amplitude leading to failure of a 24-hour cosine fit to describe a circadian rhythm. This can be seen from the second row of P-values for a brief span at the end of day 4. The remainder of the P-values are at or below the one percent level, indicating the persistence of a prominent circadian rhythm. At the bottom the time (days post-shift) is given and presented by vertical lines. In the fifth row the number of data points available for analysis is shown as a continuous line.

Table 1 Spectral classification of human rhythms.

Domain*	Ultradian	Circadian	Infradian**
Frequency, f (cycles per 24 hours)	86400 8640 1440 24 1.2	1	.86 .14 3×10^{-2} 27×10^{-4} 27×10^{-6}
Scales	Logarithmic	Linear	Logarithmic
Period, τ ($f=1/\tau$)	0.1(s) 1 1 1 (20 h) Second(s) Minute(m) Hour(h) <20 h	1(24 h) Day(d) $20\text{ h}\leq\tau\leq28\text{ h}$	(28 h) 1 1 1 1 Week(w) Month(M) Year(y) Century(c) >28 h
Rhythms in Health:	Elecroencephalogram Electrocardiogram Respiration Peristalsis REM, Rest-activity	Sleep-wakefulness Responses to drugs Blood constituents Urinary variables Metabolic processes, generally	Menstruation 17-ketosteroid and norepinephrine excretion.
Grossly apparent morbidity and mortality rhythms	cf. Conventional textbooks of medicine.	· Convulsive disorders, intermittent catatonia. · Morbidity and mortality from certain respiratory, Cardiovascular and other diseases, etc.	

* Domain [named according to frequency (f) by analogy to the physical classification of ultraviolet or ultrasound and infrared or infrasound, respectively, but delineated according to period (τ), of function approximating rhythm].

** Infradiam domain can be subdivided into several regions, as follows: Circaseptan ($\tau\sim7$ d); Circavigintan ($\tau\sim20$ d); Circatrigintan ($\tau\sim30$ d); and Circannual ($\tau\sim1$ yr). Regions in the Ultradian and Circadian domains have not as yet been delineated, except for the "specific range" of the EEG (cycles/second): delta (<1–3.5); theta (3.6–7); alpha (7.1–12); beta (12.1–30).

Several variables examined thus for exhibit statistically significant components in all three spectral domains, as does the EEG. 17-ketosteroid and norepinephrine rhythms with circadian frequency also are quite prominent.

16. CIRCAVIGINTAN: relating to biologic variation or rhythms with a frequency of 1 cycle in 21 ± 3 days.

17. COMPUTATIVE ACROPHASE, ϕ: acrophase (q.v.) referred to an arbitrarily chosen date and clock hour (e.g., midnight local time, Fig. 1).

18. CONFIDENCE INTERVAL: inferential statistical interval for a single parameter investigated.

Note: If two estimates E_1 and E_2 of the lower and upper limits of the parameter E satisfy the relation $P\ (E_1<E<E_2)=1-\alpha$, where α is some fixed probability (e.g., 0.05), the interval between E_1 and E_2 is called (e.g., 95%) confidence interval for the unknown parameter E. The assertion that E lies in this interval will be true, on the average, in a proportion $1-\alpha$ of the sampling cases when the assertion is made. The values E_1 and E_2 are called the confidence limit.

19. COSINOR: statistical summary with display on polar coordinates of a biologic rhythm's amplitude and acrophase relations by the length and the angle of a directed line, respectively, shown with a bivariate statistical confidence region (q.v.) computed (at chosen trial period) (1) to detect a rhythm (by a confidence-region not overlapping the pole) and (2) to estimate conservative confidence intervals (q.v.) for the rhythm parameters.

Note: Cosinor procedures are of several kinds:

MEAN COSINOR, cosinor-M: the original cosinor procedure applicable to 3 or more biologic series from an individual or a group for assessing the rhythm characteristics, if any, of an entire set (The Cellular Aspects of Biorhythms, H. von MAYERSBACH, Ed. Berlin, Springer-Verlag, 1967, pp. 20–48).

Inputs of cosinor-M are imputations (q.v.) consisting of amplitudes and acrophases from each individual series. Cosinor-M is applied when the mesors from individual series are different but the amplitudes are similar and the number of data points from each series is approximately equal. SINGLE COSINOR, cosinor-S: a cosinor procedure applicable to single biologic time series or to a set of series (from an individual or group) which all have similar mesors and similar amplitudes (Physiology Teacher *1*: 1–11, 1972). NUMBER-WEIGHTED MEAN COSINOR (PONDERATUS), cosinor-P: a cosinor procedure weighted with the number of observations in each series, applicable to 2 or more biologic series from one or more individuals when the mesors, amplitudes and particularly the number of data points in each series are quite different (TONG, LEE and HALBERG, International Journal of Chronobiology, in press).

20. CONFIDENCE REGION: a generalization of the confidence interval to the case where two or more parameters are being considered, i.e., the region in the parameter space to which is assigned the probability $1-\alpha$ that the parameters lie within.

21. CYCLE: the whole of consecutive states and/or specifiable changes or events recurring in a physiologically integrated fashion with a recognizable frequency, e.g., cardiac cycle, adrenal cycle, (environmental) light cycle, etc.

22. DESYNCHRONIZATION: state of two or more previously synchronized rhythmic variables that have ceased to exhibit the same frequency and/or the same acrophase relationships and show changing time relations (Fig. 3).

23. DIAN: relating to biologic variations or rhythms with a frequency of 1 cycle in precisely 24 h (±0.1 h).

24. DIURNAL: relating to biologic variations or events occurring between sunrise and sunset or during illuminated fraction of a near-daily schedule of alternating artificial light and darkness.

25. DYSCHRONISM: time structure (including rhythm) alteration associated with demonstrable physical, physiologic or mental deficit, if not disease (Fig. 4).

Note: Dyschronism, a form of ecchronism (alteration of one or several rhythm characteristics), may be reserved for ecchronism demonstrated to be associated with a deficit in physical, physiologic or mental performance, with increased risk of disease or overt illness or with life span shortening.

26. EXTERNAL ACROPHASE, φ (small Greek phi): acrophase (q.v.) referred to a point on the synchronizing environmental cycle (SYNCHRONIZER q.v.) (see Fig. 1).

27. EXTERNAL DESYNCHRONIZATION: desynchronization (q.v.) of a biologic rhythm from an evnrionmental cycle.

Note: The disappearance of 24-h-synchronized circadian periodicity in blinded mice, the rhythm in rectal temperature assuming free-running (q.v.) frequencies with 1 cycle in 23.2–23.8 h, constitutes one of many examples (Fig. 3).

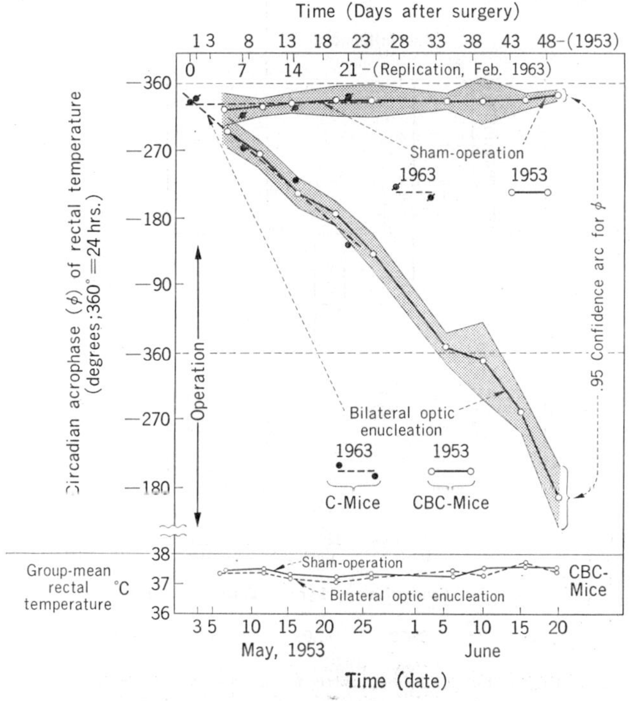

Fig. 3 Reproducibility of circadian desynchronization following blinding in studies on CBC mice and C mice done 10 years apart with different sampling techniques. A study on CBC's in 1953 is summarized by open circles connected with continuous lines. Note that results on ϕ obtained in 1963 are within the 95 per cent confidence interval computed for results on ϕ obtained a decade earlier. Note also from the bottom of the graph that the mesors, i.e., rhythm-determined averages, of body temperature (1953 data only are shown) are roughly the same in both blinded animals and controls. Thus, rhythmometry detects a clear, quantifiable and reproducible alteration of rhythm after blinding in a mammal, whereas a more classical rhythm-unqualified study fails to detect an effect. In all studies, light daily from 0600 to 1800 alterated with darkness.

28. FOURIER ANALYSIS: study of representing functions of a time series as a combination of trigonometric functions, e.g., $F(t_i)=\sum_{j=1}^{p}\cos[\phi_j+(2\pi/\tau_j)t_i]i=1,\ldots,n$ where p is the number of cosine terms, ϕ_j the acrophases, τ_j the periods, t_i the time and $F(t_i)$ the representative function.

Note: Method more general than harmonic analysis (q.v.).

29. FREE-RUNNING: desynchronized in the sense of exhibiting a continually (and systematically) changing phase relation to the schedule of a habitual (known) synchronizer following, e.g., removal of synchronizing stimuli or of their organismic transducer(s) or other major interference with the signal or its reception.

30. FREQUENCY f: the number of occurrences of a given type of event or the number of members of a population falling into a specified class. In a study of periodicity it is the number of cycles occurring per unit time, i.e., f is the reciprocal of the period (τ), $f=1/\tau$.

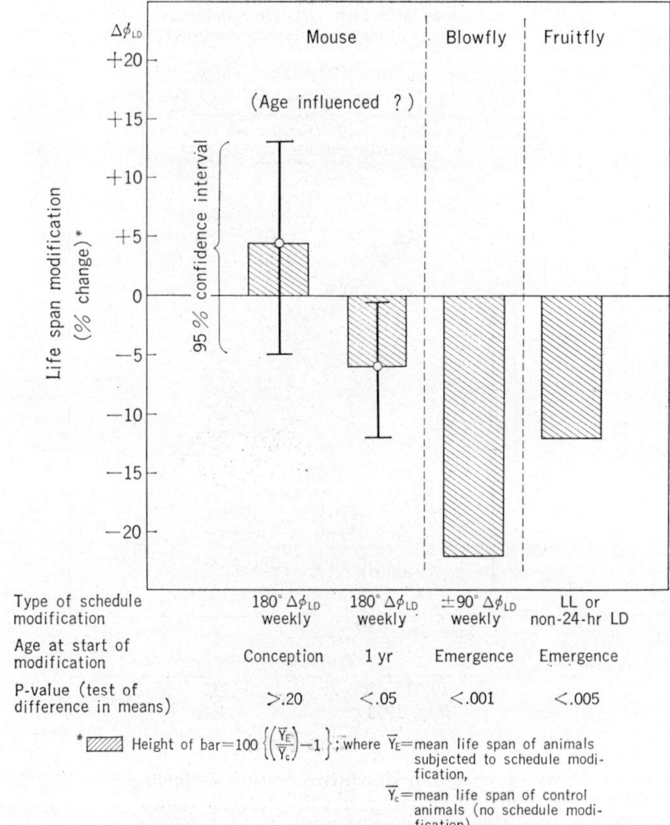

Fig. 4 Life span of Bagg albino mice apparently shortened by phase-shifts begun at about 1 year of age ($p<.05$) but not when animals (of the same inbred strain) were exposed to phase-shifts during pregnancy and their offspring thus exposed to weekly lighting regimen inversions from conception until death (HALBERG and NELSON cited in LUCE, G.G., USPHS No. 2088, page 134, 1970). Life span shortening also shown for insects (ASCHOFF, J., Naturwissenschaften, *11*: 574, 1791 and PITTENDRIGH, C. S. and D. H. MINIS, Proc Nat. Acad. Sci. .S.A., *69*: *1537,* 1972).

31. HARMONIC ANALYSIS: study of function approximating a time series as a combination of trigonometric functions with periods in harmonic order, i.e., $F(t_i)=\sum_{j=1}^{p}\cos\,[\phi_j+(2\pi j/\tau)t_i]$ where symbols are referred to Fourier Analysis (q.v.).

32. IMPUTATION: tentative estimation from limited or insufficient evidence, such as a procedure for deriving from a short time series endpoints used in further analyses such as a number-weighted or number-unweighted mean cosinor (q.v.).

33. INFRADIAN: relating to certain biologic variations or rhythms with a frequency lower-than-circadian (q.v.) (Table 1).

34. INTERNAL ACROPHASE, Φ (Greek capital phi): acrophase (q.v.) referred to acrophase of another rhythm with the same frequency in the same entity (Fig. 1).

ESTIMATING RHYTHM PARAMETERS BY
LEAST-SQUARES FITTING OF COSINE FUNCTION*
Abstract Example with 24-h Cosine Function. Y(t) [Continuous Curve],
Fitted to 2-hourly Data, yi[◉], Obtained During Wakefulness Span.

Degrees (360°≡24 h)

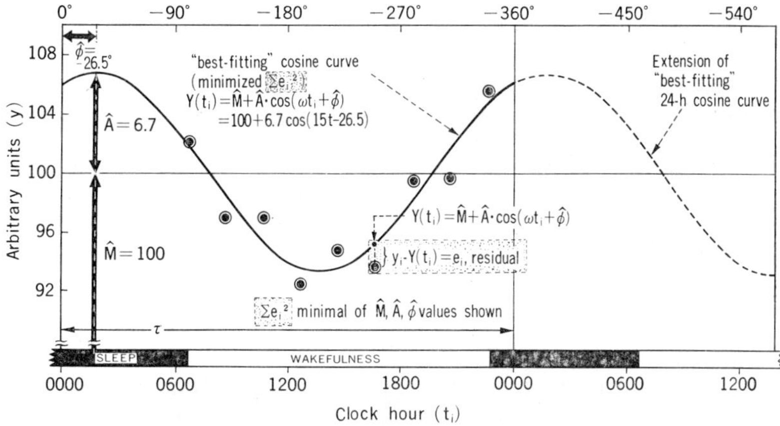

* For single series of observations or of mean volues.
τ=period=1/frequency=duration of cycle in units of time, here=24 h; frequency =1/τ=cycles per unit time, here=1 cycle per 24 h; M=Mesor; A=amplitude; ω=angular velocity=360°/τ, here 360°/24 h=15°/h; t=time (h after 0000); = computative acrophase (degree from 0000).

Fig. 5 Estimation of rhythm parameters—M=mesor, A=amplitude and ϕ=acrophase—by least squares method, i.e., by quantifying parameters—denoted as M, A and ϕ— at which $\sum_{ei}^{z}$ is minimal, where $e_i = y_i - [M + A \cos(\omega t + \phi)]$. Estimates by least squares method are the Best Linear Unbiased Estimates (BLUE) according to the Gauss-Markoff theorem. The best-fitting cosine curve, drawn as a continuous line after parameter estimation, is extended to indicate use in prediction.

35. INTERNAL DESYNCHRONIZATION: desynchronization (q.v.) of two or more rhythms from each other (in the same entity).

Note: DNA-formation rhythm desynchronizes from other cellular circadian rhythms in the liver of certain mice kept in continuous light.

36. LEAST SQUARES METHOD: estimation technique for determining quantities by minimizing the error (or residual) sum of squares. This method produces the best linear unbiased estimate (BLUE) in terms of variance (Fig. 5).

Note: Two types of least squares methods are considered: linear or nonlinear (see note in PARAMETER).

37. LIGHTING REGIMEN: schedule of light and/or darkness to which an organism is exposed.

Note: For many life-forms, lighting regimens are synchronizers (q.v.) and hence they must be given in pertinent experimental protocols. The following abbreviations are recommended: L=light; D=dark; LL=continuos light; DD=continuous dark; L_f—fluorescent lamps; L_i=incandescent lamps. Figures in parentheses after L or D may be used to indicate intensity in lux (original values measured in footcandles multiplied by 10.8). Figures in brackets after L or D may indicate either duration or span of clock hours in local time—e.g., L [12 h] : D [12 h]=a cycle of 12 h of light alternating with 12 h of darkness: L [0600–1800] : D [1800–0600] (50:0)=a cycle of light at an intensity of 50 lux from 0600 to 1800 alternating with complete darkness from 1800 to 0600. More generally, in fashion similar to suggestions by Aschoff et al. (in Circadian Clocks. North Holland, Amsterdam, 1965) one may use the notation:

$LD\,[x_1:x_2]\,(y_1:y_2)$ light-dark cycle

with

x_1=light time as duration in hours or as span of clock hours in local time
x_2=dark time, as above
y_1=intensity of illumination in L in lux
y_2=intensity of illumination in D in lux

It is pertinent that the intensity of 'illumination' of D in LD need not necessarily be unmeasurably low but can simply be lower than that in L. Thus, LD [16:8] (100: 0.1) denotes a cycle of light and less light composed of 16 h light with 100 lux and 8 h less light with 0.1 lux.

38. LINEAR LEAST SQUARES METHOD, LLS: least squares method applied to a model linear in parameters.

Note: The single cosine model $y_i=M+A\cos[\phi+(2\pi/\tau)t_i]+e_i$ is linear if the period is a priori fixed for analysis (whether or not it is known or unknown) in order to estimate the parameters M, A and ϕ.

39. LINEAR-NONLINEAR LEAST SQUARES METHOD, L-NLS: computational procedure for assessing sets of rhythm characteristics for several frequencies (unknown, unfixed) characterizing the data by applying first linear least squares (q.v.) followed by non-linear least squares, NLS (q.v.).

Note: LLS method provides a set of initial values for NLS which yield final results after an appropriate number of iterations.

40. MACROSCOPIC (analysis): analysis based solely upon inspection of original data or of averages and dispersion indices plotted as a function of time.

41. MESOR, M: rhythm-determined average, midway between the highest and lowest values of function used to approximate a rhythm (Fig. 1).

Note: M is equal to the arithmetic mean for equidistant data covering an integral number of cycles.

42. MICROSCOPIC (analysis): analysis for detecting in a signal, e.g., in biologic time series [usually containing biologic noise (q.v.)] any temporal characteristic (useful part of a signal) and for obtaining objective numerical endpoints thereof.

Note: Inferential statistical methodology using electronic computer helps to discover may important details of biologic time structure, just as the microscope serves to examine the spatial structure of tissues and cells.

43. NONLINEAR LEAST SQUARES METHOD, NLS: least squares method applied to a model nonlinear in parameters.

Note: If the period is not fixed in a single cosine model (see LLS) this model is nonlinear and the period is a parameter. This model considers a set of parameters as a vector in 4 dimensions. For computation, one chooses a set of initial values—a vector—and moves this in parameter space by iteration guided by the principle of least squares. In Fourier's Analysis, NLS is needed for assessing several cosine terms, each with a set of 3 rhythm characteristics, in addition to the overall mesor.

44. PACEMAKER: organismic entity controlling or influencing rhythmic activity.

Note: Most commonly, the term 'pacemaker' refers to the heart's rhythmic centers, e.g., the sinus node. The meaning of the term has been broadened to include any mechanisms generating rhythms from within a given system.

45. PARAMETER: in statistics the quantity characterizing the state of nature in frequency distributions.

Note: In a linear model for fitting a straight line $y_i = M + Bt_i + e_i$ M and B are parameters, while time t_i is an independent variable which can be present and an observation y_i is dependent on t_i, whereas e_i denotes random error. The making of inferences about the state of nature (parameters) is called 'statistical inference'. When say that a model is linear or non-linear, we are referring to the linearity or nonlinearity in the parameter.

46. PERCENT ERROR, PE: percentage of variability not accounted for by a fitted function.

Note: PE equals 100—PR [percent rhythm (q.v.)], i.e., PE+PR=100.

47. PERCENT RHYTHM, PR: percentage of variability accounted for by a fitted function.

48. PERIOD, τ (Greek tau): duration of (one complete cycle in) a rhythmic variation.

49. PERIODIC FUNCTION, $f(t) = f(t + \tau)$: function whose values recur in regular temporal or spatial intervals corresponding to a period τ (q.v.).

50. PERIODOGRAM: display, as a function of an abscissa linear in period, of amplitude, residual error, variance or some other characteristic investigated by harmonic or other analysis.

51. PHASE: instantaneous state of a rhythm within a cycle.

Note: For interpreting an instantaneous state of a variable as representing an instantaneous phase, information on the acrophase (q.v.) and other rhythm characteristics, as well as their limits, is required (see note in PHASE ANGLE [q.v.]).

52. PHASE ANGLE: a time point in a periodicity considered in relation to another specified time point—acrophase and phase angle are interchangeable but acrophase or some comparable term defined as to zero phase is preferred to phase angle in chronobiology.

Note: In chronobiology, a meaningful acrophase reference is essential and the nature of this reference can be made obvious by the notation of one of several (computative, external and internal) acrophases (q.v.). Use of the term 'phase' in biology may thus be limited at best to consideration of macroscopic data. Since the term is unspecific with respect to reference point, it should be replaced by 'acrophase' in any discussion of microscopic analyses.

53. PHASE-SHIFT, $\Delta\phi$: single abrupt or gradual displacement of a periodicity along the time scale.

Note: A delaying $\Delta\phi$ is denoted as $-\Delta\phi$ and involves the later occurrence of a phase marker, e.g., acrophase; an advancing $\Delta\phi$, denoted as $+\Delta\phi$ involves the earlier occurrence of a phase marker (Figs. 2 and 3).

54. PLEXOGRAM: display of original data covering spans longer than the period of a rhythm investigated along a abscissa of a single period, irrespective of time order of collection, e.g., as a function of a single conventional or other time unit—such as a day, irrespective of calendar data and/or subject.

55. POINT ESTIMATION: specification of a single value for an unknown parameter.

56. RHYTHM: a periodic component of (biologic) time series, with objectively quantified characteristics [e.g., a frequency, f, acrophase, ϕ, amplitude, A, mesor, M, and/or waveform, W] demonstrated by inferential statistical means.

> **Note:** Rhythms thus include any set of biologic changes recurring systematically according to an (algorhythmically) formulatable pattern or waveform which is validated in inferential statistical terms. Mathematically, more or less sinusoidal rhythms can be described by the use of approximating funcitons such as those of a form:
>
> $F(t)=M+A\cos(\omega t+\phi)$
>
> where ω is the angular frequency and t=time. Confidence intervals also should be estimated for rhythm parameters.

57. RHYTHMOMETRY: detection of rhythm by inferential statistics and point-and-interval estimation of characteristics such as mesor, amplitude, acrophase, period (or frequency) and/or waveform.

> **Note:** Autorhythmometry (q.v.) is a special case of rhythmometry.

58. SHIFT-SPAN (of a rhythm): $T(\Delta\phi_R)=t(\Delta\phi_R)-t(\Delta\phi_S)$: the time elapsed between the onset of a (usually sudden) shift in a synchrohizer schedule $t(\Delta\phi_S)$ and completion of the resulting shift in a rhythm's acrophase at $_t(\Delta\phi_R)$.

59. SPECTRAL WINDOW: tabulation or display, as a function of an abscissa (linear) in frequency of amplitude, percent error, variance or some other characteristic investigated by some kind of harmonic or other analysis.

> **Note:** This term is broader that the "chronobiologic window" (q.v.).

60. SPECTRUM OF BIOLOGIC RHYTHMS: the entirely of biologic rhythms with different frequencies ranging from extremely high-frequency event (e.g., 1000 cycles/sec discharges of certain electric fishes) to low-frequency population rhythyms (with one cycle in many years).

> **Note:** The biologic rhythm spectrum may be subdivided into three domains and the latter into a number of regions, as indicated in Table 1.

61. SYNCHRONIZATION: state of system when two or more variables exhibit periodicity with the same frequency and phase angle or with frequencies that are integer multiples or submultiples of one another.

> **Note:** Synchronization is possible among physiologic processes in the absence or presence of environmental cycles acting upon them. See INTERNAL and EXTERNAL SYNCHRONIZATION. Synchronization requires the operation of a primary periodic agent—called a synchronizer (q.v.) when it is an environmental factor, or a pacemaker (q.v.) when it consists or an organismic mechanism.

62. SYNCHRONIZER: environmental periodicity determining the temporal placement of a given biologic rhythm along an appropriate time scale by impelling the rhythm to assume synchronization, i.e., its frequency or an integer multiple or submultiple of its frequency.

> **Note:** Also called Zeitgeber, time-giver, entraining agent, clue or cue.

63. TEMPORARY ANTIPHASE: transient 180°±15° difference in timing of two rhythms with slightly different frequencies.

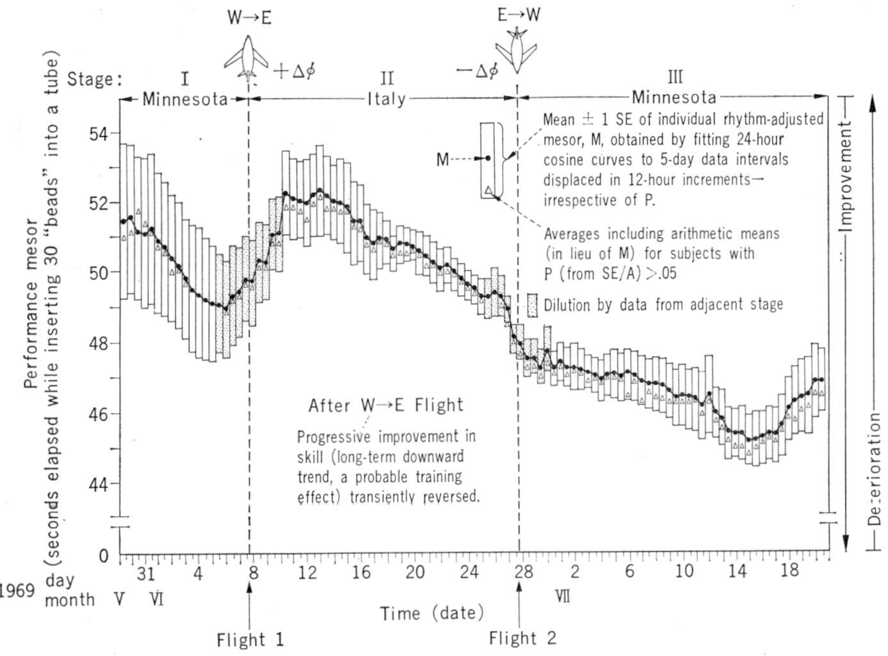

Fig. 6 The proverbial "straw that might break the camel's back." Performance decrement after certain transmeridian flights may be slight, yet it is demonstrable and its net result need not be negligible when either the top performance of extremely talented individuals or the over-all performance of large groups of individuals are cases in point. For the four male and three female adults here summarized, transmeridian dyschronism is evident in the adjustment following West-to-East flight but not following flight in the opposite direction.

Note: Macroscopic (q.v.) peak of one rhythm transiently coincides with trough (q.v.) in the other and center-line (q.v.) crossings transiently coincide, with the two rhythms' values diverging one to either side of the center line. A temporary-anti-phase-test on a desynchronized individual or group in comparison with a synchronized control validates by limited sampling (see Fig. 3) a predicted difference in the rhythm's period.

Microscopic (q.v.) transient acrophase difference of $180° \pm 15°$.

64. TIME SERIES, $(y_i,\ t_i)$: chronologic sequence of paired values, one of which is time, t_i the other a quantitative characteristic of an individual or population (y_i) at the time t_i.

Note: Biologic time series may consist of continuous, or discrete but equidistant or of unequidistant observations in time.

65. TRANSMERIDIAN DYSCHRONISM: rhythm alteration with malaise or performance decrement following transmeridian travel (Fig. 6).

66. ULTRADIAN: relating to biologic variations or rhythms with a frequency higher than-circadian.

67. VARIATION, COEFFICIENT OF: standard deviation of a distribution divided by the arithmetic mean (sometimes multiplied by 100).

Unit: Dimensionless.

Note: Serves to compare variabilities of samples or populations and is independent of units and magnitudes of means, but is sensitive to errors in the means.

68. WAVEFORM: the complete pattern of a periodic variation.

Note: Waveform can be microscopically (q.v.) represented, e.g., by multiple cosine terms. In this case, $W=g\{(A_1\phi_1)..., (A_n,\phi_n)\}$, where A_i and ϕ_i are amplitudes (q.v) and acrophases (q.v.), respectively. Cosines need not be in harmonic order. *Inter alia,* analyses including mathematical splines may also serve for approximating w.

INTRODUCTION

Chairman: Franz HALBERG

CHRONOBIOLOGY: PAST ACCOMPLISHMENTS AND FUTURE RESPONSIBILITIES

JOHN E. PAULY

Department of Anatomy, University of Arkansas Medical Center Little Rock, Arkansas, U.S.A.

The study of biological rhythms dates from the dawn of man; for certainly, ancient man noted the changes in foliage characteristic of the different seasons, the sleep-wake cycles and human menstrual periods.

The earliest recorded data predate the birth of Jesus Christ, but there were relatively few reports in the literature until the turn of the century. By 1935 interest in biological rhythms was sufficient to establish the International Society for the Study of Biological Rhythms in Roneby, Sweden.

It has been said that there are more scientists living today than the total of all who have lived before our time. A very large percentage of these workers are biologists, and they have produced an explosion in scientific literature that our forefathers would not have dreamed possible. Since 1935 there has been an almost exponential growth in reports dealing with chronobiology.

Most of the early workers assumed that biological rhythms were caused entirely by environmental cycles such as the alternating periods of light and darkness, i.e. the movement of the leaves of plants is driven by the sun rising in the morning and setting at night. There were those who believed that the temperature of man is higher during the day only as a result of his activity.

Then it was discovered that certain rhythms continue to operate in the absence of such obvious external time cues as artificial or natural light-dark cycles. Many so-called daily rhythms were observed to possess frequencies that differed from the exact 24-hour day be virtue of the fact that they would advance or delay a certain fixed period each day when maintained in presumably constant conditions.

For these reasons it was suggested that the capacity for rhythmic change is an inherent characteristic of living things and that the control is endogenous rather than exogenous. It was suggested further that these rhythms in nature are maintained within exact periods by external synchronizers such as the light-dark cycles. It became necessary to replace the usual concept of homeostasis; for living things are not constant, rather they are changing, dynamic, interrelated biological systems.

In the early fifties biologists began intensive investigations into many areas of chronobiology in attempts to better understand the mechanisms of rhythms. The one frequency that received the most attention during this period was the circadian; this was explored in both unicellular and multicellular forms of plants and animals. From these studies many generalizations about circadian rhythms were formulated. Concepts of freerunning, temperature independence, entrainment, etc. were defined and rigorously investigated with the net result being a better concept of the temporal organization of living organisms. About the same time, the practical aspects of rhythms in medicine began to be explored, and from these studies we now find ourselves at a stage where the information we have can begin to be applied by the physician.

There were those who recognized that much of the information being collected and reported was not being properly quantified. They set out to find new ways of collecting, measuring and analyzing the wealth of new data so that it could be classified and used as building blocks for future studies. Several methods are in use, including highly efficient computerized techniques such as the cosinors. Which one is best; what are the advantages or shortcomings of each? These must be discussed here in Little Rock.

While these advances were being made, many other biological scientists either payed little or no attention to this new field, or dismissed the findings as random variations with little relevance to their own work. Perhaps one can understand the neophyte who fails to recognize the significance of rhythms with small amplitudes; but such an oversight is difficult to understand when these so-called random variations represent 50%, 100%, even up to 1200% changes within a 24-hour period—and reoccur day after day with remarkable regularity. Can such misconceptions be tolerated in research when it has been shown that the same experiment performed at three different times of the day will give three different results, all statistically significant? Finally some of our colleagues in biology recognize the fact that rhythms occur, but they believe they are unimportant as long as experiments are done at the same time each day. They fail to take into consideration the seasonal changes, lighting routines and many of the complexities of cyclic activity such as variations in wave forms.

There are physicians practicing medicine today who say that biological rhythms are interesting but esoteric—having little to do with the practical problems of medicine. These physicians know the temperature of man varies about two degrees each day; but some of them pay little more attention to elevations (above the so-called normal temperature of 98.6°) during the early morning hours, when the temperature is below the mean, than in the early evening, when it is at its daily high. Some attempt to evaluate the state of health of their patients on the basis of an annual physical examination, a part of which is a single blood pressure determination. Such makes it difficult (perhaps impossible) to detect those transient elevations in blood pressure occuring at the time of the daily peak which probably are the first signs of impending hypertensive heart disease. Here failure to take periodicity into account may result in a lost opportunity to prevent, cure or control a disease.

Other practical applications to be discussed during this symposium include: the effect of manned space or of rapid transcontinental flight on body rhythms; the differential effect of drugs or insecticides when administered at different phases of the circadian system; and the effect of phase shifting by a non-lethal single dose of a drug, especially when it can be demonstrated that the degree of this phase shift is circadian phase dependent. With the advent of circadian profiles for man, perhaps we stand on the threshold of an entirely new concept of health care delivery.

Why have we assembled here in Little Rock? Why are any international meetings necessary, when modern communications and the best libraries in the world are at our disposal? We have convened to define or redefine our accomplishments, to report our latest findings, to agree on definitions and terminology so our own words do not cloud our meanings, to bring together people from such diverse backgrounds as mathematics, physics, chemistry, medicine, psychology—all with a common interest in chronobiology for the purpose of exchanging ideas, learning more about each others interests, synthesizing concepts, identifying the new horizons, and finding the paths to new discovery.

Chronobiology has "come of age". As a result of this new discipline, I submit that we are going to witness not just an evolution, but a real revolution in the biological sciences. Much of the old research needs to be reevaluated on the basis of time structure. Many are going to be surprised by different results; all will be impressed by the new information discovered. Perhaps in the future all experimental biology and medicine will be considered in the light of chronobiology.

I further submit we have responsibilities to discharge. We need to critically review some of our own data. There are obvious weaknesses in reports based on measurements taken at only two or three time points in the 24-hour day unless circadian profiles clearly have been established. Infrequent sampling when analyzing for rhythms is poor technique that leads to poor quantification. We need to advance our discipline from the macro to the micro phase, possible with modern computer technology. I hope all of us will keep this in mind during these three days—while we attend this symposium on quantitative chronobiology.

PERIODICITY AND MEDICINE

FREDERIC C. BARTTER
Endocrinology Branch, National Heart and Lung Institute
Bethesda, Maryland, U.S.A.

In speaking before this group about Chronobiology as it relates to medicine, I am somewhat in the position of a man asked to speak about motherhood before the age of Women's Liberation. In brief, I am for it. In any event, it is incumbent upon me to tell you why I am for it.

If I may make two generalizations about the relationship of periodicity to the clinician, I should say first that its *existence* is very generally accepted as fact, even though the suggestion that any new field requires treatment in terms of periodicity is accepted grudgingly. Secondly, the *meaning* of periodicity, when it refers to the control of one rhythm by another is often reasonably clear. The meaning of the primordial, first-cause rhythm is as obscure as it ever was. (In this context, the analogy of oscillations is inevitable.)

Practically, what are medical people doing about rhythms? First, even the disbeliever organizes his data in such a way as to get rid of them. His paper will not be accepted for publication if the controls were not collected at least ostensibly with the same time pattern as the experimental values. This approach might be called the trivial approach. Secondly, it is quite fashionable to assign one rhythm to another, "causal" rhythm even when the cause of *that* rhythm is quite obscure. (Here the analogy of coupled oscillators inevitably arises.) For example, if ACTH is secreted in rhythmic fashion, it is (almost) inevitable that plasma cortisol should show a dependent rhythm.

Our subject today concerns areas in which the consideration of rhythm is indispensable to good medicine. That is to say, we are concerned with the areas in which rhythms *must* be considered for diagnosis—the separation of the normal from the abnormal—and for treatment. Treatment, of course, because if the disease is periodic it is hard to avoid the conclusion that treatment must also be periodic.

I. I said rhythms are accepted as fact. For example, any intern recognizes that plasma cortisol at 8 a.m. has an entirely different meaning from plasma cortisol at 8 p.m. He may know that a value normal for 8 a.m. could represent Cushing's disease at 8 p.m. He may know that a value normal for 8 p.m. could represent Addison's disease at 8 a.m. But what would he say if he were asked to apply this same logic to blood pressure? Here I can do no better than to quote an eminent authority on hypertension to whom I presented, a very few years ago, the evidence that blood pressure is periodic. He told me he had never seen a hypertensive whose blood pressure oscillated throughout the 24-hour period. Presented with the same evidence this year, he reported that hypertensive patients with fluctuant blood pressure are so important as to require a separate category, and the title "labile".

II. As regards the ultimate meaning of periodicity, I hope this conference will enlighten us. Inevitably, one returns to the minimum requirement for periodicity, as for example in *Gonyaulax polyedra,* in which some of our eminent colleagues have shown that nucleus and cytoplasm alone are enough to establish circadian periodicity. My favorite example is *acetabularia,* however, in which ingenious experiments such as that of SCHWEIGER and HAMMERLING [1] (Fig. 1) demonstrate that the nucleus contains the essential information. Here the nucleus and cytoplasm have been given opposite phases of circadian information by lighting schedules. After transplantation, the nuclear rhythm clearly persists. If the essentials for periodicity are contained in the nucleus of a single cell, it is not difficult to envisage the enormous complexity that follows the development of a circulatory system, a central nervous system, and a system of hormones of internal secretion, all capable of carrying messages from one cell to another.

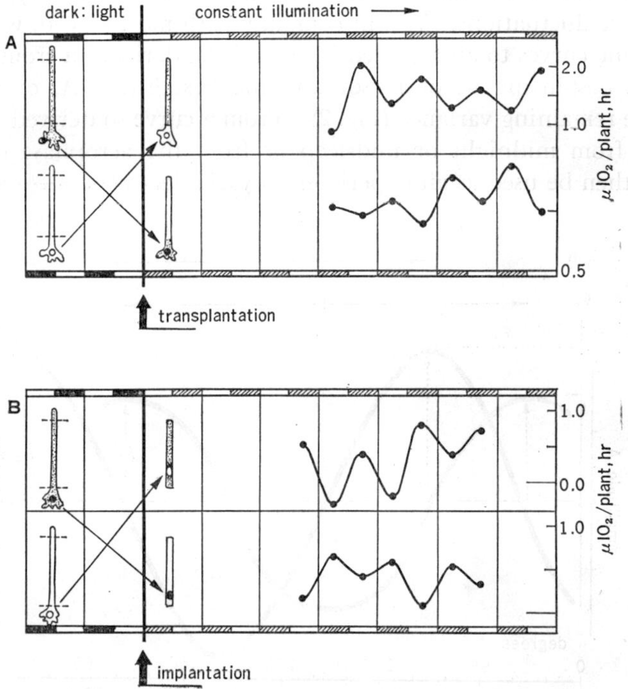

Fig. 1 The nucleus and cytoplasm of acetabularia have been given opposite phases of circadian information by lighting schedules. This information is carried by the transplanted nucleus. (Reprinted with permission from SCHWEIGER et al. [1])

III. *Methods of study*

Before I proceed to consider chronobiology in medicine, I should like to mention briefly, what is clearly a work of supererogation for this group, methods available for study of these rhythms, a number of which we shall introduce in a few moments for the study of a rhythm in medicine.

1) The classical method of removing all known synchronizers and observing the change of period or Tau with free-running is beautifully illustrated in the classical experiments of HALBERG with blinded mice [2]. Such studies convinced almost everyone that free running rhythms, in fact, exist.

2) The change of one rhythm with a change of a "causal" rhythm represents the classic method of endocrinology, and there are innumerable examples. If you remove the pituitary, the ACTH rhythm vanishes and with it, the cortisol rhythm; the urinary potassium rhythm is vastly modified or eliminated.

3) A change of phase for the whole organism, as easily produced by trans-meridian flight, allows detailed analysis of the rate of change of rhythms: if there is more than one rate (or direction) of change, there must be more than one "primary" rhythm.

4) A change of period of Tau for the whole organism, as in the use of the 21-hour day of Spitsbergen, is of value in allowing separation of rhythms that can be entrained to a new schedule from those that stay at the old schedule, or free-run. The task of analyzing the effects of such change of Tau has been very difficult.

5) In any event, our problem begins with the detection of a rhythm as opposed to non-periodic fluctuations. We, in company with most of you, will explore the fitting of cosine curves to an assumed Tau, and the derivation from such a cosine curve of the best-fitting M or mesor and the best fitting A, or amplitude, to minimize the remaining variance (Fig. 2). From a curve so derived, the acrophase as it differs from midnight or midsleep or from the acrophase of some other rhythm can then be used as an experimental variable.

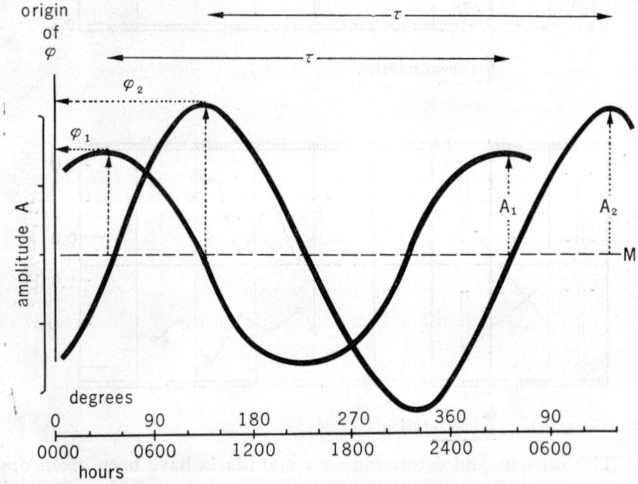

Fig. 2. Schema depicting the fitting of cosine curves to assumed Tau's. (Reprinted with permission of Dr. Alain Reinberg.)

I spoke of rhythms whose study is indispensable for good medicine. I shall consider only two, that of the blood sugar and that of the blood pressure. I suggest that failure to take these rhythms into account may result in failure to make a diagnosis, failure to find the best treatment, and ultimately failure to keep the patient alive.

Recently, a group of physicians reported that some patients may show a glucose tolerance test which is frankly "diabetic" in the afternoon, even when their morning glucose tolerance test is normal [3]. This group of physicians, having no concept of periodicity, presented their results (Fig. 3) showing convincingly

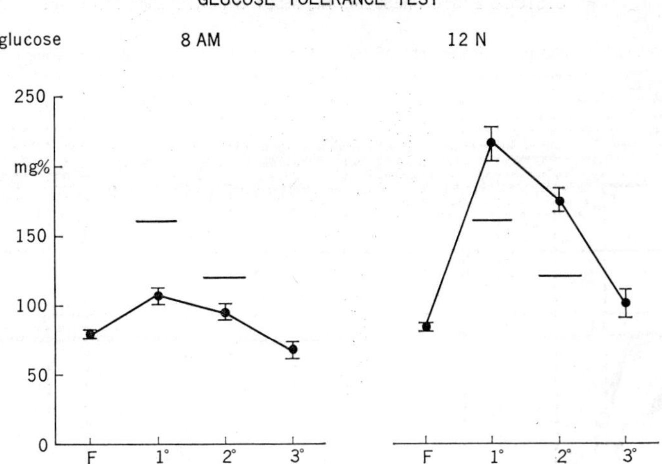

Fig. 3 The results of glucose tolerance tests given in the morning and in the afternoon. The horizontal lines at one and at two hours represent the normal values for these times in 8 a.m. to noon tests. Curves plotted from data of BOWEN et al. [3].

that their 25 patients had diabetes in the afternoon and would have escaped diagnosis with standard glucose tolerance tests in the morning. The omissions from this study are even more impressive than the results. For example, no control patient was studied! Accordingly, the results leave us no evidence with which to conclude that you and I are not also "afternoon diabetics". The implications are obvious.

The remainder of my remarks concern the question of whether periodicity is important for hypertension: I shall adopt the second view of the eminent authority quoted earlier. Fig. 4 shows, in a 61-year-old gentleman admitted for study to the NIH, the blood pressure as determined "around the clock". By conventional standards, this patient is clearly normotensive every morning. Yet the blood pressure determined each day at 6 in the afternoon provides equally convincing evidence that this patient is a hypertensive. What, then, is needed to establish a

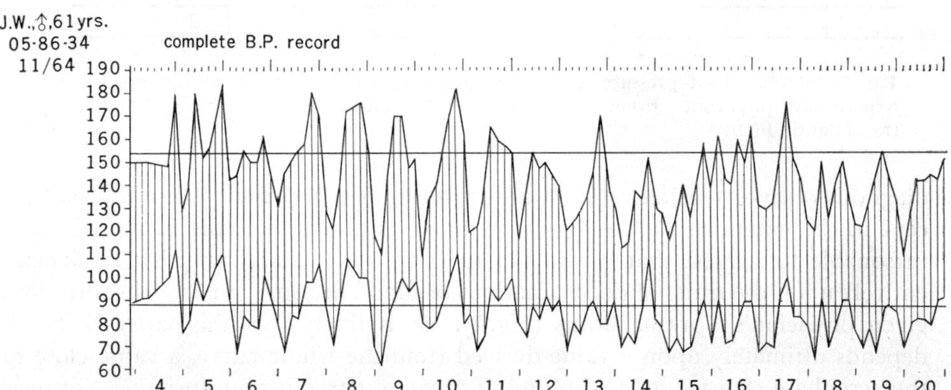

Fig. 4 Blood pressure measurements taken six times each day for 17 days in a 61-year-old patient.

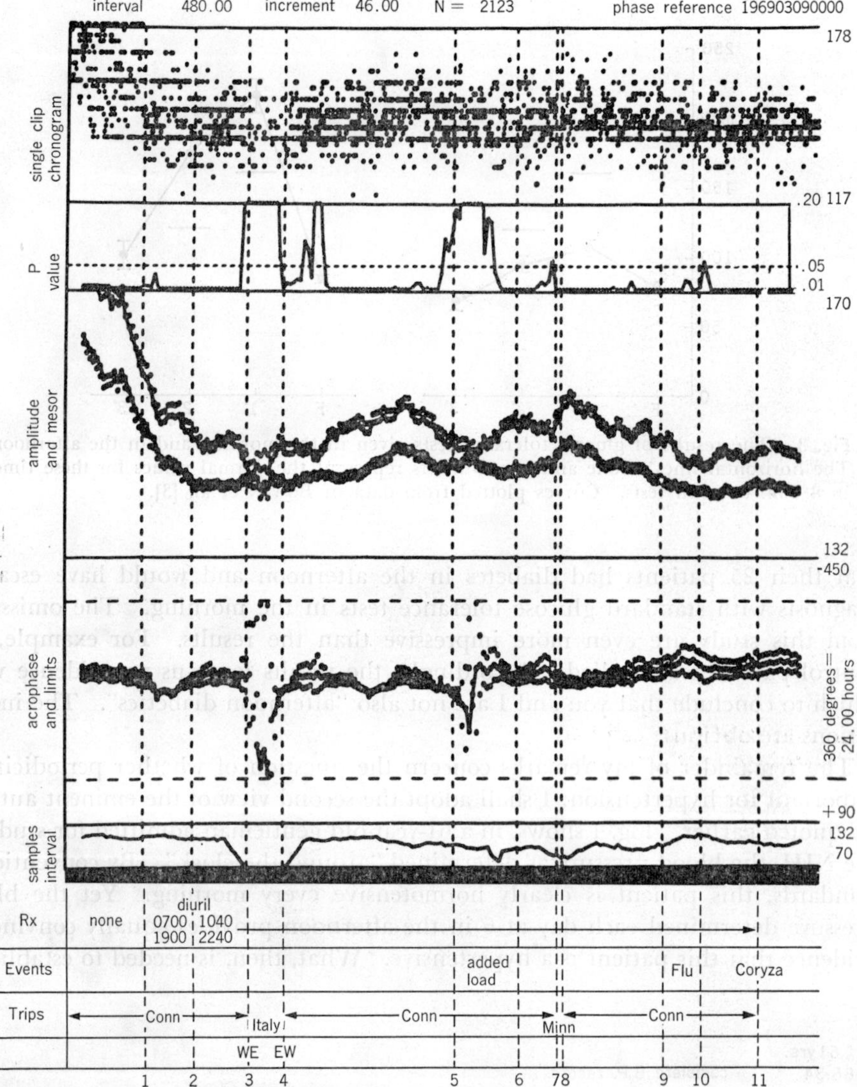

Fig. 5 Systolic blood pressure, and running averages for A and M and acrophase in a hypertensive physician. Event lines, 1–4, show start and time change for medication, and transatlantic flights.

diagnosis of hypertension? (I can assure you that this patient is not unusual except in the extent of the swings of his blood pressure.) In the first place, it is reasonable to suggest that we do not need a *mean* value, for this will depend entirely upon the times of determinations of blood pressure, and this in turn on a given physician's working hours (Fig. 2). It is likely that the patient's health depends ultimately upon a value derived from the whole curve, a value close to the acrophase or a value of M-plus-A if a cosine curve is appropriate. Not only may this be the most important determinant of the patient's state of health, but it may be the most important determinant of treatment. A given patient with hypertension may be well-treated if the *peaks* of pressure, which offer the greatest threat to his cardiovascular system, can be eliminated. Accordingly, a treatment program should also take into account a curve, be it a cosine curve or some better one. Finally, the cosine curve or the like, in which the parameters of periodic behavior are measured, may be used to study the rhythms in a given patient and again to investigate the dependence of one rhythm upon another.

I would like to illustrate a number of these points by reference to a series of meticulous studies on (and largely by) Dr. Howard LEVINE who has determined his blood pressure and a number of other variables several times a day for over two years. Fig. 5 shows the effect on the A and M of his blood pressure, of a change in rhythm-adjusted treatment [4]. At event line 2 only the hour of treatment (chlorothiazide) was changed. At event line 3 (Fig. 5) he flew to Europe (phase advance, hours or degrees); at event line 4 he returned.

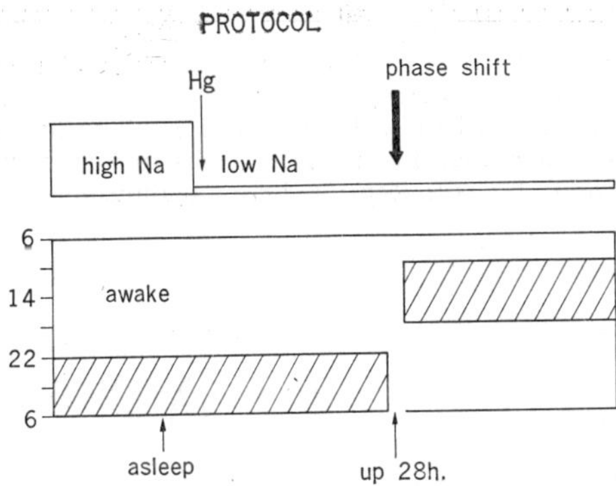

Fig. 6 Protocol of study on a patient with hypertension. Data are seen in Figs. 7 and 9.

Dr. LEVINE then entered the hospital for a protracted study according to the protocol shown in Fig. 6 [5], q.v. This allowed us to study the effect of sodium intake on the A and M of blood pressure and a number of other variables. Finally, we studied the effect of a phase shift of the whole organism, undertaken in a hospital ward by retarding his total activity and bed-time by 12 hours, on the acrophase and the rate of phase shift of a number of physiologic indices, urinary constituents, and salivary constituents. Fig. 7 illustrates one tentative observation derived from such a study. By the method of rolling averages, or serial-section for consecutive five-day periods, the acrophase for urinary calcium, phosphorus and chloride was followed. The *advance of phase to midnight* in the case of phosphorus and the *retardation to midnight* in the case of chloride suggests that these two rhythms must clearly depend upon different controlling factors.

Here I must introduce a note of warning concerning the enormous complexity of the physiological rhythms under your program discussion. Fig. 8a suggests a *minimum* number of factors that can control urinary chloride, considered as a resultant cyclic variable. Thus, at least the glomerular filtration rate, the extracellular fluid volume, and aldosterone secretion, all known to be periodic functions with a circadian period, have as their resultant the urinary chloride. Similarly, the urinary phosphorus is a variable of perhaps even greater complexity (Fig. 8b). Thus, at least the plasma phosphorus, the rate of ingress to plasma of

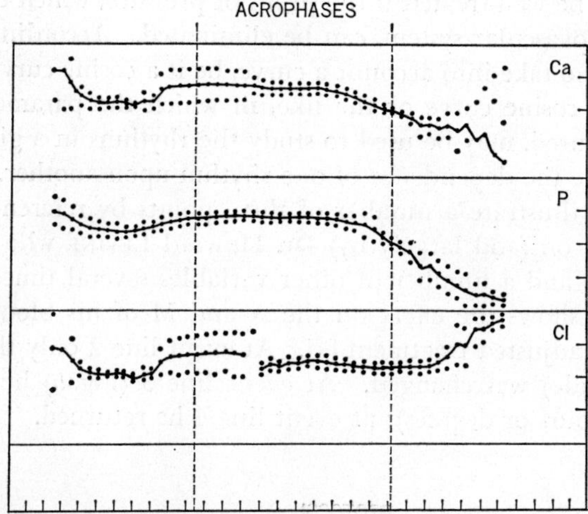

Fig. 7 Acrophases for urinary calcium, phosphorus and chloride as determined by the method of rolling averages or serial-section for consecutive five-day periods in the patient with hypertension. The first vertical dotted line corresponds to a change in diet from high- to low-sodium, the second to a phase shift of 12 hours. (Reprinted with permission from MEYER et al. [5]).

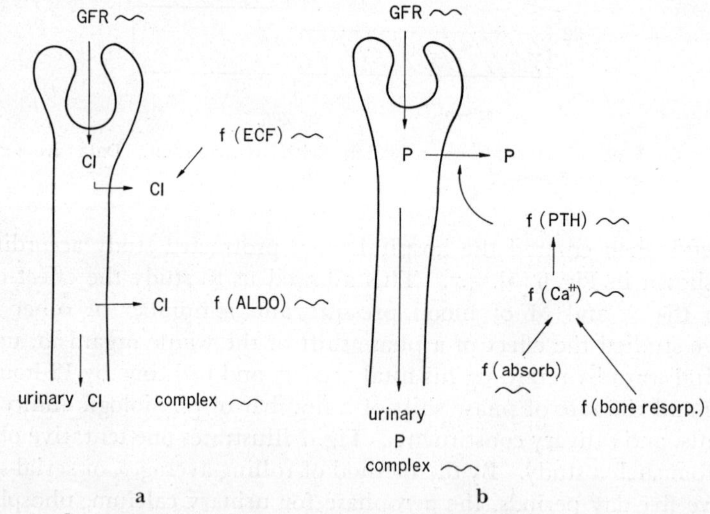

Fig. 8

a: Schema showing some variables affecting urinary chloride.
b: Schema showing some variables affecting urinary phosphorus.

phosphorus from cells or gastrointestinal tract, and the plasma parathyroid hormone control the urinary phosphorus. In addition, the analysis cannot stop here, since the function of parathyroid hormone is controlled at least by the plasma calcium, itself a periodic function.

For the patient under discussion, Fig. 9 shows the mean blood pressure as a function of time on the three regimens, high-salt, low-salt and low-salt with phase shift. My plea today is that information contained in such curves become a routine minimal amount of information accepted for the description of a patient's

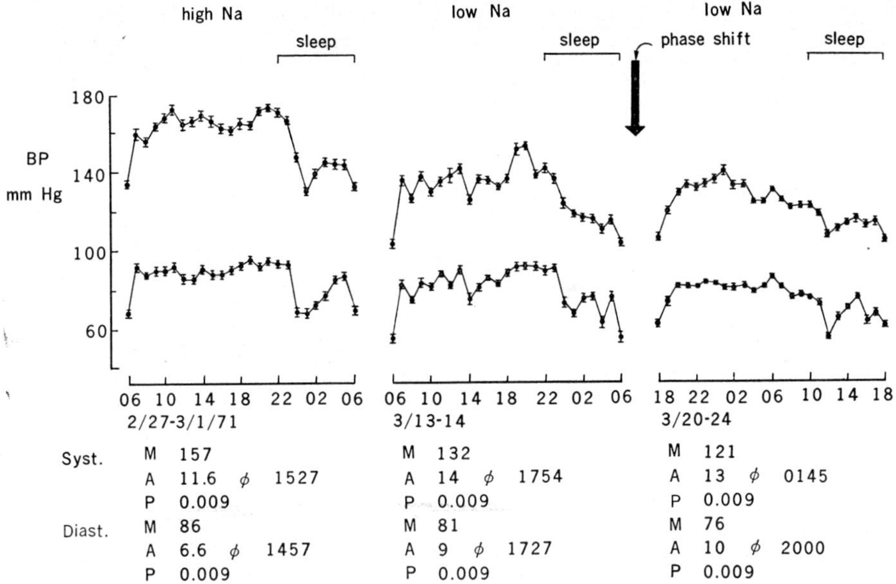

Fig. 9 The mean blood pressure of the patient with hypertension as a function of time on the three regimens: high-sodium, low-sodium and low-sodium with phase shift. (Reprinted with permission from MEYER et al. [5]).

blood pressure. The analysis of this information by cosinor should become a routine. It is essential that enough information be collected to allow objective characterization of a periodic phenomenon, to wit, an estimate of M as given for the three statuses in this patient, an estimate of A itself, and finally an estimate of acrophase, ϕ. In this way, a patient can be compared with himself at another time, or under another treatment, and the patient can be compared with a normal or with another patient.

REFERENCES

1. SCHWEIGER E., WALLRAFF H. G. and SCHWEIGER H. G. (1964): Endogenous circadian rhythm in cytoplasm of acetabularia: influence of the nucleus. Science, *146*: 658–659.
2. HALBERG F. (1960): American J. Mental Deficiency, *65*: 156–171.
3. BOWEN A. J., REEVES R. L. and NIELSEN R. L. (1968): Afternoon diabetes. J. American Medical Women's Association, *23*: 261–264.
4. LEVINE H. and HALBERG F.: Circadian rhythms of the circulatory system: Literature review; computerized case study of transmeridian flight and medication effects on a mildly hypertensive subject. Special Report, U.S.A.F. School of Aerospace Med. 1972, (in press).
5. MEYER W. J., DELEA C. S., LEVINE H., HALBERG F. and BARTTER F. C. (1974): A study of periodicity in a patient with hypertension: relations of blood pressure, hormones and electrolytes. *In*: Chronobiology. Proc. Symp. Quant. Chronobiology, Little Rock, 1971, (L. E. SCHEVING, F. HALBERG and J. E. PAULY, eds.), pp. 100–107, Igaku Shoin Ltd., Tokyo.

CHRONOBIOLOGY AT THE CELLULAR LEVEL

Chairman: Erhard HAUS

DIURNAL RHYTHMS OF HEPATIC ENZYMES FROM RATS ADAPTED TO CONTROLLED FEEDING SCHEDULES

P. Roy WALKER* and R. Van POTTER

McArdle Laboratory, The Medical School, University of Wisconsin Madison, Wisconsin, U.S.A.

INTRODUCTION

Biological rhythms at the molecular level in mammals should be examined not merely in terms of varying light and darkness in a constant chemical environment as in a chemostat but we should bear in mind the long evolutionary history of the species. Not only were circadian patterns of food gathering and activity evolving, but the constraints on the *capacity* of the organism to store energy and to draw upon alternative energy reserves were evolving to match the circadian and other rhythms in food supply and temperature in the supporting environment. Thus the capacity of the liver to store glycogen is limited, and enzyme patterns in the adult rat liver reflect the energy movement to and from the fat depots and the demands placed upon the liver depending on whether carbohydrate is constantly available, restricted in supply, restricted in availability, or both. The biological rhythms of certain adaptive enzymes vary widely in absolute amounts and may undergo wide variations in the daily oscillations both as to periodicity and amplitude, *even with no alteration in the composition of the diet fed,* merely by altering the availability of the diet. These adaptive changes must be looked upon as the various heroic measures that must be instituted to maintain an integrated system of basic homeostatic mechanisms, e.g., the regulation of blood glucose. The supporting components include a variety of hormones: the glucocorticoids, insulin, glucagon and others, and must be reflected in ketosteroid and electrolyte excretion. All of these parameters may be looked upon as molded by and exploitative of the circadian rhythm because the organism has been built to maintain reasonable homeostasis throughout the exigencies of each 24-hour day, *one way or another.* We have undertaken to describe some of the concomitant changes in liver in rats that are on various controlled feeding schedules and are also maintained under controlled lighting conditions [1, 6, 7, 11, 12].

EXPERIMENTAL

All the animals used in these experiments were adult male Sprague-Dawley rats which were trained on the "8+16" or "8+40" feeding schedules [6, 11] as required. The animals were kept in air-conditioned rooms under constant temperature, noise and humidity levels. Tyrosine aminotransferase (TAT) was assayed as described by Watanabe et al. [11], ornithine decarboxylase (ODC) as

* Present address: Department of Biochemistry, University of Sheffield, Sheffield S10 2TN, England.

described by FAUSTO [3], glucokinase (GK) by the method of WALKER and PARRY [10] and hydroxymethylglutaryl CoA (HMG CoA) reductase by the method of GOLDFARB and PITOT [4]. Basal values refer to enzyme levels at the end of the light period which is a point 16 hours after the end of an 8 hour feeding period.

RESULTS AND DISCUSSION

Fig. 1 shows changes in the activities of several enzymes during the "8+16" feeding cycle. Fig. 1a shows the changes in the activity of *HMG CoA reductase* which is an important regulatory enzyme of cholesterol biosynthesis in the liver. A similar pattern was also observed for *ornithine decarboxylase,* which catalyses the first step in a series of reactions leading to the biosynthesis of the polyamines-

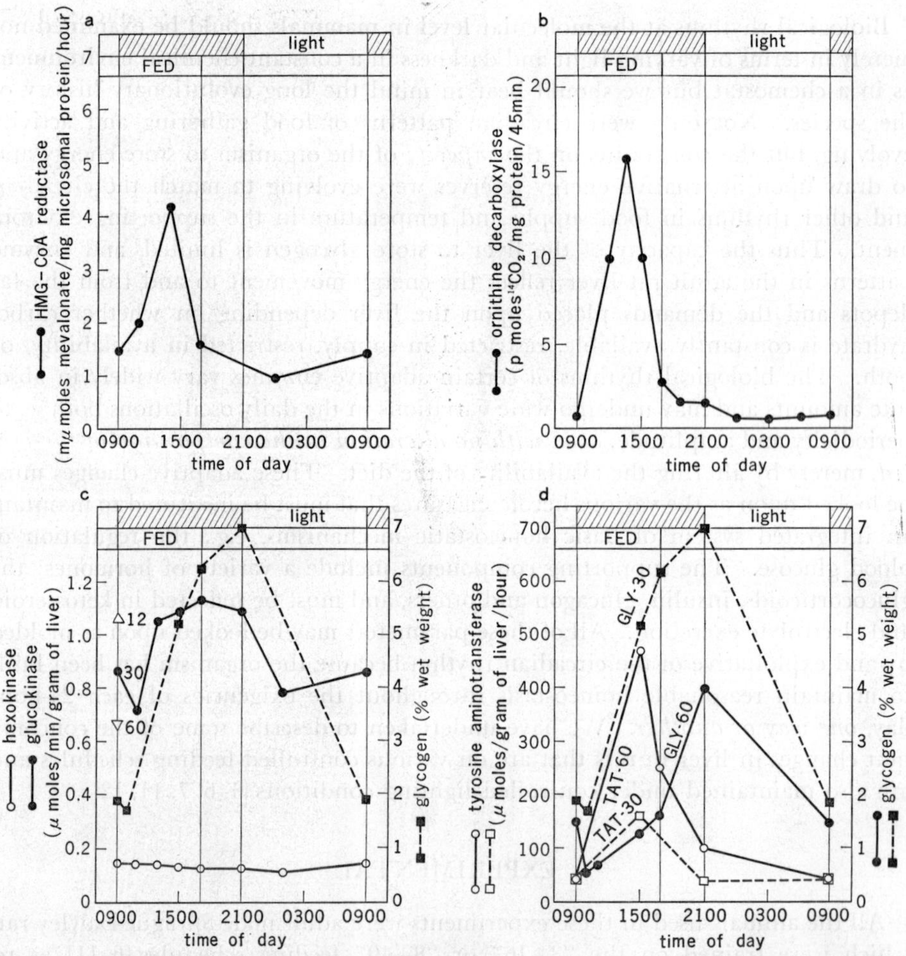

Fig. 1 Diurnal activity patterns of several hepatic enzymes and glycogen from rats trained on the "8+16" feeding schedule. (a) HMG CoA reductase data from GOLDFARB and PITOT [5]; (b) Ornithine decarboxylase data front BARBIROLI (unpublished); (c) Hexokinase, glucokinase and glycogen data from WATANABE et al. [11]; the numbers 12, 30 and 60 refer to the percentage of protein in the diet; and show the basal activities in the case of glucokinase; (d) glycogen and tyrosine aminotransferase data from WATANABE et al. [11] for two groups of animals fed either 30% or 60% protein.

putrescine, spermidine and spermine (Fig. 1b). There was a sixteenfold increase in enzyme activity from basal levels at the onset of feeding to a maximum at four and a half hours of feeding followed by an equally rapid decrease in enzyme activity back to basal levels by the end of the feeding period.

A different pattern was observed for *glucokinase* during the same 24-hour period as shown in Fig. 1c. GK is the initial enzyme of glucose metabolism in the hepatocytes and is thus of key importance in energy metabolism. The enzyme activity reached a peak at four hours after the onset of feeding and remained high until about four hours after the end of the feeding period. The changes in liver glycogen during this time are also shown in Fig. 1c and there is a close parallelism between the changes in GK and the increase in one of the products of glucose metabolism, glycogen, which accumulated during the twelve hours following the onset of feeding. The glycogen was then gradually depleted during the remaining twelve hours of fasting and GK returned to the basal level. The basal values for GK at different levels of dietary protein (12%, 30% and 60% with glucose at 79%, 61% and 31% respectively) are also given and it may be seen that the nature of the food also affects the nature of the response of an enzyme. This is more clearly demonstrated in Fig. 1d which shows changes in the activity of TAT at two levels of dietary protein. TAT is the initial enzyme in the catabolism of tyrosine and is extremely sensitive to the levels of protein and carbohydrate in the diet. At a protein level of 60% there was a rapid increase in the activity of the enzyme commencing at the onset of feeding and reaching a peak after six hours of feeding. There was then a rapid decline in activity and the basal level was essentially reached by the end of the dark period. If the protein content of the diet was lowered to 30% the magnitude of the change in enzyme activity was greatly reduced although a similar periodicity was observed. Similarly, changes in glycogen are also affected by the dietary level of carbohydrate. The amount of glycogen stored by the animal was found to be proportional to the level of carbohydrate in the diet [11].

In addition to these four enzymes many other examples of diurnal rhythms on this feeding regimen are known [1, 7, 11, 12]. Some enzymes, however, show no change during the "8+16" feeding cycle; examples include hexokinase (Fig. 1c), aldolase and pyruvate kinase (Walker and Potter, unpublished observations). In general the enzymes at the beginning of a particular metabolic pathway tend to show the greatest diurnal changes. Thus there is a great deal of variation in the response of various hepatic enzymes to daily changes in lighting and food. The particular pattern of a given enzyme must be influenced by the metabolic function of the enzyme, the rate of synthesis and halflife of both the enzyme and its corresponding mRNA, and the nature of the feedback controls on the enzyme. The composition of the diet also greatly affects the response of any given enzyme.

The importance of the feeding period in setting up and maintaining these rhythms is shown in Fig. 2. In this experiment, rats were trained on the "8+16" feeding schedule and assays were carried out during the normal feeding cycle and also during the following day when the rats received no food. Typical diurnal responses were obtained for ODC, TAT and GK during the normal feeding period. However, during the 24-hour period when food was not available ODC and TAT showed no change in activity with values remaining at the basal level throughout. There was a small response with glucokinase which showed a slight increase by four hours after the onset of darkness but a peak was not established and the

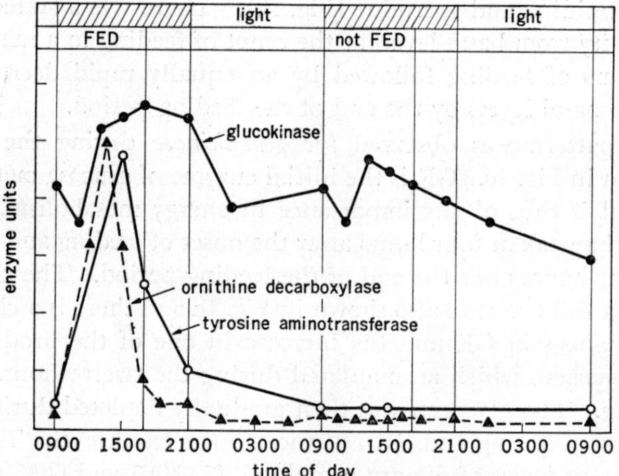

Fig. 2 Effect of fasting on the diurnal rhythms of ODC, TAT and GK in livers of rats trained on the "8+16" feeding schedule. The enzyme activities are plotted on the same scales as shown in Fig. 1.

enzyme activity then decreased gradually throughout the rest of the period studied. Thus rats trained to a "8+16" feeding schedule establish characteristic diurnal rhythms in hepatic enzymes to metabolize the dietary components adsorbed during each feeding period. When food is not available the response of some enzymes is abolished. The fact that there was a response with GK suggests that although the food is instrumental in setting up the characteristic rhythms other factors may be involved in maintaining these rhythms. Thus a humoral or neural signal anticipating food may trigger an increase in GK, although the increase is not maintained in the absence of food.

The involvement of hormones in maintaining diurnal changes in hepatic TAT is suggested by the data described in Fig. 3. The chart shows two successive 24-hour periods for five different groups of rats. One group was on the normal "8+16" feeding schedule and showed a characteristic peak on both days. The second group was also on "8+16" but was not fed the second day, thus these rats showed a peak on the first day and no peak on the second day. A third group was trained on a "8+40" feeding schedule—i.e. the rats were "fasting adapted", i.e. *trained* to receive only one meal in a two day period. This group showed a different pattern from the "8+16" group which were unfed on the second day in that there was a peak on the day without food. After the first peak during feeding the enzyme level did not return to the basal level, but remained at a higher value and then increased to give a second peak at about the onset of the dark period, after which time the activity decreased to the true basal level. Thus there was an increase in enzyme activity in the absence of dietary stimuli. The fourth and fifth groups of rats were trained on the "8+16" schedule and fed the first day but not the second day and instead were injected with glucagon (plus theophylline) at 9:00 on the second day [7]. This produced a second peak of activity. In the group receiving the lower dose of theophylline the peak coincided with the normal "8+16" peak, whereas with the larger dose there was an overshoot to give a peak about two hours later, followed by a decrease back to basal levels in both groups. WATANABE et al. [11] have shown that adrenalectomy abolishes the second peak of activity on the "8+40" schedule, whilst the first peak

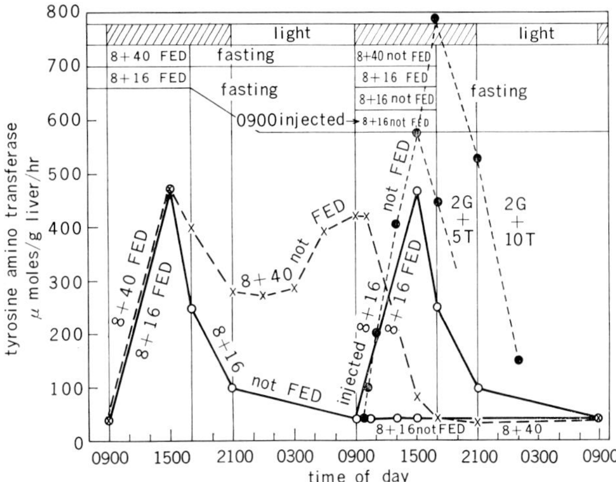

Fig. 3 Diurnal patterns of hepatic TAT as affected by changes in the feeding schedules and hormonal status of the animals. The data shown are from WATANABE et al. [11] and SCOTT et al. [7. 8].

"8+16" feeding schedule

"8+16" feeding schedule (showing data from animals not fed on the second day ± glucagon and theophylline)

"8+40" feeding schedule

2G+5T and 2G+10T refer to injection doses (mg per 100 g body weight) for glucagon and theophylline (see SCOTT et. al. [7]) given at 0900 in the absence of food.

was not affected. The influence of hormones on hepatic TAT rhythm has also been studied extensively and reviewed by BLACK and AXELROD [2].

These results show that diurnal rhythms in hepatic enzymes are established in response to training rats to particular feeding schedules. With a constant photoperiod the enzyme patterns are influenced by the length and also the frequency of the feeding period. SZEPESI and FREEDLAND [9] have shown that several enzymes show diurnal changes in response to training rats on a "1+23" schedule, including pyruvate kinase which showed no change on the "8+16" feeding schedule. The nature of the diet is also important as evidenced by the differences observed with different levels of dietary protein and carbohydrate. The studies on the effects of glucagon on TAT suggest that hormonal factors are also involved in maintaining certain diurnal rhythms. The extent of the influence of the hormones seems to depend on the nature of the external dietary stimuli. Thus several factors are involved in establishing and maintaining diurnal rhythms in hepatic enzymes and different combinations of factors are involved for any given enzyme.

ACKNOWLEDGEMENTS

The authors with to thank Drs. Bruno BARBIROLI and Stanley GOLDFARB for supplying the ODC and HMG CoA reductase data, respectively and to Drs. David SCOTT and Robert REYNOLDS for unpublished TAT data. This work was supported in part by a grant (CA-07175) from the National Cancer Institute. Roy WALKER is a recipient of a Damon Runyon Memorial Cancer Research Fund Postdoctoral Fellowship.

REFERENCES

1. BARIL E. F. and POTTER V. R. (1968): J. Nutr., *95*: 228–237.
2. BLACK I. B. and AXELROD J. (1970): *In*: Biochemical Actions of Hormones, (G. LITWACK, ed.), Vol. 1, pp. 135–155, Academic Press, New York.
3. FAUSTO N. (1971): Biochim. Biophys. Acta., *238*: 116–128.
4. GOLDFARB S. and PITOT H. C. (1971): J. Lipid Res., *12*: 512–515.
5. GOLDFARB S. and PITOT H. C. (1971): Cancer Res., *31*: 1879–1882.
6. POTTER, V. R., BARIL, E. F., WATANABE, M. and WHITTLE, E. D. (1968): Fed. Proc., *27*: 1238–1245.
7. SCOTT D. F., REYNOLDS R. D., PITOT H. C. and POTTER V. R. (1970): Life Sciences (Part II), *9*: 1133–1140.
8. SCOTT D. F., REYNOLDS R. D., PITOT H. C. and POTTER V. R.: Unpublished results.
9. SZEPESI B. and FREEDLAND R. A. (1971): Can. J. Biochem., *49*: 108–118.
10. WALKER D. G. and PARRY M. J. (1966): *In*: Methods in Enzymology, (S. P. COLOWICK and N. O. KAPLAN, eds.), Vol. 9, pp. 381–388, Academic Press, New York.
11. WATANABE M., POTTER V. R. and PITOT H. C. (1968): J. Nutr., *95*: 207–227.
12. WHITTLE E. D. and POTTER V. R. (1968): J. Nutr, *95*: 238–246.

CIRCADIAN VARIATIONS IN RAT LIVER MITOCHONDRIAL ACTIVITY

Karel M. H. PHILIPPENS

Department of Anatomy, School of Medicine, Hannover, Germany

A vast body of literature has appeared on circadian biology, but relatively few investigations have been made on the liver, the central organ in metabolism. Most studies indicate that time-order is a very important principle underlying the metabolic management, by which the liver is able to perform at optimal rates a diversity of vital functions.

Against the background of fluctuating activities, it seems likely that in the liver there are at some times prolonged periods of large energy consumption (e.g. by increased bile production, glycogen synthesis, synthesis of structural-, enzyme- and serum proteins) [1–6] and periods, during which the requirements for energy production at the cellular level are relatively low.

Chance and Williams, Klingenberg and Schollmeyer [7] described a mechanism controlling mitochondrial activity and energy production. By this mechanism any lowering of the $ATP/(ADP \times P_i)$ quotient will be compensated by an instant increase of the oxidative phosphorylation. However, our own studies suggest, that along a 24-hour period the liver cell system prepares itself for longer lasting time spans of increased energy consumption by enlarging the "respiratory capacity" of the mitochondrial apparatus. By this the liver will be able to sustain spans of higher metabolic activity. Such a system will prevent a possible exhaustion of the oxidative phosphorylation even under elevated metabolic strains. In the past we have been puzzled by the varying patterns and intensity obtained with histochemical reactions for succinic dehydrogenase in liver. As all technical factors were highly standardized, these variations could not be due to the hazards of the histochemical method. Systematic studies have revealed a circadian rhythmicity in these variables. These histochemical findings were confirmed by quantitative measurements, performed on the same material with a biochemical method developed by us, which is based on the histochemical NBT-indicator system according to Nachlas et al. [8, 9].

As reported before [10, 11] biorhythmic studies were then carried out with common biochemical techniques (succinate-cyt. c reductase and succin-oxidase determination) [12, 13, 14]. The results obtained were compared with a new method developed by us for the kinetic measurement of succinate-NBT reductase activity. All these methods revealed a circadian fluctuation of mitochondrial reactivity to succinate stimulation (Fig. 1).

The graph shows that the 24-hour excursions of mitochondrial and homogenate activity are very similar. But still, the different methods show small differences between the time points of maxima and minima. Furthermore the circadian fluctuations are expressed to a lesser extent with the NBT-indicator system than they do with cytochrome c reduction and oxygen consumption.

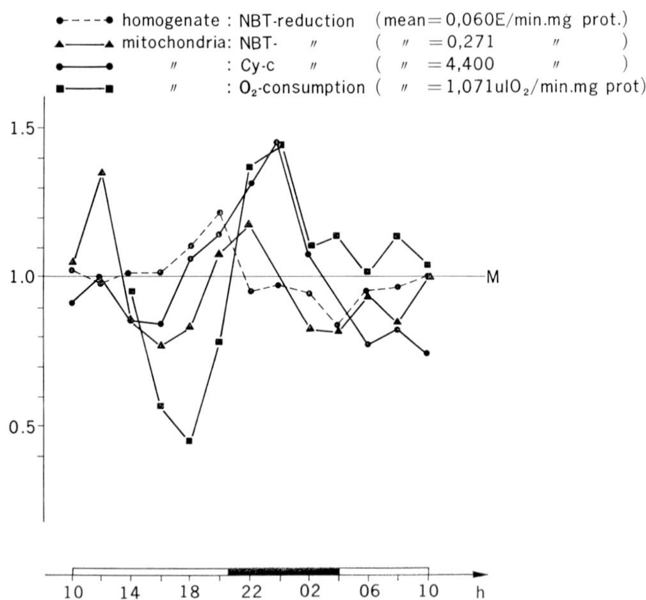

Fig. 1 Daytime dependency of rat liver succinic dehydrogenase system. Mean values of groups of 4 male rats (Wistar, AF-Han.) at each time are represented as relative values of the 24-hours mean (M).

These two aspects probably reflect the different properties of the final electron acceptors in respect to their site of reaction as well to their reactivity with the respiratory chain [15]. It hardly can be assumed that mitochondria operating as respiratory units may show heterogenous fluctuations of their constituting elements. However, the term "reactivity to succinate stimulation" expresses, that under standard incubation conditions the reduction of the final acceptors is solely limited by the enzyme system taking part in the electron transport. It permits no conclusion about the nature of the mechanism that enables mitochondria to function with varying intensity.

In other experiments, in which male Wistar rats of different inbred strains were kept on different nutritional conditions, the circadian fluctuation in mitochondrial respiratory capacity also was evaluated. It was shown, that these fluctuations successfully can be studied with the histochemical indicator system with added PMS. For the study of the biological aspects of a varying mitochondrial function in the living animal the inhibition of mitochondrial respiration by Antimycin-A was chosen. This poison is known to block electron transport at a crucial point i.e. at the cytochrome b level. Antimycin acts in vivo almost exclusively on liver succinoxidase, whilst the measurable effects on the activity in the heart, brain and muscle are minimal [16]. After Reif and Potter [17] the LD_{50} is 0.81 mg/kg body weight when given intraperitoneally to young female rats. These figures served as reference standard for the doses used to poison male albino (Wistar) rats with body weights ranging from 210 to 240 g. During the first 24-hours experiment every three hours groups of eight animals received Antimycin-A (1.5 LD_{50}) and the death ratio was recorded. At the same times four untreated animals were killed and there livers and blood taken for histochemical and biochemical analysis, so as to compare toxicological effects with the metabolic state of the normal liver. The results of this and the following experiments are represented in Fig. 2, and Table 1.

The maximal lethality occurred at 2200 h and 2400 h, when the animals are very active and the mitochondrial respiration with succinate and NADH is high.

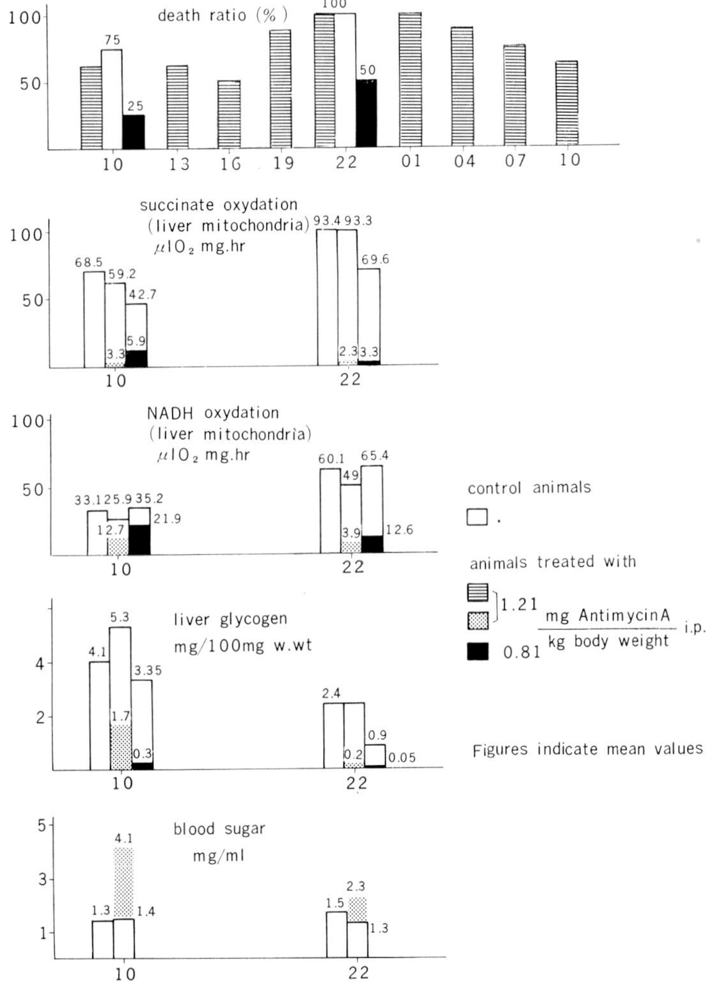

Fig. 2 Circadian variation of Antimycin A toxicity in the living animal (male albino rats, Wistar, Ivanovas inbred strain). Light period: 0700–1900 h. All biochemical parameters were analysed in groups of four untreated or treated animals.

The number of deaths decreases with decreasing mitochondrial activity when the animals are at rest. For practical reasons further experiments were performed at 2000 h and 2400 h only. In the last experiment 0.81 mg/kg was injected for intraperitoneal Antimycin-A intoxication instead of the higher doses used before.

At each time point the lethality of Antimycin-A was recorded. Furthermore, groups of four rats were injected with the same amount of Antimycin-A and sacrificed one hour later. The mitochondrial activity of their livers was measured and compared with that of untreated controls killed at the same times. In this

Table 1 Comparative representation of the effect of Antimycin A on mitochondrial respiration and glycogen content of the liver, and on blood sugar. Data based on the results in Fig. 2, second and third experiment.

		1000 h	2200 h
Inhibition of O_2- consumption (%)	NADH	44.6±9.8	84.1±5.3
m±s.e. (n=8)	Succ.	90.3±2.8	96.4±1.6
Glycogen depletion (%) m±s.e. (n=8)		80.4±16.8	83.4±1.1
Blood sugar increase (%) m±s.e. (n=8)		301.0±16.5	180.2±11.7

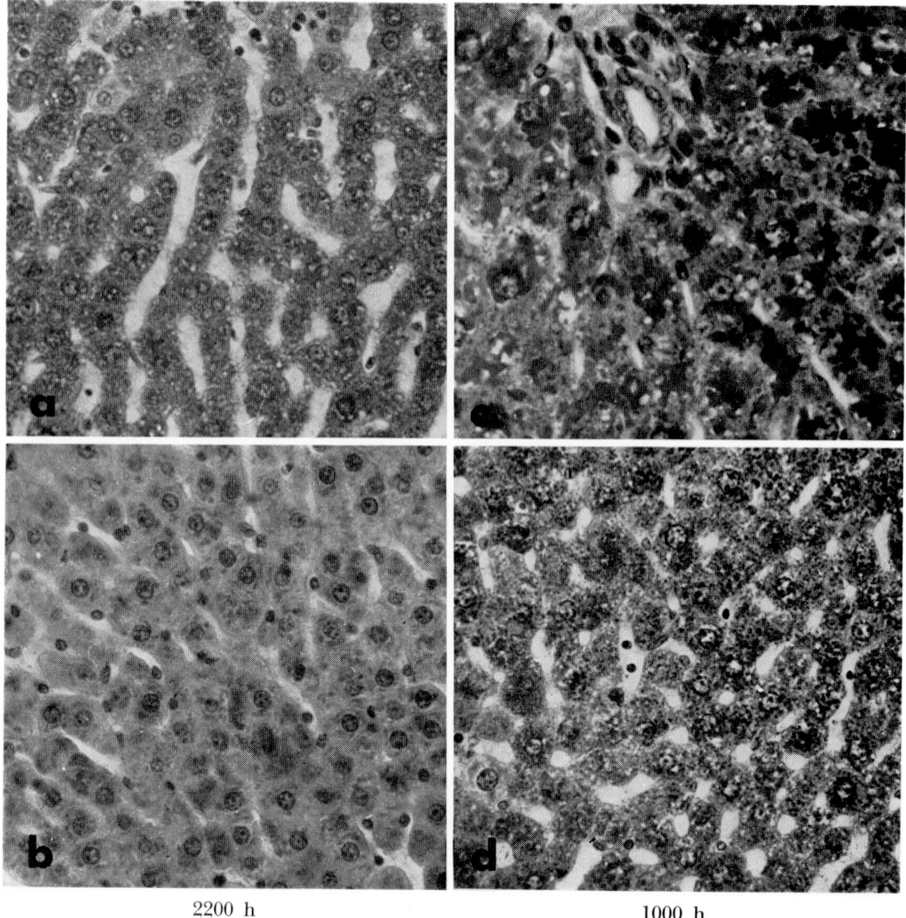

2200 h 1000 h

Fig. 3 Liver glycogen (PAS reaction) in controls (a, c) and one hour after Antimycin A (1, 5 LD_{50}, i.p.) treatment (b, d) at different times of the day. For comparison with biochemical data on glycogen, see also Fig. 2, second experiment.

experiment the same daytime dependent variations of toxicity were noticed, irrespective of the amount of Antimycin-A given.

These findings indicate that in general the animal organism is more resistant to Antimycin-A during light (1000 h) then at dark (2000 h). In the morning hours the mitochondrial activity in untreated animals is lower than at night.

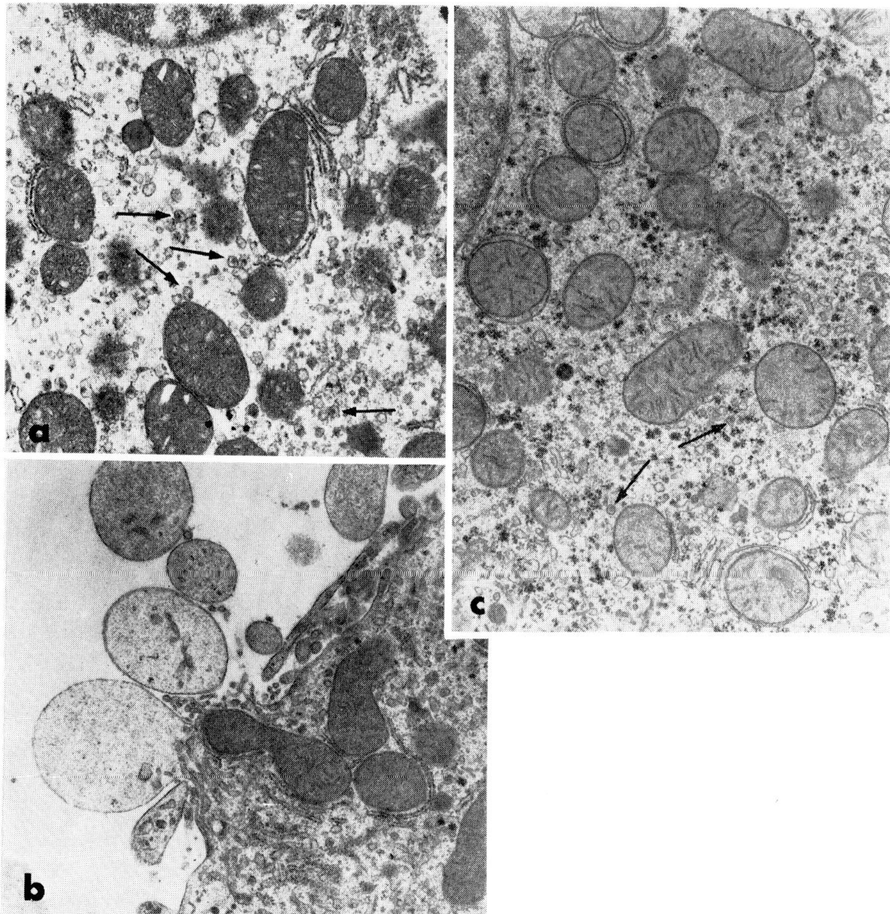

Fig. 4 Cell destructions in rat liver one hour after i.p. injected of Antimycin A (1, 5 LD_{50}). 2200 h (a, b): mitochondria are marked by a highly condensed matrix and extreme dilatation of the cristae. Some of them are incompletely surrounded by a dilatated ER. The cytoplasm is almost devoid of glycogen and shows numerous vesicles containing lipid droplets (arrows): Presumably these represent accumulations of chylomicrons resulting from an increased food intake at dark hours together with their lowered or completely inhibited metabolic processing due to the mitochondrial damage. However, they may also originate from the breakdown of lipid membranes because of an increased autophagic activity in the cell. Some cells show severe destruction with cytoplasmic protrusions that are cut off from the cell surface and give rise to intrasinusoidal, almost structureless cytoplasmic droplets (b). 1000 h (c): mitochondria occasionally show slightly dilatated cristae and have a normal density of the matrix. Many of them are encircled by an ER, which is marked by some dilatation and is partly free of ribosomes. Though glycogen has not disappeared, no glycogen areas characteristic for the liver cell at this time are seen. Vesicles are not so numerous and only few contain lipid droplets (arrows).

At both times the oxygen consumption with succinate is remarkably higher than with NADH. The activity of succinoxidase is inhibited to a much greater extent than the NADH-oxidase by the Antimycin-A intoxication of the animals. There is a marked time dependency in the sensitivity of both enzymes towards the respiratory inhibitor. At dark, when the mitochondria function at higher rates, the inhibitory effect is maximal whereas in the morning the lesser active mitochondria show lower inhibition.

As a consequence it seems reasonable to expect that metabolic damage of the liver induced by Antimycin-A will be most severe at 2200 h. At that time the greater vulnerability of the cellular respiratory system coincides with an increased requirement for its elevated activity due to an increase in endergonic metabolism. This seems to be expressed by an almost total depletion of liver glycogen under the influence of Antimycin-A (Figs. 2, 3a and b). At the same time there is an increase of blood sugar. The electron microscopical investigation of the same livers demonstrates heavy destruction of the hepatocytes (Fig. 4a–b).

These effects are much reduced after application of Antimycin-A at 1000 h. Corresponding to the lower inhibition of the mitochondrial activity, the glycogen depletion is deminished. Also the structural disorder of the cell is of a lower degree as seen ultramorphologically (Figs. 2, 3c and d, 4c).

There exists an increasing list of evidence of the fundamental capability of mitochondria to differentiate functionally in answer to changes in the organ specific and cell specific metabolic organization [18–23].

Our results add to this concept of a dynamic reacting system the important aspect of a circadian fluctuation in the constitution of the mitochondrial apparatus in the liver.

REFERENCES

1. Hamprecht B., Nüssler C. and Lynen F. (1969): FEBS Letters, *4*: 117
2. Mayersbach von, H. (1967): *In*: The Cellular Aspects of Biorhythms, (H. von Mayersbach, ed.), p. 87, Springer Verlag, Berlin.
3. Sestan N. (1964): Naturwiss., *51*: 371.
4. Lebouton A. V. and Handler S. D. (1970): FEBS Letters, *10*: 78.
5. Barnum C. P., Jardetzky D. and Halberg E. (1958): Am. J. Physiol., *195*: 301.
6. Phillips L. J. and Berry L. J. (1970): Am. J. Physiol., *218*: 170.
7. Lehninger A. L. (1965): The Mitochondrion, W. A. Benjamin Inc., New York.
8. Pearse A. G. E. (1960): Histochemistry. Theoretical and applied. 2nd edition, Little, Brown & Co., London.
9. Philippens K .M. H. (1970): Abhandl. Dtsch. Akad. Wissensch. Berlin, (herausgeg. von. L.-H. Kettler), S. 607, Akademie Verlag, Berlin.
10. Philippens K. M. H. (1968): IIIrd Int. Congr. Histochem. Cytochem., New York.
11. Philippens K. M. H. (1971): Acta histochem., (in press).
12. Kadenbach B. (1965): Biochem. Z., *344*: 49.
13. Hagihara B. (1961): Biochim. Biophys. Acta, *46*: 134.
14. Lessler M. A., Molloy E. and Shuwab C. M. (1966): Symposium on YSI Biological Oxygen Monitor, Shandon Scientific Co., London.
15. Oda T., Okazaki H. and Seki S. (1958): Acta Med. Okayama, *12*: 302.
16. Reif A. E. and Potter van R. (1953): Cancer Res., *13*: 49.
17. Reif A. E. (1953): Arch. Biochem. Biophys., *47*: 396.
18. Buno W. and Germino N. I. (1958): Acta anat., *33*: 161.
19. Kitzing W. und Schumacher H. H. (1961): Z. Zellforsch., *54*: 443.
20. Nielson R. R. and Klitgaard H. M. (1961): Am. J. Physiol., *201*: 37.
21. Gear A. R. L. (1970): Biochem. J., *120*: 577.
22. Shug A. L., Ferguson S., Shrago E. and Burlington R. F. (1971): Biochim. Biophys. Acta, *226*: 309.
23. Bohr H. P., Hundstad A.-C., Bianchi L. und Eckert H. (1970): Acta anat., *76*: 102.

CHARACTERIZATION OF THE CIRCADIAN RHYTHM OF MITOSIS IN THE CORNEAL EPITHELIUM OF THE IMMATURE RAT

Peggy J. GOODRUM, John G. SOWELL* and Sergio S. CARDOSO

Department of Pharmacology, University of Tennessee Medical Units Memphis, Tennessee, U.S.A.

The study and identification of many rhythmic patterns and variations of physiological phenomena within a 24-hour period have yielded a wealth of information concerning the importance of biological clocks in the daily existence of living organisms. Most of these studies have dealt with the adult organism. A few have dealt with the young, immature animal.

The results of many of these studies have established that for some physiological phenomena a definite circadian oscillation is present at birth [1, 2]. However, other studies have shown for some parameters that distinct differences between circadian patterns of the adult and young animal do exist [2–4].

Some physiological events such as the circadian mitotic rhythm have undergone more intensive investigation than others [5–9]. A paucity of information concerning circadian patterns of mitosis in newborn and immature animals remains to the present. Much of the original work suggests that a rhythm does exist from earliest life [2, 10, 11]. The limited amount of work in this area along with the inadequate sampling in some cases prompted us to investigate further this problem.

For this purpose experiments involving the corneal epithelium were designed (1) to determine whether a circadian mitotic rhythm was present or absent in the immediate days following birth, and (2) to establish the timing of a possible transition from an undetected to a fully developed circadian mitotic rhythm as seen in the adult animal.

METHODS AND MATERIALS

Pregnant female rats of the Holtzman strain were kept in our animal quarters at least seven days prior to the delivery of their pups. Each female was singly housed in the animal quarters which were on a 12:12 lighting system with light maintained from 0600 h to 1800 h. The temperature was maintained at 23°± 2° C. Food and water were made available *ad libitum*. Only those animals which were born between the hours of 0600 and 1800 on the day of delivery were selected for our experiments.

Each litter was maintained at the size of seven to eight animals with the remaining animals sacrificed the day after birth. No attempt was made to identify the sex of the pups.

Sacrifice was accomplished by decapitation. At sacrifice both eyes were re-

* Present address: Department of Pharmacology and Ophthalmology University of Alabama Medical Center, Birmingham, Alabama 35233.

moved from the animals and placed in ALFAC (a mixture of 85 ml of 80% ethanol, 10 ml of formalin, and 5 ml of glacial acetic acid). Whole cornea were stained by the Feulgen technique [6]. One hundred microscopic fields were scanned in each cornea, and the number of mitotic figures were recorded.

RESULTS AND DISCUSSION

The first series of experiments were designed to establish the presence or absence of a circadian variation of mitosis in the corneal epithelium of the young rat. Both ten day old and thirty day old rats were examined. Sacrifice in both cases began at 0700 on the respective days and was followed by the sacrifice of additional groups at four-hour intervals over the ensuing twenty-four hour period. Each time group contained five to six animals. The compiled data obtained from these experiments is contained in Fig. 1.

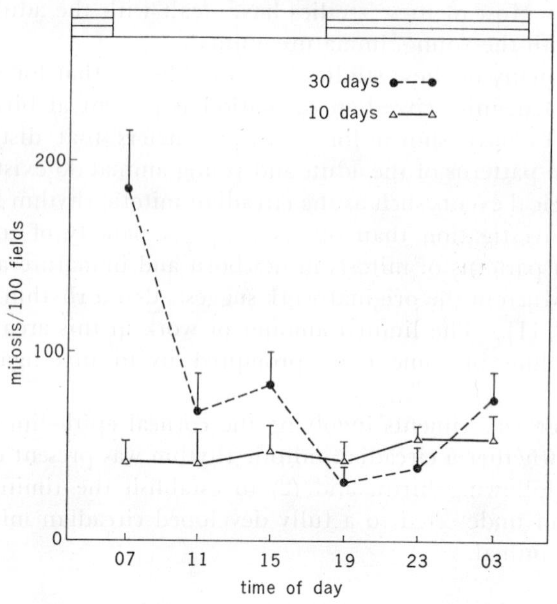

Fig. 1 Each point represents the mean value of mitosis/100 fields ± S.E. for a minimum of five animals.

As seen in the graph, thirty day old animals possess a definite circadian pattern characteristic of the adult rat [5, 6]. The crest of activity with a mean of 185 mitotic figures occurred at 0700 while the trough of activity occurred at 1900 with a mean value of 32 mitotic figures observed. In sharp contrast the cornea of the ten day old animal showed no significant oscillation of mitosis throughout the day.

Additional work was directed toward the determination of the day on which the circadian mitotic rhythm first occurred. As the thirty day old animals showed maximum and minimum levels of mitoses respectively at 0700 and 1900, groups of animals were sacrificed at these hours on selected days from day ten to day forty. Days fourteen and fifteen were especially selected as possible indicators of maturation processes that take place along with the opening of the eyes which

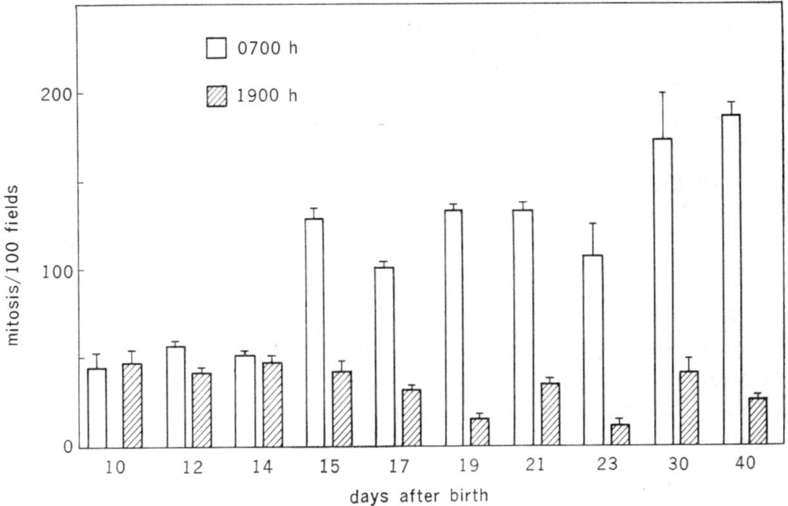

Fig. 2 Each bar represents the mean value of mitosis/100 fields ± S.E. for a minimum of five animals.

occurs in practically all the young rats by day fifteen. Light which is thought to be an important external cue could much more readily influence the circadian mitotic pattern following the opening of the eyes on day fifteen or shortly thereafter.

The results of this work showed that a steady rate of mitotic activity was maintained from day ten through day fourteen after birth with only minor differences between values found for the 0700 and 1900 on these days. From the values obtained on day fifteen it was surmised that this was the point at which a circadian fluctuation first occurred. The values found on day fifteen were comparable with those found in the older animals. Furthermore, it could be projected from these data that from day fifteen the rhythm maintained.

To verify the preceding results and to characterize further the circadian distribution of mitosis as is revealed in the cornea, a more thorough investigation was conducted. Animals were sacrificed on days fourteen, fifteen, and sixteen

Table 1 Circadian distribution of mitosis in the immature rat.

	Number of mitoses/100 Microscopic fields		
Sacrifice hours	Day 14	Day 15	Day 16
0700	64±9.8 (5)	128±11.1 (5)	188±20.5 (5)
1100	36±4.4 (4)	63± 3.1 (4)	74± 8.4 (4)
1500	50±4.7 (5)	56± 9.0 (5)	58±11.4 (5)
1900	29±3.0 (5)	28± 5.8 (5)	34± 6.5 (5)
2300	47±4.3 (4)	52± 6.1 (5)	51± 6.3 (5)
0300	49±1.2 (4)	86± 7.7 (5)	60± 5.3 (5)

± Standard error
( ) Number of animals in each group
Each value represents the mean value of mitosis/100 fields ± S. E. The number of animals in each groups is showed in parentheses.

after birth. Sacrifice was begun at 0700 on day fourteen and was continued at four-hour intervals over the next seventy-two hours. The results of this work are seen in Table 1.

On day fourteen there is not a definite daily variation. The values for day fifteen indicated a circadian rhythm with a peak value at 0700 and the trough value at 1900. The presence of the circadian mitotic rhtyhm on days fifteen and sixteen of these experiments along with the above discussed work indicated that once established the rhythm is maintained.

SUMMARY

Data have been presented here to indicate a) that in the corneal epithelium of the young rat, no circadian fluctuation of mitosis is present before day fifteen after birth and b) that once established the rhythm parallels that which is found in the adult.

REFERENCES

1. Honova E., Miller S. A., Erenkranz R. A. and Woo A. (1968): Science, *162*: 999.
2. Cooper Z. K. (1939): J. Invest. Dermatol., *2*: 289.
3. Franks R. C. (1967): J. Clin. Endocrinol., *24*: 75.
4. Allen C. and Kendall J. W. (1967): Endocrinol., *80*: 926.
5. Scheving L. E. and Pauly J. E. (1967): J. Cell Biol., *80*: 677.
6. Cardoso S. S. and Ferreira A. L. (1967): Proc. Soc. Exp. Biol. Med., *125*: 1254.
7. Liozner L. D. and Sidorova V. F. (1958): Byull. Eksp. Biol. Med. (transl.), *48*: 1532.
8. Golobova M. T. (1958): Byull. Eksp. Biol. Med. (transl.), *46*: 1143.
9. Golobova M. T. (1959): Byull. Eksp. Biol. Med. (transl.), *47*: 358.
10. Carleton A. (1934): J. Anat., *68*: 251.
11. Fortuyn-Van Leyden D. (1916): Proc. Soc. of Sciences, Amsterdam, *19*: 38.

DEMONSTRATION OF A CIRCADIAN RHYTHM OF NUCLEAR DNA IN LIVER*

JOHN R. RUBY,[1] LAWRENCE E. SCHEVING,[2] S. BRUCE GRAY[1] and KEN WHITE[1]

[1]*Departments of Anatomy, Louisiana State University Medical Center New Orleans, Louisiana, and*
[2]*University of Arkansas Medical Center Little Rock, Arkansas, U.S.A.*

Using animals that were maintained under strict controlled conditions, it firmly has been established that many chemical constituents of cells fluctuate in absolute amounts over a 24-hour period. A circadian rhythm of DNA also has been suggested [5, 7, 9, 12]. These latter findings have not been widely accepted because it is commonly believed that the DNA of a tissue whose cells are not dividing is stable and any suggestion to the contrary is thought to be influenced by technical factors. The present investigation reported herein was an attempt to determine the DNA content of liver tissue over a 24-hour cycle.

All animals used were adult male Sprague-Dawley rats which were maintained in a carefully controlled environment [17].

Since the experimental protocol was restrictive in allowing only a limited amount of time to isolate the nuclei, weak solutions of citric acid were used. Livers were homogenized in 0.05 M citric acid, filtered and washed numerous times in 0.015 M citric acid.

Prior to chemical analysis, aliquoits of the nuclear suspensions were counted with a Coulter electronic counter [16]. Adjustments were made so that all particles with a diameter greater than 2 micra were counted.

All DNA determinations were done by the diphenylamine technique of BURTON [3].

All results were analyzed by an inferential statistical method designed to estimate objectively a number of rhythmic parameters by fitting a 24-hour cosine curve to the data [8].

The results presented in this report are from four different experiments: one in June, one in July, one in November, and one in March. In the first two experiments, eight animals were sacrificed at each time point; in the last two experiments, ten animals were utilized at each time period. In the latter instances, five of the animals were injected intraperitoneally with thymidine-methyl-^{3}H (0.6 μC/g of body weight) exactly one hour prior to sacrifice.

Fig. 1 reveals the normal variation in rat liver weight over a 24-hour span during June and July. The acrophases occurred at $-132°$ (0848) and $-129°$ (0836) respectively. Similar curves were obtained in the November and March experiments; their acrophases occurred at $-107°$ (0708) and $-124°$ (0816) respectively. It is clearly shown that the weight of the liver can vary as much as 10% over the 24-hour period. It also has been noted that mild types of stress such as merely handling the animals, may produce an additional 10% drop in weight

* A more detailed description of this work has been reported previously [15].

condition of experiment	P	noise-to signal SE/A	mesor, M M±SE	amplitude, A A±SE	acrophase (.95 confidence arc) degrees
L06-18 D18-06 June 1969	.02	$.28_6$	10.67±.18	.90±.26	−132(−100, −164)
L06-18 D18-06 July 1969	.00	$.15_8$	10.74±.08	.71±.11	−129(−111, −147)

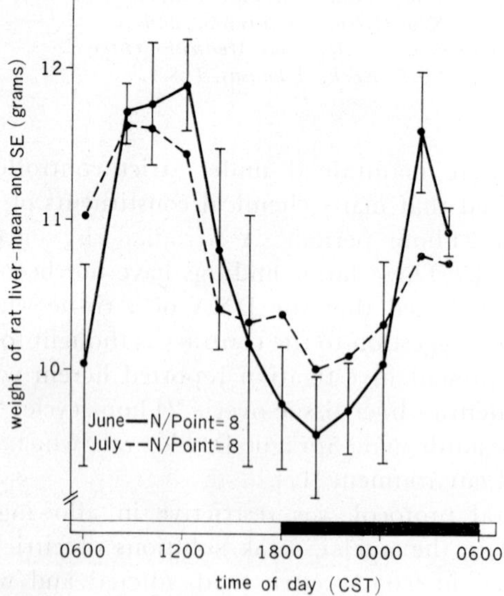

Fig. 1 Demonstrates the circadian fluctuation in rat liver weight. At the top of the graph is the summary of a fit of the data to a 24-hour cosine curve. In both instances there was a significant fit. (The acrophase reference was local midnight.)

fifteen minutes after the stress (unpublished results). Such variation in weight does question the reliability of analyses of rat liver components which are based on the amount per g or mg of liver wet weight. The results of the analyses demonstrating the fluctuation in the average amount of DNA per nucleus in suspension for the November and March experiments are recorded in Fig. 2. The acrophases of both studies occurred at similar times, −274° (1816) and −288° (1912) respectively. The differences of the high points in the late afternoon and low points during the early morning are statistically significant. The mean DNA content of the March experiment is 1.36 $\mu\mu$g higher than the experiment done in November. It is assumed this increase is due to the larger and older animal used in March. The average content per nucleus in both instances, however, remains within the range of values reported by others [6, 16].

The specific activity of DNA was determined in those animals which had received an injection of labelled thymidine. The isolated nuclei were washed twice in 0.5 N perchloric acid (acid-soluble fraction) and the DNA was subsequently extracted in hot perchloric acid. Aliquoits were taken for scintillation counting and DNA determinations. No significant activity was detected in the acid-soluble fraction which is probably due to the isolation technique. Thymi-

condition of experiment	P	noise-to signal SE/A	mesor, M M±SE	amplitude,C A±SE	acrophase (.95 confidence arc) degrees
L06-18 D18-06 Nov. 1969	.02	$.28_1$	9.09±.08	.40±.11	−274(−242,−305)
L06-18 D18-06 March 1970	.01	$.16_9$	10.45±.11	.90±.15	−288(−268,−308)

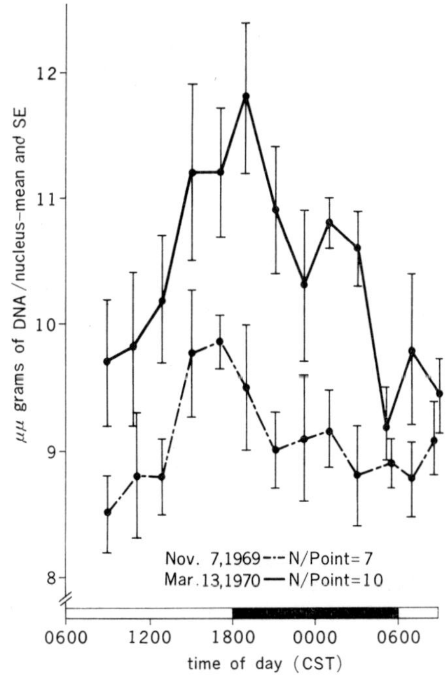

Fig. 2 Circadian fluctuation in the DNA of adult rat liver. The data in both showed a significant fit to a 24-hour cosine curve.

dine incorporation into DNA was noted at all points during the 24-hour cycle with a significant increase between 0200 and 0400 hours as indicated in Fig. 3. The precise acrophases were −84° (0536) in November and −71° (0444) in March. The sharp peak of incorporation also has been reported by others in adult livers [5], in livers of growing rat [1], and following partial hepatectomy [1, 4, 12]. In the latter two instances, this peak of incorporation can be attributed to replication of DNA prior to mitosis. In the adult liver, however, the mitotic index is very low [10]. Counts done during the present investigation on livers from the November and March experiments revealed maximum mitotic indices of 0.05% and 0.03% respectively which occurred between 0600 and 0800 hours and is approximately four hours following maximum thymidine incorporation. Since autoradiographic studies on these livers reveal some fully labelled parenchymal cell nuclei during the hours of peak incorporation, it is assumed that some of the activity indicates DNA replication prior to mitosis. It is doubtful, however, that the few nuclei undergoing mitosis can account fully for the high level of activity.

condition of experiment	P	noise-to signal SE/A	mesor, M M± SE	amplitude,A A±SE	acrophase (.95 confidence arc) degrees
L06-18 D18-06 Nov. 1969	.23	$.54_0$	21.53±.92	2.43±1.31	−84
L06-18 D18-06 March 1970	.05	$.34_9$	20.17±1.12	4.62±1.61	−71(−34,−109)

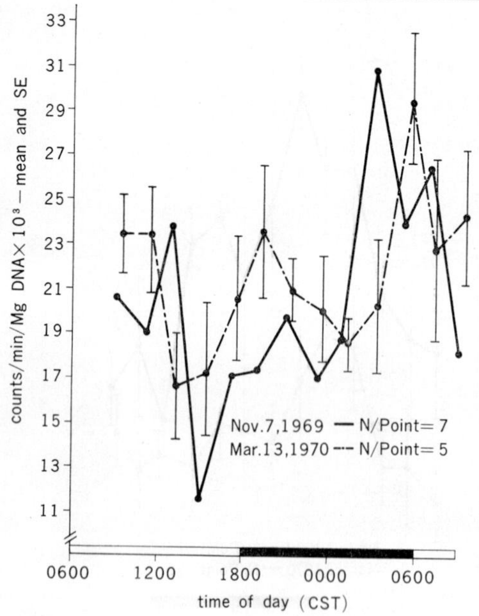

counts/min/Mg DNA× 10³ — mean and SE
33
31
29
27
25
23
21
19
17
15
13
11
Nov.7,1969 — N/Point= 7
Mar.13,1970 --- N/Point= 5
0600 1200 1800 0000 0600
time of day (CST)

Fig. 3 Circadian variation in ^{3}H-thymidine incorporation into DNA. Only the March data demonstrated a good fit to a 24-hour cosine curve.

One possible mechanism which could produce the inverse relationship between DNA content and ^{3}H-thymine incorporation would be a cyclic nuclease activity. A low nuclease activity during the light hours followed by decreased activity during the dark period would cause the variation in DNA content which we have reported. The peak incorporation of thymidine during the dark period is suggestive of DNA turnover. In this regard, these results tend to support the hypothesis of a "metabolic" DNA [13, 14]. The turnover-time of DNA in normal livers has been estimated to be 20–30 days at a rate of 3–5 per cent per day.

One other mechanism which could influence the average DNA content per nucleus is "ploidy shift" over the 24-hour period [2], although histological studies done during this investigation have not supported such an occurrence.

In addition, it is possible that there are cyclic variations in nuclear fragility, a factor which has not been fully explored in this investigation. In this case the results would be a reflection of non-random sampling of nuclei due to breakage. This would effect the average content of DNA per nucleus especially since the liver contains a large percentage of polyploid nuclei.

REFERENCES

1. BARBIROLI B. and POTTER V. R. (1971): Science, *172*: 738.
2. BUCHER O. and SUPPAN P. (1967): *In*: The Cellular Aspects of Biorhythms, (H. von MAYERSBACH, ed.), p. 126, Springer-Verlag, Berlin.
3. BURTON K. (1968): *In*: Methods in Enzymology, Vol. *12*, part B. p. 163, Academic Press, New York.
4. ECHAVE-LLANOS J.: Personal Communication.
5. ELING W. (1967): *In*: The Cellular Aspects of Biorhythms, (H. von MAYERSBACH, ed.), p. 105, Springer-Verlag, Berlin.
6. FALZONE J. A., BARROWS C. H. and YIENGST M. J. (1962): Exptl. Cell Res., *26*: 552.
7. HALBERG F., BARNUM C. P., SIEBERT J. and BITTNER J. (1958): Proc. Soc. Exp. Biol. Med., *97*: 897.
8. HALBERG F., TONG Y. L. and JOHNSON E. A. (1967): *In*: The Cellular Aspects of Biorhythms, (H. von MAYERSBACH, ed.), p. 20, Springer-Verlag, Berlin.
9. HORVATH G. (1963): Nature, *200*: 261.
10. JACKSON B. (1959): Anat. Rec., *134*: 365.
11. LOONEY W. B., CHANG L. O. and BANGHART F. W. (1967): Proc. Natl. Acad. Sci., *57*: 972.
12. MAYERSBACH H. v. (1967): *In*: The Cellular Aspects of Biorhythms, (H. von MAYERSBACH, ed.), p. 87, Springer-Verlag, Berlin.
13. PELC S. R. (1965): Exp. Geront., *1*: 215.
14. ROELS H. (1966): Int. Nat. Rev. Cytol., *19*: 1.
15. RUBY J. R., SCHEVING L. E., GRAY S. B. and WHITE K. (1973): Exptl. Cell Res., *76*: 134.
16. SANTEN R. J. (1965): Exptl. Cell Res., *40*: 413.
17. SCHEVING L. E. and PAULY J. E. (1967): J. Cell Biol., *35*: 677.

CIRCADIAN RHYTHM OF ORNITHINE DECARBOXYLASE ACTIVITY IN MOUSE LIVER; EFFECT OF ISOPROTERENOL AND CLELLAND'S REAGENT*

JOSEPH E. STONE, LAWRENCE E. SCHEVING, E. ROBERT BURNS and MAURICE GRAHAM

Departments of Pharmacology and Anatomy, University of Arkansas Medical Center Little Rock, Arkansas, U.S.A.

INTRODUCTION

Changes in liver ornithine decarboxylase activity have been associated with growth, RNA, and protein synthesis in that organ [1, 2]. The susceptibility of change in the activity of this enzyme in response to a variety of conditions and agents [3] suggested to us that rhythmic activity would be a possibility. This work was undertaken in conjunction with research involving isoproterenol (IPR), which is reported elsewhere [4].

MATERIALS AND METHODS

All mice were rigidly standardized for one week prior to and during the experiment. This consisted of food and water, *ad libitum* and a light-dark cycle (light 0600–1800 CST). Ninety animals conditioned in this manner were injected with 7.5 mgm of Isoproterenol (IPR) in 0.75 ml of distilled water or with only saline at 0900 and then divided into 18 subgroups of five mice each. The first subgroup was killed at 2100 (12 hours post injection) and thereafter control and IPR subgroups were killed every four hours throughout a 64-hour span. Thirty minutes prior to killing each animal was injected with 10 μCi of tritiated thymidine (Tdr-22 Ci/m mole).

The livers from the first 12 subgroups (saline or IPR) were removed and placed in ice cold saline weighed rapidly on a tissue balance and 20% homogenates were made with glass-Teflon Potter-Elvehjem homogenizers. The homogenates were centrifuged at $20{,}000 \times g$ for 30 minutes and the supernatants were decanted and stored at $-20°C$.

Ornithine decarboxylase activity was determined by the measurement of $^{14}CO_2$ released from I-^{14}C ornithine produced by incubation with the supernatants. The reaction mixture contained the following ingredients at pH 7.4 and a volume of 1.4 ml: Tris base, 200 μMoles; pyridoxal phosphate, 2 μMoles; DL-(I-^{14}C ornithine) HCl (specific activity 4.7 mC/m mole, New England Nuclear Corp.), 0.4 μCi; and 20% supernatant of mouse liver in 0.25 M sucrose 0.6 ml. In one series 0.1 ml of 0.01 M Clelland's reagent was added to the incubation mixture.

Before addition of the ^{14}C ornithine the mixture was preincubated for 15 min-

* Supported by a grant from the Dreyfus Medical Foundation and USPHS (12389, Phy).

utes in 25 ml center well flasks. The radioactive ornithine was added, the flasks tightly stoppered and the mixtures were incubated at 37°C for 30 minutes. The $^{14}C_2$ was trapped by 0.25 ml of 0.1 M Hyamine in methanol in the center well. The reactions were terminated by the injection of 0.5 ml of 10% trichloracetic acid through the stopper, and the flasks were allowed to stand at room temperature for one hour to trap residual CO_2. The contents of the center wells were removed and placed in vials containing 10 ml of scintillation fluid and counted in a Nuclear Chicago Mark II liquid scintillation system.

A 24-hour cosine curve was fitted by the method of least squares to obtain the following rhythmic parameters: a) Mesor, M—the rhythm-adjusted computer-determined, overall 24-hour average value (equal to the arithmetic mean if the data are equidistant and cover an integral number of periods; b) Amplitude (A—the distance from the mesor (M) to the peak of the cosine curve expressed in the same units as the variable analyzed; c) Acrophase (ϕ)—the delay in hours or degrees from the phase reference [the middle of the daily light span (1200) CST] of the peak of the 24-hour cosine curve best approximating all data. Since 360° is equated to 24 hours, 15°=1 hour. The error estimates for each of these parameters were obtained by the electronic computer methods used [5].

RESULTS AND DISCUSSION

The results of these studies with the supernatants are presented as a chronogram in Fig. 1. From examination of this chronogram, it would appear that a well defined circadian rhythm existed in ornithine decarboxylase activity with peak activities at 2100 being over 300% greater than those found at the trough (0900). By conventional statistical methods these peak and trough differences were significant ($P<0.01$) for all three sets of data.

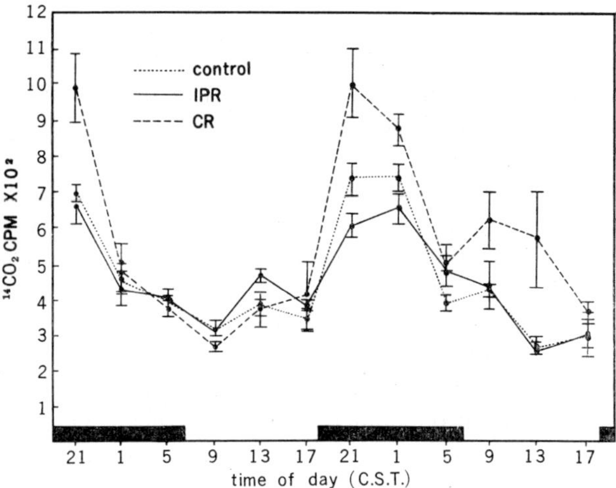

Fig. 1 Mean $^{14}CO_2$ production and the standard error of the mean by control (dotted line); **IPR** (solid line), and control supernatant and Clelland's reagent (dashed line) supernatants. Each point represents five mice (10 determinations). The differences between the Clelland's reagent and the other groups (at peak times) was statistically significant, ($P>0.01$) at the second 2100 period, and approached significance at the first 2100 and second 0100 period samples ($0.10>P>0.05$).

Table 1 Rhythmometric summary.

Ornithine decarboxylase activity in mouse liver — Cosine curve fit at: 24.0 hours

Data summary				Analysis summary using least squares fit of single cosine 24-hour period						
Variable*	Lowest & highest value	90% Range	Coef var	P	PR	Mesor M±SE	Amplitude A±SE	Acrophase ϕ Deg±SE	CK	HR(.95CA)
1. Control	257 732	294 732	38	.017	.59	445 35.07	178 49.60	−348 15	2312	2107 117
2. Clel-lands reagent	252 977	351 977	45	.092	.40	556 61.73	218 87.31	−333 22	2215	
3. Isopro-terenol	267 656	278 656	30	.061	.46	448 31.68	124 44.80	−356 20	2348	

Series Nos. 1 to 3 have 12 time points

Start time for series 1971 2721. Documented span in hours is 44.0

PR=variability ratio=percent of total variability contributed by fitted curve=(sum of squared deviations from mean (SS), of values derived, from fitted cosine curve at sampling times/SS of data themselves) ×100.

P=result of testing zero amplitude=probability of obtaining estimated parameters is sinusoidal rhythm with started period were absent.

* See text for abbreviations of variables.

When each group of data was fitted to a 24-hour cosine curve the following information was obtained: (1) The control data showed a good fit to a 24-hour cosine curve ($p<0.01$) with the acrophase occurring at $-348°\pm15°$ (translated into local clock time this would be 2312 with confidence intervals between 2107 and 0117) (Table 1). This significant fit of the data would indicate that ornithine decarboxylase activity is not fluctuating in a random fashion in the normal untreated animal. (2) Neither the IPR data or the data obtained using Clelland's reagent fit to a 24-hour cosine curve. The effect of IPR was to prevent the full exclusion of the circadian rhythm at the time of normal acrophase, at all other phases of the rat's circadian system there appeared to be no significant effect of IPR on ornithine decarboxylase activity. The use of Clelland's reagent in the chemical analysis causes a significant increase only at those times when ornithine decarboxylase activity was highest, this suggests that there may be a physical difference involving sulfhydryl groups between the enzyme found at one phase of the mouse arcadian system as compared to other phases. On the other hand this may merely demonstrate that the reagent can modify the results of the reaction.

Some of the future directions that should be taken are almost self evident. These should include study at peak and trough of enzymes from animals pretreated with actinomycin, puromycin, and other antimetabolites, as well as purification and kinetic studies of the enzymes from differing periods of activity. Such studies would be invaluable and might give insight into the basic mechanism of action of enzymatic circadian phenomena. They further demonstrate the pitfalls that await one who fails to recognize the basic rhythmicity of this enzyme.

REFERENCES

1. Russell D. and Synder S. H. (1968): Amine synthesis in rapidly growing tissues: ornithine decarboxylase activity in regenerating rat liver, chick embryo and various tumors. Proc. N. A. S., *60*: 1420–1427.
2. Fausto N. (1969): Studies of ornithine decarboxylase activity in normal and regenerating livers. Biochim. Biophys. Acta, *190*: 193–201.
3. Fauso N. (1971): The control of ornithine decarboxylase activity during liver regeneration. Biochim. Biophys. Acta., *116*: 128.
4. Burns E. R. and Scheving L. E. (1974): Circadian rhythm of mitoses in mouse corneal epithelium: alterations produced by isoproterenol. *In*: Chronobiology. Proc. Symp. Quant. Chronobiology, Little Rock, 1971, (L. E. Scheving, F. Halberg and J. E. Pauly, eds.), pp. 209–212, Igaku Shoin Ltd., Tokyo.
5. Halberg F., Tong Y. L. and Johnson E. A. (1967): Circadian system phase—an aspect of temporal morphology; procedures and illustrative examples. *In*: The Cellular Aspects of Biorhythms, (H. von Mayersbach, ed.), pp. 20–48, Springer-Verlag, Berlin.

THE HANDS OF THE CLOCK: CHEMISTRY AND CONTROL*

J. W. HASTINGS and LAURA McMURRY**

The Biological Laboratories, Harvard University
Cambridge, Massachusetts, U.S.A.

If an *in vivo* circadian rhythm has an extractable biochemical correlate which itself exhibits a circadian rhythm, it is then possible to investigate the nature of this biochemical rhythm. With a given molecular species one can study how its activity, which may be viewed as a "hand of the clock," is controlled by the rhythmic mechanism, the "clock"; one might thereby deduce something about how the mechanism itself operates.

The enzyme luciferase from the marine dinoflagellate *Gonyaulax* provides an excellent biochemical correlate for the *in vivo* rhythm of bioluminescence capacity [1]. In extracts made during the night its activity is about ten-fold greater than in similar extracts made during the day.

In the experiments summarized here we attempted to find whether the daily rise and fall of extractable luciferase is caused by daily *de novo* synthesis and destruction of the enzyme molecule. The results, though not absolutely conclusive, indicated that such synthesis and destruction does not occur. A number of other possible explanations for the activity change were also investigated and ruled out. An alternative mechanism involving the activation and inhibition of the luciferase molecule is being sought.

To detect *de novo* synthesis, a method similar to that of FILNER and VARNER [2], involving the use of heavy isotopes, was used. The luciferase was prelabeled by growth for three generations in a medium containing $C^{13}O_3=$ (25% atom) and $N^{15}O_3^-$ (99% atoms), and then transferred into a medium containing only $C_1{}^2O_3=$ and $N^{14}O_3^-$. The density of the luciferase from samples taken at the time of transfer was compared to that of the luciferase from the cells 19 hours and 42 hours after transfer to the light medium.

Assuming rapid equilibration of the C^{12} and N^{14} in the new medium with that in protein precursor pools in the cells, all luciferase molecules synthesized *de novo* after the transfer should contain only C^{12} and N^{14}. This change in isotopic content of luciferase would result in a change in its density and thereby a change in its sedimentation velocity in a sucrose gradient. No purification of the luciferase protein is required for this method.

Two molecular weight forms of luciferase have recently been reported to occur: A, M.W.$\sim$140,000 (inactive at pH 8) and B, M.W.$\sim$35,000 (active at pH 8) [3, 4]. The possibility that an *in vivo* interconversion between these two forms might figure prominently in the rhythm appears to have been ruled out, since, in parallel extracts, both A and B exhibit similar circadian rhythms.

For technical reasons the B-luciferase was used in the analyses of isotopic

* Supported in part by a grant from the National Science Foundation. This paper was taken in part from a thesis presented by Laura McMURRY, in partial fulfilment of the requirements for the Ph.D. degree, 1971, Harvard University.

** Predoctoral trainee, U.S. Public Health Service.

labeling. The experiments showed that after 19 hrs much of the heavy label remained; there appeared to be only about a two-fold increase in number of luciferase molecules due to *de novo* synthesis, not a ten-fold increase. Therefore, the occurrence of ten times as much luciferase activity in night extracts as in day extracts appears not to be caused by *de novo* synthesis and destruction of luciferase molecules. Details of these results are presented elsewhere [5] along with reasons why the conclusion is still considered tentative for several reasons.

Several alternative possibilities have been examined [5, 6] but none of these appear to provide an explanation. For example, it does not appear that different quantities of luciferase are extracted at different times of day. Several different extraction procedures were used, varying both mechanical and chemical parameters. Even extraction in 5 M guanidine hydrochloride followed by renaturation still resulted in a luciferase rhythm.

A second type of possibility is that the concentration of some hypothetical activator or inhibitor varies with time of day. Mixing experiments, ammonium sulfate precipitation, dialysis, and sucrose velocity gradient centrifugation showed that no such (dissociable) compound was present.

A possibility which has not yet been rigorously tested is that luciferase may be modified by a chemical moiety which attaches and detaches, possibly covalently, during each circadian cycle. Several specific models are available for this mechanism, especially ones involving the portion of the molecule presumed to be absent in B-luciferase.

REFERENCES

1. Hastings J. W. and Bode V. C. (1962): Ann. N.Y. Acad. Sci., *98*: 876.
2. Filner P. and Varner J. E. (1967): Proc. Nat. Acad. Sci., *58*: 1520.
3. Krieger N. and Hastings J. W. (1968): Science, *161*: 586.
4. Fogel M. and Hastings J. W. (1971): Arch. Bioch. Bioph., *142*: 310.
5. McMurry L. (1971): Ph.D. Thesis, Harvard Univ., Cambridge, Mass.
6. McMurry L. and Hastings J. W. (1972): Biol. Bull., *143*: 196.

CIRCADIAN PARAMETERS OF THE INFRADIAN GROWTH MODE IN CONTINUOUS CULTURES: NUCLEIC ACID SYNTHESES AND OXYGEN INDUCTION OF THE ULTRADIAN MODE

CHARLE F. EHRET, JAMES H. BARNES* and KENNETH E. ZICHAL**

Division of Biological and Medical Research, Argonne National Laboratory Argonne, Illinois, U.S.A.

Asynchronous populations of free-living eukaryotic cells can exhibit either of two distinctive and mutually exclusive logarithmic growth modes during continuous culture: the ultradian (or *fast* exponential) mode of growth and the infradian (or *slow* exponential) mode. The ultradian includes average generation times (GT) considerably shorter than one day; the infradian (sometimes loosely called the "stationary phase") includes GTs that range from slightly to considerably longer than a day. When infradian cells are synchronized, regardless of their GT the population of cells to which they belong displays a spectrum of temporally characteristic *circadian* properties [1]. Two of these properties, the synthesis of nucleic acids, and the capacity to be induced to reenter the ultradian mode have been observed in large-volume continuous cultures of *Tetrahymena pyriformis* W and are the subject of the present paper.

EXPERIMENTAL

Methods for continuous cultures. A 40 l carboy (CCC, Fig. 1) on a water bath shaker (WBS) at 27±0.1°C holds 15 l of cells in enriched proteose peptone (EPP) [2]. Fluorescent lights (L) were on from 0800–1800 (Fig. 2) or from 0800–2000 (Fig. 3) giving *LD, 10*:14 or *LD, 12*:12 light-dark cycles. Air is introduced through membrane filters (MF), microvalves (MV), flowmeters (FM) and cotton filters (CF) at minimal flow rates. An EPP aliquot enters the CCC from the carboy input Mariotte (CIM) each time the Zenith timer (ZT2) activates a solenoid (S1). On signal from a master clock (ZT1) a timer in the program clock complex (PC) activates the peristaltic pump to deliver 5 aliquots (5 ml) of cells and fluid from the CCC to the fraction collector (FR). The sample injector housing (SIH) is then heat sterilized. Sample 1 is discarded, 2 and 3 were fixed in 0.75 N PCA and Coulter-counted; 4 and 5 were pulse labeled (0.5 ml [5-^{3}H] uridine, 25.9 Ci/mM, or [5-^{3}H] thymidine 6.7 Ci/mM, NEN) for 15′, then fixed in 0.075 N PCA on signal from the PC to S2 and S3, serving the fixative Mariotte (FM) and fixative inputs (FI). Batches of labeled cells were moved to 4°C, washed 5× in 0.1 N PCA, collected on membrane filters, washed 4× with H_2O, and combusted in a carbon-tritium combustion apparatus (Packard Tri-Carb, Model 305). Residues collected in Scintisol were counted in a Beckman LS-250 liquid scintillation counter.

* Present address: Washington University Medical School, St. Louis, Missouri.
** Present address: The Medical School, Iowa State University, Iowa City, Iowa.

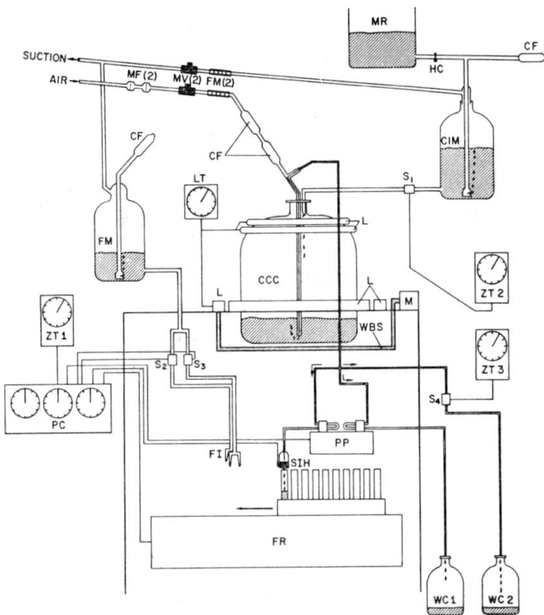

Fig. 1 Continuous culture apparatus.

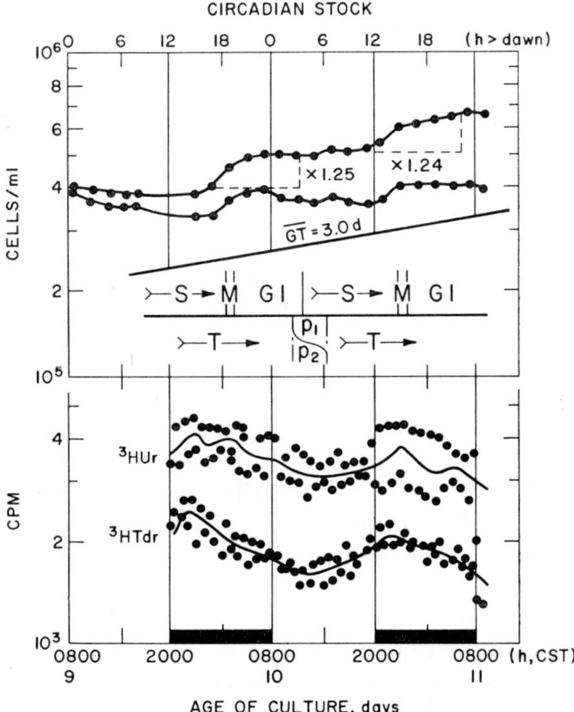

Fig. 2 Top: Graph of cell titer (lower) during a 36 hour period, and cell titer corrected for dilution (upper). This same section can be seen in the last two days of culture 2 (Fig. 3). Bottom: Graph of the isotopic uptake of ^{3}HTdR and ^{3}HUr.

RESULTS AND DISCUSSION

1. *Circadian rhythms of DNA and RNA synthesis in a photoentrained culture of Tetrahymena.* Fig. 2 shows isotope incorporation into an infradian culture (GT=3d) in which about 25% of the cells divided synchronously each day (a 36 hour section from curve 2, Fig. 3). The peaks correspond with the general trends of nucleic acid metabolism already measured in *Tetrahymena* in the ultradian mode, in which predivisional accumulation of DNA precedes that of RNA [3]. But instead of the entire cycle occurring in three hours, it is expanded here to a 24-hour period, with dramatic increases in *S* (duration of DNA replication) and in *T* (period of rapidly accelerating RNA transcription). Correlations of DNA peaks wtih the *LD* and cell division cycles also resemble those already reported in *Euglena* [4].

2. *Minimum duration of DNA S-period in circadian cells.* If one assumes that a circadian S phase encompasses *at least* the period of rapidly accelerating rates of incorporation of thymidine, then the minimum duration of S is ~12 h (Fig. 2, the span from a trough at ~1045 to peak at ~2245). Peaks on days 9 and 10 correlate well with an early period of the respective "steps" in cell titer: viz., compare the ^{3}HTdr curve with the actual plot of cell titer (upper middle) and the growth curve corrected for dilution (upper). Because S phase may continue into M ("mitosis", or better "cell division") G2 is not shown here, and M is diagrammed as a window ~1 h wide.

3. *Relation of circadian phase to RNA properties.* The curve for uridine incorporation shares the trends of that for thymidine, except that uridine troughs and peaks occur later by several hours. A suggestion of bimodality in the peaks recalls earlier measures of circadian syntheses in *Paramecium* [5, 6] in which RNA peaks were significantly bimodal during the dark period. The period of rapidly accelerating RNA synthesis (T) is seen to rise from a trough at 1300–1400, and it's cyclical occurrence may be regarded as the analogue for bulk RNA's to the G1→S→(G2)→M cycle: P1→T→P2 [7], whose minimal rates of synthesis occur at pretranscriptional (P1) and post transcriptional (P2) nodes of the cycle. The latter trough corresponds to the time at which circadian stocks RNA C6 were made, the poorest circadian competitors known to date in molecular hybridization studies of temporally characteristic transcripts [8]. Circadian stocks C12 and C18 [8, 9] (approximately equivalent to C2 and C3 [10]) have shown strength in short term annealing; this and other evidence of rRNA dominance in competition hybridizations (cross-overs, pre-annealing influences of rRNA) suggest that the peak(s) for T in Fig. 2 are dominated by ribosomal syntheses. On the late declining side of the peak(s), the stocks CO=C24 have been the strongest of all competitors in moderately long-term annealing reactions [8, 9] suggesting a minimal contribution by rRNA, and a major one by less redundant species.

4. *Characteristics of the ultradian→infradian transit in continuous cultures.* When corrected for dilution (as in Fig. 3, culture 1, top curve) growth curves of continuous cultures resemble those of batch cultures. After innoculation, a lag of ~3 h precedes entry of cells into the ultradian mode. At a peak titer the cells then switch over to the infradian mode; but unlike batch cultures, the early infradian includes an extended period of drastically diminished cell division during which the peak titer is reduced by washout: cf, growth and dilution curves

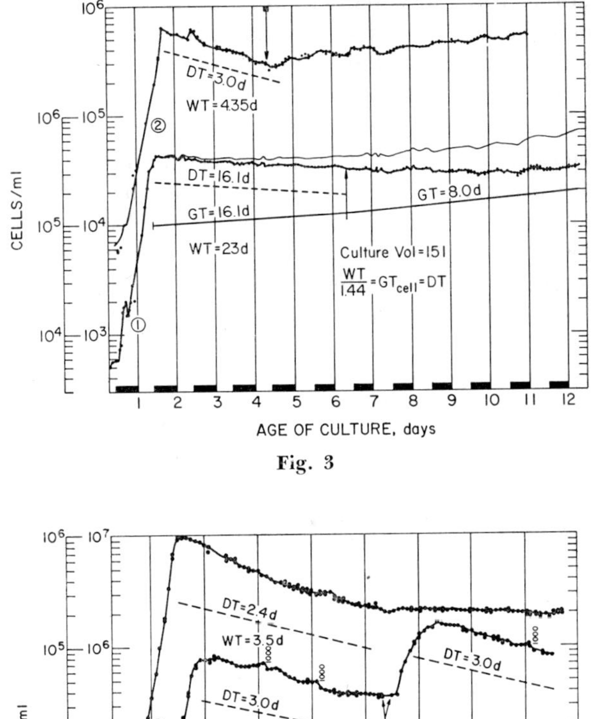

Fig. 3

Fig. 4

Figs. 3 and 4 Growth curves four continuous cultures. Arrows: at (1) switch from DR= 16.1d to DT=8.0d: at (2) no shaking for three hours; at (3) —6h vigorous shaking.

1–3, Figs. 3 and 4. Duration of this near-zero growth transit period is roughly proportional to the dilution time (DT) or washout time (WT) of the culture. After this, a culture may resume asynchronous (but infradian) cell division (curve 4), or respond to entrainment by some external zeitgeber (*LD*, all curves; decreased DT, arrow curve 1; hypoxia for 3 h, arrow, curve 2) or may be dramatically restored to the ultradian mode by hyperventilation (curve 3, arrow). Measures of dissolved oxygen in infradian cultures indicate a *relatively* anaerobic environment ($\sim 5 \times 10^{-3}$ Atm O^2), which may represent a dominant property of the infradian mode, and hence a prerequisite for the circadian oscillation. If so, then the enigmatic impotency of respiratory inhibitors vs. the circadian clock is not at all unexpected.

5. *Oxygen induction of the ultradian mode during infradian growth.* Circadian-infradian cells were collected from a CCC (WC1, Fig. 1) and then exposed

to vigorous aeration either in 250 ml bubbler-bottles, or in 250 ml Erlenmeyer flasks shaken on a New Brunswick Environmental Shaker at 220 rpm. The former system was automated to deliver aliquots every half hour for Coulter cell counts (Fig. 5). The latter system employed Erlenmeyer flasks equipped with side arms, and readings were made of light scatter every half hour with a Thorpe micronephelometer (Figs. 6–8). Microscopic observation of aerated infradian cells (GT 3d) showed a dramatic burst of cell division about three and a half hours after aeration, with about 1/3 of the cells showing the fission furrow stage. The cell count data (Fig. 5) confirms this, showing (1) a latent period of 2.5–3 h, and (2) 1.5–2.0 fold increase in 7–8 h after induction. The immediate rise in light scatter post induction (Fig. 6) corresponds to correlated cell darkening and en-

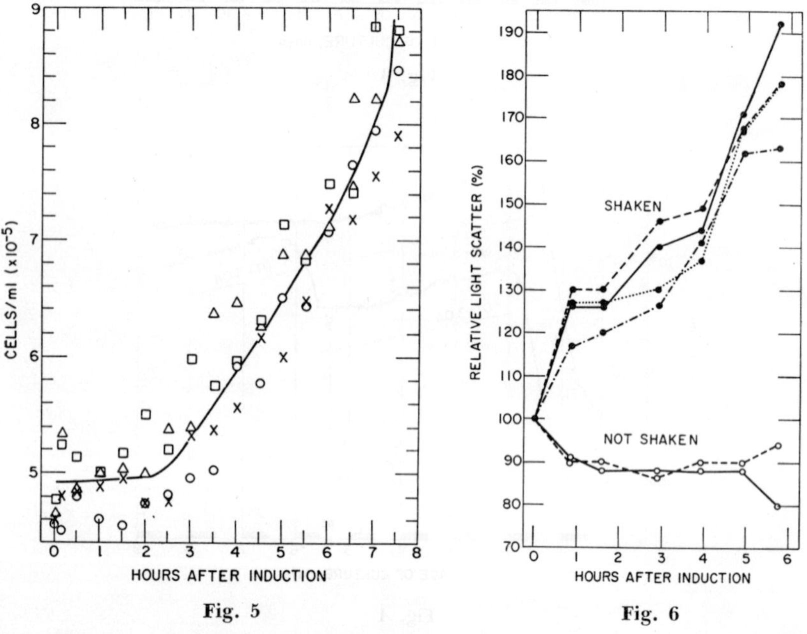

Fig. 5 **Fig. 6**

Fig. 5 Induction of ultradian cell division following aeration during P1 and early T of the circadian cycle; lag phase is 2.5–3 h.

Fig. 6 Immediate increase in light scatter following onset of aeration; unshaken cultures appear like those shaken in N_2 or CO_2-NO_2.

largement seen microscopically. The secondary rise corresponds to that seen in cell count, Fig. 5. Infradian cells not shaken, or shaken in N_2, or in CO_2+N_2 (5%+95) mixtures are not induced to divide by the manipulations. The phenomenon is slightly reminiscent of the use of hypoxic shocks to obtain synchrony in ultradian cells [12], and of the Pasteur affect in anaerobic yeast.

6. *Inhibition of oxygen induction of ultradian growth; Role of porphyrins.* Classical inhibitors of aerobic respiration, including CO, amytal, and rotenone, and also light [13], are each effective in blocking the primary and secondary rises in light scatter seen in Fig. 6. Rotenone added at zero time blocks cell division, and causes a decrease in light scatter without significant cell death (0.5×10^{-5} M); added at one and a half hours after induction a dramatic decrease in light scatter is seen, with considerable cell death. Concomitant measures of ^{59}Fe incorpora-

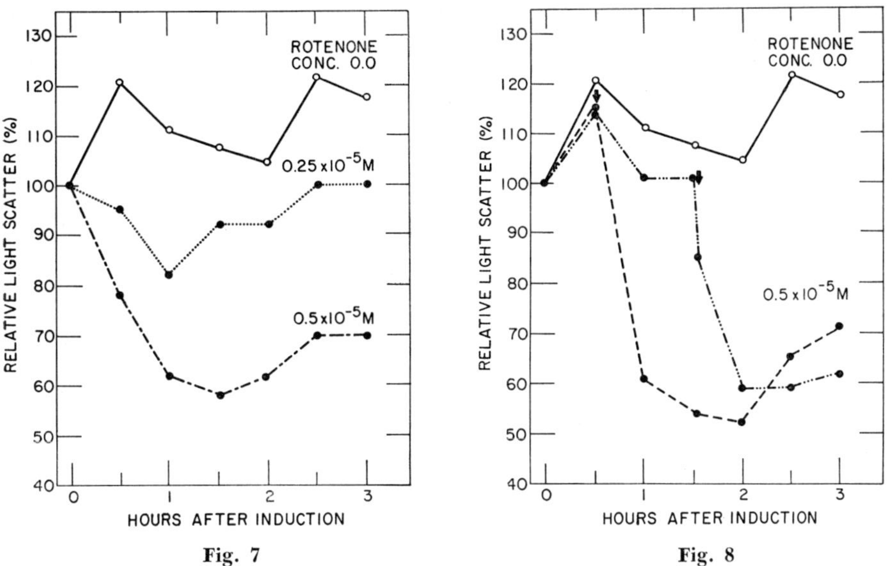

Fig. 7 **Fig. 8**

Figs. 7 and 8 Inhibitory effects of Rotenone on oxygen induction of cell division.

tion during induction suggest that the neo-synthesis of Fe-porphyrin enzymes may play an important role in the induction process.

7. *Circadian rhythm in inducibility of the ultradian mode.* Work in progress with an entirely automated sampling, aeration, and interrogation (nephelometry) system gives clear evidence of a rhythm in efficiency of inducibility—the latter falling to a minimum during the late S phase of the circadian-infradian cell cycle. Results to date suggest that the two fundamental eukaryotic growth modes are another regulatory correlate of the Pasteur effect; ultradian cells never show endogenous circadian rhythms; respiratory inhibitors should switch such cells into the infradian mode, and hence into susceptibility to circadian synchronization by nucleic acid antimetabolites and base analogues. It is of especial interest to note that the above predicts the barbiturates to have distinct and separate actions for the two modes.

ACKNOWLEDGEMENTS

We are pleased to acknowledge the excellent technical assistance of Rita Januszyk and Arlene Dobra, and fruitful discussions of the work with Drs. Audrey Barnett, Eugene McArdle, and Gregory Antipa. This work was supported under the auspices of the U. S. Atomic Energy Commission.

REFERENCES

1. Ehret C. F. and Wille J. J. Jr. (1970): *In*: Photobiology of Microorganisms, (P. Halldal, ed.), pp. 370–416, John Wiley & Sons, New York.
2. Szyszko A. H., Prazak B. L., Ehret C. F., Eisler W. J. and Wille J. J. (1968): J. Protozool., *15*: 781.
3. Cameron I. L., Padilla G. M. and Elrod L. H. (1966): *In*: Cell Synchrony, (I. L. Cameron and G. M. Padilla, eds.), pp. 269–280, Academic Press, New York.

4. Edmunds L. N. Jr. (1965): J. Cell Comp. Physiol., *66*: 147.
5. Ehret C. F. (1959): Federation Proceedings, *18*: 1232.
6. Ehret C. F. (1960): Cold Spring Harbor Symp. Quant. Biol., *25*: 149.
7. Ehret C. F. and Trucco E. (1967): J. Theoret. Biol., *15*: 240.
8. Barnett A., Ehret C. F. and Wille J. J. Jr. (1971): *In*: Biochronometry, (M. Menaker, ed.), pp. 637–651, NAS Washington.
9. Barnett A., Wille J. J. and Ehret C. F. (1971): Bioch. Biophys. Acta, *247*: 243.
10. Ehret C. F., Wille J. J. and Trucco E. (1971): *In*: Biological Oscillations, (B. Chance, K. Pye and B. Hess, eds.), Academic Press, New York, (in press).
11. Wille J. J. Jr. and Ehret C. F. (1968): J. Protozool., *15*: 785.
12. Rooney D. and Eiler J. J. (1969): J. Cell Biol., *41*: 145–153.
13. Sulkowski E., Guérin B., Defaye J. and Slonimski P. (1964): Nature, *202*: 36–39.

GENETIC ALTERATIONS MODIFYING PERIOD LENGTH OF THE SORBOSE-INDUCED HYPHAL-BRANCHING RHYTHM IN *NEUROSPORA CRASSA*

JERRY F. FELDMAN and MARIAN N. HOYLE

Department of Biological Sciences, State University of New York at Albany Albany, New York, U.S.A.

Cellular and biochemical analysis of biological clocks has yielded only vague clues about the mechanisms generating the circadian oscillations observed in many organisms [1–3]. The unique, and at times startling, properties of these oscillations [4, 5] suggests that control processes as yet unelucidated or predicted by modern biochemistry or molecular biology may underlie circadian oscillations or perhaps that currently understood mechanisms are organized in a unique pattern which generate properties unpredicted by our limited knowledge of the component steps.

One of the most powerful tools open to the biologist to unravel complex processes is the genetic and biochemical analysis of single-gene mutants in which one step in the process is altered in some defined way. For this reason, we have undertaken a search for biological clock mutants in the fungus *Neurospora crassa,* for which a vast backlog of genetic and biochemical data and techniques is available.

We have previously reported the isolation of several mutations affecting the conidiation rhythm of the *patch* strain of *Neurospora* [6]. These particular mutants, while useful in demonstrating the possibility of isolating clock mutants, for technical reasons have not yielded much additional information. Furthermore, the isolation procedure used in obtaining these mutants was extremely slow and tedious.

We have recently developed a new system for mutant isolation which seems more promising. The remainder of this paper will describe this system and the results so far obtained.

Neurospora exhibits two distinct types of morphological rhythms as well as a rhythm of CO_2 production. One of the morphological rhythms, periodic production of conidia (asexual spores), is typically circadian—the rhythm is entrained by light cycles and its period length is temperature-compensated [7, 8]. The other morphological rhythm, alternation between monopodial and dichotomous branching of the hyphae, is not entrained by light-dark cycles and the Q_{10} of its period length varies from 1.2–2.0 depending on the medium on which it is grown [9, 11]. Most previous work has been done on mutants (*patch, band,* and *clock*) which express one or the other rhythm, since wild-type strains typically do not show rhythmicity. However, WOODWARD and SARGENT [12] have recently shown that many wild-type strains are capable of expressing a conidiation rhythm if the appropriate medium and aeration conditions are used, and SUSSMAN et al. [9] showed that at least certain wild-type strains could be induced to express the hyphal branching rhythm by addition of sorbose to the medium. Our procedure for mutant isolation uses these observations and one other important fact. First, by combining and slightly modifying these two procedures, we can now induce either the hyphal-branching rhythm or the conidiation rhythm *in the same strain* —i.e., we can phenocopy both the *clock* mutant (by growth on sorbose) or the *band* and *patch* mutants (by aeration) in the same strain. Second, since *Neurospora* grows in tight colonies in the presence of sorbose, it is possible to monitor the hyphal-branching rhythm after plating colonies on a petri dish. In this way we can screen rapidly up to 50 colonies/plate for an abnormal number of bands after several days of growth.

RESULTS AND DISCUSSION

Conidia from a culture of *patch* were mutagenized with ultraviolet radiation and plated out on sorbose-complete medium [13]. After several conidial reisolations several strains were obtained which expressed a clear hyphal-branching rhythm on sorbose and a conidiation rhythm on Vogel's minimal medium with aeration. The period lengths of the two rhythms in these strains and several others are shown in Tables 1 and 2. To measure period length cultures were grown in growth tubes in triplicate in constant dark at 25°C. Experiments were run for 7–10 days and the growth tubes were marked daily in red light, which does not affect either rhythm [8, 9].

Two important results are clear. First, strains 8 and 9 have a significantly longer period length of the hyphal-branching rhythm than any of the others. However, the period length of the conidiation rhythm is normal in these two strains. Second, there is no correlation on these media between growth rate and period length of these rhythms. Genetic analysis of strains 8 and 9 has been hampered by poor germination of the ascospores, but both long and normal period isolates have been recovered from crosses with wild-type.

In addition to continuing the genetic analysis we have begun preliminary physiological studies to determine the nature of the mutations in these strains.

Table 1 Hyphal-branching rhythm on 0.5% sorbose-complete medium.

Strain	Period length (hours)	Growth rate (cm/day)
74 A (wild type)	No rhythm	
band	23.9±4.1	0.70±0.16
patch	24.0±1.3	0.28±0.05
Mutant 8	33.2±3.7	0.29±0.09
Mutant 9	35.0±5.7	0.30±0.12

Table 2 Condiation rhythm on sucrose-minimal medium with aeration.

Strain	Period length (hours)	Growth rate (cm/day)
74 A (wild type)	22.2*	9.43
band	21.8±0.4	4.45±0.01
patch	21.4	4.26
Mutant 8	22.0±0.3	4.99±0.11
Mutant 9	22.1±0.2	5.92±0.09

* Certain experiments were not repeated enough times to justify statistical analysis.

These strains grow normally on complete medium [14] and sucrose-minimal media [15]. However, their growth on acetate-minimal is significantly slower than normal. We are hopeful this will provide a clue to the biochemical nature of the genetic lesions in these strains.

We were struck of course, by the fact that although the period length of the hyphal branching rhythm is altered, the conidiation rhythm appears normal. Whether this indicates that the two rhythms are under completely separate control remains to be determined. Related to this speculation is the fact that the conidiation rhythm expressed by the *patch* and *band* mutants is sensitive to light and relatively insensitive to temperature while the hyphal-branching rhythm expressed by the *clock* mutant is insensitive to light and somewhat more sensitive to temperature. The fact that the two rhythms have these different properties seems to support the idea that the two rhythms are independent and controlled by different mechanisms. However, we have observed with our phenocopies that when the same strain can express the two different rhythms, the properties of these rhythms remains different—i.e. the conidiation rhythm is still light sensitive, temperature-insensitive while the hyphal-branching rhythm is light-insensitive, temperature-sensitive. The two-mechanism model would therefore require such a strain to have two different 24 hour clock mechanisms—one to control the conidiation rhythm and one to control the hyphal-branching rhythm. While this may be possible, it may also be that the manner in which the rhythm is expressed has some control over the properties of the rhythm itself. Such a feedback model is at variance with most thinking about circadian rhythms, in which there is a distinct separation between the clock and the "hands" of the clock and a lack of any feedback to the clock from its hands [4, 16]. These results with the phenocopies may indicate that the distinction may not be as clear as was previously thought. It may also be possible that different components of the clock mechanism—e.g., light sensitivity and temperature compensation—are coupled in different ways to different rhythms. Additional clock mutants in *Neurospora,* or perhaps mutants similar to those recently isolated in *Chlamydomonas* [17] or *Drosophila* [18], will be necessary to distinguish among these alternatives.

SUMMARY

Two mutants of *Neurospora crassa* have been isolated which show a lengthened period in their sorbose-induced hyphal-branching rhythm. Their circadian rhythm of conidiation appears normal. The observation that the properties of the two rhythms in the same strain are different (i.e., responses to light and temperature) suggests either that there are two separate clock mechanisms controlling the two rhythms, that the coupling between the rhythm and various components of the clock are different, or that the rhythm itself exerts some feedback control on the clock.

REFERENCES

1. Karakashian M. and Hastings J. W. (1962): Proc. Nat. Acad. Sci., *48*: 2130.
2. Feldman J. F. (1967): Proc. Nat. Acad. Sci., *57*: 1080.
3. Ehret C. F. and Trucco E. (1967): J. Theor. Biol., *15*: 240.
4. Pittendrigh C. S. (1960): Cold Spr. Harb. Symp. Quant. Biol., *25*: 159.

5. ASCHOFF J. (1960): Cold Spr. Harb. Symp. Quant. Biol., *25*: 11.
6. FELDMAN J. F. and WASER N. (1971): *In*: Biochronometry, (M. MENAKER, ed.), Nat'l. Acad. Sci., p. 652, Washington, D.C.
7. PITTENDRIGH C. S., BRUCE V. G., ROSENZWEIG N. S. and RUBIN M. L. (1959): Nature, *184*: 169.
8. SARGENT M. L., BRIGGS W. R. and WOODWARD D. O. (1966): Plant Physiol., *41*: 1343.
9. SUSSMAN A. S., LOWRY R. J. and DURKEE T. (1964): Am. J. Bot., *51*: 243.
10. BERLINER M. and NEURATH P. (1965): J. Cell Comp. Physiol., *65*: 183.
11. FELDMAN J. F. and STEVENS S. B. (1973): *In*: Behavior of Microorganisms, (A. PEREZ-MIRAVETE, ed.), p. 297, Plenum Press, London.
12. WOODWARD D. O. and SARGENT M. L. (1973): *In*: Behavior of Microorganisms, (A. PEREZ-MIRAVETE, ed.), p. 282, Plenum Press, London.
13. MALLING H. V. (1966): Neurospora Newsletter, *9*: 13.
14. RYAN F. (1950): Meth. in Med. Res., *3*: 51.
15. VOGEL H. J. (1956): Microb. Gen. Bull., *13*: 42.
16. HASTING J. W. (1960): Cold Spr. Harb. Symp. Quant. Biol., *25*: 131.
17. BRUCE V. G. (1973): *In*: Behavior of Microorganisms, (A. PEREZ-MIRAVETE, ed.), p. 257, Plenum Press, London.
18. KONOPKA R. and BENZER S. (1971): Proc. Nat. Acad. Sci., *68*: 2112.

CIRCADIAN VARIATIONS IN THE SUBSTRUCTURE OF THE CHLOROPLASTS OF ACETABULARIA

THÉRÈSE VANDEN DRIESSCHE and RENÉ HARS*

Laboratoire de cytologie et embryologie moléculaires
Département de Biologie moléculaire, Université Libre de Bruxelles
Bruxelles, Belgium

INTRODUCTION

The unicellular alga *Acetabularia mediterranea* displays several circadian rhythms in chloroplastic functions: photosynthesis, chloroplast shape, RNA synthesis, polysaccharide content, number of plastidial units and ATP content [1]. The degree of dependency of the various rhythms on one another is not known. The rhythms of photosynthesis and chloroplast shape are correlated since both of them are expresed or not, depending on the external light conditions [2] or the presence of biologically active substances in the culture medium; when the rhythms are expressed, the relative amplitude of the two rhythms also is comparable [3]. It was therefore of interest to investigate if there are daily changes in substructure.

METHODS

The algae were cultivated at 20°C in a LD cycle of 12:12 except when otherwise stated (daylight tubes "Phytor"). The light intensity was 1200–1400 lux. The culture medium consisted of enriched sea-water [4]. The algae were fixed with glutaraldehyde, surfixed with OsO_4 and embedded in epon according to [5]. The sections were prepared with an ultramicrotome Reichert and examined with an AEI Electron microscope 6B. Only the stalk of the alga has been considered. The sections were transversal to the long axis.

RESULTS

Striking differences have been found between the chloroplasts at 0900, onset of the light period, 1500 or 1800 and 0300. At 1500 and 1800, the chloroplasts are almost similar.

At 090, the thylakoids are peripheral, displaying pseudograna comparable to those described by BOLOUKHÈRE [6] for chloroplasts of algae kept in constant darkness. The thylakoids are not inflated. The chloroplasts usually contain one polysaccharide granule. At 1500 (acrophase of the rhythm in photosynthesis) and 1800 (acrophase of the rhythm in polysaccharide content), the thylakoids are distributed all over the organelle, some of them being diagonals. There is usually no more pseudograna. The thylakoids are inflated. There are two or three polysaccharide granule at 1500 and even more at 1800. At the latter time, some

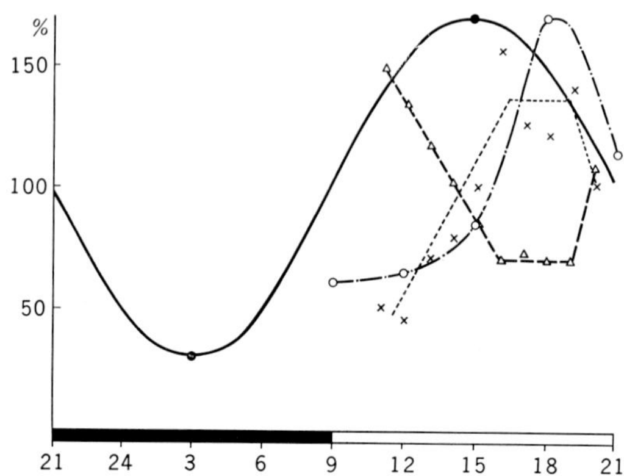

Fig. 1 Time-map of four variables of *Acetabularia*. Photosynthesis (solid line with black circles). Carbohydrate content (broken line with open circles). Number of chloroplastic granules: chloroplasts with one granule (broken line with open triangle); chloroplasts with three (or more) granules (dashed line with x).

chloroplasts appear dumbbell-shaped, which is in good agreement with the results of PUISEUX-DAO and GILBERT [7]. Finally, at 0300, the thylakoids again line up the outer part of the chloroplast, often associated in doublets and already, few pseudograna are formed. They are no more inflated. The polysaccharide granules are small and variable in number.

The whole substructure of the chloroplasts of *Acetabularia* is thus subjected to circadian variation.

0900 has been the time selected in order to check the fact that the organelles display or not in constant light (the last dark period has been skipped) the appearance that they assume at the beginning of the light period. Clearly, the maintenance in constant light does not prevent the manifestation of the circadian variation in substructure. The thylakoids are peripheral, although in a lesser extent than in the choloroplasts of L:D algae. They form a smaller number of pseudograna. They contain more polysaccharide, which is in agreement with previous findings [8].

DISCUSSION AND CONCLUSIONS

The substructure of the chloroplasts of *Acetabularia* varies with the time of the day in respect to the ordering of the thylakoids as well as to the number and size of the polysaccharide granules.

The division rhythm of the chloroplasts, as expected, also is apparent. On the one hand, the generation time of the chloroplasts is about seven days [9]; on the other hand, the diagonal lamellae delineate, according to PUISEUX-DAO [10], new plastidial units and, consequently, they are a prerequisite to chloroplast division. The results presented here demonstrate that the change in ordering is a circadian variation.

The *Acetabularia* used in this study displayed a rhythm in photosynthesis, of which the acrophase is observed at 1500. At that time, the thylakoids are inflated and situated both at the periphery and in the central part of the organelle.

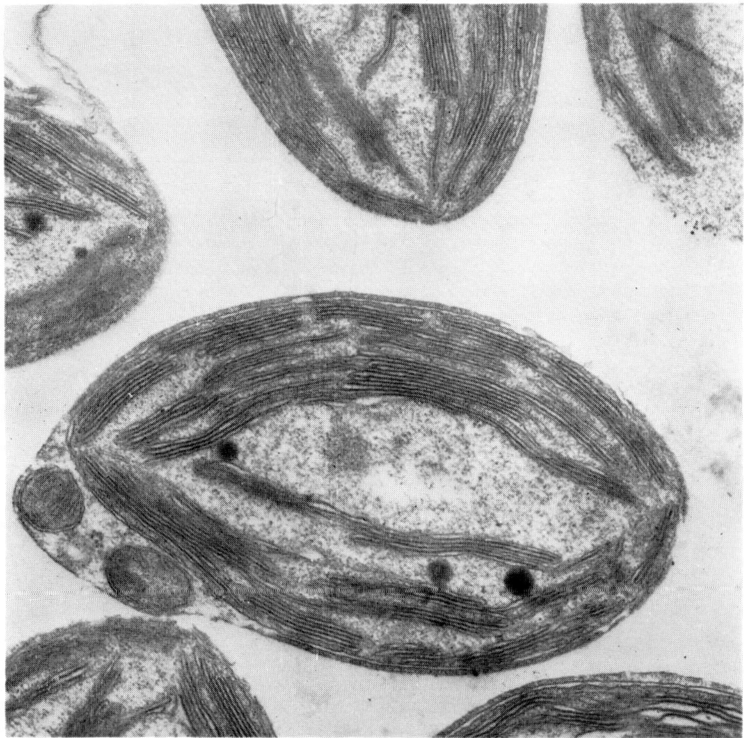

Fig. 2 Chloroplast of *Acetabularia* at 0900, beginning of the light period. ×43000.

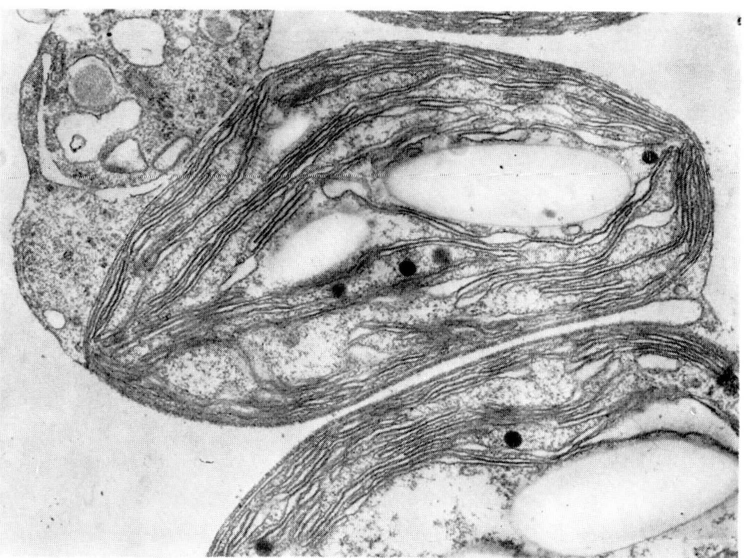

Fig. 3 Chloroplast of *Acetabularia* at 1500, middle of the light period. ×34000.

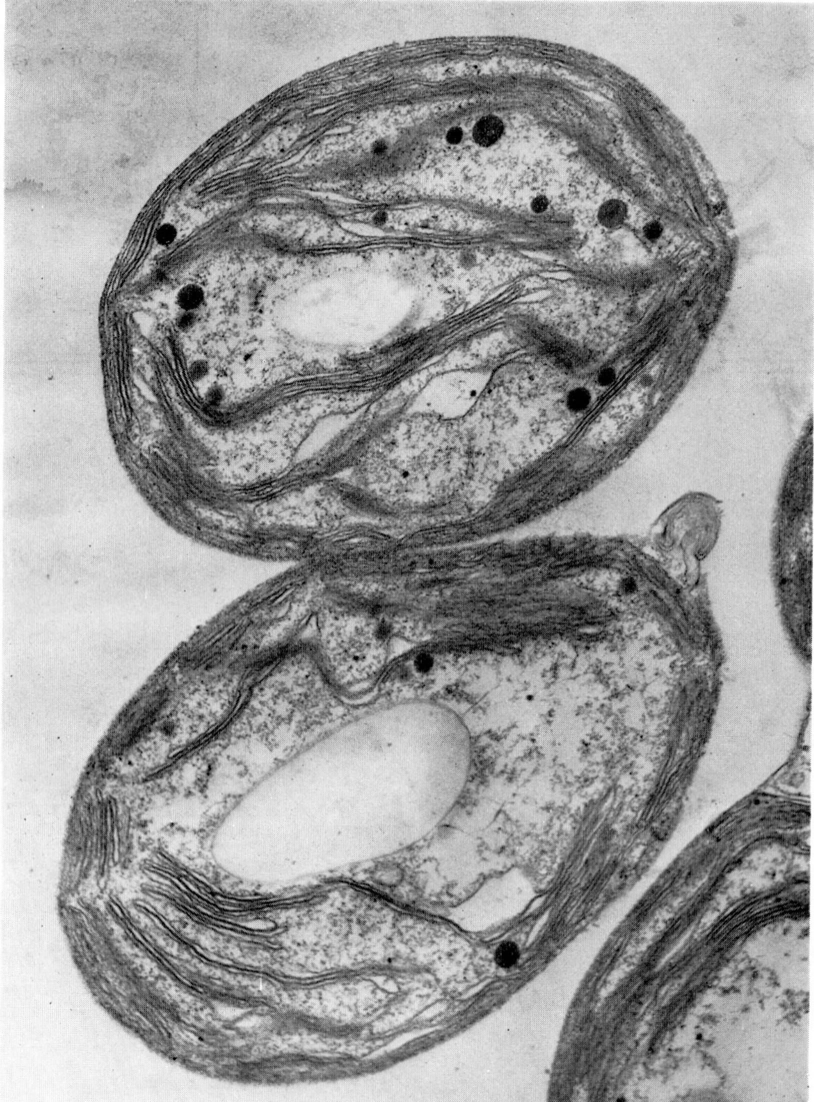

Fig. 4 Chloroplasts of *Acetabularia* at 1800, three hours after the middle of the light period. ×36000.

At 0900 h in L:L, the thylakoids are somewhat less orderly arranged than in the L:D controls. A direct influence is therefore exerted by light on the substructure of the chloroplast. Another experimental series (unpubilshed results) demonstrated that there is a drop in photosynthesis in the L:L algae as compared with the L:D ones at any time of the day. Thus, *at the level observed with the electron microscope,* no close correlation can be established between substructure and photosynthesis rate. The presence of variable amounts of polysaccharide probably influences the substructure of the organelle.

The substructural organization of the chloroplasts of *Acetabularia,* varying in a circadian way, integrates several changes, all varying with the time of the day.

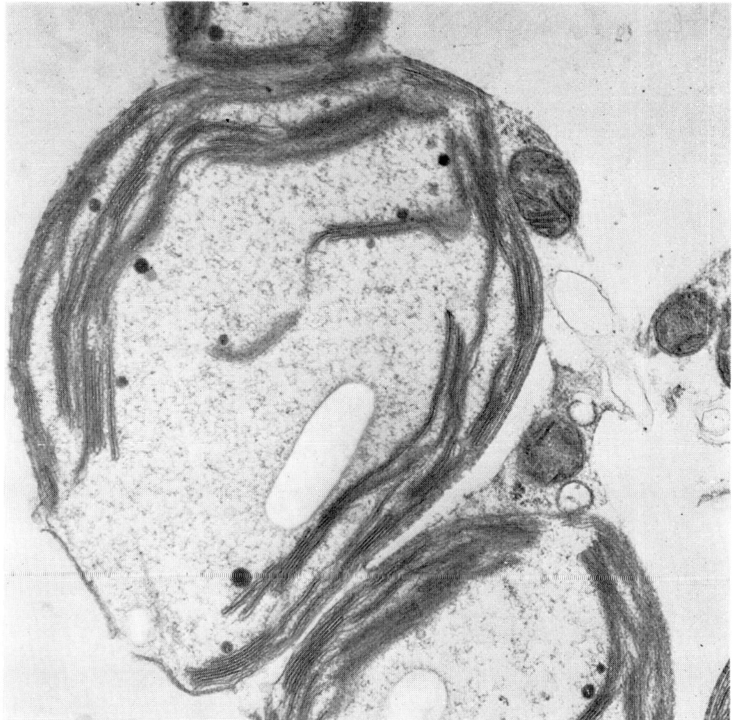

Fig. 5 Chloroplast of *Acetabularia* at 1500, middle of the dark period. ×31000.

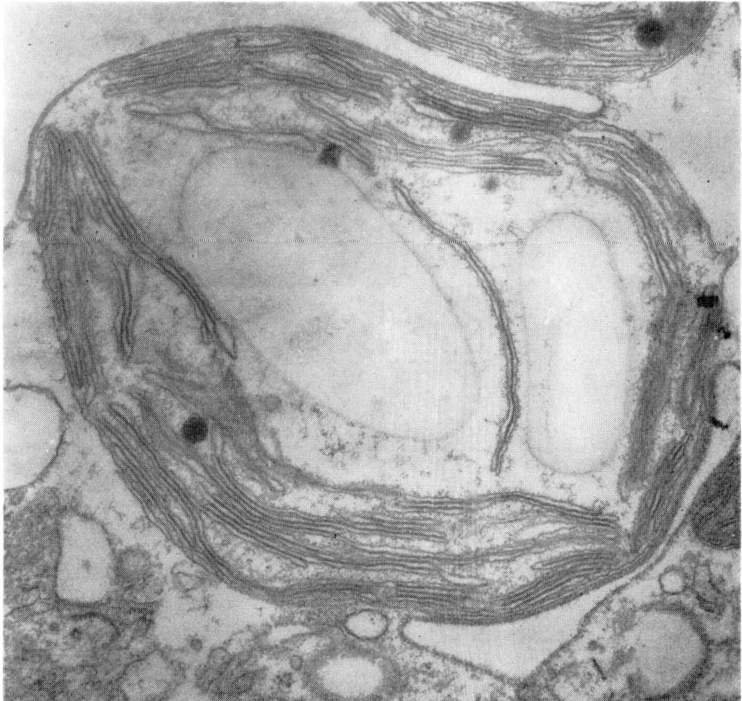

Fig. 6 Chloroplast of *Acetabularia* at the beginning of the subjective light period: the alga has been kept in continuous light for 24 hours. ×43000.

REFERENCES

1. VANDEN DRIESSCHE T. (1973): The chloroplasts of Acetabularia. The control of their activities. Sub-cell. Biochem., *2*: 33–67.
2. VANDEN DRIESSCHE T. (1966): Exptl. Cell Res., *42*: 18–30.
3. VANDEN DRIESSCHE T. (1966): Biochim. Biophys. Acta, *126*: 456–470.
4. LATEUR L. (1963): Rev. Alg., *1*: 26–37.
5. BOLOUKHÈRE M. (1970): *In*: Biology of Acetabularia, (J. BRACHET and S. BONOTO, eds.), pp. 145–175, Academic Press.
6. BOLOUKHÈRE M. (1972): J. Microscopie, *13*: 401–416.
7. PUISEUX-DAO S. et GILBERT A.-M. (1967): C. R. Acad. Sci. Paris, *265*: 870–873.
8. VANDEN DRIESSCHE T. (1969): *In*: Progres in Photosynthesis Research, (H. METZNER, ed.), Vol. 1, pp. 450–457.
9. SHEPHARD D. (1965): Exptl. Cell Res., *37*: 93–110.
10. PUISEUX-DAO S. (1966): Sixth Internat. Congr. Electron Microsc. Kyoto, pp. 377–378.

CIRCADIAN OSCILLATIONS IN ENZYME ACTIVITY IN *EUGLENA* AND THEIR RELATION TO THE CIRCADIAN RHYTHM OF CELL DIVISION

LELAND N. EDMUNDS, Jr., F. M. SULZMAN* and W. G. WALTHER**

Division of Biological Sciences, State University of New York Stony Brook, New York, U.S.A.

Circadian temporal organization is not restricted to multicellular organisms: overt persisting circadian (*sensu strictus*) rhythms have been documented and characterized in *Gonyaulax, Euglena, Acetabularia, Chlamydomonas, Tetrahymena,* and *Paramecium*—all eukaryotic, unicellular algae or protozoa—and range from bioluminescence to mating type reversal [1–4]. Indeed, in each of these six forms, *several* different rhythms have been observed concurrently, in some cases in individual cells as well as in synchronous populations [5]. These well-defined, relatively less complex, versatile systems provide, therefore, attractive experimental material for elucidating the nature of the underlying biological clock(s).

Characteristics of the rhythm of cell division in Euglena. For the past 10 years our laboratory has investigated the rhythm of cell division in the algal flagellate, *Euglena gracilis* Klebs (strain Z) and in certain of its mutants [6, 7]. Cell division can be phased, or synchronized, in cultures of this unicell (as well as in populations of numerous other microorganisms) by appropriately chosen 24-hr cycles of light or temperature so that the population approximately doubles every 24 hr. The actual mechanism, however, by which cell division is synchronized is still poorly understood.

We have postulated that an endogenous, circadian biological clock is at least partially responsible for this division rhythmicity. Appropriate *Zeitgeber* would thus entrain the division rhythm of individual cells of the population, and under "free-running" conditions divisions would be timed, or "gated", by this oscillation operating through biochemical cell cycle controls. This hypothesis is supported by our findings: that (I) synchronous division in autotrophically batch-cultured wild type *Euglena* can be precisely entrained to a 24-hr period by repetitive light-dark cycles (LD) having a driving period (T) of 24 hr (e.g., *L*D:*10*,14), although not all cells necessarily divide during any given cycle [8, 9]; that (II) entrainment by LD cycles having $T \neq 24$ hr (e.g., *L*D: *10*,10 or *L*D: *14*,14) may also occur within certain limits [9]; that (III) "skeleton photoperiods comprising the framework of normal "full-photoperiod" cycles (e.g., *L*D: *3*,6,*3*:12) will also entrain the rhythm to a precise 24-hr period [9]; that (IV) appropriate temperature cycles (e.g., *18°*/25°C: *12*,12) will entrain cell division in cultures maintained in LL [10]; that (V) the heterotrophically cultured, semiachlorophyllous P_4ZUL mutant of *Euglena,* incapable of photosynthesis, can also be synchronized by LD cycles [11];

* Present address: The Biological Laboratories, Harvard University, Cambridge, Massachusetts 02138, U.S.A.

** Present address: Biology Department, Bates College, Lewiston, Maine, U.S.A.

that (VI) rhythmic cell division will persist for a number of days with a *circadian* period ($\tau \sim 24$ hr) in both the autotrophically grown wild type batch-cultured under dim LL [12] and the mutant heterotrophically batch-cultured in either DD or bright LL [6, 11]; and that (VII) high-frequency LD cycles (e.g., *LD*: *1*,3) and even "random" illumination regimes induce circadian division periodicities [9, 13]. Persisting division rhythms have also been reported for *Gonyaulax* [14], *Chylamydomonas* [15], *Tetrahymena* [16], and *Paramecium* [17, 18], as well as in bleached *Euglena* [19].

Long-term persistence of the division rhythm. The development of a semi-continuous culture and monitoring system for light- (or temperature-) synchronized *Euglena* [20] has enabled us to overcome the limitations inherent in batch culture where the cell concentration is constantly increasing (and consequently, the nutrient and light levels are always changing). By this technique, aliquots of the master culture are automatically removed at one- or two-hour intervals for cell number determinations and are replaced by a slow, siphon-regulated flow of fresh medium through a side port from a supply bottle. Since the rate of addition of fresh medium is thus fixed to a value predetermined by the sampling volume and withdrawal rate, and since the ratio of the total sampling volume to the culture volume can be varied, the resultant *dilution rate* is given by the expression:

$$D = \frac{\ln 2}{g} = k$$

where D is the dilution rate (expressed as the fraction of the culture replaced per hour), g is the generation time of the culture in hours, and k is the growth constant in hours. Thus, one can calculate g of a semi-continuous culture from the known dilution, or "wash-out" rate, and more importantly, one can effectively offset the increase in cell number by choosing the appropriate dilution rate so as to achieve a series of identical cycles at the same cell concentration.

Using this semi-continuous system, we have demonstrated long-term persistence (in some cases two months or even longer) of the cell division rhythm in the P_4ZUL mutant grown heterotrophically on a low pH glutamate-malate medium at 18–19°C in DD [7]. A typical segment of one of these trains of oscillations ($\tau = 24.2$ hr) is shown in Fig. 1-A (top). The culture was first entrained by *LD*: *10*,14 before it was placed in DD. In this particular experiment, the generation time (g) was about 35 hr (the so-called infradian growth mode); it is not possible to elicit the rhythm when g is less than 24 hr (i.e., the ultradian mode), which is consistent with (but it does not demand) EHRET and WILLE's [2] "Circadian–Infradian Rule." The problems raised by the non-decay of the rhythm in the culture (which presumably can be considered as a population of imperfect, non-identical oscillators)—with the attendant implication of intercellular interaction or communication—has been discussed elsewhere [6, 7].

Most recently, we have further modified our system to monitor cell number every hour (rather than every two hours) and have examined the division rhythm of the mutant under continuous bright illumination (~9,000 lux) at 18°C. Typical results are illustrated in Fig. 1-B (bottom). Here τ_{FR} was approximately 23.7 hr over the 9-day LL segment shown. The individual peaks are even sharper than

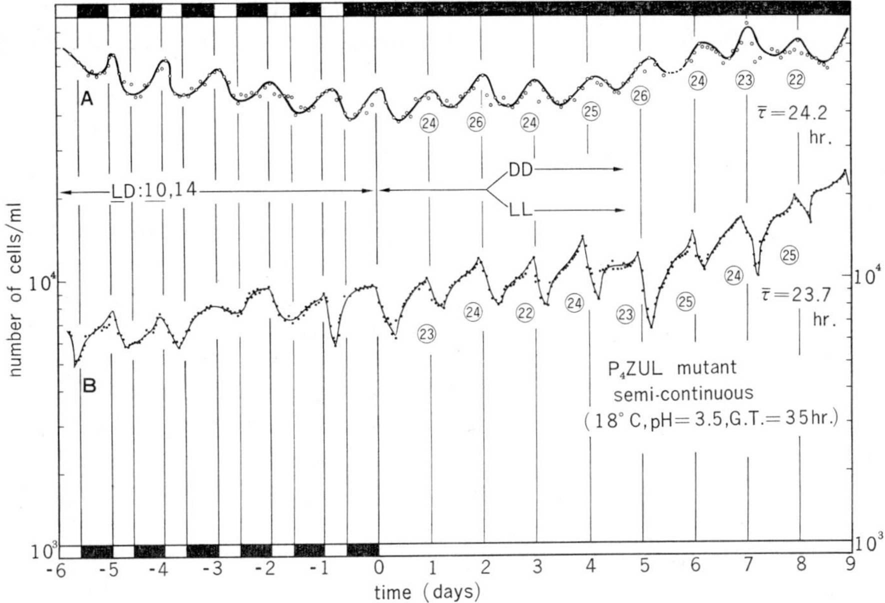

Fig. 1 Long-term, persisting circadian rhythm of cell division in two different semi-continuous cultures of the P_4ZUL photosynthetic mutant of *Euglena* grown on low pH glutamate-malate medium at 18°C. The cultures were first entrained by *LD*: *10*,14 (six monitored days shown) and then placed either in DD (top, Curve A) or in LL (bottom, Curve B); the first nine cycles under constant conditions are indicated. The overall generation time (G. T.) of both cultures was calculated from the known dilution rate to be about 35 hr. Successive period lengths are encircled just below each "free-running" cycle and the average period ($\bar{\tau}$) is given to the right for each curve.

in DD, although this might be due in part to the superposition of the circadian rhythm of cell "settling" that we have previously described [21].

Periodic changes in enzyme activity during the cell cycle. The general patterns of biosynthesis of gross parameters across the cell cycle have been mapped for our *Euglena* system synchronized by LD cycles [22]. More recently, we have discovered periodic changes in enzyme activity during synchronous growth of the wild type autotrophically batch-cultured in *LD*: *10*,14. Thus, deoxyribonuclease activity (on a per aliquot basis) remains at a constant level for about 5–6 hr after the onset of light and then sharply increases until the onset of the dark period some five hours later, at which time it levels off [23]. This discontinuous increase parallels that of total DNA [22, 24], suggesting that the two events may be causally related. Likewise, ribulose-1,5-diphosphate carboxylase and both NAD- and NADP-dependent triose phosphate dehydrogenase exhibit periodic increases in activity during the cell cycle [25]. Presumably, these enzymes and others would continue to oscillate under constant conditions of illumination (or darkness) and temperature inasmuch as the cell division rhythm persists (as well as the rhythms of phototaxis, cell settling, and—for at least several LL cycles—photosynthetic capacity).

Circadian oscillations in enzyme activity in stationary cultures. Although the cell division rhythm has provided a fruitful clock system, it also has its disadvantages, especially for a molecular approach. Thus, although one might conclude that periodic enzyme synthesis (or activation) might underlie the cell division (and other) rhythms, the situation is complicated by the fact that the cell cycle itself can act as a driving force: replication of successive genes on the chromosomes would lead to transcription, and transcription to translation and an ordered, temporally differentiated expression of enzyme activities. Of course, one is still confronted with the problem of the persisting cell cycles in the first place. GOODWIN [26] has proposed that DNA replication and oscillations in a closed feedback circuit can lock together, or couple, at the same frequency when they interact via their common element, mRNA. These mutually entrained oscillations would persist under chemostat conditions providing that the cellular oscillators interacted by some chemical species. It is thus possible to explain both the overt rhythms and the oscillations in enzyme activity that we have observed in our *Euglena* system in terms of known biochemical mechanisms.

At the same time, we have recently attempted to simplify our system by disentangling autogenous enzyme oscillations from those directly generated by the cell cycle in light-synchronized, heterotrophically batch-cultured, wild type *Euglena* by using cells that had reached the *stationary* growth phase where essentially *no net change in cell number occurs* [27]. The cultures were previously grown and then subsequently maintained in *LD*: *10*,14 at 19°C. Alanine and lactic dehydrogenase, glutamic-6-P-dehydrogenase, L-serine and L-threonine deaminase, and acid and alkaline phosphatase were assayed; all showed large-amplitude oscillations in activity (with maxima usually attained during the light period) which were entrained to a 24-hr period by the imposed LD cycle. These rhythmic changes in activity, therefore, were divorced from the cell cycle and periodic replication of the genome.

Even more interesting, however, was our finding that the activity of alanine dehydrogenase (ADH) *continues to oscillate for at least seven days in DD* in these non-dividing cultures. A typical experiment is shown in Fig. 2. Note that ADH was first monitored for several days in *LD*: *10*,14 (only two cycles are shown here) and then for three days in DD. After a two-day interval, the enzyme was monitored for two more days; the peaks during this time were still in phase, strongly implying that the oscillation was continuing throughout the entire seven-day segment. Viability checks gave no indication of a differential cell death-and-replacement process. It seems unlikely, therefore, that the observed periodic increases in enzyme activity were due merely to the progression of a subpopulation of cells (which would have just divided to replace dying cells) throughout their respective cell cycles.

These findings, then, indicate that a common, "kitchen variety" enzyme such as ADH can exhibit endogenous, circadian oscillations in activity under conditions of constant illumination and temperature—i.e., a typical biological clock—that is independent of the cell cycle. We expect this to be the case for other enzymes also. The question obviously arises as to the genesis of these oscillations. As we have already mentioned, periodic changes in enzyme activity such as we have observed may be the result of end-product repression, i.e., the enzyme and the end-product form a system in which the concentration of one determines the rate of synthesis of the other, as a consequence of which the intracellular concentrations of both parameters oscillate [26, 28, 29, 30]. Likewise, the cyclic changes in enzyme activity could result from the sequential transcription of the long, polycistronic "chronon(s)" hypothesized by EHRET and TRUCCO [31] in their

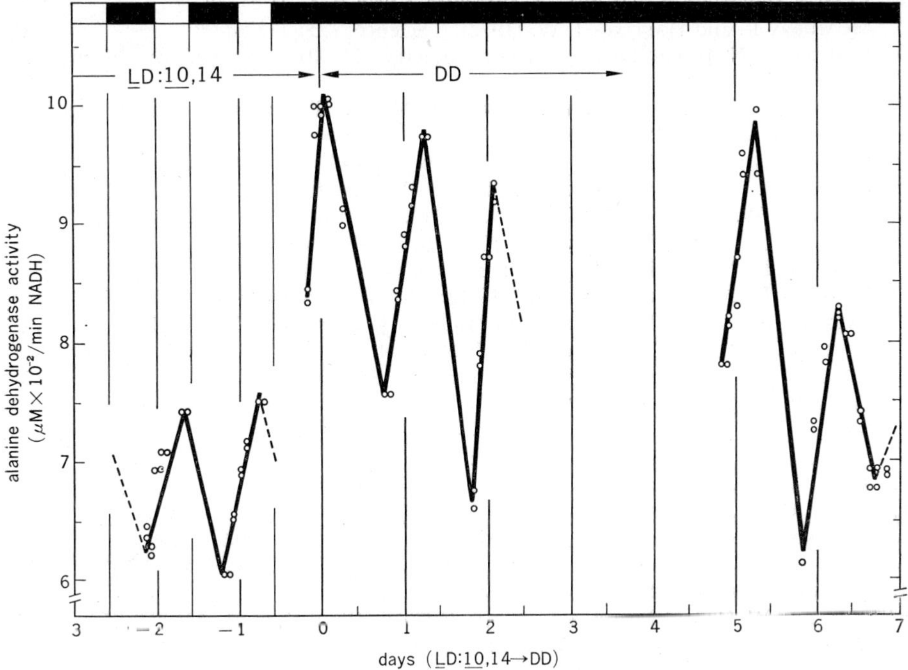

Fig. 2 Persisting circadian oscillation in the activity of alanine dehydrogenase (as measured by following the oxidation of NADH at 340 nm at 25°C) in non-dividing, stationary cultures of *Euglena* (wild type) previously grown and then maintained on low pH glutamate-malate medium. The culture was first synchronized by *LD*: *10*,14 (two monitored cycles shown) and then placed in DD at 19°C (seven days of which are shown; enzyme assays were conducted on the first three and the last two days of this time span).

molecular model for a circadian clock. These two explanations are neither mutually exclusive nor jointly exhaustive, but they do offer possible experimental approaches toward a further understanding of circadian timing in *Euglena* and substitution other microorganisms [32, 33].

ACKNOWLEDGEMENTS

The work reported represents the concerted efforts of our laboratory; particular acknowledgement should be made of the contributions of Mr. Roy R. Funch, Mr. Robert M. Jarrett, and Dr. Orville W. Terry. We thank Mrs. Lin-Whei Chuang and Mrs. Lilly Y. W. Bourguignon for their expert technical assistance. These efforts were supported by NSF research grants #GB-4140, #GB-6892, and #GB-12474 and State University of New York/Research Foundation Grant-in-Aid #31-7150A to L. Edmunds. Dr. F. Sulzman was the recipient of a NIH Predoctoral Research Fellowship (#5-FO1-GM45476-02).

REFERENCES

1. Bruce V. G. and Pittendrigh C. S. (1957): Amer. Natur., *92*: 294–306.
2. Ehret C. F. and Wille J. J. (1970): *In*: Photobiology of Microorganisms, (P. Halldal, ed.), pp. 369–416, John Wiley & Sons, New York.
3. Hastings J. W. (1959): Annu. Rev. Microbiol., *13*: 297–312.

4. VANDEN DRIESSCHE T. (1970): J. Interdisc. Cycle Res., *1*: 21–42.
5. MCMURRY L. and HASTINGS J. W. (1972): Science, *175*: 1137–1139.
6. EDMUNDS L. N. Jr. (1971): *In*: Biochronometry, (M. MENAKER, ed.), pp. 594–611, Nat. Acad. Sci., U.S., Washington, D.C.
7. EDMUNDS L. N. Jr., CHUANG Y., JARRETT R. M. and TERRY O. W. (1971): Interdisc. Cycle Res., *2*: 121–132.
8. EDMUNDS L. N. Jr. (1965): J. Cell. Comp. Physiol., *66*: 147–158.
9. EDMUNDS L. N. Jr. and FUNCH R. (1969): Planta, *87*: 134–163.
10. TERRY O. W. and EDMUNDS L. N. Jr. (1970): Planta, *93*: 106–127.
11. JARRETT R. M. and EDMUNDS L. N. Jr. (1970): Science, *167*: 1730–1733.
12. EDMUNDS L. N. Jr. (1966): J. Cell. Physiol., *67*: 35–44.
13. EDMUNDS L. N. Jr. and FUNCH R. (1969): Science, *165*: 500–503.
14. SWEENEY B. M. and HASTINGS J. W. (1958): J. Protozool., *5*: 217-224.
15. BRUCE V. G. (1970): J. Protozool., *17*: 328–334.
16. WILLE J. and EHRET C. (1968): J. Protozool., *15*: 785–788.
17. VOLM M. (1964): Z. vergl. Physiol., *48*: 157–180.
18. BARNETT A. (1969): Science, *164*: 1417–1419.
19. MITCHELL J. L. A. (1971): Planta, *100*: 244–257.
20. TERRY O. W. and EDMUNDS L. N. Jr. (1969): Biotechnol. Bioeng., *11*: 745–756.
21. TERRY O. W. and EDMUNDS L. N. Jr. (1970): Plants, *93*: 128–142.
22. EDMUNDS L. N. J. (1965): J. Cell. Comp. Physiol., *66*: 159–182.
23. WALTHER W. G. and EDMUNDS L. N. Jr. (1970): J. Cell Biol., *46*: 613–617.
24. EDMUNDS L. N. Jr. (1964): Science, *145*: 266–268.
25. WALTHER W. G. and EDMUNDS L. N. Jr. (1973): Plant Physiol., *51*: 250–258.
26. GOODWIN B. C. (1966): Nature, *209*: 479–481.
27. SULZMAN F. M. and EDMUNDS L. N. Jr. (1972): Biochem. Biophys. Res. Commun., *47*: 1338–1344.
28. GOODWIN B. C. (1963): Temporal Organization in Cells. Academic Press, London.
29. MASTERS M. M. and DONACHIE W. D. (1966): Nature, *209*: 476–479.
30. DONACHIE W. D. and MASTERS M. M. (1969): *In*: The Cell Cycle. Gene-Enzyme Interactions, (G. M. PADILLA, G. L. WHITSON and I. L. CAMERON, eds.), pp. 37–76, Academic Press, New York.
31. EHRET C. and TRUCCO E. (1967): J. Theoret. Biol., *15*: 240–262.
32. BARNETT A. (1973): *In*: Biochronometry, (M. MENAKER, ed.), pp. 637–651, Nat. Acad. Sci., U.S., Washington, D.C.
33. MCMURRY L. and HASTINGS J. W. (1972): Biol. Bull., *143*: 196–206.

A CIRCADIAN RHYTHM FROM A POPULATION OF INTERACTING NEURONS IN THE EYE OF *APLYSIA*

JON W. JACKLET

Department of Biological Sciences, State University of New York at Albany Albany, New York, U.S.A.

In recent years the brain has been increasingly implicated as the source of a circadian clock that regulates many rhythmic functions. This is true for insects [1, 2], mollusks [3–5], and vertebrates [6]. One fundamental question concerning neuronal oscillators that show circadian rhythms, and circadian oscillators in general, is whether the observed oscillations are caused by a single master oscillator or by interactions within a population of oscillators. This paper describes the neuronal interaction that produce the circadian rhythm in the eye of *Aplysia* and the organization of that circadian system.

The isolated eye and optic nerve of the marine gastropod *Aplysia* can be maintained in a culture medium for up to seven days while continuous records are made of the optic nerve activity. Under constant darkness and temperature conditions the optic nerve impulse frequency shows a circadian rhythm [5]. The maximum frequency occurs just after projected dawn of the first day after isolation and at successive circadian periods (27.5 hours) thereafter. The phase of the rhythm can be controlled by light *in vivo* and *in vitro* [7]. The latter observation shows that the basic circadian oscillators are in the isolated eye and not in the central nervous system. The optic nerve impulses, compound action potentials (CAP), are the naturally synchronized firing of many individual neurons [8]. Since the CAP represents the firing of a population of neurons, it is not surprising to find that the amplitude of the CAP, which is an index of the number of cells firing together, varies with a circadian periodicity. In fact, the CAP frequency, the CAP burst frequency, and the CAP amplitude all are circadian and all are in phase [9].

There are three subpopulations of cells in the eye: receptors, support cells, and secondary neurons. There are about 3600 receptors in the eye. They show a graded receptor potential when illuminated, but do not produce action potentials during illumination or during spontaneous optic nerve activity [8]. The support cells are closely interdigitated with the receptors and do not show any known electrical activity. There are about 950 secondary cells in the eye. These neurons are endogenously active and it is their activity that is seen in the CAP of the optic nerve [9, 10]. Simultaneous recording from the optic nerve and from a single secondary neuron show a burst of action potentials in the secondary cell precedes each CAP. However, a single action potential in the secondary cell does not evoke a CAP. Apparently a number of secondary cells must fire repeatedly before a CAP is evoked. Other records show that each secondary cell receives post-synaptic potentials from other cells. The population has inhibitory, excitatory, and electrotonic connections among its members.

* Supported by NIH Grant NS 08443 and SUNYA Res. Found. Grants.

If a number of neurons (oscillators) must fire together to produce the CAP and the circadian rhythm in CAP frequency, then reducing the number of oscillators might change the rhythmic output of the eye. Experiments were performed to test this proposition [9]. Removing the lens from the eye does not affect either the CAP waveform or the circadian rhythm. Progressively removing distal portions of the retina does progressively change the period and range of the circadian rhythm. Histological analysis of the remaining retinal populations allowed the number and kind of cells remaining in the eye to be correlated with the recorded circadian rhythm activity. Fig. 1 shows that the period and range of the oscillations are progressively reduced by reducing the population of secondary neurons from 950 to 190 or 20%. Below 20% of the population, ultradian frequencies are expressed as well as a circadian remnant. A Fourier frequency analysis was performed on some of the data shown in Fig. 1 using a Univac 1108 computer. This analysis shows that whole eyes have a dominant mode of oscillation at periods of 26–28 hours and very little energy at shorter periods. The 0.2 eye (Fig. 1) has a dominant mode at 25 hours and a minor mode at 20–22 hours. The 0.1 eye has a dominant mode at 21 hours and minor modes at 10 and 7.5 hours. The 0.02 eye has a major mode at 7.5 hours and minor modes at 12–16 hours and 18–20 hours. The Fourier analysis substantiates our other estimates of the period and demonstrates the shift of the mode of oscillation to shorter periods as the eye population is reduced. Some eyes were reduced to less than 0.02 in which all the receptor cells were removed and only a few tens of secondary cells remained. These eyes continued to show an oscillatory output.

There are at least three alternative models for the organization of the endogenously active neurons of the eye: A, a population driven by a master circadian oscillator; B, a population of circadian oscillators; C, a population of non-circadian oscillators that together produce a circadian rhythm. A master oscillator does not appear to be present since the period and range change gradually and not abruptly as one would expect when the master oscillator was cut away. If a master oscillator is present, it receives feedback from its followers and thus population interactions occur. A choice between the latter two models is more difficult to make. Fortunately both models have been studied theoretically. Winfree [11] studied a population of weakly coupled circadian oscillators. His model predicts that the period should get longer and other periods might appear as the number of oscillators is reduced. A population of non-circadian oscillators that have strong inhibitory coupling has been studied by Pavlidis [12]. This model predicts that the period should get shorter as the number of oscillators is reduced since the period of oscillation depends upon the number of oscillators and the coupling between the oscillators. This model comes close to fitting the findings from the *Aplysia* eye since the period is shortened as the number of oscillators is reduced. The question of whether each oscillator in the population is circadian or non-circadian is difficult to answer at this time. It is clear, however, that the eye of *Aplysia* has a population of interacting neuronal oscillators that produce a circadian rhythm.

The output from the eye goes to the cerebral ganglion via the optic nerve. The CAP frequency response to light [8] is proportional to the intensity of the light. This response is phasic; that is, light pulses of tens of minutes are faithfully represented but longer pulses are not because of adaptation of the system. The endogenous or tonic activity of the optic nerve shows a circadian rhythm in

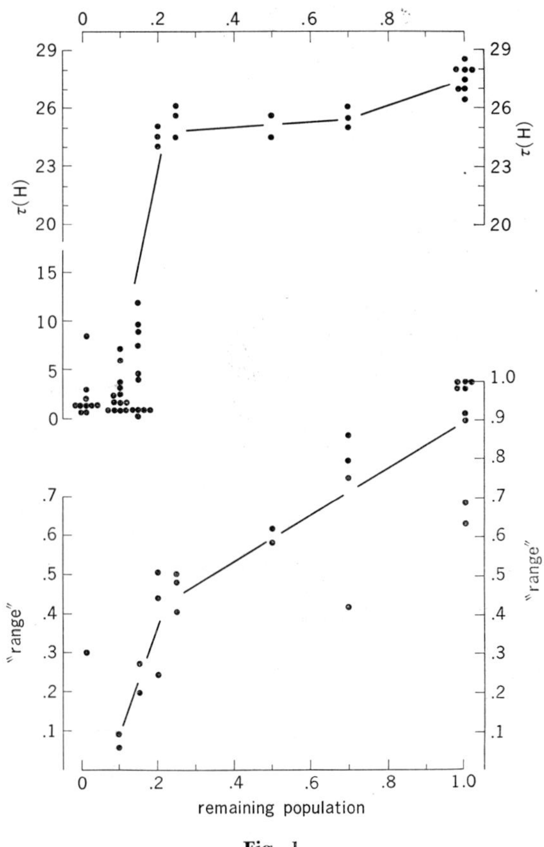

Fig. 1

either constant darkness or constant light (shorter period) and also during 15 minute pulses of light every two hours over 24 hours [7]. Thus there is a tonic system that expresses a circadian rhythm and that system may be modulated by pulses of light. If the pulses of light are of sufficient duration and intensity, they will control the phase of the circadian rhythm. The eye of *Aplysia* is unusual in that the output of the system is a naturally synchronized compound action potential. The consequence of such a system is that most of the information about form vision is lost in the synchronization. A synchronized system must have positive feedback and coupling among the units. This has the effect of pulling some units into activity that would not be active without the positive feedback coupling and tends to increase the firing of all the members of the population. If increased firing produces an increased secretion of chemicals from the axon terminals, then synchronization is useful in increasing the secretory efficiency of the system. Electron microscope studies [13] show that the secondary fibers of the retina contain large, 1000 Å, dense vesicles. The retina and optic nerve show a green-yellow fluorescence after paraformaldehyde treatment and dopamin has an excitatory effect on the eye-optic nerve system. These three facts strongly suggest that the eye is secreting biogenic amines. The systm appears to be well organized for efficient secretion.

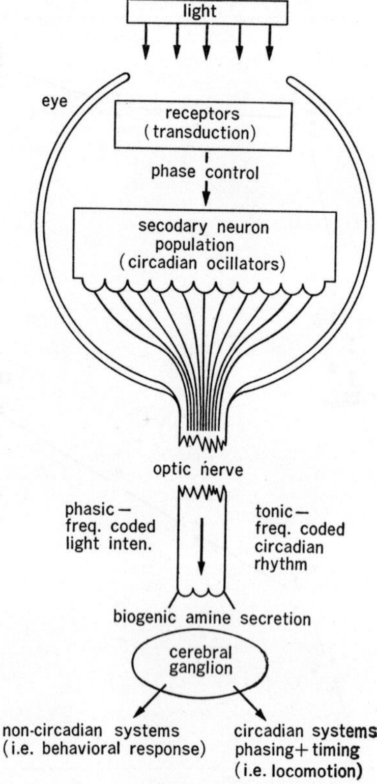

Fig. 2

The foregoing information about the eye-optic nerve circadian system has been incorporated into the diagram in Fig. 2. The usefulness of this system to the animal is two-fold; it mediates immediate (phasic) behavior responses [14] and it mediates circadian timing and phasing to other systems [10, 15].

REFERENCES

1. Brady J. (1969): How are insect circadian rhythms controlled. Nature, *223*: 781–784.
2. Truman J. and Riddiford L. (1970): Neuroendocrine control of ecdysis in silkmoths. Science, *167*: 1624–1626.
3. Strumwasser F. (1965): The demonstration and manipulation of a circadian rhythm in a single neuron, *In*: Circadian Clocks, (J. Aschoff, ed.), pp. 442–462, North Holland Publishers, Amsterdam.
4. Strumwasser F. (1967): Neurophysiological aspects of rhythms, *In*: The Neurosciences: A Study Program, (G. Quarton, T. Melnechuk and F. Schmitt, eds.), pp. 516–528, Rockefeller University Press, New York.
5. Jacklet J. W. (1969): Circadian rhythm of optic nerve impulses recorded in darkness from isolated eye of *Aplysia*. Science, *164*: 562–563.
6. Black I., Axelrod J. and Reis D. (1971): Hypothalamic regulation of the daily rhtyhm in hepatic tyrosine transaminase activity. Nature, New Biology, *230*: 185–187.
7. Jacklet J. W.: A circadian rhythm in optic nerve impulses from an isolated eye in darkness. *In*: Symposium on Biochronometry, Washington, D.C., 1971, (M. Menaker, ed.), National Academy of Sciences, (in press).

8. JACKLET J. W. (1969): Electrophysiological organization of the eye of *Aplysia*. J. Gen. Physiol., *53*: 21–42.
9. JACKLET J. W. and GERONIMO J. (1971): Circadian rhythms: Population of interacting oscillators. Science, *174*: 299–302.
10. JACKLET J. W.: Neuronal population interactions in a circadian rhythm in *Aplysia*. Proc. Sym. Invertebrate Neurobiology and Rhythms. Tihany, Hungary, (in press).
11. WINFREE A. (1967): Biological rhythms and the behavior of populations of coupled oscillators. J. Theoret. Biol., *16*: 15–42.
12. PAVLIDIS T. (1969): Populations of interacting oscillators and circadian rhythms. J. Theoret. Biol., *22*: 418–436.
13. JACKLET J. W., ALVAREZ R. and BERNSTEIN B.: Ultrastructure of the eye of *Aplysia*. J. Ultra. Res., (in press).
14. JAHAN-PARWAR B. (1970): Conditioned response in *Aplysia* californica. Am. Zool., *10* (3): 287.
15. STRUMWASSER F.: The cellular basis of behavior in *Aplysia*. J. Psych. Res., (in press).

LIGHT ENTRAINED CIRCADIAN OSCILLATIONS OF GROWTH RATE IN THE YEAST, *CANDIDA UTILIS*

John J. WILLE, Jr.

Department of Biological Sciences, University of Cincinnati
Cincinnati, Ohio, U.S.A.

In previous work it was reported that light synchronization and entrainment of a circadian rhythm of cell division could be readily obtained in slowly dividing (infradian) populations of *Tetrahymena* maintained in continuous culture in a nephelostat [1]. A thoroughgoing reappraisal of the role of cell growth and division in circadian oscillations in eukaryotic microorganisms suggested a composite view, namely that those organisms displaying a capacity for circadan regulation of cell functions are invariably capable of circadian outputs in the infradian mode of growth, as exemplified by the findings for *Tetrahymena,* and emphatically realized from the knowledge that this was also the case for *Gonyaulax* and *Euglena.* This generalization was dubbed the G-E-T effect (Gonyaulax, Euglena, and Tetrahymena) [1], and later restated as a corollary, the circadian-infradian rule, of a larger principle regarding circadian rhythms, the circadian-eukaryotic principle [2].

In this paper evidence obtained with the yeast, *Candida utilis* for a growth rate determining role in the circadian organization of cell growth will be presented. These experiments were all carried out with yeast cells grown in the modified Eisler-Webb Nephelostat described earlier [1], under different light-dark conditions. *Candida utilis* is a pseudo-mycelia, budding yeast which has been employed in a variety of studies on microbial growth [3] in continuous culture.

Growth rate of C. utilis during the transition from the ultradian to the infradian mode of growth.

When a culture of *C. utilis* is inoculated into the nephelostat growth tube from a parent batch culture in rapid exponential growth at a cell density (5×10^6 cells per ml) near the gain set (130) at which the batch culture would ordinarily enter the slow growth phase, the culture goes through approximately one doubling in cell number (cell generation time of 7.3 hours) and then all growth ceases for a day, as described above in Fig. 1. Fig. 2 shows the growth rate of *C. utilis* in constant dark (DD) immediately after the one day cessation of growth brought about the inherent limitation of growth in the transition of cells from the rapid (ultradian) to slow (infradian) mode of growth. One observes that upon resumption of growth, the growth rate does not remain constant, but continually accelerates until it regains the steady-state cell generation rate (GT cell=7.3 hours) characteristic of the ultradian mode of growth. This return required more than seven days in this particular experiment, but in other experiments, the transition may be of shorter duration. In the transition from the slow growth mode (infradian) to rapid steady-state mode of growth, there lies a region of growth rates

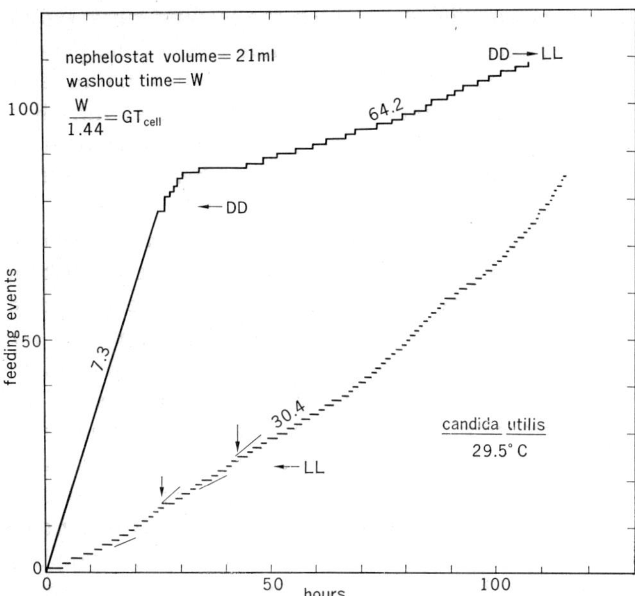

Fig. 1 Effect of visible light on the growth rate of *Candida utilis* in a nephelostat. Upper curve: Growth rate of *C. utilis* in constant dark (DD, dim red light, $\lambda > 670$ nm). At zero time the gain was reduced from 300 to 270. Each feeding event delivers 0.65 ml of nutrient. Lower curve: Growth rate of *C. utilis* in continuous light (LL 3100 Lux. λ_{max} 462 nm). At zero time the culture shown in upper curve was exposed to continuous light (DD→LL a "switch-up"). The nephelostat volume is indicated; the cell generation time (GT cell) for different periods of continuous culture are indicated above the feeding curve. The slope lines drawn in the lower curve derived by extending the linear portions of the feeding curve and fitting those which were of approximately of the same slope to a common parallel slope.

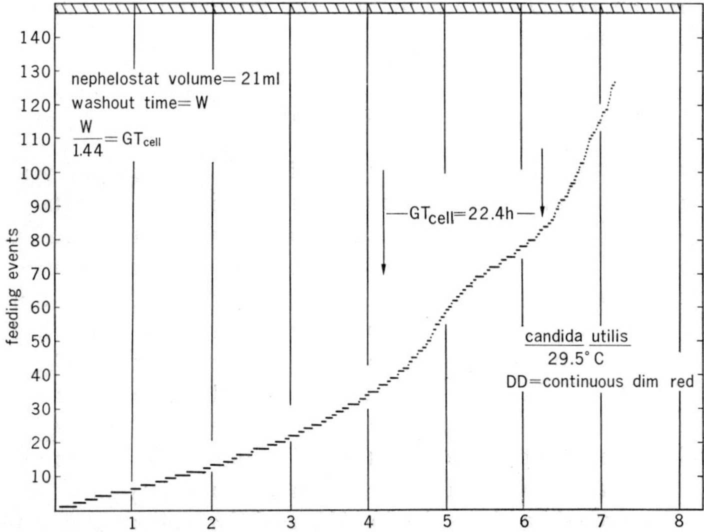

Fig. 2. Growth rate of *C. utilis* in a nephelostat in constant dark (DD, dim red light), when cell density is above 5×10^6 cells per ml. The arrows over the feeding curve delimit the period of continuous culture for which the greatest excursion in the growth rate oscillations occurs in the region of circadian dilution rates (GT cell=22.4 hours).

(bounded by the arrows in Fig. 2) in which the feeding curve typically displays a fairly large amplitude oscillation which extends over a time period of about two days, and during which the cell generation time is circadian-infradian (GT cell= 22.4 hours). This large amplitude oscillation of circadian-infradian frequency is followed by a series of small amplitude oscillations of increasingly higher frequency until a new steady-state of growth is reached in the ultradian mode of growth.

Light-dark entrainment of growth rate oscillations in a nephelostat.

Fig. 3 shows that a circadian rhythm of growth rate oscillations can be entrained by an *LD:12*,12 light-dark cycle when the yeast culture is maintained in the circadian-infradian mode of growth (GT cell=29 hours). When this light entrained culture was subjected to a "switch-down" in the level of irradiance (*LD:12*,12→DD) as shown in Fig. 4 (at zero time) a series of highly damped oscillations with a free-running period (τ) of about 22 hours was observed (as depicted by the extended slope lines above and below the feeding curve, and the arrows demarcating the period of the oscillation). Further evidence for a light entrained circadian oscillation in growth rate of *C. utilis* in a nephelostat is presented in Figs. 5 and 6. Fig. 5 shows that a circadian oscillation in growth rate is entrained by an *LD:12*,12 light-dark cycle over an 8 day period when the cell generation time in the circadian-infradian mode of growth is 36.7 hours, and that a "switch-up" (LD:12,12→LL) may or may not be effective in eliciting an endogenous rhythm of growth rate oscillations as only two very damped oscillations occurred with a period of about 20 hours. Fig. 6 shows that a circadian rhythm of growth rate oscillations can also be entrained by an *LD:10*,14 light-dark cycle over a period of six days when the cell generation time in the circadian-infradian mode was 43.3 hours. Following a "switch-down" in the level of irradiance (*LD:10*,14→DD), the growth rate oscillations had a free-running period (τ) of about 22 hours for at least three cycles (indicated by the arrows over the feeding curve in the DD portion of Fig. 6), but the oscillations appear to be highly damped and of a non-persistant nature.

DISCUSSION

A circadian rhythm of growth rate oscillations in the budding yeast, *Candida utilis* can be light entrained in continuous cultures in a nephelostat. Blue light of low intensity appears to be slightly growth stimulating under the experimental conditions used. This is readily seen both by the fact that the cell generation time in continuous blue light is about half as long as that in dim red light (Fig. 1) and by the approximately 2:1 ratio in feeding rate slopes in light versus the dark periods of cultures experiencing light-dark entrainment (Figs. 3, 5 and 6). This is consistent with the recent observations of a blue light effect on the growth and metabolism of the yeast, *Saccharomyces cerevisiae* [4–9]. These investigations have provided considerable evidence that blue light can have both inhibitory effects and stimulatory effects depending on the intensity (low intensities are stimulatory for growth and respiration [6], and for endogenous respiration of aerobically adapted cells [9], and depending on the mode of growth (slow growth is more sensitive blue light inhibition of aerobic adaptation in *S. cerevisiae* [4]

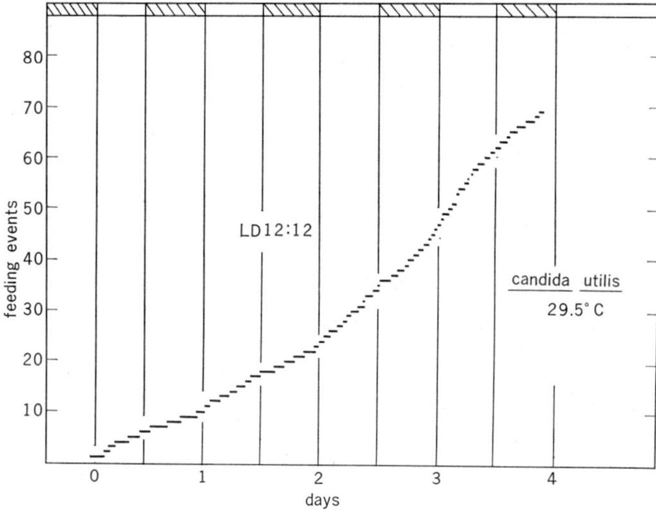

Fig. 3 Growth rate of *C. utilis* in a nephelostat on an *LD: 12,* 12 (3100, dim red light; see circadian vocabulary, Aschoff, *et al.* [11] light-dark entrainment cycle. Nephelostat volume was 21 ml; washout time, W=96 hours; and cell generation time, GT cell=29 hours.

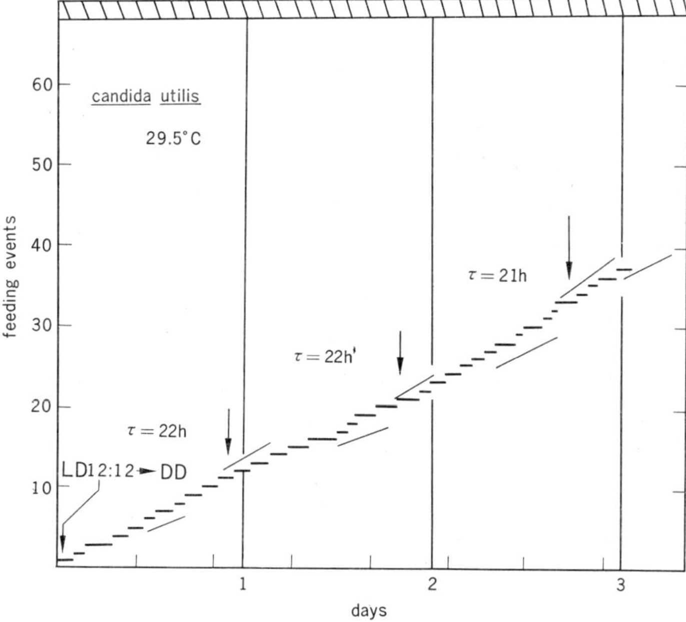

Fig. 4 Growth rate of *C. utilis* in a nephelostat following a sudden change in the level of irradiance to constant dark (DD, dim red light) of a previoutly light-dark entrained culture (*LD*: *12*,12). The free-running period (τ) of the oscillation over the three cycles shown is indicated above the feeding curve between the arrows. The period of the oscillation was obtained by the extended slope line method (see legend to Fig. 1). Th nephelostat volume was 21 ml; the cell generation time in the freerun period, GT cell= 42.4 hours.

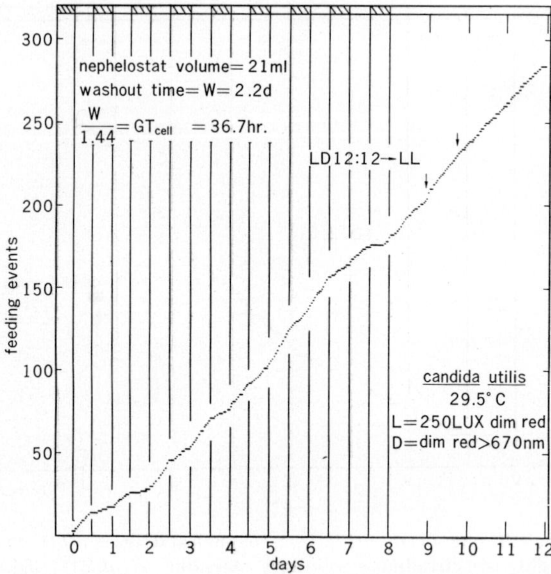

Fig. 5 Growth rate of *C. utilis* in a nephelostat on an *LD: 12,* 12 light-dark cycle, and in continuous light (LL) following a sudden change in the level of irradiance (DD→LL, a "switch-up"). The arrows indicate abrupt growth rate changes with a circadian oscillation of about 22 hours. Nephelostat volume, washout time, W, and cell generation time, GT cell, are all indicated.

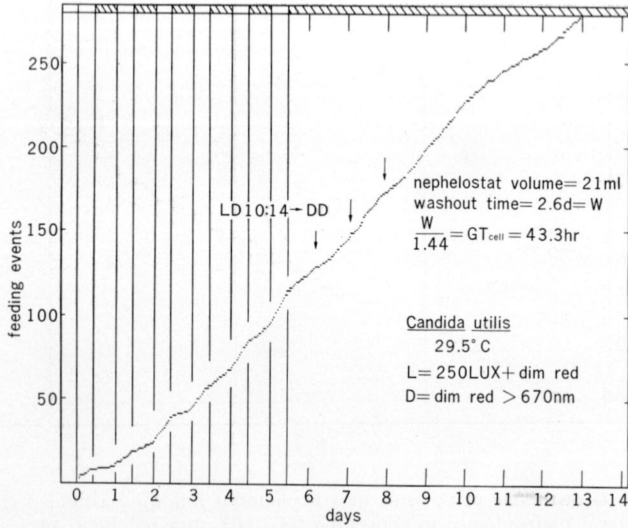

Fig. 6 Growth rate of *C. utilis* in a nephelostat on an LD: *10,* 14 light-dark cycle, and in constant dark (DD) following a sudden change in the level of irradiance (LL→DD, a "switch-down"). The arrows over the feeding curve indicate the periods of three circadian oscillations in growth rate in constant dark. Nephelostat volume, washout time, W, and cell generation time, GT cell are all indicated.

and to inhibition of glucose stimulated respiration in aerobically adapted yeast [8]. Ehrenberg [10] has convincingly shown that enhancement of glucose fermentation is the result of two antoganistic effects of blue light which is mediated by a partial reduction of the Pasteur effect.

The light entrained oscillation in growth rate seen in the nephelostat experiments reported above were obtained with well aerated cultures, and presumably do not involve anaerobic to aerobic transitions of the kind which are inhibited by blue light, but they could involve the repression and derepression of the regulatory mechanisms associated with the Pasteur effect in yeast. If the regulatory mechanisms associated with aerobic respiration and the Pasteur effect are operative primarily in yeast cells in the ultradiam mode of growth and in the transition to the infradian mode of growth, then the observations of a light entrained circadian rhythm of growth rate oscillation in yeast in the circadian-infradian mode of growth may find its ultimate explanation in the fact that a cell in switching from the rapid exponential growth mode to the slow or infradian mode of growth is capable of circadian outputs, because this switch always involves derepression of metabolic circuits collectively repressed via the feedback interactions, classically called the Pasteur effect.

REFERENCES

1. Wille J. J. and Ehret C. F. (1968): Light synchronization of an endogenous circadian rhythm of cell division in *Tetrahymena*. J. Protozool., *15*: 785–789.
2. Ehret C. F. and Wille J. J. (1970): The photobiology of circadian rhythms in protozoa and other eukaryotic microorganisms. *In*: Photobiology of Microorganisms, (P. Halldal, ed.), p. 369, John Wiley, London.
3. Tempest D. W. and Herbert D. (1965): Effect of dilution rate and growth-limiting substrate on the metabolic activity of *Torula utilis* cultures. J. gen. Microbiol., *41*: 143–150.
4. Sulkowski E., Guerin B., Defaye J. and Slonimski P. (1964): Inhibition of protein synthesis in yeast by low intensities of visible light. Nature, *202*: 36–39.
5. Guerin B. and Sulkowski E. (1966): Photoinhibition de l'adaptation respiratoire chez saccharomyces cerevisiae, I. Variations de la sensibilite àl'inhibition. Biochim. et biophys. acta, *129*: 193–200.
6. Ehrenberg M. (1966): Wirkungen sichtbaren Lichtes auf Saccharomyces cerevisiae, I. Einfluss verschiedener Faktoren auf die Hohe des Lichteffektes bei Wachstum und Stoffwechsel. Archiv für Mikrobiol., *54*: 358–373.
7. Guerin B. and Jacques R. (1968): Photoinhibition de l'adaptation respiratoire chez saccharomyces cerevisiae, II. Le spectre d'action. Biochim. et biophys. acta, *153*: 138–142.
8. Ehrenberg M. (1968): Die Hohe des Lichteffektes auf Wachstum und Stoffwechsel von Saccharomyces cerevisiae in Abhangigkeit vom Phasenstatus der Vorkulturzellen. Archiv für Mikrobiol., *61*: 20–29.
9. Ninnemann H., Butler W. L. and Epel B. L. (1970): Inhibition of respiration in yeast by light. Biochim. et biophys. acta, *205*: 499–506.
10. Ehrenberg M. (1966): Wirkungen sichtbaren Lichtes auf Saccharomyces cerevisiae, II. Die Garung unter dem Einfluss von Pasteur-Effekt und Licht-Effekt. Archiv für Mikrobiol., *55*: 26–30.
11. Aschoff J., Klotter K. and Wever R. (1965): The circadian vocabulary. *In*: Circadian Clocks, (J. Aschoff, ed.), North-Holland.

CHRONOBIOLOGY, ENDOCRINES, NEUROHUMORS AND REPRODUCTION

Chairman: Alain REINBERG

Co-Chairman: Wilbur B. QUAY

CHRONOBIOLOGICAL STUDY ON GROWTH HORMONE SECRETION IN MAN: ITS RELATION TO SLEEP-WAKE CYCLES AND TO INCREASING AGE

R. D'AGATA, R. VIGNERI and P. POLOSA

Department of Medicine and Division of Endocrinology
Catania University School of Medicine
Catania, Italy

Spontaneous, important rises in human plasma growth hormone (GH) have been demonstrated recently in relation to the onset of sleep [1, 3, 8, 11, 13]. This GH release seems to be related to synchronized sleep (EEG stages 3 and 4 or SWS, Slow Waves Sleep) [9, 12]. Therefore, in adult subjects, GH levels follow spontaneous, periodical variations [3] closely related to sleep-wake cycles [6, 14, 15].

In the present report we would like to present some observations which may give additional information about GH secretion rhythmicity, its relation to age and some neuroendocrine aspects of its regulation.

MATERIALS AND METHODS

Our study was undertaken as follows:

1) Subjects checked only for the difference between GH values during wakeful and sleeping state: a) 41 infants up to 1 year old, b) seven children 3–11 years old, c) 30 adults 21–51 years old, and d) 19 old subjects aged 61–81 years. Group a) (subjects in which wake-sleep cycles do not correspond yet to day-time-night cycles) was examined only in relation to sleep (blood samples were drawn after at least one hour of waking, fasting state and 30 min after the beginning of sleep). The other three groups were examined one hour after morning arousal (basal fasting morning value) and one hour after night-sleep onset.

2) A longitudinal study was carried out in three young men (18, 19 and 29 years old) and in four old subjects 56, 59, 66 and 72 years old) by blood sampling 5-12 times a day for 3-5 consecutive days.

3) A transversal study was carried out in: a) 26 children 3–12 years old, b) 11 adults 18–48 years old and c) eight old subjects aged 61–72 years. Blood samples were drawn 7–14 times in one single day. 4) a glucose load (100 g per os) was administered to 17 adult subjects 30 min before their usual starting time of night-sleep, after previously controlling the baseline night GH value.

All the patients, except two young adults who kept their usual life routine, were hospitalized at least 10 days before our investigation. Light-on and light-out hours are recorded for each single group in Table 1. The dark span did not strictly coincide with sleep span. The sleep onset and the waking time of each group extended into a short period of about 30 min (see Table 1). Meals were given as following: breakfast at 0830, lunch at 1230 and supper at 1830 to children, at 2000 to adults and old subjects.

Table 1 Transversal study data analysis by the "Cosinor" method (95% confidence interval).

Subjects Site of study: Catania Italy	No of subjects (No of days) Δtm, hrs	Confidence interval rhythm detection	Synchronizer light on: light out (hrs, min)	Sleep onset: Wakefulness (hrs, min)	Amplitude C (ng/ml)	Acrophase φ in degree Phase-ref: mid-sleep span	Acrophase φ in hrs, min Phase-ref: local midnight
Children 3–6 years	10 (1) 4.8	95%	0700 2145	(2130–2200) (0645–0715)	2.0 (0.9–3.1)	−331.8 (−307.5 − 5.3)	0029 (2225 0243)
Children 7–12 years	9 (1) 4.8	95%	0645 2200	(2200–2230) (0630–0700)	3.6 (0.4–6.8)	−317.3 (−277.5 −37.5)	2339 (2100 0500)
Adults 18–48 years	11 (1) 2–6	95%	0615 2300	(2300–2330) (0600–0630)	2.5 (1.3–3.8)	−325.8 (−302.2 −352.5)	0028 (2254 0215)
Old subjects 60–72 years	8 (1) 2–4.8	95%	0615 2300	(2300–2330) (0600–0630)	No statistically significant circadian rhythm detected		

Blood samples were taken by venipuncture in infants and children, and through an indwelling intravenous cannula in adults. Night blood samples were drawn 30–90 min after sleep onset on the basis of behavioral evaluation, without using EEG to determinate its depth and its course. It was necessary to awaken the subject for blood sampling. The lack of the EEG control make it possible to miss the peak of night GH value. Anyway it must be stressed that GH rises persist longer and permit the sampling of the rise to vary as much as 30–40 min before and after the time peak.

Plasma GH was determined by radioimmunoassay, using ^{125}I-HGH and a double antibody precipitation method, as previously described [17]. Our data were analyzed on an electronic computer according to Halberg's "Cosinor" programs for transversal studies and to least-squares spectrum for longitudinal ones [5].

RESULTS

First appearance of GH secretion related to sleep in infancy. Only in infants older than 3 months does a statistically significant difference between GH levels during sleeping and waking periods appear. This difference is due to a clear fall in the hormone levels during waking periods (Fig. 1).

Presence of a circadian rhythm in GH plasma levels in childhood. 17 of the 26 children examined had a night GH value higher than 5 ng/ml. GH values observed during night-time (8.7±5.9) are significantly higher than daytime values (3.7±4.4 ng/ml). This behavior is shown fairly well by chronogram analysis (Fig. 2). Data analysis, carried out with Halberg's "Cosinor" program, shows that a circadian rhythm of GH plasma levels is present, with a nocturnal phase related to sleep (see Table 1). Spontaneous GH peaks during daytime are present in this age group at a greater elevation and a greater frequency than in adults.

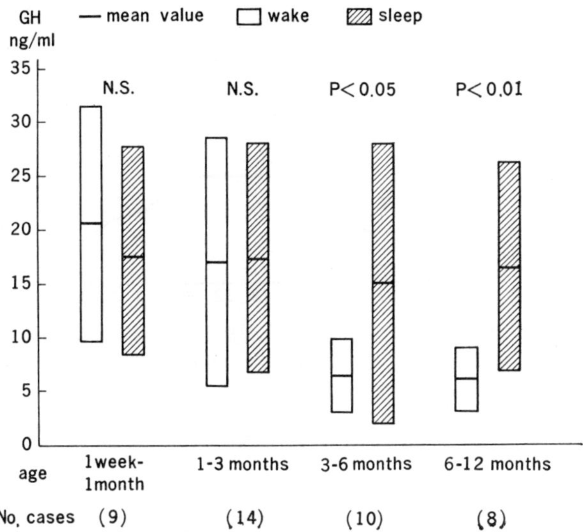

Fig. 1 In infants over three months a clear fall of GH levels (mean ±SD) is evident during wakeful state.

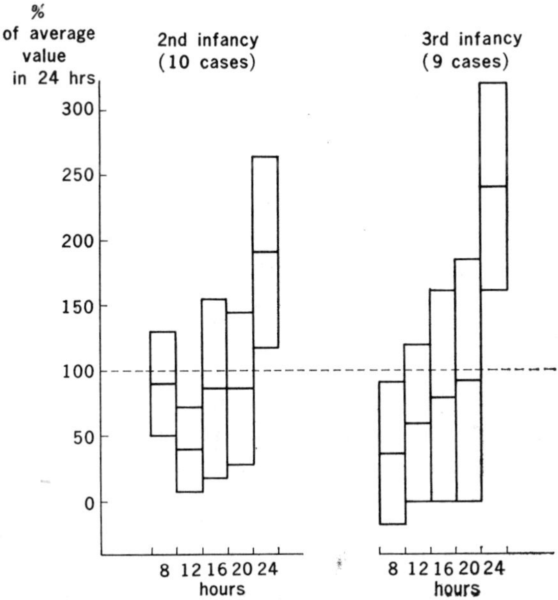

Fig. 2 In the chronogram analysis of children, in which GH values are represented as per cent of the average value in 24 hours, GH levels (mean±1 SD) are significantly higher during the night.

Presence and characteristics of GH rhythmicity in adult subjects. Twenty of the 30 adult subjects examined only for GH night peak had a post-sleep GH value greater than 5.0 ng/ml. In these 30 subjects mean GH value rose significantly ($p<.01$) from 1.3±0.82(SD) ng/ml of the basal morning sample to 5.6±3.7 ng/ml of the nocturnal sample.

In all three adults on whom a longitudinal study was carried out, our data

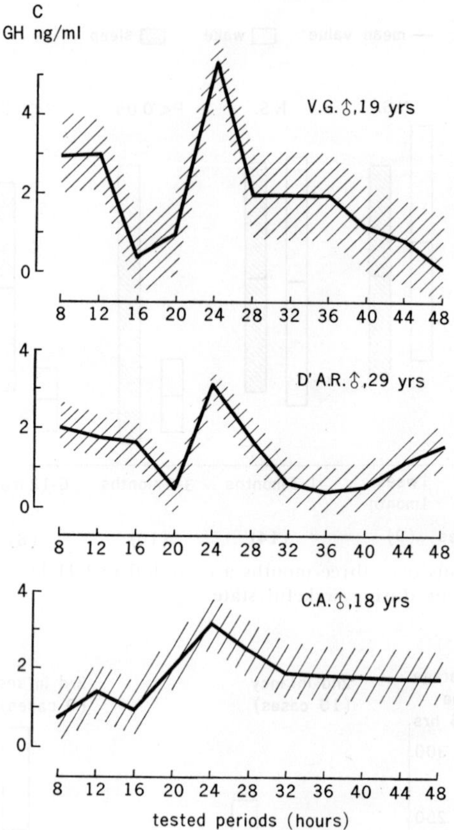

Fig. 3 In these three young male subjects a longitudinal study was carried out for five consecutive days and data analyzed by least squares method. The amplitude (C) is represented as a function of the various periods tested. The shaded area indicates ± one standard error. A single, evident peak with a clear difference in nearest points and other fluctuations, suggests the presence of a pre-eminent period for τ=24 hrs. in all the three cases.

analysis reveals that the 24-hour period seems to be significantly pre-eminent among all the other periods tested (Fig. 3). This finding is corroborated by the observation that in all three cases the standard error calculation shows its lowest value exactly in connection with a 24-hour period. Therefore in adult subjects, GH seems to follow a rhythmical pattern with a circadian period; transversal studies show the rhythm acrophase (ϕ) one hour 13 min after the average sleep onset time (Table 1). At the same night hour, seven subjects who had a GH night peak >5.0 ng/ml showed no rise in their GH level if the sleep onset was postponed until the blood sampling time.

Alteration of GH secretion in the elderly. GH secretion related to sleep seems to be altered, at least partially, in the elderly. In fact, seven of the 19-yr old subjects checked only for night GH peak, had a plasma GH value greater than 5.0 ng/ml. In these 19 subjects the difference between mean GH morning value (1.8±2.3 ng/ml) and mean night level (4.1±5.3 ng/ml) statistically were not significant. Both transversal and longitudinal studies indicate the loss of GH rhythmicity in these subjects, as no clear night peak and no pre-eminent period is observed (Table 1 and Fig. 4).

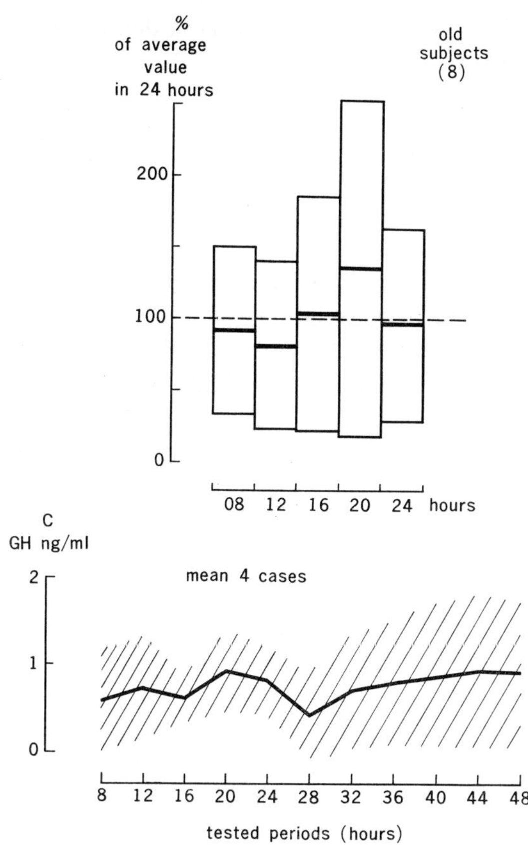

Fig. 4 In four old subjects the longitudinal study analysis does not show the presence of a pre-eminent period in GH secretion pattern.

Non-suppressibility of GH night peak by acute hyperglycemia. None of the 17 subjects to whom a glucose load was administered per os before the sleep onset showed a clearly depressed GH night peak, if referred to a previous baseline night value. In fact, no significant difference is present between the former (5.9±3.0 ng/ml) and the latter (6.1±3.5) mean value, even if a clear rise in blood sugar was evident after the glucose load.

DISCUSSION

Our results allow the following considerations:

The correlation between the GH secretion and sleep-wake cycles takes place after the third month of life, even if a certain individual variability is evident as related to chronological age. This finding seems to be closely related to EEG activity behavior at this age. In fact, until 3–4 months of age a slow waves activity in the theta frequency band prevails during wakefulness [4]; only after this age does a decrease in these waves appear during waking periods, because of a progressive acceleration in their frequency, so that at the fifth month we may observe an aspect similar to alpha waves [4]. The close relationship demonstrated in adults between GH release and Slow Waves Sleep allows one to think that the cerebral activity of the newborn, with its aspect of depressed cortical tone may be the cause of the high GH values in this age.

From childhood a circadian periodicity in GH secretion develops in man. Longitudinal and transversal studies give evidence of this rhythmical pattern of GH serum levels in physiological conditions [7, 18]. Spontaneous daytime bursts of GH are more frequent and of a greater elevation in children in respect to adults [18], as if in the latter "noise stimuli" find a higher threshold for GH release during daytime.

GH rhythmicity is present in adults, and it is closely correlated to sleep rhythmicity. We can interprete this correlation as due to a specific neural activity, also peculiar to synchronized sleep and monitoring GRF secreting structures. Alterations of normal sleep-wake cycles move and change GH rhythmicity characteristics immediately [6, 15]; these seem therefore closely and uniquely related to this strong endogenous synchronizer. In fact, the rhythmicity in nervous center activity determinates sleep-wake cycles: these give the GH rhythm its characteristics (night-time phase and circadian period). Besides, another finding must be underlined: it is the non-suppressibility of night GH discharge by hyperglycemia obtained by glucose infusion [16] or by oral glucose load.

We might call GH release related to sleep the "neural phase" of GH secretion, in respect to another phase that is suppressible by hyperglycemia ("metabolic phase"). This latter phase is not related to a periodic nervous stimulation, and therefore it is not rhythmical ("noise stimuli"). Many factors, such as metabolic and endocrine substrate variations and physical exercise, can influence this phase of GH secretion. All these stimuli, with their discontinuous, occasional influence, might be responsible for spontaneous GH bursts during daytime, not easily reproducible in the same subject at the same hour, on different days. These daytime bursts might disguise rhythmicity of the GH "neural phase", making it difficult to give evidence of a clear rhythm in GH secretion pattern.

—GH release related to sleep seems to be altered in the elderly, so that rhythmicity is often lost. Sleep modifications are suggested to determine the last finding: the percentage of EEG stages 3 and 4 of sleep decreases with increasing age [19] and consequently GH night peak, related to SWS, decreases too.

As our (unpublished data) and others' previous observations [2] in old subjects demonstrate a GH response to insulin hypoglycemia and to arginine within the normal range, we might suggest that paths of GH release related to metabolic stimulation are still efficient in the elderly, while paths for GH release related to sleep seem generally less active. The experimental observation of Pecile et al. [10] about the diminished GRF content of the hypothalamus in old rats seems to confirm these findings.

REFERENCES

1. Buckler J. M. H. (1970): Acta Endocr., *65*: 342.
2. Cartlidge N. E. F., Black M. M. Hall M. R. P. and Hall R. (1970): Geront. Clin., *12*: 65.
3. D'Agata R. and Vigneri R. (1971): Ann. Endocr., *32*: 381.
4. Ellingson R. J.: Progr. Brain Res., *9*: 26, 1964.
5. Halberg F., Engeli M., Hamburger C. and Hillman D. (1965): Acta Endocr., Suppl., *103*: 5.

6. Honda Y., Takahashi K., Takahashi S., Azumi M., Irie M., Sakuma M., Tsushima T. and Shizume K. (1969): J. Clin. Endocr., *29*: 20.
7. Hunter W. M. and Rigal W. M. (1966): J. Endocr., *34*: 147.
8. Parker D. C., Sassin J. F., Mace J. W., Gotlin R. W. and Rossman L. G. (1969): J. clin. Endocr., *29*: 871.
9. Parker D. C. and Rossman L. G. (1971): J. Clin. Endocr., *32*: 65.
10. Pecile A., Muller E., Falconi G. and Martini L. (1965): Endocr., *77*: 241.
11. Quabbe H. J., Schilling E. and Helge H. (1966): J. Clin. Endocr., *26*: 1173.
12. Quabbe H. J., Helge H. and Kubichi S. (1971): Acta Endocr., *67*: 767.
13. Sassin J. F., Parker D. C., Mace J. W., Gotlin R. W., Johnson L. C. and Rossman L. G. (1969): Science, *165*: 513.
14. Sassin J. F., Parker D. C., Johnson L .C., Rossman L. G., Mace J. W. and Gotlin R. W. (1969): Life Sc., *8*: 1299.
15. Takahashi Y., Kipnis D. M. and Daughaday W. H. (1968): J. Clin. Invest., *47*: 2079.
16. VanderLaan W. P., Parker D. C., Rossman L. G. and VanderLaan E. F. (1970): Metabolism, *19*: 891.
17. Vigneri R., Cataland F. and Papalia D. (1970): La Ricerca Clin. Lab., *10*: 20.
18. Vigneri R., D'Agata R., Morelli A., Di Bella D. and Papalia D. (1970): Folia Endocr., *5*: 504.
19. Williams R. H. (1970): J. Clin. Endocr., *31*: 461.

PERSISTING CIRCADIAN RHYTHM IN INSULIN, GLUCAGON, CORTISOL ETC. OF HEALTHY YOUNG WOMEN DURING CALORIC RESTRICTION (PROTEIN DIET)

Alain REINBERG, Marian APFELBAUM, Roger ASSAN and Dimitri LACATIS

Equipe de Recherches de Chronobiologie Humaine (CNRS, n° 105)
Fondation A. de Rothschild, rue Manin, Paris
Hôpital Bichat 170, blvd Ney, Paris
Hôtel-Dieu – Place du Parvis de Notre-Dame, Paris
Paris, France

FOOD DEPRIVATION (IN EXPERIMENTAL ANIMAL) AS WELL AS A MARKEDLY HYPOCALORIC DIET (IN MAN) DOES NOT ALTER CIRCADIAN VARIATIONS IN A LARGE NUMBER OF PHYSIOLOGIC FUNCTIONS (CHOSSAT PHENOMENON)

Chossat's early data (published in 1843) [1] demonstrate that significant circadian periodicity persists in certain birds until the day of their death from starvation and dehydration. Statistically significant within-day differences in cloacal temperature of birds completely deprived of food and water as well as in controls (eating and drinking *ad libitum*) were computed [2–4] from his data. One hundred and thirty years later Halberg et al. [2, 3, 5] demonstrated that circadian rhythms in rectal temperature, pinnal mitosis, hepatic glycogen, corticosterone content of serum and adrenal, and ACTH activity in the pituitary persist in mice deprived of food and water (Table 1). In honor of a pioneer in this field, Halberg has proposed to call this effect the Chossat phenomenon.

Circadian rhythms in hormone metabolites [6] as well as in $\dot{V}O_2$ and in R (respiratory quotient): $\dot{V}CO_2/\dot{V}O_2$ [7] were found to persist in healthy human subjects when submitted to severe caloric restriction (Table 1). The $\dot{V}CO_2$ circadian rhythm of rats persists during starvation and dehydration (M. Stupfel, unpublished data. Personal communication).

SUBJECTS, MATERIAL AND METHODS

Several groups (8 to 26 subjects) of healthy young women (18 to 25 years of age) with mild obesity were hospitalized and submitted to caloric restriction: 220 cal./24 h as protein exclusively, during 15 to 21 days. Four isocaloric meals were consumed at fixed hours: 0800, 1200, 1600, 2000. Each meal: calcium caseinate, 14 g; desalted water, 400 ml; potassium (bitartrate), 500 mg. In these conditions loss of weight was 8.02 ± 0.52 kg.

As usual in transverse circadian studies, the routine of the subjects in each involved group had to be standardized for rhythmometry [10, 11]. All the subjects were synchronized with light on at 0730 and light out at 2330. Therefore the period, τ, of their circadian rhythm is, on the average, 24 h.

A set of physiologic variables have been measured 4 times a day at fixed hours

Table 1 Persistance of physiologic circadian rhythm during food deprivation and hypocaloric diet. Chossat phenomenon.

Subjects	Synchronizer schedule	Type of diet	Studied circadian physiologic variations	References
Birds (pigeon etc.)	Natural LD cycle	Complete deprivation of food & water until death	Cloacal temperature ($\Delta t = 12$ h)	CHOSSAT [1]
766 ♀ C mice 4,5 months of age	L (0600–1800) D (1800–0600)	No food, no water No food but water	Liver glycogen ($\Delta t = 4$ h)	HAUS HALBERG [5]
192 ♀ C mice 3–4 months of age	L (0600–1800) D (1800–0600)	No food no water during a 36 hrs span of time	Rectal temperature, pituitary adrenocorticotropic activity, serum corticosterone, pinnal mitosis ($\Delta t = 4$ h)	HALBERG et al. [2] GALICICH et al. [3]
7 healthy human adults 3 ♀–4♂ 18 to 45 years of age	L (0745–2300) D (2300–0745)	Complete 36-hour bedrest with 4-hourly hypocaloric meals during sampling (336 cal/24 h as tomato juice)	Heart rate, blood pressure, urinary potassium, 17-hydroxycorticosteroid, adrenaline, noradrenaline, vanilmandelic acid ($\Delta t = 4$ h)	REINBERG, GHATA et al. [6]
8 healthy obese human females 18 to 25 years of age	L (0730–2330) D (2330–0730)	15 to 20 days of caloric restriction (220 cal/24 h as protein exclusively)	Oxygen consumption $\dot{V}O_2$ Respiratory quotient R ($\dot{V}CO_2/\dot{V}O_2$) ($\Delta t = 0.5$ h)	APFELBAUM, REINBERG et al. [7]
6 healthy human ♂ 18 to 25 years of age	L (0700–0000) D (0000–0700)	Low protein diet (30 cal/kg body w/24 h: egg protein)	Tyrosine and other amino acids concentration in plasma ($\Delta t = 4$ h)	WURTMAN et al. [8,9]

(0600–1200–1800 and midnight) and equal intervals ($\Delta t = 6$ hr), during 24 hr for each of the subjects: a) while on a spontaneously selected diet (before caloric restriction) and b) again after 14 and/or 21 days of protein diet.

Plasma cortisol (method of J. BEGUE, a modification of the De MOOR et al. procedure: [12]) as well as urinary nitrogen (Kjeldahl) were determined chemically. Plasma insulin and glucagon (peripheral venous blood samples) were determined radioimmunologically according to YALOW and BERSON's principles [14, 15]. $\dot{V}O_2$ and $\dot{V}CO_2$ were measured "continuously", ($\Delta t = 30$ min.) without a mask, during 24 h, in an open circuit metabolic chamber [13]. Medical thermometers in °C were used for rectal temperatures.

Time series thus obtained were analyzed microscopically (cosinor and related method: HALBERG et al., [16]).

RESULTS (Fig. 1—Table 2)

a) Statistically significant circadian rhythms ($P < .05$) were detected in the following physiologic variables: plasma insulin, glucagon and cortisol, respiratory quotient, rectal temperature, both before and during caloric restriction and protein diet.

b) Caloric restriction alters neither the acrophase (Fig. 1, Table 2) nor the amplitude (Table 2) of the studied variables.

Table 2 Circadian rhythms in hormonal and other physiologic variables of healthy young women before and during protein diet and caloric restriction (transverse study).

Variable (unit of measurement)	No of subjects	Caloric restric. (No of weeks)	P Rhythm detection	C_0 ±SE	Amplitude C	Acrophase ϕ in hrs min. ϕ ref.: midnight	Acrophase φ in degrees φ ref.: midsleep
					(95% Confidence limits)		
Insulin	8	no	<.05	28.6±4.1	6.5 (0.2 to 11.3)	1407 (0932 to 1841)	−159° (−90 to −227)
(μu/ml)		yes (3)	<.05	17.9±3.2	4 (0.1 to 7.1)	1611 (1058 to 2125)	−190° (−112 to −268)
Glucagon	8	no	<.05	379±20	35 (1 to 72)	1817 (1240 to 2355)	−221° (−137 to −306)
(pg/ml)		yes (3)	<.05	375±18	37 (2 to 72)	1818 (1515 to 2122)	−222° (−176 to −268)
Cortisol	11	no	<.05	8.9±1.2	4 (0.2 to 7.9)	0608 (0100 to 1115)	−39° (−322 to −116)
(μg/100 ml)		yes (2)	<.025	14.9±1.5	7.2 (2.3 to 12.2)	0733 (0548 to 0917)	−60° (−34 to −86)
		yes (3)	<.05	9.3±1.2	4.9 (0.2 to 9.6)	0643 (0302 to 1024)	−48° (−19 to −76)
$\dot{V}CO_2/\dot{V}O_2$	8	no	<.01	733±10.3	26 (9 to 43)	1424 (1248 to 1559)	−163° (−139 to −187)
(ratio. 10^3)		yes (2)	<.05	746±14.8	22 (4 to 39)	1147 (0748 to 1546)	−124° (− 64 to −184)
$\dot{V}O_2$	8	no	<.10	19.80±.074	.641	1732	−210°
(l/hr)		yes (2)	<.10	16.77±.077	.942	1724	−208°
Rectal	26	no	<.005	37.06±.05	0.3 (0.2 to 0.4)	1624 (1458 to 1743)	−192° (−172 to −213)
temperature		yes (2)	<.005	37.06±.07	0.3 (0.2 to 0.4)	1449 (1401 to 1536)	−169° (−157 to −181)
(°C)		yes (3)	<.005	37.00±.05	0.3 (0.2 to 0.4)	1502 (1423 to 1540)	−173° (−162 to −182)
Urinary	22	no	>.05	3.35±.38			
nitrogen		yes (2)	<.01	2.97±.11	.82 (.26 to .138)	1249 (0928 to 1610)	−142° (−92 to −193)
(g)		yes (3)	<.025	2.66±.10	.48 (.14 to .83)	1329 (0940 to 1718)	−152° (−95 to −209)

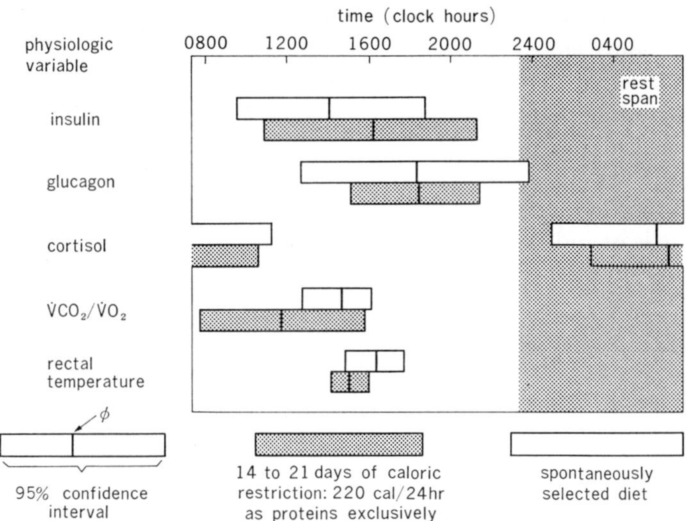

Fig. 1 Circadian acrophase in a set of physiologic variables of healthy young women before and during protein diet and caloric restriction.

c) During caloric restriction and protein diet, a statistically significant decrease occurs in C_0* of certain rhythms only: insulin, respiratory quotient, oxygen consumption.

COMMENTS

1) Circadian ϕ, C and C_0 values in insulin, cortisol and rectal temperature rhythms of healthy young women while on the diet presented here are in agreement with results obtained by other authors [4, 16–20] using similar methodology and means of investigation.

2) Caloric restriction (men) as well as starvation (mice) do not alter the acrophase and the amplitude of each of the studied physiologic variables. This is a new example of the Chossat phenomenon.

3) For certain physiologic rhythms—i.e., in hepatic glycogen of mice [5], in catecholamine urinary excretion [6], in $\dot{V}O_2$, in respiratory quotient [7] and in insulin of man, a statistically significant decrease in C_0 (24-hr mean) occurs.

Circadian rhythms in plasma insulin and glucagon persist during caloric restriction and glucose deprivation; nevertheless, insulin-C_0 decreases while glucagon-C_0 remain unchanged.

4) From the point of view of energy metabolism, some additional comments on these results may be useful. The 24-h mean of $\dot{V}O_2$ is reduced by about 15% after caloric restriction. These results demonstrate the adaptation of the organism's energetic expense to the calories supplied. The detection of statistically significant circadian rhythms in respiratory quotient, insulin and glucagon both on a free and on a restricted diet (Table 2) leads to postulating a metabolic rhythm in the utilization of nutrients. The organism (mice, men) deprived of

* When determinations are done at equal intervals along the 24-h scale and for an integral number of cycles, the 24-h mean is equal to C_0, the rhythm adjusted level.

food would be able to favor at a certain hour the gluconeogenesis and to favor the glucolysis 12 hours later or earlier.

As far as circadian rhythms in the organism's susceptibility to a wide variety of chemical (including drugs) and physical agents have been demonstrated [21–23], one has to consider also the possible existence of circadian rhythms in energy effects of sugar, fat and protein intakes.

In this connection, several studies on men (healthy individuals as well as diabetics) are at least indirectly pertinent: circadian changes in (oral) glucose tolerance tests have been reported [24–28].

ACKNOWLEDGMENTS

The authors are indebted to Pr. Agrégé G. TCHOBROUTSKY (Hôtel-Dieu Paris) for insulin determinations, and to Pr. Agrégé J. LENORMAND (Hôpital Bichat Paris) for cortisol determinations. Special thanks are due to Ch. ABULKER and J. DUPONT for their help in computer analyses of time series.

REFERENCES

1. CHOSSAT Ch. (1843): Recherches expérimentales sur l'inanition. Mem. Acad. Royale des Sciences, *8*: 438–640 (cf. particularly pp. 532–566).
2. HALBERG F., GALICICH J. H., UNGAR F. and FRENCH L. A. (1965): Circadian rhtyhmic pituitary adrenocorticotropic activity, rectal temperature and pinnal mitosis of starving, dehydrated C mice. Proc. Soc. Exp. Xiol. and Med., *118*: 414–419.
3. GALICICH J. H., HALBERG F. and FRENCH L. A. (1963): Circadian adrenal cycle in C mice kept without food and water for a day and a half. Nature, *197*: 811–813 (February).
4. HALBERG F. et REINBERG A. (1967): Rhythmes circadiens et rhythmes de basses fréquences en physiologie humaine. J. Physiol. (Paris), *59*: 117–200.
5. HAUS E. and HALBERG F. (1966): Persisting circadian rhythm in hepatic glycogen of mice during inanition and dehydration. Experientia, *22*: 113.
6. REINBERG A., GHATA J., HALBERG F., GERVAIS P., ABULKER Ch., DUPONT J. et GAUDEAU CL. (1970): Rythmes circadiens du pouls, de la pression artérille, des excrétions urinaires en 17-hydroxycorticosteroides, catécholamines et potassium chez l'homme adulte sain, actif et au repos. Ann. Endocrinol. (Paris), *31*: 277–287.
7. APFELBAUM M., REINBERG A., LACATIS D., ABULKER Ch., BOSTSARRON J. et RIOU F. (1971): Rythme circadien de la consommation d'oxygène et du quotient respiratoire de femmes adultes jeunes en alimentation spontanée et après restriction calorique. Rev. Europ. Etude Clin. Biol., *16*: 135–143.
8. WURTMAN R. J., CHOU C. and ROSE C. M. (1967): Daily rhythm in thyrosine concentration in human plasma: persistance on low-protein diets. Science, *158*: 660.
9. WURTMAN R. J., ROSE C. M., CHOU C. and LARIN F. F. (1968): Daily rhythms in the concentration of various amino-acids in human plasma. New England J. Med., *279*: 171–175.
10. HALBERG F. (1959): Physiologic 24-hour periodicity; general and procedural considerations with reference to the adrenal cycle. Z. für Vitamin-Hormon—und Fermentforschung, *10*: 225–296.
11. REINBERG A. (1971): Methodologic considerations for human chronobiology. J. Interdisc. Cycle Research, *2*: 1–15.
12. De MOOR P., STEEN O., RASKIN M. and HENDRIKX A. (1960): Fluorimetric determination of free plasma 11-hydroxycorticosteroids in man. Acta Endocrinol., *33*: 297–307.
13. APFELBAUM M., LACATIS D., AMOURIQ P., HUOT D., JOLIFF M. et BARCON F. (1971): Montage permettant des mesures prolongées de la consommation d'oxygène (VO_2)

du quotient respiratoire (R) et de laperspiration insensible chez l'Homme. J. Physiol. (Paris), *63*: 91–95.

14. Assan R., Tchobroutsky G. and Derot M. (1971): Glucagon radioimmunoassay. Technical problems and recent data. Hormone Metab. Res., *3*: 82–90.
15. Tchobroutsky G. (1969): Les dosages radio-immunologiques. Ann. Med. Interne., *120*: 717–728.
16. Halberg F., Tong Y. L. and Johnson E. A. (1967): Circadian system phase—an aspect of temporal morphology; procedures and illustrative examples. Proc. International Congress of Anatomists. *In*: The Cellular Aspects of Biorhythms (Symposium on Biorhythms), Springer-Verlag, pp. 20–48, Book Chapter.
17. Halberg F. (1969): Chronobiology. Ann. Rev. Physiol., *31*: 676–725.
18. Deschamps I., Heilbronner J., Lestradet H. et Canivet J. (1969): Les variations de l'insuline chez le sujet normal au cours du nycthémère. Ann. Endocrinol. (Paris), *30*: 589–594.
19. Haus E. and Halberg F. (1970): Circannual rhythm in level and timing of serum corticosterone in standardized inbred mature C-mice. Env. Research, *3*: 81–106.
20. Assenmacher I. et Boissin J. (1970): Rythmes circadiens et circannuels du fonctionnement cortico-surrénalien et thyroïdien en relation avec le reflexe photo-sexuel. *In*: Neuroendocrinologie. Colloque du CNRS, Paris, septembre 1969, pp. 405–423, CNRS Pub. (n° 927), Paris.
21. Halberg F. (1962): Physiologic 24-hour rhythm. A determination of response to environmental agents. *Ins* Man's Dependence on the Earthly Atmosphere, (K. E. Schaefer, ed.), pp. 48 69, MacMillan, New York.
22. Halberg F. (1964): Organisms as circadian systems: temporal analysis of their physiologic and pathologic responses, including injury and death. *In*: Walter Reed Army Institute of Research Symposium, Medical Aspects of Stress in the Military Climate, April, 1964, pp. 1–36.
23. Reinberg A. and Halberg F. (1971): Circadian chronopharmacology. Annual Reviews of Pharmacology, *2*: 455–492.
24. Bowen A. J. and Reeves R. L. (1967): Diurnal variations in glucose tolerance. Arch. Int. Med., *119*: 261.
25. Jarett R. J. and Keen H. (1969): Diurnal variation of oral glucose tolerance: a possible pointer to the evolution of diabetes mellitus. Brit. Medic. J., *2*: 341–344.
26. Roberts H. J. (1968): The value of afternoon glucose tolerance testing in the diagnostic prognosis and rational treatment of "early chemical diabetes". Acta Diabetol. Lat., *5*: 532.
27. Lestradet H., Labram C. et Alcalay D. (1970): Diabète du soir: espoir ou l'importance de l'horaire dans l'interprétation d'une épreuve d'hyperglycémie provoquée. Presse Méd., *78*: 1481–1482.
28. Alcalay D., Labram C., Deschamps I. et Lestradet H. (1971): L'importance de l'horaire de l'épreuve pour l'interprétation d'une courbe d'hyperglycémie porovoquée. Le Diabète, *19*: 27–30.

A STUDY ON THE POSSIBLE PRESENCE OF A THYROTROPIN SERUM LEVELS RHYTHMICITY IN MAN

R. VIGNERI and R. D'AGATA

Department of Medicine and Division of Endocrinology
Catania University School of Medicine
Catania, Italy

Lemarchand-Béraud [5], Blume [2] and Nicoloff [6, 7] have given evidence of a fluctuation in TSH serum levels in man. A similar observation has been reported by Bakke [1] and Singh [9] in the rat. Other studies by Utger [10], Odell [8] and Hershman [4] have stated that, even if slight fluctuations were present, no diurnal variations in TSH values appeared to occur in euthyroid man.

None of these authors carried out a real chronobiological study on TSH secretion; in particular some of them did not monitor hormone levels frequently during the time of observation, and their data were collected for a too short period (24 hrs). Furthermore they did not apply to their results any mathematical elaboration that could give objective and statistically significant evidence whether TSH serum levels follow a rhythmical pattern or not. Therefore their conclusions are affected by a subjective, individual judgement.

Since all the previous research does not give definite evidence about the presence of a rhythmical pattern in TSH secretion, we thought it would be interesting to investigate this aspect with a "longitudinal" study in normal subjects with the analysis of our results by least squares method.

MATERIALS AND METHODS

Seven normal subjects 18–66 years old (5 men and 2 women) were examined in this study. It lasted 5 days in 2 subjects, 4 days in 2 and 3 days in the other 3.

Two of them were allowed to keep their normal life routine (G.V and D'A.R, see Fig. 2); the other five all were hospitalized at least 10 days before the study and were kept resting during the investigation; light-on time was at about 0630; light-out at about 2300; the dark span roughly coincided with the sleep span. Meals were given at about 0830, 1230 and 2000. Venous blood samples were drawn through an indwelling plastic cannula at irregular intervals 5–13 times a day for every single day.

Serum TSH was measured by a double antibody radioimmunoassay [11] using rabbit anti-HTSH serum (NIH, Bethesda) and purified HTSH labelled to ^{125}I by Sorin (Saluggia, Italy). Results are expressed in terms of HTSH Research Standard A (NIMR, London). Our data were analyzed with an electronic computer according to Halberg's program of least-squares spectrum [3].

RESULTS*

Macroscopic analysis: TSH values observed in the 189 samples examined ranged from <1.0 to 14.5 μU/ml. The mean values for each three hr period are

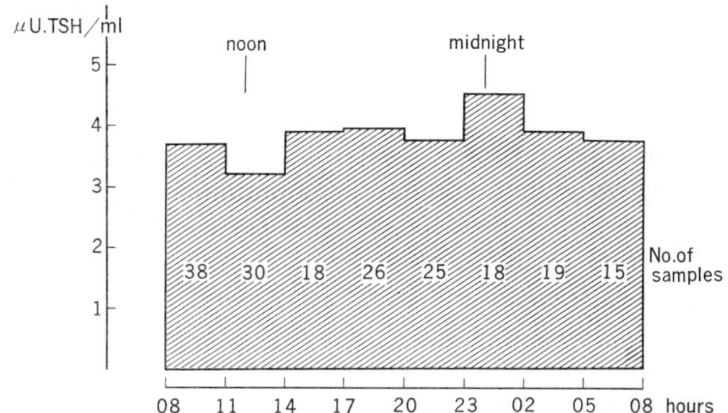

Fig. 1 Mean TSH value of samples obtained in each 3 hr period shows a 2300–0200 zenith level and a 1100–1400 nadir level. No statistically significant difference is present among these values.

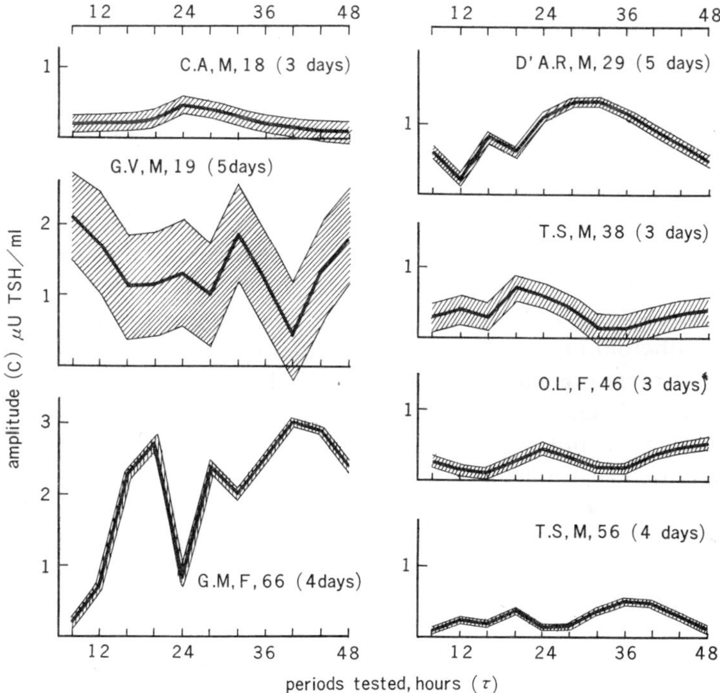

Fig. 2 Longitudinal analysis of our data following least squares spectrum. The shaded area represents $\pm\sqrt{\frac{\sum(y-yy)^2}{N}}$.

* To permit a mathematical elaboration of our data, non-detectable values (<1.0μU/ml) were arbitrarily recorded as 0.5 μU/ml when their radioactivity was still 5% over the radioactivity of 1.0 μU/ml standard; as 0.1 μU/ml when their radioactivity was over this percentage.

represented in Fig. 1: the highest mean value was observed from 2300 to 0200, the lowest one from 1100 to 1400. The difference among all these values was not statistically significant.

No relation was observed between TSH levels and sleep or meals.

Microscopic analysis: The least-squares spectrum analysis of our data showed no pre-eminent period in any of the 7 subjects examined as to the range tested (8–48 hrs). In fact, no amplitude is clearly higher than other fluctuations.

DISCUSSION

Our results confirm the presence of slight variations in TSH levels of the same subject both during the same single day and from day to day. Macroscopic analysis shows that the mean zenith value occurred between 2300 and 0200, even if a considerable individual variability is present. In our experience the values we obtained in the late afternoon and night are generally slightly higher than the ones obtained in the morning. These data are in good agreement with similar observations by Lemarchand [5], Nicoloff [6, 7] and Hershman [4].

In no case did our data analysis by least-squares spectrum show the presence of a clearly pre-eminent period. The absence of a circadian periodicity in TSH secretion is rather surprising; since a circadian periodicity may be a general property of the neurohypophyseal system. Besides, it contrasts with macroscopic chronogram analyses made by ourselves and others which seem to reveal the presence of a diurnal fluctuation in TSH levels.

The most logical explanations for this discrepancy might be: 1) the limited number of cases of our longitudinal study; but this observation contrasts with the fact that a pre-eminent 24 hour period was not observed in any case. 2) the limited number of samples drawn every day from our subjects; however it must be noticed that the human TSH half-life is about one hour or longer [4], and therefore, such permits a blood sampling 60 min before or after the zenith time without completely losing any important peak. Of course, more precise data could be obtained by a continuous blood sampling system. 3) the insufficient sensitivity of the radioimmunoassay now available; it does not allow a precise discrimination in the low euthyroid range. In our assay, values under 0.8–1.0 μU/ml cannot be detected, and therefore important falls in TSH serum levels might be missed. 4) the insufficient length of our study; the time may have been too short to detect a circadian rhythm; but, in contrast to TSH, in these subjects, the same study showed fairly well rhythmicity in GH and cortisol.

We conclude that further, longer and more complete studies which we already have begun can overcome points 1, 2 and 4; but the possibility should be considered also that there may not be a circadian rhythm in plasma TSH, even if we cannot find at the present any clear explanation and purpose.

REFERENCES

1. Bakke J. L. and Lawrence N. (1965): Metabolism, *14*: 841.
2. Blume A. S., Greenspan F. S. and Magnum J. (1968): Exc. Med. Int. Congr. Ser., No. 157 (Abst. 34).
3. Halberg F., Engeli M., Hamburger C. and Hillman D. (1965): Acta Endocr., *5* (Suppl. 103).
4. Hershman J. M. and Pittman J. A. (1971): Ann. Int. Med., *74*: 481.

5. LEMARCHAND Th. and VANNOTTI A. (1969): Acta Endocr., *60*: 315.
6. NICOLOFF J. T. (1970): J. clin. Invest., *49*: 1912.
7. NICOLOFF J. T., FISHER D. A. and APPLEMAN M. D. (1970): J. clin. Invest., *49*: 1922.
8. ODELL W. D., WILBER J. F. and UTIGER R. D. (1967): Recent Progr. Horm. Res., *23*: 47.
9. SINGH D. V., PANDA J. N., ANDERSON R. R. and TURNER C. W. (1967): Proc. Soc. Exp. Biol. Med., *126*: 553.
10. UTIGER R. D. (1965): J. clin. Invest., *44*: 1277.
11. VIGNERI R., PAPALIA D. and MOTTA L. (1969): J. Nucl .Biol. Med., *13*: 151.

GASTRIC FUNCTION IN A GROUP OF VAGOTOMIZED PATIENTS WITH DUODENAL ULCER AND IN A CONTROL GROUP: A TEMPORAL STUDY

M. CAGNONI, B. TARQUINI, R. ORZALESI and S. ROMANO

Medical School, University of Florence
Florence, Italy

A temporal study on gastric function in controls and in patients with duodenal ulcer has been carried on in our Laboratory of Chronophysiopathology.

This study was designed to investigate the daily variations of gastric titratable acidity.

Six controls (1 man and 5 women) with nonpeptic disorders (Cholecystectomy) and five patients with duodenal ulcer (3 men and 2 women) were used in this study. All subjects were operated and a small gastrostomy was performed for a better postoperative course. Complete vagotomy was performed in all patients with duodenal ulcer.

The study was started, in all subjects, at the seventh postoperative day. Patients and controls were kept in the same controlled conditions, with night rest from 2100 h to 0500 h and meals served at 0600, 1100, 1700.

Samples of gastric juice were obtained from gastrostomy at 0600, 0800, 1000, 1200, 1400, 1600, 1800, 2000, 2200, 2400, from a minimum of two to a maximum of four days, and titratable acidity was tested by neutralization with 10 N NaOH: results were expressed as ml 10 N NaOH per cent of gastric juice.

Microscopic analysis was carried on by the Cosinor method, applying to "hybrid sample" procedure [1].

Table 1 shows results obtained.

Table 1 Cosinor summary of circadian studies on gastric titratable acidity expressed as ml. of 10N NaOH per cent of gastric juice in a group of vagotomized patients with duodenal ulcer and in a control group.

Subjects investigated° (Length of study in days for entire group)	Circadian Amplitude C	Series Average (Co)	Acrophase Above: from mid-D (360°=24 h) Below: from local midnight
6 (1 ♂; 5 ♀) adult controls aged 32–70 (20)	16.7±13.90	68.1±8.02	−285° (−255 to −2) 1903 (1704 -0019)
5 (3 ♂; 2 ♀) vagotomized with peptic ulcer aged 25–64 (16)	10.1±19.26	73.7±4.90	−307° (−261 to −45) 2048 (1743 -0305)

* Synchronizer schedule: L 0500 to 2100 alternating with darkness.

It is apparent that in controls a circadian rhythm of titratable acidity is detectable at a probability of 95%. The confidence arc for the circadian acrophase extends clockwise from 1704 to 0019. The circadian crest is at 1903 (−285° from midnight).

It also is evident that vagotomized patients with duodenal ulcer displayed a circadian rhythm of gastric titratable acidity. The 95% confidence arc for circadian acrophase extends clockwise from 1743 to 0305. The circadian crest is at 2048 (−307° from midnight).

Temporal studies on gastric function are extremely rare. As far as we know only MOORE and ENGLERT [2] carried on, in 1970, a research specifically performed to investigate this topic. They obtained gastric acid samples from 27 subjects, of whom nine had a duodenal ulcer, eight had a variety of nonpeptic disorders and 10 were healthy volunteer subjects. Each was fasted for 12 h before the study and given continuously intravenous infusion of dextrose. Gastric juice was obtained by a nasogastric tube. From their data, macroscopically analized, MOORE and ENGLERT evinced a circadian periodicity in gastric acid secretion with the greatest rate of secretion in the evening and the smallest in the morning between 0500 and 1100.

Although designed with similar purposes, our study is only in part comparable with theirs. In fact MOORE's data were obtained from the group of normal subjects and from the whole experimental group which included non-operated patients with duodenal ulcer. Anyhow, under such a point of view, a remarkable similarity is detectable between our control group and their normal group. Such a similarity seems worthy of consideration; in fact our subjects were fed at various hours of the day, whereas MOORE's control group was fasted and fed intravenously with a continuous infusion of dextrose.

We could not fast our subjects because of their postoperative condition; but we believe that, if possible, it would not have been sufficient to explain whether the circadian periodicity is or is not a conditioned response to meals. In fact the fasting period could not be protracted for a sufficient time.

Results show that patients with duodenal ulcer and complete vagotomy, as well as controls, display a circadian periodicity of gastric secretion.

We like to emphasize that the results obtained might support the view that temporal changes of titratable acidity are not conditioned by parasympathetic innervation.

That is rather surprising considering the role of the parasympathetic system in regulating gastric acid secretion.

REFERENCES

1. HALBERG F., TONG Y. L. and JOHNSON A. (1967): Circadian system phase. An aspect of temporal morphology; procedures and illustrative examples. *In*: The Cellular Aspects of Biorhythms. Symposium on Rhythmic Research. Sponsored by the VIIIth Int. Congress of Anatomy, Wiesbaden, August 1965, pp. 20–48, Springer-Verlag, Berlin.
2. MOORE J. G. and ENGLERT E., Jr. (1970): Circadian rhythm of gastric acid secretion in man. Nature, *226*: 1261.

A STUDY OF PERIODICITY IN A PATIENT WITH HYPERTENSION: RELATIONS OF BLOOD PRESSURE, HORMONES AND ELECTROLYTES

WALTER J. MEYER, CATHERINE S. DELEA, HOWARD LEVINE,
FRANZ HALBERG and FREDERIC C. BARTTER

Endocrinology Branch, National Heart and Lung Institute
Bethesda, Maryland, U.S.A.
New Britain General Hospital, New Britain, Conn., and University of Minnesota, Minneapolis, Minnesota, U.S.A.

In this study we have elected to examine the circadian rhythms of a single subject, a patient with hypertension. In addition to performing the clinical studies necessary to evaluate the existence and etiology of his hypertension, we sought to answer the following questions: Is blood pressure a periodic function? What will be the effect of changing sodium intake on his blood pressure, on its periodicity, on the circadian rhythms of related variables such as serum and urinary sodium, potassium, aldosterone, chloride, phosphorus, calcium, magnesium, and 17-hydroxycorticosteroids? Can these rhythms, if they exist, be changed by having the patient undergo a 12-hour phase shift? Can they be dissociated by phase-shifting? How long will it take his rhythms to reestablish their periods on the new time schedule? The following study is a preliminary report of our results.

The patient, a 55 year-old physician, was admitted to the Clinical Center of the National Institutes of Health for evaluation and study. The initial clinical data and diagnostic tests are related in detail in Table 1. On the basis of the low serum potassium concentration, the low plasma renin activity and the elevated aldosterone secretion rate seen in this patient while he was taking a high sodium diet (240 mEq/d), we decided that his hypertension was related to the hyperfunctioning of the adrenal cortex. This syndrome is commonly called primary aldosteronism.

As can be seen from Fig. 1, the study was divided into three parts: a period of high-sodium intake (240 mEq/d), a period of low-sodium intake (9 mEq/d) and a period of low sodium intake during which a phase shift was imposed. The phase shift was initiated by a period of 28-hour wakefulness followed by the routine of sleeping during the usual waking hours. Each day of the study the patient ate the same metabolic diet (to which sodium chloride tablets were added at mealtimes to achieve the high-sodium intake), stayed within the air-conditioned constant environment and had a definite set routine which he followed each day and shifted appropriately with the 12-hour phase shift. The period of low sodium was initiated with a mercurial diuretic 2 ml intramuscularly which rapidly depleted the patient of sodium.

Urine was collected in 4-hour periods during the entire thirty days and analyzed for sodium, potassium, chloride, calcium, phosphorus, magnesium, 17-hydroxycorticosteroids and aldosterone. On seven separate days, venous blood was collected every four hours and analyzed for sodium, potassium, chloride, bicarbonate, cal-

Table 1

H.L. 55 y. M

Clinical data		
Examination: BP 190/95; heart murmur of aortic insufficiency		
Cardiac Fluoroscopy: normal heart size with mid-diastolic collapse of ascending aorta.		
Electrocardiogram: left ventricular hypertrophy, flattened T waves		
Urinalysis: specific gravity 1.027, pH 5.5; negative protein, sugar, acetone, microscopic examination		
Intravenous pyelogram and renogram: normal		
Serum thyroxine (T_4): 4.3 ug/100 ml		
Special studies		
	249 mEq Na^+ diet	9 mEq Na^+ diet
Aldosterone secretion rate	1534 μg/day	762 μg/day
Serum potassium	3.0 mEq/L	4.0 mEq/L
Plasma renin activity as Angiotensin I		
Supine	0 ng/ml/hr	.266 ng/ml/hr
Upright	.266 ng/ml/hr	.532 ng/ml/hr
Normals:		
Aldosterone secretion rate	62±28 (S.D.)	1107±763 (S.D.)
Plasma renin activity as Angiotensin I		
	110 mEq/Na	
Supine	1.02±0.21 (S.D.)	2.16±0.45
Upright	1.75±0.43	5.92±0.40
Diagnosis: Primary Aldosteronism		

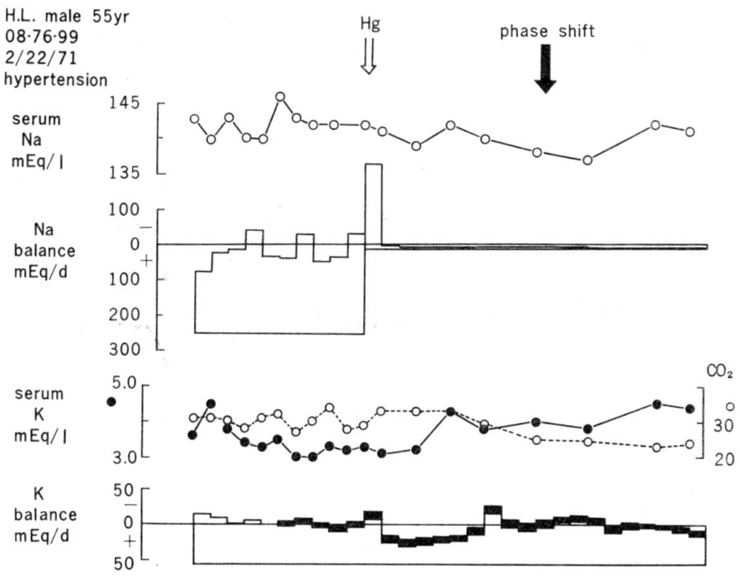

Fig. 1 The effects of high- and low-sodium intakes and of phase shift on serum sodium concentration (Na), sodium balance, serum potassium concentration (K) and potassium balance in a patient with hypertension.

cium, phosphorus, magnesium and plasma renin activity. Stools were collected in four-day pools and analyzed for sodium, potassium, calcium, phosphorus and magnesium. The blood pressure was measured at least six times a day by the patient but was also measured every ten minutes for several days continuously during each of the three periods by an automatic blood pressure-measuring instrument made by the Godart Company of Holland.

The circadian data were analyzed in two ways: 1) the means and standard errors for each 4-hour period for at least four days in each span of study were determined, and 2) for the entire body of data the best fitting cosine curve was found, and amplitude, mesor and acrophase were determined for each variable for each span of study. The best-fitting period (τ) between 23 and 25 hours was determined [1].

An overview of the balance data for sodium and potassium can be seen in Fig. 1. By day 3 of the period of high sodium intake, the patient was excreting as much sodium as he was taking in; by day 5 the patient's serum potassium concentration had fallen below the normal value of 3.5 mEq/l; it continued to fall, to reach 3.0 mEq/l, and remained low all the days of high-sodium intake. The potassium balance became slightly negative during this period. The period of low-sodium intake was initiated by the mercurial diuretic (Hg) which promptly caused a large excretion of sodium and a weight loss of two kilograms. Potassium balance initially became strikingly positive after the diuretic and then became slightly negative, as in the previous period. The serum potassium promptly rose and continued to rise during the period of the low-sodium intake.

The effect of sodium intake on the blood pressure as recorded on the Godart instrument can be seen in Fig. 2. All of the values during the period of low-sodium intake, as a whole, are lower than those during the period of high-sodium

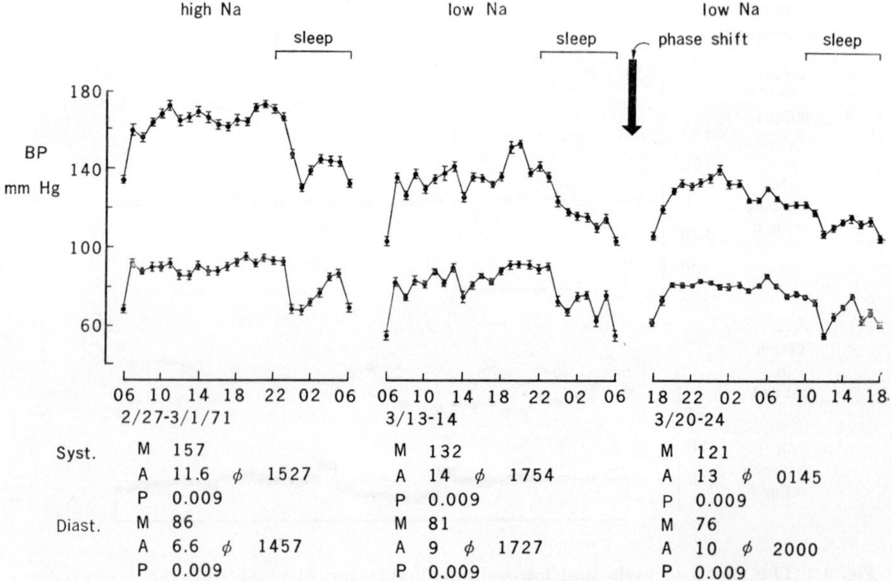

Fig. 2 The effects of high- and low-sodium intakes and of phase shift on systolic and diastolic blood pressure in a patient with hypertension. The data on the bottom of the figure are the mesor (M), the amplitude (A), the acrophase (ϕ) and P value for each curve at the top of the figure.

intake. It is noteworthy that the lowest figures in each curve are those during sleep. With the phase shift, both the systolic and diastolic blood pressures continue to fall but since the patient was still on an intake low in sodium, it is impossible to attribute this striking phenomenon to either factor alone. Each point represents a mean of all the data collected per hour for three days, two days and four days during the first, second and third regimen. Although the blood pressures are consistently lower during the period of sleep, they appear to fit a cosine function poorly, and suggest a step function. The data at the bottom of Fig. 2 illustrate the mesor (M), the amplitude (A), and the acrophase (ϕ) for each of the curves above it from the best-fitting cosine curve, statistically derived. Another point to note is that the blood pressure has completed its phase shift by the fifth day.

The plasma renin activities are of interest because of the role of renin in the etiology of hypertension. In this patient with primary aldosteronism, almost all the renin values were low [2]. However, even in this disorder after prolonged sodium deprivation the plasma renin activity rose into the normal range. In five normal subjects, plasma renin activities peaked four hours after they arose. On five out of six days in this patient, this was also observed.

The circadian data for the other variables studied are presented in Figs. 3, 4 and 5. In Fig. 3 are plotted all the serum and urinary data for sodium and potassium. Measurements of the serum values were not done frequently enough to derive statistical inferences concerning the occurrence of any rhythms; in this paper, they will be ignored. A look at the 4-hour excretion of sodium reveals no apparent pattern. However, the means of each 4-hour period during the last four days on the high-sodium intake reveal a peak excretion between 2200 and 0200 hours, which was eight hours later than the expected peak (1500–1800) as recorded in the literature for normal subjects [3]. This discrepancy might be explained by the continuous expansion of extracellular fluid volume, which caused a "flooding" of the normal excretory mechanisms for sodium, and thereby obliterated the usual circadian rhythm. The values for sodium after the mercurial diuretic were too low to be measured. To be noted on this figure are the strikingly regular patterns for urinary K and 17-OHCS during the period of high-sodium intake. The K rhythm was lost for six days following the diuretic; this may reflect the lack of sodium available for exchange with K in the renal distal tubule. This explanation does not account for the reinstitution of a rhythm for K while the Na output continues to be zero. The loss in K rhythm corresponded in time with a positive K balance and a rise in serum K concentration. The effect of changes in sodium or potassium balance in relation to circadian rhythms has not been studied.

Figs. 4 and 5 show the same data plotted as means and standard errors for potassium, 17-hydroxycorticosteroids, chloride and magnesium. The times for peak excretion on high and low sodium intakes are listed for each variable in Table 2. As others have reported [3], the urinary potassium follows the 17-hydroxycorticosteroid excretions by only a few hours, these two values peaking at 1200 and 1600 respectively. As expected, the chloride rhythm changes during the period of high-chloride intake in the same fashion (and presumably for the same reasons) as does the sodium rhythm. It is interesting to note, however, that on the low-sodium intake, the chloride rhythm is similar to that of potassium. A factor that must be kept in mind continually throughout this paper is that the

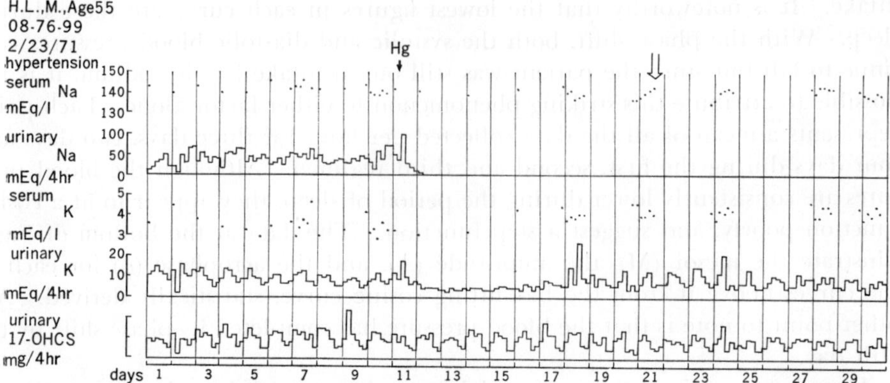

Fig. 3 The effects of high- and low-sodium intakes and of phase shift on serum sodium, urinary sodium, serum potassium, urinary potassium, and urinary 17-hydroxycorticosteroids, in a patient with hypertension. Hg indicates the administration of a mercurial and the open arrow indicates the beginning of the phase shift.

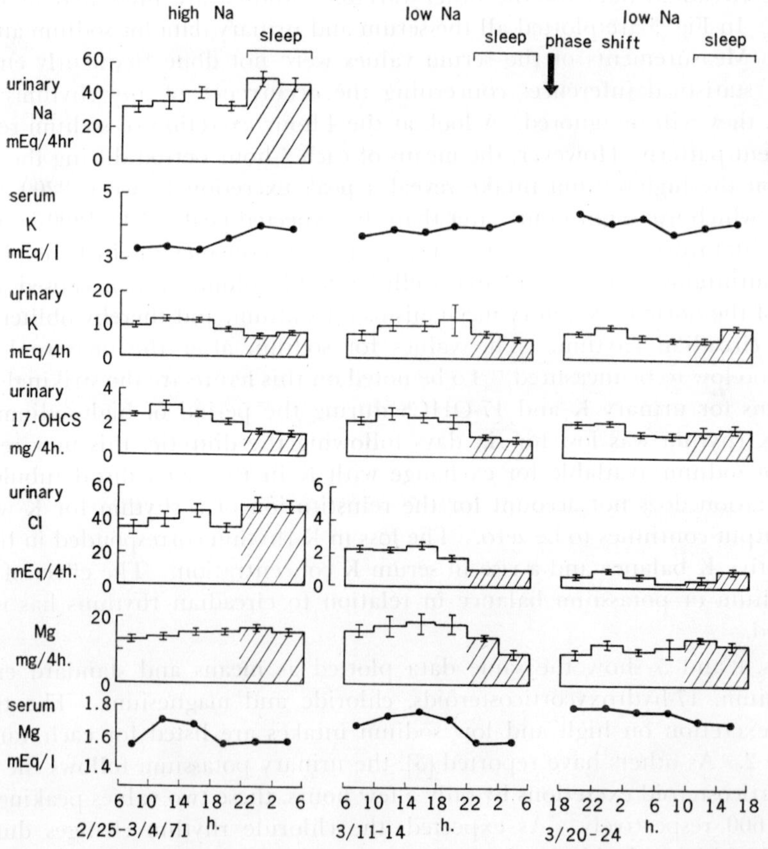

Fig. 4 The effects of high- and low-sodium intakes and of phase shift on urinary sodium, serum potassium, urinary potassium, urinary 17-hydroxycorticosteroids (17-OHCS), urinary chloride (Cl), urinary magnesium (Mg) and serum magnesium in a patient with hypertension. Data are plotted as means and standard errors for each four-hour period.

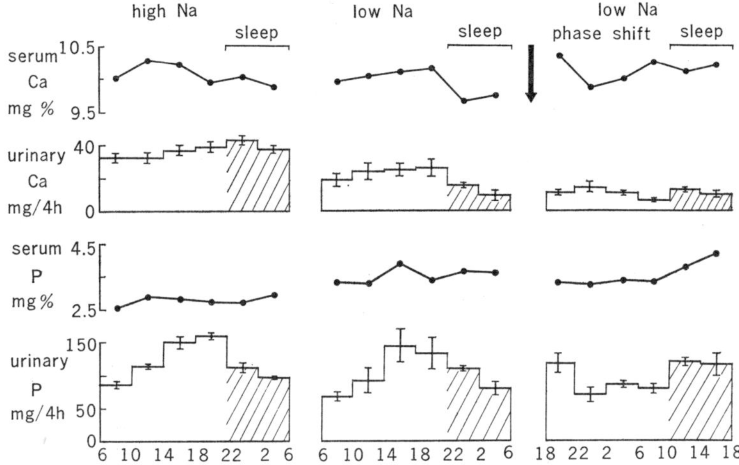

Fig. 5 The effects of high- and low-sodium intakes and of phase shift on serum calcium (Ca), urinary calcium, serum phosphorus (P) and urinary phosphorus in a patient with hypertension. Data are plotted as means and standard errors of each four-hour period.

Table 1 Peak excretions of urinary constituents.

	High Na	Low Na	Literature	
Sodium	2200–0200	–	1500–1800	[3,8]
potassium	1000–1800	1800–2200	1500–1800	[3]
17–OHCS	1000–1400	1000–1400	1200–1500	[3]
chloride	2200–0600	1400–1800	day	[6,7,8]
magnesium	2200–0600	1400–1800	0300–0600	[4,6]
calcium	2200–0200	1000–2200	night afternoon	[5]
phosphorus	1400–2200	1400–2200	1500–1800	[3]

paient has been shown to have primary aldosteronism. Therefore, peak times in this patient may not agree with those in normal subjects.

There is a suggestion in these data that the excretion of one ion can shift the peak excretion time of another ion. Urinary magnesium and calcium (Figs. 4 and 5), whose excretion is known to be influenced by the sodium intake, show peak times of excretion at night on the high-sodium intake and in the late afternoon on the low-sodium intake; the data reported in the literature agree with the peaks on the high-sodium intake. Because of the closeness of the observed values between the various time periods, it is not clear whether there is a complete loss of rhythm or a shift in peak time. Urinary phosphorus throughout the study demonstrated the expected peak times of 1600 to 2000.

Although none of the ions seemed to complete the phase shift, the direction of phase shift can be examined.

Figs. 6 and 7 present the acrophase in serial section (running 5-day average) for the urinary variables as estimated by the cosinor method of analysis. These data are most interesting during the period of phase shifting. The patient was awake for 28 hours—that is, his sleep was retarded. The acrophases for urinary volume, magnesium, calcium and phosphorus advance, whereas the acrophases for potassium, 17-hydroxycorticosteroids, and chloride retard during the march of the acrophase towards a completed phase shift.

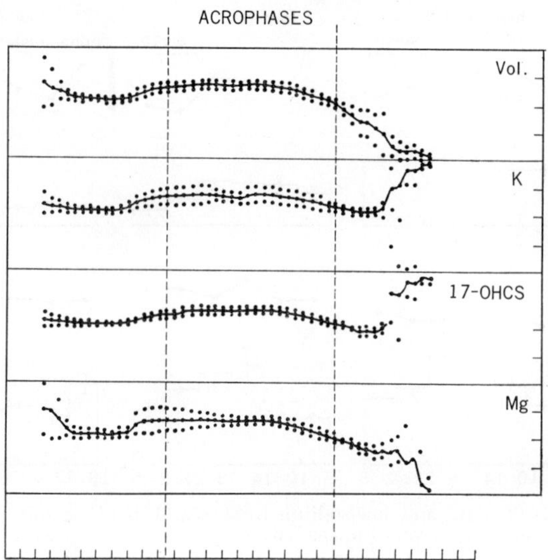

Fig. 6 Acrophases for urinary volume, potassium, 17-hydroxycorticosteroids, and magnesium in a patient with hypertension.

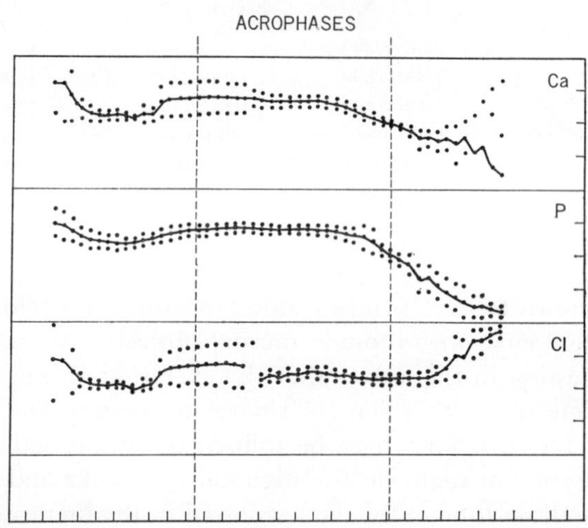

Fig. 7 Acrophases for urinary calcium, phosphorus and chloride in a patient with hypertension.

The discrepancy between these two groups of variables may represent a separation of adrenal from parathyroid function. If so, then the predominant determinant of magnesium excretion is parathyroid hormone rather than aldosterone in this patient.

SUMMARY

A patient with hypertension was found to have primary aldosteronism. A study of circadian periodicity of serum and urinary hormones and electrolytes revealed the following:

1. The blood pressure was effectively lowered by the imposition of a low-sodium diet. With an imposed phase shift of 12 hours it continued to fall: this fall may be attributed to the phase shift and the extension of the effect of the low-sodium diet.

2. A high-sodium intake appeared to "flood" the excretory mechanisms and obliterate the usual rhythm for urinary sodium. Other ions, whose excretion is known to be influenced by sodium, namely, chloride, calcium and magnesium, had different rhythms on high- and low-sodium intakes.

3. During the phase shift, the acrophases for urinary volume, magnesium, calcium and phosphorus advanced, while those for potassium, chloride and 17-hydroxycorticosteroids were retarded.

ACKNOWLEDGEMENT

Numerical analyses supported by USPHS (5-K6-GM-13,981), NSF-GW 7613, NASA and Connecticut Regional Medical Program; Central Chapter, Connecticut Heart Association.

REFERENCES

1. Halberg F. (1969): Chronobiology. Ann. Rev. Physiol., *31*: 675.
2. Bartter F. C., Meyer W. J., Levine H. and Delea C. S. (1972): Circadian aspects of hormone and electrolyte metabolism in hypertension. *In*: Proceedings of the International Symposium on the Renin-Angiotensin-Aldosterone-Sodium System in Hypertension, Montreal, 1971, "Hypertension 1972" (J. Genest, ed.), Montreal.
3. Bartter F. C., Delea C. S. and Halberg F. (1962): A map of blood and urinary changes related to circadian variations in adrenal cortical function in normal subjects. An. N.Y. Acad. Sci., *98*: 969.
4. Doe H. P., Vennes J. A. and Flink E. B. (1960): Diurnal variation of 17-hydroxy-corticosteroids, sodium, potassium, magnesium and creatinine in normal subjects and in cases of treated adrenal insufficiency and Cushing's syndrome. J. Clin. Endo. Metab., *20*: 253.
5. Aschoff J. (1965): Circadian rhythms in man. Science, *148*: 1427.
6. Fiorica V., Burr M. J. and Moses R. (1968): Contribution of activity to the circadian rhythm in excretion of magnesium and calcium. Aerospace Med., 714.
7. Stanbury S. W. and Thomson A. E. (1951): Diurnal variations in electrolyte excretion. Clinical Science, *10*: 267.
8. Wesson L. G. and Lauler D. P. (1961): Diurnal cycle of glomerular filtration rate and sodium and chloride excretion during responses to altered salt and water balance in man. J. Clin. Invest., *40*: 1967.

CIRCADIAN RHYTHMS IN URINARY EXCRETION OF 17-HYDROXYCORTICOSTEROIDS, DEHYDRO-EPIANDROSTERONE, ANDROSTERONE AND ETIOCHOLANOLONE OF TWO HEALTHY MALE SUBJECTS

A. CAVALLERI[1], N. MONTALBETTI[2] and A. REINBERG[3]

[1] *Instituto di Medicina del Lavoro, Università di Pavia, Italy*
[2] *Centro Malattie Endocrine e Metaboliche—Ospedale Civile di Magenta, Italy*
[3] *Equipe de Recherche de Chronobiologie Humaine CNRS, Paris, France*

Several authors [1], after PINCUS [2], have studied the rhythmic variations of urinary excretion of 17-ketosteroids, finding circadian and other frequency rhythms [3].

17-Ketosteroids, includes 11-oxo–17-ketosteroids, of exclusive adrenal origin, and 11-deoxo–17-ketosteroid metabolites, either from adrenal and testicular steroids.

Recently TOCCAFONDI et al. [4] have studied the rhythms of excretion of individual 11-deoxy–17-ketosteroids, but the analysis of time series has been limited to a macroscopic inspection, and now available computer programs for the detection of rhythms and evaluation of their parameters with confidence limits have not been employed.

SUBJECTS AND METHODS

Two male subjects, students at an Institute of the University of Pavia, 23 years of age, collected urines at 4 hour intervals for six days.

For a month before and during the whole study, the sleeping span was ~0045–~0800.

Urines were preserved at 0–4°C for no more than 5 days before being assayed for 17-hydroxycorticosteroids (17-OHCS) with the method of SCHOLLER et al. [5], and for dehydroepiandrosterone, androsterone and etiocholanolone with the gas chromatographic method by CAVALLERI et al. [6] which includes enzymatic hydrolysis, alcohol-ether extraction, washing with NaOH 1 N and water, formation of trimethysilyl-ethers and gas liquid chromatography on NGS 2% at 220°C. The time series have been analyzed with a computer program for the Cosinor method developed by HALBERG [7].

RESULTS

The results are summarized in Table 1. Statistically significant circadian rhythms in urinary 17-OHCS, dehydroepiandrosterone, androsterone, and etiocholanolone have been detected ($P<0.005$).

The level (C_0), amplitude (C) and acrophase (ϕ) of the circadian rhythms of 17-OHCS are similar to those referred to in the literature [8].

Table 1 Circadian rhythms of urinary 17-hydroxycorticosteroids, dehydroepiandrosterone, androsterone, etiocholanolone.

Physiologic variable	Rhythm detection P	Rhythm adjusted level* C_0 ±1 ES	Amplitude C	Acrophase** ϕ minutes in hours ϕ Ref. local midnight (Confidence limits 95 per cent)	φ in degree (360°=24 h) Ref. mid-sleep span
17-hydroxy-corticosteroids	<0.005	0.885± 0.104	0.427 (0.233 to 0.621)	1220 (0944 to 1455)	−131° (−92° to −170°)
Dehydroepi-androsterone	<0.005	0.162± 0.061	0.084 (0.028 to 0.140)	1053 (0848 to 1259)	−109° (−78° to −141°)
Androsterone	<0.005	0.229± 0.126	0.078 (0.030 to 0.128)	1029 (0707 to 1350)	−103° (−53° to −144°)
Etiocholanolone	<0.005	0.351± 0.100	0.089 (0.023 to 0.155)	0837 (0457 to 1217)	−75° (−20° to −130°)

Time series analyzed by cosinor method.
Subjects two healthy males 23 years of age.
Synchronization: light-on∼0800—light-out∼0045; mid-sleep: 0337 or–54°
Duration of each study (T): 6 Days. Sampling interval (Δt): 4 hrs.

* C and C_0 are expressed in mg. In these experiments C_0 corresponds to the mean of samples collected 6 times a day at equidistant (Δ t=4 hrs) intervals. For example: at the acrophase (1220) the excretion of 17-OHCS will reach as a mean 1.312 mg (0.885+0.427) with a upper limit at 1.506 mg; 12 hrs later or earlier this excretion will fall around 0.458 mg, with a lower limit at 0.264 mg.

** Acrophase: peak of the sine function used to approximate the biorhythm.

Table 2 Circadian rhythms of several hormones and hormone metabolites.

Phisiologic variable	Acrophase[a] or peak	Authors
Urines		
dehydroepioandrosterone	1000—1400	Toccafondi et al. 1970[b]
androsterone	1000—1400	Toccafondi et al. 1970[b]
etiocholanolone	0600—1000	Toccafondi et al. 1970[b]
testosterone	1200—1600	Okamoto et al. 1971[c]
epitestosterone	1200—1600	Okamoto et al. 1971[c]
Plasma		
testosterone	1400	Okamoto et al. 1971[c]
	1130	Dray et al. 1965[d]

[a] Dray et al. have analyzed their time series by cosinor method which furnishes a value for acrophase. Other authors have done only a macroscopic analysis of their time series, therefore only the span of maximum excretion, or time of maximum level are referred to.

[b] 7 males; no information on synchronizer schedule.

[c] 12 males, 20–23 years of age; no information on synchronizer schedule.

[d] 5 males, 20–42 years of age; sleep span 2200/2400 0700.

The value of C compared with the level C_0 is very similar for 17-OHCS and dehydroepiandrosterone, while for androsterone and etiocholanolone the relative value of C is inferior.

The acrophases of the individual 11-deoxy–17-ketosteroids occur somehow earlier than the acrophase of 17-OHCS, particularly for etiocholanolone.

DISCUSSION

The results of our "microscopic" [9] analysis are in favor of a circadian rhythm of individual 11-deoxy–17-ketosteroids, thus confirming the macroscopic analysis of Toccafondi et al. [4].

The relative amplitude of dehydroepiandrosterone, the prevalent metabolite of adrenal steroids, is similar to the relative amplitude of 17-OHCS.

The acrophase of etiocholanolone occurs earlier than other 11-deoxo–17-ketosteroids in accordance with Toccafondi et al.; no explanation for this fact has been found.

As the acrophases of urinary testosterone and epitestosterone occur later (1200–1600) [10] and the acrophases of plasma testosterone occur earlier 1130 [11], 1000–1400 [10], it seems that no correlation exists between the rhythms of individual 11-deoxo–17-ketosteroids and the rhythms of testosterone.

It must be emphasized that a correct comparison can be done only with the data of Dray et al. for which the synchronizer schedule is referred.

SUMMARY

The cosinor analysis has detected circadian rhythms of urinary excretion of dehydroepiandrosterone, androsterone and etiocholanolone in healthy male subjects, 23 years of age in which the circadian rhythms of urinary 17-OHCS was normal.

From the timing of the acrophases it seems that no correlation exists between the above mentioned rhythms and the rhythm of testosterone referred to in the literature.

REFERENCES

1. Review in: Conroy R.T.M.L. Human circadian rhythms I. & A. Churchill, London, 1970.
2. Pincus G. (1943): A diurnal rhythm of the excretion of urinary ketosteroids by young men. J. Clin. Endocr. Metab., *3*: 195.
3. Halberg F., Engeli M., Hamburger C. and Hillman D. (1965): Spectral resolutions of low frequency, small amplitude rhythms in excreted ketosteroids: probable androgen induced circaseptan desynchronisation. Acta Endocr. Suppl., 103, vol. 50.
4. Toccafondi R., Bianchi G., Rolandi E. and Madeddu G. (1970): Circadian variations of urinary 11-deoxy–17-ketosteroids in normal subjects. Ann. Endocrinol., *31*: 1087.
5. Scholler L., Busigny M. and Jayle M. F. (1962): Méthode rapide de dosage des 17-hydroxycorticostéroïdes urinarires totaux. *In*: Analyse des stéroïdes hormonaux, (M. F. Jayle, ed.), p. 137, Vol. II, Masson & C., Paris.
6. Cavalleri A., Salvadeo A. and Favino A. (1964): Analisi gascromatografica degli 11-desossi–17-chetosteroidi urinari come trimetilsilileteri. Folia Endocrinol., *17*: 646.
7. Halberg F., Tong Y. L. and Johnson E. A. (1967): Circadian system phase—an aspect of temporal morphology; procedures and illustrative examples. *In*: The Cellular Aspects of Biorhythms, Springer-Verlag, Berlin.

8. Halberg F., Reinhardt J., Bartter F. C., Delea C., Gordon R., Reinberg A., Ghata J., Halhuber M., Hofman H., Günther R., Knapp E., Perra J. C. and Garcia Sainz M. (1969): Agreement in endpoints from circadian rhythmometry on healthy human beings living on different schedules. Experientia, *25*: 106.
9. Halberg F. (1966): Le pouvoir de résolution des calculateurs électroniques en chronopatologie. Une analogie avec la microscopie. Science, *101*: 1.
10. Okamoto M., Setaishi C., Nakagawa K., Horiuchi Y., Moriya K. and Itoh S. (1971): Diurnal variations in the levels of plasma and urinary androgen. J. Clin. Endocr., *31*: 846.
11. Dray F., Reinberg A. and Sebaoun J. (1965): Rhythme biologique de la testosterone libre du plasma chez l'homme adulte sain: existence d'une variation circadianne. C. R. Acad. Sc. Paris, *261*: 573.

CIRCADIAN RHYTHM IN PLASMA ACTH IN HEALTHY ADULTS*

PH. VAGUE, CH. OLIVER and J. Y. BOURGOIN

Clinique Endocrinologique de la Faculté de Médecine de Marseille
France

The circadian rhythmicity of the hypothalamopituitary-adrenal axis is well known in animals and human beings.

While data on variations of plasma ACTH levels in man throughout the day have been published [2], no conclusive evidence for a rhythmic variation has been offered. This probably is related to the fact that ACTH appears to be secreted in intermittent bursts. Frequent sampling is necessary if one is to detect a rhythm.

SUBJECTS AND METHODS

Five healthy adults, four women and one man aged 20 to 31 were studied while on their habitual diurnal activity (meals were taken at 0800, 1200, 1900 and the dark sleep period extended from 2300 to 0700).

After the insertion of an indwelling needle in the antecubital vein, hourly samples were obtained for 26 hours starting at 0800 in three and at 0900 in two of the subjects. Sampling was painless and did not awaken the subjects during the dark period.

Plasma ACTH was assayed by a radioimmunoassay method previously described [5, 7]. Its sensitivity allows detection of less than 5 micromicrogr/ml (pg/ml). The precision and the reproducibility are ±10% for a 95% degree of confidence in the range 5–400 pg/ml. In the system used, the entire sequence of the ACTH molecule is detected; the recognition by the antiserum of fragments of the molecule is insignificant.

Electronic computer analyses by the Cosinor method [3] have been used:
—to detect the rhythm probability of a cosine curve best approximating all data.
—to estimate its amplitude and its acrophase.

RESULTS (Fig. 1)

A circadian rhythm has been detected at a statistical level of significance $p<.05$.

The amplitude (with 95% confidence interval) is: 39.4 pg/ml (10.5–89.3).

The acrophase (with 95% confidence interval) is at $-96°$ ($-5°$ to $-178°$) with $0°$ taken at 0.00.

* Supported in part by a research grant of the Caisse Nationale d'Assurance Maladie des Travailleurs Salariés.

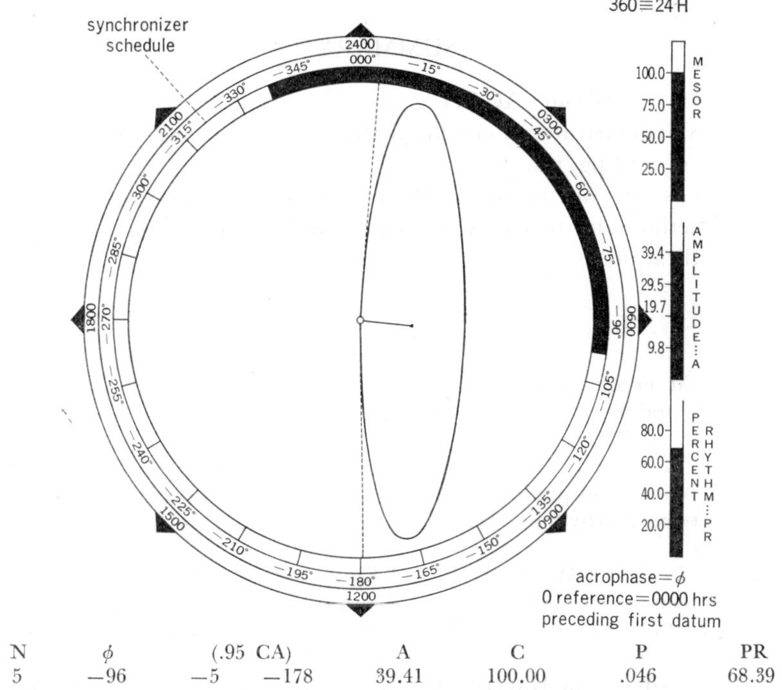

N	φ	(.95 CA)		A	C	P	PR
5	−96	−5	−178	39.41	100.00	.046	68.39

Fig. 1 Circadian rhythms in plasma ACTH of healthy human adults. Cosinor summary: Chronobiology Laboratories, University of Minnesota. Because of wide confidence arc for φ we deal with a candidate rhythm.

DISCUSSION

The existence of a circadian rhythm of plasma ACTH in man is not surprising, and macroscopic examinations already have led to its demonstration [2].

It must be noted that in this study the 95% confidence interval of the acrophase is large enough in comparison to the one of 17-OHCS in plasma or in urine in other studies [1, 6]. This may be accounted for by the small number of subjects tested but more probably by the fact that ACTH is secreted in bursts which faithfully are reflected in the plasma levels owing to the short half life of the hormone (5–10′). Even with hourly sampling peaks of ACTH levels may have been missed, and this may account for the lower degree of statistical significance of the estimated cosine curve. Furthermore the individual variations in the time of appearance of the burst of secretion may not have been evaluated sufficiently with unfrequent sampling. In these conditions the description of a rigid timing for the acrophase could be an artifact. Nevertheless, in this study the acrophase occurs at 0630, half an hour before awakening. As expected it slightly precedes the acrophase of 17-OHCS in plasma which in other studies happens just after awakening on macroscopic examination [4] and at 0730 with data processed by the Cosinor method [1].

SUMMARY

Determinations of plasma ACTH levels at hourly intervals for 26 hours in five healthy adults demonstrated a circadian rhythm with an acrophase at −96° and an amplitude of 39.4 pg/ml.

We wish to express our thanks to Pr. F. Halberg for processing the data with the Cosinor method and to Dr. A. Reinberg for his interest in this work and his help in its realization.

REFERENCES

1. Apfelbaum M. et al.: (in press).
2. Berson S. A. and Yalow R. S. (1968): Immunoassay of ACTH in plasma. J. Clin. Invest., *47*: 2725–2751.
3. Halberg F., Tong Y. L. and Johnson E. A. (1967): Circadian system phase—an aspect of temporal morphology; procedure and illustrative examples. *In*: The Cellular Aspects of Biorhythms, (H. von Meyersbach, ed.), pp. 20–48, Springer-Verlag, Berlin.
4. Krieger D. T., Allen W., Rizzo F. and Krieger H. P. (1971): Characterization of the normal temporal pattern of plasma corticoid levels. J. Clin. Endocr., *32*: 266–284.
5. Oliver Ch. (1971): Le dosage radioimmunologique de l'ACTH plasmatique. Thèse de Méd., Marseille.
6. Reinberg A., Ghata J., Halberg F., Gervais P., Abulker C. H., Dupont J. and Gaudeau Cl. (1970): Rythmes circadiens du pouls, de la pression artérielle, des excrétions uinaires en 17 OH corticostéroïdes, catécholamine et potassium chez l'homme adulte sain, actif et au repas. Ann. Endocr. Paris, *31*: 277–287.
7. Vague Ph., Oliver Ch., Jaquet Ph. and Vague J. (1971): Le dosage radioimmunologique de l'ACTH plasmatique. Résultats chez les sujets normaux. Rev. Eur. d'Et. Clin. Biol., *16*: 485–493.

CIRCADIAN SUSCEPTIBILITY-RESISTANCE CYCLE OF BONE MARROW CELLS TO WHOLE BODY X-IRRADIATION IN BALB/C MICE*

ERHARD HAUS[1)], FRANZ HALBERG[2)] and MERLE K. LOKEN[3)]

[1)]*Department of Anatomic and Clinical Pathology*
St. Paul-Ramsey Hospital, St. Paul, Minnesota
[2)]*Chronobiology Laboratories, Department of Laboratory Medicine and Pathology*
University of Minnesota
[3)]*Division of Nuclear Medicine, Department of Radiology*
University of Minnesota, Minneapolis
Minnesota, U.S.A.

A circadian susceptibility-resistance cycle characterizes mouse bone marrow cells. At the time of high radioresistance of the animals, higher numbers of surviving nucleated cells are present four days after exposure to 350 r whole body x-irradiation. This radiosensitivity cycle is amenable to a 180° phase shift within two weeks after a 180°-shift in synchronizer phase (a change from L0600–1800, D1800–0600 to L1800–0600, D0600–1800).

INTRODUCTION

Circadian (about-24-hour) rhythms, almost ubiquitous phenomena [1–4], characterize cellular variables from DNA synthesis to mitosis in rodent liver [5–9], gastric mucosa [10, 11], epidermis [12], and bone marrow [13, 49]. Human bone marrow [14, 15] and epidermis [16, 17] as well as counts of circulating lymphocytes, eosinophils and other formed elements in blood [10, 13, 46] from man [18–22] and mice [13, 23, 25] also exhibit circadian rhythms.

In bone marrow from diurnally active human beings, large numbers of mitoses are seen in the evening, mitotic activity being at a low in the morning. The opposite pattern is observed for nocturnal animals like rats and mice kept on a regimen of light during the daytime alternating with darkness during the night hours. The ensemble of these rhythms plays a critical role in the organism's responses to stimuli, including resistance to potentially harmful agents [26–32].

Cellular radiation sensitivity is cell-cycle dependent [33] and mammalian radiation lethality is a sequel of cell lethality. If stem cell populations are partially synchronous, mean lethal doses should depend upon the phase of the circadian system at the time of radiation exposure. This hypothesis has been investigated in rats [34–38] and mice [39, 40]: survival or death as endpoints showed lower radiosensitivity during the daily light span. Some seemingly contradictory results [41, 42] could largely be traced to techniques inadequate for circadian rhythmometry [41, 42] as discussed by MENAKER [43].

* Supported in part by USPH (5-K6-GM-13,981), NASA (NGR-24-005-006), NCI IROI-CA-14445-01 and the St. Paul Ramsey Medical Research and Education Foundation.

Exploring the mechanisms involved in the susceptibility-resistance cycle to whole body irradiation, Ueno [44] and Vacek and Rotkovska [45] studied the number of endogenous spleen colonies of hematopoietic elements in mice ten days after irradiation. These investigators observed a higher number of colonies present in the spleen of animals irradiated during the time of greater radio-resistance as compared with those irradiated at times of increased radio-suscept-ibility as indicated by a lesser 30-day survival rate [45]. Based upon these findings, Ueno [44] as well as Vacek and Rotkovska [45] suggest that a higher number of surviving hematopoietic stem cells may be the cause of the differences in the mortality of mice irradiated at different circadian system phases.

MATERIAL AND METHODS

Inbred Bagg Albino (C) strain mice maintained in the Chronobiology Laboratory at the University of Minnesota, 13–17 weeks of age were used.

Three weeks prior to irradiation, a total of 406 mice were transferred to two relatively sound-deadened "periodicity rooms". They were housed one animal per cage measuring 20×15×15 cm at a room temperature of 24±1°C, in 12-h light (L) alternating with 12-h darkness—LD12:12. Fox chow and tap water were provided *ad libitum* throughout the study. For a first week of standardization, both rooms were kept on the same illumination regimen—L0600–1800, D1800–0600. On the eighth day of standardization, for two additional weeks, the lighting cycle in one room only was shifted by 180°, to L1800–0600, D0600–1800.

The animals were exposed to whole body x-ray irradiation, three weeks after housing and two weeks after the 12-hour shift of synchronizer phase in one of the rooms. On the days of treatment, at 4-hour intervals, from 0800 of one day to 0800 of the next, subgroups of 11 to 12 mice were removed from each room and were irradiated in a plastic cage holder with 200 KVp x-rays at 15 mA with a half-value-layer of 0.85 mm copper. The dose rate was 45.1 r/min in air—corresponding to 66.1 rads/min at mouse midline. A single dose of 350 r was delivered (on measurements of roentgens in air). Midway through each radiation exposure the plastic box containing the mice was rotated 180° to equalize inhomogeneities in the radiation field.

After exposure, the mice were returned to their individual cages and to mouse rooms with the same lighting conditions as before x-irradiation, regular or "inverted", and were kept on these same schedules until killing. Groups of five mice on the "regular" regimen and seven mice on the inverted one were sham irradiated at 0800, 1600 and 0000: they were processed like the irradiated animals but the x-ray machine was not turned on while the plastic box was standing on the irradiation table. On the fourth day after irradiation or sham irradiation of each subgroup, the mice were killed in two sessions. A morning session lasted from 0700 until 1100, an afternoon session from 1700 to 2100. The mice were removed alternatingly from the two mouse rooms, a mouse from the room on a regular regimen being always followed by one from the room on the "inverted" one and vice versa.

In an adjacent laboratory the rectal temperature was measured. Tail blood was obtained for a total white blood cell count and blood smear. Thereafter the mouse was killed with ether. The thymus, spleen and the right groin lymph node were removed "quantitatively" and fixed in formalin, to be later carefully dissected free from adherent fat tissue and weighed on a Roller Smith balance. A femur was dissected and both ends were clipped off. The length of the shaft was measured and the bone marrow was squirted out with 0.2 ml of bovine serum albumin into a small (5 ml) flask. Larger pieces of marrow were broken up with a bacteriologic loop and 2 ml of white cell counting fluid was added. The bone marrow cell suspension was then mixed for five minutes on an automatic shaker. An aliquot was then filled into a Fuchs-Rosenthal chamber and counted in duplicate. Throughout both sessions, the same procedures were carried out by the same ten individuals in a production line fashion.

RESULTS AND DISCUSSION

Whole body x-irradiation with a single dose of 350 r leads in the young adult male C-mouse on the fourth day after exposure to a reduction in the number of nucleated cells in the bone marrow to about one tenth of those found in the non-irradiated controls. This reduction was found in the numbers of nucleated bone marrow cells for mice killed both in the morning and in the afternoon, a statistically significant difference between the two groups not being detected; hence data were pooled for analysis.

Fig. 1 shows the number of nucleated bone marrow cells per mm femur shaft

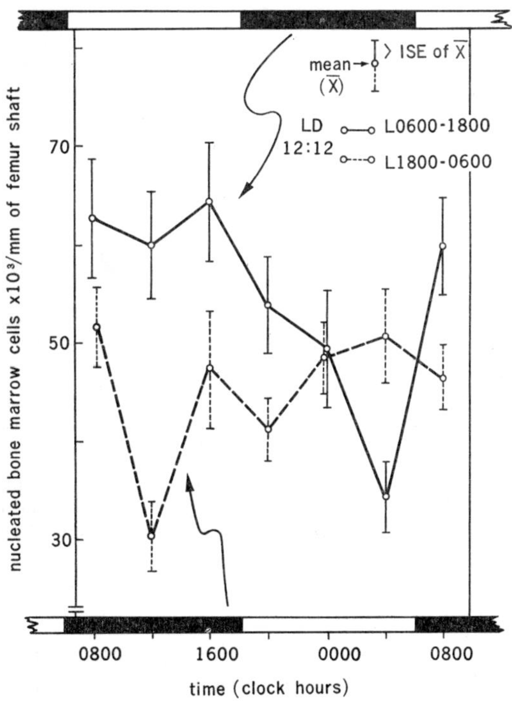

Fig. 1 Number of nucleated bone marrow cells per mm femur shaft in mice kept under two different lighting regimens [LD12:12 with L 0600–1800 (continuous line) or L 1800–0600 (broken line)] and exposed to whole body x-irradiation (350 r) at different circadian system phases.

* At the time of exposure of adult male C mice to single dose (350 r).

of the irradiated animals as a function of circadian system phase at time of exposure—four days prior to killing. The reduction in the number of nucleated bone marrow cells after exposure to 350 r of whole body x-irradiation is neither equal nor random in these animals exposed at different circadian system phases: a time effect for both the animals on a "regular" as well as on an "inverted" lighting regimen is significant below the 1% level (analysis of variance). By forming from the bone marrow cell numbers of the first animals irradiated at each of the seven time points a first series of values and by forming additional series from the following mice, the second, third, etc. animals at each time point, a set of admittedly artificial series can be obtained for analysis by cosinor. Thus a rhythm is detected for both experimental groups, Fig. 2. This bone marrow depression rhythm is nearly in antiphase for mice on the regular and the reversed regimens. In both

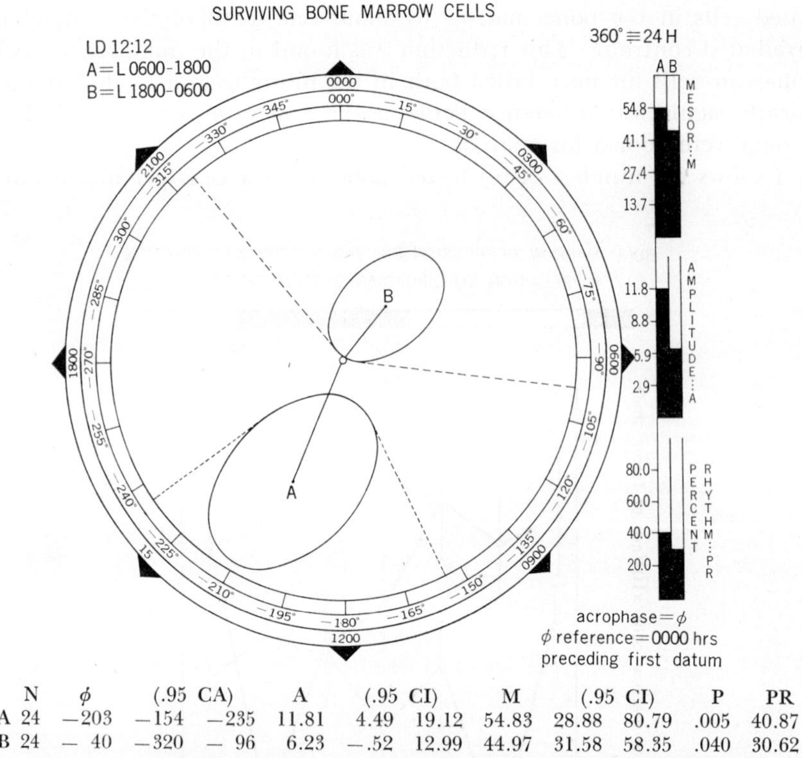

	N	φ	(.95 CA)		A	(.95 CI)		M	(.95 CI)		P	PR
A	24	−203	−154	−235	11.81	4.49	19.12	54.83	28.88	80.79	.005	40.87
B	24	− 40	−320	− 96	6.23	−.52	12.99	44.97	31.58	58.35	.040	30.62

Fig. 2 Cosinor analysis of Fig. 1 data detects circadian susceptibility resistance cycle of bone marrow cells to whole body x-irradiation and demonstrates phase shift of susceptibility rhythm in response to single dose at 14 days after shift in lighting regimen.

groups, the largest number of nucleated bone marrow cells at four days after exposure is found during the second half of the light span. This time corresponds to that of highest radioresistance reported previously for mice of the same inbred strain, sex and age [40], as gauged in Fig. 3 by the LD 50. Mice exposed to doses of 400, 450 or 500 r of whole body x-irradiation were treated as in the study summarized in Figs. 1 and 2, except that they were not killed but observed for mortality during two months after irradiation. Differences in mortality as a function of exposure time were ascertained by a method for the determination of crests in time series [47]. However, the Fig. 3 data were not obtained in the same studies as those in Figs. 1 and 2. Seasonal variation in circadian rhythms of the susceptibility-resistance cycle to x-irration have been described [37, 38] and may persist even on an artificial LD12:12 regimen [48].

Circulating leucocytes in the peripheral blood of irradiated animals on the fourth day after exposure to 350 r appeared to be too low in number for reliable counting. The weights of spleen, thymus and lymph nodes showed no significant differences in weight as a function of time of exposure four days previously. This result obtained four days after exposure on the spleen differs from that of Vacek and Rotkovska [45] who found a circadian cycle in splenic weight ten days after irradiation, parallel to the numbers of colonies of hematopoietic cells present at

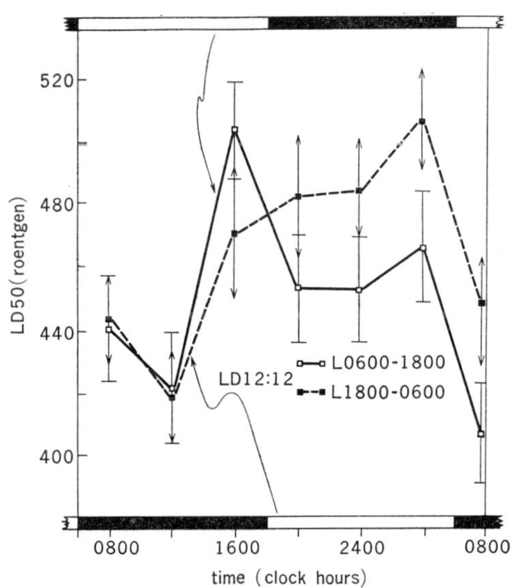

Fig. 3. Circadian variation of LD50 of whole body x-irradiation in mice on two lighting regimens, irradiated at different circadian system phases. Peak and trough of susceptibility of the animals to whole body x-irradiation correspond to high and low numbers of surviving bone marrow cells shown in Fig. 1.

that time. Increased regeneration of hematopoietic elements may take place ten days after irradiation, whereas on the fourth day after exposure, immediate damage rather than regeneration will be prominent. Young adult (10–20 weeks old) male Balb/C mice exposed to 400–500 r whole body x-irradiation begin to die about four days after exposure, maximal deaths occurring between days 10 and 14 after irradiation. At that time infection may be a frequent complicating factor.

The reaction of mammals to mid-lethal exposure, including death or survival, is influenced by the surviving number of hematopoietic stem cells and by the rate at which the stem cells and their progeny proliferate and repopulate the hematopoietic system [33, 45]. In addition to a difference in growth potential of hematopoietic stem cells of intact non-irradiated animals as function of circadian system phase, suggested by VACEK and ROTKOVSKA [45], a difference in number of cells present in itself may contribute to the circadian susceptibility-resistance cycle to whole body x-irradiation. At the time of high radioresistance, larger numbers of bone marrow stem cells may survive the irradiation and provide more viable cells for seeding and repopulating the hematopoietic system. In view of the thus far unavoidable damage inflicted on human bone marrow in the course of certain forms of radiotherapy, the clinical implications of these findings deserve early attention and investigation in man.

REFERENCES

1. HALBERG F. (1960): The 24-hour scale: A time dimension of adaptive functional organization. Perspectives in Biology and Medicine, *3*: 491–527.
2. HALBERG F. (1960): Temporal coordination of physiologic function. Cold Spring Harbor Symposia on Quantitative Biology, *25*: 289–310.
3. HALBERG F. (1969): Chronobiology. Annual Review of Physiology, *31*: 675–725.
4. HALBERG F. (1969): Chronobiologie; rythmes et physiologie statistique. *In*: Theoretical Physics and Biology, (M. MAROIS, ed.), pp. 347–393. Discussion remarks pp. 339–341 and 394–411, North-Holland, Amsterdam.
5. HALBERG F., HALBERG E., BARNUM C. P. and BITTNER J. J. (1959): Physiologic 24-hour periodicity in human beings and mice, the lighting regimen and daily routine. *In*: Photoperiodism and Related Phenomena in Plants and Animals, (R. WITHROW, ed.), pp. 803–878.
6. POTTER Van R., BARIL E. F., WATANABE M. and WHITTLE E. D. (1968): Systematic oscillations in metabolic functions in liver from rats adapted to controlled feeding schedules. Federation Proceedings, *27/5*: 1238–1245.
7. RUBY J., SCHEVING L. E., GRAY S. B. and WHITE K. E.: Daily fluctuation in protein and DNA content of cell nuclei isolated from rat liver. Proc. IX Int. Cong. Anatomists, Leningrad, August, 1970. Experimental Cell Research, (in press).
8. HALBERG F. (1959): Physiologic 24-hour periodicity; general and procedural considerations with reference to the adrenal cycle. Ztsch. Vit. Horm. Fermentforsch., *10*: 225–296.
9. JACKSON B. (1959): Time-associated variations of mitotic activity in livers of young rats. Anat. Rec., *134*: 365–377.
10. CLARK R. H. and BAKER B. L. (1962): Effect of adrenalectomy on mitotic proliferation of gastric epithelium. Proc. Soc. Exptl. Biol. Med., *111*: pp. 311–315.
11. CLARK R. H. and BAKER B. L. (1963): Effect of hypophysectomy on mitotic proliferation in gastric epithelium. Amer. J. Physiol., *204*: 1018–1022.
12. BLUMENFELD C. M. (1943): Studies of normal and abnormal mitotic activity. II. The rate and the periodicity of the mitotic activity of experimental epidermoid carcinoma in mice. Arch. Path., *35*: 667–673.
13. CLARK R. H. and KOST D. R. (1969): Circadian periodicity of bone marrow mitotic activity and reticulocyte counts in rats and mice. Science, *166*: 236–237.
14. MAUER A. M.: (1965): Diurnal variation of proliferative activity in the human bone marrow. Blood, *26*: 1–7.
15. KILLMAN S. A., CRONKITE E. P., FLIEDNER T. M. and BOND V. D. (1962): Mitotic indices of human bone marrow cells. I. Number and cytologic distribution of mitoses. Blood, *19*: 743–750.
16. SCHEVING L. E. (1959): Mitotic activity in the human epidermis. Anat. Rec., *135*: 7–14.
17. FISHER L. B. (1968): The diurnal mitotic rhythm in the human epidermis. Br. J. Derm., *80*: 75–80.
18. ELMADJIAN F. and PINCUS G. (1946): A study of the diurnal variations in circulating lymphocytes in normal and psychotic subjects. J. Clin. Endocrinol., *6*: 287–294.

19. Halberg F., Visscher M. B., Flink E. B., Berge D. and Bock F. (1951): Diurnal rhythmic changes in blood eosinophil levels in health and in certain diseases. Journal-Lancet, *71*: 312–319.
20. Halberg F. (1962): Circadian temporal organization and experimental pathology. *In*: VII Conferenza Internazionale della Societa per lo Studio dei Ritmi Biologici, Siena, Italy (Sept. 1960), pp. 20.
21. Bartter F., Delea C. S. and Halberg F. (1962): A map of blood and urinary changes related to circadian variations in adrenal cortical function in normal human subjects. Ann. N.Y. Acad. Sci., *98*: 969–983.
22. Malek J., Suk K., Brestak M. and Maly V. (1962): Daily rhythm of leukocytes, blood pressure, pulse rate and temperature during pregnancy. Ann. N.Y. Acad. Sci., *98*: 1018–1041.
23. Halberg F., Visscher M. B. and Bittner J. J. (1953): Eosinophil rhythm in mice: Range of occurrence; effects of illumination, feeding and adrenalectomy. Am. J. Physiology, *174*: 313–315.
24. Panzenhagen H. and Speirs R. (1953): Effect of horse serum, adrenal hormones, and histamine on the number of eosinophils in the blood and peritoneal fluid of mice. Blood, *8*: 536–644.
25. Brown H. E. and Dougherty T. F. (1956): The diurnal variation of blood leucocytes in normal and adrenalectomized mice. Endocrinol., *58*: 365–375.
26. Halberg F. (1962): Physiologic 24-hour rhythms: A determinant of response to environmental agents. *In*: Man's Dependence on the Earthly Atmosphere, (K. E. Schaefer, ed.), pp. 48–89, The Macmillan Co., New York.
27. Haus E. (1964): Periodicity in response and susceptibility to environmental agents. Ann. N.Y. Acad. Sci., *117*: 292–315.
28. Halberg F., Johnson E. A., Brown B. W. and Bittner J. J. (1960): Susceptibility rhythm to *E. coli* endotoxin and bioassay. Proc. Soc. Exp. Biol. and Med., *103*: 142–144.
29. Scheving L. E., Vedral D. F. and Pauly J. E. (1967): A daily (circadian) rhythm in rats to D-Amphetamine sulfate: Effect of continous illumination on the rhythm. Nature, *210*: 621–622.
30. Pauly J. E. and Scheving L. E. (1964): Temporal variations in the susceptibility of white rats to pentobarbital sodium and tremorine. Intern. J. Neuropharmacol., *3*: 651–658.
31. Reinberg A. (1967): The hours of changing responsiveness or susceptibility. Perspect. Biol. Med., *11*: 111–126.
32. Halberg F. (1971): Chronopharmacology. Ann. Rev. Pharmacology, *11*: 455–492.
33. Bond V. P., Fliedner T. M. and Archambeau J. O. (1965): Mammalian Radiation Lethality. Academic Press, London.
34. Pizzarello D. J., Witcofski R. L. and Lyons E. A. (1963): Variations in survival time after whole-body radiation at two times of day. Science, *139*: 349.
35. Peters K. (1963): Veränderung der Überlebenszeit von Ratten nach Ganzkörperbestrahlung zu verschiedenen Tageszeiten. Strahlentherapie, *122*: 554–557.
36. Hellwig O. and Rosenkranz J. (1968): Über circadiane Schwankungen der Starhlensensibilität von Ratten. Strahlentherapie, *135*: 220–222.
37. Fochem K., Michalia W. and Picha E. (1967): Über die pharmakologische Beeinflussung der tagesrhythmischen Unterschiede in der Strahlenwirkung und zur Frage der jahreszeitlich bedingten Unterschiede der Strahlensensibilität bei Ratten und Mäusen. Strahlentherapie, *133*: 356–361.
38. Fochem K., Michalica W. and Picha E. (1968): Über tages- und jahrenszeitliche Unterschiede der Strahlenwirkung bei Versuchstieren und die Möglichkeiten einer pharmakologischen Beeinflussung. Strahlentherapie, *135*: 223–226.
39. Pizzarello D. J., Isaak D., Chua K. E. and Rhyne A. L. (1964): Circadian Rhythmicity in the sensitivity of two strains of mice to whole-body radiation. Science, *145*: 286–291.
40. Haus E., Halberg F., Loken M. and Kim Y. S. (1971): Circadian Rhythmometry of Mammalian Radiosensitivity. *In*: Space Radiation Biology, (C. Tobias, ed.), Chapter 9, (in press).

41. STRAUBE R. L. (1963): Examination of diurnal variation in lethally irradiated rats. Science, *142*: 1062.
42. RUGH R., CASTRO V., BALTER S., KENNELLY E. V., MAROSDEN D. S., WARMUND J. and WOLLIN M. (1963): X-Rays: Are there cyclic variations in radiosensitivity? Science, *142*: 53–56.
43. MENAKER M. (1964): X-Rays: Are there cyclic variations in radiosensitivity? Science, *143*: 597.
44. UENO Y. (1968): Diurnal rhythmicity in the sensitivity of haematopoietic cells to whole-body irradiation of mice. Int. J. Radiat. Biol., *14*: 307–312.
45. VACEK A. and ROTKOVSKA D. (1970): Circadian variations in the effect of X-irradiation on the hematopoietic stern cells of mice. Strahlentherapie, *140*: 302–306.
46. HALBERG F., TONG Y. L. and JOHNSON E. A. (1967): Circadian system phase: an aspect of temporal morphology; procedures and illustrative examples. *In*: The Cellular Aspects of Biorhythms, (H. von MAYERSBACH, ed.), 8th Int. Congr. Anat., pp. 20–48, Springer-Verlag, Berlin.
47. HAUS E. and HALBERG F. (1970): Circannual rhythm in level and timing of serum corticosterone in standardized inbred mature C-mice. Environmental Res., *3*: 81–106.
48. SAVAGE I. R., RAO M. M. and HALBERG F. (1962): Test of peak values in physiopathologic time series. Exp. Med. and Surg., *20*: 309–317.
49. PIZZARELLO D. J. and WITCOFSKI R. L. (1970): A possible link between diurnal variations in radiation sensitivity and cell division in bone marrow of male mice. Radiology, *97*: 165–167.

CIRCADIAN RHYTHM OF ACTH AND GROWTH HORMONE IN HUMAN BLOOD; TIME RELATIONS TO ADRENOCORTICAL (BLOOD AND URINARY) RHYTHMS*

David J. Lakatua, Erhard Haus, Ernest M. Gold and Franz Halberg

Department of Anatomic and Clinical Pathology
Department of Medicine, St. Paul-Ramsey Hospital, St. Paul, Minnesota
and the Chronobiology Laboratories
Department of Laboratory Medicine and Pathology
University of Minnesota, Minneapolis
Minnesota, U.S.A.

As part of a summer program in chronobiology sponsored by the Twin Cities Institute for Talented Youth, six male and seven female high school students 13 to 17 years of age participated in a study of various blood and urinary variables measured at six time points over a 24-hour span. The students carried out self measurements of oral temperature, systolic and diastolic blood pressure, short-term memory and 1-minute estimation and two types of eye-hand skill measurements for several weeks prior to the 24-hour blood and urine study and for a similar span thereafter. Analyses of these physiologic and clinical functions [1] as well as of several hematologic and other biochemical variables will be discussed separately [2]. In this communication the circadian rhythms of serum ACTH levels, serum cortisol, growth hormone and glucose and the urinary rhythms of corticosteroid, sodium and potassium excretion studied simultaneously in a group of adolescents are presented and wherever feasible their rhythm parameters are quantified by cosinor analysis [3].

SUBJECTS AND METHODS

Six boys, mean age 15.5 years (range 13–17), and seven girls, mean age 15.3 years (range 14–16), spent a 30-hour span beginning at 1000 July 19, 1971, at St. Paul-Ramsey Hospital, St. Paul, Minnesota. During daytime they visited different sections of the hospital and attended a number of short presentations on health science related subjects. Overnight they were housed in the nurse's dormitory. They had free choice of hospital food. On July 19 lunchtime was at 1230 and dinner at 1700. A nighttime snack was provided which was eaten sometime between 2000 and midnight. On July 20, 1971, breakfast was at 0820 and lunch at 1200. Blood was obtained by venipuncture on July 19 at 1130, 1530, 2000 and midnight and on July 20, at 0400 and 0800. The sampling sessions lasted about 30 minutes each.

The students voided as completely as possible after the first blood sampling session at 1200. The urine was discarded. From 1200 on July 19 until 1200 on

* Supported by the St. Paul-Ramsey Medical Education and Research Foundation and by grants from the U.S. Public Health Service (5-K6-GM-13,981) NSF-GW 7613 and NASA (NGR-24-005-006).

July 20 all urine was collected and measured at 4-hour intervals and aliquots of about 100 ml were frozen and stayed in that state.

The health of each student was explored by a physical and laboratory examination, including a complete blood count and differential, SMA 12 battery, erythrocyte sedimentation rate, chest x-ray, and electrocardiogram.

The following laboratory examinations were performed: serum glucose by the cupric-phenanthroline chelate method on the Auto-Analyzer [4]; serum growth hormone by the Rosselin modification [5] of the radioimmunoassay method described by Glick et al. [6]; serum ACTH by radioimmunoassay according to Berson and Yalow [7] and plasma cortisol by the fluorometric method of Smith and Muehlbaecher [8]. In urine, free 11-hydroxycorticosteroids were determined fluorometrically [8] and total 17-hydroxycorticosteroids after enzymatic hydrolysis by a modification of the phenylhydrazine-sulfuric acid method as "Porter-Silber Chromogens" [9]. Sodium and potassium were measured by flame photometry with an IL flame photometer.

For quantitative evaluation of the rhythm parameters, cosine curves were fitted with the help of a Varian 620 electronic computer to the data of each variable studied in each subject and the best fitting curve determined by the least squares method [3]. The results obtained for the group of students were summarized by the mean cosinor method [3].

RESULTS

The circadian rhythms of the blood variables studied are shown in Fig. 1 on the top as chronograms and at the bottom of the figure as a cosinor graph printed by the electronic computer. The cosinor validated and quantified the circadian rhythms in ACTH (A), Cortisol (C) and glucose (B). Serum growth hormone showed a marked variability of interindividual hormone levels and its error ellipse overlapped the point of origin. For this reason a one-way analysis of variance was done and showed a significant time effect ($p<.01$), thus validating a circadian variation also for this parameter.

In the urine (Fig. 2) the cosinor described a circadian rhythm for sodium (A), potasium (B) and free cortisol (C) excretion. A time factor in the excretion of Porter-Silber Chromogens was significant ($p<.01$) in a one-way analysis of variance although this otherwise well-demonstrated rhythm could not be quantified by mean cosinor.

DISCUSSION

A frequency structure of rhythms characterizes endocrine and metabolic events in man and in experimental animals. Endocrine rhythms of pituitary-adrenocortical function as well as of pituitary-gonadal function have been described for the ultradian frequency range [10–16], for the circadian [17–20] and for several infradian frequencies [21–28]. The frequency of the menstrual cycle is, of course, well recognized. Circadian variations of ACTH levels have been shown in clinically healthy, adult individuals [14, 29, 30] and in patients with adrenal insufficiency [31]. Higher frequencies in both serum ACTH and in serum corticosteroid levels have recently gained interest [14, 10–16]. In these studies, as in the one here reported, ACTH levels usually paralleled those of the simultaneously studied

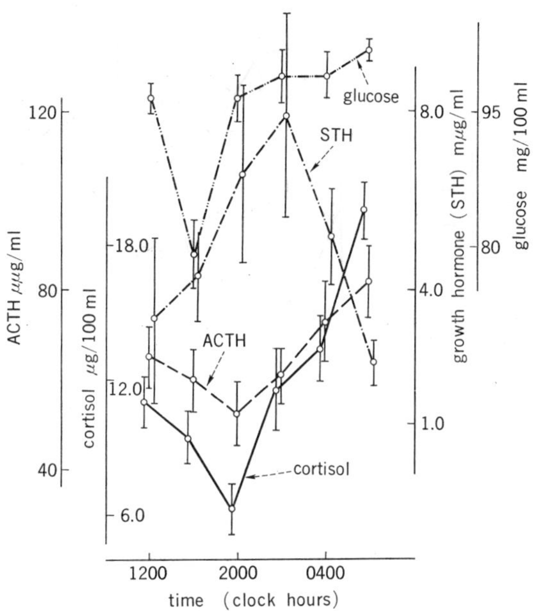

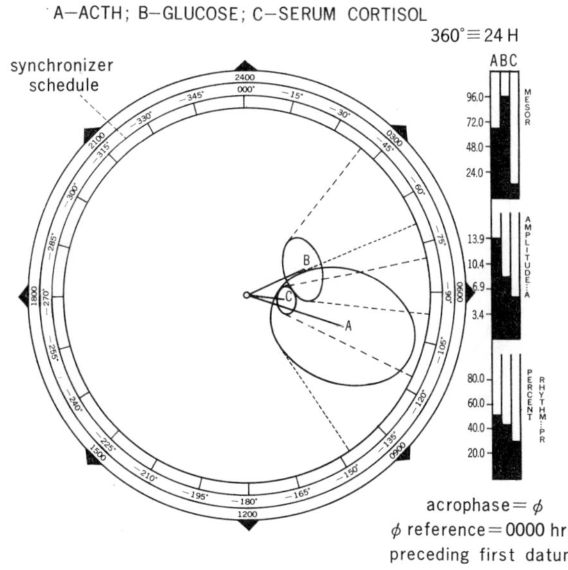

N	φ	(.95 CA)		A	(.95 CI)		M	(.95 CI)		P	PR
A 12	−108	−68	−146	13.98	5.39	22.58	65.20	33.80	96.60	.001	51.16
B 12	− 66	−39	− 96	8.50	6.32	10.69	96.00	83.99	108.02	.001	42.66
C 11	− 98	−79	−115	5.62	4.54	6.69	11.86	4.43	19.28	.001	29.90

Fig. 1 Circadian variations of ACTH, cortisol, growth hormone and glucose in serum shown as chronogram (top). The rhythms of ACTH, cortisol and glucose are quantified by mean cosinor (bottom).

* 6 male and 7 female high school students, 13–17 years of age.

CIRCADIAN URINARY RHYTHMS IN CLINICALLY HEALTHY ADOLESCENTS *

11-OHCS, 17-OHCS, sodium, and potassium

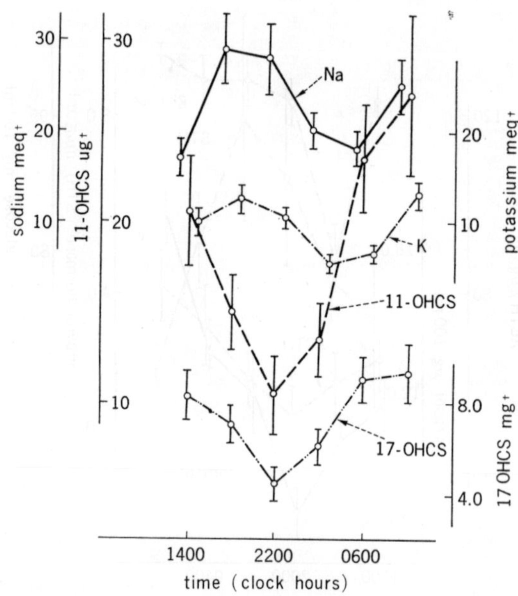

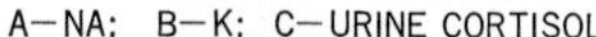

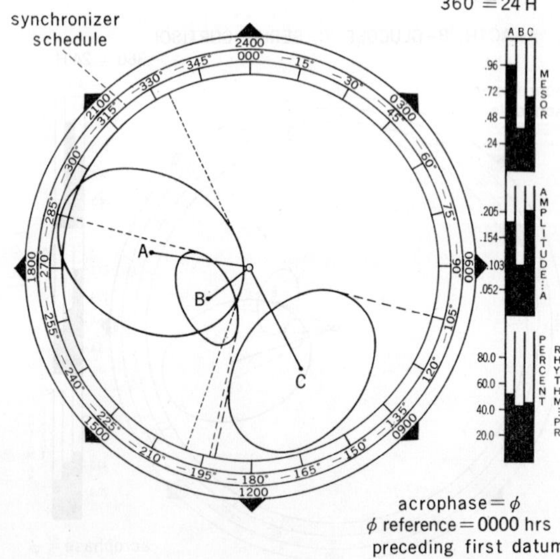

	N	ϕ	(.95 CA)		A	(.95 CI)		M	(.95 CI)		P	PR
A	13	−278	−200	−334	.18	−.01	.37	.96	−.04	1.97	.036	51.38
B	13	−235	−192	−284	.09	.05	.13	.37	.09	.66	.001	42.23
C	13	−152	−104	−191	.20	.10	.30	.67	.15	1.18	.001	44.30

Fig. 2 Circadian variations of urinary excretion of free 11-hydroxycorticosteroids, total 17-hydroxycorticosteroids (Porter-Silber Chromogens), sodium and potassium shown as chronogram (top). The rhythms of sodium, potassium and 11-hydroxycorticosteroids are quantified by mean cosinor (bottom).

* 6 male and 7 female high school students, 13–17 years of age.

† per 4 hours.

corticosteroids. However, with some notable exceptions most reports on high and low frequency rhythms of endocrine functions as well as those on simultaneously measured rhythms of ACTH and corticosteroids are based on the observation of chronograms and a mathematical analysis other than that of quantitative rhythmometry.

A circadian variation in radioimmunoassayable growth hormone (STH) levels with a sharp nocturnal rise has been observed in fasting adult males and females [32]; in children 8–15 years of age a circadian variation has been found with consistently higher levels at night [33, 34]. A group of healthy young adults, two men and four women maintained on four different feeding regimens, were sampled at 30-minute intervals over a 24-hour span [35]: growth hormone levels were found to be consistently higher at night than during the day on all four feeding regimens, including fasting and a half-hourly equicaloric liquid diet. The circadian variation of plasma growth hormone was thus shown to be independent of glucose concentration and unrelated to meal times, with maximum values observed between 1800 and 0300. The highest growth hormone values in our series occurred at 2000 in day-active individuals and thus are in keeping with these earlier results. The lack of rhythm detection and quantification by cosinor is due to the great variability between individuals, part of which may be caused by superimposed higher frequencies. The rhythm detection in the excretion of free 11-hydroxycorticosteroids and the lack of simultaneous detection of the otherwise widely and reliably demonstrated rhythm in 17-hydroxycorticosteroid (Porter-Silber Chromogen) excretion [19] appears to be due to the larger relative amplitude of the rhythm of the free steroid excretion. As reference function the free 11-hydroxycorticosteroids in urine may thus be preferable to the widely used Porter-Silber chromogens, also, because of the faster and simpler fluorometric method which is amenable to automation.

The results presented quantify by methods of inferential statistical rhythmometry the circadian rhythm of serum ACTH in a group of clinically healthy adolescents and define this rhythm in its phase relations to plasma cortisol and to urinary 11-hydroxycorticosteroid, sodium and potassium excretion. A circadian variation of serum growth hormone is also found but its quantification by cosinor will require more extensive data.

SUMMARY

Thirteen clinically healthy adolescents, aged 13–17, were studied over a 24-hour span. Circadian rhythms of serum ACTH, cortisol and glucose and of urinary 11-hydroxycorticosteroid and electrolyte excretion were detected and quantified by cosinor analysis. Circadian variations in serum growth hormone and of urinary 17-hydroxycorticosteroids were shown by conventional statistical methods but the data available were not sufficient for reliable rhythmometric quantification of their rhythm parameters.

REFERENCES

1. Halberg F., Haus E., Halberg E., Kühl J. F. W., Lucas R., Gedgaudas E., Leong J., Ahlgren A., Strobel H. and Angellar A. (1974): Blood pressure self-measurement for computer-monitored health assessment and the teaching of chronobiology in high

school. *In*: Chronobiology. Proc. Symp. Quant. Chronobiology, Little Rock, 1971, (L. E. Scheving, F. Halberg and J. E. Pauly, eds.), pp. 372–378, Igaku Shoin Ltd., Tokyo.

2. Haus E. and Lakatua D. J.: Time qualified normal ranges of hematologic and biochemical parameters in clinically healthy adolescents. Human Pathology (*In*: Preparation).
3. Halberg F., Tong Y. L. and Johnson E. A. (1967): Circadian system phase: an aspect of temporal morphology; procedures and illustrative examples. *In*: The Cellular Aspects of Biorhythms, (H. von Mayersbach, ed.), 8th Int. Congr. Anat., pp. 20–48, Springer-Verlag, Berlin.
4. Brown M. E. (1961): Ultra-micro sugar determinations using 2,9-dimethyl-1,10-phenanthroline hydrochloride (neocuproine). Diabetes, *10*: 60–62.
5. Rosselin G., Assan R., Yalow R. S. and Berson S. A. (1966): Separations of antibody-bound and unbound peptide hormones labelled with iodine-131 by talcum powder and precipitated silica. Nature, *212*: 355.
6. Glick S. M., Roth J. Yalow R. S. and Berson S. A. (1963): Immunoassay of human growth hormone in plasma. Nature, *199*: 784.
7. Berson S. A. and Yalow R. S. (1968): Radioimmunoassay of ACTH in plasma. J. Clin. Invest., *47*: 2725.
8. Smith E. K. and Muehlbaecher C. A. (1969): A fluorometric method for plasma cortisol and transcortin. Clin. Chem., *15*: 961.
9. Glenn E. M. Nelson D. H. (1953): Chemical method for the determination of 17-hydroxycorticosteroids and 17-ketosteroids in urine following hydrolysis with B-glucuronidase. J. Clin. Endocr. and Metab., *13*: 911.
10. Halberg F. (1961): The Adrenal Cycle. *In*: Circadian Systems, Report of 39th Ross Conference on Ped. Res., pp. 57–60, Ross Laboratories, Columbus, Ohio.
11. Galicich J. H., Haus E., Halberg F. and French L. A. (1964): Variance spectra of corticosteroid in adrenal venous effluent of dogs. Ann. N.Y. Acad. Sci., *117*: 281–291.
12. Hellman L., Weitzman E. D., Roffwarg H., Fukushima D. K., Yoshida K. and Gallagher T. F. (1970): Cortisol is secreted episodically in Cushing's syndrome. J. Clin. Endocr., *30*: 686–689.
13. Weitzman E. D., Fukushima D., Nogeire C., Roffwarg H., Gallagher T. F. and Hellman L. (1971): Twenty-four hour pattern of the episodic secretion of cortisol in normal subjects. J. Clin. Endocr., *33*: 14–22.
14. Krieger D. T., Allen W., Rizzo F. and Krieger H. P. (1971): Characterization of normal temporal pattern of plasma corticosteroid levels. J. Clin. Endocrinol., *32*: 266–284.
15. Nankin H. R. and Troen P. (1971): Repetitive luteinizing hormone elevations in serum of normal men. J. Clin. Endocr., *33*: 558–560.
16. Boyar R., Perlow M., Hellman L., Kapen S. and Weitzman E. (1972): Twenty-four hour pattern of luteinizing hormone secretion in normal men with sleep stage recording. J. Clin. Endocr., *35*: 73–81.
17. Halberg F. (1969): Chronobiology. Ann. Rev. Physiology, *31*: 675–725.
18. Halberg F. (1967): Ritmos y corteza suprarenal. IV Simposio Panamericano de Farmacología y Terapéutica. Excerpta Medica Int. Congress Series No. 185, pp. 7–39.
19. Halberg F., Reinhardt J., Bartter F., Delea C., Gordon R., Reinberg A., Ghata J., Hofman H., Halhuber M., Günther R., Knapp E., Pena J. C. and Garcia Sainz M. (1969): Agreement in endpoints from circadian rhythmometry on healthy human beings living on different continents. Experientia, *25*: 107–112.
20. Faiman C. and Winter J. S. D. (1971): Diurnal cycles in plasma FSH, testosterone and cortisol in men. J. Clin. Endocr., *33*: 186–192.
21. Halberg F. and Hamburger C. (1965): Electronic computer techniques for the study of endocrine rhythms. Acta Endocr. (Kbh) Suppl., *100*: 170.
22. Halberg F., Engeli M., Hamburger C. and Hillman D. (1965): Spectral resolution of low-frequency, small amplitude rhythms in excreted 17-ketosteroids; probable androgen-induced circaseptan desynchronization. Acta Endocr. (Kbh) Suppl., *103*: pp. 54.

23. Soule J. and Assenmacher J. (1966): Mise en evidence d'un cycle annuel de la function cortico-surrenalienne chez le canard male. C. R. Acad. Sci. (Paris), *263*: 983–985.
24. Okamoto M., Kohzuma K. and Horiuchi Y. (1964): Seasonal variations of cortisol metabolites in normal men. J. Clin. Endocr., *24*: 470–471.
25. Watanabe G. I. (1964): Seasonal variation of adrenal cortex activity. Arch. Environ. Health, *9*: 192–200.
26. Haus E. and Halberg F. (1970): Circannual rhythm in level and timing of serum corticosterone in standardized inbred mature C-mice. Environmental Res., *3*: 75–90.
28. Engle E. T. and Shelesnyak M. D. (1934): First menstruation and subsequent menstrual cycles of pubertal girls. Human Biol., *6*: 431–453.
29. Ney R. L., Shimizu N., Nicholson W. E., Island D. P. and Liddle G. W. (1963): Correlation of plasma ACTH concentration with adrenocortical response in normal human subjects, surgical patients, and patients with Cushing's disease. J. Clin. Invest., *42*: 1669–1677.
30. Vague Ph., Oliver Ch. and Bourgoin J. Y. (1973): Circadian rhythms in plasma ACTH of healthy human adults. *In*: Chronobiology. Proc. Symp. Quant. Chronobiology, Little Rock, 1971, (L. E. Scheving, F. Halberg and J. E. Pauly, eds.), Igaku Shoin Ltd., Tokyo.
31. Graber A. L., Givens J. R., Nicholson W. E., Island D. P. and Liddle G. W. (1965): Persistence of diurnal rhythmicity in plasma ACTH concentrations in cortisol-deficient patients. J. Clin. Endocr., *25*: 804–807.
32. Quabbe H. J., Schilling E. and Helge H. (1966): Pattern of growth hormone secretion during a 24-hour fast in normal adults. J. Clin. Endocr., *26*: 1173–1177.
33. Hunter W. M. and Rigal W. M. (1966): The diurnal pattern of plasma growth hormone concentration in children and adolescents. J. Endocrin., *34*: 147–153.
34. Vigneri R., D'agata R., Morelli A. et al. (1970): Daily variations of blood somatotropin in children. Folia Endocrinol. (Roma), *23*: 504–512.

A REVIEW OF ANNUAL ENDOCRINE RHYTHMS IN CATS WITH BRAIN STEM LESIONS*

Walter RANDALL and Virginia PARSONS
Department of Psychology, University of Iowa
Iowa City, Iowa, U.S.A.

Cats with lesions in the ventro-lateral pontile tegmentum have modified annual rhythms in adrenalin and glucocorticoid excretion. We have been studying these abnormal rhythms in order to determine the physiological bases of an abnormal behavior that is also present in these cats [1, 2]. The present report compares the annual endocrine rhythms of normal and lesioned cats in order to specify the primary endocrine dysfunction. The main conclusion is that abnormal endocrine rhythms are produced by temporal uncoupling of variables that normally have orderly and systematic interrelationships.

METHODS

Adult, male cats (*Felis catus* L.) were individually caged in an air conditioned room (72°F±2°). Windows provided natural lighting, and the artificial lights in the room were on only during daylight. Six of the cats received stereotaxic, DC lesions of the ventral lateral brain stem at the level of the pons. The exact locus of the lesions has been illustrated in previous reports [3]. Several weeks after the lesion, when the temporary effects (e.g., motor and eating debilities) were no longer present, the home cages of both normal and lesioned cats were converted to metabolism cages by exchanging the floor of the cage with a sloping tray with a spout and collecting bottle. For 96 hours the urine was collected fresh by frequent monitoring and frozen and stored at −25°C. This collection procedure was repeated once a month for a year. Body weight and caloric intake data were collected also for the same year. Then adrenalin and noradrenalin concentrations were determined by the method of von Euler and Lishajko [4], and 11-hydroxycorticoid concentration was determined by the method of Silber, Busch and Oslapas [5]. To maximize the chances of detecting a rhythm, the 12 monthly samples for a cat were analyzed on the same day in one single procedure. Thus any day-to-day variation in the method could not obscure the rhythm because differences among cats in overall level are not relevant in the rhythm analysis. Least-squares fitted curves to the endocrine data were obtained with the model

$$y_t = c_0 + c_1 t + \sum_{i=1}^{j}\left(a_i \cos\frac{2\pi it}{k} + b_i \sin\frac{2\pi it}{k}\right).$$

where y is the derived value at time t, c_0, c_1, a_i and b_i are the least-squares derived coefficients, k is the number of repeated measures (k=12 for this study) and j, the last harmonic of the Fourier series, is $[k/2]$. Statistical evaluation of

* This research was supported by Grant MH15402-04 from the National Institute of Mental Health, United States Public Health Service.

the fitted curves was accomplished with an analysis of variance. The linear term was used in order to test for any systematic effects of storage of the urines. In every case, no linear trend was present. The model then reduced to the familiar Fourier series and to the analysis of variance model for periodic regression developed by BLISS [6]. A computer program [7] performed all the mathematical and statistical operations for the combined algebraic and trigonometric model. Details and illustrations of the periodic regression technique are presented by BLISS [8]. Differences between normal and lesioned groups were tested with a Type I analysis of variance (a two-factor mixed design with repeated measures on one factor, LINDQUIST [9], p. 267), and the Pearson product moment correlation coefficient was used to test for interrelationships among the variables within a group. A step-wise regression procedure was employed (the "forward selection procedure" [10]) to determine the best predictor of the adrenalin rhythm in each group.

RESULTS

The periodic regression analyses detected annual rhythms in adrenalin and noradrenalin excretion in both normal and lesioned cats, with a difference between normal and lesioned cats in the adrenalin rhythm but not in noradrenalin (Figs. 1 and 2). Differences in trend between normal and lesioned cats were found in the adrenalin data (Type I analysis of variance, $p<.05$) but not in the noradrenalin data ($p>.2$). A prominent annual rhythm in 11-hydroxycorticoid excretion was detected for the normal cats, and multiphasic variations during the year existed in the lesioned group (Fig. 3). The Type I analysis of the 11-hydroxycorticoid data revealed trend differences between normal and lesioned groups ($p<.05$). Fig. 4 exhibits two abnormal rhythms in the lesioned group, the abnormal rhythm in 11-hydroxycorticoid excretion and an abnormal rhythm in adrenalin per body weight. No rhythm was detected in the normal group with the periodic regression technique when the body weight transformation was made. Table 1 presents a summary of the periodic regression analysis.

One outstanding feature of the difference between normal and lesioned cats was the correlations among the variables as determined with Pearsons correlation coefficients (r). In general no significant correlations were found in the lesioned group, but in the normal group all the variables exhibited significant correlations. For example, some correlations in the normal group were adrenalin and body weight ($r=+.72$, $p>.01$), adrenalin and noradrenalin ($r=-.85$, $p<.001$), 11-hydroxycorticoids and caloric intake ($r=+.76$, $p<.01$), and body weight and food intake ($r=+.42$, $p<.05$), but none of these correlations were significant ($p<.05$) in the lesioned group, except for a negative correlation between food intake and body weight ($r=-.44$, $p<.01$).

To further determine group differences in the relationships of adrenalin excretion to the other variables, and especially to determine if the separate correlations of the different variables with adrenalin in the normal group were independent, a step-wise regression equation was derived for both groups with adrenalin as the dependent variable. The independent variables were noradrenalin, 11-hydroxycorticoids, body weight and food intake. The equation for the normal group was Adrenalin=3.35−.16 Noradrenalin. Even though the other variables of the normal group were correlated significantly with adrenalin, they were not admitted

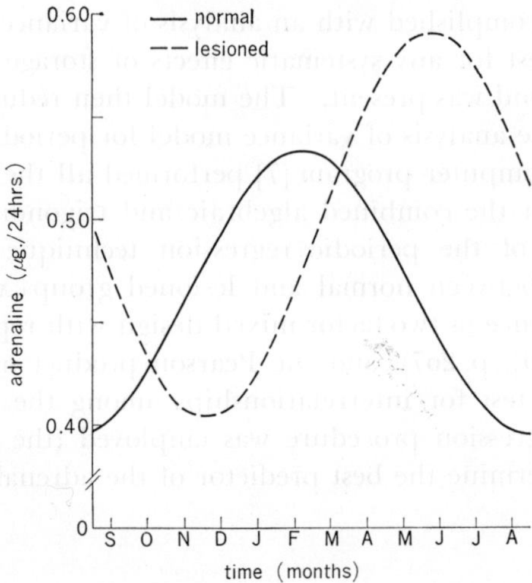

Fig. 1 The least-squares-fitted curves, using the trigonometric fitting functions of the Fourier series, to the adrenalin data for eight normal cats and six lesioned cats. An annual harmonic accounted for all the systematic trend in both sets of data. Table 1 shows the least-squares derived coefficients and the outcomes of the F-tests.

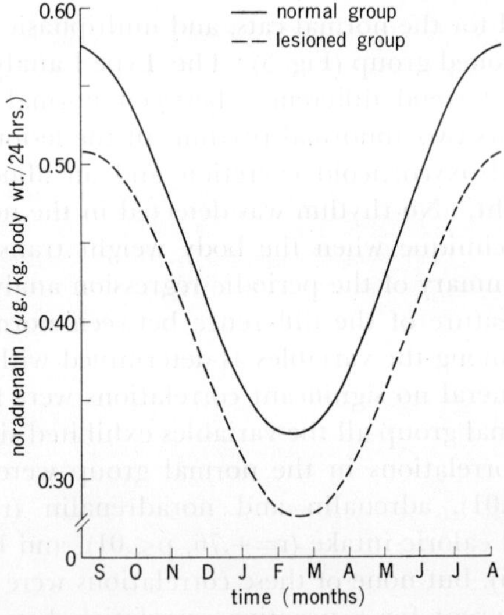

Fig. 2 The least-squares-fitted curves to the noradrenalin data for the same cats. The annual harmonic accounted for all the trend in both sets of noradrenalin data.

to the regression equation because they were all accounting for the same variance as noradrenalin. Thus in the normal group a highly integrated, interdepedent system is indicated, a kind of phase angle integration. For the lesioned group, all variables were rejected from the regression equation, indicating an independency and dissociation of the annual adrenalin rhythm from the other variables.

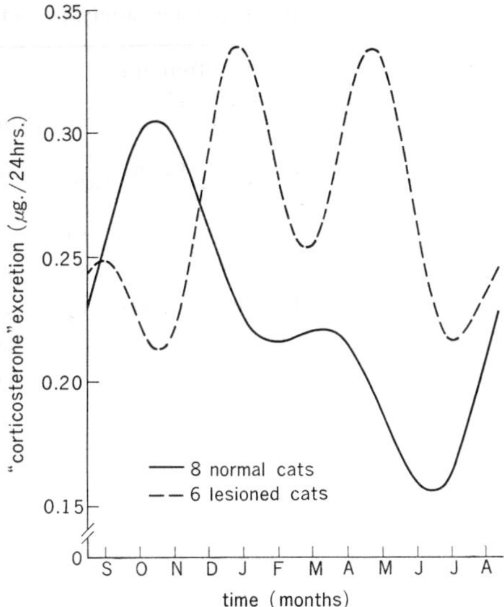

Fig. 3 The least-squares-fitted curves to the 11-hydroxycorticoid data. In the normal group the annual harmonic accounted for about 75% of the trend, but an additional harmonic was added in order to account for the remainder of the trend. Three sine fitting functions were required to account for the trend in the data of the lesioned group, with the annual and four-month harmonics each accounting for approximately 50% of the trend. F-test results and the least-squares derived coefficients are presented in Table 1.

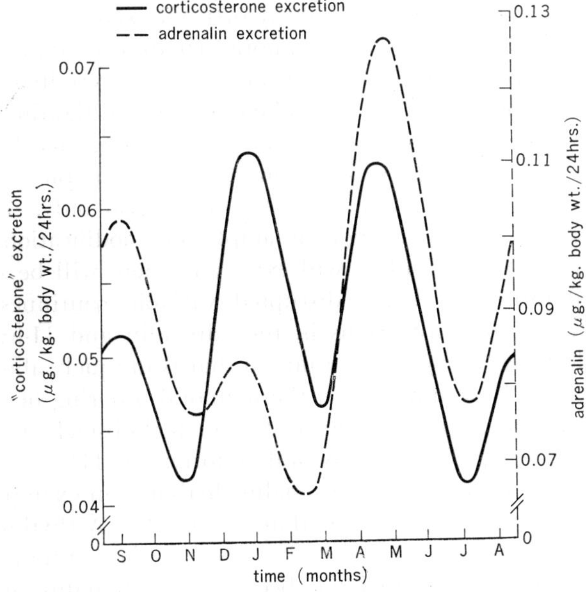

Fig. 4 Two abnormal rhythms in the lesioned group illustrated in terms of per kilogram body weight. The 11-hydroxycorticoid rhythm is the same as illustrated in Fig. 3, indicating that body weight is out of phase with the 11-hydroxycorticoid rhythm. No adrenalin/kg. body weight rhythm was present in the normal group because these two variables were in phase. In cats with brain stem lesions, the interaction of modified and uncoupled rhythms results in an abnormal rhythm in adrenalin/kg. body weight that is similar to the abnormal 11-hydroxycorticoid rhythm.

Table 1 Summary of the results of the regression analyses coefficient and p-value.

Fitting Functions	Adrenalin				Noradrenalin				11–Hydroxycorticoids			
	Normal Group		Lesioned		Normal Group		Lesioned		Normal Group		Lesioned	
	Coeff.	p-values	Coeff.	p-values	Coeff.	p-values	Coeff.	p-values	Coeff.	p-values	Coeff.	p-values
1	0.45	.01	.497	.05	.458	.01	.400	.01	.225	NS	.269	.05
t		NS		NS		NS		NS		NS		NS
cos π t/6	−.068	.05*	.034	.05*	.118	.01*	.106	.01*	−.007	.05	−.036	.05
sine π t/6	−.001		−.362		.024		.036		.055		−.018	
cos π t/3									−.017	.05*	−.011	NS
sine π t/3									.029		−.015	
cos π t/2											.009	.05*
sine π t/2											.037	

* The remaining trend was not significant ($p > .05$) after adding this harmonic.

DISCUSSION

The integration of the endocrine system during a year consists in part of orderly phase angle relationships among annual rhythms. Furthermore, orderly temporal relationships exist among the endocrine rhythms and the food intake and body weight rhythms. For example, in the normal cat in the fall of the year, increases in adrenalin excretion, 11-hydroxycorticoid excretion, body weight, and food intake occur, along with decreases in noradrenalin excretion. These changes occurred simultaneously in all the normal cats, cats who were living in a constant-temperature environment but with normal environmental lighting. The features of the normal physiological milieu, then, vary in an integrated fashion with the seasons of the year. Thus at any given time there is a complex set of endocrine and metabolic values that is normal for that time of year only. If one rhythmic variable is disrupted by phase angle or amplitude modifications, then for all times, an abnormal internal milieu will exist, and there will be systematic variations in the abnormality as the disrupted variable continues its abnormal rhythm. The temporal dysfunctions in the adrenalin and 11-hydroxycorticoid rhythms in the lesioned cats provide a milieu that is uniquely abnormal for each time of the year. Such complicities in the temporal ordering of complex sets of endocrine and metabolic variables constitutes our real obstacle in elucidating the physiological bases of a periodic behavioral abnormality [11].

The lesion-induced change in the adrenalin rhythm is an example of a dysfunction in temporal integration of two rhythmic variables. A rhythm was found in adrenalin/kg. body weight in the lesioned cats only. No rhythm in adrenalin/kg. body weight existed in the normal cats because the adrenalin and body weight rhythms were in phase. BERNARD [12, p. 132] warns against dividing by body weight because "all sorts of tissues foreign to the phenomenon in question are included." But in longitudinal studies, where rhythms in body weight are generally found, the question of temporal integration arises, i.e., whether or not the endocrine rhythms are concomitants of or responses to the changes in body weight. In addition BERNARD's warning has new meaning when applied to logitudinal studies because dividing by body weight may transform a nonrhythmic variable into a rhythmic one, with the rhythm solely attributable to the variation in body weight.

REFERENCES

1. Randall W. and Parsons V. (1972): Rhythmic dysfunctions in 11-hydroxycorticoid excretion after midbrain lesions and their relationship to an abnormal grooming behavior in cats. J. interdis. cycle res., (in press).
2. Parsons V. and Randall W.: The relationship of grooming fragments to catecholamine excretion in cats, (in preparation).
3. Randall W. and Liittschwager J. (1967): The relationship between cyclic changes in thyroid function and behavior of cats with brain stem lesions. J. Psychiat. Res., *5*: 39–58.
4. Von Euler H. and Lishajko F. (1961): Improved technique for the fluorimetric estimation of catecholamines. Acta physiol. Scand., *51*: 348–356.
5. Silber R., Busch R. and Oslapas R. (1958): A practical procedure for the estimation of corticosterone or hydrocortisone. Clin. Chem., *4*: 278–285.
6. Bliss C. (1958): Periodic regression in biology and climatology. Bulletin 615, The Connecticut Agricultural Experimental Station.
7. Shama K. (1968): A generalized program for treatment of periodic data. Unpublished Master's thesis. University of Iowa, Iowa City, Iowa.
8. Bliss C. (1970): Statistics for biologists, Vol. 2, McGraw-Hill, New York.
9. Lindquist E. (1953): Design and analysis of experiments in psychology and education, p. 267, Houghton Mifflin, Boston.
10. Draper N. and Smith H. (1966): Applied regression analysis, p. 169, John Wiley and Sons, Inc., New York.
11. Rogers W., Parsons V. and Randall W. (1971): Consummatory grooming fragments: A model for periodic behaviors. Psychon. Sci., *23*: 375–376.
12. Bernard C. (1957): An introduction to the study of experimental medicine, 1865. Translated by Henry Copley Green, Dover Publications, Inc., New York.

ENDOGENOUS MECHANISMS FOR THE SYNCHRONIZATION OF CORTICOSTEROID PRODUCTION AND SEASONAL RHYTHMS IN HIBERNATING MAMMALS

M. G. KOLPAKOV, S. G. KOLAEVA and M. G. POLYAK

Laboratory of Endocrinology, Institute of Cytology and Genetics Siberian Branch of the Academy of Sciences of the U.S.S.R. Novosibirsk, U.S.S.R.

This study was undertaken on a hibernating mammal, the ground squirrel (*Citellus erythrogenys*). Animals were caught in nature at the most active period of their life. The hibernation took place in a special animal house at a temperature of 2–3°C. The body temperature was 3°C.

The annual cycle of activity in the ground squirrel (*C. erythrogenys*) inhabiting the steppe regions of Western Siberia is characterized by a short period of intense activity alternating with a prolonged period of hibernation [1]. There are reasons to think that the synchronization of vital processes is influenced greatly by seasonal rhythms in corticosteroids which may act as a metronome [2].

At the beginning of the active period (May), the production of hydrocortisone and aldosterone is highest. An increase of nuclear and nucleolar volumes in the fasciculata and glomerulosa cells of the adrenal cortex at this time is combined with an increase in incorporation of labelled uridine and leucine; the latter may be indicative of an intensified functioning of the genetic apparatus in the cell nucleus which in turn causes an intensification of the protein-synthetic and secretory activity. In June–July the size of the adrenal cortex decreases sharply, the secretion of all corticosteroids falls, and nuclear and nucleolar volumes diminish to a great extent.

It is essential that at the time of greatest activity the secretory rhythm of the adrenal z. glomerulosa is the same as the rhythm of functional activity of the renal endocrine apparatus. The renin content in kidneys decreases from May to July five times, and three times in the JGI [3]. In the latter case the specific height of MD cells diminishes; this suggests that the signal changes which enters the juxtaglomerular granulated cells from the part of renal tubules that determines the sodium status of the body. The possibility of such a change in this status is suggested by an increase in plasma sodium concentration.

According to our data, angiotensin II stimulates selectively the *in vitro* aldosterone production by adrenals of *C. erythrogenys.* It is noteworthy that the reaction of the adrenals to the same dose of angiotensin II in May and July is different. In July the increase in aldosterone production is one and a half times less than in May.

The greatest response of the glucocorticoid-producing tissue to ACTH was found in May, and the reaction to the same dose in July was one and a half times less [4].

Taking into consideration the existence of a circadian rhythm in the *in vitro* response of adrenals to ACTH [5] all experiments with the incubation of adrenals have been made at the same morning hours.

The change in the response of the adrenal cortex to regulatory stimuli found in our laboratory must be attributed to the main mechanisms responsible for the formation of a rhythmic seasonal activity of the adrenal which can be interpreted as a reflection of the intrarenal clock movement synchronized with rhythms of other parts of the endocrine system by means of angiotensin II and ACTH.

The study of corticosteroid production a few months after initiation of hibernation has shown that the adrenals have started to prepare for the activity phase. In hibernating animals the production of aldosterone in March rises by 108%, over that of January, and the production of hydrocortisone by 150%. In March sharp increases in all indices of adrenal function were found. The state of the adrenal cortex, as those of other glands in hibernating mammals [6], is a dynamic process with quite different levels of functional activity at the beginning and at the end of the period of torpor; undoubtedly this suggests an endogenous control of the seasonal rhythm.

The results of our experiments may be considered as a confirmation of PENGELLY's suggestion [7] about the existence of an "annual" cycle in the life activity of hibernating mammals which is not dependent on the environment.

REFERENCES

1. SLONIM A. D. (1971): Ecological Physiology of Animals. Moscow.
2. HALBERG F. (1960): Temporal coordination of physiologic function. *In*: Biological Clocks. Cold Spring Harbor Symposia on Quantitative Biology, 25, 4.
3. KOLAEVA S. G., LUTSENKO N. D. and SOKOLOVA G. P. (1969): Seasonal morphofunctional changes in adrenal z. glomerulosa and renal JGA in *Citellus erythrogenys*. 5th Conf. Europ. Compar. Endocrinol., Utreht (Abstr. No. 82).
4. KOLPAKOV M. G., KRASS P. M. and POLYAK M. G. (1971): Mechanisms of seasonal rhythms of corticosteroid regulation in the ground squirrel. 6th Conf. Europ. Compar. Endocrinologists, Montpelier (Abstr. No. 111).
5. UNGAR F. and HALBERG F. (1962): Circadian rhythm in the *in vitro* response of mouse adrenal to adrenocorticotropic hormone. Science, *137*: 1058–1060.
6. KAYSER Ch. and PETROVIC A. (1958): Rôle du cortex surrénalien dans le méchanisme du sommeil hivernal. Comp. Rend. Soc. Biol., 152, *3*: 519–528.
7. PENGELLY E. T. and ASMUNDSON S. J. (1971): Annual biological clocks. Scientific American, 224, *4*: 72–79.
8. KOLPAKOV M. G. and SAMSONENKO R. A. (1970): Adrenal reactivity in *Citellus erythrogenys* at different periods of the life activity. Proc. Acad. Sci. of USSR, *191*: 1424–1426.

CORTICOSTERONE RHYTHMOMETRY AT A SINGLE TIME-POINT—UTILIZING ANTIPHASE TEST*

FRANZ HALBERG[1)], GEORGE S. KATINAS[1,2)], ERNA HALBERG[1)] and ERHARD HAUS[3)]

[1)]*Chronobiology Laboratories, Department of Pathology*
University of Minnesota
Minneapolis, Minnesota, U.S.A.
[2)]*Institute for Experimental Medicine*
Academy of Medical Science
Leningrad, U.S.S.R.
[3)]*Department of Pathology, St. Paul-Ramsey Hospital*
St. Paul, Minnesota, U.S.A.

ABSTRACT

High values in serum corticosterone (rhythm) of one group of mice can be made to coincide with low corticosterone levels in another group—by manipulating a regimen of alternating 12-h light and 12-h darkness, as well as by several other precautions. Antiphase-tests carried out on this basis allow the study of statistically and physiologically different states at the same times of day.

In the follow-up on an earlier study of skin-grafts [1], it seemed of interest to schedule transplantations at both the circadian high and low in serum corticosterone of certain recipient mice. In view of earlier work [2–4], these times were assumed to be both predictable and amenable to manipulation by the lighting regimen. Accordingly, a study was planned in such a fashion that the predicted corticosterone high and low chosen for both pertinence and operational convenience occurred at the same time each day.

The graft recipients used were of several inbred strains (C57 Black subline 6 stock and Dilute Brown, subline 2 males), as were the donors involved in these operations. This information, as well as the use of both males and females from one strain, is given in Table 1; thus possible strain differences had to be considered, along with a sex difference also predicted from earlier work [5].

For several preoperative weeks and for several months after grafting, the total number in the several different recipient animal groups (Table 1) was kept in darkness for 12 hours, alternating for one-half with light from 0900 to 2100, for the other half with light from 2100 to 0900. Assuming that the corticosterone rhythms in all recipient subgroups of mice can be similarly synchronized by a regimen of light (L) and darkness (D) alternating at 12-hour intervals (LD12:12), and knowing that such lighting regimens were 180° out of phase with each other in the study on hand, it followed that at two time points during a 24-hour span the crest of one rhythm coincided with the trough of the other. Of these two antiphases on the above specified LD12:12 regimens, one occurred in the morning at a convenient time for all the personnel involved in this study. Thus, from earlier

* Supported by the U.S. Public Health Service (5-K6-GM-13,981) NASA (NGR-24-005-006) and the St. Paul Ramsey Research and Education Foundation.

Table 1 Predicted statistically significant difference in serum corticosterone of mice kept on different lighting regimens but sacrificed at a fixed time*.

Light (in LD12:12)	Serum corticosterone in mg % (N of mice)					Overall Mean±SE	t P
	Recipient	C57Bl/6		DBA/2	C57Bl/6**		
	Donor	Balb/C	DBA/2	Balb/C	C57Bl/6		
L 0900–2100		11.5 (4)	13.5 (4)	15.5 (4)	18.0 (3)	11.5±.72	
		10.5 (4)	10.0 (4)	14.0 (4)	11.5 (3)		
		7.0 (4)	11.0 (4)	15.0 (4)	17·5 (3)		
		9.5 (3)	8.0 (3)	12.5 (4)			
		9.5 (3)	8.5 (3)	11.0 (3)			
		6.0 (3)	8.5 (3)	14.0 (3)			
	Mean±SE	9.0±0.86	9.9±.85	13.7±.68	15.7±2.09		4.8 <.001
L 2100–0900		17.0 (4)	20.0 (4)	10.0 (4)	24.5 (3)	17.4±1.59	
		14.5 (4)	18.0 (4)	23.0 (4)	22.0 (3)		
		14.0 (4)	13.5 (4)	18.5 (4)	27.5 (3)		
		13.0 (4)	14.0 (4)	16.0 (4)			
			15.0 (3)	17.0 (4)			
			15.0 (3)	18.5 (3)			
	Mean±SE	14.6±.85	15.9±1.04	17.2±1.74	24.7±1.59		

* 0630 to 0749 on two consecutive days. Purina laboratory chow and tap water freely available.

Single time-point sampling also suffices to confirm predicted sex difference.

** Female mice (all others are male).

work, a high corticosterone level for the subgroup kept in L2100 to 0900 and a low level for the subgroup in L0900–2100 could be predicted to occur around 0700±1 hour. Also anticipated from results of earlier sampling was that the females would have higher serum corticosterone values than the males [5].

All these assumptions and predictions were validated by the limited data available in Table 1. These were obtained from animals on each of the two lighting regimens randomly selected for killing between 0630 and 0749. Killing was done on two consecutive days (July 28 and 29, 1971) in order to limit the extent of disturbance as well as the length of the sampling time on any one day [6]. Table 1 confirms the anticipated lower corticosterone values for animals kept in light from 0900 to 2100 as compared to those from animals kept in light from 2100 to 0900. A difference in corticosterone values shown in the top and bottom halves of Table 1 for mice on each of the two lighting regimens, and the t-test shown in the last column of this table, validate the predicted antiphase relation of the two rhythms; thus assumptions concerning *two periodicities* are validated at a *single-time-point* spotcheck.

The sex difference is a significant one below the 1% level on either of the two lighting regimens, as well as in a pool of data, as may be seen from Table 2. At the chosen sampling time a statistically significant strain difference also is found in data from one of the two lighting regimens—at the low in the corticosterone rhythms of two strains. This finding is missed at the other circadian system phase, corresponding to the circadian high of the same corticosterone rhythms,

Table 2 Statistical significance of differences investigated.*

Factor compared	Group compared	Lighting	Difference Mean±SE	t	P
Sex M vs. F	C57 Black/6, M versus C57 Black/6, F	0900–2100	6.2±1.5	4.1	<.010
	C57 Black/6, M versus C57 Black/6, F	2100–0900	5.9±1.7	3.6	<.010
	C57 Black/6, M versus C57 Black/6, F	0900–2100& 2100–0900	8.0±1.9	4.2	<.001
Strain	C57 Black/6 recipient, Balb/C donor vs. Dilute Brown/2 recipient, Balb/C donor	0900–2100	2.5±2.3	1.1	>.500
	C57 Black/6 recipient, Balb/C donor vs. Dilute Brown/2 recipient, Balb/C donor	2100–0900	4.7±1.1	4.3	<.010
	C57 Black/6 recipient, Balb/C donor vs. Dilute Brown/2 recipient, Balb/C donor	0900–2100& 2100–0900	4.2±1.5	2.8	<.020

* Predicted sex difference in serum corticosterone of mice is consistently detected; strain difference also is detected as statistically significant for all mice as well as for the subgroups kept on one of two different lighting regimens, but at the same fixed clock hour (cf. Table 1); it is not detected for mice kept on another lighting regimen.
Purina laboratory chow and tap water freely available.

although the inter-strain difference is in the same direction as that at the corticosterone low; and hence it is significant as well in a pool of data from each strain, irrespective of lighting.

In summary, without chronobiologic considerations, even though the "time of day" is controlled, certain differences may be missed, whereas with rhythmometry even during a short and convenient span of the day, many statistically as well as biologically significant findings may be made. The approach here utilized—an antiphase test—may have yet broader utility in planning intergroup comparisons on rhythmic variables at a single convenient time point.

ACKNOWLEDGEMENT

The invaluable help of Miss Linda Cadotte, Chronobiology Laboratories, University of Minnesota, Minneapolis, is gratefully acknowledged.

REFERENCES

1. Halberg J., Halberg E., Runge W., Wicks J., Cadotte L., Yunis E., Katinas G., Stutman O. and Halberg F. (1974): Transplant chronobiology. *In*: Chronobiology. Proc. Symp. Quant. Chronobiology, Little Rock, 1971, (L. E. Scheving, F. Halberg and J. E. Pauly, eds.), pp. 320–328, Igaku Shoin Ltd., Tokyo.
2. Halberg F., Halberg E., Barnum C. P. and Bittner J. J. (1959): Physiologic 24-hour periodicity in human beings and mice, the lighting regimen and daily routine. *In*: Photoperiodism and Related Phenomena in Plants and Animals, (R. B. Withrow, ed.), pp. 803–878, Ed. Publ. No. 55 of Amer. Assoc. Adv. Sci., Washington, D.C.
3. Halberg F., Barnum C. P., Silber R. H. and Bittner J. J. (1958): 24-hour rhythms at several levels of integration in mice on different lighting regimens. Proc. Soc. Exp. Biol. and Med., *97*: 897–900.

4. Halberg F., Peterson R. E. and Silber R. H. (1959): Phase relations of 24-hour periodicities in blood corticosterone, mitoses in cortical adrenal parenchyma and total body activity. Endocrinology, *64*: 222–230.
5. Halberg F. and Haus E. (1960): Corticosterone in mouse adrenal in relation to sex and heterotopic pituitary isografting. Am. J. Physiol., *199*: 859–862.
6. Halberg F., Albrecht P. G. and Bittner J. J. (1959): Corticosterone rhythm of mouse adrenal in relation to serum corticosterone and sampling. Am. J. Physiol., *197*: 1083–1085.

CIRCADIAN SYSTEMS IN THE HORMONAL REGULATION OF AMPHIBIAN METAMORPHOSIS

M. M. JOSEPH

Department of Zoology and Physiology
Louisiana State University
Baton Rouge, Lousiana, U.S.A.

The role of thyroid hormone in the induction of amphibian metamorphosis is well documented. Recent evidence from a number of laboratories suggests that a prolactin-like hormone may also be involved in the timing of metamorphosis. Treatment of anuran tadpoles with preparations of mammalian prolactin stimulates larval growth and inhibits metamorphosis [1–4]. *In vitro* studies support the view of BERN et al. [3] that prolactin inhibits the responsiveness of the tissues to thyroxin [5, 6].

Another interesting effect of prolactin is its ability to promote migration to water in the immature terrestrial stage (red eft) of *Notophthalmus viridescens* [7, 8]. This return to water is, in fact, a second metamorphosis [9]. The first metamorphosis is comparable to that occurring in anurans. GRANT and COOPER [10] refer to thyroxin as the hormone responsible for "land-drive" in contrast to prolactin, the hormone responsible for "water-drive." Studies of amphibian metamorphosis have led to the concept that normal growth and development are regulated by a balance between a prolactin-like hormone and thyroid hormone, favoring larval growth on the one hand and metamorphosis on the other [1].

Recent studies in our laboratory have demonstrated that circadian systems are involved in the effects of prolactin on amphibian metamorphosis. In premetamorphic tadpoles (*Rana pipiens*) on 14-hour daily photoperiods, daily injections of prolactin administered two hours after the onset of light completely inhibited metamorphosis, whereas injections given at midday or late in the photoperiod had little, if any, effect in preventing it [11]. There are also daily variations in the effectiveness of prolactin to induce the water drive in red efts (*Notophthalmus viridescens*) [12]. Injections given late in a 16-hour photoperiod are more effective (155 hours to migrate to water) than injections given early (389 hours) or at midday (341 hours).

The daily variations in the effectivenes of prolactin indicate that the photoperiod synchronizes a daily oscillation of an endogenous system that sets a biological rhythm of tissue responsiveness to prolactin. Both the thyroid and the interrenal gland have been implicated as parts of the system that mediates the photoperiodic effect. Thyroxin [13] and hydrocortisone [14] injections set circadian rhythms of fattening response to prolactin that persist for many days in fish under conditions of continuous light. Corticosterone can also phase or drive circadian rhythms of fattening responses to prolactin in a number of other vertebrate species [14, 15]. In the light of these findings, we tested the ability of corticosterone to set up a circadian rhythm of water-drive responsiveness to prolactin in red efts [12]. It was found that the injections of prolactin given 8 hours after the time of corticosterone injections were most effective (100 hours) in promoting migration to water, while prolactin injections given 16 hours after corticosterone were least effective (245 hours). Intermediate values were obtained in the efts that received injections of prolactin and corticosterone at the same time (146 hours) and in the group that received prolactin alone (171 hours) (Fig. 1).

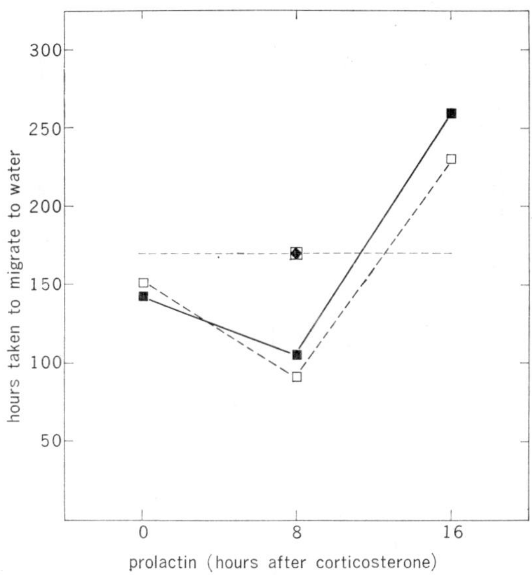

Fig. 1 Corticosterone phases the circadian variations in water-drive response to prolactin in *Notophthalmus viridescens*. Efts held on continuous light were injected with corticosterone (1 μg/day) either at 0800 (■) or at 2400 (□) starting on 6 October, 1970, and continued on alternate days. Each corticosterone group received prolactin injections (2 μg/day) at 0800, 1600, or 2400 starting on 8 October, and continued daily until the efts migrated to water. One group received prolactin alone (◆). Each group consisted of four to five individuals. According to a least-squares analysis of variance, the temporal differences in responses to prolactin for both corticosterone-injected groups were significant at the 1% level.

Breaux and Meier [11] have postulated that combinations of a stimulatory (thyroxin) and an inhibitory factor (prolactin) in various time relations might offer a basic mechanism for a graded and controlled sequence of metamorphic events. To test the validity of this hypothesis, premetamorphic tadoples (weighing 1.5 to 3 g) of *Rana pipiens* held on continuous light were injected daily for 6 days with ovine prolactin (10 μg/day) in various time relations (0, 4, 8, 12, 16 and 20 hours) following the time of thyroxin (10 μg on alternate days beginning one day prior to prolactin treatment) or saline injections. The hormones were administered intraperitoneally in 0.01 ml 0.65% saline. There were considerable differences in the effectiveness of prolactin to inhibit thyroxin-induced metamorphosis (Fig. 2). Prolactin injections given 20 hours after the injection of thyroxin were most effective. Tadpoles that received prolactin at 0 and 4 hours after thyroxin injections underwent metamorphic changes in parallel with the controls, which received thyroxin but no prolactin. Prolactin did not completely block the thyroxin-induced metamorphosis since all the tadpoles eventually metamorphosed regardless of the times of injections. However, the dosage of thyroxin used was quite high. It is conceivable that the amounts of thyroxin (40 μg) administered overwhelmed the blocking action of prolactin and that lower dosages more in line with physiological levels might have revealed a complete inhibition by prolactin.

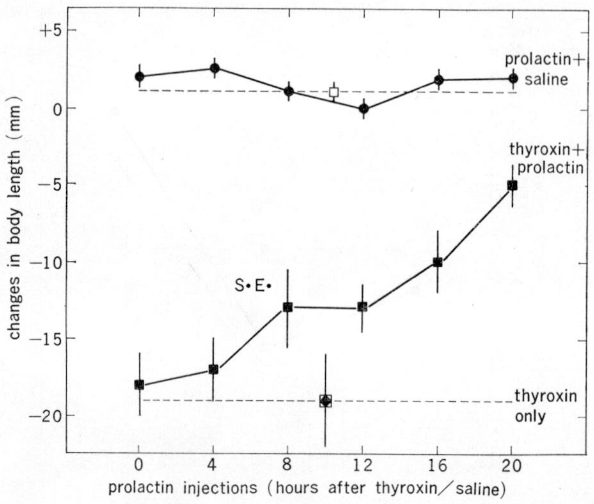

Fig. 2 Effect of prolactin on thyroxin-induced metamorphosis in *Rana pipiens* tadpoles. Body length changes were measured in tadpoles held on continuous light and injected daily for six days with ovine prolactin (10 μg/day) in various time relations following the time of thyoxin (■) or saline (●) injections. Thyroxin (10 μg) or saline injections were made on alternate days beginning one day prior to prolactin treatment. Group which received thyroxin alone =◈; uninjected controls □. Each group consisted of five individuals. S.E. is the standard error of the mean.

Conflicting results have been published concerning the ability of prolactin to inhibit thyroxin-induced metamorphosis. In premetamorphic bullfrog tadpoles, injections of prolactin did not alter the metamorphic changes induced by thyroxin [2]. Bern et al. [3], however, found that the tail reabsorbing influence of thyroxin added to the aquarium water could be inhibited by injections of prolactin in bullfrog tadpoles. In a similar experiment with *Rana pipiens* tadpoles, injections of prolactin did not inhibit metamorphosis induced by thyroxin [4]. It seems apparent from our studies that there is a "right time" so far as the function of prolactin is concerned, and the temporal relations of the hormones might account for the discrepancies reported.

Available evidence indicates that hormones are produced and released periodically. Plasma levels of adrenocortical hormones exhibit circadian oscillations that are synchronized by the photoperiod [16]. Circadian rhythms of pituitary prolactin have been reported in mammals [17, 18] and birds [19]. The content of thyroid-stimulating hormone in the rat hypophysis and serum also varies during the day [20, 21]. Whether or not these rhythms are present in tadpoles is unknown. However, the widespread occurrence of daily rhythms of these hormones suggests that they may also be present in amphibians. Our results indicate that such rhythms should be considered.

The system that could emerge as the regulator of amphibian growth and metamorphosis might be one similar to that described for the regulation of body fat in a variety of vertebrates [14, 15]. In the white-throated sparrow, the time of the daily rise in plasma corticosterone [22] and the time of the daily release of prolactin [19] changes from season to season in a manner that accounts for the seasonal fluctuations in body fat stores. For example, in May when the birds are fat, the interval is 10 to 14 hours; whereas in August when the birds are lean, the interval is five to nine hours. When injections of the hormones are made, gains in body fat result from daily injections of prolactin that follow injections of corticosterone by 12 hours, and losses in body fat result from prolactin injections that follow corticosterone by eight hours [15]. The interval between the daily release of thyroxin, corticosterone, and prolactin may change during amphibian development, resulting in a graded and controlled sequence of metamorphic events.

ACKNOWLEDGMENTS

The author is deeply grateful to Dr. Albert H. Meier for constant encouragement and guidance. This report was supported by a grant from the National Science Foundation (GB-20913) to Albert H. Meier. The ovine prolactin (NIH-P-S8: 1 mg=28 IU) was a gift of the Endocrinology Study Section of the National Institutes of Health.

REFERENCES

1. Etkin W. and Gona A. G. (1967): Antagonism between prolactin and thyroid hormone in amphibian development. J. Exp. Zool., *165*: 249–258.
2. Gona A. G. (1967): Prolactin as a goitrogenic agent in amphibia. Endocrinology, *81*: 748–754.
3. Bern H. A., Nicoll C. S. and Strohman R. C. (1967): Prolactin and tadpole growth. Proc. Soc. Exp. Biol. Med., *126*: 518–521.
4. Brown P. S. and Frye B. E. (1969): Effects of prolactin and growth hormone on growth and metamorphosis of tadpoles of the frog, *Rana pipiens*. Gen. Comp. Endocrinol., *13*: 126–138.
5. Derby A. (1968): An *in vitro* quantitative analysis of the response of tadpole tissue to thyroxine. J. Exp. Zool., *168*: 147–165.
6. Etkin W., Derby A. and Gona A. G. (1969): Prolactin-like antithyroid action of pituitary grafts in tadpoles. Gen. Comp. Endocrinol. Suppl., *2*: 253–259.
7. Chadwick C. S. (1940): Identity of prolactin with water drive factor in *Triturus viridescens*. Proc. Soc. Exp. Biol. Med., *45*: 335–337.
8. Grant W. C. and Grant J. A. (1958): Water drive studies on hypophysectomized efts of *Diemyctylus viridescens*. Part 1. The role of lactogenic hormone. Biol. Bull., *114*: 1–9.
9. Grant W. C. (1961): Special aspects of the metamorphic process—Second metamorphosis. Am. Zool., *1*: 163–171.
10. Grant W. C. and Cooper G. (1965): Behavioral and integumentary changes associated with metamorphosis in *Diemyctylus*. Biol. Bull., *129*: 510–522.
11. Breaux C. B. and Meier A. H. (1971): Diurnal periodicity in the effectiveness of prolactin to inhibit metamorphosis in *Rana pipiens* tadpoles. Amer. Midl. Natur., *85*: 267–271.
12. Meier A. H., Garcia L. E. and Joseph M. M. (1971): Corticosterone phases a circadian water-drive response to prolactin in the spotted newt, *Diemyctylus viridescens*. Biol. Bull., *141*: 331–336.

13. Meier A. H. (1970): Thyroxin phases the circadian fattening response. Proc. Soc. Exp. Biol. Med., *113*: 1113–1116.
14. Meier A. H., Trobec T. N., Joseph M. M. and John T. M. (1971): Temporal synergism of prolactin and adrenal steroids in the regulation of fat stores. Proc. Soc. Exp. Biol. Med., *137*: 408–415.
15. Meier A. H. and Martin D. D. (1971): Temporal synergism of corticosterone and prolactin controlling fat storage in the White-throated Sparrow, *Zonotrichia albicollis*. Gen. Comp. Endocrinol., *17*: 311–318.
16. Halberg F. (1969): Chronobiology. Ann. Rev. Physiology, *31*: 675–725.
17. Clark R. H. and Baker B. L. (1964): Circadian periodicity in the concentration of prolactin in the rat hypophysis. Science, *143*: 375–376.
18. Kent G. C., Jr., Turnbull J. G. and Kirby A. C. (1964): A daily rhythm in prolactin secretion of release in hamsters. Assoc. Southeast. Biol. Bull., *11*: 48 (Abstr.).
19. Meier A. H., Burns J. T. and Dusseau J. W. (1969): Seasonal variations in the diurnal rhythm of pituitary prolactin content in the White-throated Sparrow, *Zonotrichia albicollis*. Gen. Comp. Endocrinol., *12*: 282–289.
20. Bakke J. L. and Lawrence N. (1965): Circadian periodicity in thyroid stimulating hormone titer in the rat hypophysis and serum. Metabolism, *14*: 841–843.
21. Singh D. V., Ponda J. N., Anderson R. R. and Turner C. W. (1967): Diurnal variation of plasma and pituitary thyrotropin (TSH) of rats. Proc. Soc. Exp. Biol. Med., *126*: 553–554.
22. Dusseau J. W. and Meier A. H. (1971): Diurnal and seasonal variations of plasma corticosterone in the White-throated Sparrow, *Zonotrichia albicollis*. Gen. Comp. Endocrinol., *16*: 399–408.

DAILY RHYTMS IN THE HORMONAL CONTROL OF FAT STORAGE IN LIZARDS

Terry N. TROBEC

Department of Zoology and Physiology
Louisiana State University
Baton Rouge, Louisiana, U.S.A.

Many animals experience seasonal fluctuations of fat stores. One such animal is the green anole, *Anolis carolinensis* [1–3], which is an iguanid lizard indigenous to the southeastern United States. Total lipid content is maximal in November and December and minimal in April and May. The seasonal pattern of abdominal fat-body weights is similar to the seasonal pattern of total lipid content. An experiment performed by Arlene Meier [4] demonstrated that prolactin injections could elicit fattening in the green anole. However, the injections caused fattening only if administered at certain times of day. Lizards receiving prolactin injections early in the photoperiod (16L:8D) in November had a total lipid index similar to that of the controls. Midday injections of prolactin caused a significant increase in total lipid stores ($p<0.02$), whereas injections of prolactin made late in the photoperiod caused a significant decrease in the total lipid stores ($p<0.02$).

Further studies of the green anole reported herein substantiate the daily fattening response to prolactin injections in this lizard (Fig. 1A & Table 1). Anoles maintained on a 12L:12D photoperiod in August at an air temperature of 30±2°C received daily injections of prolactin (1 μg/g body wt) for eight days at 0, 4, 8, 12, 16, or 20 hours after the onset of light. The animals were killed on the ninth day, 24 hours after the last injection, and the fat bodies were dissected out and weighed. Weights of fat bodies in lizards that received prolactin injections approximately 16 hours after the onset of light were three-fold larger than the weights of fat bodies in the untreated controls. Injections of prolactin administered at four to eight hours after the onset of light suppressed fat-body weights. The time of day when prolactin caused fattening in this experiment is somewhat different from that reported previously [4]. The difference may be attributed to the lack of strict control of air temperature in the experiment by Meier. Under conditions of continuous light, a thermoperiod can also entrain a daily pattern of fattening response to prolactin injections, yet this pattern is not the same as the pattern entrained by a comparable photoperiod (Trobec and Meier, unpublished). Also, endogenous physiological rhythms that influence fattening may have varied with the time of year (November versus August).

A similar experiment was performed on the granite night lizard, *Xantusia henshawi,* maintained at 30±2°C on a 16L:8D photoperiod (Fig. 2A and Table 1). The lizards received daily injections of prolactin (1 μg/g body wt) for eight days at 0, 3, 6, 9, 12, 15, 18 or 21 hours after the onset of light. They were killed on the ninth day, and the total lipids were extracted by ether in Soxhlet equipment. Injections of prolactin administered at the onset of light stimulated a two-fold increase in total lipid content as compared to the lipid content of the

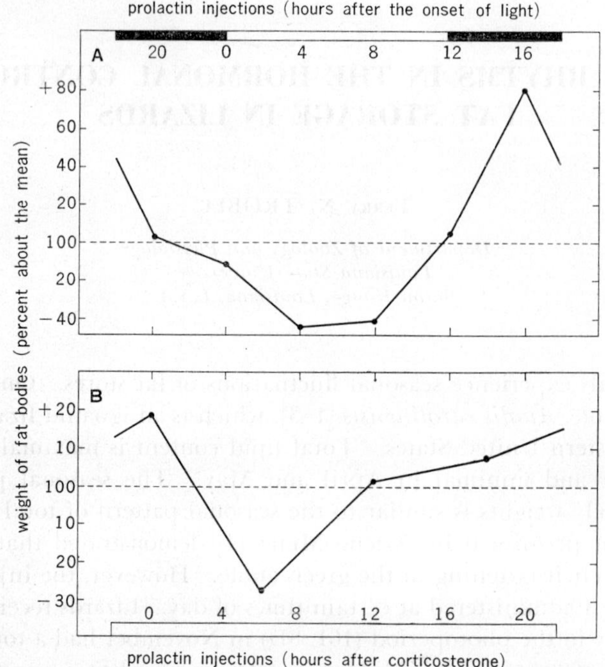

Fig. 1 Daily variation in responses to prolactin injections in the lizard *Anolis carolinensis.* (A) In this experiment, daily prolactin injections were administered to groups of lizards held on a 12L:12D light regimen and at an air temperature of 30±2°C. (B) In another experiment, daily prolactin injections were administered at a specified number of hours after daily corticosterone injections to groups of lizards held on continuous bright light and at an air temperature of 30±2°C. Both patterns of variation (A and B) are statistically significant at the 5% level by the least squares analysis of variance.

untreated controls, whereas injections administered 15 hours later had no effect.

In both the green anole and the granite night lizard prolactin injections given at specific times of day stimulated large gains in fat stores, while prolactin injections given at other specific times of day suppressed fat stores. These daily variations in fattening responses to prolactin could account for the fact that Licht [5, 6] has been unable to demonstrate a fattening effect of prolactin in the green anole.

The daily variations in the fattening responses to prolactin suggest that another system sets the time of tissue responses to this hormone. Because the adrenocortical hormones have pervasive influences throughout the vertebrate body, and because the plasma levels exhibit circadian oscillations that are synchronized by the photoperiod, it seemed possible that a daily rhythm of plasma adrenocortical hormones may entrain or drive rhythms of fattening sensitivity to prolactin. Consequently, experiments were performed to test whether the interval between daily injections of prolactin and of corticosterone, the principal adrenal steroid in some lizards [7], is important with respect to the levels of body fat stores in *Anolis carolinensis* [8] and *Xantusia henshawi* (as reported herein). The experiments were performed in continuous bright light to remove photoperiodic information that could complicate interpretation of the results. The lizards were given an abundance of food and water and were maintained in incubators at an air temperature of 30±2°C. They were divided into two groups that received daily in-

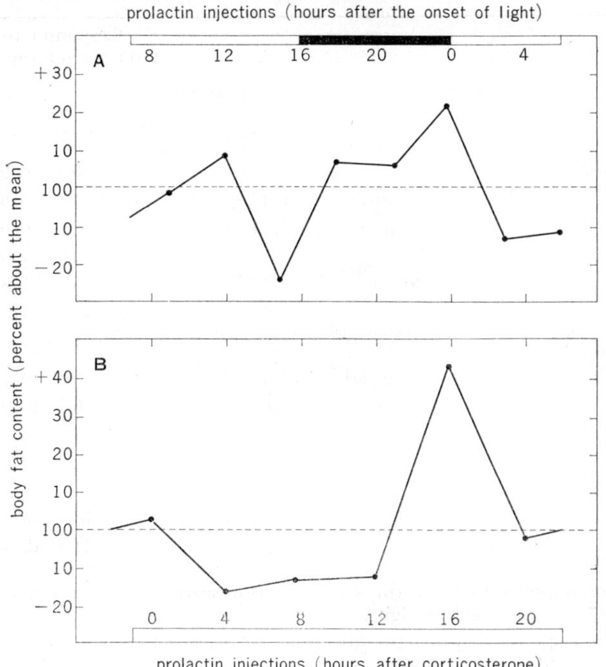

Fig. 2 Daily variation in responses to prolactin injections in the lizard *Xantusia hensawi.* (A) In this experiment, daily prolactin injections were administered to groups of lizards held on a 16L:8D light regimen and at an air temperature of 30±2°C. (B) In another experiment, daily prolactin injections were administered at a specified number of hours after daily corticosterone injections to groups of lizards held on continuous bright and at an air temperature of 30±2°C. While group variation precluded experimental verification of the pattern of response in experiment A, the response pattern of experiment B is statistically significant at the 5% level by the least squares analysis of variance.

jections of corticosterone; one group received the hormone at 0600 hours and the other received the hormone at 1800 hours. In the experiment with the green anole,both groups that received corticosterone were further divided into four subgroups that received daily injections of prolactin at 0600, 1200, 1800 or 2400 hours, while in the experiment with the granite night lizard, both groups that received corticosterone were divided into six subgroups that received daily injections of prolactin at 0400, 0800, 1200, 1600, 2000 or 2400 hours. This method allowed comparison of two groups that received daily injections of prolactin at various time intervals after corticosterone treatment, and provided a simple means of determining whether the variations in the responses to prolactin were set by the time of injection of corticosterone. Both hormones were administered via subcutaneous injections at a dosage of 1 μg/g body weight. Daily injections of prolactin were initiated five days after the beginning of daily injections of corticosterone. Both hormones were injected daily for nine more days, which made a total of 14 days of corticosterone injections and nine days of prolactin injections.

In the green anole, injections of prolactin administered six hours after corticosterone treatment inhibited fat-body growth regardless of whether the adrenal steroid was administered at 0600 or 1800 hours, whereas injections of prolactin administered 18 to 24 hours after corticosterone treatment stimulated fat-body

Table 1 Daily variations in fattening responses to prolactin injections in lizards.

Species		Controls		Experimental		
		Mean±S.E.	(N)	Time	Mean±S.E.	(N)
*Anolis carolinensis**				0	0.90±0.21	(6)
	initial	1.17±0.36	(5)	4	0.64±0.28	(6)
	final	0.71±0.33	(5)	8	0.68±0.27	(4)
				12	1.05±0.23	(7)
				16	1.84±0.26	(7)
				20	1.04±0.42	(6)
*Xantusia henshawi***				0	15.2±1.5	(9)
	initial	13.4±0.6	(8)	3	10.7±2.1	(9)
	final	8.6±1.2	(9)	6	10.9±1.4	(10)
				9	12.2±1.2	(9)
				12	13.3±1.8	(9)
				15	9.4±1.5	(9)
				18	13.2±1.0	(9)
				21	12.5±1.7	(6)

Responses to prolactin (Hours after the onset of light)

* The fattening response to prolactin injections is measured in *A. carolinesis* by the weight of fat bodies as % wet body weight.

** The fattening response to prolactin injections is measured in *X. henshawi* by the body fat content as % dry body weight.

Table 2 Temporal synergism of corticosterone and prolactin in the regulation of fat storage in lizards.

Species	Corticosterone injection time	Responses to prolactin (Hours after corticosterone) Time	Mean±S.E.	(N)
*Anolis carolinesis**	0600	0	2.06±0.14	(7)
		6	1.30±0.14	(6)
		12	1.54±0.18	(7)
		18	2.20±0.34	(7)
	1800	0	2.62±0.46	(7)
		6	1.52±0.18	(5)
		12	2.46±0.22	(7)
		18	2.01±0.20	(7)
*Xantusia henshawi***	0600	0	7.80±2.03	(5)
		4	8.17±1.57	(6)
		8	10.54±1.25	(7)
		12	7.17±1.68	(6)
		16	13.33±1.59	(6)
		20	8.80±1.19	(6)
	1800	0	9.51±2.50	(7)
		4	6.03±1.13	(6)
		8	4.46±0.54	(7)
		12	7.85±0.95	(6)
		16	11.27±1.00	(7)
		20	9.16±0.79	(7)

* The fattening response to prolactin injections is measured *A. carolinesis* by the weight of fat bodies as % wet body weight.

** The fattening response to prolactin injections is measured in *X. henshawi* by the body fat content as % dry body weight.

growth (Fig. 1B and Table 2). However, in the granite night lizard, injections of prolactin administered four hours after corticosterone treatment suppressed lipid content, whereas injections of prolactin administered 12 hours later increased lipid content (Fig. 2B and Table 2).

The temporal relationship between corticosterone and prolactin is of considerable importance in the regulation of fat storage in both the green anole and the granite night lizard. The temporal synergism of these two hormones elicits a similar pattern of fattening response in both genera of lizards tested. Points during the respective photoperiods which simulate the time of injection of corticosterone can be visualized if the peaks and troughs of Fig. 1A are superimposed on 1B and those of Fig. 2A onto 2B. These points would be approximately 20 hours after the onset of light in the green anole and approximately eight hours after the onset of light in the granite night lizard. The location of these points is of interest in that the green anole is diurnally active whereas the granite night lizard is nocturnally active. Studies on the green anole are presently being done in this laboratory to determine plasma levels of corticosterone and their probable circadian fluctuation. It is hoped that such information will clarify the regulatory role of the temporal synergism of corticosterone and prolactin on lipid storage in lizards.

ACKNOWLEDGMENTS

The author is an NSF Predoctoral Fellow and wishes to acknowledge the financial support of experimentation by NSF Research Grant GB-20913 awarded to Dr. Albert H. Meier. I would like to thank Dr. P. E. Schilling and the personnel of the Computer Center at Louisiana State University for their generosity in making the statistical evaluations. I also thank Maureen Trobec for technical assistance with the experiments and for editing the manuscript. Special thanks go to Dr. Albert H. Meier for his advice and encouragement.

REFERENCES

1. Dessauer H. C. (1953): Hibernation of the lizard, *Anolis carolinensis.* Proc. Soc. Exptl. Biol. Med., *82*: 351–353.
2. Dessauer H. C. (1955): Seasonal changes in the gross organ composition of the lizard, *Anolis carolinensis.* J. Exptl. Zool., *128*: 1–12.
3. Dessauer H. C. (1955): Effect of season on appetite and food consumption of the lizard, *Anolis carolinensis.* Proc. Soc. Exptl. Biol. Med., *90*: 524–526.
4. Meier A. H. (1969): Diurnal variations of metabolic responses to prolactin in lower vertebrates. Gen. Comp. Endocrinol. Suppl., *2*: 55–62.
5. Licht P. (1967): Interaction of prolactin and gonadotropins on appetite, growth, and tail regeneration in the lizard, *Anolis carolinensis.* Gen. Comp. Endocrinol., *9*: 49–63.
6. Licht P. and Jones R. E. (1967): Effects of exogenous prolactin on reproduction and growth in adult males of the lizard, *Anolis carolinensis.* Gen. Comp. Endocrinol., *8*: 228–244.
7. Phillips J. G., Chester Jones I. and Bellamy D. (1962): Biosynthesis of adrenocortical hormones by adrenal glands of lizards and snakes. J. Endocrinol., *25*: 233–237.
8. Meier A. H., Trobec T. N., Joseph M. M. and John T. M. (1971): Temporal synergism of prolactin and adrenal steroids in the regulation of fat stores. Proc. Soc. Exptl. Biol. Med., *137*: 408–415.

CIRCADIAN RHYTHM AND PHASE-SHIFTING IN RUNNING ACTIVITY BY FERAL WHITE-FOOTED MICE (*PEROMYSCUS*): EFFECTS OF DISTAL PINEALECTOMY*

WILBUR B. QUAY**

Department of Zoology, University of California
Berkeley, California, U.S.A.

INTRODUCTION

GASTON and MENAKER [1] reported that the circadian rhythm in locomotor activity by house sparrows (*Passer domesticus*) was lost following pinealectomy. Further studies by this group support the conclusion that the activity of these pinealectomized birds was arrhythmic. However, in laboratory rats (*Rattus norvegicus*) it has been shown that the circadian rhythm in locomotor (running) activity remains after pinealectomy both in static photoperiod (LD) conditions and in continuous illumination (LL) [2–5]. It has been shown also that laboratory rats that have been pinealectomized still follow 'Aschoff's rule', in that illumination of reduced intensity leads to reduction in circadian cycle length (τ) (e.g. $\tau LL > \tau DD$) [5]. Recent results suggest that the primary defect in circadian mechanisms in the pinealectomized rat concerns control of phase-shift rate rather than either maintenance of rhythmicity or of attainment of entrainment [6–10]. Stabilization of circadian phase-shift rate appears to be a life-long function of the rat pineal organ [9]. Nevertheless, the vectorial characteristics of this pineal circadian control mechanism change with age and are responsive to characteristics of the environmental illumination [11].

At the present time there is no direct or clear evidence that the pineal circadian mechanism outlined above for the rat might not be true as well for other mammalian species, including man. It remains both possible and desirable to clinically test phase-shift rates in patients suspected of pineal lesions or other neurological or neuroendocrine disorders [12]. On the other hand it might be argued on hypothetical grounds that the laboratory rat is not representative of feral species or of other mammals in pineal-circadian relations. Therefore, it is important to extend the investigation of circadian phase-shift and rhythmic mechanisms to other and feral species of mammals. Such studies in comparison with those on laboratory rats will be handicapped by having to deal with heterogeneity in the subject animals, in genetic composition and in life history.

* This work was supported by a research grant (NS-06296) from the National Institutes of Health, U.S. Public Health Service, aid from Biomedical Sciences Support grant funds, and the Miller Institute for Basic Research in Science, University of California, Berkeley. I am especially grateful to Helen SHERRY for extensive field, laboratory and office assistance on this project. Additional excellent help by Joe WONG and Edith CARRIGAN is also gratefully acknowledged.

** Present address: Waisman Center on Mental Retardation and Human Development, University of Wisconsin, Madison, Wisconsin, U.S.A.

It is suggested that white-footed mice (genus, *Peromyscus*) are ideal for research on pineal-circadian mechanisms in wild, small rodent species. It was shown earlier that the distal or posterior division of the pineal organ in a species (*P. leucopus*) of this genus, responds cytologically to changes in photoperiod [13]. This species, and at least many of its congeners, is nocturnal, polyestrous and is affected in its annual cycle of reproductive activity by changes in day-length [14]. Preliminary results obtained over the past year with species of white-footed mice show that circadian rhythmicity in running activity is maintained following pinealectomy and that pineal-phase-shift control is probably similar to that demonstrated in the laboratory rat.

METHODS

Mice of both sexes (*P. californicus, P. truei* and *P. maniculatus*) were collected at two localities (Lafayette, Contra Costa County and Strawberry Canyon [University of California, Berkeley], Alameda County, California) and at two seasons (November–December, 1970, and April, 1971). They were caged individually in a windowless, temperature-controlled and isolated room with an automatically timed, daily photoperiod (LD 12:12). The general conditions and methods of activity recording, handling and analysis were as described previously for studies on pinealectomized rats [5, 6].

The same experiment plan was followed for the two series of mice, first from November, 1970, to April 1–5, 1971 (9 animals) and second from April 11 to September 14, 1971 (18 animals). In both plans two reversals of photoperiod were imposed before operation and two afterwards, followed by several weeks of continuous weak red light (maximum illumination <1 foot candle) until the time of autopsy. At operation, mice in alternate or adjacent cages were pinealectomized (p) or sham-operated (s) [15] while under Nembutal (sodium pentobarbital, 5–7 mg i.p./100 g body wgt.) anesthesia. The one animal that died, possibly through hemorrhage or the effect of surgery, had been sham-operated. At autopsy all animals were examined for pathology (none found) and their central brain and pineal region was fixed as a block in Bouins fluid for microscopic study.

RESULTS AND DISCUSSION

The analysis of the results has progressed sufficiently by this date to show that: (1) all animals, whether pinealectomized (p) or sham-operated (s) had free-running circadian rhythms when in continuous weak red light, and (2) individual variation in phase-shift rate both before and after operation does not yet allow a statistically significant difference to be demonstrated between p and s white-footed mice.

As was noted earlier in the case of laboratory rats [5] the mice studied here gradually went out of phase with each other over the several weeks of free-running circadian activity. This is evidence that the animals were not following potential cues provided by the animals in adjacent cages. Furthermore, mice in adjacent cages, whether p or s, usually differed in: (1) the relative length of the free running cycle, some being >24 hours, some $\simeq$24 hours and some <24 hours; and (2) the number of days following the start of continuous conditions before which free-running occurred.

The apparent contradiction in results from pinealectomized sparrows and small rodents (rats and white-footed mice) is probably most wisely viewed as a presumptive species difference or as a possible difference between birds and mammals, until results from more species are obtained. The survival in all pinealectomized laboratory rats and white-footed mice of circadian rhythmicity in running activity under constant conditions suggests that the pineal gland is not 'the biological clock' in these species. Other studies, on phase-shifting in pinealectomized rats, indicate that the pineal does play a role in circadian mechanisms in these mammals, but that it is more subtle in its effects, although possibly no less significant physiologically. The gathering of additional results from more species should be a primary concern at this stage in the development of our knowledge concerning pineal contributions to circadian mechanisms.

REFERENCES

1. Gaston S. and Menaker M. (1968): Pineal function: the biological clock in the sparrow. Science, *160*: 1125–1127.
2. Richter C. P. (1964): Biological clocks and endocrine glands. Excerpta Medica, *83*: 119–123.
3. Richter C. P. (1965): Biological Clocks in Medicine and Psychiatry. Charles C. Thomas Publisher, Springfield, Illinois.
4. Quay W. B. (1965): Photic relations and experimental dissociation of circadian rhythms in pineal composition and running activity in rats. Photochem. Photobiol., *4*: 425–432.
5. Quay W. B. (1968): Individuation and lack of pineal effect in the rat's circadian locomotor rhythm. Physiol. Behav., *3*: 109–118.
6. Quay W. B. (1970): Physiological significance of the pineal during adaptation to shifts in photoperiod. Physiol. Behav., *5*: 353–360.
7. Quay W. B. (1970): The significance of the pineal. *In*: Hormones and the Environment, (G. K. Benson and J. G. Phillips, eds.), pp. 423–444, Cambridge University Press, Cambridge.
8. Quay W. B. (1970): Precocious entrainment and associated characteristics of activity patterns following pinealectomy and reversal of photoperiod. Physiol. Behav., *5*: 1281–1290.
9. Quay W. B. (1972): Pineal homeostatic regulation of shifts in the circadian activity rhythm during maturation and aging.
10. Kincl F. A., Chang C. C. and Zbuzkova V. (1970): Observations on the influence of changing photoperiod on spontaneous wheel-running activity of neonatally pinealectomized rats. Endocrinol., *87*: 38–42.
11. Quay W. B. (in press): Behavioral effects of the mammalian pineal gland: Quantitative analysis and elicitation by environmental and intracranial factors. *In*: Pineal Workshop, (D. C. Klein, ed.), Raven Pres, New York.
12. Quay W. B. (1970): Diagnosis of destructive lesions of the pineal. The Lancet, No. 7662: 42.
13. Quay W. B. (1956): Volumetric and cytologic variation in the pineal body of *Peromyscus leucopus* (Rodentia) with respect to sex, captivity and day-length. J. Morphol., *98*: 471–495.
14. Whitaker W. L. (1940): Some effect of artificial illumination on reproduction in the white-footed mouse, *Peromyscus leucopus noveboracensis*. Jour. Exp. Zool., *83*: 33–60.
15. Quay W. B. (1965): Experimental evidence for pineal participation in homeostasis of brain composition. Prog. Brain Res., *10*: 646–653.

EVIDENCE FOR A SEASONAL RHYTHM IN PINEAL GLAND FUNCTION

Russel J. REITER

Department of Anatomy, The University of Texas
Medical School at San Antonio
San Antonio, Texas, U.S.A.

Mammalian reproductive cycles are very carefully controlled so that insemination, pregnancy and parturition occur within closely defined periods of the year. These events are synchronized with environmental changes to ensure that the progeny are born and nurtured during the period of the year most conducive to the survival of individuals and, thus, of the species. Characteristically, the annual cycle of reproductive competence alternates with periods of pituitary-gonadal dormancy. Several environmental factors, e.g., day-length, ambient temperature, rainfall, etc., are known to influence the reproductive states of mammals. How these external factors are evaluated and summated by the animal is open to question, but it is possible that part of the mechanism involves the pineal gland.

ANATOMICAL EVIDENCE

The possibility of a seasonal antagonistic action of the pineal on reproductive functions was suggested by Quay in 1956 [31] and was more strongly suggested by the work of Pfiugfelder [2] and Mogler [2]. These authors reported that darkness caused morphological changes within the pineal gland that were indicative of increased pineal activity and that these were accompanied by depression in the reproductive system. Pflugfelder [1] and Mogler [2], in fact, observed an inverse relationship between the size of the pinealocyte nuclei and the nuclei of the testicular interstitial cells in the hamster (*Mesocricetus auratus*). Even more convincing than the previously discussed studies are those of Něsić [3]. He determined that the annual periods of sexual activity and anestrus in the ewe are correlated with characteristic appearances of the pinealocytes suggesting a physiological nexus between the pineal and the reproductive system.

In light of these findings it might be expected that species which inhabit the Arctic and Antarctic regions would possess pineal glands with unique structural features. These species are, of course, exposed to great variations in photoperiodic length. If the pineal is influenced by environmental illumination and if its secretory activity determines the functional status of the gonads, then investigations on species living at the extremes of latitude should be informative. Indeed, it has been found that the walrus (*Odobenus rosmarus*) [4], the penguin (*Aptenodytes patagonica*) [5], the Weddell seal (*Leptonichotes weddelli*) [6], and the northern fur seal (*Callorhinus ursinus*) [7] all have inordinately large pineal glands. For example, the Weddell seal, a carnivore of the suborder Pinnipedia, has a pineal gland five to ten times as large as that of an equivalent sized dog which is a carnivore of the suborder Fissipedia.

Furthermore, CUELLO and TRAMEZZANI [6] observed conspicuous seasonal differences in the histological and histochemical character of the pineal of the Weddell seal. Thus, pineals of seals captured in May and June, when the daily light cycle was very short, were found to contain large amounts of lipid while pineals obtained during the period of maximal photoperiodic stimulation, i.e., December and January had very little stainable lipid material. QUAY [8] has found the lipid content of the rat pineal to show a similar decrease in response to increased amounts of light. On the basis of their results, CUELLO and TRAMEZZANI [6] concluded that there is a relationship between the rhythmic activity of the pineal gland and the annual sexual cycle of the animal. The conclusions of ELDEN and colleagues [7] in reference to the northern fur seal were very similar. They were "convinced that the pineal gland of the northern fur seal is an important functional organ and that it accounts for the marked photoperiodic nature of the fur seal's life cycle". In essence, they feel that the pineal serves as a "biological clock" which is impelled by the predominating photoperiodic conditions.

PHYSIOLOGICAL EVIDENCE

The Syrian hamster (*Mesocricetus auratus*) has been widely studied to determine the interplay of ambient lighting conditions, pineal activity and reproductive functions [9]. In this regard, the relationships of the pineal and the gonads to photoperiod have been relatively thoroughly studied while the influence of lighting intensity and spectral distribution of the light source have been only sparingly investigated.

One consequence of maintaining normally seasonally breeding rodents under laboratory conditions is the spreading of their reproductive capability through the year. However, hamsters raised under laboratory conditions still exhibit residual annual reproductive cyclicity such that they are less sexually competent during the winter than during the summer. In an attempt to circumvent the problems of breeding hamsters in the winter, CUSICK and COLE [10] increased the length of the artificial photoperiod within the animal rooms and found that this ameliorated their breeding difficulties.

What CUSICK and COLE [10] had accomplished by increasing the length of the photoperiod was the depression of pineal antigonadotropic activity. Hence, the reproductive organs were maintained in a functionally mature state and the animals were capable of reproducing. This was experimentally tested several years later. CZYBA and colleagues [11] noted that adult male hamsters displayed grossly regressed gonads and involuted germinal epithelia during the months of January and February. By pinealectomizing the hamsters in November they were able to prevent the usual involution of the testes and secondary sex organs which occurred six to eight weeks later. The details of the seasonal cycle in the fecundity of male and female hamsters have been described [12–15].

The dependency of reproductive organs on photoperiod has also been studied. Unless hamsters are exposed to 12.5 or more hours of light each day the gonads atrophy [16]. Conversely, hamsters that have been subjected to pinealectomy never exhibit dark-induced gonadal involution regardless of how they are deprived of light, i.e., by exposure of the animals to "short days" or by surgical removal of the animals' eyes [17, 18].

The involuted reproductive organs of hamsters kept in "short days" have several noteworthy characteristics. Firstly, once the gonads are regressed the pineal gland must be left intact or the sex organs immediately begin to recrudesce [19]. Secondly, after a period of quiescence, the reproductive organs *spontaneously* regrow to the mature adult condition [18, 20, 21] even though the animals are still in darkness and their pineals are intact. The length of gonadal dormancy (20–25 weeks) is roughly equivalent to the duration of the winter period. Hence, this experimental scheme may depict what normally happens to hamsters in nature. They enter hibernation with atrophic sexual organs in the autumn and exit from their burrows the following spring with an active reproductive system.

Recent findings from the author's laboratory illustrate further the probable role of the pineal gland in the sexual cycle of the hamster [22]. If male hamsters were maintained in light:dark (LD) cycles of 1:23 (in hours) for 30 weeks, the testes and accessory sex organs underwent degeneration and subsequent spontaneous regeneration. At this time the hamsters were placed in LD cycles of 14:10 ("long days") for either 1, 10, or 22 weeks after which they were again subjected to short daily photoperiods. Only those hamsters that were exposed to "long days" for 22 weeks (after the initial 30 week period of LD 1:23) experienced a second gonadal regression when they were returned to LD cycles of 1:23 (Fig. 1).

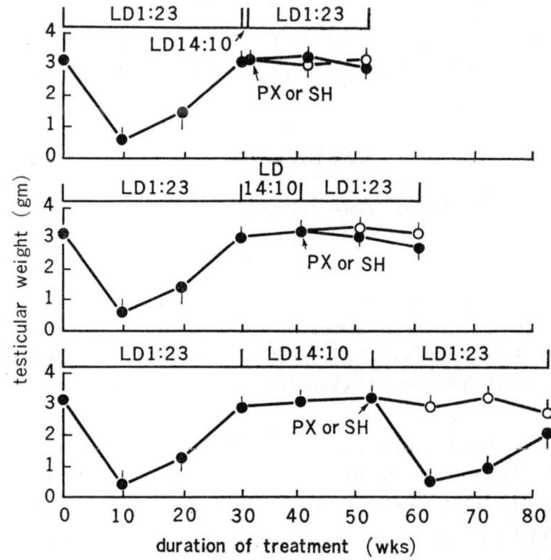

Fig. 1 Responses of the testes of adult male golden hamsters to photoperiodic changes. The animals were exposed to either "short" (LD 1:23) or "long days" (LD 14:10) for variable lengths of time. Solid dots indicate mean testicular weights in intact or sham operated (SH) hamsters. Hollow dots indicate testicular weightts in pinealectomized (PX) hamsters.

The total time required (i.e., 52 weeks) for these responses may coincidentally correspond to the normal length of the sexual cycle of hamsters in nature. The ovaries or testes of pinealectomized hamsters never undergo atrophy in response to reduced photoperiodic length [9].

Some of the most compelling evidence that the pineal is linked to annual changes in the sexual cycle of rodents comes from the work of Rust and Meyer

[23] on the weasel (*Mustela erminea*). Summer captured weasels possessed enlarged testes and were ostensibly capable of reproducing. The weekly subcutaneous implantation of melatonin-beeswax pellets into these animals caused the gonads to become infantile within 8 weeks. Similarly, weasels captured and brought into the laboratory in the winter displayed testicular recrudescence unless they received melatonin-beeswax pellets subcutaneously. Melatonin is considered to be one of the pineal antigonadotropic factors. These findings are obviously strongly suggestive of the pineal substance, melatonin, being involved in determining the breeding cycle of the short-tailed weasel.

CONCLUDING REMARKS

There are at least two mechanisms by which the pineal gland may regulate seasonal reproductive cycles in mammals. The pineal could merely synthesize and release greater quantities of its antigonadotropic substance during certain seasons. Alternatively, pineal activity may not fluctuate with seasonal changes in the photoperiod, but rather, the pituitary-gonadal system may become seasonally more sensitive to the pineal antigonadotropic influence. In either case, gonadal regression would ensue. It may be well to consider that light is not the exclusive impeller of pineal biosynthetic activity. Certainly environmental temperature [24], nutritional factors [25, 26], status of olfactory stimulation [27, 28], and possibly even auditory information [29] may alter the ability of the pineal to influence reproduction. In conclusion, it should be pointed out that, in humans as well, photoperiod, pineal activity and reproduction may not be unrelated [30].

REFERENCES

1. Pflugfelder O. (1956): Deut. Zool. Gesell. Verh., *50*: 53–75.
2. Mogler R. K.-H. (1958): Z. Morphol. Oekol. Tiere, *47*: 267–308.
3. Něsić L. (1962): Acta Anat., *49*: 376–377.
4. Turner W. (1888): J. Anat. Physiol., *22*: 300–303.
5. Watzka M. and Voss H. (1967): Verh. Anat. Gesell., *120*: 177–183.
6. Cuello A. C. and Tramezzani J. H. (1969): Gen. Comp. Endocr., *12*: 154–164.
7. Elden C. A., Keyes M. C. and Marshall C. E. (1971): Amer. J. of Vet. Res., *32*: 639–647.
8. Quay W. B. (1961): Gen. Comp. Endocr., *1*: 211–217.
9. Reiter R. J. and Sorrentino S. Jr. (1970): Amer. Zoologist, *10*: 247–258.
10. Cusick F. J. and Cole H. (1959): Texas Rep. Biol. Med., *17*: 201–204.
11. Czyba J. C., Girod C. et Durand W. (1964): C. R. Soc. Biol., *152*: 742–745.
12. Smit-Vis J. H. and Akkerman-Bellaart M. A. (1967): Experientia, *23*: 844–845.
13. Czyba J. C. (1968): C. R. Soc. Biol., *162*: 113–116.
14. Vendreley E., Guerillot, Basseville C. and Da Lage C. (1970): C. R. Soc. Biol., *164*: 1442–1448.
15. Frehn J. L. and Liu C.-C. (1970): J. Exper. Zool., *174*: 317–323.
16. Gaston S. and Menaker M. (1967): Science, *158*: 925–928.
17. Hoffman R. A. and Reiter R. J. (1965): Science, *148*: 1609–1611.
18. Reiter R. J., Sorrentino S. Jr. and Hoffman R. A. (1970): Int. J. Fertil., *15*: 163–170.
19. Reiter R. J. (1969): *In*: Progress in Endocrinology, (C. Gual., ed.), pp. 631–635, Excerpta Medica, Amsterdam.
20. Hoffman R. A., Hester R. J., and Towns C. (1965): Comp. Biochem. Physiol., *15*: 525–533.
21. Reiter R. J. (1969): Gen. Comp. Endocr., *12*: 460–468.

22. Reiter R. J. (1971): Anat. Rec., *169*: 410.
23. Rust C. C. and Meyer R. K. (1969): Science, *165*: 921–922.
24. Miline R., Dećeverski V., Šijacki and Krstić R. (1970): Hormones, *1*: 321&331.
25. Negus N. C. and Berger P. J. (1971): Experientia, *27*: 215–216.
26. Sorrentino S. Jr., Reiter R. J. and Schalch D. S. (1971): Neuroendocriology, *7*: 105–115.
27. Miline R., Dećeverski V. and Krstić R. (1963): Ann. d'Endocr., *24*: 377–379.
28. Reiter R. J. and Sorrentino S. Jr. (1971): *In*: The Pineal Gland, (G. E. W. Wolstenholme and J. Knight, ed.), pp. 329–344, J. and A. Churchill, London.
29. Miline R., Dećeverski V. and Krstić R. (1969): Acta Anat., (Suppl. 56), *73*: 293–293–300.
30. Timonen S. and Carpen E. (1968): Annal. Chir. Gyn. Fenn., *57*: 135–138.
31. Quay W. B. (1956): J. Morph., *98*: 471–495.

PHARMACOLOGY, TOXICOLOGY AND TIMED TREATMENT

Chairman: Alejandro ZAFFARONI

SERENDIPITY AND CHRONOBIOLOGY IN PHARMACOLOGY

ALEXANDER H. FRIEDMAN

Department of Pharmacology and Therapeutics
Loyola University Stritch School of Medicine
Maywood, Illinois, U.S.A.

On the 50th anniversary of the publication in *Pflüger Archiv* of Otto LOEWI's discovery of chemical transmission of nerve impulses [1] I had the opportunity to consider some of the events that ultimately resulted in his sharing the Nobel Prize in Physiology and Medicine with Henry DALE in 1936. An awareness of circadian and other periodic influences on biological phenomena suggested that these played a heretofore unsuspected role in the success of his initial experiments and also explained the basis for the subsequent difficulties arising in reproducing his momentous experiment some years later [2]. Dr. LOEWI had the first of his two famous dreams some 51 years ago on the evening before Easter Sunday in 1920 [3]. In that dream he devised the experiment that would prove that nerve impulses could be transmitted chemically from synapse to effector. Unfortunately he could not recall the details of the experiment when he awoke the next morning. That night the dream recurred. To avoid the disappointment of the previous day he arose promptly and sped to his laboratory at 0300 hours and performed the experiment in which he demonstrated that the effluent from one frog heart obtained after vagus nerve stimulation produced identical slowing in a recipient heart. The coincidence of several chronobiological phenomena including the responsiveness of Ranidae at that time of day and season probably assured the success of this now classic work. Had Dr. LOEWI merely written down his experimental design as he conceived it and proceeded to his laboratory in the morning for a normal working day, he might have encountered a less sensitive preparation and might have been diverted into other approaches to the solution of the problem of transmission in the nervous system. At any rate, propitious timing undoubtedly was a potent factor in the serendipity which manifested itself in Professor LOEWI's life.

The thesis which I am expounding is not that chronobiology will present one with the opportunity of achieving a place in the Hall of Scientific Greats but that when one focuses attention on rhythmic aspects of biological events one gains insights that probably would not be possible otherwise. In this regard I would like to review some of the work we have accomplished in our pharmacological and toxicological studies in recent years.

I was led into a consideration of the biological clock by my own skepticism with regard to quality control at one of the country's major drug firms. After using pentobarbital sodium to produce sleep in rats I discovered that my predictions with regard to the duration of effect which I could achieve with a fixed dose per kilogram were poor. I called Dr. R. K. RICHARDS of Abbott Laboratories and in my naiveté suggested that there might be considerable variation from batch to batch of this hypnotic agent. He suggested that I consider my animals more carefully and concern myself with *their* altered responsiveness. As a result of this suggestion and after further stimulation from a seminar of Dr. SCHEVING (then at The Chicago Medical School), I began the investigations that have produced insights I might not otherwise have attained had I continued my studies with a more usual approach to scientific experimentation.

We learned in rats that pentobarbital sleep had its longest onset and shortest duration during the dark phase of a programmed illumination cycle when motor activity was maximal [4]. A reverse relationship was observed during the light phase when our nocturnal animals are normally quiescent. In addition to the 24 hour pentobarbital rhythm we have studied biogenic amine rhythms of rat brain and their phase relationships [5, 6]. A number of established and putative neurotransmitters of the CNS, viz., norepinephrine, serotonin, acetylcholine, dopamine, homovanillic acid, histamine, gamma aminobutyric acid and glycine have been examined in our laboratories from the standpoint of chronobiology [4, 6–11]. In most instances we have detected rhythms that help to clarify our ideas regarding physiological function, behavior and the responsiveness to drugs. The toxicity of cholinergic drugs has been shown to vary on a circadian or ultradian basis [4, 6]. Drug types include cholinomimetics (e.g., pilocarpine, oxotremorine, and carbaminoylcholine), anticholinergics (e.g., atropine sulfate, scopolamine bromide and atropine methylnitrate), and cholinesterase inhibitors (e.g., physostigmine sulfate and neostigmine methyl sulfate). The first study demonstrating the possibility of cholinergic rhythms was reported by HAUS and HALBERG [12] in 1959, and most recently HANIN, MASSARELLI and COSTA [13] corroborated our finding [4, 6] of 24 hr rhythms in rat brain acetylcholine.

The variability one encounters in the toxicity of drugs acting at cholinoceptive sites is displayed in Table 1. When LD50's for cholinergic drugs are estimated in mice over a 24 hour period statistically-significant shifts in the per cent toxicity can be demonstrated in every case (as much as 56.5% over control in the case of neostigmine). In general such drugs are more toxic during the dark phase of the illumination cycle. In the case of the anticholinergic compounds, the reverse is true. Such studies not only tell us something about drug efficacy as a function of the time of administration but also allow us to consider the processes that alter responsiveness and endanger the drug recipient. Improved drug regimens can then be planned.

Recently, WALKER, SPECIALE and FRIEDMAN [14] used electron microscopic techniques to examine the ultrastructure of the rat hypothalamus and caudate nucleus at times when maximum differences in body temperature, motor activity and neurotransmitter levels could be anticipated. They were able to demonstrate for the first time that: 1) there were quantifiable changes in the level of the granular and agranular synaptic vesicles of these tissues which correlate to some degree with biogenic amine levels and with gross motor and behavioral activity, and 2) that these were modifiable by drugs that mobilize or alter the concentration of neurotransmitter amines.

Many of the studies here mentioned interdigitate with those of Drs. SCHEVING and PAULY, corroborating or extending some of theirs. Occasionally we have advanced along independent lines.

Our experimental design in these studies has always included an adaptation period for our animals of at least three weeks. This period of adaptation subsequently was shown to be *critical* in obtaining stable responses in rats [15]. The

Table 1 LD_{50} values for cholinergic and anticholinergic drugs in mice as per cent of minimum value.

Compound	Time of day (hours) 0600	1200	1800	2400
Cholinergic				
Acetylcholine Cl	104.0	120.8**	114.5	100.0
Carbachol Cl	100.0	151.4**	108.1	151.4
Pilocarpine HCl	100.0	116.1	117.2*	102.5
Oxotremorine	106.6	136.6*	126.7	100.0
Anticholinesterase				
Physostigmine SO_4	139.2**	123.5	115.7	100.0
Neostigmine CH_3SO_4	100.0	121.7	156.5**	126.0
Anticholinergic				
Atropine SO_4	127.0	100.0	116.9	127.3**
Scopolamine HCl	113.4	100.0	128.1*	112.6
Atropine CH_3NO_3	109.2	100.0	112.8**	110.5
Cholinomimetic				
(After 25 mg/kg Atropine CH_3NO_3)				
Acetylcholine Cl	100.0	123.9	127.7**	104.0
Physostigmine SO_4	116.3	117.4*	112.8	100.0

n=9–11 mice/dose

* Difference between minimum value (expressed as 100%) and maximum value significant at $P<0.05$

** Difference between minimum value (expressed as 100%) and maximum value significant at $P<0.01$

lighting program was automatically-timed lasting from 0800 to 2000 hours on our "normal" cycle and from 2000 hours to 0800 hours on our "reverse" cycle. The lighting used in the majority of these studies was full-spectrum fluorescent Vita-Lite* which reproduces 93% of the spectrum of daylight. Initially we had used incandescent lamps in which wavelengths toward red are heavily-represented and in which the near UV is absent. Food and water was received *ad libitum* and cage cleaning was scheduled at random intervals to avoid entrainment.

The number of points examined during any 24 hour period was from 4 to 8. The usual number of animals was 6/point for the biogenic amine studies; 18–22/point for the toxicity studies and 60/point in the case of a body temperature study. Previous experience had shown that because of the unequal distribution of brain amines [16] some information was lost if whole brain was homogenized for the analyses. Therefore, we examined such CNS parts as the brain stem, caudate nucleus, corpora quadrigemina, hypothalamus, thalamus, medulla, pons, cervical, thoracic and lumbar spinal cord. Fluorescence techniques [17–22] were used in most of the neurochemical studies except for GABA which was analyzed by an enzymatic technique [23] and acetylcholine which was quantified by bioassay [24] (CROSSLAND, 1967).

* Kindly supplied by Mr. H. A. ANDERSON, Duro-Test Corporation, North Bergen, New Jersey.

In our studies of rat brain amines we were able to demonstrate an inverse relationship between norepinephrine and serotonin in rat brain stem. The studies of VALZELLI and GARATTINI [25] help to explain this phenomenon which might be likened to a switching action in the CNS from a state of alertness to one of sleep. Norepinephrine has been implicated in alerting responses and serotonin in sleep so that the patterns obtained over a 24 hour period follow a logical sequence.

Serendipity enters the picture when we consider the 24 hour patterns obtained for central histamine. Measurements of this or any amine at a single point during the usual laboratory day tells us very little beyond the concentration obtained (much in the manner that a still photograph tells us of a dynamic event). But when the pattern is examined (as in a cinematographic analysis) we see a correlation between the normal waking-sleeping cycle. Histamine levels are directly related to waking and inversely related to sleeping. When this finding is coupled with those from recent studies of MONNIER [26–28], who showed that histamine could produce an alert cortex after central or systemic administration, that of AKCASU [29] who showed that morphine withdrawal signs were ameliorated by histamine depletion with 48/80 and the work of PFEIFFER et al. [30] who demonstrated the involvement of histamine in mental states, we are led to consider that histamine has an important role in CNS function—quite possibly in the maintenance of the awakened or alert state or as a modulator of transmission. Furthermore, this information provides one with a possible explanation of the sedative action of antihistamines which is a common side-effect of this drug type [31]. Of course this interpretation is not the only one, since coincident with the waking state one finds in the rat that acetylcholine levels [4, 6], dopamine [9], GABA [10] and spermidine levels (FRIEDMAN and RODICHOK [33]) are also at their maxima in the brain, although not entirely in phase with each other. Furthermore they are not equally responsive to changes in light programming.

In the spinal cord, glycine levels (depending on species—APRISON and WERMAN, 1965) correlate with the degree of neuronal traffic of the motor area subserved by a particular segment of the cord. When examined over a 24 hour period under the controlled conditions described earlier, we find [10, 32, 34] that peak levels of this inhibitory transmitter occur in the cervical and lumbar enlargements during maximal motor activity at a time when greater regulation of cord activity would be necessary. Glycine levels are minimal when the animals are quiescent or asleep. The toxicity of two glycine-antagonist convulsants, strychnine and allylglycine, is maximal when spinal cord glycine levels are minimal (PIEPHO and FRIEDMAN, unpublished data). These findings support the hypothesis that glycine is a neurochemical of prime importance in inhibitory regulation of spinal cord activity.

To summarize the theme of this paper succinctly—one gains considerably more understanding of functional processes and pharmacological activity when one considers them from the standpoint of chronobiology and investigates patterns and their interrelationships.

REFERENCES

1. Loewi O. (1922): Pflügers Arch., *189*: 239.
2. Friedman A. H. (1971): Pflügers Arch., *325*: 85.
3. Loewi O. (1960): Perspect. Biol. Med., *4*: 3.
4. Friedman A. H. and Walker C. A. (1969): J. Physiol., *202*: 133.
5. Friedman A. H. and Walker C. A. (1968): J. Physiol., *197*: 77.
6. Friedman A. H. and Piepho R. W. (1970): The Pharmacologist, *12(2)*: 236.
7. Friedman A. H. and Walker C. A. (1969): Fed. Proc., *28(2)*: 447.
8. Piepho R. W. (1972): Ph.D. Thesis.
9. Piepho R. W. and Friedman A. H. (1969): The Pharmacologist, *11(2)*: 255.
10. Piepho R. W. and Friedman A. H. (1971): Life Sciences, *10(1)*: 1355.
11. Speciale S. G. Jr. and Friedman A. H. (1971): The Pharmacologist, *13(2)*: 305.
12. Haus E. and Halberg F. (1959): J. Appl. Physiol., *14*: 878.
13. Hanin I., Massarelli R. and Costa E. (1970): Science, *170*: 341.
14. Walker C. A., Speciale S. G. Jr. and Friedman A. H. (1971): Neuropharmacology, *10*: 325.
15. Massarelli R., Hanin I. and Costa E. (1970): Fed. Proc., *29(2)*: 417.
16. Hornykiewicz O. (1966): Proc. 2nd Symposium Parkinson's Disease and Information Res. Center.
17. Shore P. A. and Olin J. S. (1958): J. Pharmacol., *122*: 295.
18. Shore P. A., Burkhalter A. and Cohn V. H. (1959): J. Pharmacol., *127*: 182.
19. Sourkes T. L. and Murphy G. F. (1961): Methods in Medical Research, *9*: 147.
20. Laverty R. and Sharman D. F. (1965): Brit. J. Pharmacol., *24*: 759.
21. Anton A. H. and Sayre D. F. (1968): Fed. Proc, *27*: 243.
22. Ansell G. B. and Beeson M. T. (1968): Anal. Bioch., *23*: 196.
23. Jakoby W. D. (1962): Methods in Enzymology, *7*: 777.
24. Crossland J. (1961): Methods in Medical Research, *9*: 125.
25. Valzelli L. and Garatinni S. (1968): Adv. Pharmacol., *6b*: 249.
26. Monnier M., Fallert M. and Bhattacharaya I. C. (1967): Experientia, *23*: 21.
27. Monnier M. (1969): 4th Int. Pharmacological Meeting Abstracts, Pergamon Press, Basel.
28. Monnier M., Sauer R. and Hatt A. M. (1970): Int. Rev. Neurobiol., *12*: 265.
29. Akcasu A. (1969): 4th Int. Pharmacological Meeting Abstracts, Pergamon Press, Basel.
30. Pfeiffer C. (1969): 4th Int. Pharmacological Meeting Abstracts, Pergamon Press, Basel.
31. Douglas W. W. (1970): *In*: The Pharmacological Basis of Therapeutics, 4th ed., (Goodman and Gilman, eds.), Chap. 29, Macmillan & Co.
32. Friedman A. H. and Piepho R. W. (1968): The Pharmacologist, *10(2)*: 160.
33. Friedman A. H. and Dodichok L. D. (1970): Fed. Proc., *29(2)*: 617.
34. Piepho R. W. and Friedman A. H. (1971): Life Sciences, *10(1)*: 1355.

TREATMENT SCHEDULES MODIFY CIRCADIAN TIMING IN HUMAN ADRENOCORTICAL INSUFFICIENCY

ALAIN REINBERG, JEAN GHATA, FRANZ HALBERG, MARIAN APFELBAUM, PIERRE GERVAIS, PIERRETTE BOUDON, CHARLES ABULKER and JEAN DUPONT

Equipe de Recherche de Chronobiologie Humaine (CNRS n° 105)
Laboratoire de Physiologie—Fondation A. de Rothschild
Paris, France
Chronobiology Laboratories, Department of Pathology
University of Minnesota
Minneapolis, Minnesota, U.S.A.

Rhythms of several physiological functions are altered as group phenomena in adrenalectomized animals [1–4] and in patients suffering from adrenal insufficiency (AI) [5–10]. The administration of cortisol—with or without aldosterone or 9-α-fluorocortisol—allows these patients to live and restores their performance. However, the question remains whether the timing of substitution treatment should be such as to restore any changed external and/or internal timing of rhythms. If so, it would follow that the timing of medication, e.g. hormone intake—in relation to a subject's social synchronization—has to be taken into consideration when one wants to normalize any biologic time structure* altered by disease, e.g. in patients with AI [11–13]. Properly timed treatment may change any ecchronism** related to the adrenal insufficiency into a condition close to euchronism. Indeed, by 1956, GHATA and REINBERG [14, 15] tried to obtain in patients suffering from AI a resynchronization of circadian rhythms in the urinary excretion of K, Na and water by administering a high dose of cortisol early in the morning. Moreover, also in 1956, KAINE, SELTZER and CONN [16] documented the re-induction of a circadian rhythm in eosinophil blood cells of patients with Addison's disease by a single timed dose of adrenal cortical hormone. It was of interest to reconsider with the help of objective means of analysis [17] and appropriate methodology [3, 10, 18] the question whether physiologically timed substitution treatment might have a demonstrable advantage with respect to a rhythm in grip strength and other variables as well.

SUBJECTS, MATERIALS AND METHODS

Circadian rhythmic rectal temperature, heart rate, grip strength and urinary excretion of 17-hydroxycorticosteroids [19], 17-ketosteroids [20], potassium and sodium were assessed in seven healthy adults and in seven adult patients with an

* The sum of non-random and thus predictable time-dependent biologic changes, including with growth, development and aging, a spectrum of rhythms and near-rhythms. Time structure characterizes any biologic entity, including ecosystems and populations as well as individual or grouped organisms, organ-systems, organs, tissues, cells and subcellular structures.

** An altered time-structure element such as a rhythm with one or more characteristics (frequency, acrophase, amplitude, mesor, waveform) outside certain statistical tolerance limits.

adrenal cortical insufficiency secondary to hypophysectomy for removal of an adenoma (two patients) or as a result of Addison's disease (five others). Subjects were standardized on a routine of diurnal activity (0700 to 2300) and nocturnal rest (2300 to 0700) for a month before testing. By such similar synchronization of subjects, group comparisons were to be facilitated. Sampling was carried out at 4-hourly intervals for 24 or 48 hours.

Oral corticosteroid treatment (Cortisol, 25 to 35 mg/24 h, with or without 50 μg/24 h of 9-α-fluorocortisol) was given in one of two ways to the patients with adrenal insufficiency. Mode A: three equal medications at 0800, 1300 and 2000 (mealtimes). Mode B: two unequal doses, 3/4 or 2/3 of the total dose at 0700 (upon awakening or getting up) and the remainder at 2300 (upon retiring). Four of the patients were studied on both medication modes.

Electronic computer analyses by the cosinor method [7] involved fitting a cosine curve (best approximating all data) to estimate rhythm characteristics such as the extent of predictable change (amplitude) and its timing (acrophase).

RESULTS

Circadian rhythms, largely synchronized in phase as well as in frequency, were detected with statistical significance by cosinor—for healthy subjects and also for two sets of series from patients with adrenal insufficiency treated, one set according to Mode A, the other set by Mode B (Table 1). The circadian amplitude and acrophase for rectal temperature and heart rate of the patients, whether treated according to A or B, were similar to those of healthy subjects, Fig. 1.

Circadian rhythms in grip strength and urinary excretion of 17-OHCS, 17-KS, K and Na in patients treated according to Mode A were delayed in acrophase, as compared to controls. By contrast, the acrophase and the amplitude of the same functions in patients treated according to Mode B were similar to those of healthy subjects.

DISCUSSION

It will be important to find out on a world-wide basis what the most common treatment schedule of AI might be and to compare this treatment with a mode of administration that takes into account the circadian rhythm in adrenal cortical secretion. Moreover, when more appropriate means of administration than tablets are available for timed treatment, one should simulate not only the circadian rhythm, as was attempted in a simplified fashion herein, but the ultradian component as well [21, 22]. Until such automated physiologic means of administration become available, the problem of a *physiologically* desirable timing will have to be weighed against its *logistic* feasibility. Certain schedules for medication may be difficult to remember, and if a patient is likely to forget altogether to take a drug on certain schedules, this intended timing will not be superior to three pills a day that are taken at the readily remembered meal-times. For this reason, the times of awaking and retiring, also readily remembered times, were chosen for the present study as an alternative to meal-times. The major focus of this report is directed upon the common-sense point that certain bioperiodic variables are best timed if they are above their overall rhythm-adjusted level during working hours, decreasing perhaps slightly during leisure times and reach-

Table 1 Circadian acrophases [ϕ and φ] and amplitudes [A] of a set of physiologic variables in healthy subjects (H) and in patients suffering from adrenal insufficiency (AI) when receiving cortisol according to different timing.

Physiologic variable	(Subjects) group (n)	Rhythm detection P	Amplitude A (% of M: Rhythm adjusted mean)	Acrophase ϕ ϕ (hrs min from midnight)	Acrophase φ φ (degrees from midsleep)
			(.95 confidence limits)		
Rectal temperature	(H) (7)	< .005	6.9 (3.8 to 9.7)	15 00 (12 20 to 17 20)	−180° (−140 to −214)
	(AI) A (5)	< .005	8.1 (6.5 to 10)	15 42 (14 49 to 16 36)	−191° (−177 to −204)
	(AI) B (5)	< .005	6.8 (4.6 to 8.9)	14 31 (12 25 to 16 36)	−173° (−141 to −204)
Heart rate	(H) (7)	< .002	10.9 (6.4 to 15.5)	15 56 (13 48 to 17 40)	−194° (−162 to −220)
	(AI) A (4)	< .005	7.5 (6.3 to 8.9)	16 16 (12 47 to 19 45)	−199° (−147 to −251)
	(AI) B (3)	< .050	12.1 (1.9 to 22.1)	17 52 (14 00 to 21 43)	−223° (−165 to −281)
Grip strength	(H) (7)	< .005	11.2 (6.0 to 16.5)	15 00 (12 55 to 17 06)	−188° (−156 to −219)
	(AI) A (5)	< .005	7.5 (3.4 to 11.7)	18 12 (14 46 to 21 38)	−228° (−177 to −280)
	(AI) B (5)	< .005	6.9 (2.6 to 11.3)	13 49 (11 01 to 16 37)	−162° (−120 to −204)
Urine volume	(H) (7)	< .010	29.5 (12.7 to 46.3)	08 50 (06 45 to 11 00)	− 88° (− 56 to −120)
	(AI) A (5)	< .025	35.1 (9.1 to 61.0)	20 13 (16 56 to 23 31)	−258° (−209 to −313)
	(AI) B (5)	< .100	22.9	00 47	−327°
17–OHCS	(H) (7)	< .005	46.4 (19.7 to 65.9)	11 56 (08 28 to 14 28)	−135° (− 82 to −172)
	(AI) A (5)	< .005	44.1 (31.3 to 56.8)	17 54 (15 46 to 20 02)	−224° (−192 to −256)
	(AI) B (5)	< .005	48.3 (28.7 to 67.8)	12 40 (10 16 to 15 05)	−145° (−109 to −181)
17–KS	(H) (7)	< .005	20.2 (10.7 to 29.8)	12 15 (10 28 to 14 04)	−139° (−112 to −166)
	(AI) A (5)	< .005	18.5 (8.2 to 28.7)	17 45 (14 08 to 21 21)	−221° (−167 to −275)
	(AI) B (5)	< .005	19.3 (9.8 to 28.8)	12 41 (10 11 to 15 11)	−145° (−108 to −183)
Potassium	(H) (7)	< .005	46.6 (29.7 to 63.6)	13 20 (11 03 to 14 52)	−155° (−123 to −178)
	(AI) A (5)	< .005	46.8 (28.7 to 64.9)	18 03 (15 46 to 20 20)	−226° (−192 to −260)
	(AI) B (5)	< .005	37.6 (21.3 to 53.9)	13 12 (11 08 to 15 15)	−153° (−122 to −184)
Sodium	(H) (7)	< .010	25.6 (9.3 to 41.9)	10 02 (07 30 to 12 48)	−107° (− 67 to −147)
	(AI) A (5)	< .025	38.4 (11.0 to 65.8)	18 31 (15 16 to 21 47)	−233° (−184 to −282)
	(AI) B (5)	< .050	14.1 (1.8 to 26.5)	11 19 (06 43 to 15 55)	−115° (− 56 to −194)

Mode A: cortisol at 0800–1300 and 2000 equal doses. Mode B: two inequal doses, 3/4 or 2/3 of the total dose at 0700 and the rest at 2300. L: 0700–2300; D: 2300–0700 Total observation span=T=24 to 48 h. Uncontrolled diet. Interval between consecutive observation =Δt=4 h.

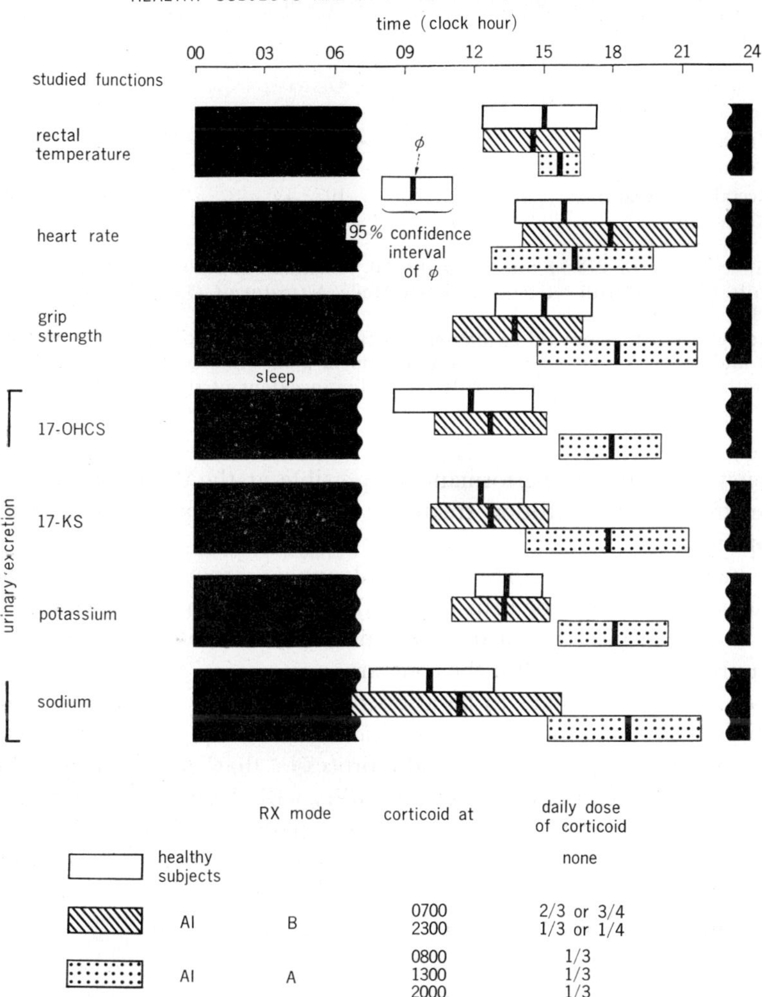

Fig. 1 Circadian acrophase chart for several physiologic variables in healthy subjects and patients with incompetent adrenals (AI), receiving Cortisol on two schedules (T= 24 to 48 h; $\Delta t = 4$ h; L from 0700 to 2300; D from 2300 to 0700; uncontrolled diet). Circadian rhythms in grip strength and urinary excretion of 17-OHCS, 17-KS, K^+ and Na^+ in patients given three equal doses of cortisol at 0800, 1300 and 2000 lag in acrophase by about 6 hours behind controls. By contrast, the acrophase of the same functions was similar for healthy subjects and patients treated 2/3 or 3/4 of the daily cortisol dose at 0700 and the remaining fraction at 2300.

ing their trough during rest and/or sleep when there is no need for top performance. Grip strength is a case in point. It does not appear to be an optimal adjustment to have the best grip strength in the late afternoon or early evening if one then rests after working with one's hands earlier in the day. For others, who are mentally active, it will be important to find out whether a test of mental performance such as adding speed and random number memory or other *pertinent* performance tests also can be influenced by timing the treatment of adrenal

Table 2 Change in timing of adrenocortical substitution treatment alters circadian computative acrophase (ϕ) of several physiologic variables.

Variable examined	DF	Change in ϕ in Rx-Mode B* Mean ± SE	t	P
Grip strength	5	+47°±20°	2.28	.05
Urinary 17–OHCS	8	+76°±19°	4.10	<.02
Urinary 17–KS	7	+78°±27°	2.94	.02
Urinary Na	8	+124°±32°	3.84	<.01
Rectal temperature	8	−10°±13°	.77	>.40
Heart rate	3	−48°±26°	1.84	>.10

* Mode B (3/4 or 2/3 of total does at 0700, remainder at 2300) induced acrophase-shift, expressed with reference to ϕ on Mode A treatment (3 equal doses at 0800, 1300 and 2000).
Acrophase-shift expressed as difference between arithmetic averages of ϕ's on Mode B and Mode A treatments; for intertreatment-mode difference in cosinor ϕ that are the same in direction of shift but differ in extent of shift, see Table 1.

insufficiency so that peak performance is possible at the desirable time. Otherwise, the timing of substitution treatment for adrenal cortical insufficiency will not be optimal.

When such obvious times of "relative advantage" from a rhythm's timing are not obvious (as they are for grip strength), the circadian acrophases of healthy subjects may be regarded as a desirable norm. This will be presumed by the authors until proof is offered to the contrary.

However, many factors are to be assessed in the future in relation to optimal substitution treatments for adrenal cortical insufficiency. Hormone breakdown and elimination are likely to be periodic processes, that change in rate along the 24-hour scale. Hence, a hormone's availability will change somewhat differently from the way it is offered. Moreover, when a patient with adrenal insufficiency works or is exposed to major loads or threats transiently or permanently at night, his schedule (one trusts) will (or should) be adjusted according to information on phase-shifting and related problems [11, 12].

CONCLUSION

With noted qualifications, circadian acrophase charts indicate that adrenal cortical insufficiency of persons on a socially-synchronized ~24-hour periodic schedule is better treated if the total daily dose of corticoid is distributed in two unequal parts, namely 2/3 or 3/4 on awakening and the rest before retiring, rather than in equal doses at each of three principal meal-times, say at 0800, 1300 and 2000.

REFERENCES

1. Halberg F., Visscher M. B. and Bittner J. J. (1953): Eosinophil rhythm in mice: range of occurrence; effects of illumination, feeding and adrenalectomy. Am. J. Physiol., *174*: 313–315.
2. Brown H. E. and Dougherty T. F. (1956): The diurnal variation of blood leucocytes in normal and adrenalectomized mice. Endocrinol., *58*: 365–375.
3. Halberg F. (1960): Temporal coordination of physiologic function. *In*: Cold

Spring Harbor Symposia on Quant. Biol., Long Island Biol. Assoc., New York, *25*: 289–310.

4. Halberg F. and Howard R. B. (1958): 24-hour periodicity and experimental medicine. Posgrad. Med., *24*: 349–358.
5. Halberg F. Visscher M. B., Berge K. and Bock F. (1951): Diurnal rhythmic changes in blood eosinophil levels in health and in certain diseases. Lancet, *71*: 312–319.
6. Halberg F. and Kaiser I. (1954): Lack of physiologic eosinophil rhythms during advanced pregnancy of a patient with Addison's disease. Acta Endocrinol. (Kbh), *16*: 227–232.
7. Azerad E., Ghata J. and Reinberg A. (1957): Disparition du rythme nycthéméral de la diurèse et de la kaliurie dans 8 cas d'insuffisance surrénale. Ann. Endocr. (Paris), *18*: 484–491.
8. Bartter F. C. Delea C. S. and Halberg F. (1962): A map of blood and urinary changes related to circadian variations in adrenal cortical function in normal subjects. Ann. N.Y. Acad. Sci., *98*: 969–983.
9. Reinberg A. (1966): Rythmes des fonctions corticosurrénaliennes et systèmes circadiens. In: Symposium de Neuroendocrinologie, pp. 75–89, L'Expansion Scientifique Française, Paris.
10. Mills J. M. (1966): Human circadian rhythms. Physiol. Rev., *46*: 128–171.
11. Halberg F. (1969): Chronobiology. Ann. Review of Physiology, *31*: 675–725.
12. Halberg F., Nelson W., Doe R., Bartter F. C. and Reinberg A. (1969): Chronobiologie. J. Europ. Toxicol., No. 6, 311–318.
13. Reinberg A. and Halberg F. (1971): Circadian chronopharmacology. Ann. Rev. Pharmacol., *11*: 455–492.
14. Ghata J. (1956): Etudes des fonction rénales et réabsorption d'eau et des électrolytes et du rythme nycthéméral des éliminations chez l'addisonien traité par la cortisone. Sem. Hôp. Paris (Pathologie et Biologie-), *32*: 84–89.
15. Reinberg A. and Ghata J. (1957): Les ryhhmes biologiques. 2[d] ed. P.U.F., Paris.
16. Kaine H. D., Seltzer H. S. and Conn J. W. (1955): Mechanisms of diurnal eosinophil rhythm in man. J. Lab. Clin. Med., *45*: 247–252.
17. Halberg F., Tong Y. L. and Johnson E. A. (1967): Circadian system phase—an aspect of temporal morphology; procedure and illustrative examples. *In*: The Cellular Aspects of Biorhythms, (H. von Mayersbach, ed.), pp. 20–48, Springer-Verlag, Berlin and New York.
18. Reinberg A. (1971): Methodologic considerations for human chronobiology. J. Interdisc. Cycle Research, *2*: 1–15.
19. Scholler R., Busigny M. and Jayle M. F. (1957): Méthode de dosage des 17-21-déhydroxy-20-cétostéroides. Sem. Hôp. Paris. Arch. Biol. Med., *33*: 2.
20. Seligson D. (1958): Method of Drekter. *In*: Standard Method of Clinical Chemitry. Vol. 2, Academic Press, Inc., New York.
21. Halberg F. (1964): Organisms as circadian systems; temporal analysis of their physiologic and pathologic responses, including injury and death. *In*: Medical Aspects of Stress in the Military Climate. Walter Reed Army Institute Symposium, April, 1964, pp. 1–25.
22. Galicich J. H., Haus E., Halberg F. and French L. A. (1964): Variance spectra of corticosteroid in adrenal venous effluent of anesthetized dogs. Ann. N.Y. Acad. Sci., *117*: 281–291.

CIRCADIAN RHYTHMS IN THE THRESHOLD OF BRONCHIAL RESPONSE TO ACETYLCHOLINE IN HEALTHY AND ASTHMATIC SUBJECTS

ALAIN REINBERG, PIERRE GERVAIS, MARTINE MORIN and CHARLES ABULKER

Equipe de Recherche de Chronobiologie Humaine (CNRS n° 105)
Laboratoire de Physiologie — Fondation A. de Rothschild
Paris, France

Circadian chronotoxicologic effects of acetylcholine (ACh) have been demonstrated in the mouse by JONES, HAUS and HALBERG [1], while circadian chronopharmacologic effects of the same drug on the isolated heart (right atrium) of rat were reported by SPOOR and JACKSON [2]. Susceptibility around the clock in human respiratory function has been noted to be rhythmic by DE VRIES et al [3].

Human bronchial response (constriction) to ACh inhaled as an aerosol can be evaluated precisely under standardized experimental conditions defined by TIFFENEAU [4, 5]. This "ACh test" now is used widely in the study of patients suffering from lung diseases, more specifically from asthma. The response threshold, however, usually is considered as a constant for a given subject [4–6]. Since a) chronopharmacologic changes have to be studied systematically [7] since b) the ACh test is conventionally done without any time (even interval clock hour) reference and since c) the bronchial response to ACh seems to play an important pathophysiologic role in asthma, it is useful to explore possible temporal relationships from a quantitative point of view.

Eight healthy adults—smokers or ex-smokers—(4 women: 28 to 47 years of age and 4 men: 24 to 48 years of age) and six patients suffering from extrinsic asthma (4 women: 11 to 53 years of age and 2 men: 27 and 35 years of age) were synchronised at least one week before the first test and thereafter with L 0700–2300 and D 2300–0700.

Serial determinations of the bronchial response to ACh chloride* (LEMATTE et BOINOT. Paris, France) were made at fixed hours (0800, 1500, 1900 and 2300) on five different days, each separated by an interval of about one week.

One measurement was made per day, but at different times of day over the 5 weeks. The order of test schedules was randomized among the subjects in this fashion to minimize the time-of-day effects attributable to psychological or other reactions from one determination to the next. The ACh test at 1500 or at 0800 has been made twice.

Each measurement session began with 3 determinations** of forced expiratory volume/1 second (FEV_1) and vital capacity (VC). Then by successive assay the smallest quantity of ACh via aerosol inhalation (particle size ~1 micron) was recorded which provoked a 15% decrease of the initial FEV_1 value recorded at the beginning of each respective session.

* ACh was in anhydrous triethylene glycol solution which could be diluted at will in saline solution (NaCl: 9 g/100 ml); the hydrolysis of ACh is very slow in this condition.

** Spiromètre Cara-Modèle B2 — Gauthier, Paris, France.

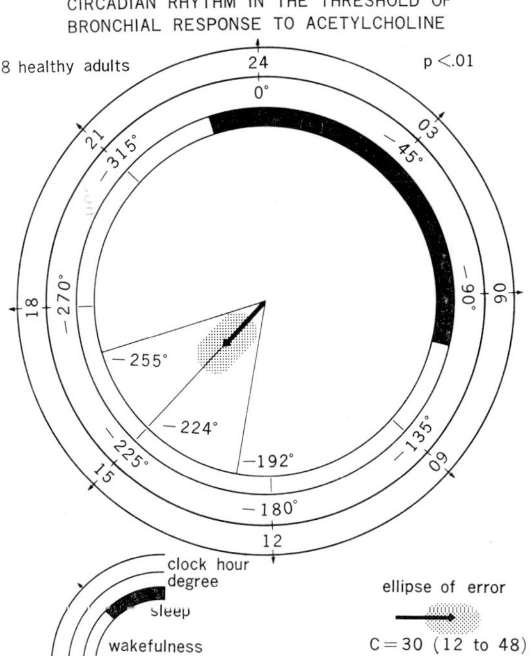

Fig. 1 Circadian rhythm in the threshold of bronchial response to acetylcholine (8 healthy adults).

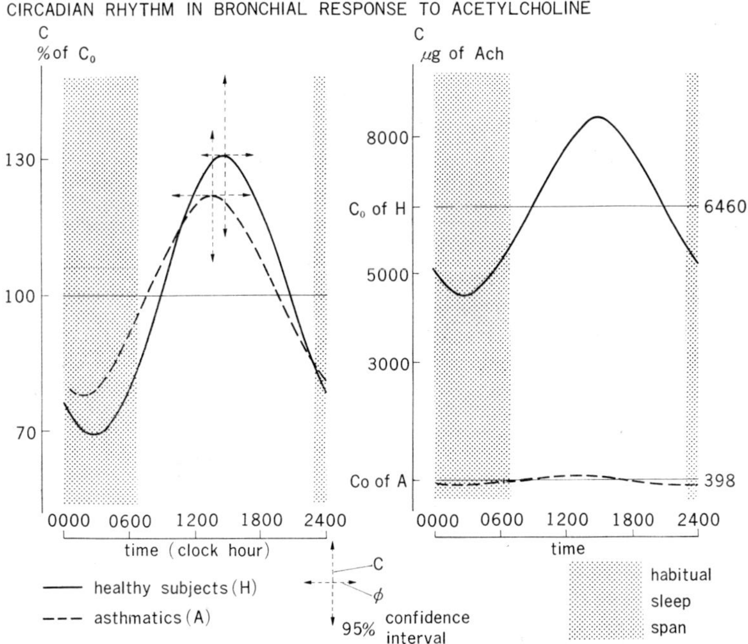

Fig. 2 Circadian rhythm in bronchial response to acetylcholine expressed as per cent of C_0 (left) and in μg of ACh (right).

Table 1 Circadian rhythms in forced expiratory volume/1 sec. (FEV_1) & in threshold of bronchial response to acetylcholine (ACh test) of 8 healthy adults (H) & 6 asthmatic patients (A) synchronized with L [0700–2300] & D [2300–0700]

Physiologic variable	Subjects	Rhythm detection P	Rhythm adjusted level $C_0 \pm SE$ [ml.]	Amplitude C [ml]	Acrophase ϕ ϕ [hrs$^{min.}$ from midnight]	Amplitude C C [% of C_0]	Acrophase φ φ [degrees from midsleep]
				....	(.95 confidence limits)	....	
FEV_1	8H	<.025	3350±25	117 (32 to 202)	1559 (1228 to 1930)	3.7 (.9 to 6.4)	−195° (−142 to −248)
	6A	<.025	2225±42	112 (21 to 203)	1556 (1102 to 2049)	5 (1 to 9.1)	−194° (−128 to −260)
			C_0 [μg ACh]	C [μg ACh]	ϕ [hrs$^{min.}$ from midn.]	C [% of C_0]	φ [degrees from midsleep]
ACh test	8H	<.010	6460±1360	1944 (762 to 3120)	1454 (1247 to 1701)	30.1 (11.8 to 48.3)	−179° (−147 to −210)
	6A	<.050	398±96	80 (24 to 131)	1336 (1036 to 1640)	20.2 (7.5 to 33)	−159° (−114 to −205)

C_0 and C are expressed in ml for FEV_1 and in μg for ACh test. C is also expressed as percent of C_0. Acrophase, ϕ, is given in hours and minutes with midnight as ϕ reference. Acrophase, φ, is aso given in degrees, with midsleep (−45°; 1 hour=15°) as φ reference.

Microscopic analysis of the time series (cosinor and related methods) yielded results shown in Table 1 and Figs. 1 and 2.

Statistically significant circadian rhythms ($P<.05$) are detected for FEV_1 and the bronchial response to ACh in both the healthy and asthmatic groups. A statistically significant circadian rhythm in CV (not represented) was detected in healthy subjects only.

The acrophases (ϕ) and amplitudes (C) in FEV_1 do not show statistically significant differences between healthy and asthmatic groups of subjects, while the rhythm adjusted level (C_0) is lower in asthmatic patients as compared with normal subjects.

The acrophase of the circadian rhythm in the ACh test of healthy subjects does not differ from that of asthmatic patients. For the same physiologic variable, there is a marked difference in C_0 of healthy subjects ($6460 \pm 1360\ \mu g$) by comparison to the C_0 of asthmatic patients ($398 \pm 96\ \mu g$). Thus, by chronobiologic methods, precision can be added to the usual expression of ACh test results.

When the amplitude of the circadian rhythm in ACh test is expressed as per cent of C_0, there is no statistically significant difference between the C value of healthy subjects in comparison to the C value of asthmatic patients (Table 1, Fig. 2). Nevertheless, when both C values are expressed in μg of ACh, a difference between the two groups becomes obvious. This fact gives a new experimen tal example of the relationship between the level and the amplitude of a biorhythm.

It is pertinent to consider the relation between FEV_1 and the ACh test from a chronobiologic point of view. The respective acrophase of the two rhythms differs only by about one hour; this difference is not statistically significant. Circadian changes in FEV_1 cannot explain circadian changes in the ACh test; the amplitude of the former is too small by comparison to the that of the latter. Therefore the study of circadian changes in the ACh test furnishes a new argument for considering bronchial response to ACh as a major phenomenon in human respiratory function.

ACKNOWLEDGEMENT

The advice of Mr. J. Bohlen of the Chronobiology Laboratories, University of Minnesota, Minneapolis, is greatly appreciated.

REFERENCES

1. Jones F., Haus E. and Halberg F. (1963): Murine circadian susceptibility resistance cycle to acetylcholine. Proc. Minnesota Acad. Sci., *31*: 61–62.
2. Spoor R. P. and Jackson D. B. (1966): Circadian rhythms: variation in sensitivity of isolated rat atria to acetylcholine. Science, *154*: 782.
3. De Vries K., Goei J. T., Booy Noord H. and Orie N. G. (1962): Changes during 24 hours in the lung function and bronchitic patients. Int. Arch. Allergy, *20*: 93.
4. Tiffeneau R. (1955): Hyperexcitabilité acétylcholinique du poumon. Critère physio-pharmacodynamique de la maladie asthmatique. Presse Med., *63*: 227–230.
5. Tiffeneau R. (1957): Examen pulmonaire de l'asthmatique, p. 245, Masson et Cie, Paris.
6. Makino S. (1966): Clinical significance of bronchial sensitivity to acetylcholine and histamine in bronchial asthma. J. Allergy, *38*: 127–142.
7. Reinberg A. and Halberg F. (1971): Circadian chronopharmacology. Ann. Rev. Pharmacol., *11*: 455–492.

THE EFFECTS OF HORMONES ON THE SKIN TEMPERATURE OF NORMAL AND CANCEROUS BREASTS*

Carl M. MANSFIELD, John D. WALLACE, Robert F. CURLEY and Ralph A. CARABASI

Thomas Jefferson University Hospital
Department of Radiation Therapy and Nuclear Medicine
Philadelphia, Pennsylvania, U.S.A.

INTRODUCTION

Various investigators [1, 3, 6] have studied the possibility of predicting the clinical response of a patient with advanced cancer of the breast to a particular therapeutic modality. In searching for such a method, we have developed a system involving the continuous monitoring of the breast skin temperature in tumorous and comparable normal anatomical areas. This approach is being used because it has been shown that there is usually a relative increased temperature over the skin of a tumorous breast when compared with the opposite normal breast [4, 7, 8]. In addition, when the skin temperature of the normal and abnormal breasts are continuously monitored and recorded for later computer analysis, it is possible, on occasion to detect a difference, not only in the temperature level, but in the temperature pattern [9, 10]. It has also been shown that the normal body temperature is circadian in nature [5]. Therefore, by continuously monitoring the breast temperature it should be possible to detect this circadian rhythm in the normal breast and any deviation from this rhythm in the abnormal breast.

METHOD AND MATERIALS

A thermogram was performed in order to locate the region of the skin that most likely represented the tumor's temperature. A thermistor was placed on this area and a second, control thermistor, was placed on normal skin on the contralateral side in a comparable anatomical area. The thermistors were connected to a small radio transmitter which sent the information to a receiver. Upon decoding, the system gave the approximate temperature sensed by each thermistor. The temperatures were sampled every five minutes. Details of this system have been described elsewhere [2]. These data have been sampled once every two to five hours to form the working data base which is presented as a 12 hour moving average. The mean temperature for the entire study period for each breast was determined, then the variations of the temperature in degrees centigrade from this mean were plotted against time. The difference between these last two values were plotted against time.

---

* Supported by N. I. H. Contract PH 43-68-1304.

* This work was performed in part in the Clinical Research Center under N. I. H. Grant #RR72.

The usual procedure was to monitor the temperature for at least two days to establish a baseline. The patient was then given Stilbesterol tablets, 15 mgm, t.i.d. for several days. Following a few days of no medication, 100 mgm of Testosterone Proprionate was given, i.m., once daily.

RESULTS

We have performed approximately 140 studies since 1966. Because of space limitation, only two graphs (Figs. 1, 2) are presented as examples of the possible effects of hormones on breast temperature.

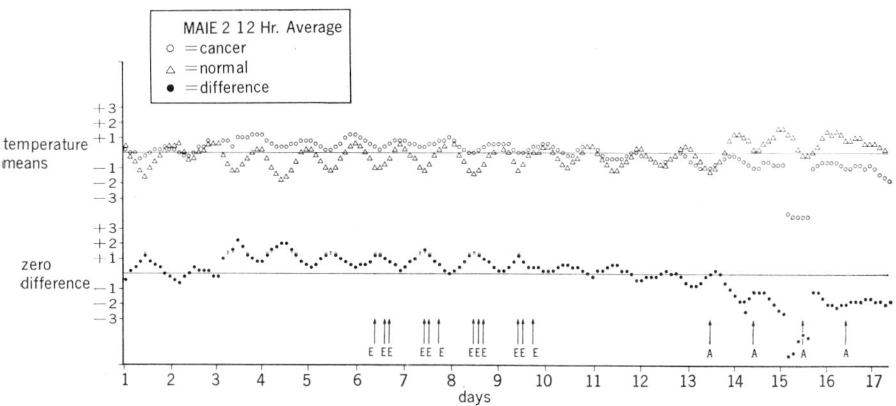

Fig. 1

1. There is a change in the relative amplitudes of the signals at the cancer and normal sites before and after the start of Stilbesterol (E).
2. After the Stilbesterol (E) the curves nearly coincided.
3. Upon the administration of Testosterone (A) the relative levels of temperature are reversed and the amplitude of the signal at the cancer site is deminished.

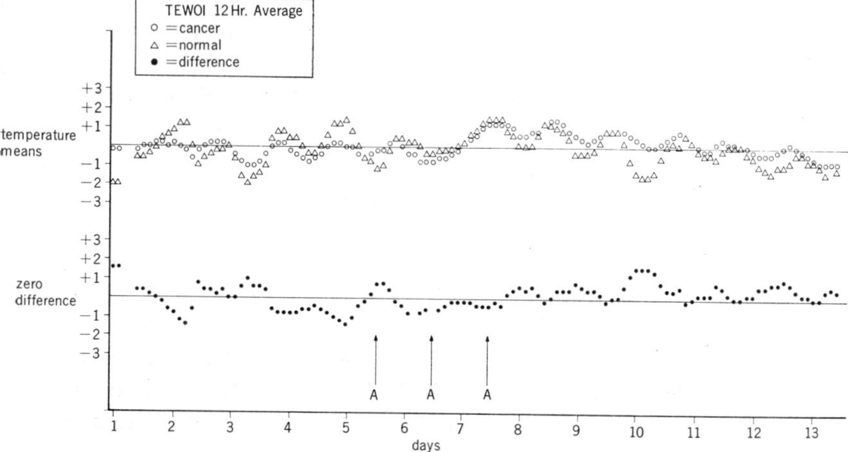

Fig. 2

Note that the cicadian rhythm is present at both cancer and normal sites. After the administration of Testosterone (A) the cancer tended to be warmer as compared to the normal.

When comparing the time intervals before and after the administration of Stilbesterol (E), there appear to be three noticeable changes in the first graph (Fig. 1). The first is a relative change in the amplitude of the temperature cycles for the cancer and normal sites. Secondly, after Stilbesterol (E), the temperature curves of the cancer and normal sites approached each other, in that the amplitudes and temperature are comparable. In addition, after the administration of Testosterone (A) it is evident that the relative levels of temperature are reversed and the amplitudes of the temperature cycle at the cancer site is diminished.

In the second graph (Fig. 2), as in the first, the circadian rhythm is also evident. However, after the administration of Testosterone (A) the cancer temperature tended to increase as compared to the normal site, as is indicated by the difference curve.

We have noted in many patient studies, the temperature patterns showed no apparent change during the entire procedure. The correlation of change or lack of change in the pattern, with the clinical response to treatment, is under investigation.

DISCUSSION

Recent reports in the literature have been encouraging. Stoll et al. [12] have reported that they are able to predict a patient's response to a hormonal agent by noting the patient's "temperature profile" before treatment and two weeks after treatment. A temperature drop of 1°C was associated with a favorable clinical response in 8 out of 14 patients. An increase in temperature or a decrease of less than 1°C was associated with no response. Gillespie et al. [13] reported that a temperature increase in patients treated with nor-ethisterone acetate was associated with a favorable clinical response in 6 out of 6 patients.

If the value of this system can be established, most tumors could be studied. Deep tumors may have to be implanted with heat sensitive probes. Many modalities of therapy could be tested and the relative merits of timed therapy could be evaluated. It will be necessary to try and associate these changes with other methods presently being used to predict a patients response to therapy. Under current investigation is an evaluation of the correlation of temperature pattern and the response to irradiation and chemotherapy. In addition, a relationship between this work and the uptake of ^{32}P by tumors will be further investigated [11].

CONCLUSION

This paper tends to confirm earlier observations that hormones can cause an alteration in the skin temperature over a tumorous and normal breast. These changes were not noted in every patient studied. It remains to be determined if there is a useful correlation between the temperature response curve and the clinical response. If such proves to be the case, this method could be a useful aid in determining, in advance, the therapeutic modality to which a tumor is more likely to respond.

REFERENCES

1. Bullen M. D., Freundlich H. F., Hale B. T., Marshall D. H. and Tudway R. C. (1963): The activity of malignant tumors and response to therapeutic agents, studied by continuous records of radioactive phosphorus uptake. Postgrad. M.D., *39*: 265–277.
2. Curley R. F., Wallace J. D., Mansfild C. M. and Marsh L. (1968): A system for the evaluation of cancer treatment. Proceed. 6th Nat. ISA Biomed Sciences Instrumentation Symposium.
3. Gerbrandy J. and Hellendoarn H. B. (1957): The diagnostic value of calciuria during hormonal treatment of metastasized mammary carcinoma. Acta Endocrinol. (Suppl. 31), *24*: 275–288.
4. Gershon-Cohen J., Berger S. M., Habberman J. D. and Barnes R. B. (1964): Thermography of the breast. Amer. J. Roentgenol., *91*: 919–926.
5. Halberg F. (1962): Physiologic 24-hour rhythms: A determinant of response to environmental agents. *In*: Man's Dependence on the Earthly Atmosphere, (K. Schaefer, ed.), pp. 48–89, The Macmillan Co., New York.
6. Handley R. S. (1962): The temperature of breast tumors as a possible guide to prognosis. Acta Un. Int. Cancer, *18*: 822.
7. Lawson R. N. and Chightai M. A. (1963): Breast cancer and body temperature. Can. Med. Assoc. J., *88*: 68–70.
8. Lloyd W. K. (1964): Pictorial heal scanning. Phys. Med. Biol., *9*: 433–456.
9. Mansfield C. M., Wallace J. D., Curley R. F., Kramer S., Southard M. and Driscoll D. (1970): A comparison of the temperature curves recorded over the normal and abnormal breasts. Radiol., *94*: 3, 697–698.
10. Mansfield C. M., Dodd G. D., Wallace J. D., Kramer S. and Curley A. F. (1968): Use of heat sensing devices in cancer therapy. Radiol., *91*: 4, 673–678.
11. Stoll B. A. and Burch W. M. (1968): Surface detection of circadian rhythm in ^{32}P content of breast cancer. Cancer (phil.), *21*: 193.
12. Stoll B. A. (1971): The thermoprofile as an early indicator of breast cancer response to hormonal therapy. Cancer (phil.), *27*: 1379–1383.
13. Gillespie P. J., Burrows B. D. and Edelystyn G. A. (1971): Possible clinical implications of therapeutically induced temperature changes in continuously monitored tumor mass. Brit. J. Cancer, *25*: 95–93.
14. Dodd G. D., Wallace J. D., Freundlich I. M., Marsh L. and Sermino A. (1969): Thermography and cancer of the breast. Cancer (N.Y.), *23*: 797–802.

CIRCADIAN RHYTHM IN DRUG ACTION: A PHARMACOLOGIC, BIOCHEMICAL, AND ELECTRONMICROSCOPIC STUDY

VELAYUDHAN NAIR

Department of Pharmacology, Chicago Medical School
University of Health Sciences
and Michael Reese Hospital, Psychiatric Institute,
Chicago, Illinois, U.S.A.

There have been several reports of circadian variation in response to drugs, both in experimental animals and man. Among those reported showing circadian variation in action are ouabain, lidocaine, endotoxin, ethanol, reserpine, chlordiazepoxide, ACTH, metyrapone, phenobarbital, pentobarbital, histamine, periactine, salicylate, halothane, and hexobarbital. The mechanisms underlying the circadian variation in drug action may be different for the different drugs and may depend on several factors; the specific type of interaction the drug has at the biochemical and morphological level; the intrinsic rhythms, if any, of the organism's biochemical and morphological components, environmental conditions such as light-dark cycle, diet, sleep-wake pattern and so forth.

In this study, we have attempted to relate the pharmacological observations in the whole animal with the biochemical findings from *in vitro* studies and ultrastructural data from electronmicroscopic studies. A circadian rhythm in drug action observed in the whole animal (rat) is shown to be related to a corresponding rhythm in the hepatic enzyme which metabolizes the drug as well as to a rhythm in hepatic endoplasmic reticulum structures. Experiments were also done to understand the factors regulating the enzyme rhythmicity. We chose the drug hexobarbital for these studies, because as shown below, it is an ideal drug for such correlative studies.

Male Sprague-Dawley rats, 70–80 days old were used throughout and were fed *ad libitum* unless otherwise indicated. All animals were allowed to acclimatize for at least two weeks before use in experiments.

Duration of hexobarbital action was determined as reported earlier [7]. In animals housed under alternating periods of light and darkness, the duration of drug action was maximal in the light period (2 PM) and minimal in the dark period (10 PM) (Fig. 1). The hepatic enzyme which metabolizes the drug, hexobarbital oxidase (HO) also showed a circadian variation in its activity. Hepatic hexobarbital oxidase was determined in a whole liver homogenate system as described in earlier papers [5, 11]. The enzyme activity was minimal in the light period (2 PM), i.e. at the time when the duration of hexobarbital action was maximal and maximal in the dark period (10 PM) when the drug action was minimal (Fig. 1). This reciprocal relationship will be expected if the duration of drug action is predominantly determined by the activity of the metabolizing enzyme. Indeed, similar correlations of enzyme activity and drug action have been shown for hexobarbital in other conditions also [6, 9], which is one of the reasons for its selection in this study.

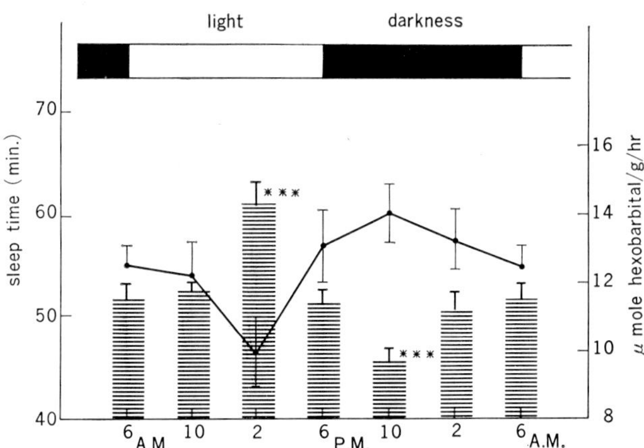

Fig. 1 Circadian variation in duration of hexobarbital action (vertical bars) compared to the circadian changes in hepatic hexobarbital oxidase activity (solid line). The sleep time at each point represents the average of 6–10 animals. Hexobarbital Sodium (150 mg/kg) was given intraperitoneally. The enzyme activity is expressed as μ moles of hexobarbital metabolized per gram tissue per hour and each point represents the mean of 6-8 animals. The vertical line represents the standard errors. ***=$P<0.001$

Experimental conditions for the electronmicroscopic (EM) studies were the same as in enzyme studies except that only two time periods, those of maximal and minimal enzyme activity, were selected for EM study. For the initial study, this was considered adequate since a more detailed examination every four (4) hours, was done for enzyme measurements. Both enzyme and EM experiments were done in the same animal and tissue from contiguous areas was used. The experimental details are described in another publication [1]. EM studies showed that the amount of smooth endoplasmic reticulum (SER) varied with the time of day. We have also observed regional differences in the distribution of SER and rough endoplasmic reticulum (RER) within the hepatic lobule. In general, within the same lobule, there appeared to be a centroportal gradient in the distribution of SER. SER was more abundant at 10 PM in the pericentral areas and to a certain extent in the midzones when compared to the same areas at 2 PM (Fig. 2). The RER in contrast showed less variability. When the enzyme rhythm was abolished as in rats rendered blind by bilateral enucleation, the circadian variation in SER was not evident [1].

By the judicious selection of a drug, whose varied aspects can be followed from the whole organism level to the level of ultrastructure, the present studies have shown the relationships among the rhythms in drug action, in the activity of the drug metabolizing enzyme in liver and in the endoplasmic reticulum of hepatocytes. No doubt, parallelism in the pattern of two parameters, by itself, does not establish cause and effect relationship. However, it can be stated that the diurnal variation in the duration of hexobarbital action is largely due to the diurnal changes in the activity of metabolizing enzyme. It is known from the work of other investigators [2–4], that in normal rat liver, DNA synthesis and mitotic activity are highest in the morning hours and lowest at night. If one assumes that these events precede the formation of ER membranes and enzyme proteins, it is consistent with our findings of an abundant SER at 10 PM and lower amount at 2 PM.

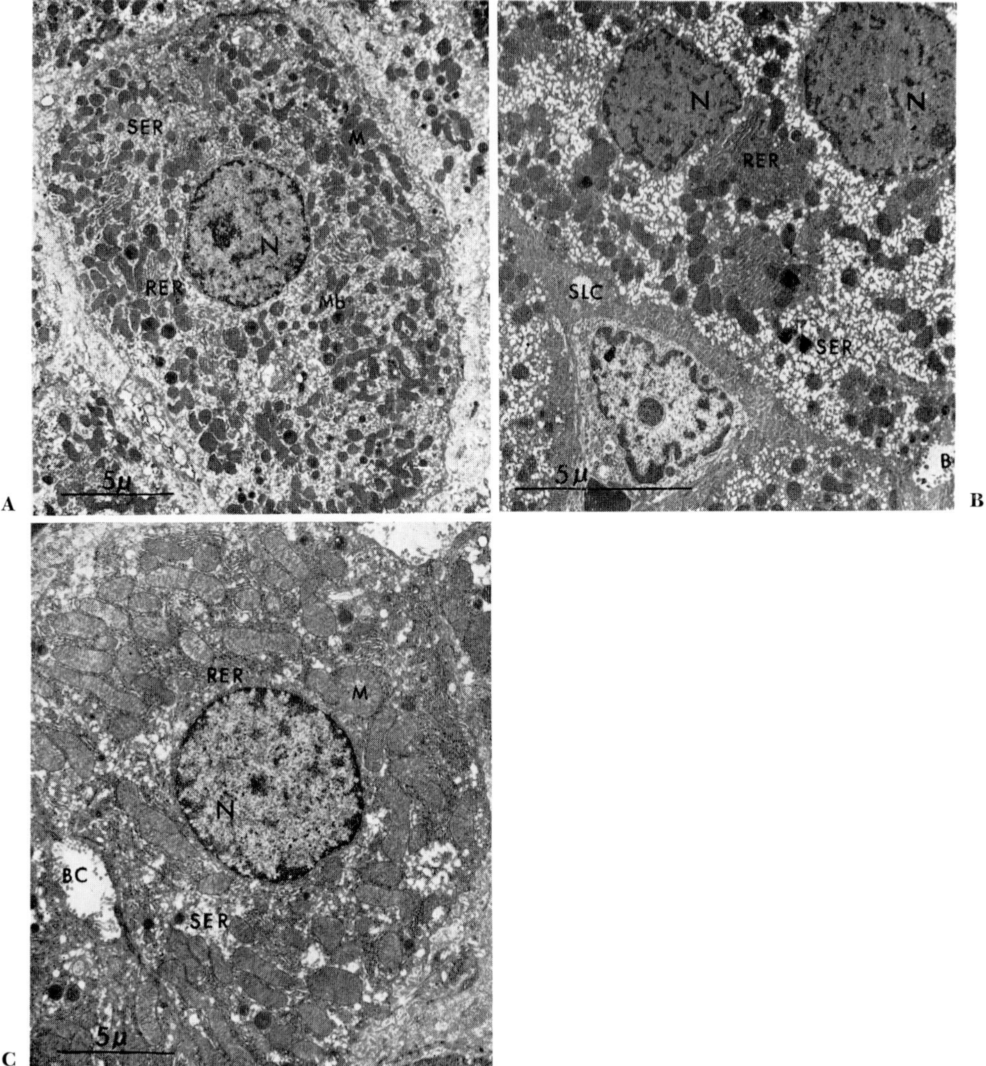

Fig. 2 Circadian changes in the endoplasmic reticulum of rat hepatocytes. (A) and (B), normal cells at 2 and 10 PM respectively from the pericentral area. (C), normal cell at 10 PM from the periportal region. No significant differences were observed between 2 and 10 PM specimens in the periportal region. N–nucleus; M–mitochondrion; SER–smooth endoplasmic reticulum; RER-rough endoplasmic reticulum; Mb-microbody; BC-bile canaliculus; SLC-sinusoid lining cell. Magnification is indicated in the bottom of the picture.

Regulation of the Diurnal Rhythm in Enzyme Activity.—In order to understand the factors regulating the rhythmicity, rats were subjected to each of the following treatments and the effects on enzyme activity examined [8].

1) Constant illumination
2) Constant darkness
3) Blinding by enucleation
4) Bilateral superior cervical ganglionectomy (SCG)

5) Hypothalamic lesions, medial or lateral and
6) Regulated hourly feeding.

Both constant illumination and constant darkness abolished the rhythmicity, but the enzyme activity was generally higher in the dark reared group than in the light exposed one. There was no enzyme rhythm in the blinded animals also and the activity was high as in those reared in total darknes. Following SCG, the rhythm was attenuated but not abolished and exhibited phase shifts. The hypothalamic lesions abolished the enzyme rhythm. Regulated feeding every hour maintained the rhythm, but produced phase shifts thus ruling out the feeding pattern as a principal "zeitgeber". When the plasma corticosterone levels were determined in the animals housed in constant light, we noted a persistence of the steroid rhythm (although with phase shifts) while the enzyme rhythm was now absent. These results suggest that 1) the enzyme rhythmicity is not determined by the feeding pattern or plasma corticosterone rhythm even though the base levels may be influenced by the plasma steroid levels. The rhythmicity is dependent on environmental illumination and appears to be regulated by the hypothalamic centers and is partially dependent on the sympathetic innervation to the pineal gland. REITER [10] studying pineal function in hamsters has suggested an interaction of pineal autocoid with hypothalamo-hypophyseo-gonadal axis in light gonadal relationships. Since the enzyme under study here can be influenced by interference at any level in the axis [6], and since we have shown an attenuation of the enzyme rhythm with SCG (by interrupting the sympathetic input to the pineal) [8] it is tempting to speculate an interaction of the type suggested by REITER. The resolution of this complex matter should await further investigation.

ACKNOWLEDGMENTS

This work was supported in part by a grant from the State of Illinois Department of Mental Health. The cooperation and assistance of Dr. A. CHEDID and Dr. R. CASPER and Messrs. D. BAU, S. SIEGEL and M. RAFFELD are gratefully acknowledged.

REFERENCES

1. CHEDID A. and NAIR V. (1972): Diurnal rhythm in endoplasmic reticulum of rat liver: an electronmicroscopic study. Science, *175*: 176–179.
2. FABRIKANT J. I. (1967): The spatial distribution of parenchymal cell proliferation during regeneration of the liver. The Johns Hopkins Med. J., *120*: 137–147.
3. HALBERG F., ZANDER H. A., HOUGLUM M. S. and MUHLEMANN H. R. (1954): Daily variations in tissue mitoses, blood eosinophils and rectal temperature of rats. Am. J. Physiol., *177*: 361–366.
4. JACKSON B. (1959): Time associated variations of mitotic activity in livers of young rats. Anat. Rec., *134*: 365–377.
5. NAIR V., BAU D. and SIEGEL S. (1968): Effects of prenatal x-irradiation: studies on the mechanism of x-irradiation-induced inhibition of microsomal enzyme development in rat liver. Radiation Res., *36*: 493–507.
6. NAIR V., BROWN T., BAU D. and SIEGEL S. (1970): Hypothalamic regulation of hepatic hexobarbital metabolizing enzyme system. Europ. J. Pharmac., *9*: 31–40.
7. NAIR V. and CASPER R. (1969): The influence of light on daily rhythm in hepatic drug metabolizing enzymes in rat. Life Sciences, *8*: 1291–1298.

8. Nair V., Casper R., Bau D. and Siegel S. (1970): Regulation of the diurnal rhythm in hepatic drug metabolism. Fed. Proc., *29*: 804.
9. Quinn G. P., Axelrod J. and Brodie B. B. (1958): Species strain, and sex differences in metabolism of hexobarbitone, amidopyrine, antipyrine and aniline. Biochem. Pharmac., *1*: 152–159.
10. Reiter R. J. (1969): Pineal function in long-term blinded male and female golden hamsters. Gen. Comp. Endocr., *12*: 460–468.
11. Yam K. and DuBois K. P. (1967): Effects of x-irradiation on the hexobarbital metabolizing enzyme system of rat liver. Radiation Res., *31*: 315–326.

CIRCADIAN RHYTHMICITY IN RESPONSE TO BARBITURATES

O. MÜLLER

Department of Anatomy, School of Medicine
Hannover, Germany

Among the many reports on a circadian rhythm in response to various chemical agents, circadian drug susceptibility for certain barbiturates also has been published [1–6]. The aim of the present study is to compare the circadian susceptibility of a long-acting and a short-acting drug of the barbiturate series. As an example of a long acting barbiturate, phenobarbital was chosen. A circadian rhythm of the mortality and sleeping time after phenobarbital administration until now has not been established. Furthermore, this drug is known to produce an alteration in the liver cell ultrastructure. Such alterations will be shown to depend on the time of the day phenobarbital is administered and will be compared with circadian structural changes in the normal liver. The other drug investigated in this report is the short acting hexobarbital, because it is known to be almost completely metabolized in the liver and thus can be related to the circadian rhythms of liver function.

In a first experiment, 72 seven month old male Wistar rats (AF-Han, 340–380 g body wt.) kept in a temperature-controlled room with natural daylight (sunrise 0430, sunset 2030) were used. Every three hours during an uninterrupted 24-hour period, eight animals received 190 mg/kg phenobarbital-sodium intraperitoneally (=LD 50).

The results are shown in Fig. 1: mortality changes between 100% and 0% and the duration of sleep of the survivals between 28 hrs and 14 hrs.

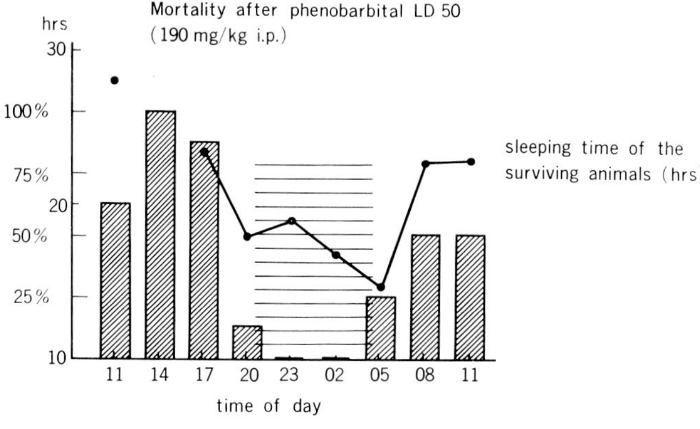

Fig. 1

In an additional experiment the circadian toxicity of the long-acting phenobarbital was compared with that of the short-acting hexobarbital and the *in vitro* hexobarbital-oxidation in the liver. This study was performed on 180 four months old male Wistar rats (Ivanovas, 220–240 g body wt.) in one uninterrupted

24-hour period. The animals were kept under standardized environmental conditions (temperature, humidity) including an artificial light-dark regimen (light period 0700–1900 hrs, dark period 1900–0700 hrs), and divided into 3 subgroups. Every 3 hours 8 animals from subgroup I received 190 mg/kg phenobarbital-sodium i.p., and the sleeping time and mortality were recorded. 8 animals of subgroup II received 150 mg/kg hexobarbital-sodium i.p., and the sleeping time was recorded. Subgroup III serves as control group from which four animals were sacrificed at the same intervals. Their livers were used for the estimation of the *in vitro* oxidation of hexobarbital and for morphologic analysis [7].

The results are given in Fig. 2: The maximum of the hexobarbital oxydation (0.36±0.015 μMol/g body wt.×30 min)* occurred at the end of the dark period, while the hexobarbital-induced sleeping time was at its minimum (20.2 min±1.2). The longest anesthesia (35.7 min±1.1) was found at the end of the light period, in the time of the lowest oxydation potency of the liver (0.23±0.015 μMol/g body wt.×30 min).

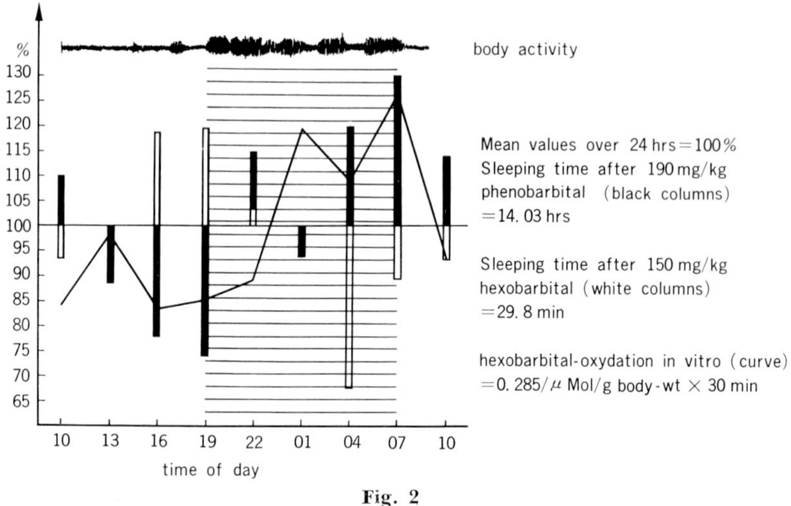

Fig. 2

Phenobarbital caused highest mortality (5 out of 8 animals) and longest sleeping time (18.2 hrs±2.7) when given in the early light period; whereas in the late light period no animal died, and the shortest sleeping time of the survivors was found (10.4 hrs±3.1).

The early dark period and the early light period showed very striking differences in liver ultrastructure. These are summarized schematically in Fig. 3. Fig. 3A demonstrates a liver cell in the early light period, with abundant glycogen (as a result of the glycogen-synthesis during the dark period), packed lamellae of granular endoplasmic reticulum and poorly developed agranular reticulum. Fig. 3B shows a liver cell in the early dark period. Glycogen has disappeared during the light period, the arrangement of mitochondria and granular reticulum has changed remarkably and smooth reticulum now is visible [8, 9].

A single dose of 80 mg/kg phenobarbital i.p. given at the beginning of the light period after seven hours produced changes in liver ultrastructure which are

* mean values ± standard error of the mean.

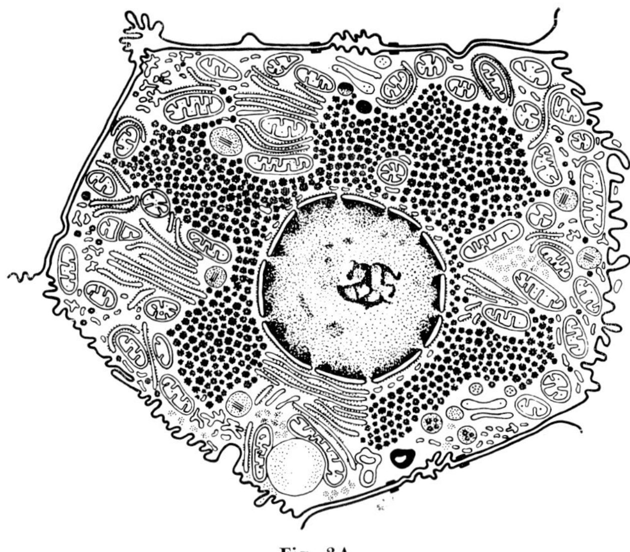

Fig. 3A

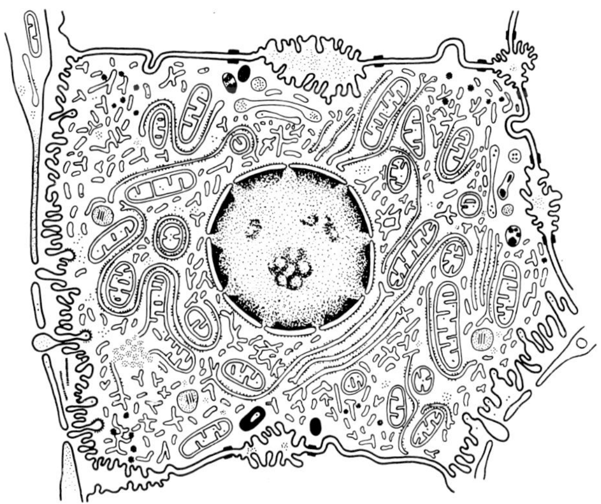

Fig. 3B

very similar to those shown in Fig. 3B and therefore do not differ from untreated control animals. The same dose given at the beginning of the dark period resulted in dilatation and partial fragmentation of the endoplasmic reticulum (Fig. 4). We never could observe such alterations during the normally occurring circadian oscillation of the liver ultrastructure. Some other parameters as liver glycogen, liver weight and respiration of isolated liver mitochondria, also showed different changes after a single dose (80 mg/kg i.p.) of phenobarbital (Fig. 5). They depend on the time of day of administration and therefore on the different structural and functional state of the liver cell.

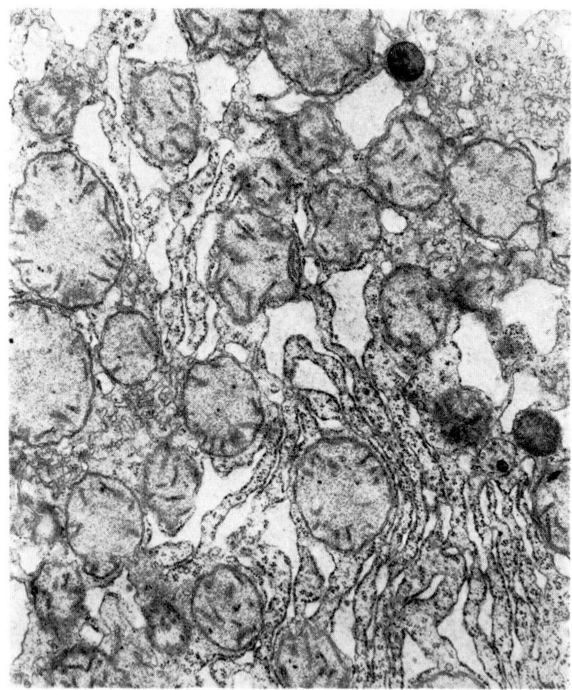

Fig. 4

Table 1

	control	control	seven hrs after Luminal at 1000	control	control	seven hrs after Luminal at 1900
	1000 hrs	1700 hrs	1700 hrs	1900 hrs	0200 hrs	0200 hrs
liver-weight %	3.92 ±0,085	3,53 ±0,055	3,38 ±0,045	3,43 ±0,03	3,53 ±0,065	3,00 ±0,05
liver-glycogen %	3,38 ±0,23	1,86 ±0,22	0,88 ±0,19	1,63 ±0,17	3,99 ±0,12	0,35 ±0,14
0_2 consumption by liver mitochondria after succinate-stimulation μ l 0_2/mg Prot. min	0,986 ±0,123	1,343 ±0,240	1,294 ±0,140	1,555 ±0,263	0,831 ±0,142	1,467 ±0,099

REFERENCES

1. Davis W. M. (1962): Experientia, *18*: 235.
2. Pauly J. E .and Scheving L. E. (1964): J. Neuropharmacol., *3*: 651.
3. Lindsay H. A. and Kullman V. S. (1966): Science, *151*: 576.
4. Scheving L. E. et al. (1968): Anat. Rec., *160*: 741.
5. Roberts P. et al. (1970): Europ. J. Pharmacol., *12*: 375.
6. Nair V. and Casper R. (1969): Life Sciences, *8*: 1291.
7. Müller O. (1971): Naunyn-Schmiedebg. Arch. Pharmak., *270*: Suppl. R 99.
8. Müller O. et al. (1966): Z. Zellforsch., *69*: 438.
9. Müller O. (1971): Acta Histochem., Suppl. X, 141.

CIRCADIAN LIVER DETOXICATION AND ACETYLCHOLINESTERASE RHYTHMICITY: TWO LIMITING FACTORS IN CIRCADIAN E 600 TOXICITY

H. von MAYERSBACH

Department of Anatomy, School of Medicine
Hannover, Germany

E 600 (Paraoxon) is an excessive poison which acts by blocking acetylcholinesterase. It is used both as an insecticide and as a powerful parasympathicomimetic drug in ophthalmology.

HOLTZ and WESTERMANN [1] found that after intraperitoneal injection, E 600 is seven times less toxic than after intramuscular application. They concluded, that paraoxon is detoxicated by the liver of mice and rats. In histochemistry and biochemistry, E 600 serves as an inhibitor substance for several other esterases, i.e. the so called nonspecific esterases. The nonspecific esterase activity in the liver is almost completely inhibited by E 600 in dilutions up to 10^{-8} M.

By histochemical means it was possible to show that the E 600 sensitive esterase (A-esterase) is diffusely located in the cytoplasm of liver cells and therefore regarded as microsomal, whilst the organophosphate resistent (B-esterase) is connected with the lysosomes. The latter enzyme is regarded as identical with the nonspecific esterase in the blood serum of rabbits found by ALDRIDGE [2] which is able to hydrolyse organophosphates like paraoxon.

MAYERSBACH and YAP [3] have shown by histochemical and biochemical means, that the nonspecific esterases in the liver exhibit a marked circadian rhythm which is influenced by sex and seasons. Using certain histochemical techniques there is a fluctuation of esterases that becomes apparent when investigating livers along a 24 hour period. The esterase reactions sometimes are confined solely to lysosomes; at other times, they are located in microsomes. Between these extremes, transitional forms are visible (Fig. 1).

If the classical LD 50 dose of E 600 is intraperitoneally given into rats at different times, a marked circadian drug susceptibility in death ratios is evident (Fig. 2). The analysis of livers revealed that at time of maximal lethality, the activity of E 600 resistent esterase is definitely lowest. As this enzyme most likely is responsible for detoxication by paraoxon hydrolysis, it is puzzling that at times of high activity (0100 and 0700 hours) the death ratios vary considerably (Fig. 5).

An explanation of this effect can be seen only in a different circadian sensitivity of its target, i.e. acetylcholinesterase. The analysis of the brain revealed that the acetylcholinesterase activity also exhibits a marked circadian rhythm (Fig. 3). Now if detoxication (expressed by the activity of E 600 resistant liver esterase) is compared with the acetylcholinesterase activity in the brain, a very good correlation can be found. The varying effectiveness of E 600 expressed as acetylcholinesterase-inhibition in the brain therefore may be correlated directly with the rate of circadian lethality (Figs. 4, 5).

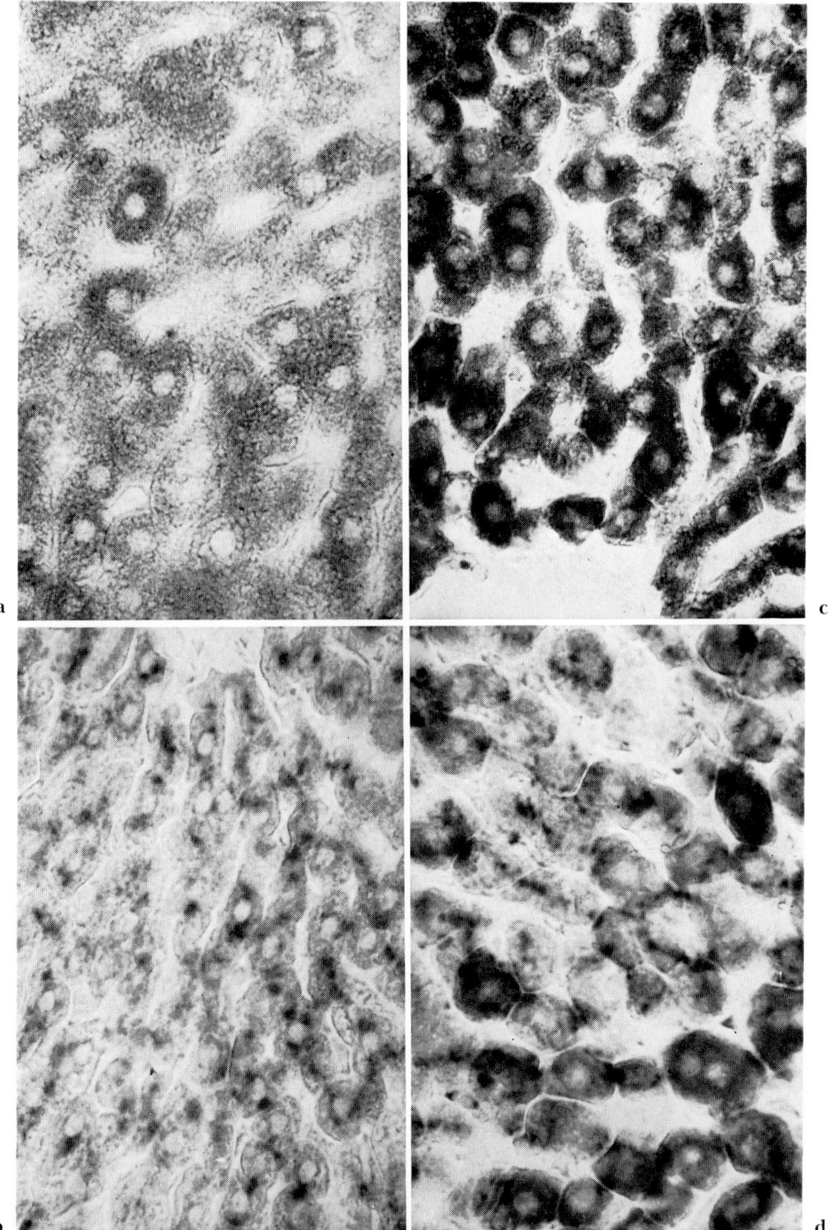

Fig. 1. Histochemical demonstration of circadian esterase activity in the normal rat liver (5-Br-4-Cl-Indoxylacetatreaction after Holt pH 7.2).

a. 0800 Pure microsomal localization
b. 0100 Pure lysosomal localization
c. 1800 Strong microsomal reaction with few lysosomes
d. 2200 Strong lysosomal and mediocre microsomal reaction

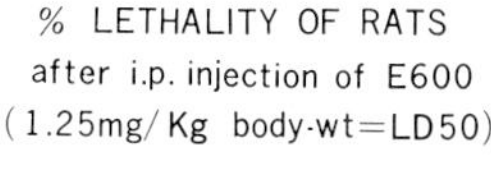

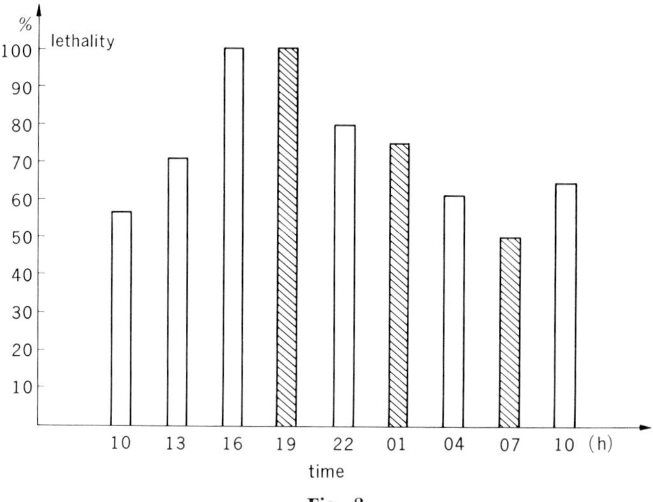

Fig. 2

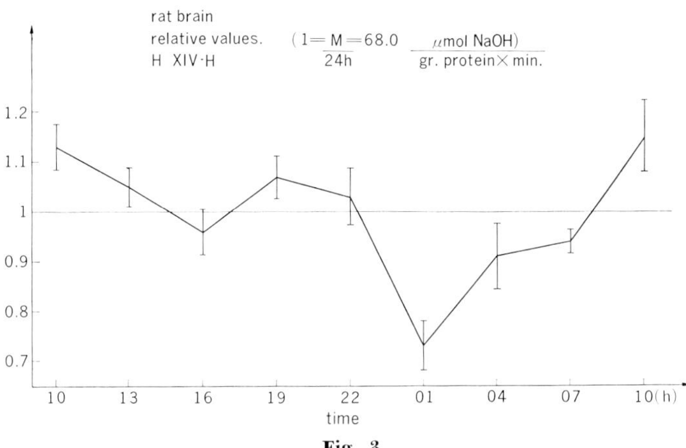

Fig. 3

The histochemical analysis of livers of E-600 treated rats showed very strikingly, that E 600 is able to stimulate the synthesis of E 600 resistent esterase in lysosomes (Fig. 6); but this effect is only visible at times in which normal lysosomal activity is high. Similarly there are marked differences in the side effects of E 600 according to the time of day. Glycogen depletion is very great at times of natural circadian glycogen dissimilation; whereas this effect is much lower at times when the liver is normally in a state of glycogen accumulation (Fig. 7). Ultramorphologically this effect is expressed by the appearance of many glycogenosomes after the administration of E 600 at times of normal glycogen degradation. Summarizing, we can conclude from these experiments:

1. A varying pharmacological effect such as lethality may not be explained only by circadian differences in drug metabolism or only by circadian dependent sensitivity of the targets.

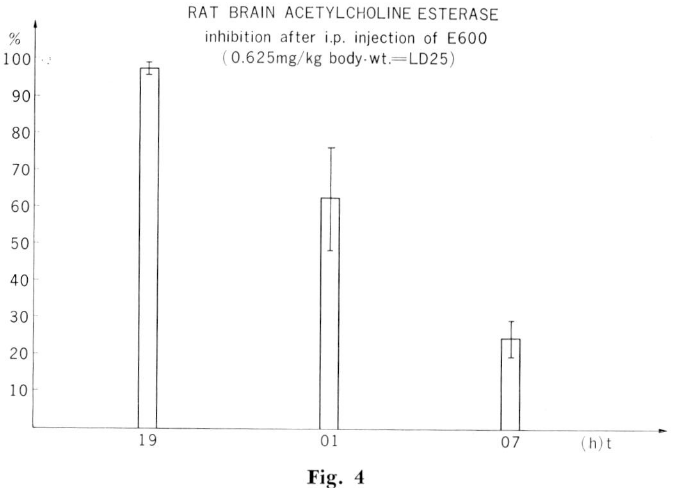

Fig. 4

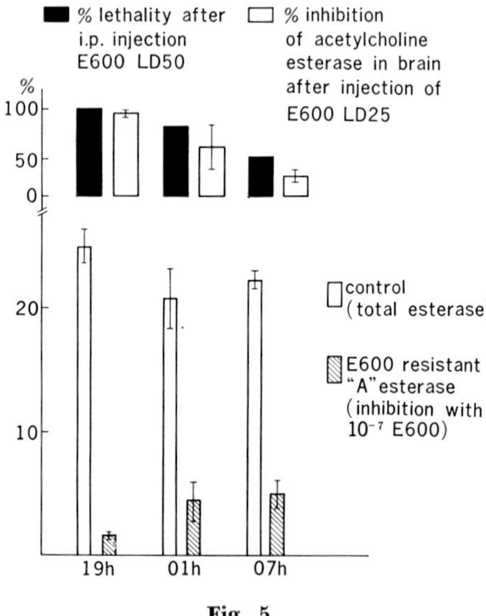

Fig. 5

In the case of E 600, the rhythmicity of detoxication by the liver must be matched with the differences of acetylcholinesterase activity.

2. Generally it may be assumed that a cell stimulation, e.g. enzyme induction, can be achieved only when the cell is prepared for this certain function by its normal rhythm.
3. The same is true for side effects that result from pharmacological influences in which the cell will act mainly in the way already provided by biological rhythms.

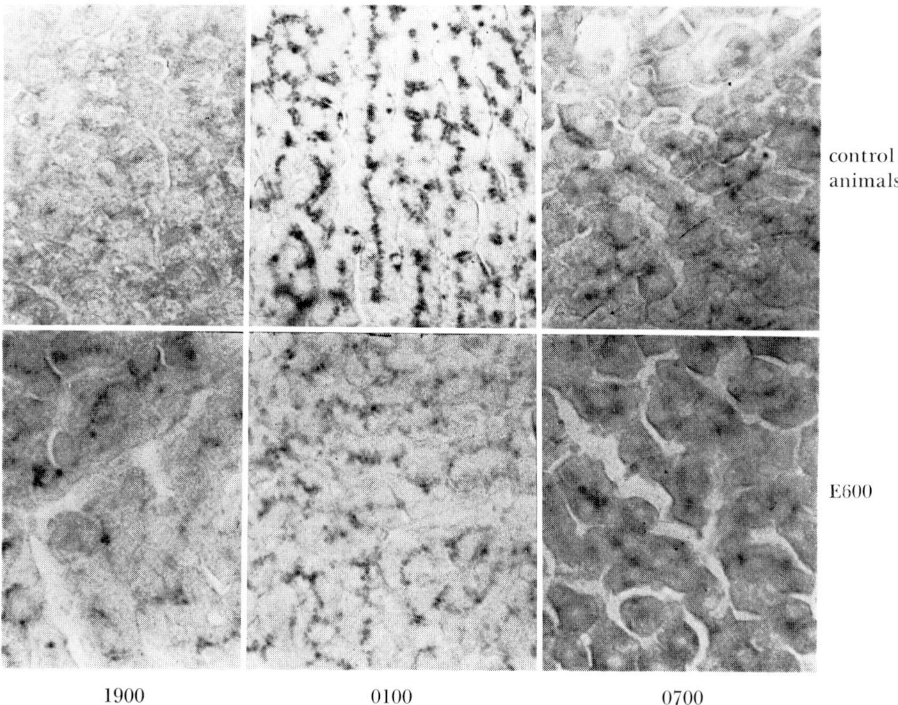

Fig. 6 Histochemical appearance of lysosomal esterase (5-Br-4-Cl-Indoxylacetat after Holt, pH 8.5).

Upper row: Control animals. Lower row: Liver, 1 hour after E 600 (LD 25=0.625 mg/kg) i. p. injection.

It clearly can be seen that after E 600 treatment lysosomal activity is stimulated, but markedly only at times when lysosomes become active during a non disturbed circadian rhythm.

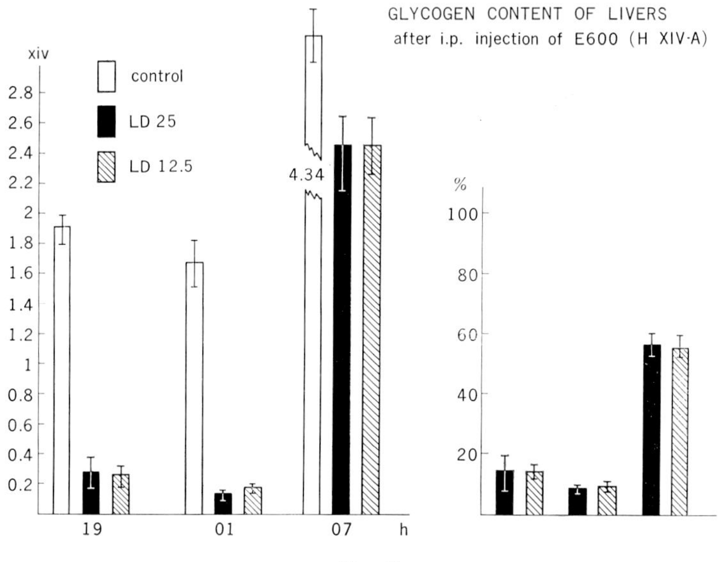

Fig. 7

REFERENCES

1. Holtz P. und Westermann E. (1959): Naunyn-Schmiedebergs Arch. exp. Path. und Pharm., *237*: 211.
2. Aldridge W. N. (1954): Biochem. J., *57*: 692.
3. Mayersbach H. v. und Yap P. (1965): Histochemie, *5*: 297.

RATIONALE FOR CIRCADIAN-SYSTEM PHASED GLUCOCORTICOID MANAGEMENT

MICHAEL S. SMOLENSKY

The University of Texas at Houston
School of Public Health
Houston, Texas, U.S.A.

INTRODUCTION

In recent years evidence has accumulated on circadian rhythms as a determinant of effect by a given agent. The stage of an organism's many rhythms—its circadian-system phase—has been demonstrated to be an important factor in the evaluation of rodent and human susceptibility to challenge as well as to the effects of pharmacological preparations [1]. Of particular interest here is the question of whether steroid-dependent side effects may be reduced by timing synthetic glucocorticoids to the circadian phase of expected high endogenous serum corticosteroid concentration as a single dose, either daily or on alternate days.

BACKGROUND

Since the initial documentation of circadian rhythms in adrenocortical secretion in humans by PINCUS in 1943 [2], many works on both humans and other species have followed. In particular, HALBERG et al. [3] summarized the microscopic findings for the circadian rhythm of 17-OHCS excretion for inhabitants of various geographic locations. For our purposes, the most interesting observation is that the timing of highest expected 17-OHCS values varies but little. Although the acrophases,* when referenced to the middle of the habitual sleep span of the respective samples, range from $-109°$ to $-188°$, the great majority occur within the limits of a 3-hour arc extending from $-125°$ to $-170°$. Thus, the attainment of peak 17-OHCS values are expected to follow about 8 to 10 hours after the middle of the habitual sleep span.

It is important to point out that rhythms do not merely represent "a time-of-day effect". Conceptualization of circadian rhythms in this manner is incorrect and can be very misleading in chronotherapy (the application of therapeutic measures according to rhythms). For example, rhythms of persons adhering to varying work-rest schedules, as observed in day versus night workers, are not expected to have comparable peak times with regard to clock hours. When rhythms are referenced to comparable phase markers, the middle of the usual sleep span or the acrophase of another circadian system of the same organism, agreement in timing between samples of subjects adhering to differing work-rest schedules is expected. This point is well illustrated by work of HALBERG and SIMPSON [4].

* In this paper, the acrophase (an approximation by least squares spectral analyses of the rhythm's peak) is expressed as a delay in timing from the chosen phase reference; thus, as a negative value. Since in the analysis of these time series a cosine of 360° is fit to 24-hour data, each 15° is equivalent to 1 clock hour.

RHYTHMS AND CHRONOTHERAPY BY GLUCOCORTICOIDS

Early work by DiRaimondo and Forsham in 1958 [5] suggested that administration of 10 mg prednisone as a single dose, for example at 0800, to diurnally active persons would result in considerably less adrenal suppression as compared to administration of the same total dosage as four separate ones, each of 2.5 mg.

Grant, Forsham and DiRaimondo [6] confirmed these findings. As summarized in Table 1, 8 mg of Triamcinolone given at 0800 for eight days resulted in no adrenal suppression; the change from the baseline value of plasma 17-OHCS was from 20.2±1.0 ($\overline{X}$±S.E.) to 19.1±1.6 μg/100 ml. When the same total daily dosage was divided equally into four dosages of 2 mg (each) the degree of adrenal suppression was considerable. Identical results were obtained in a second investigation on three of the same subjects similarly studied. Also included in Table 1 is a summary of results obtained by Scalabrino and Pasquariello [13] and Schulz and Retiene [14] which indicate comparable findings.

Animal work provides additional support for a rational (chronotherapeutic) approach to management of steroid-dependent disorders with glucocorticoids. Tarquini et al. [7] reported minimal adrenal atrophy and corticosterone suppres-

Table 1 Glucocorticoid administration and adrenal suppression: Comparison of single daily versus divided doses.

Study No. (Duration)	Glucocorticoid Dose, Schedule	Subjects No., Sex	Plasma/ Urine	17–Hydroxycorticosteroids Baseline ($\overline{X}$±S.E.)	Final
I[1] (8 days)	8 mg daily at 0800	6 ♂	Plasma (μg/100 ml)	20.2±1.0	19.1± 1.6
	8 mg daily 2.0 mg/divided)	,,	,,	21.5±1.4	*9.0± 2.1
II[1] (8 days)	8 mg daily at 0800	3 ♂	Urine (mg/24 hr.)	10.2±0.4	9.9± 1.1
	8 mg daily 2.0 mg/divided)	,,	,,	8.7±0.2	*4.7± 0.0
III[2] (9 days)	1.5 mg daily at 0800	3 ♂	Plasma (μg/100 ml)	15.9±2.8	16.1± 1.6
	1.5 mg daily 0.5 mg/divided)	,,	,,	15.7±1.8	10.7± 3.8
	1.5 mg daily at 0800	,,	Urine (μg/24 hrs)	+351.0±73.	+315.0±42.
	1.5 mg daily 0.5 mg/divided)	,,	,,	+372.0±26.	+278.0±92.
IV[3] (weeks)	20–50 mg Morning)	5	Plasma (μg/100 ml)	17.1.....	17.7± 1.8
	20–50 ng/divided)	7	,,	17.7.....	8.1± 6.0

1) Data from Grant *et al.* (6) for Triamcinolone

2) Data from Scalabrino *et al.* (13) for Dexamethasone

3) Data from Schulz *et al.* (14) for Prednisone

* Statistical Significance between baseline and final 17–OHCS level, P<.01.

\+ Urinary 17, 21–OHCS.

sion in rats following several once-a-day administrations of Dexamethasone given around the expected daily crest of corticosterone concentration. Identical treatment at a time of reduced serum hormone concentration resulted in considerable adrenal atrophy. RETIENE et al. [8], using the same drug, confirmed these findings. REINDL, ULLNER and HALBERG [9] demonstrated that methylprednisolone timed to the expected crest of adrenal activity resulted in far less adrenal suppression as compared to identical treatment during the span of reduced adrenal activity.

NICHOLS et al. [10], carried out comparable work on diurnally active healthy adult male volunteers. A dose of 0.5 mg Dexamethasone at 0800 or 1600 resulted in slight suppression of cortisol secretion. The same dose given at 0000 produced virtually complete suppression of cortisol production during the ensuing 24-hour span.

A study by SEGRE and KLAIBER [11] confirms and extends the findings by NICHOLS et al. [10]. In this study, three healthy, diurnally active male volunteers received Flumethasone daily during a four-day span of study. Tests of adrenal response were carried out in separate studies at 0800, 1600 or 0000. The mean adrenal suppression, resulting from Flumethasone, was dependent on the circadian-system phase of treatment. Administration of Flumethasone at 0800 resulted in, as compared to control values, a 43% decrease in 17-OHCS concentration. Flumethasone at 1600 resulted in a 61% decrease; at 0000 nearly an 83% decrease in 17-OHCS concentration occurred. The difference between the resulting adrenal suppression for daily single doses at 0800 versus such at 0000 is statistically significant, $P<0.01$.

Based upon the circadian rhythmic pattern of pituitary-adrenal interaction, HARTER et al. [12] developed an acceptable approach to management of steroid-dependent disorders. By choice of a proper glucocorticoid such as prednisolone, which is characterized by activity of intermediate duration and strength, adrenal suppression can be minimized appreciably by administrations on alternate days around the expected circadian acrophase of adrenocortical activity.

Investigations by this author [15] indicate other undesired side effects of prolonged glucocorticoid management, besides adrenal suppression, may be minimized by proper choice of circadian-phased treatment. Results of studies on rodents, which are summarized in Fig. 1, suggest the inhibitory effects on weight gain in immature male C-mice by methylprednisolone can be reduced by selection of proper circadian phase of administration. In this figure, daily body weight changes from methylprednisolone relative to control groups are summarized for each of six circadian phases of administration (corresponding to specific clock hours of 0000, 0400, 0800, 1200, 1600 and 2000) by means of an index (DCI). A DCI value of zero indicates no difference between body weight changes of experimentals relative to controls; a negative value of great magnitude indicates great inhibition of weight gain by experimentals relative to controls studies at the same time. The data summarized in Fig. 1 show that either daily or alternate day treatment at 1600 [a time corresponding to highest expected adrenocortical activity for animals exposed in L (0600 to 1800): D (1800 to 0600)] results in lowest reduction in weight gain in these immature mice. Treatment at other times produced a greater degree of weight gain inhibition depending on the circadian phase of methylprednisolone administration. For example, methylprednisolone at 0400 or 0800 results in upto a 1.5 to 1.8 gm weight gain inhibition of experimental relative to corresponding time point controls after 20 days of treatment.

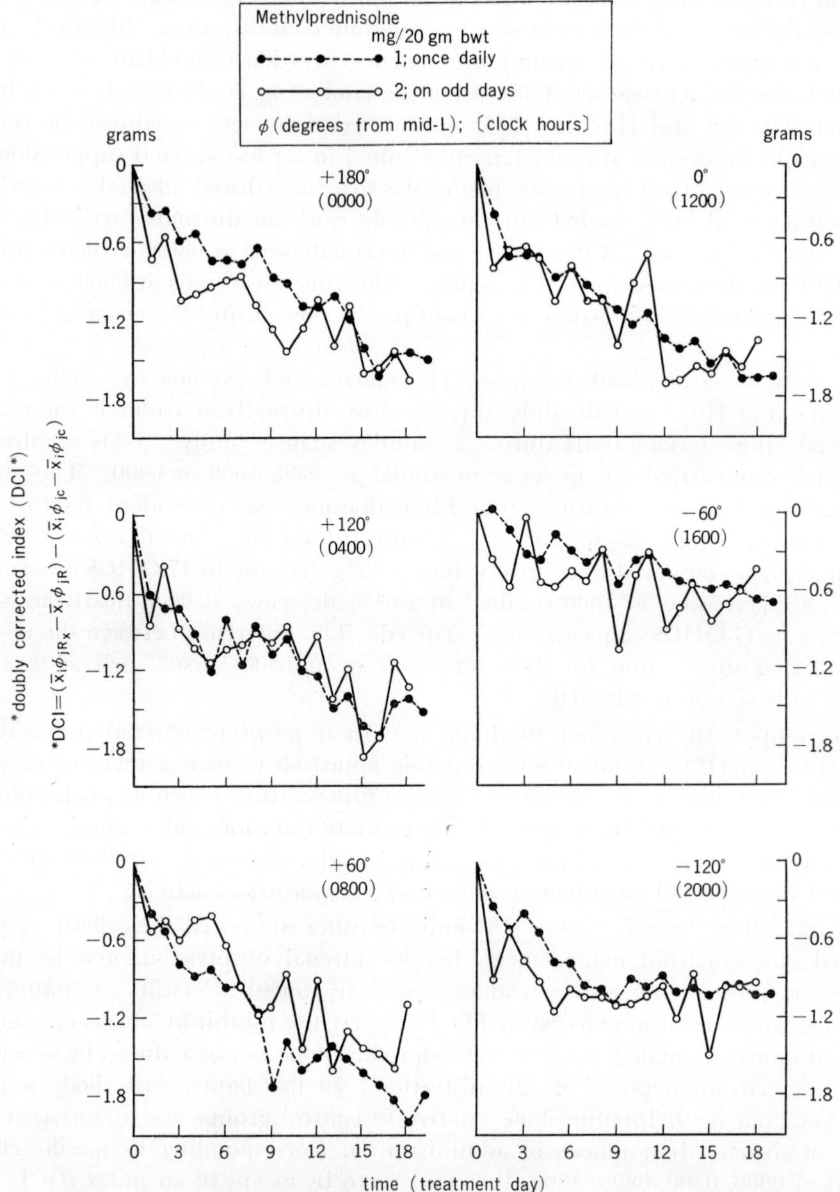

Fig. 1 Weight gain difference between experimental and control groups indicated by means of an index-DCI (See text). Methylprednisolone for 20 consecutive days given either daily or on alternate days at the circadian crest of adrenal cortical function results in minimal effects upon weight gain. Identical treatment at other circadian phases results in considerably greater effect.

CONCLUSION

In managing steroid-dependent illnesses, reduction of such iatrogenic effects as adrenal inhibition and growth impairment in children may be moderated by proper choice of circadian phase of administration. Specifically, findings indicate reduction of undesired side effects of synthetic corticosteroids may be achieved by single daily or alternate day treatments phased to the circadian crest of adreno-cortical secretion, which is expected to coincide by clock hour to the habitual time of awakening following sleep. Future studies are planned to evaluate the role of circadian rhythms as a factor affecting the magnitudes of both desired as well as undesired drug effects.

REFERENCES

1. Reinberg A. and Halberg F. (1971): Annual Review Pharmacology, *11*: 455.
2. Pincus G. (1943): J. Clin. Endcr., *3*: 195.
3. Halberg F., Reinhardt J., Bartter F., Delea C., Gordon R., Reinberg A., Ghata J., Hofmann H., Halhyker M., Gunther R., Knapp E., Pena J. C. and Garcia Sainz M. (1968): Experientia, *25*: 107.
4. Halberg F. and Simpson H. (1967): Human Biol., *39*: 405.
5. DiRaimondo V. C. and Forcham P. H. (1958): Am. J. Med., *21*: 321.
6. Grant S. D., Forsham P. H. and DiRaimondo V. C. (1965): New England J. Med., *273*: 21.
7. Tarquini B., Fantini F. and Cagnoni M. (1966): Lo Sperimentale, *116*: 373.
8. Retiene K., Schulz F. and Marco J. (1967): Rass. D. Neurol. Veg., *21*: 217.
9. Reindl et al. (1966): Unpublished findings.
10. Nichols T., Nugent C. A. and Tyler F. H. (1965): J. Clin. Endcr., *24(1)*: 343.
11. Segre E. J. and Klaiber E. L. (1966): Calif. Med., *104*: 363.
12. Harter J. G., Reddy W. J. and Thorn G. W. (1963): New England J. Med., *269*: 591.
13. Scalabrino R. and Pasquariello G. (1967) Reumatismo, *18(2)*: 81.
14. Schulz F. and Retiene K. (1971): Klin. Wschr., *49*: 100.
15. Smolensky M. H., Pitts G. C., Stubbs S. S., Sothern R. B., Nelson W. L., Mason P., McHugh R. B., Cadotte and Halberg F. (1971): Fed. Proc., March.

DISSECTION OF CAUSAL RELATIONS BETWEEN HORMONAL AND METABOLIC ULTRADIAN RHYTHMS

JOHN URQUHART and NANCY KELLER

Biodynamics Laboratory, ALZA Corporation
Palo Alto, California, U.S.A.

The large literature on the circadian rhythms of many physiological variables [1, 2] contrasts with the much less extensive work on ultradian rhythms, particularly those whose periods range between 0.25 and 4 hours. This unbalanced emphasis on different parts of the physiological spectrum is exemplified by the field of adrenocortical physiology: a sizeable diurnal rhythm in plasma cortisol concentration was discovered about 15 years ago [3], and has been extensively discussed since (cf. 2), but only within the past three years has it been learned that there is an almost equally large ultradian rhythm in the adrenocortical system, manifested by large variations in plasma cortisol concentration that cycle with a period of 2½ to 4 hours [4–9]. It has, of course, required the sampling of plasma at least as often as every 20 to 30 minutes to demonstrate these ultradian cycles, and that is a sampling rate which had rarely been applied to the adrenocortical system and never previously for long enough to show a rhythmic pattern. It was BERSON and YALOW's extensive work on the radioimmunoassay of ACTH in plasma [4] that first combined the necessary frequency and duration of sampling for both ACTH and cortisol that revealed the clear-cut ultradian rhythm. Subsequent work [5–9] has amply confirmed its existence.

There is a limited amount of evidence that other physiological variables also cycle with periods in the range of 0.25 to 4 hours. These include heat production [10], oxygen consumption and CO_2 production [11], various ventilatory parameters and arterial pO_2 and pCO_2 [12], and blood glucose concentration [13]. Each of these is a significant factor in metabolism or its regulation, which makes one puzzle at the paucity of confirmatory, and the absence of contradictory, evidence. It is almost as if physiologists were selectively deaf to frequencies in the range 0.25 to 4 cycles per hour. Nevertheless, we would like to propose an experimental strategy for endocrine and metabolic physiology that can be exploited from ultradian rhythms in the 0.25 to 4 cycles per hour range.

One of the long-standing questions that this strategy may provide at least partial answers for is the role of the various endocrine systems in metabolic regulation. The relatively primitive state of our understanding of the physiological roles played in metabolic regulation by the secretions of the thyroid, adrenal cortex, and pituitary is masked by a voluminous literature that emphasizes effects observed either in the absence of one or more of these glands' secretions, or at pathological extremes of hormonal dosage. To use the adrenocortical system as an example, consider the question: what physiological role does the circadian rhythm in plasma cortisol concentration play? It is not an unreasonable question, because the ratio between maximum and minimum values is 2–5. Despite a large literature, spanning 20 years, on circadian variations in adrenocortical function, there is no satisfactory answer to this question, indicating how little we know about the roles played by the adrenocortical secretions in physiological regulation.

The strategy we propose is, first, to take a straightforward spectroscopic approach to the various endocrine systems, utilizing sampling regimens that can expose whatever ultradian frequencies may exist. The same spectroscopic approach should also be made to various metabolic parameters, particularly those of fat metabolism, which is the major source of metabolic energy. It will be of great interest to measure both endocrine and metabolic variables simultaneously in the same experimental subject or animal. The temporal relations between various metabolic and endocrine cycles will suggest hypotheses about causality. The same approach has been taken to circadian rhythms, but the significance of emphasizing ultradian cycles is that they involve actions that occur on a time scale short enough to permit their dissection in acute experimental preparations, notably isolated perfused organs or tissues. The value of the perfused organ is that it gives the experimentor both control over the concentration of one or more substances in the organ's arterial inflow, and the ability to measure the organ's rates of production or uptake of various metabolites. The chief limitation of the perfused organ as an experimental preparation is that its functional lifetime is almost invariably limited to a few hours. While perfusion technology may improve and lengthen this functional lifetime, it is in general adequate to permit the exploration of ultradian cycles. Thus, the specific strategy we propose is to reproduce experimentally the precise time course of an ultradian hormonal cycle in the arterial blood of an isolated, perfused organ and to look for its effects on the organ's uptake or production of one or more metabolites.

We would emphasize that an essential element in the design of these experiments is that the hormonal concentration changes induced in the perfused organ's arterial supply ought to be of no greater magnitude than those actually observed in the systemic blood of the intact organism. This approach obviously calls for techniques of organ perfusion that can provide the necessary experimental control without crippling the organ and making it unresponsive to its normal physiologic signals, within physiological concentration ranges. Elsewhere, we have described two general methods of organ perfusion that appear largely to meet this criterion [14, 15].

This strategy will undoubtedly involve many iterations around a cycle which consists of measurements on intact animals, hypothesis formulation, and hypothesis testing in the perfused organ, but a vigorous experimental approach to ultradian rhythms in endocrine and metabolic systems ought to enrich our understanding of their regulation, and incidentally provide new rationales for therapy of endocrine and metabolic disorders.

REFERENCES

1. Sollberger A. (1965): Biological Rhythm Research. Elsevier, Amsterdam.
2. Conroy R. T. W. L. and Mills J. N. (1970): Human Circadian Rhythms. Churchill, London.
3. Bliss E. L., Sandberg A. A., Nelson D. H. and Eik-Nes K. (1953): J. Clin. Invest., *32*: 818.
4. Berson S. A. and Yalow R. S. (1968): J. Clin. Invest., *47*: 2725.
5. Hellman L., Nakada F., Curti J., Weitzman E. D., Kream J., Roffwarg H., Ell-

MAN S., FUKUSHIMA D. K. and GALLAGHER T. F. (1970): J. Clin. Endocrinol. Metab., *30*: 411.
6. HELLMAN L., WEITZMAN E. D., ROFFWARG H., FUKUSHIMA D. K., YOSHIDA K. and GALLAGHER T. F. (1970): J. Clin. Endocrinol. Metab., *30*: 686.
7. WEITZMAN E. D., FUKUSHIMA D., NOGEIRE C., ROFFWARG H., GALLAGHER T. F. and HELLMAN L. (1971): J. Clin. Endocrinol. Metab., *32*: 14.
8. TOURNIAIRE J., ORGIAZZI J., RIVIERE J. F. and ROUSSET H. (1971): J. Clin. Endocrinol. Metab., *32*: 666.
9. KRIEGER D. T., ALLEN W., RIZZO F. and KRIEGER H. P. (1971): J. Clin. Endocrinol. Metab., *32*: 266.
10. IBERALL A. (1960): Trans. ASME, J. Basic Eng., *103*: 96.
11. GOODMAN L. (1964): IEEE Trans. Biomed. Electr., *11*: 82.
12. LENFANT C. (1967): J. Appl. Physiol., *22*: 675.
13. ANDERSON G. E., HILLMAN R. W., van ELK I. F. A. and PERFETTO A. J. (1956): Am. J. Clin. Nutr., *4*: 673.
14. URQUHART J. (1970): The Physiologist, *13*: 7.
15. URQUHART J. and KELLER N. (1972): Acta Endocrinol., Suppl. *158*: 9.

IMPLICATIONS OF BIOLOGICAL RHYTHMS IN BRAIN AMINE CONCENTRATIONS AND DRUG TOXICITY

CHARLES A. WALKER

Department of Physiology and Pharmacology
School of Veterinary Medicine, Tuskegee Institute
Tuskegee, Alabama, U.S.A.

Circadian patterns for rectal temperature and biogenic amine levels in the midbrain and caudate nucleus were established in rodents maintained under stable laboratory conditions for at least three weeks on an automatically-timed light-dark cycle. The light phase lasted from 0800 to 2000 hours daily. Levels of norepinephrine, histamine and acetylcholine in the rat were maximal during the dark phase of the illumination cycle coinciding with maxima for rectal temperature and motor activity and were minimal during the light phase when these nocturnal animals are quiescent [1]. The circadian pattern for serotonin was twelve hours out of phase with the other parameters measured. Bilateral adrenalectomy did not alter these patterns qualitatively although quantitative changes were observed, e.g., peak rectal temperature was significantly lowered. Pentobarbital sodium reversed the rectal temperature pattern significantly [2]. Minima occurred during the dark phase of the illumination cycle. Significant increases in norepinephrine maxima in midbrain and caudate nucleus and in caudate nucleus histamine were observed after pentobarbital sodium. The serotonin pattern however was decreased. The onset and duration of sleep induced by pentobarbital sodium exhibited a circadian pattern. Onset time was significantly longer and duration significantly shorter during the dark phase coinciding with acetylcholine, norepinephrine, histamine, rectal temperature and motor activity maxima. Peak blood histamine levels were noted at the end of the light phase and the circadian patterns were unaltered by pentobarbital sodium. Blood glucose patterns were bimodal in nature. A primary peak occurred at the end of the dark phase toward the end of the normal feeding period, while a secondary peak occurred at the end of the light phase. These patterns were unaltered by pentobarbital sodium.

Norepinephrine levels in rat hypothalamus were significantly higher at 0300 hours than at 1500 hours, reflecting similar changes observed in midbrain and caudate nucleus [3]. Significant reductions and elevations were obtained by pretreatment with reserpine or dl-DOPA and pargyline (an MAO inhibitor), respectively.

An examination of the ultrastructure of the hypothalamus and caudate nucleus showed that circadian changes were demonstrable when synaptic vesicles in tissue removed at 0300 and 1500 hours were counted. The number of granular vesicles (presumably containing norepinephrine) were significantly higher at 0300 hours in the anterior hypothalamus and caudate nucleus, but not in the posterior hypothalamus and caudate nucleus, but not in the posterior hypothalamus. The number of agranular vesicles (presumably containing norepinephrine) were significantly higher at 0300 hours in the anterior hypothalamus and caudate nucleus but not in the posterior hypothalamus. Granular vesicle counts from tissues of rats pretreated with reserpine or dl-DOPA and pargyline were similar to those of the control study in that they were higher at 0300 hours than at 1500 hours. A more variable pattern was obtained with the agranular vesicle counts. The most significant finding was the greater number at 1500 hours in the tissues from reserpinized animals. A similar finding was obtained in the anterior hypothalamus after dl-DOPA and pargyline administration.

Circadian patterns for the toxicity of cholinergic compounds were demonstrated in mice [4], LD50's [5] for acetylcholine, pilocarpine and oxotremorine were minimal (greatest toxicity) during the dark phase of the illumination cycle. The circadian pattern for atropine toxicity was essentially a mirror image with maximum toxicity (minimal LD50) during the light phase. A quaternary anticholinesterase, neostigmine, was most toxic during the late dark phase, while a tertiary anticholinesterase, physostigmine, was least toxic at that time and became progressively more toxic during the late light and early dark phase of the illumination cycle. Atropine methyl nitrate decreased the toxicity of acetylcholine and physostigmine but the circadian pattern was essentially unchanged. An ultradian toxicity pattern having two peaks, one at midday, the second at midnight, was obtained with carbachol.

The greatest changes in toxicity were 56 and 51 percent for neostigmine and carbachol respectively. Atropine sulfate showed a maximum change of 27 percent while maximum for acetylcholine was 21 percent. The smallest percent difference for the cholinomimetics was a maximum 17 percent for pilocarpine and a 17.4 percent for physostigmine following atropine methyl nitrate pretreatment.

Twenty-four hour LD50 rhtyhms for sympathomimetics [6] were determined for mice adapted to a light-dark cycle as previously described. The greatest toxicity for the directly acting (norepinephrine, epinephrine and isoproterenol) and the indirectly acting (ephedrine and amphetamine) sympathetic stimulants occurred at a time when endogenous brain norepinephrine levels are elevated and during a period of increased motor activity (dark phase). Minimum toxicity for these compounds occurred during the light phase. The biorhythmic toxicity pattern for the alpha adrenergic blockers (phenoxybenzamine and yohimbine), the beta adrenergic blockers (dichlorisoproterenol and propranolol) and the neuronal blocker-catecholamine depletor (reserpine) were inversely related to the patterns observed for adrenergic stimulants. Differences between peaks and troughs in all cases were significant.

The influence of biorhythms on dl-amphetamine toxicity in isolated and aggregated mice was recently investigated. The greatest toxicity for both isolated and aggregated mice occurred during the dark phase while least toxicity was noted during the light phase (Table 1). For isolated animals, peak toxicity was observed at D-4 and D-7 (4th and 7th hour of darkness) during the months of June and January respectively. The differences between peak toxicities at different times of the year were significant ($P<0.05$). For aggregated animals *dl*-amphetamine was approximately 6 to 10 fold more toxic than for isolated animals during the same period of the year. The maximum changes in the twenty-four hour toxicity for isolated mice were 16 to 21 percent for the months of June and January respectively while a maximum 97 percent change in toxicity was observed

Table 1 Twenty-four hour seasonal LD50 values for DL-amphetamine in mice.

Time of day	Isolated				Aggregated	
	January LD50(mg/kg)±S.D.	% Change	June LD50(mg/kg)±S.D.	% Change	June LD50(mg/kg)±S.D.	% Change
0300	112.4± 8.10**	0.00	132.4±8.00	9.95	9.7±0.97**	0.00
0600	121.3±13.90	7.91	126.9±2.35	6.19	14.4±2.03	48.18
0900	128.9± 3.55	14.67	139.0±3.55**	16.31	19.1±2.85**	97.01
1200	134.5± 5.90**	20.55	134.9±5.50	12.88	16.6±2.35	70.80
1500	129.6±10.95	15.30	133.2±9.90	11.46	14.0±1.84	44.59
1800	128.4± 5.35	14.23	139.0±3.55**	16.31	16.7±3.32	71.88
2100	132.4± 8.00	17.79	127.5±8.40	6.69	13.6±5.63	39.85
2400	117.8± 7.40	4.80	119.5±1.80**	0.00	14.1±2.99	45.10

* Difference between maximum and minimum values significant at $P<0.01$.
** Difference between maximum and minimum values significant at $P<0.05$.

for aggregated mice. The suggested seasonal changes in addition to the daily biorhythmic pattern for isolated animals might be related to seasonal variations in specific body amine concentrations. A diurnal phase relation was noted in our laboratory for norepinephrine and serotonin [7] and GARATTINI and VALZELLI [8] reported a seasonal rhythm for serotonin and 5-hydroxy indole acetic acid in mouse brain tissue. These authors reported seasonal serotonin peaks during the summer months of June, July and August.

The great maximum percent difference in the twenty-four hour LD50 value for aggregated mice when compared to isolated mice might be related to endogenous norepinephrine stores. BLACK et al. [9] indicated a circadian rhythm in the ability of reserpine to deplete brain norepinephrine and suggested a more rapid turnover rate for brain norepinephrine during the dark period than during the light period. Nerve impulse frequency apparently increases during the dark period due to increased motor activity. Crowded conditions probably augments the sensory output resulting in a greater release and utilization of specific neurotransmitter amines at a time when the norepinephrine storage and synthesis capacities are maximal.

These studies indicate that drug responses can probably be influenced by biological rhythms involving physiological functions and body chemistry. The objectives of these investigations were to study some of the commonly used neuropharmacological compounds and determine the effects of the time of day on their potency and toxicity which might have implications in terms of greater drug safety and therapeutic effectiveness.

REFERENCES

1. FRIEDMAN A. H. and WALKER C. A. (1968): Circadian rhythms in rat midbrain and caudate nucleus biogenic amine levels. J. Physiol. (London), *197*: 77–85.
2. FRIEDMAN A. H. and WALKER C. A. (1968): Rat brain amines, blood histamine and glucose levels in relationship to circadian changes in sleep induced by pentobarbital sodium. J. Physiol. (London), *232*: 133–147.
3. WALKER C. A., SPECIALE S. G. and FRIEDMAN A. H. (1971): The influence of nor-

epinephrine levels and ultrastructure of rat hypothalamus and caudate nucleus during a programmed light-dark cycle. J. Neuropharm., *10*: 325–334.
4. Friedman A. H. and Walker C. A. (1969): Circadian rhythms in central acetylcholine and the toxicity of cholinergic drugs. Fed. Proc., *28*: 447.
5. Litchfield J. T. and Wilcoxon F. (1949): A simple method of evaluating dose-effect experiments. J. of Pharm. and Exp. Therap., *96*: 99.
6. Walker C. A. and Friedman A. H. (1970): Twenty-four hour toxicity rhythms of sympathomimetics in mice. Pharmacologist, *12*: 198.
7. Walker C. A. and Friedman A. H. (1967): Circadian rhythms of biogenic amine levels in the brain stem and caudate nucleus of the rat. Pharmacologist, *9*: 236.
8. Garattine S. and Valzelli L. (1965): Serotonin, Elsevier Publ. Co., Amsterdam.
9. Black I. B., Parker L. and Axelrod J. (1969): A daily rhythm in the rate of depletion of brain norepinephrine by reserpine. Biochem. Pharmacol., *18*: 2688–2691.

CIRCADIAN RHYTHM OF MITOSIS IN MOUSE CORNEAL EPITHELIUM: ALTERATIONS INDUCED BY ISOPROTERENOL*, **

E. Robert BURNS and Lawrence E. SCHEVING

Department of Anatomy
University of Arkansas Medical Center
Little Rock, Arkansas, U.S.A.

The mouse parotid gland and kidney are stimulated to undergo DNA synthesis after a single injection of isoproterenol (IPR) [1, 2]. The uptake of tritiated thymidine (^{3}H-TdR) by the parotid gland and kidney, as well as that of the duodenum, after a single injection of isoproterenol, varied with circadian frequency as did the uptake of ^{3}H-TdR by each of these organs in saline treated animals [3]. The effect of IPR has on the uptake of ^{3}H-TdR into mouse kidney and duodenum is circadian phase dependent; at one phase it may stimulate, at another it may inhibit or at still another phase there may be no significant effect. In the parotid gland, IPR significantly stimulates the uptake of ^{3}H-TdR at all phases of the mouse circadian system. However, the magnitude of the positive response of the parotid gland to a single injection of IPR was characterized by a significant circadian rhythm [3].

Because of IPR's effect on the uptake of ^{3}H-TdR and the marked influence the mouse circadian system has on this effect, an experiment was designed to test the effect of a single injection of IPR on the mitotic index of mouse corneal epithelium.

Mice were carefully standardized and kept in a light-dark cycle (light from 0600–1800 C.S.T.) for one to two weeks prior to and throughout the duration of the experiments. Food and water was available *ad libitum*. In the first experiment, 36 mice were injected intraperitoneally with 7.5 mg of IPR in 0.75 ml saline in subgroups of six mice each at 0900, 1300, 1700, 2100, 0100 and 0500. Identical subgroups were injected with 0.75 ml saline. Exactly 28 hrs after each injection of IPR or saline, the subgroups were killed and whole mounts of the cornea prepared. The mitotic index of the corneal epithelium was determined by counting 5000 cells/cornea and expressing the index as mitoses/1000 cells.

In the second experiment, groups of 85 to 90 standardized mice were injected with 7.5 mg IPR at 0900 or 2100. Identical groups of mice were injected with 0.75 ml saline and served as controls. Animals were killed in subgroups of five mice each every four hrs for 68 consecutive hours and the corneal mitotic indices were determined.

In the first experiment (IPR or saline injected every four hrs for one 24 hr period and the mice killed exactly 28 hrs later), the saline treated animals demonstrated a circadian rhythm in mitotic index of the corneal epithelium with a peak

* This research was supported by a grant from the Damon Runyon Memorial Fund for Cancer Research (E.R.B.) and a grant from the National Institutes of Health, AM 12389 (L.E.S.).

** The complete details of this work has been submitted for publication in the *Journal of Cell Biology*.

mitotic index at 0900 and the trough at 2100. This rhythm essentially was identical to that reported for the rat cornea [4, 5]. In the subgroups of mice treated with IPR (one subgroup every four hrs for 24 hrs) and killed 28 hrs later, the corneal epithelium mitotic indices at the six different time points were low and statistically similar to the trough level of the controls (2100). Thus in this particular experimental situation it *appeared* as if the normal circadian rhythm in mitotic index of corneal epithelium was completely abolished by a single injection of IPR.

The second experiment was designed to test for the duration of the "apparent" inhibition of mitosis by a single injection of IPR. This involved injection of IPR at 0900 or 2100 and following each at four hr intervals for 68 hrs. When IPR was injected at 0900 and subgroups of mice killed every four hrs for 68 consecutive hrs, the circadian rhythm in corneal epithelium mitotic index was present, but the peak of the rhtyhm was advanced by eight hrs on the first day when compared to the control peak (Table 1). On the second and third days after IPR injection, the peak occurred four hrs prior to the control peaks at 0900 (Table 1). The times of occurrence of the troughs, however, were in phase with the control troughs, but were prolonged for up to 16 hrs (Table 1). When IPR was injected at 2100, the mitotic index showed a phasing on the first two days after injection similar to controls, however, on the third day in the IPR treated animals, the peak was advanced by four hrs when compared to the controls (Table 1). All of the peaks in the control as well as the IPR treated groups were sharp when compared to the troughs which were prolonged in the IPR treated animals (Table 1).

Table 1 Times of peak and trough mean mitotic indices during the three days after a single injection of isoproterenol.

Treatment	Days		After		Injection	
	First		Second		Third	
	Peak	Trough	Peak	Trough	Peak	Trough
Control	0900* (6.9)**	1700–2100 (2.2–1.7)	0900 (5.64)	1700–2100 (0.89–1.96)	0900 (5.05)	1700 (0.60)
IPR at 0900	0100 (6.08)	0900–2100 (2.64–1.32)	0500 (10.8)	1900–0100 (2.0–2.0)	0500 (9.5)	1700 (0.68)
IPR at 2100	0900 (6.56)	1700–0100 (1.05–0.67)	0900 (8.20)	2100 (1.40)	0500 (5.95)	1300–1700 (1.36–0.56)

* Time of day
** Mitotic index (mitosis/1000 cells counted).

Table 2 is a rhythmometric summary of the fit of the data to a 24 hr cosine curve; a very significant fit was obtained for all three experimental conditions. It is emphasized that this summary does not adequately demonstrate the day to day phase shifts described above since each rhythmic parameter represents only the averages. It probably is because of the averaging that the confidence intervals of each acrophase slightly overlap. The summary also does not reflect any day by day changes in levels or amplitude, however, even when an analysis of the levels of the mitotic index were analysed by conventional statistical methods, for each day there was no statistically significant difference between injected and control

Table 2 Rhythmometric summary* of data obtained from corneal epithelial mitoses/1000 cells in two groups of animals treated with isoproterenol at 0900 and 2100 and a control group; the data was fit to a 24–hour cosine curve. The documented span in hours is 64.0.

Variable**	Low & high values		90% Range	Coef. Var.	P	PR	Mesor +S.E.	Amplitude +S.E.	Acrophase		
									Degress +S.E.	Clock hour	(.95 CA)
Control	0.5	7.6	0.8– 7.6	66.79	.009	0.70	3.18±.29	2.44±.41	−113± 9	0733	0617–0849
IPR at 0900	0.6	10.7	0.6–10.7	86.79	.009	0.53	3.31±.49	2.88±.69	− 77±13	0510	0322–0658
IPR at 2100	0.5	8.1	0.6– 8.1	79.09	.009	0.48	2.93±.45	2.39±.66	−144±14	0939	0742–1137

* Kindly provided by Professor Franz Halberg, Chronobiology Laboratories, University of Minnesota.

** mitotic index=mitoses/1000 cells counted.

P=Result of testing zero amplitude=probability of obtaining estimated parameters if sinusoidal rhythm with stated period were absent.

PR=Variability ratio=percent of total variability contributed by fitted curve=sum of squared deviations from mean (ss), of values derived, from fitted cosine curve at sampling times/ss of data themselves ×100.

Mesor=Rhythm-determined average.

animals. We conclude that the principal effect of IPR on the mitotic rhythm is to phase shift the normal circadian rhythm in mitotic index in corneal epithelium. The data presented in Table 2 serve to quantify the three circadian parameters of the mitotic index in mouse corneal epithelium.

This investigation clearly demonstrates the danger of single time point sampling in this circadian system when studying IPR and, most assuredly, other pharmaceuticals as well.

REFERENCES

1. Baserga R. (1970): Induction of DNA synthesis by a purified chemical compound. Fed. Proc., *29*: 1443–1446.
2. Malamud D. and Malt R. A. (1971): Stimulation of cell proliferation in mouse kidney by isoproterenol. Lab. Invest., *24*: 140–143.
3. Burns E. R., Scheving L. E. and Tsai T. (1972): Circadian rhythms in the total uptake of tritiated thymidine into parotid gland, kidney and duodenum in saline and isoproterenol treated mice. Science, *175*: 71–73.
4. Scheving L. E. and Pauly J. E. (1967): Circadian phase relationships of thymidine-H^3 uptake, labeled nuclei, grain counts, and cell division rate in rat corneal epithelium. J. Cell Biol., *32*: 677–683.
5. Scheving L. E. and Pauly J. E. (1967): Effects of adrenalectomy, adrenal medullectomy and hypophysectomy on the daily mitotic rhythm in corneal epithelium of the rat. *In*: The Cellular Aspects of Biorhythms, (H. von Mayersbach, ed.), pp. 167–174, Springer-Verlag, Berlin.

VARIATION IN SUSCEPTIBILITY OF MICE TO THE CARCINOSTATIC AGENT ARABINOSYL CYTOSINE*

L. E. SCHEVING[1], S. S. CARDOSO[2], J. E. PAULY[1], F. HALBERG[3] and E. HAUS[4]

[1]*Department of Anatomy, University of Arkansas Medical Center Little Rock, Arkansas*
[2]*Department of Pharmacology, University of Tennessee Medical Center Memphis, Tennessee*
[3]*Chronobiology Laboratories, University of Minnesota Minneapolis, Minnesota.*
[4]*Department of Anatomic and Clinical Pathology, St. Paul-Ramsey Hospital and Medical Center St. Paul, Minnesota, U.S.A.*

INTRODUCTION

The antimetabolite, arabinosyl cystosine (ara-C), currently in very extensive clinical use [1] was administered in a single daily dose at a presumably fixed circadian system phase to BDF_1 mice for either five or six consecutive days to determine any circadian and other changes in susceptibility.

METHODS

Five experiments are summarized here (Fig. 1): the first three were conducted while the senior author was at the Louisiana State University Medical School in New Orleans, Louisiana, and another two studies after he had moved to the University of Arkansas Medical Center at Little Rock.

For the first three experiments, male BDF_1 mice were obtained from Southern Animal Farms of Prattville, Alabama. The animals were shipped just after weaning and averaged 12 g at the time of arrival. For seven days prior to the beginning of each study the mice were singly housed in artificial light from 0600 to 1800 alternating with darkness. Food and water were available *ad libitum.* During the standardization span the animals gained weight and averaged about 16 g at the time of the initial injection.

The *first toxicity experiment* (I) was conducted in March, 1970. Beginning at 0830, March 15, 1970, fifteen animals were injected intraperitoneally with 400 mg/kg ("low-dosage") of arabinosyl cytosine (ara-C), and a comparable group was injected with 800 mg/kg ("high dosage"). This procedure was repeated on different subgroups of fifteen mice each at 1300, 1830, 2300 and 0300. Thus ten subgroups received "low" or "high" dosage of ara-C at defined circadian system phases. Beginning on the second day at 0830, the subgroup injected on the previous day at the same time was given one half the initial dosage (200 or 400) for four more consecutive days. Exactly the same procedure was followed for the

* Supported by grants from the U.S.P.H.S. (5-K6-13,981), (AM, 12389), NSF-GW 7613 and NASA (NAS 9-12338) and by the St. Paul-Ramsey Medical Research and Education Foundation. Mr. S. Poiley, National Cancer Institute, National Institutes of Health, Bethesda, Maryland, kindly contributed the mice used.

1300, 1830, 2300 and 0300 subgroups of mice. (The animals on the high dosage of ara-C thus received a total of 2400 mg/kg and those on the low dosage a total of 1200 mg/kg divided in five injections over a five-day span.) The number of deaths was recorded each day for 13 days after the last injection when all survivors looked perfectly healthy and had regained their initial weight. No animals died after the ninth day following the initial injection.

The *second toxicity experiment* (II) was conducted in April, 1970. The procedures described in the first experiment were followed except that only one dose level was used. The initial dosage at each time point was 400 mg/kg and at all subsequent time points it was 200 mg/kg. The injections began at 1300 on April 11, 1970, and were continued through the sixth day rather than the fifth as in the first experiment. (The total dose of ara-C in this study thus was 1400 mg/kg.) No deaths occurred after the eleventh day following the initial injection.

The *third toxicity experiment* (III) began May 10, 1970, and was carried out exactly as the initial experiment except that the drug again was administered for 6 rather than 5 days. The same two daily dose levels were used leading to a total dose of 1400 mg/kg and 2800 mg/kg.

Toxicity experiments four (IV) and *five* (V) were begun on January 23, 1971, and February 23, 1971, respectively. The same procedure as described in the first experiment was followed except that the initial injection was given at 1830 rather than 0830; again, the duration of treatment was six days. The total dose of ara-C given was thus 1400 mg/kg and 2800 mg/kg respectively. (The BDF_1 mice for these studies were obtained from Rawley Farms of Plymouth, Michigan, and were slightly larger than those used in Experiments I, II and III, weighing at the time of the initial injection 25 g in Experiment IV and 21 g in Experiment V.)

The same person administered injections in all five experiments. In all cases, they were given in the lower right intraperitoneal space, with saline being routinely added to the drug just prior to injection. The same drug lot was used throughout the five experiments. At night, injections were made in dim light.

In all studies, animals were weighed repeatedly, and weight losses between the initial and final injections were recorded.

Mortality data were fitted to a 24-hour cosine curve by the method of least squares to obtain the following parameters: a) Mesor (M)—the rhythm-adjusted, computer-determined, over-all 24-hour average value (equal to the arithmetic mean if the data are equidistant and cover an integral number of periods); b) Amplitude (A)—the distance from the mesor (M) to the peak of the cosine curve expressed in the same units as the variable analyzed; c) Acrophase (ϕ)—the delay in hours or degrees from the acrophase reference (middle of the daily light span [1200 CST]) of the peak in the 24-hr cosine curve best approximating all data. Since 360° is equated to 24 hours, 15°=1 hour. The error estimates for each of these parameters were obtained by the electronic computer methods used.

RESULTS AND DISCUSSION

The toxicity of ara-C at different circadian system phases of young adult BDF_1 mice was gauged by the percentage of animals dying during or after administration of the drug. The results expressed in percent mortality for each time point of injection are summarized in Fig. 1. It becomes apparent that the mortality after ara-C injections varies in all six studies and for both dosages tested as a

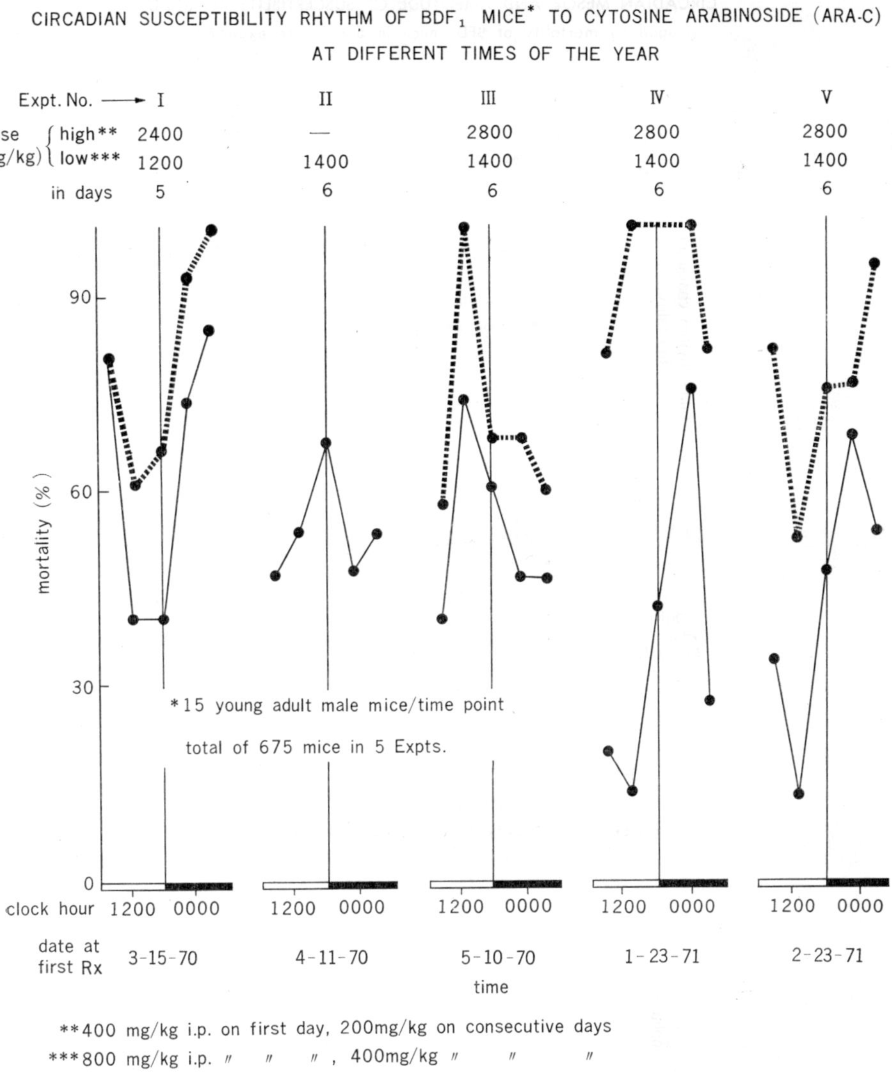

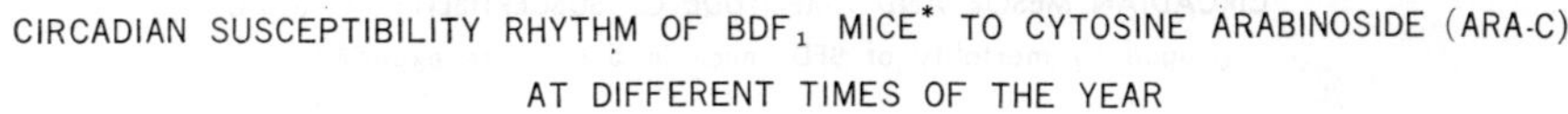

Fig. 1

function of drug administration time. However, the times of high and low mortality after injection of the same dose of ara-C are not the same in all studies.

In the first experiment conducted in March, mortality from the lower dosage was high (86%) for animals injected at 0300 and 2300; mortality was lower (40%) for the mice injected at 1300 and 1800 (Fig. 1). With the higher dosage the same general trend was seen, with a 100% and 93% mortality at 0300 and 2300 and 60% and 66% mortality at 1300 and 1800.

In the second experiment, about one month later in April, with only one dose level, and in the third experiment, another month later in May, with two dose levels, the highest mortality (67% and 73% respectively) occurred in the mice injected at 1830 and at 1300, the minimum (40%) at 0830 (Fig. 1). With the higher dosage in the third experiment, the highest mortality of 100% occurred also at 1300, the lowest (57%) at 0830.

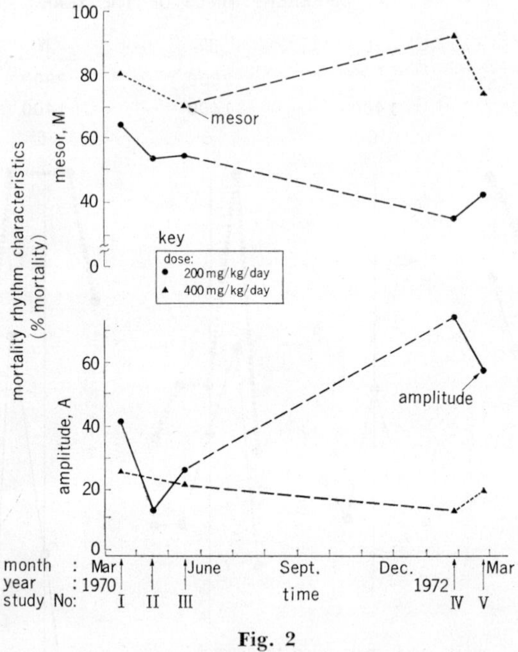

Fig. 2

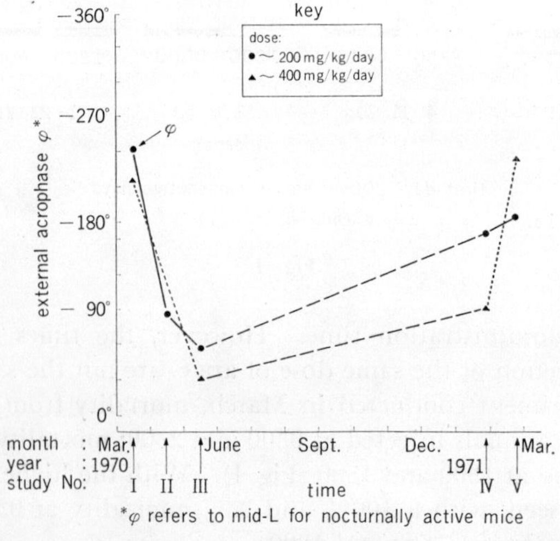

Fig. 3

In the fourth experiment conducted in January and in the fifth conducted during the last week of February, the lowest mortality was found at 1300 (Fig. 1). With both dosages the highest mortality occurred during the dark span at 2300 or at 0300 respectively, thus resembling the timing found in the first experiment.

In viewing the first five experiments as a whole, a 4-gram weight loss of the treated animals (from the initial to the final injection) was the average for all time points of injection.

In examining the mortality of the animals injected at different times along the 24-hour scale, it becomes obvious that the considerable differences in circadian timing observed among the studies done at different times of the year cannot be ignored and that a pooling of the results of these studies for evaluation would thus be misleading.

A rhythmometric summary, obtained by the least squares fit of a 24-hour cosine function to each series of the six experiments is shown in Figs. 2 and 3. The difference in the mesor (rhythm-adjusted average) expresses the difference in mortality between low and high dosage. From one study to the next, the relative amplitudes vary greatly, particularly for the low dosage (Fig. 2). The acrophase, shown in Fig. 3, represents the mathematically determined times of higher mortality, namely, the lag from the acrophase reference of the peak in the cosine function best fitting all data; it is expressed in degrees from the midpoint of the light span (1200 CST) as zero reference (0°). This acrophase of the circadian susceptibility cycle of the BDF_1 mice to ara-C changes drastically in the course of the time span during which the five toxicity experiments were conducted. As yet, we cannot ascertain whether this change in timing occurs at random or follows an about-yearly and/or other low-frequency pattern. A follow-up on these findings over several years will be necessary, if indeed a circannual rhythm in timing of susceptibility to ara-C occurs, in order to define its synchronization or lack of synchronization with the seasons of the calendar year. However, the possibility remains that some unknown variable other than an about-yearly rhythm contributes to the obtained results.

SUMMARY

A variation in susceptibility of BDF_1 mice to 1200 to 2800 mg of arabinosyl cytosine (ara-C) given at defined circadian system stages, divided over 5 or 6 consecutive days, was found in five separate studies. Differences in timing between the experiments conducted during different times of the year indicate effects from superimposed factors, including, possibly, a circannual susceptibility rhythm.

REFERENCE

1. Skipper H. E., Schabel F. M. Jr. and Wilcox W. S. (1967): Cancer Chemotherapy Reports, No. 3, 51.

CIRCADIAN AND CIRCANNUAL MAPS

Chairman: Lawrence E. SCHEVING

Co-Chairman: Günther HILDEBRANDT

CHRONOBIOLOGY

LAWRENCE E. SCHEVING

Department of Anatomy, University of Arkansas Medical Center
Little Rock, Arkansas, U.S.A.

During the past two decades the science of chronobiology has emerged as a distinct subdivision of biology and has as its objective the study of the temporal aspects of plants and animals. Although a comparatively young science it is documented by an impressive body of scientific data which already has yielded new insight into fundamental biology.

The facts clearly demonstrate that practically every physiological or behavioral variable amenable to measurement is characterized by being rhythmic, at least in the healthy organism. The spectrum of rhythms is broad. The one frequency that has received the greatest attention in the past few years is that which approximates the frequency of the rotation of the earth, or the circadian. It is this frequency, primarily in the mammalian organism, upon which I will comment. Fluctuations of most physiological variables are not apparent to us in the same sense that the respiratory or menstrual rhythm is; they only become overt when they are properly measured at frequent intervals over the 24-hour time scale. Because of their somewhat "invisible" nature there is a tendency, on the part of some investigators, to slight or to ignore them in their experimental design. In spite of all that is known they simply are not being accorded the attention they deserve; this undoubtedly is in large part due to the fact that the science is young. It is my considered opinion that the following attitudes characterize what some scientists believe regarding this phenomenon:

(1) Some still rationalize that they represent no more than minor fluctuations around a daily mean and consequently do not warrent the additional amount of work and expense required to properly explore for their significance. Those who believe this, simply are ignoring the scientific facts available to them. Admittedly oral or rectal temperature may change only one or two degrees over the 24-hour time scale; there are, however, few physiologists who would minimize the effect that this small amplitude rhythm has on metabolic or physiological activity. On the other hand the corticosterone concentration in the plasma of rats may vary three or four-fold; the same applies to many enzymes in the liver and plasma. When 5-hydroxytryptamine levels in the whole brain are measured at frequent intervals over a 24-hour period, the difference between the lowest and highest recorded mean may represent only an 18% change, but the same substance in the pineal may vary as much as 900%. When similarly compared the cell division rate in a number of mitotically labile tissues, such as the cornea, may represent over a 1200% increase. Generally fluctuations in man are equally as great as in the rat; the evidence to support the above statements is extensively documented. Because the biological system is continually changing, the organism is a different biochemical entity at different phases of its circadian system. If the organism is subjected to a potentially noxious agent at one time, it may succumb to the agent,

whereas at another phase of the circadian system it may have very little effect. This circadian differential in response has led to the concept of the *Hours of Changing Resistance.* The overwhelming evidence that supports this concept necessitates total rejection of the idea that circadian time structure represents no more than minor (and hence unimportant) fluctuations around a daily mean. Furthermore, the facts render completely untenable the concept of homeostatic balance as is presently taught in many freshman medical school physiology courses.

(2) Others feel that circadian rhythms represent nothing more than day-night changes, or that they are some kind of direct response to the ingestion of a meal, and that if one simply samples once during the day and once during the night he will have in some vague way taken care of this nuisance variable. This attitude toward rhythmicity simply cannot be defended, but it must be recognized that it still is held by a few investigators. The facts demonstrate that oscillations are responsive to changes in the environment but are endogenous and inherited. Teleologically their properties provide the mechanisms that enable the organism to adjust to geophysical changes.

(3) A large number of workers sample each day at the same time of day, ostensibly to avoid or minimize rhythms. In view of what has been learned about their properties, such as the endogenous nature, the ability to phase shift with lighting changes or to freerun in continuous light or darkness, it is this widespread practice that is most difficult for the chronobiologist to accept as the means for avoiding the effects of rhythms. All that this practice does is to assure an investigator that he is sampling in the trough, on the incline, at the peak or on the decline; he can only be sure of this if he has first gone to the trouble of mapping his rhythm under controlled conditions, i.e., if synchronizing cycles such as those of light and darkness on a fixed alternating schedule are carefully controlled; even then the circadian system phase will shift with the changes in season so that sampling at 0900 June 1, will not be the same as sampling at 0900 Jan. 1. This practice can lead the investigator into pitfalls.

(4) There are many excellent investigators who have at least an intuitive appreciation of the importance of rhythms and would explore for their effects in depth, but find that proper control of the animal quarters simply cannot be obtained. Most animal facilities are geared more to housekeeping than to carefully controlling lights, etc. I believe this is a major reason for staying away from this type of investigation; it certainly is the most frequent excuse given. Recently I attended a lecture by a young investigator (successful enough to be funded by three agencies) who was reporting on the neural control of ovulation. To my amazement he admitted that he did not pay much attention to the circadian time structure of his female rats, but felt that to do so might be beneficial. When I urged him to seriously consider such an approach he shrugged his shoulders and stated "I have enough trouble with the animal care people now; I could never get them to control lights, etc." This scientist did his research in a prestigious institution that is supported by large federal appropriations. I could not help but feel that this young man was working against himself by ignoring the circadian and other rhythmic frequencies *known* to characterize the system he was exploring, with the important objective of elucidating the mechanism of neural control of ovulation.

An attitude prevails among a few scientists that the only significant or relevant work in chronobiology is that which revolves around locating the controlling mechanism or the so-called "biological clock." I readily admit that explaining the mechanism of rhythmicity would be as big a scientific breakthrough as if one were to come up with an adequate explanation to the mechanism of cell division; to date no significant progress has been made along this line. Pharmacologists exploring the mechanism of drug action might have great success if they considered the concept of *hours of changing resistance*; again there has been little effort in this direction. Cancer chemotherapy has not attempted to utilize timed therapy to improve the action of drugs with potent side effects; this certainly should be attempted.

I would predict that the property of an organism to oscillate eventually will come to be considered as fundamental a property of all life as is irritability at the present time. Also I predict that the control of this variable in scientific investigation will become a *sine qua non* for serious research. If this is done chronobiology could play a significant role in unraveling some of the yet unexplained mysteries of biology. Although still very inadequate, the present body of knowledge on temporal biology does not justify the ignoring of this phenomenon. It should be given highest priority when considering the variables to be controlled in research planning.

The question arises as to what can be done to enable chronobiology to to gain the recognition and status it deserves. I offer the following suggestions:

1. First the leadership in the biological scientific community must critically examine the body of knowledge upon which chronobiology presently is founded. I assume that if this were done it would lead one to enthusiastically recognize the potential that this science has to aid the understanding of fundamental biology. Once this was recognized they would have to emphasize the importance of including in proposed research protocol the design necessary for carefully controlling and exploring the possible effect of rhythms on the system being investigated—with such an approach on a large scale, the spin-off could be rewarding.

2. There must be frequent conferences to permit exchange of ideas. In the past, conferences, especially in the U.S., have been sporadic and frequently restricted in scope and to a selected few because of limited funds. Chronobiologists should come together at least annually in much athe same manner as do cell biologists and biochemists.

3. Those working in the field must adopt a unified methodology which would enable one to qualify rhythms. I believe that adequate methods already are available and, although documented in the literature, they are not yet widely adopted.

4. An official journal which can serve to educate other scientists regarding progress in the field is essential.

CIRCADIAN AND CIRCANNUAL MAPS FOR HUMAN SALIVA*

Colin Dawes
Department of Oral Biology, Faculty of Dentistry
University of Manitoba
Winnipeg, Canada

In most studies of rhythms in human salivary flow rate or composition, the results have not been subjected to the type of statistical analysis which gives quantitative information about the presence or characteristics of rhythms.

In addition, many factors such as flow rate, duration of stimulation, the method of saliva collection, nature of the stimulus and serial dependency of sampling can influence salivary composition. Inadequate standardization of such variables will reduce the possibility of rhythm detection.

The only established salivary circannual rhythm is one for the flow rate of unstimulated parotid saliva [1]. Saliva was collected by a well standardized procedure on one occasion, before breakfast from 3,868 healthy, male, U.S. Air Force recruits. The collections were spaced fairly evenly throughout the 12 months of the year and Table 1 shows the results obtained by a least-squares cosine fit of the mean flow rates for each month. The minimum flow rate in July was attributed by Shannon to the fact that the subjects, situated in Texas, may have been dehydrated at that time of year. No objective evidence was available to support this possibility.

The best macroscopically established circadian rhythm in human saliva is that for the sodium concentration in unstimulated whole saliva, maximum values occurring early in the morning [2].

The results shown for 8 subjects in Table 1 are mean values derived by a least-squares cosine fit to data obtained over time spans of between 4 and 26 days. Five min samples of unstimulated and ten min samples of stimulated parotid saliva were collected about five times per day, prior to meals. The parotid saliva was collected at a constant flow rate of 1.0 ml/min and only the final 5 ml were obtained for analysis.

Statistically significant rhythms, for the group, as tested by the cosinor method [3], could not be detected for unstimulated whole saliva protein, potassium, calcium or phosphate concentrations or for stimulated parotid phosphate concentration. The constituents for which a significant circadian rhythm was detected by cosinor are shown in Table 1.

The failure to detect any rhythm by cosinor does not imply that rhythms do not occur in a given set of time series analyzed but it could be that any rhythms that might characterize the data are not synchronized between individuals in period and/or phase or they differ markedly in amplitude [4]. Moreover, unstimulated whole saliva is derived from a number of glands and it is possible that

* This work was supported by Grant MA 3610 from the Medical Research Council of Canada.

Table 1 Circadian and circannual rhythm parameters for saliva from presumably healthy human adults.

Rhythm investigated	Type of saliva	Components (Units)	No. & sex of subject	Level C_0 ±S.E.	Amplitude C ±S.E.	Acrophase ϕ (in degrees) ±S.E.	C.V.† ±S.E.
Circadian		Oral Temp (°C)	5♂, 3♀	36.7±0.07	0.38±0.07	−250±4	34±4
	Unstimulated Whole Saliva	Flow Rate (ml/m)	〃	0.49±0.09	0.19±0.03	−231±11	34±5
		Na (mM)	〃	6.2 ±1.2	4.3 ±1.2	−70 ±3	46±7
		Na/K	〃	0.29±0.05	0.18±0.06	−67 ±3	46±7
		Cl (mM)	〃	17.4±1.4	4.8 ±0.9	−74 ±6	42±4
	Stimulated parotid saliva	Protein (mg%)	〃	270±29	81±14	−238±7	32±7
		Na (mM)	〃	29.0±4.2	4.2±0.7	−76±9	23±6
		K (mM)	〃	22.0±0.8	1.4±0.2	−260±9	22±4
		Na/K	〃	1.38±0.23	0.28±0.04	−78±8	31±6
		Ca (mM)	〃	1.02±2.8	0.04±0.01	−288±14	7±2
		Cl (mM(	〃	17.1±2.8	2.0±0.05	−76±8	16±4
		Corticoids (μg/%) (E+F)	5♂	—	—	−107	—
Circannual	Unstimulated Parotid Saliva	Flow Rate (ml/m/gland)	3868 ♂	0.0381± 0.0006	0.0079± 0.0009	−17*±6	90

† =% of total variability accounted for by the rhythm.
* =Phase referenced to January 1st.
Values for C_0, C, ϕ & C.V. are the means ±S.E. of values obtained for each individual by a least-squares fit of a 24 h cosine curve to the individual time series.

different glands, although they may be producing saliva rhythmically, are not synchronized in their frequency and/or phase of production. With respect to the failure to detect a significant potassium rhythm in unstimulated whole saliva, whereas a potassium rhythm was detectable in stimulated parotid saliva, the added fact may be considered that parotid saliva usually contains more potassium than does submandibular saliva but the proportional contributions from these glands to whole saliva may not be constant.

The circadian rhythm in the flow rate of unstimulated whole saliva is of possible clinical significance. It is well known that during sleep the salivary flow rate is very low [5] and it would seem that the most important time of day to carry out oral hygiene procedures would be before going to bed. During sleep the protective effects of saliva, such as its buffering capacity, will be largely absent.

The acrophases for sodium and potassium in parotid saliva are almost exactly 180° out of phase with each other and could be due to the known rhythm [6], in the plasma concentration of aldosterone. The Na/K ratio is known to be very low in patients with primary aldosteronism [7] and returns to normal after successful removal of aldosterone-secreting tumors. The rhythm in sodium concentration is absent in patients with Addison's disease [8].

The acrophase for salivary corticoids (E+F) derived from the data of Katz and Shannon [9], agrees with published values for plasma corticoids, which supports Katz and Shannon's suggestion that salivary corticoid levels tend to reflect those in plasma.

The chloride acrophase is similar to that for sodium and this is presumably due to the fact that chloride reabsorption in the salivary ducts tends to follow passively that of sodium.

The cause of the prominant rhythm in stimulated parotid protein concentration is uncertain but the physiological advantage may be that the acrophase is close to the time at which the main meal of the day is normally consumed.

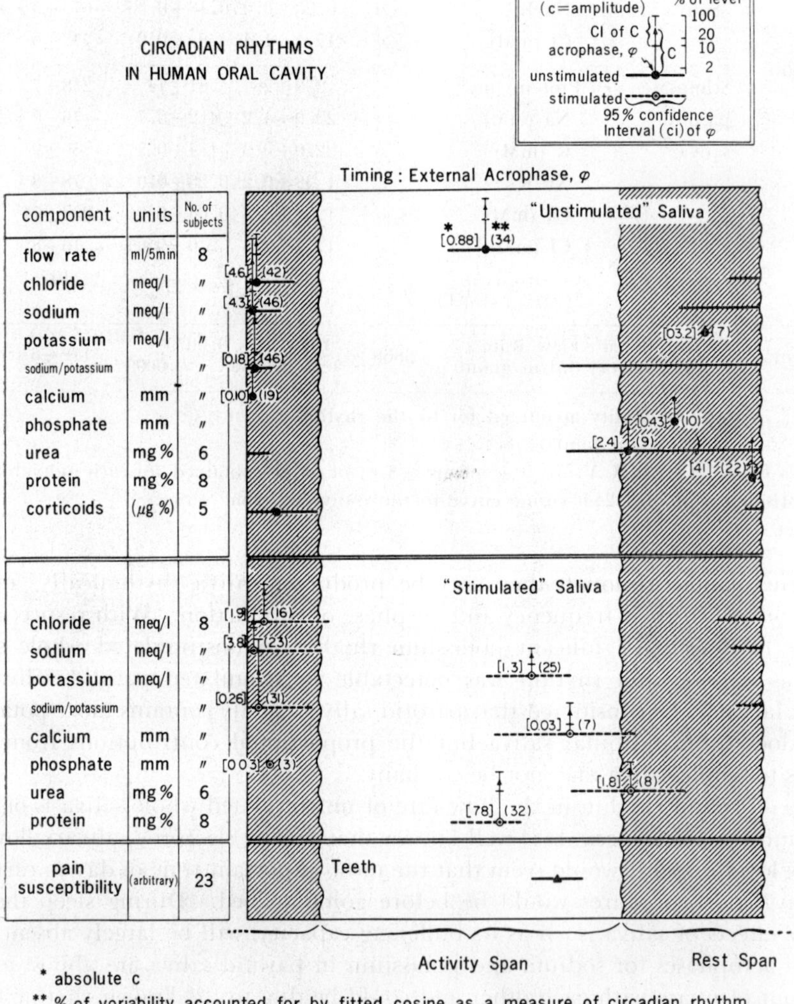

Fig. 1

Fig. 1 depicts as a map amplitudes and acrophases of rhythms in the oral cavity, including a behavioral one (pain threshold to electrical stimulation of teeth; see Reinberg et al. [10] for analysis and Jores (1937) for the original data).

REFERENCES

1. Shannon I. L. (1966): Archs. oral Biol., *11*: 451–453.
2. Grad B. (1954): J. Geront., *9*: 276–286.
3. Halberg F., Tong Y. and Johnson E. (1967): *In*: The Cellular Aspects of Biorhythms, (H. von Mayersbach, ed.), pp. 20–48, Springer Verlag, Berlin.
4. Halberg F. (1966): Scientia, *101*: 412–419.
5. Schneyer L. H., Pigman W., Hanahan L. and Gilmore R. W. (1956): J. dent. Res., *35*: 109–114.
6. Wesson L. G. Jr. (1964): Medicine (Baltimore), *43*: 547–592.
7. Wotman S., Goodwin F. J., Mandel I. D. and Laragh J. H. (1969): Archs. intern. Med., *124*: 477–480.
8. Pawan G. L. S. (1955): Biochem. J., *60*: xii.
9. Katz F. and Shannon I. L. (1969): J. Clin. Invest., *48*: 848–855.
10. Reinberg A., Zagulla-Mally Z. W., Ghata J. and Halberg F. (1967): Proc. Soc. exp. Biol. Med., *124*: 826–832.

COSINOR MAPPING OF PHYSIOLOGIC AND PSYCHOLOGIC VARIABLES IN 10 HEALTHY MEN BEFORE AND DURING BALNEOTHERAPY*

ROBERT GÜNTHER, EDWIN KNAPP, ERHARD HAUS and FRANZ HALBERG

Department of Medicine, University of Innsbruck, Innsbruck, Austria
Department of Anatomic and Clinical Pathology
St. Paul-Ramsey Hospital, St. Paul, Minnesota, U.S.A. and
Chronobiology Laboratories, University of Minnesota
Minneapolis, Minnesota, U.S.A.

INTRODUCTION

Circadian rhythms characterize most physiologic and psychologic functions commonly studied in clinical medicine. Thus, arises the need to compare any simple casual sample with a range of "usual" values qualified for an individual's physiologic time. Moreover, from multiple samples, rhythm parameters can be quantified by inferential statistical methods of rhythmometry, applied with the help of an electronic computer; rhythm characteristics serve as new endpoints for the study of human adjustment and response to environmental stimuli as well as for the early detection of occult disease.

The following study was undertaken to (1) map with the help of cosinor analysis a number of circadian rhythms of physiologic and psychologic functions in a group of 10 clinically healthy young male volunteers living under relatively standardized conditions and (2) to study the possible effects on circadian rhythm parameters of a hydrotherapeutic treatment used during about 20 centuries for a variety of disorders by local patients and tourists in one of the spas of Central Europe. Methodologic details and some results of this study were the subject of several previous communications [1–3].

SUBJECTS AND METHODS

Data from ten clinically healthy male medical students over two consecutive 25-day spans will here be considered. The average age of the group was 23 years, with a range of 22 to 26 years of age. Prior to the study the individuals had a conventional physical and laboratory examination, with a chest x-ray, electrocardiogram, complete blood count, urinalysis and an erythrocyte sedimentation rate—all found to be within the "normal limits".

The subjects followed a diurnal activity and nocturnal rest pattern. They registered the time of rising and the time of retiring as physiologic reference functions for a time span extending from May 9 to November 30, 1966. On July 15, 1966 the students traveled to Badgastein, Austria, without crossing any time

* Work supported by USPHS (5-K6-GM-13,981 and 1RO1-CA-14445-01), NSF GW-7613, NASA and St. Paul Ramsey Hospital Medical Research and Education Foundation.

zones. There they were housed in the Badehospiz, a minimum-care medical facility designed for patients undergoing treatment in the spa.

Among the numerous functions studied the following variables will be considered in this report:

1. Oral temperature, measured with calibrated fever thermometers kept sublingually for eight min for each measurement.

2. The systolic and the diastolic blood presure. The blood pressure was measured with a calibrated mercury-hemosphygmomanometer using a 12-cm wide cuff applied with its lower end about 6 cm above the elbow in a sitting position. The cuff was inflated to about 30 mm of mercury above the expected systolic blood pressure value and then was slowly deflated. The systolic pressure was registered at the first appearance of Korotkoff sound and the diastolic pressure was recorded at the moment of disappearance of the sounds. The mean of three consecutive measurements was used as the value of a given test time in this rhythmometric evaluation.

3. Urine was collected in four portions per day. The sampling spans were 0700 to 1300, 1300 to 1600, 1600 to 2200, and 2200 to 0700. Volume was measured and the pH determined with a pH-meter.

4. Sodium and potassium in urine were determined in duplicate aliquots by the flame photometric micromethod of Stamm and Herrmann [4]. The excretion of these electrolytes was expressed in mEq/h. The Na/K quotient was calculated. The pH determination and all chemical measurements were done in duplicates.

5. A subjective estimation of mood and of vigor was registered by each student at the time of each sampling session. A scoring system with seven steps from 1 to 7 was used, 1 being the least active or the lowest mood rating, resp., and 7 being the score for highest activity or best mood. The first score after rising in the morning served as point of reference for the estimation of mood and of vigor on any given day. Grip strength was measured with a JAMAR-DYNAMOMETER [5] in standing position with arm and hand extended and freely hanging. Three consecutive measurements were performed using the right hand; the highest of the three values was recorded for evaluation in this study.

Sampling was begun on July 16, 1966. During the first 25 days of the study the subjects rose at 0645. Measurements were taken immediately after rising and at 0830, 1000, 1130, 1300, 1600, 1730, 1900 and 2030. Breakfast was scheduled for shortly after 0700, lunch at 1200 and dinner at 1930, and the subjects were to retire after 2200, yet there was recorded individual variation in schedules—analysed to obtain an objective mid-sleep time, shown as the synchronized time in the graphs. The sampling was limited to the waking hours; the nightly sleep span of the subjects remained undisturbed.

Beginning on August 10, 1966, until September 3, 1966, the subjects were exposed daily with the exception of Sunday beginning at 0500 for a duration of exactly 20 min to the thermal water of the spa. The water temperature in the bath tub was 37°C. The thermal water used contained at the time of the treatment an average of 10×10^{-9} Curie (c) of radon per liter. The relation of radon to radium A, B and C in the water was 1:0.8:0.4:0.3. The thoron content in the water was estimated at about 10 picocurie (pc) per liter [6].

The content of radon in the open air in Badgastein is about 2.6 pc per liter (as compared to levels of 0.1 pc in other parts of Austria). In the building where the students were housed the radon content of the air was an average of 4 pc per liter, rising in the therapy rooms to levels of 7–8 pc per liter.

The results obtained during the two 25-day time spans were analyzed separately. The rhythm parameters of each function in each individual were estimated by the least squares fit of a cosine function for a quantification and inferential statistical evaluation for the entire group. Results were obtained with the help of a Varian 620 electronic computer by cosinor analysis [7, 8].

RESULTS AND DISCUSSION

By the mean cosinor technique, the occurrence of a circadian rhythm could be verified for all functions investigated during both 25-day spans.

A statistically significant circadian rhythm was described both for systolic and for diastolic blood pressure. This inference is based upon a lack of pole-overlap by the error ellipses, Figs. 1 and 2. An acrophase shift related probably to the earlier time of rising can be detected for the circadian rhythm of the diastolic blood pressure but not for that of the systolic pressure—an effect that can be obscured by variations with a frequency-lower-than circadian, including circannual and other infradian rhythms.

Body temperature, grip strength, mood and vigor show the expected circadian rhythms without any detectable effect of the thermal baths (Table 1). A slightly higher mesor (rhythm-determined mean) of grip strength in the second time span is interpreted as training effect.

The evening acrophases of both systolic and diastolic blood pressure in this group are similar to those observed in other studies on mesornormotensive [9, 15] and mesorhypertensive [10] individuals.

All urinary functions studied showed significant circadian rhythms. Circadian renal periodicity has been recognized early and has been observed in many studies as summarized by Conroy and Mills [11]. Quantitative data on rhythm parameters, however, are considerably more limited [12–15]. The rhythm characteristics for the two 25-day spans are similar, for all urinary rhythms studied. The ealier acrophases during the second 25-day span represent largely a phase-shift due to the earlier rising of the subjects taking the thermal baths at 0500. The intergroup differences in acrophases in Fig. 2 and Table 1 are for most functions of less than two hours (30°) and could indicate a gradual rather than an instantaneous phase-shift, quite apart from the effect of infradian [16] rhythms. Since the second time span was analyzed as a whole the shift time and the completeness of the phase shift are not here assessed. There is no indication of any effect of the daily thermal baths upon the urinary rhythms studied.

SUMMARY

The circadian rhythms of oral temperature, systolic and diastolic blood pressure, of a subjective mood and vigor rating and of urinary volume, urinary pH, sodium and potassium excretion and of the grip strength of the right hand were mapped in 10 clinically healthy male students of an average age of 23 years (range 20–26) under standardized living conditions during two consecutive 25-day spans. By mean cosinor analysis, statistically significant rhythms were reproducible for all functions studied during both time spans and by the same method rhythm

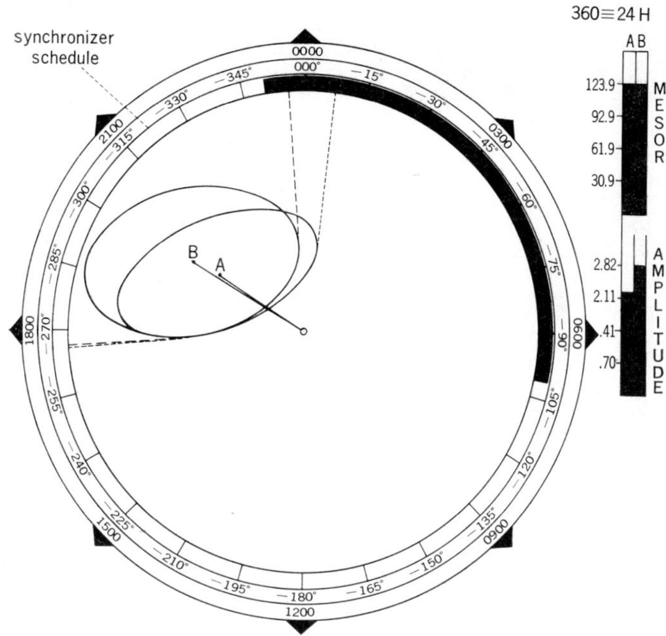

Fig. 1 Mean cosinor summary of carcadian rhythm in systolic blood pressure of ten healthy male medical students.
A: during 25 days without hydrotherapy and
B: during a susequent 25-day span with hydrotherapy.

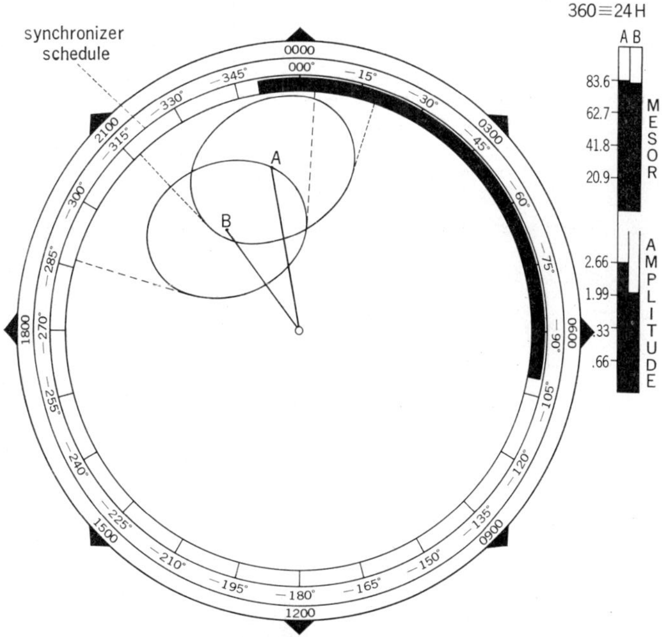

Fig. 2 Mean cosinor summary of circadian rhythm in diastolic blood pressure of ten healthy make medical students.
A: during 25 days without hydrotherapy and
B: during a susequent 25-day span with hydrotherapy.

Table 1 Rhythm parameters of variables investigated by autorhythmometry (10 clinically healthy medical students).

Variable investigated	Unit	Treatment*	P	Mesor, M (.95 CI of M)	Amplitude, A (.95 CI of A)	Acrophase ϕ (.95 CA of ϕ)
Blood pressure	Systolic mm Hg	A	.003	123 (107 139)	2.24 (1.11 3.38)	−301 (−265 — 7)
		B	.005	123 (107 140)	2.82 (1.29 4.35)	−300 (−266 −355)
	Disatolic mm Hg	A	.001	83 (65 101)	2.66 (1.69 3.63)	−350 (−316 — 18)
		B	.002	81 (61 102)	2.05 (1.10 2.99)	−323 (−286 — 3)
Oral	Temperature °C	A	.001	36.7 (36.4 37.0)	.26 (.20 ·31)	−265 (−249 −281)
		B	.001	36.8 (36.5 37.1)	.22 (.14 .29)	−265 (−244 −290)
Dynamometer	(right hand) lb	A	.001	102 (76 128)	3.71 (2.16 5.26)	−244 (−231 −263)
		B	.002	113 (83 143)	5.14 (2.69 7.59)	−244 (−225 −263)
Mood	arbitorary	A	.006	3.9 (3.2 4.5)	.53 (.23 .82)	−243 (−200 −254)
		B	.021	3.7 (2.3 5.0)	.71 (.02 1.39)	−237 (−226 −246)
Vigor	arbitorary	A	.014	3.9 (2.9 5.7)	.52 (.05 .99)	−241 (−203 −263)
		B	.011	3.6 (2.7 5.5)	.70 (.10 1.30)	−234 (−225 −252)
Urine	Vol ml	A	.002	66 (38 94)	15.0 (8.1 22.48)	−254 (−224 −266)
		B	.022	52 (31 73)	9.0 (0.2 19.52)	−236 (−143 −264)
	pH	A	.001	6.06 (5.84 6.29)	.52 (.37 .68)	−211 (−198 −226)
		B	.001	5.99 (5.75 6.24)	.58 (.43 .72)	−197 (−180 −211)
	Na meq/hr	A	.001	11.12 (7.57 14.67)	4.59 (3.73 5.46)	−241 (−223 −257)
		B	.001	9.83 (6.18 13.48)	3.89 (2.99 4.79)	−235 (−213 −252)
	K meq/hr	A	.001	3.22 (1.72 4.73)	1.77 (1.25 2.29)	−200 (−184 −220)
		B	.001	2.89 (1.86 3.91)	1.53 (1.11 1.96)	−184 (−170 −208)
	Na/K	A	.001	4.16 (2.65 5.68)	1.24 (.19 2.29)	−342 (−300 — 13)
		B	.008	4.10 (2.22 5.98)	1.35 (.55 2.14)	−314 (−248 −337)

* A–without hydrotherapy
B–with hydrotherapy

parameters were quantified. Physical therapy in the form of a "thermal bath" of 20-minute duration on six mornings of the week in water of a temperature of 37°C containing 10×10^{-9} c radon/kg of water is compatible with the demonstration of prominent rhythm parameters in various body functions.

REFERENCES

1. GÜNTHER R., KNAPP E. and HALBERG F. (1969): Z. angew. Bäder- und Klimaheilk, *16*: 33–38 and *16*: 123–153.
2. GÜNTHER R., HALBERG F. and KNAPP E. (1969): Kurverlaufs and Kurerfolgsbeurteilung, (W. TEICHMAN, ed.), Symposium II, Sanitas Verlag, Bad Wörishofen.
3. GÜNTHER R., HALBERG F. and KNAPP E. (1971): Z. Phys. Med., *2*: 180–197.
4. STAMM M. and HERRMANN R. (1965): Z. klin. Chem., *3*: 193–197.
5. JAMAR-Dynamometer by Asimow Engineering Company, Los Angeles, California, U.S.A.
6. Analysis by Forschungsinstitut Gastein, Badgastein, Austria, 1966.
7. HALBERG F., ENGELI M., HAMBERGER C. and HILLMANN D. (1965): Acta Endocr. (Kbh.) Suppl. *103*: 54.
8. HALBERG F., TONG Y. L. and JOHNSON E. A. (1967): *In*: The Cellular Aspects of Biorhythms, (H. von MAYERSBACH, ed.), pp. 20–48, Springer-Verlag, Berlin.
9. HALBERG F., HAUS E., AHLGREN A., HALBERG E., STROBEL H., ANGELLAR A. and KÜHL J. F. et al. (1974): *In*: Chronobiology. Proc. Symp. Quant. Chronobiology, Little Rock, 1971, (L. E. SCHEVING, F. HALBERG and J. E. PAULY, eds.), pp. 372–378, Igaku Shoin Ltd., Tokyo.
10. LEVINE H. and HALBERG F. (1972): U.S.A.F. School of Aerospace Medicine Aerospace Medical Division AFSC. April, 1972. AFM-TR-72-3.
11. CONROY R. and MILLS J. (1970): Human Circadian Rhythms. J. & A. Churchill, London.
12. DOE R., VENNES J. and FLINK E. (1960): J. Clin. Endocrinol. Metabol., *20*: 253–265.
13. KRIEGER D. and KRIEGER H. (1967): Metabolism, *16*: 815–823, (table 1, p. 817).
14. HALBERG F., REINHARDT J., BARTTER F., DELEA C., GORDON R., REINBERG A., GHATA J., HALHUBER M., HOFMAN H., GÜNTHER R., KNAPP E., PENA J. and GARCIA SAINZ M. (1966): Experientia, *25*: 107–112.
15. REINBERG A., GHATA J., HALBERG F., GERVAIS P., ABULKER C., DUPONT J. and CAUDEAU C. (1970): Ann. d'Endocrin. (Paris), *31*: 277–287 (Fig. 1, p. 281).
16. KÜHL Y., LEE Y. K., HALBERG F., HAUS E., GÜNTHER R. and KNAPP E.: Circadian and lower-frequency rhythms in male grip strength and body weight. *In*: Biorhythms and Human Reproduction. Proc. Intl. Inst. for the Study of Human Reproduction Conference, (M. FERIN, F. HALBERG, R. RICHART and R. V. de WIELE, eds.), Wiley, N.Y., in press.

CIRCADIAN VARIATIONS OF THERMOREGULATORY RESPONSE IN MAN

GÜNTHER HILDEBRANDT

Institut für Arbeitsphysiologie und Rehabilitationsforschung der Universität Marburg/Lahn W. Germany

Through the daily course of the body temperature it long has been known that temperature regulation also is subjected to circadian variation. Recent investigations have shown that spontaneous changes in heat production are slight under resting conditions, so that the course of the body temperature must be ascribed mainly to changes in heat loss [1, 2].

Under constant ambient conditions, skin temperature in the extremities shows a daily course which runs counter to that of the body temperature. This was first discovered by HEISER and COHEN [3], and was later confirmed by our studies [4] and by others [2]. Recent climate-chamber studies carried out in our institute by DAMM and DÖRING [5] with heat-clearance measurements of skin blood-flow showed that the contrary course taken by acral skin temperatures in comparison with the daily course of the core temperature is matched by considerable changes in skin blood-flow (Fig. 1). If the core temperature rises in the morning, peripheral skin blood-flow in the hand and foot decreases; vice versa, the afternoon values increase when body temperature sinks. Skin blood-flow in the forehead, however, takes the same course as the core temperature, with slighter amplitude, as is already known to be true for skin temperature. Likewise, the daily variation of muscle blood-flow in the extremities runs, as might be expected, contrary to that of skin blood-flow, as was shown by the recent investigations of KANEKO et al. [6].

Thus, the daily course of body temperature is the result of two different thermoregulatory adjustments: warming-up in the morning and cooling-down in the afternoon (cf. Fig. 1). The transition from the one functional direction to the other generally occurs, in our experience, around 0300 and 1500, local time, when the core temperature reaches its minimum and maximum, respectively.

Practical experience had already indicated that response to thermal stimuli is influenced by the phase of thermoregulatory adjustment, and thus by the time of day [4, 7]. For this reason, we systematically examined the effect of cold and hot test stimuli in a daily course. Our findings were as follows:

Cold stimuli applied during the morning warming-up phase lead to much stronger and longer-lasting reactions. Subjects in cool baths develop cold-shivering faster, and the rewarming time of the acral skin is considerably prolonged after cool hand and foot baths. Fig. 2 shows as an example the daily course of the rewarming time in the fingers after a cold hand bath in three subjects who were examined at 2-hour intervals under constant external conditions. In all three cases, the constrictive reactions increase in the morning, and decrease in

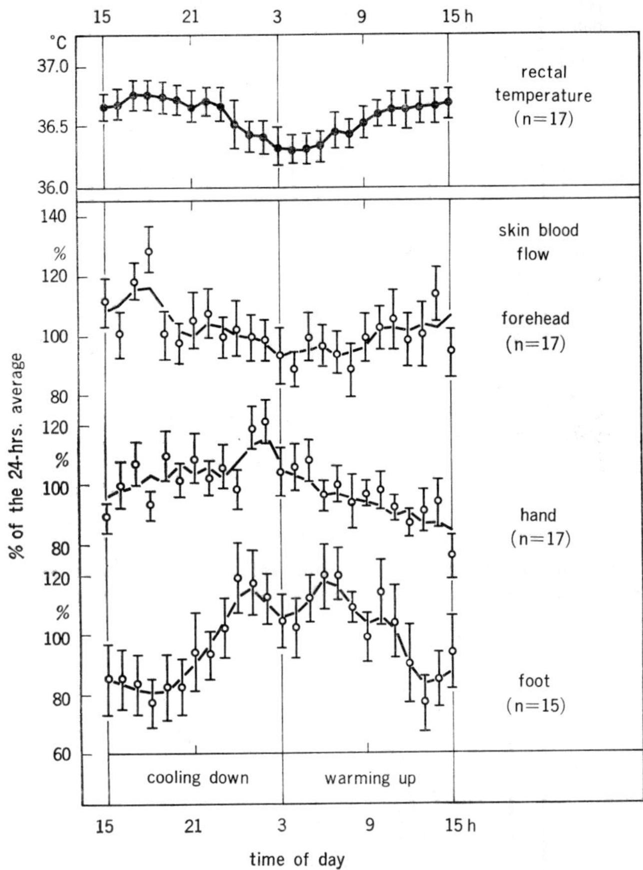

Fig. 1 The mean daily course of rectal temperature and skin blood flow of different regions in healthy subjects under constant ambient conditions (After DAMM and DÖRING [5]).

the afternoon. Likewise, consensual vasoconstrictions are stronger and longer-lasting in the morning, regardless of whether they were caused by cold stimulation of the extremities or of the trigeminal region [4, 8–13].

In contrast, response to heat stimuli is stronger and longer in the afternoon cooling-down phase of the circadian thermoregulatory rhythm. Consensual vasodilatations can be evoked more easily [4]; circulatory loading is considerably higher in hyperthermal treatment [7]. In particular, sweating reactions are considerably increased. In our investigations we administered a defined amount of diaphoretic tea at a given temperature to healthy subjects at 3-hour intervals under constant external conditions, and then continuously recorded sweat evaporation on the forehead. Fig. 3 shows the superimposed curves from this experiment. Here it can be seen that it is often impossible to evoke a sweating reaction in the morning during the warming-up phase, whereas the same stimulus applied during the afternoon, that is, during an increased cooling-down tendency, immediately produces an abundant outburst of sweat [14].

All in all, the organism is thus more sensitive to cold stimuli during the circadian warming-up phase, and more sensitive to heat stimuli during the cooling-

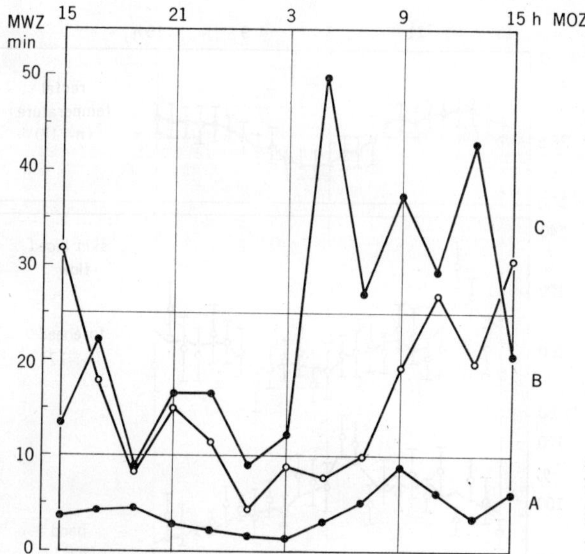

Fig. 2 The daily course of the rewarming time (MWZ) in the fingers after a cold hand bath (15°C; 5 min) in three subjects examined at 2-hour intervals under constant external conditions (After HILDEBRANDT [9]).

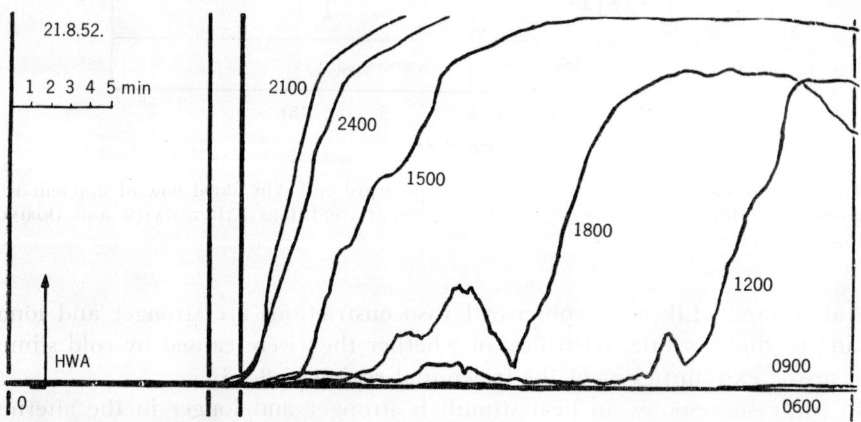

Fig. 3 Superimposed curves of sweat evaporation on the forehead (HWA) from an experiment, in which a defined amount of diaphoretic tea (55°C) was given to a healthy subject at 3-hour intervals (After HILDEBRANDT et al. [14]).

down phase. This means that in each case there is a particularly strong response to those stimuli which cause reactions that increase the predominant phase direction of thermoregulation. Recently, SMITH [15] likewise found indications of circadian fluctuations in thermal reactions. In the field of physical therapy, practical experience in the application of such knowledge has been available for some time now (for a survey of the literature, see HILDBRANDT [11]).

According to our findings, average maximal sensitivity to cold and heat stimuli does not occur at the time of maximal or minimal body temperature, but rather approximately in the middle of the two circadian phases, at the time at which body temperature is changing most rapidly. Fig. 4 shows this in regard to warming-up reactions following cold stimulation, and to the sweating reaction following heat stimulation. From this it can be concluded that sensitivity to the reaction is not controlled exclusively by the varying initial situation, but that there is also a dynamic dependence upon the direction of the phase, for instance upon the rate of change in the core temperature [9].

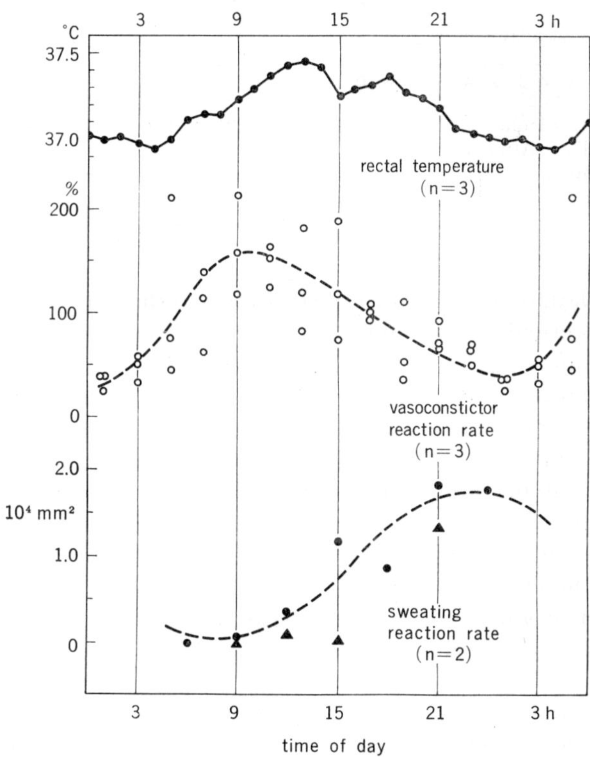

Fig. 4 Daily course of rectal temperature, duration of vasoconstrictor reaction after cold hand-bath, and planimetric magnitude of sweating reaction on the forehead after drinking diaphoretic tea. (Data from HILDEBRANDT et al. [14]; HILDEBRANDT [9]).

There is also ample documentation for fluctuations in thermal sensitivity during the annual biological rhythm in man. According to the extensive investigations of SCHULZ [16], the annual course of subjective sensitivity thresholds for "warm", "mild", "cool", and "cold" has an amplitude of almost 15°C (Fig. 5). The annual maximum for sensitivity to cold occurs in August, the minimum in February. In agreement with this, HENTSCHEL and SCHIRGEL [17] found the extremes of acral rewarming time after cold hand baths, as measured at the beginning and end of climatic therapy in large groups of patients.

As to the significance of thermal stimuli as a Zeitgeber for circadian systems, it is generally considered to be slight for light-active homoiotherms. This is in agreement with the fact that the ambient thermal rhythm is less precise than the light-dark cycle, because the former is dependent upon advective atmospheric processes. However, from the physiological point of view, the decisive fact is

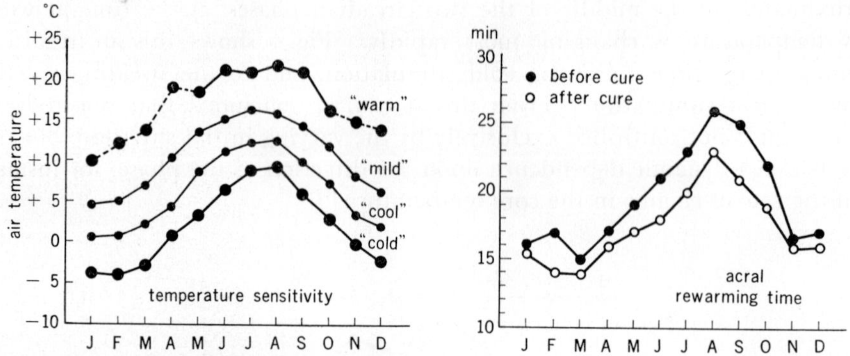

Fig. 5 Left part: Annual course of subjective sensitivity thresholds for "warm", "mild", "cool", and "cold". (After Schulz [16]).
Right part: Annual course of acral rewarming time after cold hand baths, measured at the beginning and end of climatic therapy in large groups of patients. (After Hentschel and Schirgel [17]).

that the phase relation of ambient thermal stimuli to the daily fluctuations found in sensitivity to thermal stimuli under natural conditions rules out the possibility of a Zeitgeber function [11]. When the ambient air temperature rises in the morning, the organism is less sensitive to heat; when this temperature declines in the afternoon, the organism is less sensitive to cold. This relation would lead us to expect more of a damping influence on the daily biological rhythm, rather than a phase-intensifying effect. In contrast, we have found in studies made with Lowes [18] that, for instance, in man the vegetative sensitivity to light stimulus ("light on") is greatest at the time of normal increase of light in the morning, and slightest in the evening. This variation thus supports, in its natural phase situation, the synchronizing effect of the natural "light on"-signal and decreases the disturbing effects of light at other times of day, thus stabilizing the phase relation.

The behavior of civilized human beings also can lead to a schedule of thermal stimuli which could support the synchronization of the biological rhythm. Theoretically, at least, leaving the climate of the bed in the morning, with subsequent cold stimulus, means an effective action which could intensity the normal morning warming-up phase. Likewise, the application of heat towards evening, and the return to the warm climate of the bed, could strengthen the normal cooling-down tendency.

In principle, each reaction system the sensitivity of which is subjected to fluctuations in circadian rhythm can be used to support and stabilize synchronization. However, there is another question here: To what extent will it be possible in practice to use thermal reactions in this sense for supporting a new synchronization after a phase-jump on the part of the Zeitgeber? This question requires further experimental clarification.

There is still another point to be considered in the use of very strong thermal stimuli. It is known that every strong thermal stimulus imposed upon the organism leads to a superimposition of more frequent periods upon the 24-hour rhythm. These periods are usually submultiples of the 24-hour rhythm, e.g. 12-, 8-, or 6-hour periods, and they have a tendency to die out in a damped manner. We have designated these stimulus-induced submultiples as "reactive periods".

According to our investigations, the organism's sensitivity to stimulation, or its responsiveness, is modified not only by the 24-hour rhythm, but also by the phases of such shorter reactive periods. This means that the facilitation and inhibition of ambient effects, which are limited by the 24-hour rhythm to one-half of the day in each case, will be distributed over the entire day in several sections by the appearance of more frequent reactive periods, as is shown schematically in Fig. 6.

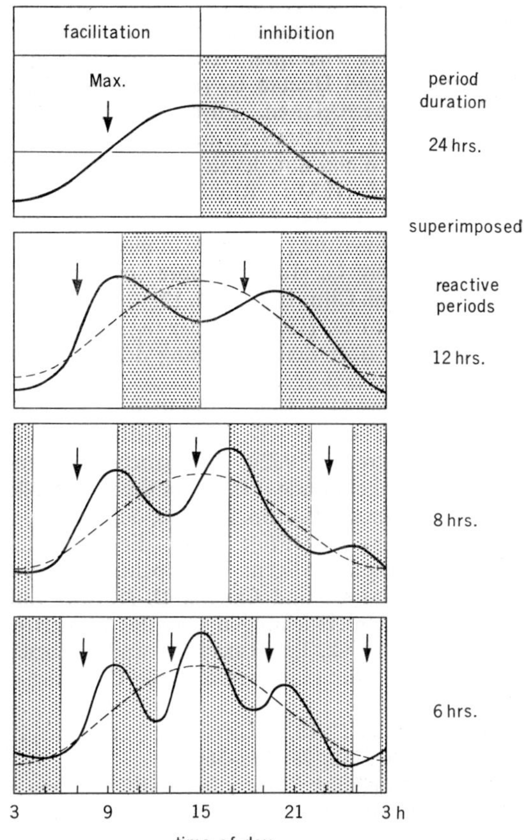

Fig. 6 See text for explanation. Max.=Maximum of facilitation.

This kind of process thus also leads to a stronger response by the organism to Zeitgeber stimuli which appear in abnormal phase positions. By this means, it might be easier for the organism to synchronize to a new Zeitgeber schedule. Here, however, we could expect that phase-jumps of a particular magnitude on the part of the Zeitgeber would be more readily accepted than others of a different magnitude. Thus, the recent observations of Haus and Halberg [19] that the rate of phase shifting in new synchronization is not uniform, but rather occurs as an alternation of fast and slow steps, could be taken also an indication of the efficacy of reactive-periodic fluctuations in sensitivity toward the Zeitgeber stimulus.

Whether this mechanism, which is necessarily nonspecific, can be applied to support the new synchronization of circadian systems is a question which requires further investigation.

REFERENCES

1. Aschoff J. (1955): Der Tagesgang der Körpertemperatur beim Menschen. Klin. Wschr., *33*: 545–551.
2. Koe F. K. Höfler W. und Lüders K. (1968): Mittlere Hauttemperatur und periphere Extremitätentemperaturen bei den tagesperiodischen Änderungen der Wärmeabgabe. Arch. phys. Ther. (Leipzig), *20*: 221–226.
3. Heiser F. and Cohen L. H. (1933): Diurnal variations of skin temperature. J. Industrial Hyg., *15*: 243–256.
4. Hildebrandt G. und Engelbertz P. (1953): Bedeutung der Tagesrhythmik für die Physikalische Therapie. Arch. phys. Ther. (Leipzig), *5*: 160–170.
5. Damm F. und Döring G. (1973): Der Tagesgang der Hautdurchblutung verschiedener Körperregionen. Phys. Med. u. Rehabil., (in print).
6. Kaneko M., Zechman N. and Smith R. E. (1968): Circadian variation in human peripheral blood flow levels and exercise responses. J. appl. Physiol., *25*: 109–114.
7. Lampert H. (1953): Rhythmische Reizbarkeitsänderung des Organismus und ihre Bedeutung für die Krankenbehandlung. Verh. III. Konf. Int. Ges. f. Biologische Rhythmusforschung 1949. Acta med. scand., Suppl., *278*: 141–144.
8. Hildebrandt G. (1952): Sobre los periodos diarios de la termoregulación física (span.). Fol. Clin. Int., *2*: H. 12.
9. Hildebrandt G. (1957): Über tagesrhythmische Steuerung der Reagibilität. Untersuchungen über den Tagesgang der akralen Wiedererwärmung. Arch. phys. Ther. (Leipzig), *9*: 292–303.
10. Hildebrandt G. (1960): Spontane Schwankungen der Thermoregulation und Physikalische Therapie. Berliner Medizin, *11*: 37–40.
11. Hildebrandt G. (1962): Biologische Rhythmen und ihre Bedeutung für die Bäder- und Klimaheilkunde. *In*: Handbuch der Bäder- u. Klimaheilkunde, (W. Amelung und A. Evers, eds.), pp. 730–785, Schattauer, Stuttgart.
12. Hildebrandt G. (1970): Über die Bedeutung einer tageszeitlichen Ordnung der Hydrotherapie. Allgem. Therapeutik, *10*: 30–36.
13. Hildebrandt G. und Crnjak D. (1970): Zur tageszeitlichen Ordnung der Hydrotherapie. Untersuchungen über das Wassertreten. Z. Phys. Med., *1*: 51–61.
14. Hildebrandt G., Engelbertz P. und Hildebrandt-Evers G. (1954): Physiologische Grundlagen für eine tageszeitliche Ordnung der Schwitzprozeduren. Z. klin. Med., *152*: 446–468.
15. Smith R. E. (1969): Circadian variations in human thermoregulatory responses. J. appl. Physiol., *26*: 554–560.
16. Schulz L. (1960): Der jahreszeitliche Gang der Temperaturempfindung des Menschen anhand einer zehnjährigen Beobachtungsreihe. Arch. phys. Ther. (Leipzig), *12*: 245–255.
17. Hentschel G. und Schirgel L. (1960): Beobachtungen über Funktionsänderungen der akralen Durchblutung als klimatherapeutischer Effekt. Arch. phys. Ther. (Leipzig), *12*: 235–240.
18. Lowes E.-M. (1973): Tagesrhythmische Empfindlichkeitsschwankungen optisch-vegetativer Funktionen. Med. Inaug.-Diss. Marburg/Lahn.
19. Haus E. and Halberg F. (1969): Phase-shifting of circadian rhythms in rectal temperature, serum corticosterone and liver glycogen of the male C-mouse. Rassegna di neurologia vegetativa, *23*: 83–112.

LOW-FREQUENCY RHYTHMS IN THE BEARD GROWTH OF A MAN*

ROBERT B. SOTHERN

Chronobiology Laboratories, University of Minnesota
Minneapolis, Minnesota, U.S.A.

Beard growth in man is a generally unheeded parameter of hormonal activity, although it reflects changes in pituitary and thyroid hormones as well as endocrine secretions of the gonads and the adrenal cortex. That the beard is a secondary sex characteristic of the human male is attested to by the observation that under normal conditions it achieves full development in mature males, being dependent upon gonadal secretions for development and in some instances for maintenance. Indeed, ROOK [1] asserts that male sexual hair (beard, ears, upper pubic triangle, body hair and nasal tip), which is dependent upon male levels of steroid hormones, is a remnant of the elaborate plummage of the mature male, once serving the purpose of sexual display.

METHOD

In order to examine quantitatively any low frequency rhythm(s) of follicular activity manifesting beard growth, weights of shavings were obtained from a 24-year old, white male in Minnesota from March 1 to October 1, 1971 (with one missing span of 21 days). The subject arose each morning at 0900 ± 1/2 hour and washed his face with soap and hot water. So as to minimize the effect of Verel's phenomenon of edema when changing from a recumbent posture [2], the time between arising and shaving was kept as constant as possible. Thirty to forty-five minutes after awakening the face was conscientiously shaved with a Schick electric razor and the shavings collected from the razor's head in a glassine packet and later weighed in a laboratory on a Sartorius balance to the nearest milligram. The 24-hour weights were then converted to growth rate in mg/hr (to correct for slight discrepencies in time elapsed between successive shavings) and then two-day moving averages were calculated to refine irregularities in shavings collection (if the total beard area was not closely shaved one day the next day usually showed a jump in beard weight). Body weight was also measured at the time of shaving and several vital signs were monitored just prior to shaving and 5–6 times during the rest of the waking span for the duration of the study (including oral temperature, pulse, respiration, right and left grip strengths, time estimation, systolic and diastolic blood pressure, eye-hand coordination, peak expiratory flow, mood and vigor ratings). All measures were then coded onto punch cards as time of collection and values and each series analyzed on a Varian 620 computer for any significant low-frequency rhythms between 40 and 3 days by fitting cosines with the method of Least Squares [3].

* Supported by U.S. Public Health Services (5-K6-GM-13,981) and NASA (NGR-24-005-006).

RESULTS

Upon inspecting the raw data plotted at the top of Fig. 1 several cyclic patterns in the beard weights can readily be seen with the naked eye. The individual 24-hour weights ranged from less than 30 mg to more than 50 mg and the analysis of the two-day moving means of the hourly beard growth rate shows a best cosine fit at 16.5 days (p<.01). This cosine is plotted through the raw data and while it describes a general pattern which is not random but somewhat predictable, it is not completely adequate in mapping the long-term, lower-frequency changes in mesor, i.e.—the tendency of lower values at the left and higher values to the right in the data plot perhaps suggesting seasonal or yearly influences.

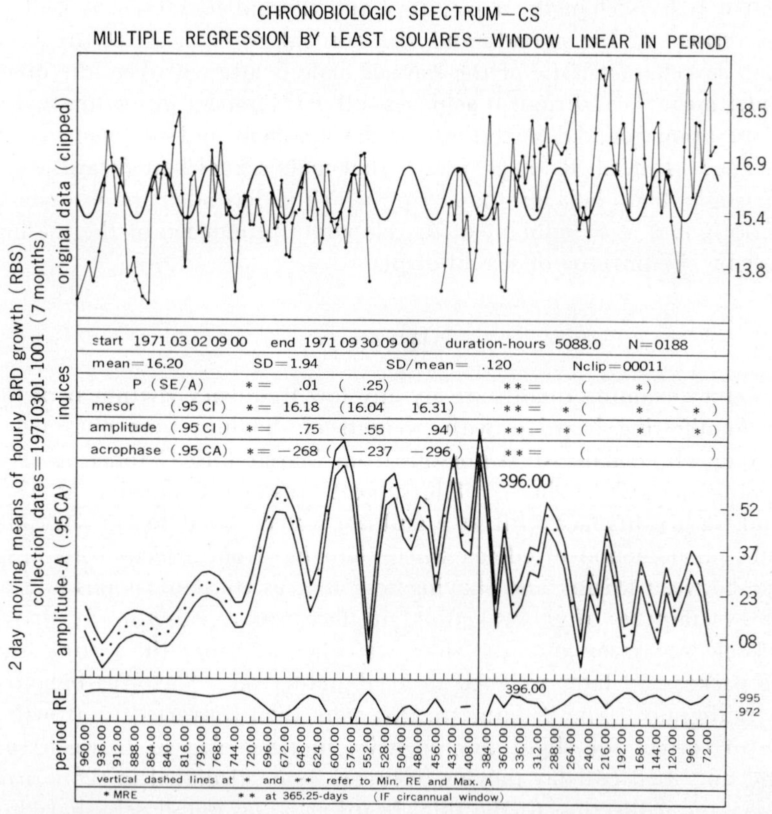

Fig. 1 Chronobiologic window—CS. Multiple regression by least squares-window linear in period.

Looking at the amplitude chart near the bottom of Fig. 1, one can see several other peaks at various trial periods, those at 24.5 and 18.0 days (in hours on chart) also having P-values less than .01 but accounting for less of the variability than the 16.5 day curve. Significant low-frequency rhythms were also detected in similar manner for each vital sign except body temperature (summarized in Table 1 for 'best fitting cosine' and best fit in the circavigintan (about 20-day) range). These results give an impression that some overall phenomenon is reflected in these functions which may or may not be related to the beard growth

Table 1 Rhythmometric summary of vital signs on healthy male over a span of 7 months (19710301–0931).

Variable		N of obser.	Best fitting period (days)*	P values	Circavigintan component (days)**	P values
Beard weight(mg/hr)		188	16.5	.01	18.0	.01
Morning body weight (lbs)		190	31.0	.01	22.0	.08
Pulse (beats/minute)		868	34.0	.01	18.0	.02
Grip strength (dynes)	right hand	869	26.0	.01	17.0	.01
	left hand	869	26.0	.01	19.0	.01
Time estimation, one-minute (seconds)		772	10.0	.04	16.0	.07
Blood pressure (mmHg)	Systolic	528	22.0	.01	16.0	.01
	Distolic fade	860	27.0	.01	22.0	.01
	Distolic disappear	860	6.0	.01	17.0	.01
Respiration (breaths/2 min)		869	28.0	.01	17.0	.04
Oral temperature (°F)		867	32.0	.12	17.0	.16
Mood (1–7 scale)		869	38.0	.01	16.0	.01
Vigor (1–7 scale)		869	7.0	.04	18.0	.27
Eye-hand coordination (seconds)		861	27.0	.01	21.0	.01
Expiratory peak flow (L/min)		869	18.0	.01		

* irrespective of period length.
** though not necessarily most prominent infradian rhythm.

rhythm. It's interesting to note, for example, that in a young man a variable as unpredictable as mood has a significant rhythm around 16 days (close to that of beard growth), when in females noticeable mood swings often accompany their menstrual periods or 'sex cycles.'

DISCUSSION

Since care was taken to shave equally carefully each day and the time between arising and shaving was similar, explanations other than measurement error must be sought for the detected beard growth rhythm of 16.5 days. All types of razors, when removing the beard will also remove small quantities of lipids and epithelial debris but it was hoped that the outer layer of epithelial cells would be mostly washed away with the soap and water and that the abrasive action of the electric razor, when used in a standardized motion each day, would collect a more or less similar proportion of these cells. BURTON, et al. [4] showed that sebum excretion exhibited a circadian rhythm with a late morning phase of $-169°$ for 19 subjects. Their previous studies suggested that the sebaceous glands responded slowly over days or weeks to internal stimuli and thus they could be a contributor to any weight differences since no attempt was made to isolate the beard from other material collected by the razor. The subject was in presumably good health throughout the study so Sims' [5] finding that the diameter of hair from well-fed individuals was constant would be applicable.

Anonymous [6] was led to suspect a resumption of sexual activity after isolation on a remote island was the stimulus for his increased beard growth, indeed even the presence of female company after separation seemed to cause an increase in growth. In exploring variations in the rate of human sexual behavior, Udry and Morris [7] found that rates of intercourse and orgasm at one point of the female menstrual cycle (for 98 females) to be 2–6 times that rates at other points, demonstrating a periodicity similarly characterizing the sexual behavior of lower mammals—the highest rate occurring at the time when ovulation is thought to occur. Kihlström [8] though, dissociated the variation of beard growth with sexual activity solely determined by the phase of the sexual partner, since his wife used an oral contraceptive (with an exact 28-day period) and his six months of 24-hour beard weights revealed a 33-day cycle. Since in the study here reported the subject refrained from coitus it seems likely that unexplored hormonal as well as social-psychological processes are involved in the mechanism of beard growth—the beard either being a target organ or simply responding to stimuli provided the body.

It is thus interesting to note that in the inferential statistical analysis of Hamburger's thirteen year collection of 24-hour 17-ketosteroid samples, Halberg [9] found a low-frequency component in the 17–21 day region (along with daily, weekly, monthly and yearly components)—all the more convincing since the time series was of such an unusual length and produced a circavigintan rhythm reproducible for 13 years. A probable connection between beard growth and male hormones is thus suggested, with further work required to isolate the human male 'sex cycle' and understand its relationship to the timing of other bodily processes, such as chemical constituents, sperm counts or potency, as well as strength, blood pressure, mood, etc.

REFERENCES

1. Rook A. (1965): Endocrine influences on hair growth. Brit. Med Journal, *5435*: 609–614.
2. Verel D. (1955): Observations on the effect of posture on the distribution of tissue fluid in the face. Journal of Physiology, *130*: 72–78.
3. Halberg F. (1969): Chronobiology. Ann. Review of Physiol., *31*: 675–725.
4. Burton J. L., Cunliffe W. J. and Shuster S. (1970): Circadian rhythm in sebum excretion. Br. J. Derm., *82*: 497–500.
5. Sims R. T. (1968): Hair growth as an index of protein synthesis. Brit. Journal of Derm., *80*: 337–339.
6. Anonymous (1970): Effects of sexual activity on beard growth in man. Nature (London), *226*: 869–870.
7. Udry J. R. and Morris N. M. (1968): Distribution of coitus in the menstrual cycle. Nature, *220*: 593–596.
8. Kihlström J. E. (1971): A monthly variation in beard growth in one man. Life Sciences, *10*: 321–324.
9. Halberg F., Engeli M., Hamburger C. and Hillman D. (1965): Spectral resolution resolution in low-frequency, small amplitude rhythms in excreted 17-ketosteroids; probable androgen-induced circaseptan desynchronization. Acta Endocrinology: suppl. 103, Periodica, Copenhagen.

CHRONOBIOLOGIC SERIAL SECTION ON 8876 ORAL TEMPERATURES COLLECTED DURING $4^1/_2$ YEARS BY PRESUMABLY HEALTHY MAN

(age 20.5 years at start of study)

Robert B. SOTHERN

Chronobiology Laboratories, University of Minnesota
Minneapolis, Minnesota, U.S.A.

Serial sections have proved indispensable to the classical microscopic pathologist. He has come to rely on them in searching for the tumor in a tissue. The chronobiologic serial section is potentially of yet greater utility, reflecting as it does changes in a time structure that may signal risk or disease and may reveal even the effects of an odd routine for experimentation in the case described herein.

The purpose of this paper is to present a new display for the chronobiologic serial section. It is intended as a first step toward obtaining a rhythm description by a concomitantly computed P-value and by providing 24-hour rhythm parameters such as amplitude, acrophase and mesor, with some dispersion indices as well [1].

Fig. 1 represents a composite serial section, covering a span from May 11, 1967 to September 30, 1971 of oral temperatures measured 5–9 times/day for four and one-half years by a presumably healthy young man. It may be seen from the top row of chronograms that the variability of body temperature around the proverbial 98.6 line covers several degrees Fahrenheit, although the scale of the chronogram is so condensed that changes within the day are not readily apparent. With few exceptions, the P-value for rhythm detection lies below the .05 level. The indices of mesor and amplitude show remarkable changes with intervals of days, weeks, months and possibly years, reflecting as they do the influences of exogenous as well as endogenous factors.

Though the naked eye suspects that infradian rhythms with frequencies much lower than a day do characterize the data, the plot of all temperatures is not optimally suited for examining this question, and if one desired to scrutinize the results for this possibility one would have to re-analyze daily averages or mesors for yet longer spans to detect and map these frequencies lower than 24-hours. The positioning of the acrophase along the 24-hour scale is remarkably stable through the years, being displaced only by phase-shifts resulting from actual or simulated transmeridian flights and by phase-drift during a span in isolation. The data thus serve to document that in the case of human body temperature we deal with a biorhythm.

More specifically, to qualify as a *rhythm* a set of biological changes must recur systematically according to a formulatable pattern or waveform which is validated by *inferential statistical means.* However, the term *biorhythm* has been reserved for cases meeting *two additional biologic criteria* [2]. A first requirement for a biorhythm (if it be amenable to frequency sychronization with an environmental

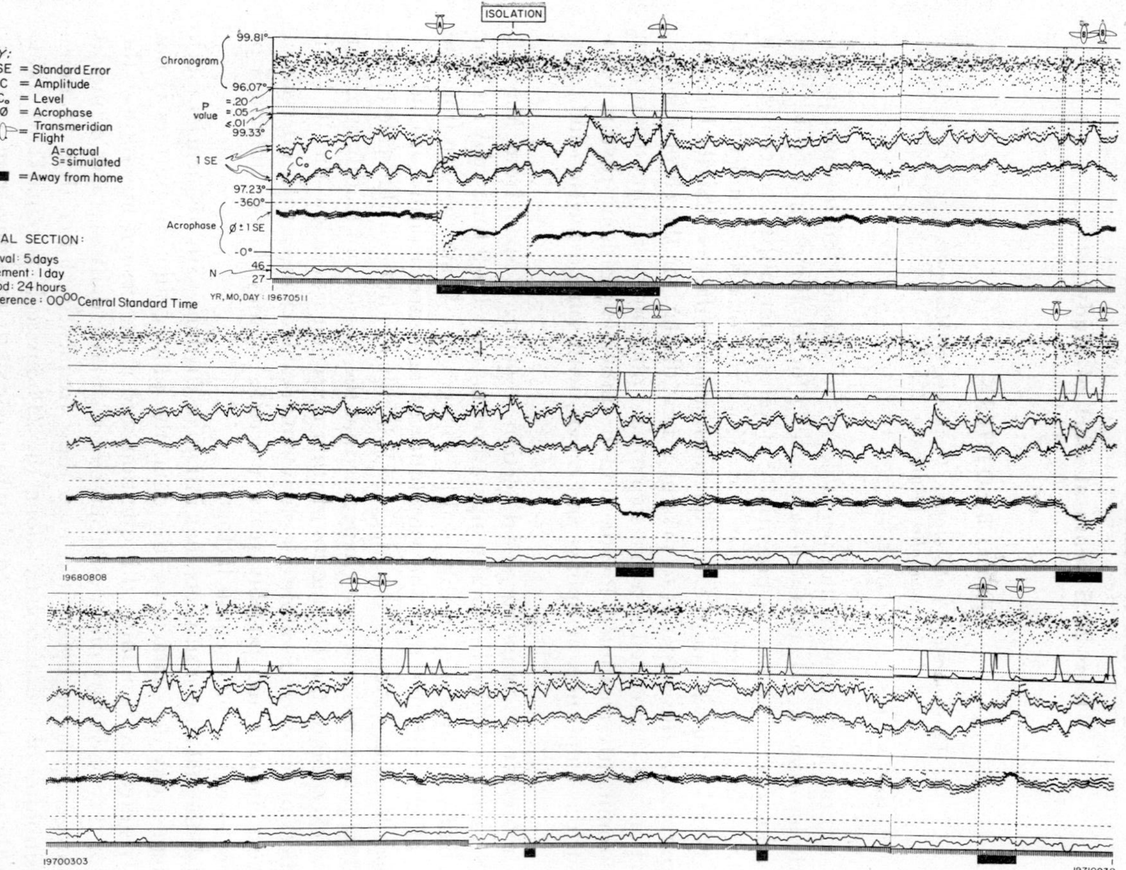

Fig. 1 Actual and simulated transmeridian flight-determined shifts and isolation-induced drifts of circadian acrophase and other characteristics of rhythms quantified by chronobiologic serial section of 8876 oral temperatures collected during $4\frac{1}{2}$ years by presumably healthy man (age 20.5 years at start of study).

Table 1

Chronobiology Laboratories—University of Minnesota — Minneapolis Minnesota 55455 USA (612)–373–2920

Rhythmometric Summary

Self-measures of a healthy male (RBS. age 22) for 3 weeks Chronobiologic window from 28.0 to 20.0 hours

	Variable	Data summary					Analysis summary using least squares fit of single cosine at															
							24-hour period											Best fit				
		Lowest & highest values		90% range		Coef. var.	P	VR	Mesor M	Mesor SE	Amplitude A	Amplitude SE	Acrophase ϕ Deg	SE	CK	HR (.95CA)		Period	P	PR	A	ϕ
1	Urine temp	96.7	99.7	97.1	99.3	.7	.009	.52	98.34	.05	.84	.07	−253	4	1653	1619	1726					
2	Mood rating	2.0	6.0	2.0	5.0	22.8	.009	.08	3.90	.08	.32	.09	−137	25	908	546	1231	23.0	.009	.13	.4	−177
3	Vigor rating	1.0	4.0	1.0	4.0	47.4	.009	.54	2.01	.07	1.23	.11	−192	4	1249	1211	1327					
4	Two min T.E.	97.0	192.0	116.0	173.0	12.3	.009	.35	144.83	1.49	18.81	2.39	− 62	5	410	330	450					
5	Pulse/min	55.0	96.0	59.0	89.0	12.2	.028	.05	72.71	.91	3.09	1.23	−180	24	1202	849	1515	23.2	.009	.12	4.4	−311
6	Syst BP	103.0	142.0	110.0	136.0	6.5	.009	.08	122.88	.80	2.98	.90	−321	25	2124	1805	43	24.2	.009	.21	5.3	−233
7	Dst 1–fades	60.0	89.0	71.0	87.0	5.9	.460	.01	78.34	.49	.68	.54	−141	67	926			24.4	.009	.10	2.3	−176
8	Dst 2–dsprs	60.0	88.0	66.0	85.0	7.6	.009	.17	75.47	.56	3.39	.74	−177	13	1152	1003	1341	23.9	.009	.20	3.6	−219
9	Fngr count	8.7	12.0	9.0	11.1	6.3	.009	.30	10.13	.05	.65	.09	− 50	5	324	242	406					
10	Add speed	28.1	51.7	29.1	41.7	11.6	.009	.21	34.94	.36	3.39	.60	− 53	6	334	239	428					
11	Oral temp	96.3	99.1	97.1	99.0	.5	.009	.61	98.07	.03	.76	.05	−252	3	1651	1625	1717					
12	Rt Dyn	108.0	155.0	119.0	152.0	7.5	.009	.45	134.32	.80	11.41	1.22	−262	5	1730	1650	1810					
13	Lft Dyn	112.0	163.0	121.0	155.0	7.4	.009	.32	137.15	.88	9.78	1.36	−260	6	1721	1630	1812					
14	Peak flow 1	555.0	634.0	583.0	626.0	2.2	.009	.07	601.26	1.34	6.60	2.16	−216	13	1425	1238	1611					
15	Peak flow 2	547.0	635.0	574.0	631.0	2.9	.009	.09	599.81	1.75	7.38	2.27	−175	20	1144	905	1423	24.1	.009	.10	7.8	−122
16	Peak flow 3	456.0	651.0	538.0	621.0	4.7	.169	.02	586.32	2.89	7.21	4.09	−188	31	1233			26.6	.028	.05	9.4	−272
67	Highest pef	574.0	651.0	590.0	634.0	2.1	.009	.09	608.70	1.28	6.56	1.96	−202	14	1331	1140	1521	24.5	.009	.11	6.6	−215

(Series No. .N) 1 120 2 119 3 120 4 116 5 119 6 117 7 117 8 117 9 120 10 119 11 117 12 120
13 120 14 120 15 120 16 120 17 120

Start time for series 1969 531 8

VR=Variability Ratio=percent of total variability contributed by fitted curve =(sum of squared deviations from mean(ss). of values derived. from fitted cosine curve at sampling times/ss of data themselves).

P=result of testing zero amplitude=probability of obtaining estimated parameters if sinusoidal rhythm with stated period were absent.

cycle—a condition met in Fig. 1, as can be seen from the stable acrophase) is that the institution of an abrupt, single-cycle shift by 90° or more in the environmental routine should be followed by a gradual rather than within-single-cycle acrophase adjustment. Indeed, at the times marked by airplanes (indicating an abrupt environmental cycle or synchronizer shift) there is usually (in at least one direction) a gradual rather than abrupt adjustment of the acrophase. A second point merging from these phase-shifts of routine is that the acrophase adjustment after the flight in one direction may differ considerably from that required after a flight in the opposite direction. This difference in speed of acrophase adjustment after two flights may be seen, e.g., in the middle of the chart on the extreme right.

Second, a biorhythm should persist as a statistically significant entity (validated by a test indicating that its amplitude is not zero with results demonstrated in the second row of this serial section) for two or more cycles after elimination of the environmental synchronizer. Indeed, during the isolation span the rhythm persisted for the 19 days with a high amplitude that did not decrease as isolation progressed and with an acrophase that had no known environmental counterpart —the drift in acrophase indicating the occurrence of a period longer than 24-hours.

Thus, the serial section complements the rhythmometric summary, Fig. 2, and awaits, inter alia, clinical applications, as discussed elsewhere [3].

REFERENCES

1. Halberg F. (1969): Chronobiology. Ann. Rev. Physiol., *31*: 675–725.
2. Halberg F., Nelson W., Runge W. J., Schmidt O., Pitts G. C., Tremor J. and Reynolds O. (1971): Plans for orbital study of rat biorhythms. Results of interest beyond the biosatellite program. Space Life Sciences, *2*: 437–471.
3. Halberg F., Johnson E. A., Nelson W., Runge W. and Sothern R. (1972): Autorhythmometry-procedures for physiologic self-measurements and their analysis. Physiology Teacher, *1*: 1–11.

THE TRANSFER AND UTILIZATION OF VITAMIN C AS INTERPRETED BY ITS HUMAN BIOLOGICAL RHYTHMS

C. W. M. WILSON

Department of Pharmacology, University of Dublin Trinity College Dublin, Ireland

In Britain and Ireland, the recommended dietary intake of ascorbic acid is 30 mg daily [1]. There is a considerable range in individual dietary intake of ascorbic acid. About one-third of old people living at home do not achieve the minimum recommended intake [2]. It has been shown that preferences for chemical substance and foods as assessed by taste threshold measurements [3], vary between species and with different compounds. These fluctuations in taste threshold have both innate and learned components [4]. The taste threshold can be affected by the blood concentrations of the substances under test, as in the case of glucose and salt [5]. The taste threshold modalities [6] to a number of different substances including alcohol [7–9] and Vitamin C [10, 11] undergo circadian rhythms.

The taste threshold to Vitamin C will be discussed in relation to ascorbic acid plasma levels to illustrate the factors which influence Vitamin C concentrations and metabolism on a circadian basis in man. Circadian variations in plasma concentrations of ascorbic acid are correlated with rhythmic variation in other blood constituents. Comparisons of these variations provide information about the transfer and storage of ascorbic acid in the tissues. Evidence can finally be obtained about the metabolic function of ascorbic acid in specific tissues by correlating rhythmic variations in ascorbic acid content with alterations in other physiological processes in these tissues [12].

METHODS

The taste threshold to Vitamin C was measured by a modification of the method described by Harris and Kalmus [13] for the measurement of the taste threshold to phenylthiocarbimide [10]. The subjects tasted increasing concentrations of ascorbic acid in water. As soon as the characteristic slightly acid taste of Vitamin C was detected, the taste threshold was confirmed by giving the subjects three solutions of this particular concentration of ascorbic acid to taste. These had to be differentiated correctly from three containers of water. Plasma, leucocyte and urinary ascorbic acid concentrations were measured by the standard procedure of extraction into tri-chloracetic acid, and spectrophotometric estimation of ascorbic acid by chemical combination with dinitrophenyl hydrazine [14, 15]. Procedures for measurement of other compounds are described in the original publications.

THE TASTE THRESHOLD TO VITAMIN C

Taste threshold to ascorbic acid has a circadian rhythm with its acrophase at 0148 hours. The plasma ascorbic acid acrophase occurred at 1102 hours (Fig. 1). The taste threshold measurements were carried out in two groups of subjects on two separate occasions in May, 1969. In one of these groups plasma ascorbic acid was simultaneously measured. Plasma ascorbic acid concentrations were measured in another group in November 1968.

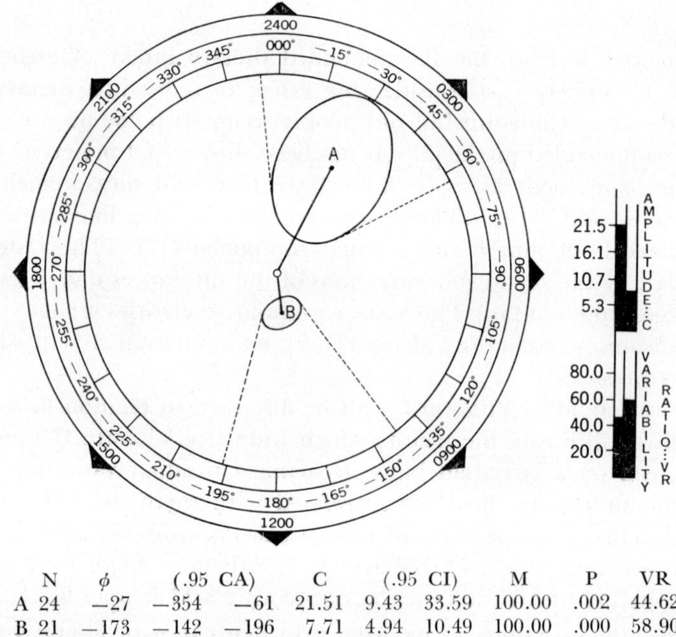

	N	ϕ	(.95 CA)		C	(.95 CI)		M	P	VR
A	24	−27	−354	−61	21.51	9.43	33.59	100.00	.002	44.62
B	21	−173	−142	−196	7.71	4.94	10.49	100.00	.000	58.90

Fig. 1 Mean levels for acrophases of ascorbic acid taste threshold and plasma levels obtained from two, and four groups of subjects respectively. The taste thresholds were measured during a period of one month, and the plasma levels were measured at an interval of six months. Amplitudes expressed as percentage of means. A: ascorbic acid taste threshold. B: plasma ascorbic acid.

The acrophases for the taste thresholds in the different groups did not differ significantly and they did not vary in the different investigations even though they were carried out during a period of four weeks. The taste threshold amplitudes were large in comparison with the amplitudes for the plasma concentrations of ascorbic acid. The acrophases of the plasma ascorbic acid in the different investigations did not differ significantly in the separate groups of subjects, and the amplitudes were similar. The plasma ascorbic acid circadian rhythms were demonstrated to be reproducible from year to year and in different student population samples. The circadian rhythms for plasma ascorbic acid levels and taste threshold to Vitamin C did not synchronize either in acrophase or amplitude. This means that when an individual was at his most sensitive phase to the taste for ascorbic acid, the level of plasma ascorbic acid was at its highest amplitude during the circadian cycle.

CIRCADIAN RHYTHMS OF ASCORBIC ACID IN THE BLOOD

Ascorbic acid undergoes a circadian rhythm in the plasma which is synchronized with the rhythm in the leucocytes (Fig. 2). Cosinor analysis demonstrated that the plasma ascorbic acid acrophase occurred at 1148 hours. The acrophases for the leucocyte ascorbic acid and for the serum iron corresponded in the cycle with that for the plasma ascrobic acid.

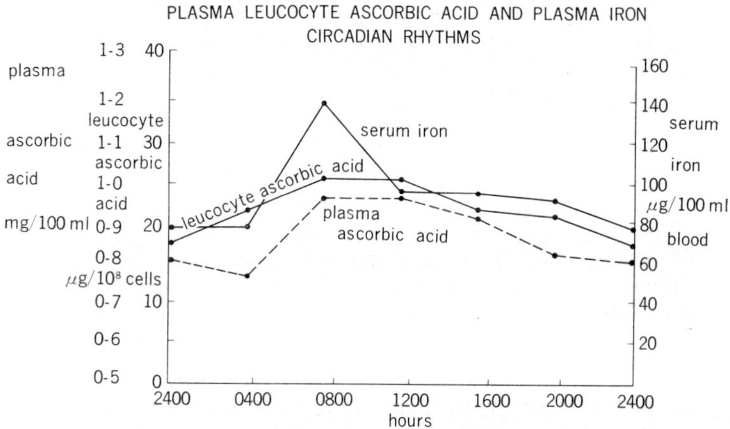

Fig. 2 Twenty-four hour circadian rhythms of plasma and leucocyte ascorbic acid concentrations and of plasma iron. Synchronised acrophases for ascorbic acid values and for iron occurred at 1148 hours.

It has been demonstrated that about 5% of the leucocyte ascorbic acid in male human beings exists in the reduced form as a fixed store in the leucocytes. The remaining 95% is available as a labile store which can be utilized for metabolic purposes during normal physiological conditions [11]. The existence of synchronized acrophases for ascorbic acid in the leucocytes and plasma indicates the extensive lability of ascorbic acid in the leucocytes. This lability is dependent upon a reversible flow of ascorbic acid between the plasma and leucocytes.

Factors in the red and white blood cells determine the extent and duration of flow of ascorbic acid from the plasma. The uptake of ascorbic acid by the leucocytes takes place through an active energy regaining process which is enhanced by the presence of ferric iron, and as a result of passive diffusion [16]. Leucocytes do not absorb dehydroascorbic acid to any significant extent, but incubation of erythrocytes in dehydroascorbic acid significantly increases the concentration of ascorbic acid in the supernatant fluid. Simultaneous incubation of erythrocytes with leucocytes in dehydroascorbic acid results in a significant increase in the ascorbic acid content of the erythrocytes (Fig. 3). A mechanism therefore exists to enable leucocytes actively to absorb ascorbic acid which originates from the dehydroascorbic acid manufactured by the erythrocytes. The two acrophases for serum iron and for leucocyte ascorbic acid occur at the same time. In consequence the rhythm for serum iron is synchronized with that for plasma ascorbic acid so that conditions are suitable for the active absorption of ascorbic acid by the leucocytes. It has been shown by Loh and Wilson [17] that the total white

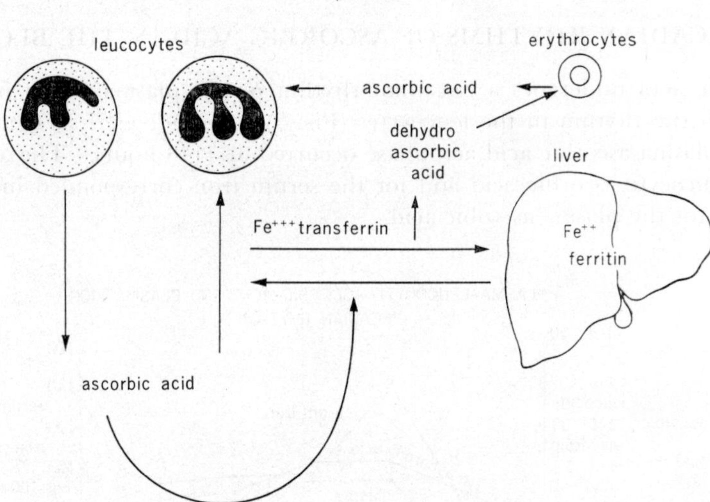

Fig. 3 Factors influencing the uptake and metabolism of ascorbic acid between the blood and tissues, illustrating the lability of ascorbic acid in relation to its transfer during the circadian rhythm.

count in the blood is inversely correlated with the concentration of ascorbic acid in the buffy coat of the blood, that is with the leucocyte ascorbic acid concentrations. It is therefore of interest to observe that the rhythm for the total white count is completely out of phase with the circadian rhythm for the leucocyte ascorbic acid. The close relationship of ascorbic acid and iron rhythms in the blood to the rhythmic change in numbers and biochemical activity of erythrocytes and leucocytes facilitates the transfer of ascorbic acid out of the plasma into cells and back again. This enables a circadian rhythm of ascorbic acid metabolic utilization and storage to take place.

THE METABOLIC UTILIZATION OF ASCORBIC ACID ON A MONTHLY BASIS

The fraction of ascorbic acid being transferred through the plasma, or the labile pool temporarily being stored in the leucocytes, or the fraction of the plasma ascorbic acid transferred into the renal tubules, individually, or when measured on isolated occasions, give little indication of the metabolic utilization of ascorbic acid in specific bodily activities. The ascorbic acid status of the patient can be evaluated by correlating two of these criteria simultaneously or by correlating one of them with a separate but physiologically related criterion which provides an adequate comparison for the ascorbic acid on a chronobiological basis.

An example of the metabolic utilization of ascorbic acid on a monthly chronobiological basis has been provided by Loh and Wilson [18] in their study of the relationship between the excretion of ascorbic acid and human ovulation. The occurrence of human ovulation has in the past been most easily determined by noting the morning rise in temperature during the monthly menstrual cycle. If women are saturated by the daily administration of 500 mg of ascorbic acid for two weeks preceding menstruation, the urinary excretion of ascorbic acid provides an indication of the metabolic utilization of ascorbic acid in the tissues. It can then be compared with the early morning variations in basal body temperature, and with the urinary excretion of luteinizing hormone (Fig. 4).

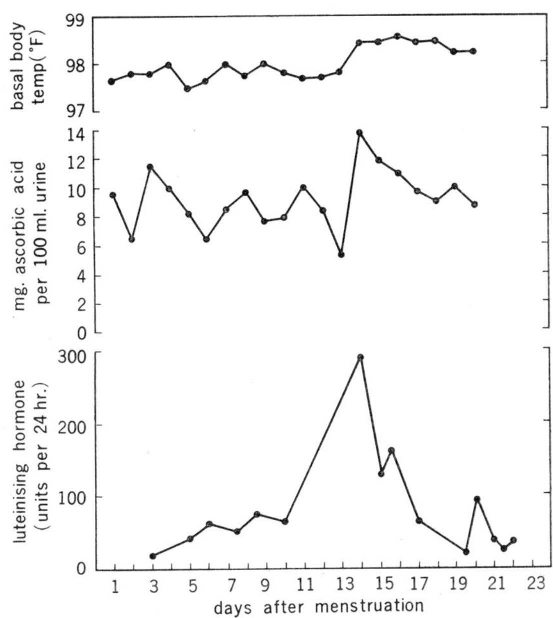

Fig. 4 Mean daily values for basal body-temperature and excretion of ascorbic and in morning specimen of urine from five female subjects superimposed on mean values for excretion of luteinising hormone in morning specimen of urine from five other comparable ovulating subjects [18].

The basal body temperature has been adjusted so that it is shown to take place on the 14th day of the cycle in all the subjects, corresponding to the time during the cycle when ovulation occurs. The excretion of luteinizing hormone begins to increase three days prior to ovulation. At the time of ovulation the concentration of luteinizing hormone in the urine reaches a peak. There are two subsidiary, but smaller, peaks in the excretion of LH during the seven days following ovulation. The excretion of ascorbic acid attains a peak three days prior to ovulation following which excretion progressively diminishes until the day preceding ovulation. On day of ovulation, ascorbic acid attains maximum excretion in the urine like L.H. Following ovulation, the excretion of ascorbic acid mimics precisely that of L.H.

CONCLUSIONS

Examination of human ascorbic acid metabolism on a chronobiological basis provides valuable information about factors influencing its consumption, its tissue transfer, and utilization. It differs from other nutritive agents in that taste threshold becomes more sensitive when plasma levels rise [27]. This is contrary to the findings with glucose [19, 20], salt [21–23] and alcohol [24, 25]. The extreme lability of ascorbic acid as shown by its passage between the plasma, erythrocytes, and leucocytes, and the demonstration that plasma concentrations of ferric iron are correlated with ascorbic acid variations, indicate the occurrence of a closely interlinked circadian rhythm in the blood. This is designed to promote the temporary storage of ascorbic acid when tissue demand is reduced, and its release at specific times for metabolic purposes. Ascorbic acid is available in the plasma at a high concentration during the 24 hour cycle when adrenal corticosteroids simultaneously achieve their acrophase. The inter-relationship of ascorbic acid utilization at the time of ovulation with secretion of luteinizing hormone demonstrates the essential role of ascorbic acid in endocrine metabolism.

The clinical ignorance about the functions and metabolism of ascorbic acid can be reduced by study of its biological rhythms in relation to many vital human mechanisms. When such studies are performed, the implications have relevance to modern entrenched conceptions about human nutrition, haemopoiesis, and reproduction. Glucose is required to provide the power in the human thermodynamic engine, but the engine also requires the agents to enable it to use this power source. Ascorbic acid is one of the exogenous agents whose rhythmic availability determines utilization of thermodynamic power by the tissues at the cellular level and enables the endocrine control mechanisms to function correctly and effectively.

ACKNOWLEDGEMENTS

The author would like to express his grateful thanks to Professor Franz HALBERG for his encouragement during the course of these investigations, and for carrying out the cosinor analysis of ascorbic acid taste threshold and plasma rhythms. He would also like to express his thanks to Dr. H. S. LOH for collaboration and assistance in the work, and to Mr Kevin FORAN and Mr Paul DEMPSEY for their technical help and assistance in many ways.

REFERENCES

1. H.M.S.O. (1969): Rep. publ. Hlth. med. subj. Londo., No. 120.
2. WILSON C W. M. and NOLAN C. (1970): The diets of elderly people in Dublin. Irish J. Med. Sci., *3*: 345–355.
3. RICHTER C. P. (1957): Production and control of alcoholic cravings in rats. *In*: Neuropharmacology: Transactions of the Third Conference, (H. A. ABRAMSON, ed.), pp. 39–146, Josiah Mary Jr., Foundation, New York.
4. JACOBS H. L. (1967): Taste and the role of experience in the regulation of food intake. *In*: The Chemical Senses and Nutrition, (M. R. KARE and O. MALLER, eds.), pp. 187–200, Johns Hopkins Press, Baltimore.
5. PANGBORN R. M. (1967): Some aspects of chemoreception in human nutrition. *In*: The Chemical Senses and Nutrition, (M. R. KARE and O. MALLER, eds.), pp. 45–60, Johns Hopkins Press, Baltimore.
6. HENKIN R. I. (1967): Abnormalities of taste and olfaction in various disease states. *In*: The Chemical Senses and Nutrition, (M. R. KARE and O. MALLER, eds.), pp. 95–107, Johns Hopkins Press, Baltimore.
7. McAIRT J. G. P., O'BRIEN C., ROLFE D. A. H. and WILSON C. W. M. (1969): A taste for drink and its control. Irish J. Med. Sci., *1*: 579–580.
8. WILSON C. W. M. (1970): The human taste threshold to alcohol and its implications. Brit. J. Nutr., *29(2)*: 29A.

9. WILSON C. W. M. (1972): The limiting factors in alcohol consumption. Biological Aspects of Alcohol Consumption, (FORSANDER O. and ERIKSSON K., eds.), Finnish Foundation for Alcohol Studies, *20*: 207–215.
10. LOH H. S. and WILSON C. W. M. (1969): Studies in ascorbic acid taste threshold circadian rhythm in relation to plasma ascorbic acid levels. Irish J. Med. Sci., *2*: 396.
11. WILSON C. W. M. (1972): The metabolic availability of Vitamin C. Vitamines, *3*: 75–97, Editiones Hoffman-La Roche, Bâle.
12. LOH H. S. and WILSON C. W. M. (1971): The relationship between leucocyte and plasma ascrobic acid concentrations. Brit. Med. J., *3*: 733–735.
13. HARRIS H. and KALMUS H. (1949): The measurement of taste sensitivity to phenylthiourea (P.T.C.). Ann. Eugen. (Lond.), *15*: 24–31.
14. DENSON K. W. and BOWERS E. F. (1961): The determination of ascorbic acid in white blood cells. A comparison of W.B.C. ascorbic acid and phenolic acid excretion in elderly patients. Clin. Sci., *21*: 157–162.
15. LOH H. S. and WILSON C. W. M. (1971): An improved method for the measurement of leucocyte ascorbic acid concentrations. International Journal of Vitamin and Nutrition Research, *41*: 90–98.
16. LOH H. S. and WILSON C. W. M. (1970): The relationship between leucocyte ascorbic acid and plasma iron. Brit. J. Pharmacol., *40*: 566–567P.
17. LOH H. S. and WILSON C. W. M. (1971): The relationship between leucocyte ascorbic acid concentrations and the total white blood cell count. International Journal of Vitamin and Nutrition Research, *41*: 253–258.
18. LOH H. S. and WILSON C. W. M. (1971): Relationship of human ascorbic acid metabolism to ovulation. Lancet, *1*: 110–112.
19. MAYER-GROSS W. and WALKER J. W. (1946): Taste and selection of food in hypoglycemia. Brit. J. Exp. Path., *27*: 297–305.
20. GOETZL F. R., AHOKAS A. J. and PAYNE J. G. (1950): Occurrence in normal individuals of diurnal variations in acuity of the sense of taste for sucrose. J. Applied Physiol., *2*: 619–626.
21. YENSEN R. (1958): Influence of salt deficiency on taste sensitivity in human subjects. Nature, *181*: 1472–1474.
22. YENSEN R. (1959): Some factors affecting taste sensitivity in man II. Depletion of body salt. Quart. J. Exp. Psychol., *11*: 230–238.
23. HENKIN R. I., GILL J. R., BARTTER F. C. and SOLOMON D. H. (1962): On the presence and character of the increased ability of the Addisonian patient to taste salt. J. Clin. Invest., *41*: 1364–1365.
24. IRVIN D. L., AHOKAS A. J. and GOETZL F. R. (1950): The influence of ethyl alcohol in low concentrations upon olfactory acuity and the sensation complex of appetite and satiety. Permonente Fdn. Med. Bull., *8*: 97–101.
25. WILSON C. W. M. (1972): The pharmacological actions of alcohol in relation to nutrition. Proc. Nutr. Soc., *31*: 91–98.
26. GOETZL F. R., AHOKAS A. J. and GOLDSCHMIDT M. (1951): Influence of sucrose in various concentrations upon olfactory acuity and sensations associated with food intake. J. Applied Physiol., *4*: 30–36.
27. LOH H. S. and WILSON C. W. M. (1973): Vitamin C: plasma and taste threshold circadian rhythms, their relationship to plasma cortisol. Internat. J. Vit. Nutr. Res., *43*: in press.

CIRCANNUAL CYCLES IN MORNING PLASMA CORTISOL LEVELS IN MAN IN FAIRBANKS, ALASKA

BETTY ANNE PHILIP[1] and KENNETH H. HANSON[2]

[1]*Arctic Health Research Center, U.S. Public Health Service, and Institute of Arctic Biology, University of Alaska College, Alaska*

[2]*Arctic Health Research Center, U.S. Public Health Service College, Alaska, U.S.A.*

The arctic is characterized by wide variations in temperature and light patterns. However little is known about changes in normally occurring daily and seasonal cycles in steroid hormone concentrations in blood and urine under these conditions. Variations in the concentration of these compounds profoundly affect the health and functioning of the individual. Thus, with the increased immigration of a temperate zone urban white population into northern regions, human response to photoperiod and temperature extremes becomes particularly interesting.

METHODS

A. Sampling: Monthly blood samples for cortisol analysis were collected for two years (1967/68 and 1969/70) from adult males residing in or near Fairbanks, Alaska, latitude 64°49′ N, longitude 147°52′ W and quarterly for one year (1969/70) from a control group in Salt Lake City, Utah, latitude 40°45′ N, longitude 111°53′ W (Table 1). Each individual was assigned a time for blood collection between 0800 and 0930. The plasma was separated immediately and frozen for later analysis. Frozen Utah samples were hand carried to Fairbanks and intermingled with the local samples for chemical analysis. A smaller number of 24-hour urine samples was also collected by Fairbanks volunteers and frozen for later analysis for 17-ketogenic steroids (Table 2).

B. Chemical Analysis: Cortisol, 6β-OH Cortisol, and Corticosterone were extracted from plasma and measured as a group by acid fluorescence using the method of DE MOOR et al. [1]. The urinary 17-ketogenic steroid levels were determined by the borohydride-metaperiodate method [2, 5].

C. Mathematical Analysis of Data: The plasma steroid levels, expressed as cortisol concentrations, were analyzed as a function of time for each individual by least squaring the percent deviation from the year's average, Δ, to fit a Cosine curve:

$$\Delta_i = \frac{yi - C_{0i}}{C_{0i}} = C_i \cos\left(\frac{2\pi\dagger}{\tau} + \phi_i\right)$$

where y_i = Plasma cortisol concentration (μg/100 ml)

C_{0i} = Invidual's yearly average cortisol concentration

C_i = Amplitude

Table 1 Morning levels of plasma cortisol, μg/100 ml (Fairbanks, Alaska).

	1967			1968								
Subject	Oct 23	Nov 27	Dec 19	Jan 3	Feb 6	Mar 5	Apr 2	May 7	May 28	Jun 25	Jul 23	Sep 17
1	22.7	24.2	—	13.0	23.0	—	14.5	20.3	15.3	14.7	—	—
2	12.9	16.0	22.2	16.3	14.2	13.6	19.8	30.2	17.9	14.8	12.3	13.6
3	13.9	13.0	20.3	13.8	16.0	15.7	13.9	10.9	16.5	14.4	19.7	17.3
4	18.8	12.2	10.2	9.0	10.0	14.2	10.3	—	25.1	22.3	12.1	11.6
5	27.8	10.1	21.8	10.2	10.7	10.5	9.9	—	9.7	9.0	15.3	10.5
6	15.5	23.6	18.8	23.5	21.4	7.2	7.9	14.1	17.0	9.4	—	13.2
7	16.7	9.8	15.8	13.2	19.8	12.6	17.7	—	—	—	—	—
8	18.0	12.5	18.4	20.7	17.9	22.4	14.1	16.1	13.4	12.0	—	11.4
9	17.7	15.8	18.0	—	20.3	18.6	—	27.7	36.2	20.7	11.5	18.0
10	11.7	16.4	16.8	21.6	14.2	7.2	9.7	13.4	19.5	20.3	20.6	16.8
11	11.2	9.5	12.1	14.9	15.2	10.0	8.5	20.5	10.3	20.9	11.8	17.1
12	30.0	16.5	17.3	16.1	18.0	13.0	17.5	14.3	17.9	—	—	—
13	10.7	13.7	13.0	10.0	12.6	—	12.3	6.9	7.0	11.6	9.4	—
14	14.2	16.4	11.1	13.8	11.6	9.3	18.3	13.2	11.0	15.3	11.1	8.2

	1969				1970								
Subject	Sep 23	Oct 21	Nov 25	Dec 23	Jan 20	Feb 24	Mar 24	Apr 21	May 19	Jun 23	Jul 21	Aug 18	Sep 22
2*	25.0	—	—	13.6	—	—	15.8	—	—	23.7	—	—	—
3	15.0	—	27.7	27.3	13.6	20.4	15.3	19.5	17.2	25.8	30.0	18.0	14.4
5	15.6	—	—	12.7	—	—	21.1	—	—	17.0	—	—	—
6	23.9	—	—	16.2	—	—	13.1	—	—	20.8	—	—	—
9	10.0	—	—	17.8	—	—	21.7	—	—	15.8	—	—	—
10	16.4	—	—	18.1	—	—	15.6	—	—	21.3	—	—	—
14	13.2	42.5	37.0	—	16.7	43.8	29.6	12.7	23.5	18.8	15.0	19.3	13.5
15	15.0	22.0	26.7	25.4	14.4	15.0	11.6	11.3	22.1	22.5	21.0	18.5	24.1
16	13.9	12.9	24.0	12.8	19.3	21.8	16.9	16.4	14.7	15.6	—	—	—
17	15.3	19.1	19.2	9.4	19.1	21.9	18.8	11.7	18.2	15.8	14.0	17.4	23.6
18	15.3	17.8	25.6	9.1	21.5	12.1	14.4	20.0	—	—	—	—	—
19	28.0	32.9	37.0	25.0	29.4	21.4	28.3	21.0	17.1	—	12.2	21.2	17.0
20	11.9	7.3	24.4	22.4	19.0	20.0	42.1	17.5	20.6	—	—	—	9.3
21	13.3	36.7	25.3	17.3	23.2	23.6	13.4	17.5	11.1	—	—	—	—
22	10.5	15.3	16.0	12.4	11.0	12.1	10.4	21.3	18.9	10.0	—	—	—
23	21.7	15.3	28.6	17.6	24.4	19.6	13.8	18.5	9.4	33.0	29.0	19.6	26.4
24	30.0	22.7	38.5	44.1	24.3	45.0	20.9	22.8	23.9	—	26.7	25.7	25.2
25	22.5	10.9	15.3	19.5	18.1	19.6	21.1	15.7	31.7	35.0	26.7	19.6	13.0
26	—	—	—	16.3	—	—	13.1	12.5	40.0	20.0	21.7	27.2	13.0
27	—	—	—	—	—	—	—	—	34.1	27.5	28.9	39.5	18.1

* Subject number repeats indicate a volunteer for all years of study.

τ=Period (Days)

†=Serial day number with Day 1=Jan. 1, 1900

ϕ_i=Phase Angle measured in counterclockwise direction

Programs based on the Halberg Cosinor procedure [3] were developed for the University of Alaska's IBM 360 computer. For each test value of τ, vectors derived from the best fit values of C_i and ϕ_i for each individual were averaged to yield the Fairbanks population vector and error ellipses were calculated at the 90, 95 and 99% confidence levels. Periods ranging from 100 to 400 days were tested at intervals of 5 days or less. The Utah blood data and Fairbanks urine

Table 2

A. Urine: 17-Ketogenic Steroids mg/24 hrs (Fairbanks, Alaska).

	1969		1970								
Subject	Nov 25	Dec 23	Jan 20	Feb 24	Mar 24	Ap 21	May 19	Jun 23	Jul 21	Aug 18	Sep 22
15	17.4	15.8	19.6	33.0	24.9	21.4	22.7	23.3	28.4	22.6	20.1
16	25.1	30.6	21.5	15.7	21.7	19.8	23.8	19.4	—	—	—
17	16.5	18.3	11.4	14.9	17.8	15.0	12.2	—	32.7	15.2	18.2
23	—	17.0	19.3	19.2	26.5	21.9	15.3	27.0	19.1	29.3	27.0
26	—	28.2	27.9	18.2	10.7	25.6	13.5	18.2	18.9	17.0	14.8
27	—	—	—	—	—	18.4	20.9	29.0	31.2	18.3	19.2

B. Morning Levels of Plasma Cortisol μg/100 ml (Salt Lake City, Utah)

Subject	Dec 12, 1969	Mar 24, 1970	June 23, 1970	Sep 22, 1970
28	14.0	17.1	25.7	12.4
29	19.3*	13.6	13.5	33.3
30	40.8	23.6	15.0	12.1
31	14.6	12.9	18.3	18.6
32	10.7*	9.5	10.0	11.7
33	17.9	25.0	16.4	40.0
34	4.3	8.9	20.0	7.1
35	24.6*	24.3	5.0	—
36	—	8.3	12.1	7.2

* Dec 13, 1969

$C_0=16.5\pm1.9$ S.E. μg/100 ml

data were treated similarly. The 1967/68 and 1969/70 series for Fairbanks were also analyzed separately.

RESULTS

A. Chemistries: To allow other statistical approaches, observed steroid concentrations are given in Tables 1 and 2.

B. Mathematical Analysis: Combined analysis of all the Fairbanks plasma data yielded 17 periods significant at the 99% level. However, when the two yearly series were analyzed separately, these were resolved into the four periods significant in 1967/68 and the seven in 1969/70 listed in Table 3. Pseudo spectra for the separated yearly series are shown in Fig. 1, which assumes that the smaller the error ellipse, the more significant the rhythm.

The Utah data yielded only one period at the 99% level:

$\tau=215$ days; $C=.12$ (.011–.22); $\phi=1.04$ radians (6.14–2.9); $C_0=16.5\pm1.9$ μg/100 m*l*.

However, this data set may be too small for meaningful comparisons.

Cycles in the urinary steroids were sinificant only at the 95% level and are given in Table 4.

DISCUSSION

Curve fitting to deviations from an average is not as satisfactory a procedure as similar operations on absolute values. However, the Fairbanks population aver-

Table 3 Fairbanks plasma Cortisols. Rhythms significant at 99% confidence level.

Period (days)	Amplitude C′ (a)	C (b) (limits)	Phase (limits) (Radians)	Area (c) error ellipse
A. 1967/68 Series				
104	1.0	.065 (.010–.12)	.55 (6.2–2.5)	.024
114	.99	.064 (.002–.13)	.62 (6.1–2.8)	.033
150	1.1	.071 (.004–.14)	1.8 (6.2–2.7)	.048
182.625	1.3	.085 (.001–.17)	1.4 (5.8–2.5)	.057
B. 1969/70 series				
110	2.4	.12 (.010–.23)	.60 (5.9–2.1)	.088
173	1.7	.085 (.001–.17)	1.95 (.50–3.4)	.073
175	2.0	.10 (.007–.20)	.58 (5.7–1.8)	.056
182.625	2.0	.10 (.006–.20)	.90 (6.0–2.25)	.073
188	2.2	.11 (.006–.21)	.95 (6.0–2.3)	.074
198	2.4	.12 (.012–.23)	.43 (5.9–1.7)	.056
215	2.5	.125 (.014–.24)	1.6 (.36–2.6)	.079
C. Combined data				
110	1.8	.099 (.032–.17)	.60 (.018–1.5)	.027
112	1.6	.090 (.014–.17)	.91 (.046–2.2)	.041
114	1.6	.090 (.020–.16)	.89 (.080–2.1)	.037
116	1.2	.070 (.003–.14)	.76 (6.2–2.6)	.029
150	8.6	.048 (.004–.091)	1.78 (.059–2.7)	.020
168	1.0	.056 (.003–.11)	.94 (.072–2.7)	.019
173	1.1	.062 (.010–.11)	2.13 (.88–3.2)	.026
175	1.5	.082 (.026–.14)	.96 (.21–1.9)	.023
182.625	1.6	.091 (.037–.14)	1.14 (.28–1.9)	.025
185	1.5	.083 (.006–.16)	.35 (6.0–1.5)	.017
188	1.7	.096 (.033–.16)	1.03 (.33–1.7)	.024
215	1.4	.076 (.015–.14)	1.66 (.51–2.6)	.031
230	1.4	.080 (.009–.15)	.97 (.085–2.6)	.041
270	1.1	.062 (.004–.12)	2.00 (.38–3.1)	.033
350	1.1	.062 (.004–.12)	1.28 (.13–2.9)	.036
360	1.1	.060 (.001–.12)	1.20 (.081–3.0)	.036
370	1.0	.058 (.000–.12)	.95 (6.1–2.9)	.032

(a) μg/100 ml plasma; (b) Dimensionless; (c) Arbitrary Units

age for plasma cortisol in the study period 1967/68, $C_0=15.5\pm0.6$ S.E. μg/100 ml. was significantly lower than that for 1969/70, 20.3 ± 1.0 $p<.001$). Therefore this procedure was used to offset the effect of a shifting baseline. To test the effect of this adjustment on the population vectors, the latter were also calculated by averaging individual vectors formed from C_i' and ϕ_i. C_i' $(=C_{0i}\times C_i)$, the amplitude in terms of absolute values of the concentration, was calculated from the best fit value of C_i and the yearly average for each individual. When so calculated, two periods in the 67/68 series. 114 and 182.625 days, were reduced in apparent significance from $p<.01$ to $p<.05$; one period, 166 days, was increased from $p<.05$ to $p<.01$ in the 69/70 series; and two periods, 116 and 370 days, also were reduced in significance from $p<.01$ to $p<.05$ when vectors were formed using the combined data of both series.

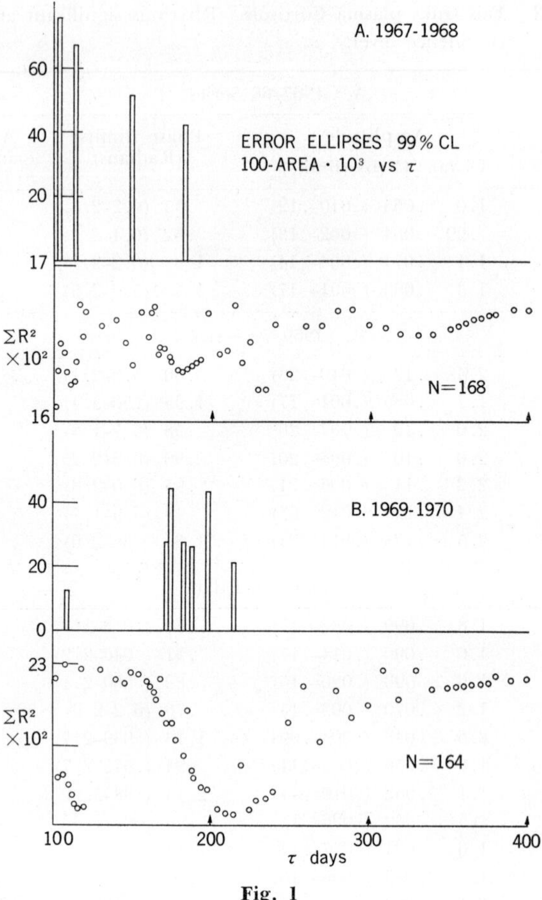

Fig. 1

Table 4 Significant Rhythms. Fairbanks, Alaska: 24 hour excretion of 17-ketogenic steroids 95% Level of confidence.

Period (Days)	Amplitude**	Limits	Phase* (Radians)	Limits
195	.074	.006–.14	6.07	3.4–.16
200	.13	.017–.24	1.2	5.5–1.8
205	.10	.010–.20	2.0	5.9–2.6
210	.11	.027–.19	1.9	6.0–2.4
215	.13	.004–.25	1.1	5.0–1.6

* Measured in conventional counterclockwise direction Day 1=Jan 1, 1900.
** $C_0=21.1\pm9$ S.E. mg/24 hrs.

The reasons for the difference between the results for 1967/68 and 1969/70 are not readily obvious. Unfortunately, urine data are not available for the earlier series. However, the urine values for the later series are higher than normally reported and are thus consistent with the observed plasma levels. In addition, although there were too few urine data to demonstrate cycles at the p<.01 level, p<.05 cycles with similar amplitudes and phases were observed for periods between 195 and 215 days.

Both plasma series yielded a significant semi-annual cycle which peaked February 11 and August 12, 1968 and January 27 and July 29, 1970 with negligible phase differences. However an equally good fit was observed for $\tau = 150$ days for the earlier series with peaks November 18, 1967; April 15 and September 12, 1968. The closely related 175 days period in 69/70 peaked December 30, 1969; June 24 and December 16, 1970. Phase and period differences here can only be resolved by further studies with more frequent sampling rates. The latter procedure could also validate the questionable high frequency cycles observed near 110 days which may be an artifact of the sampling schedule. To check for self-consistency in the data analysis and to test the feasibility of using hospital records for cycle studies, each series was analyzed as if it were one individual with the results shown in Fig. 1. Note that the observed minima in the sums of squares of deviations from fitted cosine curves ($\sum R^2$) correspond very well to significant period for each of the study years.

Possible mechanisms for semiannual cycles in plasma steroid levels include (1) Seasonal shifts in daily secretion rates, (2) Phase shifts in circadian rhythms or (3) Dampling of circadian cycles. The magnitude of the observed changes is consistent with (2). However, the urine data suggests an integrated change in level assuming that cortisol catabolism itself is not on an appreciably different cycle. The data is insufficient to make clear choices.

Physical mechanisms for the observed yearly differences are equally difficult to assess. Although photoperiod changes are known to affect seasonal plasma corticosterone levels in mice by changing circadian phase [4], the annual light/dark cycle is relatively constant from year to year. The possibility exists that in man, a combination of environmental factors may act not only to shift the circadian phase but also to change the daily secretion rate.

REFERENCES

1. De Moor P., Steeno O., Raskin M. and Hendrix A. (1960): Acta Endocrinologica, *33*: 297–307.
2. Few J. D. (1961): Journal Endocrinology, *22*: 31–46.
3. Halberg F., Tong Y. L. and Johnson E. A. (1967): *In*: The Cellular Aspects of Biorhythms. Symposium on Rhythmic Research, VIIIth International Congress of Anatomy, Wiesbaden, August 1965, pp. 20–48, Springer-Verlag, Berlin.
4. Haus E. and Halberg F. (1970): Environmental Research, *3*: 81–106.
5. Metcalf M. (1963): Journal Endocrinology, *26*: 415–423.

CHRONOPATHOLOGY, IMMUNOLOGY AND CANCER

Chairman: SERGIO S. CARDOSO

Co-Chairmen: FREDERICK URBACH
DIMITRIJE VULOVIĆ

AUTORHYTHMOMETRY IN RELATION TO RADIOTHERAPY: CASE REPORT AS TENTATIVE FEASIBILITY CHECK*

K. CHARYULU, F. HALBERG, E. REEKER, E. HAUS
and H. BUCHWALD

Chronobiology Laboratories, Department of Pathology, and the Departments of Radiology and Surgery University of Minnesota, Minneapolis, Minnesota, U.S.A. and the Department of Radiation Therapy University of Miami, Miami, Florida, U.S.A.

INTRODUCTION

A patient described elsewhere [1] continued self-measurements while receiving ("timed") radiotherapy administered with consideration of circadian system phase at time of exposure. Early work on experimental animals and related findings on human susceptibility rhythms [2] suggest the desirability of timing radiotherapy as a function of circadian system phase. The LD_{50} from whole body radiation is a circadian rhythmic index for inbred C mice. In the rat, the mean survival time and even the weight loss from partial body irradiation also depend upon the stage of the circadian system. For a critical discussion of the extensive literature on the subject of rhythms and radiosensitivity, the interested reader can be referred to a review [3].

The question whether the radiosensitivity rhythm is amenable to phase shifting has thus far been tested only in the mouse by manipulation of the lighting regimen. Since the murine radiosensitivity rhythm's phase-shift was incomplete at two weeks after the lighting reversal [3], the possible interaction of other synchronizers with the lighting regimen in determining the timing of this rhythm deserves further study. Moreover, since results from phase-shift studies as yet are incomplete even in experimental animals, no attempt was made in this study to phase-shift the subject's rhythms, nor was the attempt made to time radiation according to a specific circadian rhythm such as that in body temperature. The choice of timing was based upon the observation that in the rat or mouse, resistance to radiation is found to be highest toward the end of the daily light span (corresponding to the daily rest span of these rodents). In the absence of an indication that the timing of other variables may be more pertinent—a topic for further study indeed—it was decided in this case to time radiotherapy toward the end of the patient's customary rest span, namely in the early morning hours.

RADIOTHERAPY

In order to achieve this aim without the possible undue disturbance of physiologic functions associated with an early morning drive to the hospital, the

* Supported by grants from the United States Public Health Service (5-K6-GM-13,981) NCI 1RO1-CA-14445-01 and NASA.

patient entered the radiotherapy room each evening at about 2200 and spent the night sleeping there. She received treatment at 0700 the following morning, leaving the hospital for the remainder of the daylight hours. Treatments were given daily Monday through Friday and were omitted not only on weekends but also on holidays (in this case, Christmas, and New Years Day, December 25, 1968, and January 1, 1969, respectively).

The patient was initiated on a course of radiation therapy on 12/11/68. Therapy was planned with an L-shaped internal mammary supra-clavicular and axillary field and two tangential fields—the medial tangential being adjacent to the parasternal mark and the lateral tangential along the mid-axillary line. The planned dose was 4000 rads at 4 cm for the L-shaped port and the same dose at the midplane for the tangential fields of the chest wall. Bolus (unit-density material) was used for the tangential fields to uniformly irradiate surface and chest wall tissues.

During therapy the patient had no significant nausea, malaise, vomiting or other symptoms. As usually occurs, a sore throat had developed by 12/21/68, rapidly becoming intense and persisting for about four days after completion of therapy on 1/2/69. Moderate erythema of the treated area was also evident at this time. She was seen again about three weeks post-therapy; at that time the mastectomy incision site was well healed. There was also a certain amount of deep erythema associated with a marked peeling of the skin of the treated areas.

SURGERY (A SECOND OPERATION)

The patient was readmitted to the hospital on April 16, 1969, with a small palpable mass in the opposite breast, the right breast. The mass measured approximately 2 cm and was superior and lateral to the right nipple. Under a general anesthesia the lesion was excised and the wound primarily closed.

Recovery from the procedure was uneventful. Immediate study of frozen sections and subsequent histological examination led to the diagnosis of a benign fibroadenoma.

MATERIALS AND METHODS

The kinds of self-measurements made, along with the units used for expressing results, are summarized as a table in a related paper [1]. The durations and sampling details of the several measurement series reveal that not all variables were evaluated in all stages. Body core temperature and heart rate were evaluated consistently, and results on these variables before and after surgery have been compared [1].

Series of determinations on blood and urine during the post-surgical observation span are available as indices for the timing of radiotherapy but were not evaluated before radiotherapy was administered. The ability to judge the passage of two minutes, included as a "tempo", rather than performance test, only during the first stage represents a side issue and serves for comparison with similar series on presumably healthy subjects.

Oral temperature was evaluated by an Ovulindex thermometer left in the mouth for five minutes. Heart rate was evaluated for one minute as was the radial pulse. The methods for electronic computer analyses have been described elsewhere [4].

Blood was taken by fingerpricks three to four times a day at intervals of at least three hours. Blood sampling varied during the course of the study from sampling three times a week, Monday, Wednesday and Friday, to sampling once weekly but on different days of the week. Urine volume was measured and separate samples were collected to analyze potassium, sodium, chloride, calcium, 17-hydroxycorticosteroids and 17-ketosteroids on the one hand, and for catecholamine determinations on the other hand. About 30 ml were collected for the former determinations and 50 to 60 ml over sodium metabisulfite for catecholamine determinations.

RESULTS

Whenever noise-to-rhythm ratios obtained by fitting a 24-hour cosine curve to the data are around or below .33 and the sample consists of 20 or more observations, a circadian rhythm is fairly acceptably described. A table in reference 1 shows that for certain variables in a patient after mastectomy and before radiotherapy, such ratios are favorable for rhythm description. Table 1 in this paper reveals during and after radiotherapy a rhythm alteration or dyschronism for several variables investigated.

The behavior of the urinary variables can be further scrutinized in Figs. 1, 2 and 3. Figs. 1 and 2 show a rather stable acrophase for the rhythms in urinary potassium and volume before, during and after radiotherapy. Fig. 3, in turn, summarizing the time course of sodium excretion, suggests a rhythm alteration. The interpretation is rendered difficult by an unfavorable noise-to-rhythm ratio. With this qualification, there seems to occur an acrophase change shortly after institution of radiotherapy, with an eventual apparent resynchronization (first with an odd acrophase and then with an anticipated acrophase) following the institution of aspirin administration while radiotherapy is continued. One can only speculate whether a possible desynchronization of the circadian rhythm in sodium excretion is a result of tissue destruction. The question also remains whether any possible partial desynchronization of the urinary sodium excretion rhythm may be a general phenomenon and, if so, whether it should be avoided. The data do suffice however to make the point that during radiotherapy it is not necessarily the case that all rhythms are consistently synchronized with the 24-hour social day. Tentatively it is concluded that autorhythmometry following a radical mastectomy, before, during and after radiotherapy, not only is possible but will have to be carried out systematically if treatment is to be timed according markers.

In the case under study, action was required promptly. We regard the present results simply as a feasibility check. Systematic work with proper design on groups of comparable patients consistently exposed to radiotherapy in the several physiologically defined circadian phases is overdue. Before selecting a reference rhythm for such work, several variables should be investigated as candidate standards.

Whether the sodium excretion rhythm is necessarily a critical one in timing treatment constitutes a different though related question also awaiting further study.

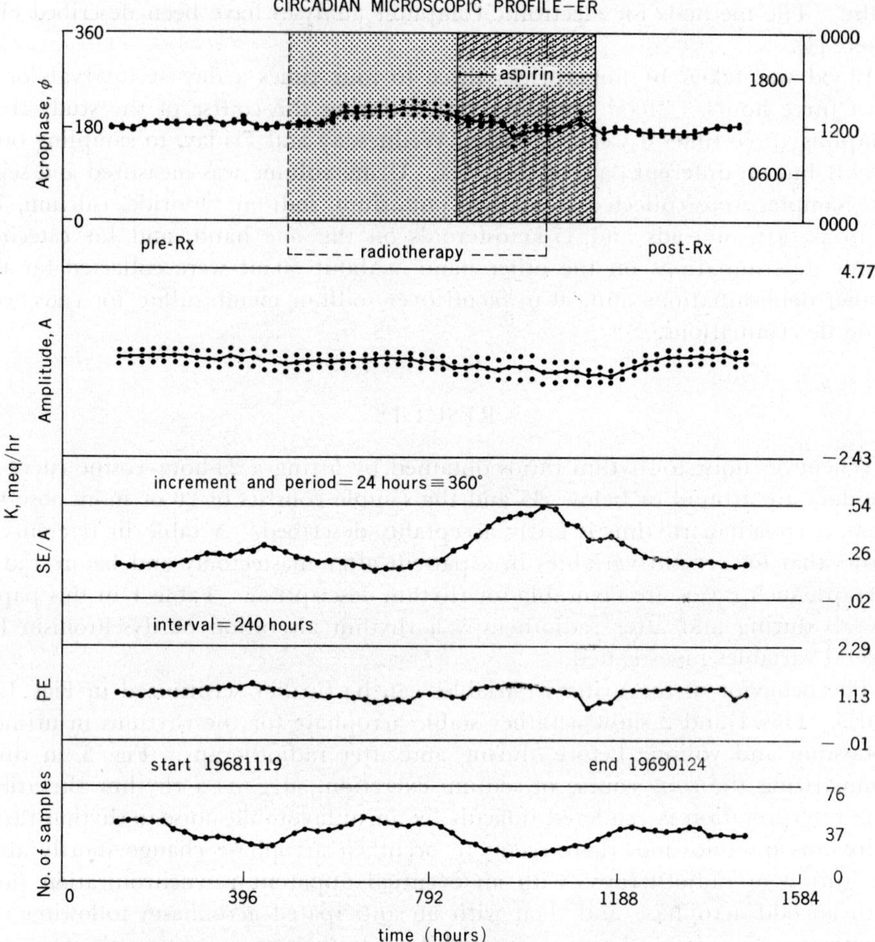

Fig. 1 Chronobiologic serial section of urinary potassium rhythm. In consecutive rows from top to bottom:
acrophase as a function of time;
amplitude;
noise-to-rhythm ratio;
minimal percent error; and
number of samples.

Note stable acrophase and amplitude with some transient worsening of noise-to-rhythm ratio toward end of radiotherapy span when aspirin also is taken.

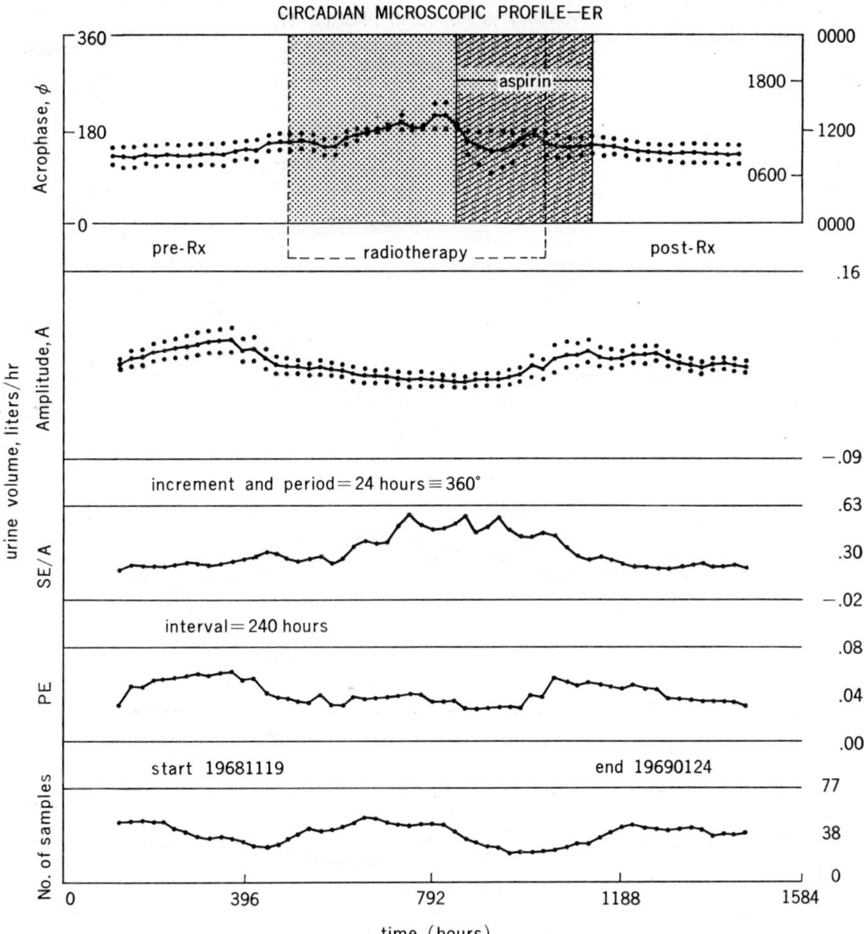

Fig. 2 Chronobiologic serial section of urinary volume excretion in a patient receiving radiotherapy after radical mastectomy (see text and legend to Fig. 1).

With proper planning, future studies may well attempt to quantify both the desired and the undesired effects. The current study, then, can no more than suggest that some circadian rhythms that are quite prominent in health persist even after mastectomy but can be radically altered during radiotherapy. Circadian rhythm assessment not only prior to timed radiotherapy but also during treatment is therefore a desideratum. In clinical work timed according to rhythms the drastic effect of radiotherapy upon certain rhythms, as shown in Table 1 and Fig. 3, herein must be kept in mind. That such timing is desirable emerges from recent findings in our laboratory that the bone-marrow-depressing effect of whole-body irradiation in the mouse also is circadian-system-phase-dependent and thus can be reduced in extent [5].

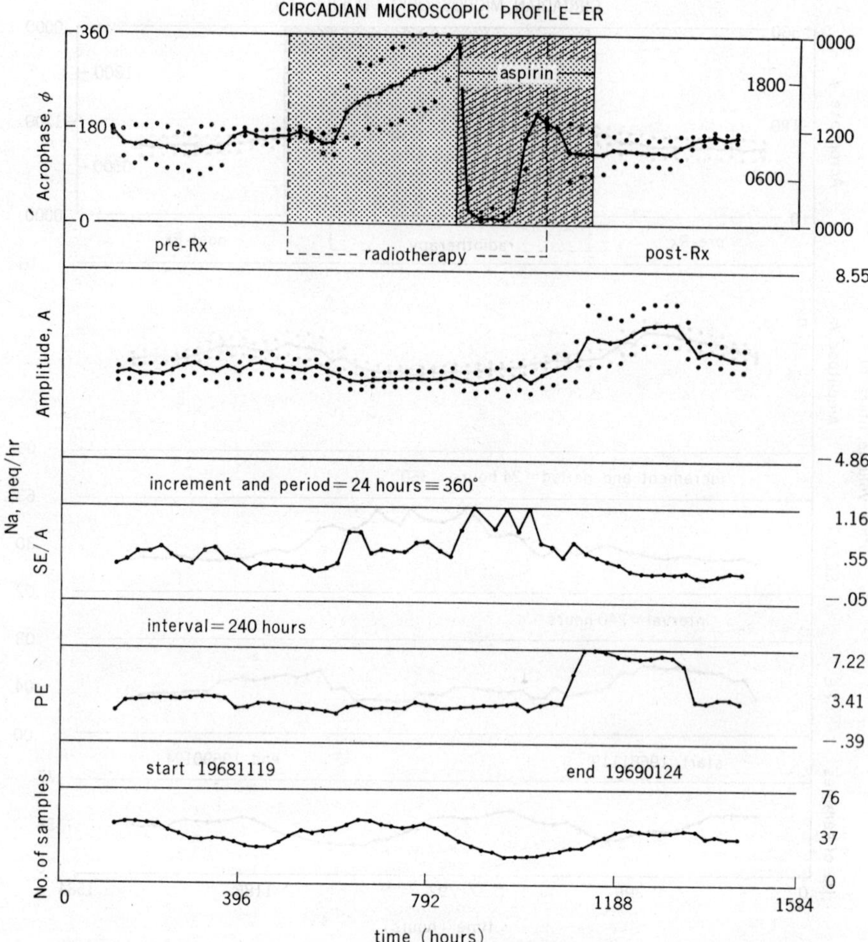

Fig. 3 Chronobiologic serial section of urinary sodium excretion in a patient receiving radiotherapy after radical mastectomy. Note drastic changes in sodium acrophase during radiotherapy and increase in rhythm's amplitude after treatment.

Table 1 Results from fitting 24-hour cosine curve to data of different stages (IIA to IIC) from a patient receiving radiotherapy.*

Subject stage*	Variable unit	N of Samples	Mesor, M M±SE	Amplitude, A A±SE	Noise-to rhythm SE/A	Acrophase (.95 confidence arc)
IIA	Urine volume,	89	70±5.0	50±7	$.14_2$	−97° (− 80 to −114)
IIB	ml/h	83	60±4.0	20±6	$.30_7$	−131° (− 98 to −164)
IIC		86	50±5.0	40±7	$.14_9$	−105° (− 89 to −122)
IIA	Sodium,	89	5.83± .30	1.46± .41	$.27_9$	−124° (− 92 to −156)
IIB	mEq/h	81	6.25± .31	.19± .43	2.24	−208°
IIC		83	6.53± .60	3.20± .84	$.26_4$	−118° (− 90 to −146)
IIA	Potassium,	89	2.76± .13	1.37± .19	$.13_9$	−143° (−128 to −158)
IIB	mEq/h	81	2.64± .13	1.14± .19	$.16_4$	−166° (−150 to −183)
IIC		83	2.30± .13	1.15± .19	$.16_2$	−145° (−128 to −162)
IIA	Chloride,	88	7.65± .39	4.13± .53	$.12_9$	−119° (−104 to −134)
IIB	mEq/h					
IIC						
IIA	Calcium, mg/h	88	6.16± .27	1.25± .35	$.28_6$	−111° (− 76 to −145)
Controls	Oral Temp					−259° (−224 to −294)
IIA		105	98.22± .05	.21± .08	$.39_3$	−136° (−101 to −171)
IIB		93	97.98± .07	.35± .12	$.33_5$	−164° (−136 to −191)
IIC		88	98.12± .06	.10± .08	$.62_5$	−151°
				A as % of M		
Controls	Polymorpho-	7		16±2		−212° (−198 to −225)
IIA	nuclear cells	21		12±3	$.29_5$	−154° (−119 to −189)
IIB		18		5	$.70_9$	−166°
IIC		36		5±3	$.34_2$	−167° (−136 to −199)
Controls	Lymphocytes	7		18±3	$.18_7$	−305° (−284 to −326)
IIA		21		14	$.61_5$	−330°
IIB		18		24	$.55_1$	−331°
IIC		36		24±7	$.28_0$	−346° (−320 to − 12)
Controls	Eosinophils	7		29±10	$.36_5$	−342°
IIA		21		34±14	$.41_7$	− 8°
IIB		17		26	1.57	−169°
IIC		35		16	$.73_9$	−332°
Controls	Monocytes	7		12± 2	$.19_6$	−339° (−319 to −359)
IIA		19		46±12	$.26_0$	−312° (−278 to −345)
IIB		18		40	$.66_8$	−123°
IIC		34		9	1.11	−148°
Controls	Basophils					
IIA		21		32± 9	$.27_8$	−279° (−242 to −316)
IIB		18		9	1.86	−323°
IIC		35		9	1.31	−352°

* Stage IIA=pre-treatment span, XI/27 to XII/9/68.

Stage IIB=radiation therapy treatment span, XII/11/68 to I/I/69.

Stage IIC=post-treatment span, I/3 to IV/11/69.

Acrophase reference=middle of average daily sleep span.

Some noise-to-rhythm ratios are favorable in Stage I; marked increase in ratio during treatment for all blood cell rhythms studied. Except for lymphocytes, circadian dyschronism of white blood cells persists in Stage III.

REFERENCES

1. Buchwald H., Halberg F., Charyulu K. and Reeker E. (1973): Autorhythmometry and cancer of the breast—a case report. *In*: Chronobiology. Proc. Symp. Quant. Chronobiology, Little Rock, 1971, (L. E. Scheving, F. Halberg and J. E. Pauly, eds.), pp. 293–298, Igaku Shoin Ltd., Tokyo.
2. Reinberg A., Zagula-Mally Z., Ghata J. and Halberg F. (1969): Circadian reactivity rhythm of human skin to house dust, penicillin and histamine. J. of Allergy, *44*: 292–306.
3. Haus E., Halberg F., Loken K. and Kim Y. S.: Problems associated with circadian rhythmometry of mammalian radiosensitivity. *In*: Space Radiation Biology, (A. Tobias and P. Todd, eds.), AIBS Publ., (in press).
4. Halberg F., Tong Y. L. and Johnson E. A. (1967): Circadian system phase—an aspect of temporal morphology; procedures and illustrative examples. *In*: The Cellular Aspects of Biorhythms, (H. von Mayersbach, ed.), pp. 20–48, Springer-Verlag, Berlin.
5. Halberg F., Haus E., Cardoso S. S., Scheving L. E., Kühl J. F. W., Shiotsuka R., Rosene G., Pauly J. E., Runge W., Spalding J. F., Lee Y. K. and Good R. A. (1973): Toward a chronotherapy of neoplasia; Tolerance of treatment depends upon host rhythms. Experimenta, *29*: 909–934.

CIRCADIAN ASPECTS OF PARKINSONISM AND NEUROSIS*

HELMUT COPER[1)], WOLGANG GIRKE[2)], FRANZ HALBERG[3)]
and HELMUT KÜNKEL[4)]

[1)]*Institute for Neuro-Psycho-Pharmacology, Free University of Berlin Germany*
[2)]*Psychiatric and Neurologic Clinic, Psychiatric Clinic I Free University of Berlin, Germany*
[3)]*Chronobiology Laboratories, University of Minnesota Minneapolis, Minnesota, U.S.A.*
[4)]*Institute for Clinical Neurophysiology and Experimental Neurology Medical Academy of Hanover, Germany*

The chronobiology and possible chronopathology of parkinsonism was explored on a limited number of patients by comparing them to subjects without organic neurologic defect yet with the psychosomatic label "neurosis". This hybrid group-study establishes by the techniques of both individual and mean cosinors the occurrence of circadian rhythms in several variables of the patients studied.

MATERIAL AND METHODS

Over a span of several months, we examined, one or two patients at a time, from two groups [each composed of 5 human males, 40–64 years of age (mean age 55.4 years)]. A history of encephalitis or neurologic disease in the family background was absent for both groups.

One group consisted of patients with parkinsonism with roughly comparable neurologic status. All of these patients received conventional physical and drug treatment before and during the observations but had not been given l-dopa.

A second group consisted of subjects matched by age (within one year) with the patients suffering from parkinsonism. These patients were being treated clinically for unspecific psychosomatic or neurotic illness, yet appeared to be intact from the viewpoint of an organic neurologic examination. For at least one week before study and during an ensuing week of actual sampling, the subjects adhered strictly to a daily schedule of getting up at 0600, retiring for sleep at 2200 and eating each day at the same clock hours. For each series, data collection started at 0600 or 0700 on a Monday and terminated at 2000 or 2100 on a Saturday, documenting a span of 134 or 135 hours. Data collection was performed at 2-hour intervals during the "light span" (0600 to 2200); thus 48 data points were obtained on each patient and variable. Table 1 lists the ten variables analyzed.

Urinary cortisol was determined according to the method of SILBER and PORTER [1] and urinary VMA according to PISANO, CROUT and ABRAHAM [2]. Data analysis was accomplished by preparing first a least squares spectral window in the region of 28 to 20 hours with .1-hour intervals between consecutive trial periods on each individual time series [3]; thereafter, results of the 24-hour trial

* Supported from USPHS (5-K6-GM-13,981 and 1RO1-CA-14445-01), NSF GW-7613, and NASA.

Table 1A Rhythms described.*

Variable	Circadian rhythms			
	Neurosis		Parkinsonism	
	Hybrid	Individual	Hybrid	Individual
Oral temperature	+	5	+	4
Pulse	—	4	—	3
Systolic blood pressure	—	0	—	2
Diastolic blood pressure	—	2	—	3
Vital capacity	+	1	—	2
2 Minute estimation	—	2	—	2
Urine volume/hr	—	4	—	3
Urine creatinine/hr	—	3	—	3
Urine VMA/hr	—	2	+	4
Urine cortisol/hr	—	5	+	5

* Probability of obtaining estimated parameters ≤ 0.05 if sinusoidal rhythm with period of 24 HR were absent. + = yes; — = no

Table 1B Group acrophases for 2 variables of patients with neurosis and parkinsonism, found to differ with statistical significance (Wilcoxon test); other rhythm characteristics are pooled.

Variable	Diagnosis	Nr.	Percent rhythm	Mesor	Amplitude	Acrophase ϕ degrees (Hr Min)	SE (ϕ)
Oral temperature (°C)	Neurosis	5	26	36.2	0.25	−256 (1704)	19
	Parkinsonism	4				−230 (1520)	12
Urine cortisol (mg/h)	Neurosis	5	29	189.89	82.83	−186 (1224)	15
	Parkinsonism	5				−161 (1044)	

period were summarized separately for each 6-day time series as an intermediate-length individual sample [4]. Finally, all series for a given variable were treated as a hybrid sample and subjected to group analysis by mean cosinor [5], separately for patients with and without parkinsonism.

RESULTS AND DISCUSSION

Tables 2 and 3 present some of the results on individual time series obtained by the spectral window. The column headed by P provides an estimate of rhythm description. The percent variability accounted for by the fit is given under PR (for percentage rhythm). The mesor and amplitude of the 24-h fit follow with their standard errors. From the column headed "Period" under "Best fit" it can be seen that for most subjects and variables a best-fitting period was near a 24-hour length and in 4 cases at that precise length. Indeed >90% of the best-fitting periods were within a range of trial periods extending from 21 to 27 hours. By the same token, the central tendency of these trial periods was similar for the two groups; a possible inter-group difference in the location of the best fitting τ was not detected.

Table 2 Rhythmometrics of parkinsonism (P) and neurosis (N).

Variable investigated	Condition	patient No.	At 24-h Trial Period (τ)						Best fit (if τ different from 24 h)			
			P	PR	Mesor	SE	Amplitude	SE	Period	P	PR	Amplitude
Oral temperature (°C)	P	1	.02	17	36.2	.1	.3	.1	24.4	.01	18	.3
		2	.01	45	36.0	.0	.3	.1	24.4	.01	48	.3
		3	.01	20	36.6	.0	.2	.1	24.4	.01	21	.2
		4	.01	27	36.1	.0	.3	.1	23.8	.01	28	.3
		5	.25	06	36.4	.0	.1	.1	22.0	.01	18	.2
	N	1	.01	19	36.0	.1	.2	.1	22.6	.01	26	.2
		2	.01	18	36.4	.1	.2	.1	26.6	.01	50	.4
		3	.01	21	36.0	.1	.2	.1	22.8	.01	24	.2
		4	.01	35	36.2	.0	.2	.1	24.2	.01	35	.2
		5	.01	40	36.1	.1	.4	.1	23.8	.01	41	.4
Pulse (beats/min)	P	1	.12	09	92	1.2	3.5	1.8	23.8	.12	09	3.4
		2	.01	63	81	1.1	8.7	1.0	23.2	.01	70	9.3
		3	.57	02	77	.5	.6	.6	22.4	.21	07	1.0
		4	.02	15	64	1.2	5.6	2.0	24.4	.02	16	5.6
		5	.01	33	67	.7	3.4	.8	23.4	.01	37	3.5
	N	1	.02	15	79	1.3	5.8	2.0	26.4	.01	18	4.2
		2	.37	04	97	1.5	2.4	2.0	26.6	.01	23	5.1
		3	.01	27	90	1.4	5.6	1.3	22.4	.01	46	7.8
		4	.05	13	81	1.7	4.2	1.7	22.6	.01	17	5.0
		5	.01	18	91	1.5	5.0	1.7	22.6	.01	32	6.6
Systolic blood pressure (mmHg)	P	1	.15	08	128	1.3	4.1	2.1	20.2	.01	28	5.0
		2	.85	01	151	1.7	1.2	2.4	26.6	.03	15	4.7
		3	.01	37	161	1.7	9.7	2.1	23.4	.01	40	9.8
		4	.95	00	128	1.5	.8	2.4	21.2	.43	04	2.1
		5	.01	20	124	1.5	5.1	1.6	25.0	.01	23	6.7
	N	1	.45	03	139	1.3	2.7	2.1	21.6	.01	36	5.7
		2	.10	10	121	1.5	3.9	2.1	23.4	.09	10	3.7
		3	.21	07	125	1.6	2.7	1.6	26.8	.01	41	7.3
		4	.45	03	123	1.3	1.9	1.7	20.2	.33	05	2.0
		5	.24	06	130	1.1	2.0	1.3	20.4	.18	07	2.2
Diastolic blood pressure (mmHg)	P	1	.01	18	90	1.2	5.6	1.9	24.0			
		2	.02	16	78	.9	3.9	1.4	23.2	.01	22	4.3
		3	.01	26	106	1.1	4.8	1.4	23.8	.01	26	4.7
		4	.45	03	89	1.3	1.5	1.2	25.4	.28	05	2.1
		5	.15	08	86	1.2	2.3	1.2	22.4	.03	14	3.6
	N	1	.05	12	93	1.1	3.5	1.6	22.8	.01	18	4.0
		2	.15	08	83	1.0	2.5	1.4	20.2	.04	13	2.3
		3	.01	24	79	1.1	4.3	1.3	26.6	.01	40	5.5
		4	.12	09	84	1.1	2.6	1.4	24.2	.12	09	2.6
		5	.34	05	89	1.2	2.4	1.8	24.2	.33	05	2.4
Vital capacity	P	1	.01	19	2.2	.0	.3	.1	24.2	.01	19	.3
		2	.56	02	3.1	.1	.1	.1	26.6	.12	09	.1
		3	.01	31	3.4	.1	.2	.1	24.4	.01	32	.3
		4	.73	01	4.4	.1	.1	.1	27.0	.01	33	.3
		5	.68	02	2.7	.1	.1	.1	20.8	.04	13	.1
	N	1	.01	19	3.7	.1	.2	.1	24.6	.01	20	.2
		2	.09	10	3.8	.1	.2	.1	22.8	.01	23	.2
		3	.53	03	4.5	.1	.1	.1	28.0	.01	53	.4
		4	.27	06	3.8	.1	.1	.1	20.2	.01	23	.2
		5	.27	06	1.9	.0	.1	.1	26.8	.01	18	.1

Table 3 Rhythmometrics of parkinsonism (P) and neurosis (N).

Variable investigated	Condition patient No.	At 24-h Trial period (τ) P	PR	Mesor	SE	Amplitude	SE	Best fit (if τ different from 24h) Period	P	PR	Amplitude
2′–Tempo (seconds)	P 1	.01	32	100.5	2.3	14.1	3.4	23.0	.01	48	15.2
	2	.31	05	108.3	.9	1.4	1.0	28.0	.01	23	3.0
	3	.01	19	135.4	2.6	12.9	4.1	24.2	.01	19	13.3
	4	.18	07	121.7	1.5	4.3	2.4	21.6	.08	10	3.7
	5	.98	00	117.1	2.2	.5	2.9	27.4	.13	09	4.6
	N 1	.27	06	124.4	1.7	3.7	2.5	23.0	.11	09	4.4
	2	.73	01	123.2	3.4	4.3	5.4	20.8	.01	24	11.5
	3	.31	05	123.6	1.6	3.0	2.2	21.4	.21	07	2.9
	4	.01	25	122.3	1.7	7.1	2.1	23.8	.01	25	7.0
	5	.01	29	117.5	3.9	17.1	4.2	25.6	.01	41	0.6
Urine – Volume	P 1	.06	13	41.9	3.3	10.4	4.2	22.8	.02	19	12.1
	2	.01	59	62.9	3.2	34.4	4.4	24.2	.01	60	34.5
	3	.15	14	38.9	4.1	10.7	5.3	25.6	.04	24	14.5
	4	.01	36	52.2	3.5	22.3	4.9	24.0			
	5	.01	44	38.4	3.0	20.2	4.2	23.8	.01	44	20.3
	N 1	.01	58	46.7	3.5	34.7	4.8	24.2	.01	59	35.2
	2	.01	40	35.8	1.7	11.4	2.1	23.4	.01	42	11.7
	3	.83	01	47.3	3.6	2.9	4.7	20.8	.18	10	8.5
	4	.02	21	42.5	3.6	14.5	4.9	26.0	.01	23	14.9
	5	.01	48	47.6	3.9	32.0	5.2	24.8	.01	52	32.5
Urine – Creatinine	P 1	.04	23	61.6	4.1	14.5	5.3	25.0	.02	27	16.2
	2	.05	14	71.6	2.1	7.2	2.9	23.6	.05	15	7.2
	3	.02	24	56.5	3.0	12.8	4.2	22.4	.01	32	14.4
	N 1	.01	21	60.8	3.8	16.8	5.4	23.4	.01	22	16.5
	2	.05	13	72.8	2.7	8.5	3.4	24.4	.04	13	8.8
	3	.50	04	77.2	2.6	4.2	3.5	21.2	.20	09	6.1
	4	.45	05	71.8	1.9	3.3	2.6	26.4	.08	14	5.5
	5	.01	25	66.1	3.6	18.3	5.0	23.8	.01	25	18.3
Urine – VMA	P 1	.09	11	200.0	22.7	69.2	30.9	23.0	.06	13	72.5
	2	.01	28	222.6	15.6	84.0	20.5	23.8	.01	29	84.6
	3	.05	22	181.7	15.7	53.2	20.2	25.0	.03	26	59.5
	4	.01	22	295.4	10.3	46.4	14.0	23.4	.01	24	48.2
	5	.01	32	220.4	12.9	69.1	18.5	22.6	.01	42	77.4
	N 1	.18	09	211.4	27.6	72.5	38.6	23.2	.15	10	73.0
	2	.03	15	255.0	10.9	39.6	14.5	25.0	.01	18	42.4
	3	.24	08	261.1	13.5	31.9	18.6	22.0	.10	13	38.5
	4	.41	05	177.5	9.1	16.8	12.4	27.0	.07	15	26.3
	5	.01	40	204.7	13.6	96.7	18.4	24.2	.01	41	96.9
Urine – Cortisol	P 1	.02	18	201.0	17.3	66.2	22.2	23.0	.01	24	74.2
	2	.01	52	190.3	12.3	115.0	16.8	24.4	.01	53	117.0
	3	.05	21	186.3	20.7	71.5	28.2	25.2	.01	30	86.1
	4	.02	19	176.8	13.5	54.5	18.4	24.8	.01	23	60.9
	5	.01	57	168.1	12.1	104.3	16.8	23.8	.01	57	104.0
	N 1	.01	28	199.1	26.9	142.4	37.7	24.0			
	2	.01	19	197.3	8.9	39.9	12.5	24.0			
	3	.04	18	202.7	13.8	51.2	18.9	22.6	.01	26	58.6
	4	.01	38	203.4	12.3	75.5	16.8	24.2	.01	38	75.9
	5	.01	19	174.0	25.0	108.0	34.6	24.6	.01	20	109.0

The significance of the deviation of most "best-fitting" periods shown in Table 2 from a precise 24-hour period remains questionable; it may be presumed that added measures for standardization and/or a longer span of life on a fixed routine prior to sampling might have resulted in a closer synchronization of individuals in the two groups.

Indeed it can be seen from Table 1 that in a group summary by mean cosinor (hybrid) a within-group synchronization of the subjects with neurosis could be found only for oral temperature and vital capacity. For the patients with parkinsonism, a mean cosinor detected a group rhythm for oral temperature, urinary VMA and urinary cortisol.

The failure in neurosis to detect a cortisol rhythm—a most consistent and prominent rhythm in health [6]—calls for comment. The differences in the acrophases at the trial period of 24 hours indeed were small for the group with neurosis, in keeping with results on the group with parkinsonism or the healthy subjects studied earlier [6]. The failure of the mean cosinor to detect a rhythm is probably not attributable to differences in acrophase but rather to differences in amplitude; it is thus perhaps a failure due more to the limitations of the analytical technique than to a true desynchronization among the subjects investigated.

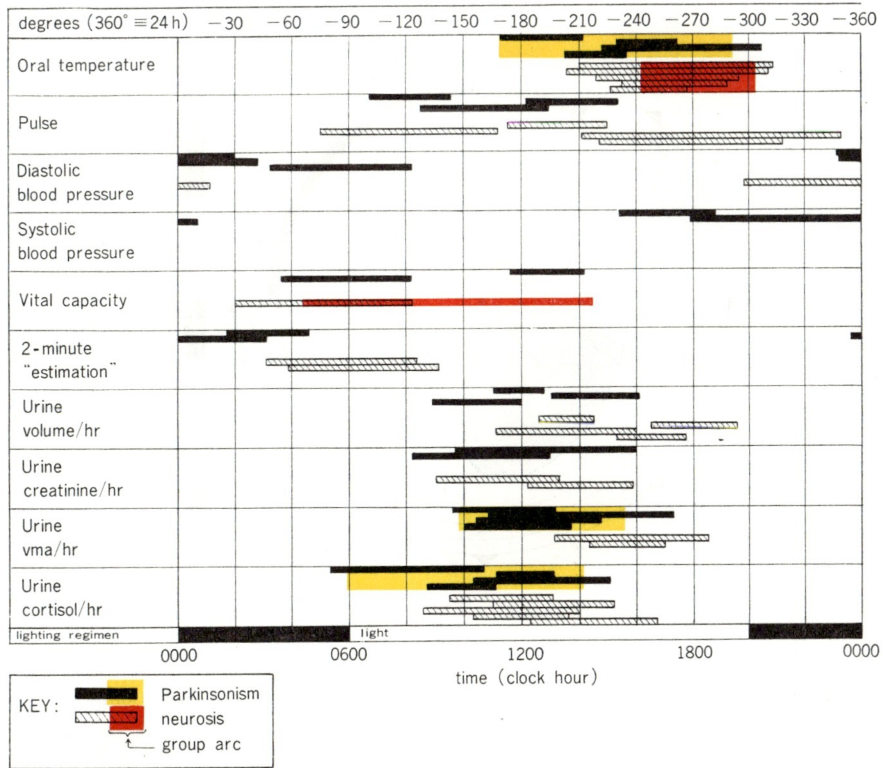

Fig. 1 95% confidence arcs for circadian acrophases in parkinsonism and neurosis.

The extent of acrophase agreement for all of the variables investigated is shown in Fig. 1. The analytical results are the confidence arcs of the circadian acrophases, shown as black bars for each patient with parkinsonism and as hatched bars for each patient with neurosis. Upon inspection of Fig. 1, it appears that for a number of variables the acrophases occur somewhat earlier in patients with parkinsonism as compared to those with a diagnosis of neurosis. The acrophase lead varies between 25° and 60°, i.e., a span from approximately 1.5 to 4 hours. Unfortunately, results that are usable because of a significant rhythm description are limited by the small sample. The acrophase lead is seen in the two cases, from each group compared, for 2′-tempo and for urine volume and VMA. The hypothesis of intergroup differences can be tested nonparametrically for the case of oral temperature and again for urinary cortisol—variables that are prominently rhythmic in the subjects investigated. A statistically significant difference thus is found between the acrophases of subjects with parkinsonism and with neurosis, the acrophase lead being on the average about 25° (100 minutes). A statistically significant difference was not detected in the PR's (percent rhythm), mesors and amplitudes of the two groups, Table 1B.

Table 1A also reveals that a statistically significant intergroup difference can not be found in a number of rhythms described for various variables in neurosis and parkinsonism. Rhythms were described for 28 time series from patients with neurosis and 31 series from patients with parkinsonism.

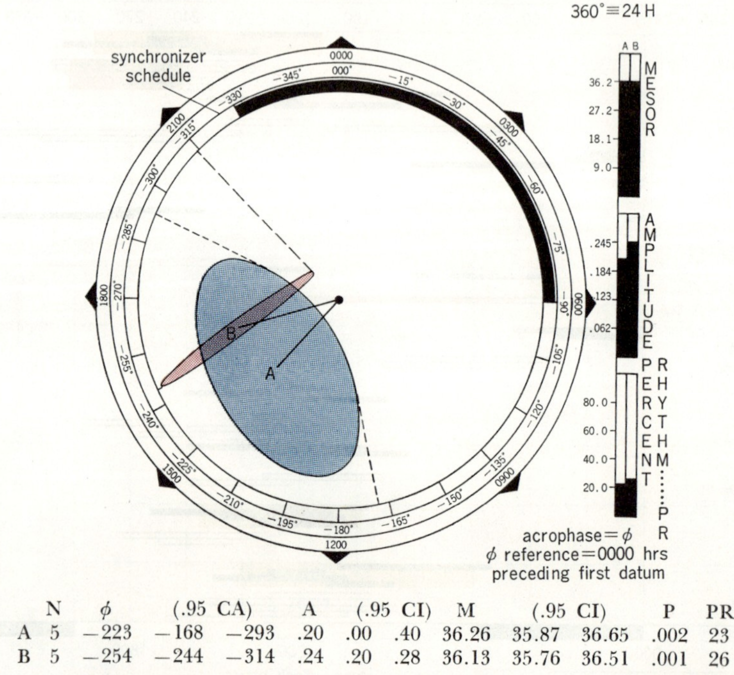

	N	ϕ	(.95 CA)		A	(.95 CI)		M	(.95 CI)		P	PR
A	5	−223	−168	−293	.20	.00	.40	36.26	35.87	36.65	.002	23
B	5	−254	−244	−314	.24	.20	.28	36.13	35.76	36.51	.001	26

Fig. 2 Circadian rhythms of oral temperature in patients with A: Parkinsonism. B. Neurosis.

Fig. 2 shows the mean cosinor representation of the oral temperature rhythm in the two groups investigated. The difference in acrophase is relatively small, i.e. 31 degrees (2 hours and 4 minutes). The amplitudes are similar in the two groups and the rhythm is detected with statistical significance in each.

SUMMARY

Circadian rhythms characterize oral temperature, pulse, systolic and diastolic blood pressure, vital capacity, two-minute estimation (or rather two-minute "tempo") as well as urinary volume, creatinine, VMA and cortisol in certain patients with parkinsonism and neurosis. A six-day time series restricted to the wakefulness hours, except for an uninterrupted collection during the sleep span in the case of urinary variables, suffices to reveal these rhythms in most though not all of the variables and patients investigated. With the short series of intermediate-length on hand and a restriction in analytical technique to linear least squares analysis a precise definition of the best-fitting period is not possible. It is the occurrence of an approximate circadian period that can be validated for a majority of the patients and variables. The question whether patients with neurosis or parkinsonism can be regarded as being "homeostatic" or circadian periodic has been answered in favor of the latter hypothesis. Further investigations have to take into account the periodicities here demonstrated as well as the sampling requirements revealed in this first rhythmometry of two medical conditions that urgently await further study.

REFERENCES

1. Silber R. H. and Porter C. C. (1954): The determination of 17,21-dihydroxy-20-ketosteroids in urine and plasma. J. of Biological Chemistry, *210*: 923–932.
2. Pisano J. J., Crout J. R. and Abraham D. (1962): Determination of 3-methoxy-4-hydroxymandelic acid in urine. Clin. Chem. Acta, 7: 285–291.
3. Halberg F., Engeli M., Hamburger C. and Hillman D. (1965): Spectral resolution of low-frequency, small-amplitude rhythms in excreted 17-ketosteroid; probable androgen-induced circaseptan desynchronization. Acta Endocr. Suppl., *103*: 54.
4. Halberg F., Johnson E. A., Nelson W., Runge W. and Sothern R. (1972): Autorhythmometry—procedures for physiologic self-measurements and their analysis. Physiology Teacher, *1*: 1–11.
5. Halberg F., Tong Y. L. and Johnson E. A. (1967): Circadian system phase—an aspect of temporal morphology; procedures and illustrative example.s *In*: The Cellular Aspects of Biorhythms, (H. von Mayersbach, ed.), pp. 20–48, Springer-Verlag, Berlin.
6. Halberg F. (1969): Chronobiology. Annual Review of Physiology, *31*: 675–725.

A STUDY OF CIRCADIAN RHYTHMS OF VARIOUS PARAMETERS IN PATIENTS WITH LEPROSY*

CARL C. ENNA[1], LAWRENCE E. SCHEVING[2], FRANZ HALBERG[3], ROBERT R. JACOBSON[4] and ALAN MATHER[5]

[1] *U.S.P.H.S. Hospital, Carville, Louisiana, U.S.A.*
[2] *Department of Anatomy, University of Arkansas Medical School Little Rock, Arkansas, U.S.A.*
[3] *Department of Pathology, Lyon Laboratories, University of Minnesota Minneapolis, Minnesota, U.S.A.*
[4] *U.S.P.H.S. Hospital, Carville, Louisiana*
[5] *Chemical Chemistry Section, Control Disease Center, Atlanta, Georgia, U.S.A.*

This study was undertaken to determine if circadian rhythms are exhibited by patients with leprosy, and if so, to determine a rhythm-adjusted mean or mesor, and also to inquire into the extent of change, i.e., the amplitude and into the timing of these rhythms, as gauged by the so-called acrophase.

The study includes variables in four categories: psychologic, vital signs, urine and blood. Ten patients with lepromatous leprosy ranging in age from 29 to 66 years were studied. Functions in the first three categories were sampled for ten days, every three hours, except for a sleep span from 2100 to 0600. Blood was collected only on the last day at three-hour intervals during 24 hours.

The psychological tests were: (a) a short-term memory test, (b) random-number addition test, (c) an eye-hand skill test, and (d) mood determinations.

The vital signs examined were: oral temperature, pulse, blood pressure, peak expiratory flow and the strength of grip in both hands.

The volume, specific gravity, and pH of urine were measured and urine temperature recorded to the nearest tenth of a degree. A 60 ml sample was collected for catecholamine determinations and a 30 ml sample collected for determination of 17 KG steroids and the electrolytes Na, K, and Cl. Fourteen blood constituents were analyzed.

All of the time series were transferred to punch cards and were processed first by the least squares fit of a 24-hour cosine function [1, 2]. Fits were prepared on each time series as a whole for each of the behavioral, vital sign, urinary and blood variables. Chronograms also were drawn to check on the gross consistency of circadian rhythms. The analyses of individual series for heart rate and temperature are shown in Table 1; the subgroup summaries are in Table 2. Table 3 shows the mean cosinor summaries for vital signs and urinary variables. Fig. 1 is a display of the acrophases for each variable measured. Also in Fig. 1, the confidence arcs of all variables having a significant fit to a 24-hour cosine curve are shown.

RESULTS

Upon analysis by mean cosinor, a statistically significant circadian rhythm was described near the 5% level of statistical significance (or below) in oral tempera-

* Supported from USPHS (5-K6-GM-13,981 and 1RO1-CA-14445-01), NSF GW-7613, and NASA.

Table 1 Summary of circadian rhythmometry on oral temperature and heart rate of patients with active and inactive leprosy.

Patient no.	Age (yrs)	P Rhythm Detection	Mesor, M M±SE	Amplitude, A A±SE	Noise-to rhythm SE/A	Acrophase, ϕ (.95 confidence arc)
				Temperature (°F)		Angular degrees
				Healthy		
(3000)	23	<.005	98.2± .1	.63± .10	$.15_2$	−259° (−239 to −279)
				Inactive disease		
2885	29	<.005	98.0± .1	1.28± .12	$.09_8$	−249° (−238 to −257)
2898	35	<.005	97.9± .1	.80± .09	$.11_3$	−238° (−228 to −248)
2921	50	<.005	98.3± .1	.74± .13	$.17_5$	−226° (−210 to −241)
2334	50	<.005	97.8± .1	.59± .10	$.17_8$	−266° (−241 to −290)
2174	57	<.005	98.1± .05	.47± .08	$.17_8$	−232° (−216 to −247)
1810	66	<.10	98.3± .1	.24± .13	$.54_1$	−232°
				Active disease		
2684	36	<.005	98.4± .1	.74± .10	$.14_0$	−253° (−238 to −268)
2453	48	<.01	97.8± .1	.51± .12	$.23_5$	−243° (−220 to −265)
2793	62	<.005	98.6± .1	.68± .10	$.15_3$	−241° (−226 to −256)
2926	64	<.005	98.3± .1	.45± .08	$.18_3$	−257° (−234 to −281)
				Heart rate		
				Healthy		
(3000)	23	<.005	86.0±1.4	8.7 ±2.0	$.22_9$	−235° (−221 to −259)
				Inactive disease		
2885	29	<.005	92.3±1.7	14.7 ±2.6	$.17_4$	−237° (−221 to −253)
2898	35	<.005	85.3±1.2	7.9 ±1.7	$.21_0$	−251° (−229 to −274)
2921	50	<.01	74.4±1.0	4.7 ±1.6	$.33_2$	−182° (−152 to −211)
2334	50	<.005	69.7± .9	3.9 ±1.0	$.26_3$	−280° (−241 to −320)
2174	57	<.10	66.3± .7	2.4 ±1.2	$.47_4$	−209°
1810	66	<.10	76.9± .6	1.7 ± .8	$.47_8$	−170°
				Active disease		
2684	36	<.005	78.0±1.2	6.9 ±1.8	$.26_1$	−249° (−222 to −276)
2453	48	<.10	71.1± .9	1.9 ±1.5	$.78_7$	− 35°
2793	62	<.005	89.2± .9	10.2 ±1.4	$.13_9$	−220° (−208 to −233)
2926	64	<.04	77.8± .8	2.4 ±1.0	$.39_8$	−263° (−209 to −316)

Table 2 Cosinor summary of data on temperature and heart rate from subjects with and without leprosy.

Status of group studied (N of subj.)	P Rhythm Detection	Mesor, M M±SE °F	Amplitude, A °F	Acrophase, ϕ (.95 confidence limits)
		Temperature		Angula degrees
Healthy (1)	<.005	98.2± .1	.63(.43, .63)	−240° (−220 to −260)
Inactive disease (6)	<. 04	98.1± .1	.67(.08, 1.26)	−223° (−168 to −254)
Active disease (4)	<. 04	98.3± .2	.59(.08, 1.11)	−229° (−190 to −282)
		Heart rate		
Healthy (1)	<.005	86.0±1.4	8.7 (4.8, 12.6)	−216° (−192 to −240)
Inactive disease (6)	>.10	77.5±4.0	5.1	−214°
Active disease (4)	>.10	79.0±3.8	4.2	−219°

Acrophase reference=middle of average sleep span (2130–0600).

Table 3 Cosinor results for the carville project—rhythmometry in leprosy
10 Patients with leprosy studied for a ten-day span.

Physiologic Variable	Mesor, M M±SE	Rhythm detection P-value	Amplitude, A A±SE	Acrophase, ϕ (95% confidence arc)
Oral temperature	98.14± .36	<.001	.56± .27	−264 (−240 to −308)
Heart rate	78± 3	.056	4.78± 4.76	−235 (−149 to −306)
Systolic BP	130± 4	<.020	3.12± 2.45	−310 (−241 to − 11)
Diastolic BP	85± 4	.100	1.74± 2.08	−315 (—)
Peak flow	468± 30	.006	23.35±15.42	−220 (−203 to −248)
Mood	4.0± .1	.050	.14± .14	−224 (−161 to −332)
Forward numbers	5.02± .1	.017	.24± .19	−220 (−185 to −255)
Backward numbers	3.18± .17	.006	.35± .23	−268 (−239 to −308)
Eye-hand skill	39.58±2.13	<.001	1.74± .77	− 47 (− 31 to − 65)
Dynamometer-right	57± 10	.007	2.78± 1.86	−245 (−219 to −296)
Dynamometer-left	49± 11	.137	2.69± 3.64	−237 (—)
Urinary Variable				
Volume-ml/hr	82± 9	<.001	28.75±10.87	−228 (−200 to −250)
pH		.480	.10± .23	−200 (—)
Specific Gravity	1.014±.001	.037	.0014± .0013	− 55 (−358 to −145)
K—Meq/hr	2.83 ±.14	<.001	.97± .34	−197 (−171 to −224)
Na—Meq/hr	6.61 ±.66	<.001	1.65± .84	−218 (−186 to −238)
Cl—Meq/hr	6.47 ±.78	<.001	1.95± .95	−191 (−164 to −216)
17-ketosteroids	.04 ±.04	<.001	.13± .06	−207 (−191 to −225)
Epinephrine—Mg/hr	.43 ±	.007	.19± .13	−206 (−159 to −233)
Norepinephrine	.88 ±	.035	.33± .31	−197 (−182 to −350)

ture, heart rate, systolic blood pressure, peak expiratory flow, and the grip strength of the right hand (this hand being regarded as the major member by all individuals in this group). The circadian variation in diastolic blood pressure was described at the 10% level of statistical significance by the fit of a 24-hour cosine (Table 3, Fig. 1).

Table 4 Circadian computative acrophases (ϕ).*

Variable \ Condition	Healthy (N=13 Subjects)	Leprosy (N=10 Subjects)
Calcium	−303 (−217 −339)	−250 (−203 −326)
Potasium	−354 (−311 − 57)	−245 (−165 −339)
Sodium	−302 (−281 −324)	−302
Chloride	−333 (−279 − 2)	− 33
Total protein	−261 (−252 −277)	−232 (−200 −267)
Urea nitrogen	−326 (−301 −345)	−313 (−230 −346)
GO-transaminase	−126 (−104 −149)	−267 (−222 −304)
Bilirubin	−150 (−107 −271)	−132 (− 80 −189)

* 24 hr≡360°; hence 15°≡1 hr; 0°=local midnight.

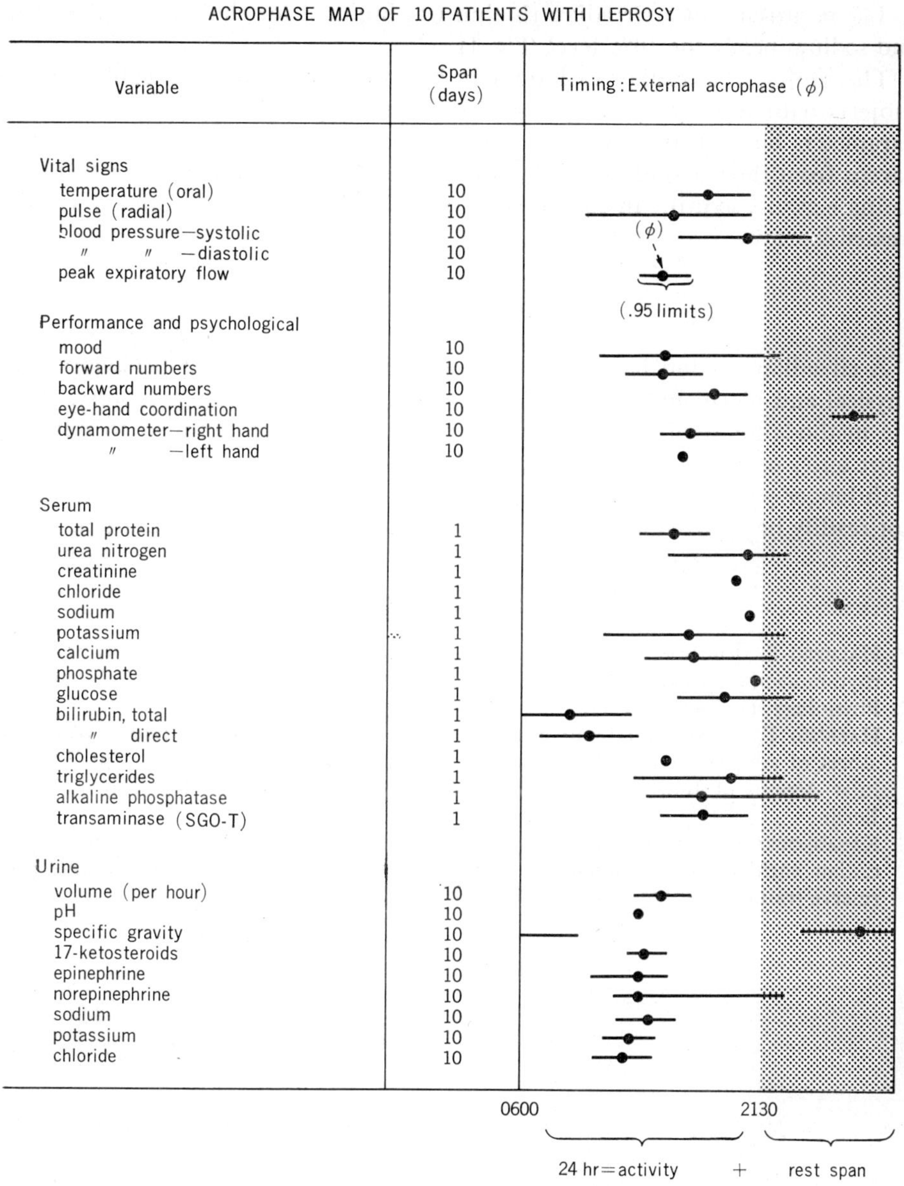

ACROPHASE MAP OF 10 PATIENTS WITH LEPROSY

Variable	Span (days)	Timing: External acrophase (ϕ)
Vital signs		
temperature (oral)	10	
pulse (radial)	10	
blood pressure—systolic	10	(ϕ)
〃 〃 —diastolic	10	
peak expiratory flow	10	(.95 limits)
Performance and psychological		
mood	10	
forward numbers	10	
backward numbers	10	
eye-hand coordination	10	
dynamometer—right hand	10	
〃 —left hand	10	
Serum		
total protein	1	
urea nitrogen	1	
creatinine	1	
chloride	1	
sodium	1	
potassium	1	
calcium	1	
phosphate	1	
glucose	1	
bilirubin, total	1	
〃 direct	1	
cholesterol	1	
triglycerides	1	
alkaline phosphatase	1	
transaminase (SGO-T)	1	
Urine		
volume (per hour)	10	
pH	10	
specific gravity	10	
17-ketosteroids	10	
epinephrine	10	
norepinephrine	10	
sodium	10	
potassium	10	
chloride	10	

0600 2130

24 hr=activity + rest span

Fig. 1 Acrophase map of ten patients with leprosy.

With regard to urinary variables, only on the data on pH did we fail to ascertain statistical significance for the description of a circadian rhythm by the mean cosinor method (Table 3, Fig. 1).

Indeed, rhythm description was feasible for urine volume, 17-ketogenic steroids, epinephrine, norepinephrine, sodium, potassium, chloride, and specific gravity.

Circadian rhythms of blood parameters were detected below 10% or at least below the 5% level of statistical significance for SGOT, total as well as direct bilirubin, triglycerides, urea, potassium, calcium, as well as alkaline phosphate, protein and glucose (Fig. 1).

The occurrence of a circadian rhythm was imputed for cholesterol, phosphate, and sodium below the 10% level (Fig. 1).

The timing of circadian variations in blood is summarized in Table 4 for 10 subjects with leprosy by comparison to 13 healthy soldiers [3]. A number of parameters but not all of them, are locked in phase with one another as is apparent from their acrophases. The difference in acrophase for serum GO-transaminase awaits scrutiny in comparisons of healthy subjects and patients with untreated leprosy.

SUMMARY

A circadian rhythm has been demonstrated for a number of behavioral, biophysical and biochemical functions in patients with lepromatous leprosy. Since the methods for resolving a circadian time structure were applied to relatively short series of data available at 3-hour intervals covering only ten days, rhythms with longer periods, as well as rhythms with a period shorter than six hours can not readily be resolved and remain beyond the scope of this presentation.

REFERENCES

1. HALBERG F., JOHNSON E. A., NELSON W., RUNGE W. and SOTHERN R. (1972): Autorhythmometry—procedures for physiologic self-measurements and their analysis. Physiology Teacher, *1*: 1–11.
2. HALBERG F., TONG Y. L. and JOHNSON E. A. (1967): Circadian system phase—an aspect of temporal morphology; procedures and illustrative examples. *In*: The Cellular Aspects of Biorhythms, (H. von MAYERSBACH, ed.), pp. 20–48, Springer-Verlag, Berlin.
3. KANABROCKI E. L., SCHEVING L. E., HALBERG F., BREWER R. and BIRD T. (1973): Circadian variation in presumably healthy men under conditions of peace-time army reserve training. Space Life Sciences, *4*: 258–270.

CHANGES IN THE CIRCADIAN RHYTHM OF BLIND PEOPLE

F. HOLLWICH and B. DIECKHUES

Augenklinik der Westfalen-Wilhelms. Universität, Münster, Germany

INTRODUCTION

There is abundant evidence that light is an important synchronizing force in the regulation of metabolism in man.

Stimulated by the many fundamental findings on circadian organization by Professor Franz HALBERG, we decided to explore the effects of ocular perception of light on various urinary constituents of the rabbit and also on constituents of urine and blood from blinded and normal-sighted man.

RESULTS AND DISCUSSION

A study on 15 rabbits longitudinally sampled at 1100, 1400, 1700 and 0800 over a 4-day period demonstrated a prominent circadian group rhythm in the following urinary constituents: chloride, inorganic phosphate, bilirubin, urobilinogen, 17-ketosteroids, uric acid, creatinine, potassium, sodium, the ratio of uric acid to creatinine and the ratio of sodium to potassium. Following the 4-day natural LD schedule the rabbits were subjected to an environment of constant darkness for 38 consecutive days. During this constant dark period with the same sampling interval the profile of the rhythms of most variables became very modified when compared to the LD colony rhythm. The tendency was for a progressive lowering of the amplitude as the rabbits continued under these conditions. After 38 days the animals were returned to the natural LD cycle; within three days the colony rhythms seen earlier under the natural LD schedule were again apparent. Unfortunately space limitations do not permit presentation of these data; they will be published elsewhere.

Several studies were conducted on blinded and normal-sighted individuals. In one study circadian rhythms in certain urinary constituents were compared between a group of 100 blind and 50 normal-sighted persons. Fig. 1 best summarizes these findings; the most notable finding is a reduction in urinary output of most variables in the blind as compared to the normal. Also the amplitudes of the rhythms of certain variables appear to be reduced in the blind.

In a study on 50 patients blind because of cataracts it was found that very little circadian variation characterized the group data when sampled at 3-hour intervals during the waking hours. However, upon removal of the cataracts and restoration of sight the group circadian rhythm became prominent (Fig. 2a and 2b).

Fig. 3 demonstrates that for a group of 300 blind patients the characteristic morning eosinopenia of normal-sighted individuals is not seen in the blind. Furthermore the pre- and post-operative cases of blindness due to cataracts were

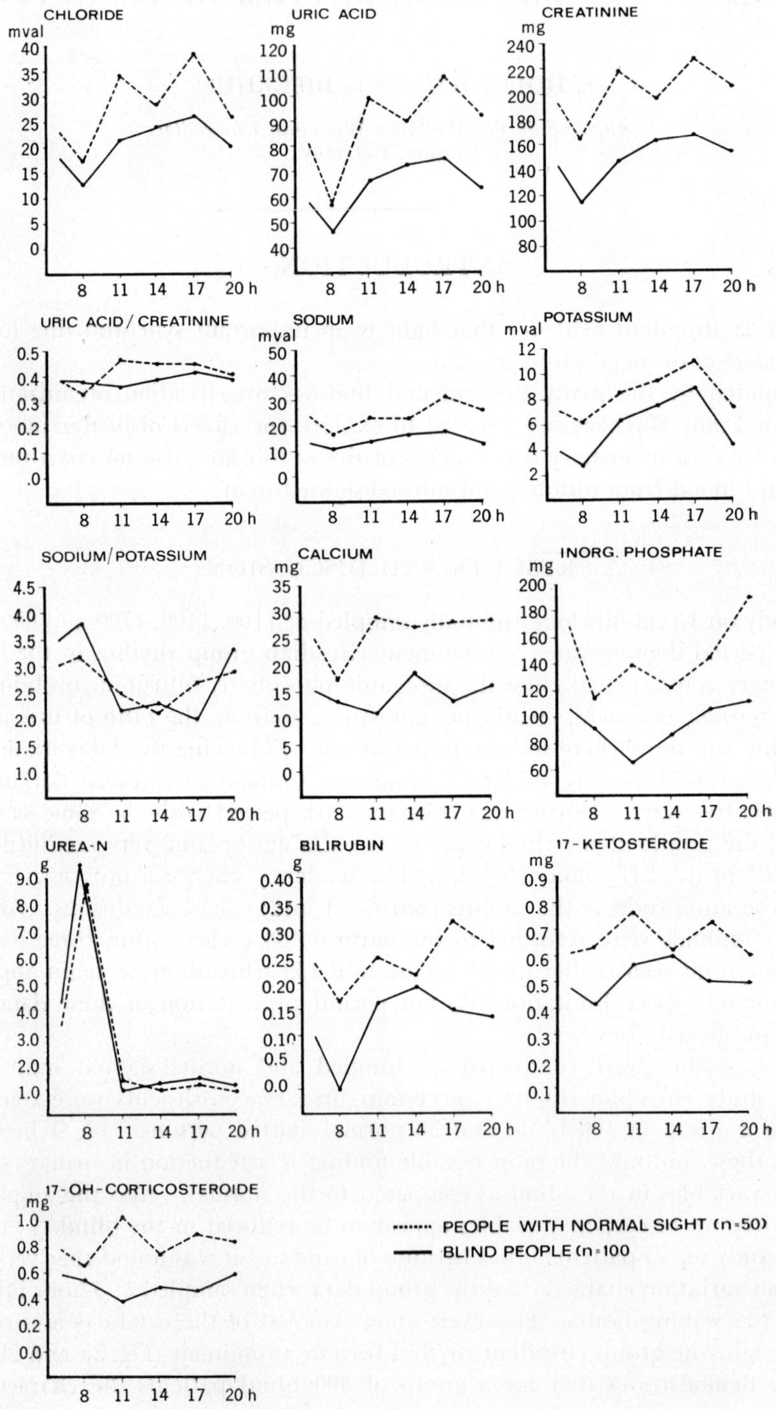

Fig. 1 Demonstrates the group mean values with sampling at 3-hour intervals during waking hours for urinary metabolic constituents. When one compares the rhythms between those with ocular light perception one finds a disturbance of the circadian rhythm in the blind; the circadian rhythm is desynchronized, e.g., the amplitude is diminished and the over-all excretion also is reduced.

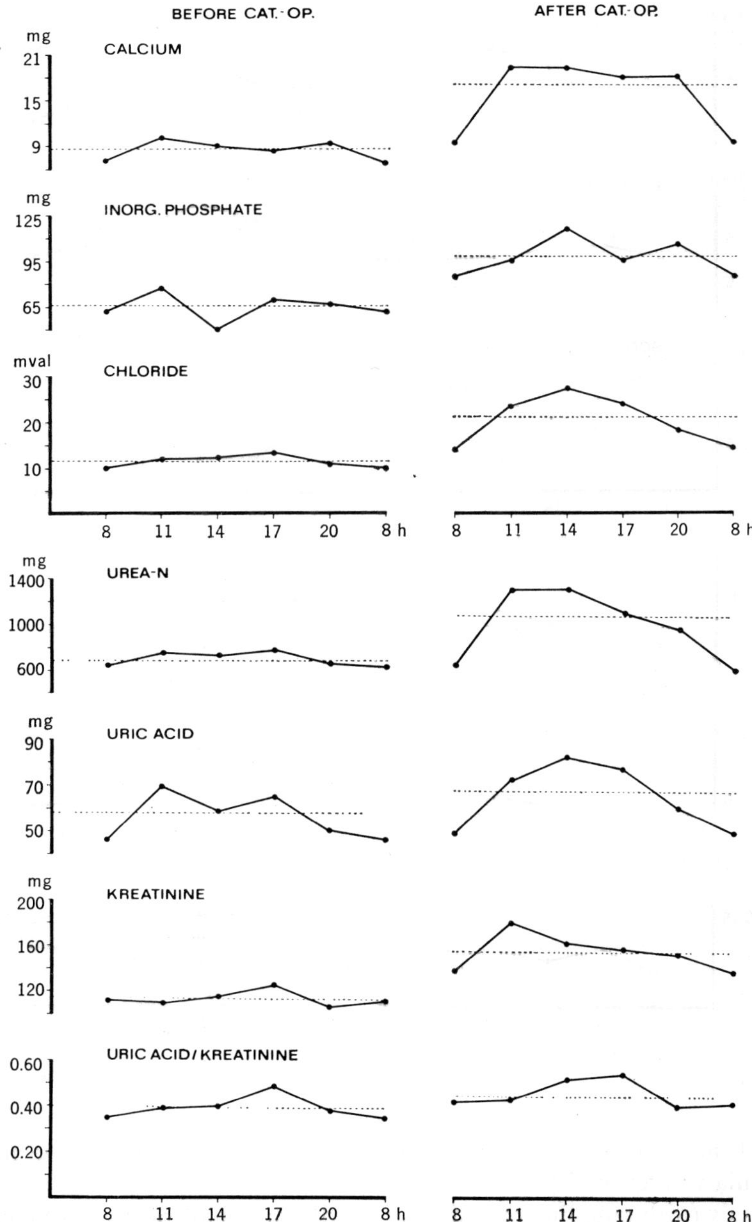

Fig. 2 Compares the rhythmic fluctuation in urinary metabolites in 50 patients (practically blind) before and after (normal ocular light perception restored) cataract operation. In the preoperative state there was a diminished corticosteriod, chloride, uric acid, phosphate, postassium, urea nitrogen and bilirubin excretion and an increased excretion of sodium. In the postoperative state there was a normalization of the rhythms of the various metapdorucbolic products.

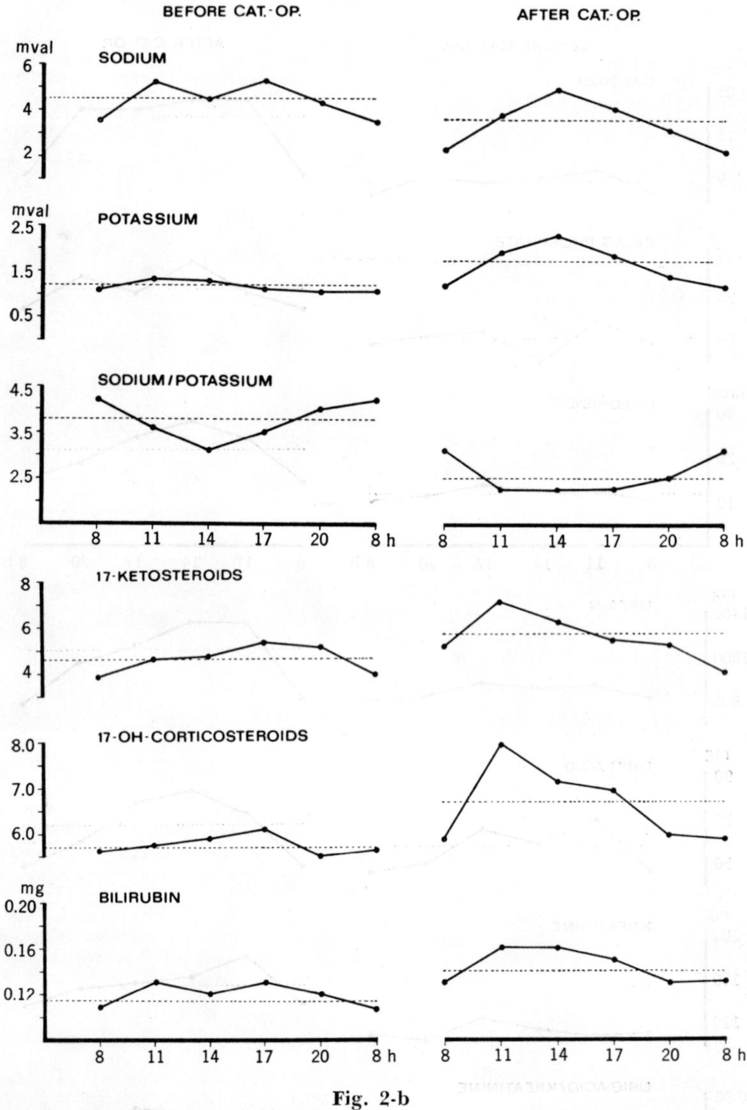

Fig. 2-b

compared, and after cataract extraction and restoration of vision the morning eosinopenia was restored (Fig. 4).

In a study (with sampling only during the diurnal phase of the circadian cycle) on 50 blind and 250 normal-sighted persons, thrombocyte levels showed remarkable differences (Fig. 5). The blind present an inverse pattern when compared to the normal-sighted individual. The characteristic morning decline between 0800 and 1100 found in the normal-sighted was absent in the blind. Thrombocytes decrease significantly in the blind and the bimodal curve seen in the normal-sighted individual is absent in the blind (Fig. 5). Fig. 6 demonstrates that the morning increase in thrombocytes is restored after cataract operation.

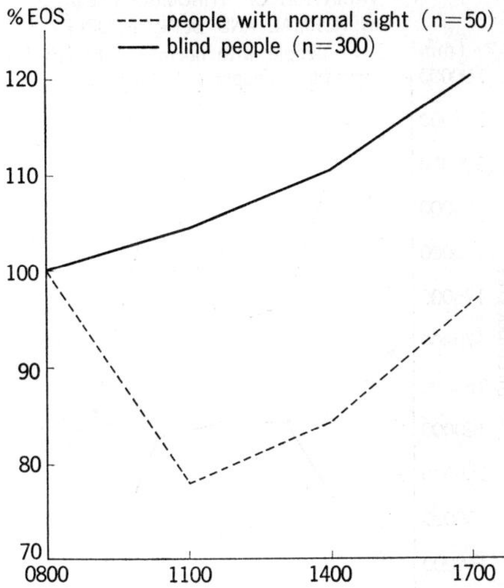

Fig. 3 Demonstrates the diurnal variation in 300 blind and 50 normal-sighted individuals. In the normal-sighted individual there is a fall in the value of eosinophils in the forenoon; this was not the case for the blind. The over-all level of eosinophils is much lower in normal-sighted individuals.

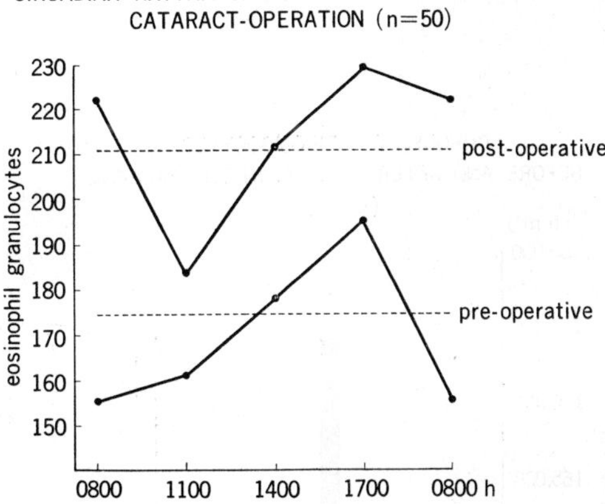

Fig. 4 Demonstrates the behavior of eosinophils before and after cataract operation. The change of eosinophils of 50 cataract patients with a vision less than 1/20 was determined. The normal morning eosinopenia was absent in the blind. After cataract operation the same patients showed a decrease in eosinophils.

When cortisol levels between blind (due to cataracts) were compared with the normal-sighted (N+10 each category), the circadian variation characterizing the group was absent in the blind (Fig. 6). With restoration of sight subsequent to cataract operation, the cortisol rhythm was restored (Fig. 7).

CONCLUSIONS

Our comparative study on individuals with normal (good) sight and blind patients, as well as examinations on experimental animals, show that the penetra-

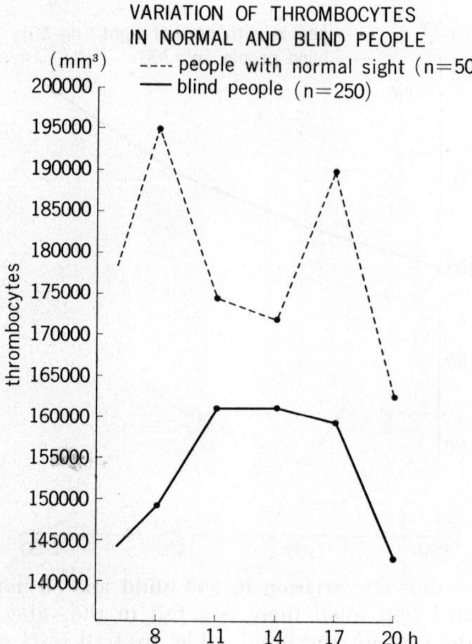

Fig. 5 Comparison of rhythm in thrombocyte between normal and blind people during the diurnal phase of the 24-hour period. Thrombocytes of people with normal vision show marked diurnal changes. In blind persons this rhythm is disturbed. The amplitude is lowered and the absolute number of thrombocytes in the blood is reduced.

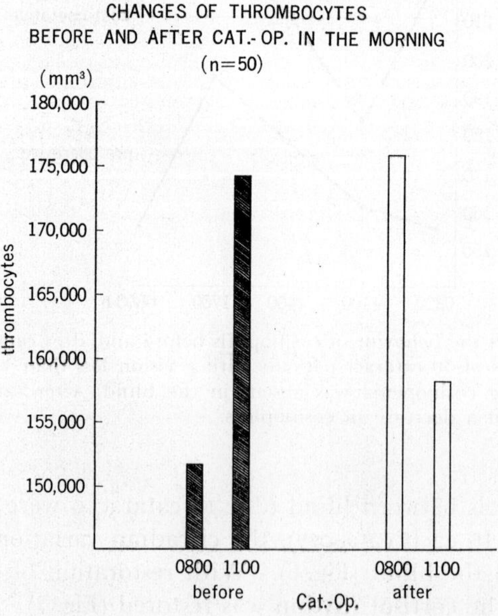

Fig. 6 Demonstrates changes in thrombocytes in the morning before and after cataract operation (N=50). The characteristic morning decrease between 0800 and 1100 found in people with normal sight is absent in cataract patients. Postoperatively there was a decrease in thrombocytes with restoration of ocular light perception.

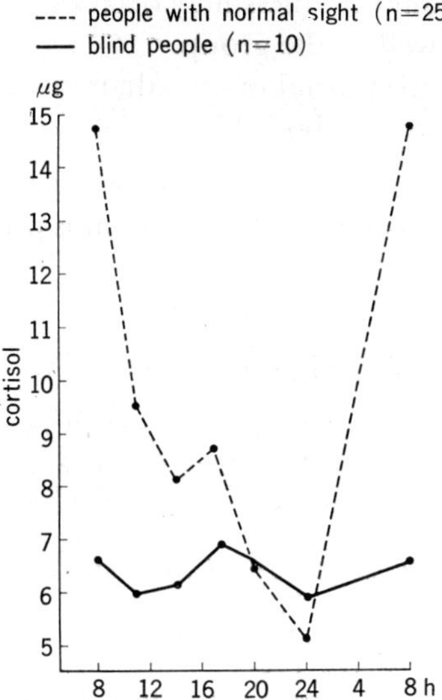

Fig. 7 Comparison of blood cortisol levels for a group of blind and normalsighted individuals. The increase of cortisol in plasma in the morning in the normal patients is less pronounced in the blind. In sighted patients the mean level is 9.2 μg/ml whereas in the blind this was reduced to 6.2 μg/ml. Sampling was at 3-hour intervals during waking hours.

CIRCADIAN VARIATION OF CORTISOL IN CATARACT PATIENTS
BEFORE AND AFTER CAT.-OP. (n=10)

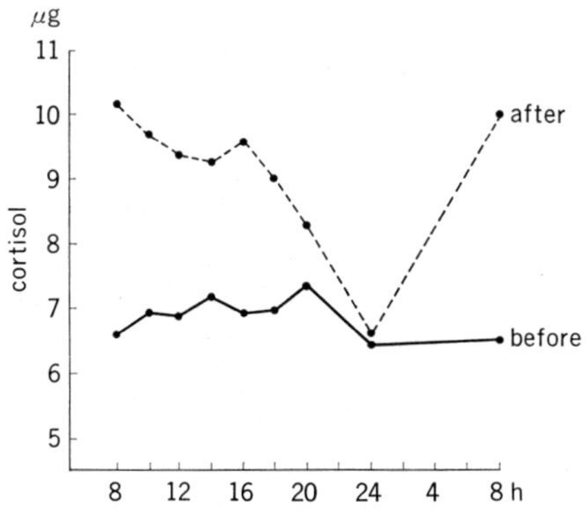

Fig. 8 Compares the difference in cortisol levels in cataract patients before and after cataract operation. After restoration of normal ocular light perception there was a marked increase in the morning plasma content of cortisol.

tion of light into the eye as well as exclusion of light have an influence on the circadian rhythms of various metabolites.

Under conditions of light exclusion due to blindness and in light deprivation in experimental animals, we have found the following significant changes:

(1) Light deprivation or blindness can alter circadian rhtyhms in constituents of both urine and blood.

(2) Light deprivation from blindness may decrease the amplitudes of rhythms in the individual as well as the group.
(3) Changes in the over-all diurnal or circadian mean may occur due to blindness or light deprivation in both blood and urinary variables.

We conclude that the influence of ocular light perception via "energetic pathways of the visual system" acts on the hypothalamus-pituitary system and causes "corticotropic activity".

REFERENCES

1. Halberg F. (1970): cited in Biological Rhythm in Psychiatry and Medicine Monography National Inst. of Mental Health, Chevy Chase, Maryland.
2. Hollwich F. (1964): The influence of light via the eyes on animals and man. Ann. N. Y. Acad. Sci., 117.
3. Hollwich F. and Dieckhues B. (1971): Circadian rhythm in the blind. J. interdiscipl. Cycle Res., Vol. 2, No. 3, pp. 291.
4. Hollwich F. and Dieckhues B. (1971): Endocrine system and blindness. German Medical Monthly, No. 3, Vol. I, 122–128.
5. Hollwich F. and Dieckhues B. (1972): Circadian rhythm of cortisol level in normal subjects and in the blind. Annals of Ophthalmology, Vol. 5.

AUTORHYTHMOMETRY AND CANCER OF THE BREAST— A CASE REPORT*

F. HALBERG, H. BUCHWALD, K. CHARYULU
and E. REEKER

Chronobiology Laboratories, Department of Pathology and the Departments of Surgery and Radiology, University of Minnesota Minneapolis, Minnesota, U.S.A. and Department of Radiation Therapy, University of Miami Miami, Florida, U.S.A.

INTRODUCTION

In this paper we document for a subject pursuing her daily routine while presumably healthy and also after surgery and during radiotherapy the feasibility of carrying out self-measurements with a frequency compatible with the demonstration of a circadian rhythm in several variables and possibly also of an alteration in one of the temperature rhythm's endpoints, namely its amplitude. The feasibility of self-measurement in such a case seems important, since in the future, possible merits of longitudinal rhythmometry as a tool for clinical care or preventive family practice [1, 2] will have to be evaluated—more often than not, from autorhythmometry [3].

Implantable transensors and related instrumentation for collecting time series amenable to rhythmometry are available for experimental animals; body core temperature as well as gross motor activity of "unharnessed" animals, moving freely though in a relatively limited space, thus may be monitored for weeks, months or years [4]. However, similar devices sufficiently low in cost, as well as dependable and innocuous, are not yet available for large-scale use on man; they may be forthcoming only after a critical need for them has been demonstrated.

Time series obtained as self-measurements by an ambulatory subject pursuing at least minimal tasks of her customary routine will be summarized herein by methods of microscopic [5] rhythmometry, such as the least squares fit of a cosine curve [3, 6], a more direct approach than the variance spectra [7–10] which earlier demonstrated a rhythm alteration in selected patients. The specificity of earlier findings as a feature of cancer chronopathology and of the results here presented, however, await further large-scale work. In this context the present report aims more at concept formulation and represents a fledgling "feasibility check" rather than a completed investigation and experimental validation.

Apart from the foregoing methodologic points, this report deals with observations of interest in their own right: the two time series which demonstrate the feasibility of collecting self-measurements under certain conditions and executing certain kinds of analyses also allow a comparison, in one and the same individual,

* Supported by grants from the United States Public Health Service (5-K6-GM-13,981) NCI 1RO1-CA-14445-01 and NASA.

of circadian rhythm parameters in body temperature and heart rate in a first series obtained about a year prior to the diagnosis of metastatic breast cancer and in a second series obtained after surgical removal of the breast and axillary nodes. However, in this case, it cannot be ascertained whether the pre-surgical series was obtained prior to the onset of a malignant process: only on subjects systematically sampled from early childhood does it seem possible to scrutinize the occurrence of any rhythm alteration prior to the onset of overt disease.

CASE HISTORY

The patient, a 44-year-old widow, was first seen in consultation in October of 1968 with the complaint of a mass in her left breast. She noticed engorgement of her left breast approximately one and a half months prior to consulting a physician. About one month before being seen, she noted a mass in the upper outer quadrant of the left breast. The mass was neither painful nor tender. There had been no discharge from the nipple, skin discoloration, or other local signs. The patient felt well and there had been no weight loss or any evidence of systemic disease.

The patient has three children, ages 17, 15 and 10. She attempted to breast feed the first two but discontinued her efforts in each case—in the first instance because of anxiety that she might not provide enough, in the second after development of mastitis and a breast abcess, treated by antibiotics. However, she breast-fed her last child for seven months. The patient had undergone a dilatation and curettage procedure in 1964 for intermenstrual bleeding; no evidence of malignancy was found. She had otherwise enjoyed good health and her past medical history was noncontributory. There was no sanguinous familial history of cancer. She had a tonsillectomy at age 6 and chicken-pox, mumps, whooping cough, measles during childhood.

At physical examination the patient's record describes her as "a well-developed lady of stated years, with a casual blood pressure of 120/80, pulse 80 and regular weight 135 pound." Head and neck, pulmonary, cardiac, abdominal, extremity, neurological, and peripheral vascular examinations were all "normal." A 1 cm freely movable firm, but not hard, lymph node was palpable in the mid-left axilla; no other lymphadenopathy was found. There was a visible mass in the upper outer quadrant of the left breast. On palpation this was firm, and again not hard, freely movable and irregular in its boundaries. The size of the mass, on physical examination, was approximately 4×6 cm. There was no evidence of erythema, skin retraction, attachment to the chest wall, or other local changes.

With a "normal" single-sample blood count and urine analysis, and a chest x-ray revealing no evidence of metastatic disease, the patient was taken to the operating room on October 23, 1968, where under local anesthesia an excisional biopsy of the above mentioned mass was attempted. The mass removed measured 3×5 cm and consisted of a poorly defined lobulated fibrous tumor. Histological diagnosis was: Infiltrating ductal carcinoma of breast.

At this stage in her therapy, a thorough search for distant disease was carried out. Specifically, a bone x-ray survey was obtained, a supraclavicular lymph node was biopsied, and liver function tests were done. These procedures all failed to reveal any evidence of distal metastatic disease.

RADICAL MASTECTOMY

On October 28, 1968, she was again taken to the operating room where a standard Halsted radical mastectomy was carried out. The residual growth was removed with the pectoralis major muscle, the pectoralis minor muscle and the axillary contents from the level of the axillary vein down. At the time of operation, one "suspicious" lymph node was palpated in the mid-axilla. The area in the high axilla appeared to be, grossly, free of disease. The skin incision was made in a line from the left axilla diagonally across the left anterior chest toward the lower sternum. Primary closure was attempted under mild tension. It was apparent on the third post-operative day that there would be an area of skin slough in the middle of the skin closure. This was allowed to slough and the wound permitted to granulate to the point where healthy, bleeding granulation tissue covered the raw area. On November 25, pinch grafts were applied to the area from a donor site on the left anterior thigh and with nearly 100 per cent take of the graft the area was rapidly covered by epithelium.

The radical mastectomy specimen showed breast tissue with the following overall diameters—22×14×4 cm and skin overlying the breast tissue measuring 17×5 cm, and containing a 4 cm long transverse incision about 2 cm above and superior to the nipple. There were nodular areas in the vicinity of the previous biopsy and they appeared to consist of residual tumor. On sectioning of these nodules, residual carcinoma was indeed found.

Under the microscope these areas showed lobules of breast tissue with proliferation of the ducts, for the most part, characteristic of a benign sclerosing adenosis. However, some of the ducts were quite dilated and formed small cystic spaces with interlobular tissue of dense fibrous appearance. In several of these areas of cellular accumulation, epithelial cells were forming small cords which were infiltrating the interlobular fibrous tissue. These cords consisted of small hyperchromatic cells characteristic of infiltrating ductal carcinoma. There were areas of nodular breast tissue distant from the primary site of tumor and these areas showed benign cystic disease but also several areas where epithelial cells were again seen to be infiltrating the interlobular fibrous tissue surrounding several round ducts. This too represented infiltrating ductal carcinoma.

Sections from the nipple and tissues subjacent showed ducts but no evidence of tumor in this area. Sections of the axillary contents showed portions of six lymph nodes from the apex of the axilla, all of which were negative for tumor. Nodes from the midportion of the axilla numbered four, of which two were positive for metastatic disease. Nodes from the base of the axilla also numbered four and consisted of two positive and two negative lymph nodes. The metastases seen were diffusely scattered throughout the lymph nodes and occasionally formed small clumps of tumor cells. Thus, the overall pathological diagnosis is: infiltrating ductal carcinoma, possibly originating in several areas of the left breast, which spread to the middle and lower planes of the left axillary lymph nodes. In addition, the patient had evidence of fibrocystic disease of the breast.

The patient made an uneventful recovery from surgery and within three weeks was able to resume her activities as a research worker. One week after take of the skin graft, she was started on radiotherapy [12], as indicated in Table 1.

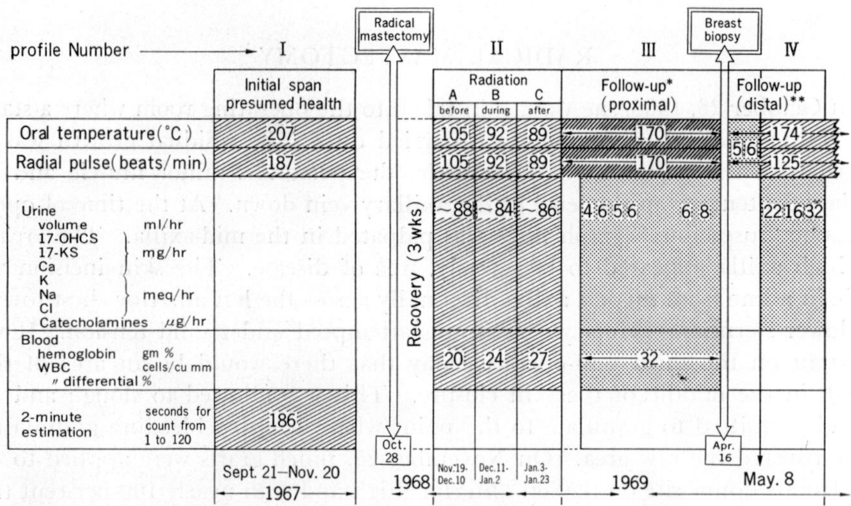

Fig. 1 Variables and time spans investigated in relation to two operative procedures and radiotherapy.

RESULTS

The mostly favorable noise-to-rhythm ratios in Table 2 suggest the occurrence of a rhythm in the time series studied. Thus, for the oral temperature measured in 1967 data, a very favorable noise-to-rhythm ratio is readily apparent. On the other hand, for the 1968 series there are changes in rhythmicity although the mesor of temperature (36.8°C) is identical with that observed in 1967. Thus, the higher noise ratio in 1968 suggests a lessor stability of the temperature rhythm. Also, the amplitude of this rhythm appears to be less in 1968 after surgery, and the temperature acrophase occurs slightly earlier during the post-surgical observation—whether all 1967 data, or only parts of the 1967 series matched to the 1968 series by the number of observations or by the number of days documented by measurements, are used.

Similar findings are not made for pulse. The circadian mesors for the two series of heart rates obtained about one year apart are almost identical; as compared to the corresponding standard error, the circadian amplitude, A, of the heart rate rhythm is large for both series. There is also a surprising similarity in the circadian acrophase of heart rate for the time series obtained in 1967 and in 1968. These estimates for the circadian rhythm in heart rate agree quite well with estimates from other subjects [11]. MPE values in Table 2 suggest that rhythms are 24-h synchronized.

DISCUSSION

Does a malignant breast tumor alter any body rhythms? A definitive answer to this question and related ones will require proper sampling on pertinent

Table 2 Results from fitting cosine curves to self-measurements during two observation spans about one year apart. Prior to detection of a mammary tumor in 1967 and immediately after radical mastectomy in 1968.

Series Variable (units) (No. investigated)	N of obs.	Noise-to rhythm SE/A	Mesor, M M±SE	Amplitude, A A±SE	Acrophase ϕ (.95 conf. arc)	τ at MPE*	M±SE	A±SE	SE/A
			November 19 – December 20 1968						
1. Heart rate (beats/minute)	105	$.13_9$	75 ±.58	5.2 ±.7	−183° (−163, −203)	24.0			
2. Oral temperature (°C) (19 days)	105	$.39_3$	36.8±.02	.08±.03	−135° (−100, −171)	24.0			
			September 21–November 20 1967						
3. Heart rate (beats/minute)	187	$.21_1$	73 ±.54	3.4 ±.7	−180° (−152, −208)	24.0			
4. Oral temperature (°C)	207	$.11_4$	36.8±.01	.18±.04	−168° (−154, −182)	24.0			
**5. Oral temperature (1st 19 days)	71	$.16_7$	36.6±.02	.17±.03	−172° (−150, −195)	23.9	36.7±.02	.17±.03	.158
**6. Oral temperature (1st 105 points)	105	$.16_3$	36.7±.004	.17±.03	−168° (−148, −189)	23.9	36.7±.02	.17±.03	.152
7. 2′-time estimation (count from 1–120)	186	$.11_8$	113 ±.64	5.9 ±.8	− 14° (−358, − 30)	24.0			

Acrophase reference=middle of sleep span.

* MPE=minimal percent error

** 1967 temperature data matched for comparison with 1968 data by N of days [5] or N of observations [6].

Note that the amplitude of the 1968 oral temperature drops to less than half that of 1967 oral temperature, whether the entire data span from 1967 is taken as reference standard, i.e., row 2 as compared to row 4, or whether the reference standard is slightly modified so that 1967 data covering the same number of days or consisting of the same number of measurements are used for comparison, i.e., row 2 compared to rows 5 or 6.

variables of populations rather than individuals. Observations on this single case are useful primarily for the design of sampling requirements for future studies. Nonetheless, they command some interest in themselves when, as in the case here reported, series of self-measurements available not only after the diagnosis of a tumor but also at a time preceding the detection, if not the inception, of a malignant growth lead to analytical results that suggests rhythm alteration. It is our ability to estimate rhythm parameters and their alteration (in the case of body temperature) that underlines the feasibility of carrying out meaningful self-measurements with sufficient frequency to allow rhythmometry in the post-surgical state and even during a span of radiotherapy while the ambulatory patient pursues at least the minimal tasks of her usual routine [12].

REFERENCES

1. Halberg F. (1970): Chronobiology and the Delivery of Health Care. *In*: A Systems Approach to the Application of Chronobiology in Family Practice, (J. O'Leary, ed.), pp. 31–96, Health Care Research Program, Dept. of Family Practice and Community Health, University of Minnesota.
2. Halberg F.: Education, biologic rhythms and the computer. *In*: Engineering, Computers and the Future of Man. Proc. Conf. on Science and the International Man: The Computer. Chania, Crete, June 29–July 3, 1970. International Science Foundation, Paris, (in press).
3. Halberg F., Johnson E. A., Nelson W., Runge W. and Sothern R. (1972): Autorhythmometry—procedures for physiologic self-measurements and their analysis. Physiology Teacher, *1*: 1–11.
4. Halberg F., Nelson W., Runge W. J., Schmitt O. H., Pitts G. C., Tremor J. and Reynolds O. E. (1971): Plans for orbital study of rat biorhythms. Results of interest beyond the Biosatellite program. Space Life Sciences, *2*: 437–471.
5. Halberg F. (1969): Chronobiology. Annual Review of Physiology, *31*: 675–725.
6. Halberg F., Tong Y. L. and Johnson E. A. (1967): Circadian system phase—an aspect of temporal morphology; procedures and illustrative examples. Book Chapter. *In*: The Cellular Aspects of Biorhythms, (H. von Mayersbach, ed.), pp. 20–48, Springer-Verlag, Berlin.
7. Halberg F., Stein M., Diffley M., Panofsky H. and Adkins G. (1964): Computer techniques in the study of biologic rhythms. Annals N. Y. Acad. Sci., *115*: 695–720.
8. Halberg F., Panofsky H. and Mantis H. (1964): Human thermo-variance spectra. Annals N. Y. Acad. Sci., *117*: 254–270.
9. Halberg F. and Panofsky H. (1961): I. Thermo-variance spectra; method and clinical illustrations. Experimental Med. and Surg., *19*: 284–309.
10. Panofsky H. and Halberg F. (1961): II. Thermo-variance spectra; simplified computational example and other methodology. Exp. Med. and Surg., *19*: 323–338.
11. Halberg F., Vallbona C., Dietlein L. F., Rummel J. A., Berry C. A., Pitts G. C. and Nunneley S. A. (1970): Human circadian circulatory rhythms during weightlessness in extraterrestrial flight or bedrest with and without exercise. Space Life Sciences, *2*: 18–32.
12. Charyulu K., Halberg F., Reeker E., Haus E. and Buchwald H. (1974): Autorhythmometry in relation to radiotherapy: case report as tentative feasibility check. *In*: Chronobiology. Proc. Symp. Quant. Chronobiology, Little Rock, 1971, (L. E. Scheving, F. Halberg and J. E. Pauly, eds.), pp. 265–272, Igaku Shoin Ltd., Tokyo.

COMPARISON OF DAILY BLOOD SUGAR PROFILES IN DIABETICS UNDER TREATMENT WITH DIFFERENT ORAL ANTIDIABETIC DRUGS

FRIEDRICH WILHELM STRATMANN

Farcharzt für Innere Krankheiten
Stuttgart, West Germany

INTRODUCTION

Sulfonylureas produced a fundamental change in the treatment of diabetes mellitus. Since their introduction, a tremendous amount of clinical experience dealing with the use of these drugs has been published. Through the effort of pharmaceutical research a few years ago a new agent, glibenclamide (HB 419, Euglucon 5) was developed. In clinical evaluation glibenclamide proved to be 200 times more efficient than the standard drug tolbutamide, with less frequent side effects. It was called the sulfonylurea of the second generation. When this drug was first used in clinical practice it was given too freely and in too high a dosage without appropriate control of the patient. As a consequence hypoglycemia was reported in a large number of cases, especially in the afternoon. Glibenclamide undoubtedly was one of the most potent hypoglycemic agents developed. Glibenclamide exerted a long lasting effect on insulin secretion. In the case of tolbutamide a high initial peak and a subsequent quick decline in insulin secretion was noticed, whereas glibenclamide produced a slowly rising curve which plateaued for hours, thereafter the curve declined slowly. This different behavior might well explain hypoglycemia in a patient occurring in the late morning after a single dose of tolbutamide, but in the early afternoon after a single morning dose of glibenclamide.

The aim of the present investigation was to determine whether varying oral antidiabetic drugs would exert different effects on blood sugar, with constant dietetic intake. It was of special interest to determine whether a tendency to hypoglycemia could be detected under treatment programs without obvious hypoglycemic reactions.

MEHODS

The records of 600 diabetic patients were statistically evaluated. Six groups with 100 patients were defined, the treatment of each group consisted of:

group 1 diet alone
group 2 sulfonylurea first generation, e.g. tolbutamide (SU)
group 3 sulfonylurea second generation, i.e. glibenclamide (GLIB)
group 4 sulfonylurea first generation plus biguanides (SU plus BIG)
group 5 sulfonylurea second generation plus biguanides (GLIB plus BIG)
group 6 insulin

On the days of control evaluation, blood glucose was determined before the three main meals. Only the profiles of the days with constant dosage of the drug were taken for comparative studies between the groups. All patients were on the same carbohydrate intake of 150 g distributed in five meals in a carbohydrate ratio 3:1:4:1:4. Thus conditions did not vary from group to group.

Two facts should be mentioned: for 16 years all the patients have been under the care of one physician, for 8 years the same technician determined blood sugar levels using the same method (reduction method). On each patient a daily profile was done every 8–12 weeks.

RESULTS

Tables 1 and 2 show the mean and the median values respectively of blood glucose determinations, and the differences between the points of measurements in the different forms of treatment. For each point of measurement (before brakfast, lunch and dinner) the different forms of treatment were compared. Parametric (BARTLETT-test, simple variance analysis with subsequent DUNCA-test) and nonparametric tests (H-test according to KRUSKAL-WALLIS with subsequent paired test according to MNN-WHITNEY-WILCOXJN) were performed to compare the measurements statistically eror probability 5%).

Table 1 Mean values of blood sugars and mean values of differences between the points of measurements in the various groups of treatment. See text for defimtion of abbreviation.

Time	Group 1 (Diet)	Group 2 (SU)	Group 3 (GLIB)	Group 4 (SU plus BIG)	Group 5 (GLIB plus BIG)
Morning	100.98	127.76	122.36	138.51	129.45
Noon	112.25	148.94	154.98	175.68	176.76
Evening	107.10	126.57	134.23	144.69	136.39
Difference between:					
Noon-morning	11.27	21.18	32.62	37.17	47.31
Evening-morning	6.12	− 1.19	11.87	6.18	6.94
Evening-noon	− 5.15	−22.37	−20.75	−30.99	−40.37
Number of values	100	100	101	101	71

Table 2 Median of blood sugar values and median of differences between the points of measurements in the various groups of treatment.

Time	Group 1 (Diet)	Group 2 (SU)	Group 3 (GLIB)	Group 4 (SU plus BIG)	Group 5 (GLIB plus BIG)
Morning	84.00	123.00	117.00	136.00	129.00
Noon	106.00	137.50	146.00	178.00	168.00
Evening	106.00	119.50	125.00	138.00	136.00
Difference between:					
Noon-morning	11.00	14.00	32.00	34.00	45.00
Evening-morning	6.00	− 2.00	9.00	4.00	9.00
Evening-noon	− 6.50	−19.00	−23.00	−38.00	−42.00
Number of values	100	100	101	101	71

At each of the three points of measurements the zero hypothesis (equality of the different distributions) was rejected with an error probability of 5%, i.e. the values do not derive from the same populations.

Especially group 1 (diet) differs significantly at all points of measurements from the other groups. Group 4 (SU plus BIG) differs significantly from group 2 (SU) and group 3 (GLIB), whereas there is no statistically significant difference between the groups 2 (SU) and 3 (GLIB), and the groups 4 (SU plus BIG) and 5 (GLIB plus BIG).

In addition, the values of the three different points of measurements were analyzed in detail. In Fig. 1, the 50% borders, the mean values and the medians of blood glucose levels in the different forms of treatment at the points of measurements (morning, noon, evening) are shown. 25% of the values are above, 25% are below the middle line. For the differences between the readings (noon-morning, evening-morning, evening-noon) the same statistical tests were performed as for each single value.

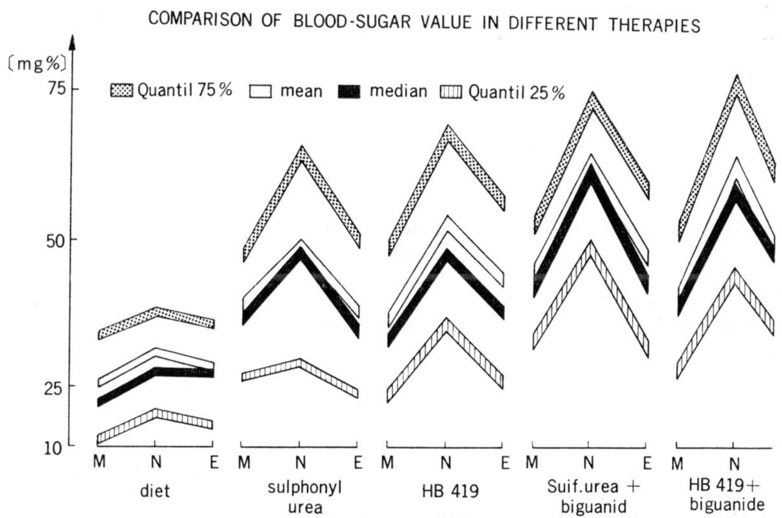

Fig. 1 Comparison of daily blood sugar profiles in persons under different oral antidiabetic drugs. The gray and red middle group of profiles represent the mean (gray) and median (red). The above group of blue profiles represents the values which are 25% above the middle values and the blue represents values that are 25% below the middle group.

Furthermore, it was proved by means of the WILCOXON matched pairs-signed-ranks test, whether the differences between the three points of measurements in the therapeutical procedures under study, were different from zero in a more than accidental way. If one calculates from an error probability of 5%, only the groups 2 (SU) and 5 (SU plus BIG) do not show a statistical difference to zero, i.e. in these groups morning and evening values do not differ more than accidentally. In all other cases, the differences between morning, noon and evening values are significant.

Table 3 indicates how frequently blood glucose levels below 80 mg% were observed in relation to the various forms of therapy, this relation being tested with a contingence table (x^2-test). Excluding group 1 (diet) there is no difference

Table 3 Number of blood sugar values below 80 mg % in the various groups of treatment at the different points of measurements.

Therapy	Number of values	Number of values below 80 mg %		
		Morning	Noon	Evening
Diet	100	20	8	16
SU	100	20	8	16
GLIB	101	9	4	7
SU plus BIG	101	3	1	1
GLIB plus BIG	71	5	0	2

between the various groups of treatment at the different points of measurements (error probability 5%).

A significant difference is noted only in the evening measurements between the group 3 (BLIB) and group 4 (SU plus BIG). When all groups are tested together, the x^2-test is only significant for group 1 (diet). This is confirmed by the exact test according to FISHER. At all points of measurements a significant difference is noted between group 1 (diet) and the other groups (error probability 5%).

DISCUSSION

These investigations show that patients treated with diet alone differ significantly from diabetics treated with oral hypoglycemic agents. The variations in the blood glucose profiles as evidenced by the differences between morning and noon, and between noon and evening, are pronounced in the sulfonylurea treated group as compared to the diet group, and are even more striking in the group treated with sulfonylureas and biguanides. This might reflect the increasing severity of diabetes. Groups with monotherapy and with combined therapy do not differ.

The more frequent occurrence of hypoglycemic states in the diet group, especially in the morning, might be due to the fact that in this group preprandial measurements reflect more likely the state of starvation than simply a fasting glucose level. Patients treated with oral agents do not exhibit any statistical difference in the incidence of hypoglycemia. A stronger hypoglycemic effect of one of the oral agents could not be detected, provided one follows an exact program of dosage and control.

The only significant difference in the frequency of blood glucose levels below 80 mg% in the evening values (Table 3),—compared is group 3 (GLIB) versus group 4 (SU plus BIG)—, might be interpreted as a shift of the optimal effect of glibenclamide to the late afternoon. However, since there is no significant difference to other groups under oral treatment, especially not to group 2 (SU), the hypoglycemic effect is thought to be dosage dependent. Whether or not the elevated blood glucose levels in the groups under oral treatment compared to the diet group demonstrate a risk factor, is not proved yet.

It is remarkable, howeevr, that even well conducted oral antidiabetic therapy is not able to avoid high alimentary peaks. These peaks do not occur to this extent in the diet group.

Although comparative investigations in a group of 100 insulin treated patients show a certain smoothing effect of insulin on the daily blood glucose profile, the variations around a mean glucose level still are higher than in the diet group. The smoothing effect seems to be more pronounced if insulin treatment is combined with biguanides. The number of patients, 29, however, does not allow yet a statistical evaluation.

SUMMARY

Long term studies were conducted in a large number of diabetic patients to ascertain, whether various antidiabetic agents exert different effects on blood sugar levels during chronic treatment. The results were compared with those obtained in groups of diabetics treated with diet alone, and with diet and insulin.

Patients treated with diet alone differed significantly from all other groups, in that (1) the mean values of blood glucose at the different points of measurements were lower, (2) the variations of blood glucose levels during the day were not as pronounced and (3) hypoglycemic values occurred more freuent.

Provided constant dietetic intake, adequate mangement and choice of the oral agent, esentially no difference in the blood sugar profiles could be detected under treatment with various oral agents, including the most potent glibenclamide.

CLINICAL IMPORTANCE OF BIORHYTHMS LONGER THAN THE CIRCADIAN

Hobart A. Reimann

Hahnemann Medical College and Hospital
Philadelphia, Pennsylvania, U.S.A.

Since 1947 I have collected reports of more than 2,000 examples of various medical disorders recurrent at weekly, fortnightly, monthly or irregular intervals in otherwise healthy people. These repetitive disturbances are far more intense and disabling than those incident to the circadian rhythm. They are heritable and, with an exception, only one of several entities afflicts a family [1].

Generally, and in current textbooks of Medicine, the proposed nosologic relation of the disparate entities is receied skeptically and the subject is confused. I probably have not made my views clear, or my essays may not be read carefully or the idea is disregarded. Each entity is different clinically, but all have in common heredity, precise or irregular periodicity of short episodes of illness for decades, overlapping features, suppression of episodes during pregnancy, occasional replacement of one entity by another, occasional amyloidosis, similar serum complement defects and they resist therapy.

The cause is unknown. Theoretically, an inherent rhythm or a feedback mechanism provokes a diencephalic discharge the effects of which are mediated autonomically as various neurovascular and cellular responses. The rhythm or rhythms probably are universal and seldom recognized. When their amplitudes are exaggerated in rare susceptible persons, overt symptoms or clinically inapparent episodes are provoked periodically. To account for the different distinct entities and the familial predilection for any one, different generally sensitive organs or tissues serve as target areas. Their reaction gives rise to the respective entities and their names.

The names applied are periodic (a) polyserositis, (b) fever, (c) synoviosis, (d) edema, (e) purpura, (f) myelodysplasia (neutropenia), (g) thrombocytopenia, (h) pancreatosis, (i) sialadenosis and others.

Periodic polyserositis and *periodic fever* often are wrongly regarded as identical. Of 15 names of the former, the least desirable is familial Mediterranean fever. It was coined because Armenians, Jews and Arabs are chiefly affected. Etiocholanolone was wrongly implicated as causal in both disorders. Periodic polyserositis is inflammatory and febrile with painful serosal involvement. No lesions are obvious in periodic fever. Amyloid ensues oftener in polyserositis.

Corticosteroid and colchicine therapy has biven temporary relief in some victims, but not in others. Estrogens reportedly relieved the symptoms of both disorders. In my experience, contraceptive therapy, which gives effects resembling pregnancy, has not been successful.

In *periodic synoviosis,* the knees chiefly are painfully swollen usually fortnightly. The disorder is related to *periodic edema* (intermittent hydrarthrosis). Both are afebrile. In the former, plasma enters the synovia; in the latter, into the skin.

In both there is a lack of the serum inhibitor of Cl esterase of the complement system [2]. One entity may replace, accompany or alternate with the other. The mortality rate from glottic edema exceeds 10 percent in periodic edema.

Report that infused normal inhibitor in donor plasma suppresses Cl and meliorates periodic edema has been questioned. The same doubt applies to the value of E-aminocaproic acid for either entity. Any therapy to be effective probably must be applied continually.

Periodic myelodysplasia is characterized by the depletion or disappearance of neutrophil cells in the marrow and blood every 21 days accompanied by mucosal lesions. Infection is a hazard [3]. Morley observed 20 victims in five afflicted families [4]. Splenectomy has not given satisfactory results. A similar disorder affects Collie dogs. Morley demonstrated 14–23 days oscillation in the numbers neutrophils [5] and 21–35 days cycles of platelets in normal persons [6], as I had predicted [1].

Fewer cases of periodic *thrombocytopenia, pancreatosis, sialadenosis* and others are on record.

REFERENCES

1. Reimann H. A. (1963): Periodic Diseases, F. A. Davis Co., Philadelphia, Blackwell Scientific Publications, Oxford, pp. 1–189.
2. Reimann H. A., Coppoea E. D. and Villegas G. R. (1970): Serum complement defects in periodic diseases. Ann. Int. Med., *73*: 737.
3. Reimann H. A. (1971): Haemocytic periodicity and periodic disorders. Postgrad. Med. Jour., *47*: 504.
4. Morley A. A., Carew J. P. and Baikie A. G. (1967): Familial cyclic neutropenia. Brit. Jour. Haemat., *13*: 719.
5. Morley A. A. (1966): A neutrophil cycle in healthy individuals. Lancet, *2*: 1220.
6. Morley A. A. (1969): A platelet cycle in normal individuals. Australasian Ann. Med., *18*: 127.

SUPPRESSION OF MITOSES IN MOUSE MAMMARY TUMOR CELLS FOLLOWING ADMINISTRATION OF CIRCADIAN DESIGNED CHEMOTHERAPY REGIMENS*

GORDON L. ROSENE**

Department of Physiology and Health Science
Muncie, Indiana, U.S.A.

INTRODUCTION

This pilot study was designed to examine the effects of administration of chemotherapeutic agent, Cytosar,*** at different time points along the 24-hour time scale. A dose regimen was sought which would yield minimum lethal effects upon the normal cell population and maximum lethal effects upon the tumor cell population.

Cytosar was observed to be an active agent in suppression of DNA synthesis [1–3]. The drug was frequently employed in animal tumor studies. The majority of these investigations involved neoplasms in the lymphoid line [4, 5].

Reports of clinical application were published [6–8]. Patients with lymphoid neoplasms were studied most often. Patient populations were treated with the drug and followed carefully during the post-treatment period. Complete and partial remissions of the leukemic symptoms were observed in a limited number of cases. The study of FREI et al. [8] differed from the other two in that the Cytosar was administered by perfusion over 24-, 48, and 72-hour periods. Patients with carcinoma, were also included in the study. Cytosar is cleared from the system quite rapidly, when injected intravenously in single doses [5]. Its perpetual presence was assured during the treatment period with intravenous perfusion, and any cell in the body undergoing DNA synthesis during these periods was in jeopardy. On the whole, the results obtained from the infusion regimens were superior to the earlier studies. However, as in dosage regimen techniques employed by other investigators, the hyperplastic segment of the normal cell population was endangered (i.e. bone marrow, lymphoid system etc.).

The present study is directed toward this problem. Circadian rhythms of mitosis were reported to exist in experimental animals as well as human subjects. The subject area was thoroughly discussed by ROSENE and HALBERG [9]. COOPER and SCHIFF [10] and SCHEVING [11] performed studies on the human epidermis. The period where maximum mitosis of the hyperplastic cells of the body occurs is predictable in each 24-hour period. The DNA synthesis phase of the replication cycle also follows circadian periodicity and is the keystone of the present experiment. Maximum destruction to the normal, hyperplastic cells will occur if cytosine arabinoside is administered during this period. Although the mitotic index of some tumors have been shown to follow circadian rhythms [9] the majority of the tumors studied do not. Time plots of mitotic indicies in tumors

---

* Supported by grant from Delaware County Cancer Society No. 2362

** Present address: Department of Physiology and Health Sciences. Ball State University, Muncie, Indiana, U.S.A.

*** Cytosine arabinoside, Upjohn

displaying circadian rhythms demonstrate much less precision than do those time plots made from non-neoplastic populations. The peaks are broader, and the overall level of the mitotic index seen during the 24-hour period is more elevated. If the periods of maximum DNA synthesis of the normal cell populations were avoided, extensive damage could be still afforded the tumor cell population and minimum effects would be felt by the normal cells. This is the premise upon which the experimental design of the present study was based.

EXPERIMENTAL PROCEDURES AND RESULTS

The work completed at the present time can be divided into two sections.

1. Lethal Dose Curve in Normal Mice

A group of 75 female A-strain mice was selected from our colony. These animals were separated into five subgroups of fifteen members each. Each animal was given daily injections of Cytosar (200 mg/kg). The regimen was continued for a six-day period. This part of the study is a duplication of work previously done by Drs. L. SCHEVING and S. CARDOSO and was suggested to us by Dr. SCHEVING. Our findings agreed closely with theirs which are reported in this symposium. In each study, maximum survival was seen during the forenoon hours. The survival ratios at various time points in our laboratory were as follows: 0330 hours, 0/15; 0830 hours, 3/15; 1230 hours, 5/15; 1830 hours, 1/15; 2330 hours, 0/15.

2. Dose Regimen in Tumor Animals

A second experiment was designed, utilizing the information obtained from the dose response study. Sixteen, female, A-strain mice were employed. Each had developed a spontaneous mammary tumor. A tumor biopsy specimen (3-day, pre-treatment biopsy) was removed from each animal at the onset of the study for histologic examination. The mice then were randomly relegated to one of two groups. Three days after biopsy, mice in Group I received a single daily injection of Cytosar (50 mg/kg) at 2400 hours. This was continued for a three-day period. Equivalent dosage was given to mice in Group II over a similar three-day period, but the drug was administered at four-hour intervals during the 24-hour period, with a fraction of the total amount given at each time point. Eighty per cent of the daily dose was administered during the previously determined period of maximum survival (0800 to 1200). The remaining twenty per cent was given during the period of minimum survival (1800 to 0300).

A second tumor biopsy specimen was obtained from each mouse, three days after the last injection of Cytosar (three-day, post-treatment biopsy). A third tumor biopsy specimen was obtained from mice of each group, 24 days after the last injection of Cytosar (24-day, post-treatment biopsy). All biopsy specimens were removed between 0900 and 1100 hours.

Mitotic incidence was expressed as the ratio of metaphase mitotic stages per oil emersion field (Fig. 1). It was found that consistent data was more easily obtainable when only metaphase stages of mitosis were counted. Considerable nuclear pycnosis, leucocyte infiltration and cellular debris made the distinction between the other phases of mitosis difficult. Slides were masked and coded during counting procedures.

The significance between differences in mitotic incidence in three-day, pre-treatment and three-day post-treatment biopsies, as well as the significance between differences in three-day, pre-treatment and 24-day, post-treatment mitotic index was determined by paired t-test statistical methods (Fig. 1).

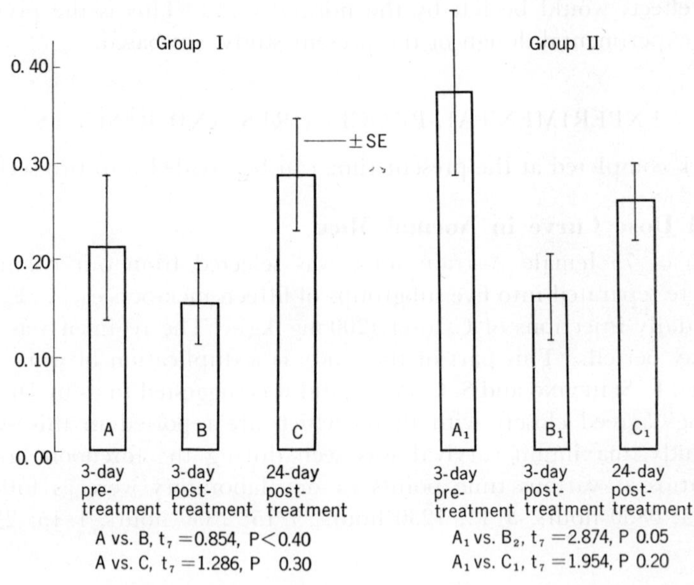

Fig. 1

RESULTS

Alterations of mitotic incidence were seen to occur subsequent to Cytosar administration.

1. Mitotic Index in Three-Day, Pre-Treatment Versus Three-Day, Post-Treatment Biopsy Specimens

Depressions of the mitotic index were observed in five of the eight animals in Group I (50 mg/kg administered at 2400 hours during a period of three days). An increase in mitotic activity was evidenced in the remaining three animals. Paired t-test evaluation did not indicate a significant depression of mitotic activity (t_7=0.854, $p<0.40$).

Depression in the mitotic index was observed in all eight animals in Group II. These depressions were seen to be significant when the data was submitted to paired t-test analysis (t_7=2.874, $p<0.05$).

2. Mitotic Index in Three-Day, Pre-Treatment Versus Twenty-Four-Day Post-Treatment Biopsy Specimens

Four of the eight Group I animals demonstrated elevations of mitotic index during the 24-day, post-treatment period (Fig. 1). A net increase in mitotic activity was observed when mean values were compared (0.081 cells in metaphase per field). This increase was not significant when data were submitted to paired t-test analysis (t_7=1.286, $p<0.30$).

In Group II, less within-group variation was seen in the animals (Fig. 1). A depression of tumor cell mitotic incidence was still evident 24 days after Cytosar treatment in all but one of the eight group members. Although mitotic incidence showed some evidence of suppression 24 days after the final Cytosar injection, paired t-test evaluation revealed no significant difference when mitotic incidence in these tumor cell biopsy specimens were compared with the three-day, pretreatment tumor tissues ($t_7=1.954$, $p>0.05$).

DISCUSSION

We have observed a more effective suppression of tumor cell mitosis (reflecting a more severe inhibition of DNA synthesis) when the drug was administered throughout the 24-hour period (Fig. 1) than when given at a single time point (Fig. 1).

Skipper et al. [5] using a transplantable leukemic neoplasm (L1210) found a similar enhancement of drug effectiveness following administration of the drug at three hour intervals throughout the 24-hour period. They reported the half life of the drug to be 10–20 minutes. This characteristic explains the enhancement of the drug seen in this study. Frei et al. [8], administering Cytosar by intravenous perfusion techniques, found enhanced antitumor effects during administration periods of 24, 48, and 72 hours.

We also have found an enhanced anti-mitosis response with dose regimens involving injections at several points of the 24-hour period. Our approach, however, differs in one very important respect. We are attempting to find a dose regimen that will protect the normal cell population by taking into account circadian fluctulations of DNA synthesis.

The protection of the normal cell population is essential to normal metabolic functions as well as to the maintenance of the reticuloendothelial system. It is well established that enhancement of neoplastic growth follows immunosupression [12]. With suppression of the hemopoietic and lymphopoietic capacities of the animal, the survival and metastatic capacities of the tumor would be enhanced [13–15].

The control of neoplastic growth depends upon the cell replication capacity of the individual tumor cells as well as the effectiveness of the antitumor response of the reticuloendothelial system. Cytosar therapy can reduce cell replication capacity. Circadian-timed dosage, as employed here, can better maintain the immune capacity of the treated individual.

REFERENCES

1. Chu M. Y. and Fisher G. (1962): A proposed mechanism of action of 1-B-D-Arabinofuranosylcytosine as an inhibitor of growth of leukemic cells. Biochem. Pharmacol., *11*: 423–430.
2. Kim J. H. and Eidenoff L. (1965): Action of Cytosine Arabinoside on the neucleic acid metabolism and viability of He La cells. Fed. Proc., *24*: 331.
3. Silagi S. (1965): Metabolism of 1-B-D-Arabinofuranosylcytosine in L cells. Cancer Res., *25*: 1446.
4. Evans J. S., Musser E., Bostwick L. and Mengel G. D. (1964): The effect of 1-B-D-Arabinofuranosylcytosine on murine neoplasms. Cancer Res., *24*: 1285–1293.
5. Skipper H., Schabel F. M. and Wilcox W. (1967): Experimental evaluation of potential anticancer drugs. Cancer Chemother. Rep., *51*: 125–165.

6. HENDERSON E. S. and BURKE P. J. (1965): Clinical experience with Cytosine Arabinoside. Proc. Am. Assoc. Can. Res., *6*: 26–29.
7. ELLISON R. R., HOLLAND J. F., WEIL M., JACQUILAT C., BOIRON M., BERNARD J., SAWITSKY A., ROSNER F. and GUSSOFF B. (1968): Arabinosyl Cytosine—A useful tool in the treatment of acute leukemia in adults. Blood, *32*: 507–523.
8. FREI E., BICKERS J., HEWLETT J., LANE M., LEARY W. and TALLY R. W. (1969): Dose schedule and antitumor studies of Arabinosyl Cytosine. Cancer Res., *29*: 1325–1331.
9. ROSENE G. and HALBERG F. (1970): Circadian and ultradian mitotic rhythms in livers and Ehrlich asites tumors in mice. Bull. All India Institute of Medical Sciences, *4*: 77–94.
10. COOPER Z. K. and SCHIFF A. (1938): Mitotic rhythm in human epidermis. Proc. Soc. Exp. Biol. Med., *39*: 323–326.
11. SCHEVING L. E. (1959): Mitotic activity in human epidermis. Anat. Res., *135*: 7–19.
12. FAIRLEY G. H. (1969): Immunity to malignant disease in man. British Medical Journal, *2*: 467–473.
13. BASERGA R. and SHUBIK P. (1954): Action of cortisone on transplanted and induced tumors in mice. Cancer Res., *14*: 12–16.
14. FISHER B. and FISHER E. R. (1959): Experimental evidence in support of the "dormant tumor cell." Sci., *130*: 918–919.
15. MARTINEZ C. and BITTNER J. J. (1955): The effect of cortisone on lung metastases production. Proc. Soc. Exp. Biol. Med., *89*: 569–570.

EFFECT OF DEXAMETHASONE UPON CIRCADIAN DISTRIBUTION OF MITOSIS IN THE CORNEA OF RATS II. THE ESTABLISHMENT OF A 48-HOUR RHYTHM

S. S. CARDOSO, J. G. SOWELL and P. J. GOODRUM

Department of Pharmacology, University of Tennessee Medical Units Menphis, Tennessee, U.S.A.

The different metabolic steps (synthesis of protein, RNA and DNA) which preceed and control cell division have been subject to intensive investigation [1]. These studies have been greatly aided by *in vitro* synchronization of cell division which can be accomplished with the use of different systems often involving the use of agents which inhibit the synthesis of nucleic acid [2]. Necessarily, however, these *in vitro* methods eliminate the participation of control mechanisms which most likely play important roles in the conditioning and/or modulation of different pathways of cell metabolism *in vivo* [2]. On the other hand, relatively limited attention has been given to circadian mitotic rhythms, a *natural* synchronization of cell division [3–6] as a tool in the study of cell division cycles. Different degrees of mitotic synchronization exist among the several tissues of the body, although the peaks of mitotic activity seem to occur at approximately the same time in most tissues thus far studied. Circadian mitotic peaks for the corneal and digestive tract epithelium of rats occur from 0600 to 1100 hours of the day in animals kept on a light/dark schedule, with lights from 0600 to 1800 hours. Their ratios of peak to trough are of the order of 10/1 to 2/1, respectively. Furthermore, the understanding of the mechanism which controls the circadian rythms would represent a significant step towards the understanding of normal growth processes as well as disturbances of growth resulting in neoplasia, since it has been reported that several experimental and human neoplasms are not responsive to these controlling mechanisms [5, 6, 9–13], whereas healthy tissue is responsive. Also, studies in this area might ultimately lead to the willful modification of circadian mitotic rhythms both in normal and neoplastic cells. Circadian synchronization of normal and neoplastic cell populations with the establishment of mitotic achrophases distinctly separated in time from each other, could permit a greater exposure of neoplastic cells to effective therapeutic agents [14–16]. Without increased injury to healthy tissue, Roentgen therapy and S phase specific chemotherapeutic agents for instance could be delivered to tumor bearing animals or patients at times in which maximum resistance to these agents could be expected from critical tissues (bone marrow and gastrointestinal epithelium) of the host [10, 15, 16].

Evidence has been presented indicating that glucosteroids seem to play an important role in the overall mechanism(s) which control circadian rhythms in the normal tissues of the host [17–20].

Results obtained over the past several months and presented in this communication [8] indicate that the circadian mitotic rhythm as found in the cornea of rats can be profoundly altered by the intraperitoneal administration of Dexamethasone. Furthermore these studies indicate that this effect is dependent upon the time of the day in which Dexamethasone is administered. Three different types of responses (circadian mitotic rhythm) are observed, following Dexamethasone administration at 0700, 1500 and 2300 hours. The responses are respectively: 1) depression, 2) synchronization, 3) establishment of a 48 hour rhythm. The selection of the different hours of the day for drug administration (0700, 1500 and 2300 hours) was based upon the animals' natural corticosterone circadian pattern. With Dexamethasone administration at 0700 hours of the day (Table 1) the overall levels of mitotic activity were found to be depressed in the immediate days following the last dose of Dexamethasone. The mitotic peaks (at 0700 hours) were significantly reduced when compared with those of untreated controls. On the other hand, the general circadian mitotic pattern in Dexamethasone treated animals was comparable with that of controls.

Table 1 The effect of dexamethasone administration at 0700 hours on alternate days upon mitosis in the cornea of rats.

Groups[a]	Mitosis per 100 field±SE[b] hour of day[c]						
	1500	1900	2300	0300	0700	1100	0700
I	74±15	74±6	31±5.5	89±10	186±186.11	106±7.7	186±11
II[d]	95±14	25±2	18±5	84±8	122±5	48±17	112±4
p[e]	>0.05	>0.05	>0.05	>0.05	<0.001	<0.02	<0.001

[a] Groups I & II respectively Controls & Dexamethasone.
[b] SE Standard error by the mean value.
[c] Hour of day in which the different groups of rats were sacrificed.
[d] Dexamethasone, 0.4 mg, administered every other day (I. P. route) at 0700 hours. (total of 5 doses)
[e] p Represents the comparison between Control Groups and Dexamethasone.

With Dexamethasone administration at 1500 hours (Fig. 1) on alternate days, a sharp synchronization of mitosis around 0700 to 1100 hours of the first day after treatment: a 30 to 35 fold rise in cell division rates is present between 0300 and 0700 hours of the day in dexamethasone treated animals. The cell division rates rise approximately 10 fold between 1900 and 0700 hours (trough to peak) in control animals. Thus an apparent enhancement of the natural circadian synchronization of cell division was obtained with dexamethasone administration at 1500 hours. In the second day following the last dose of the steroid, the mitotic peak was delayed and depressed. This reduction of the mitotic peak is statistically significant ($p<0.005$). Additional experiments, with the sacrifice of animals at hourly intervals, will be needed to establish more closely the extent of the above delay. We feel that this delay might be related to the return of the adrenal-pituitary function following dexamethasone administration [21, 22].

With dexamethasone administration at 2300 hours, on alternate days, pronounced changes also occurred. These changes seen in Fig. 2 were as follows: a) delay from 0700 to 1200 hours of the first mitotic peak to occur after the last dose of dexamethasone, with maintenance of high levels of mitosis at 0700 hours;

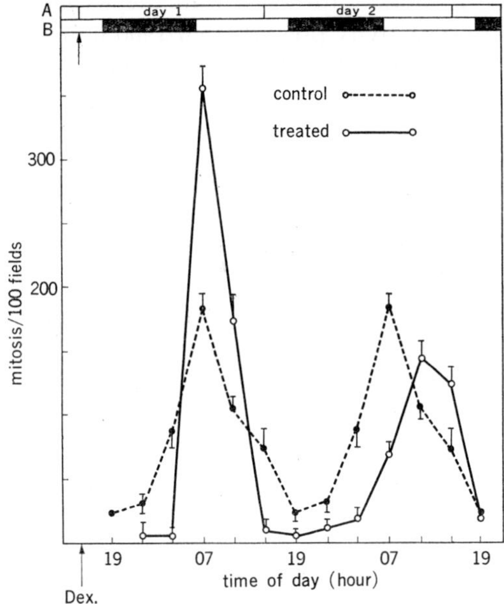

Fig. 1 The effect of Dexamethasone administhation at 1500h, on alternate days upon mitosis in the cornea of rats. Control ●---● and Dexamethasone treated rats ○—○ were sacrificed at four hour intervals, with the first treated group being sacrificed eight hours after the fifth dose of dexamethasone (DEX and arrow in the (figure). Drug administration was made on alternate days at 1500 hours by the I.P. route. A minimum of six rats was sacrified to obtain each of the means and Standard Errors of the means shown in the figure.

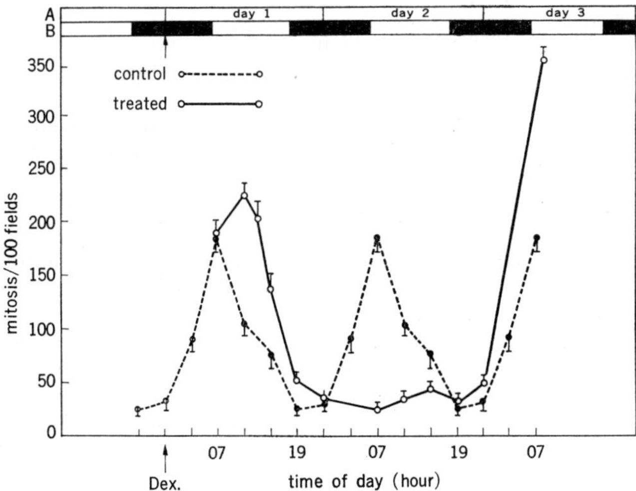

Fig. 2 The effect of Dexamethasone administration at 2300 hours on alternate days, upon mitosis in the cornea of rats. Control ●---● and dexamethasone treated rats ○—○ were sacrified at four hour intervals, with the first treated group being sacrificed 8 hours after the last dose of dexamethasone (DEX abd arrow in the figure). Drug administration was made by the I.P. route on alternate day at 2300 hours (total of five doses of dexamethasone 0,4. mg per dose). A minimum of six rats was sacrificed to obtain the means shown in the figure.

b) suppression of the 24 hour mitotic cycle in the second day after treatment; c) trough levels of mitotic activity present from 1900 hours of the immediate day after treatment to 2300 hours of the second day after treatment; d) increased levels of mitotic activity at 0700 hours of the third day after treatment. In effect, the changes produced by dexamethasone treatment at 2300 hours, resulted in establishment of a 48 hour mitotic cycle.

SUMMARY

Data are presented to indicate that Dexamethasone administration to rats produces dramatic changes in the circadian distribution of mitotic activity. They also indicate that the type of response observed is related to the time of the day (circadian system phase) of Dexamethasone administration. Thus, either depression, sharp synchronization or the establishment of a 48 hour mitotic rhythm were observed, following the administration of the steroid at 0700, 1500 and 2300 hours of the day, respectively.

REFERENCES

1. Muller G. C. (1969): Biochemical events in the animal cell cycle. Federation Proceedings, *28*: 1780.
2. Petersen R. A., Tobey R. A. and Anderson C. (1969): Synchronously dividing mammalian cells. Federation Proceedings, *28*: 1771.
3. Halberg F. (1960): Temporal coordination of physiologic function. Cold Spring Harbor Symp. Quant. Biology, *25*: 289.
4. Halberg F. (1969): Chronobiology. Annual Review of Physiology, *31*: 675.
5. Cardoso S. S., Ferreira A. L., Camargo A. C. M. and Bohn G. (1968): The effect of partial hepatectomy upon distribution of mitosis in the cornea of rats. Experientia, *24*: 569.
6. Gololobova M. T. (1958): Changes in mitotic activity in rats in relation to the time of the day or night. Bulletin of Experimental Biology and Medicine (USSR), *46*: 1143.
7. Messier B. and Leblond C. P. (1960): Cell proliferation and migration as revealed by radioautography after the injection of thymidine H3 into male rats and mice. American Journal of Anatomy, *106*: 247.
8. Clark R. H. and Korst D. R. (1969): Circadian periodicity of bone marrow activity and reticulocyte counts in rats and mice. Science, *166*: 236.
9. Blumenfeld C. M. (1943): Studies of normal and abnormal mitotic activity. II. The rate and the periodicity of experimental epidermoid carcinoma in mice. Archives of Pathology, *35*: 667.
10. Garcia-Sainz M. and Halberg F. (1966): Mitotic rhythms in human cancer, reevaluated by electronic computer programs. Evidence for chronopathology. Journal of the National Cancer Institute, *37*: 279.
11. Maini M. M. and Stich H. F. (1962): Chromosones of tumor cells. III. Unresponsiveness of precancerous hepatic tissues and hepatomas to mitotic stimulus. Journal of the National Cancer Institute, *28*: 753.
12. Watanabe M., Potter Van R., Pitot H. C. and Morris H. P. (1969): Systematic oscillations in metabolic activity in rat liver and hepatomas: effect of hydrocortisone, glucagon and adrenalectomy. Cancer Research, *29*: 2085.
13. Potter Van R., Watanabe M., Pitot H. C. and Morris H. P. (1969): Systematic oscillation in metabolic activity in rat liver and hepatomas. Survey of normal diploid and other hepatoma lines. Cancer Research, *29*: 55.
14. Bertalanfy F. D. and Gibson M. H. (1971): The *in vivo* effect of Arabinosylcyto-

sine on cell proliferation of murine B_{16} mellanoma and Erlich Acites tumor. Cancer Research, *31*: 66.

15. Cardoso S. S. and Carter J. R. (1969): Circadian mitotic rhythm as a guide for the administration of antimetabolites. Proceedings for Experimental Biological Medicine, *131*: 1403.
16. Cardoso S. S., Scheving L. E. and Halberg F. (1970): Mortality of mice as influenced by the hour of the day drug administration. Pharmacologist, 12. (Abst.)
17. Cardoso S. S. and Ferreira A. L. (1967): Effect of adrenalectomy and of Dexamethasone upon circadian distribution of mitosis in the cornea of rats. Proceedings of Soc. Experimental Biology and Medicine, *125*: 1254.
18. Cardoso S. S., Goodrum P. and Sowell J. (1971): Synchronization of cell division and the establishment of a 48 hour rhythm *(in vivo)* by Dexamethasone. Federation Proceedings, *V30*: 679.
19. Radzialowski F. M. and Bousquet W. F. (1968): Daily rhythmic variation in hepatic drug metabolism in the rat and mouse. Journal of Pharmacology and Experimental Therapeutics, *163*: 229.
20. Reinberg A. and Halberg F. (1971): Circadian chronopharmacology. Annual Review of Pharmacology, *11*: 455.
21. Liddle G. W. (1960): Tests of pituitary adrenal suppressibility in diagnosis of Cushing Syndrome. Journal of Clinical Endocrinology, *20*: 1560.
22. Gibbs F. P. (1970): Circadian variation of ether-induced corticosterone secretion in the rat. American Journal of Physiology, *219*: 288.

CIRCADIAN VARIATION OF ULCER-SUSCEPTIBILITY IN THE RAT*

Steven D. BROWN and Frank W. FINGER
Department of Psychology, University of Virginia
Charlottesville, Virginia, U.S.A.

Ader [1, 2] reported that immobilization produced gastric ulcers in a greater proportion of rats restrained during the hours of predicted maximal activity than among those similarly treated during that portion of the day previously characterized by lesser running in the activity wheel. Consistent with his guiding hypothesis, it might reasonably be argued that this outcome reflected the differential stressfulness of preventing free movement during the opposite phases of the activity cycle. In order to demonstrate circadian fluctuation in susceptibility to ulceration, it seems essential to employ a stressor the intensity of which is less likely to vary as a function of time of day. Our choice is unpredictable and inescapable shock to the unrestrained animal.

SUBJECTS

The subjects were male Long-Evans rats, about 100 days old, habituated to a regimen of LD 12:12 (lights on from 0600 to 1800). Two groups of 14 each were constituted for the major experiment, Group L to be subjected to the stress of shock during light (0800 to 1400) and Group D to be stressed during dark (2000 to 0200). Assignment was random, except for the interchange of one pair to equate the groups initially with respect to mean body weight. Twelve similar animals served as subjects in the secondary experiment.

APPARATUS

To permit simultaneous treatment of two subjects, duplicate experimental chambers were provided, wired in series. Each was a plywood box, 12×12×12 in., with hinged Plexiglas top. Footshock was delivered through brass floor grids, spaced ½ in. apart. The constant-current scrambled-shock source (Grason-Stadler E6070B) was controlled by a Gerbrands tape programmer.

Illumination during the light 12 hr of the day was by incandescent bulbs, with intensity of 32 foot candles at the floor of the experimental chambers. A red fluorescent tube (General Electric F40R) was continuously on, making it possible for the experimenter to work during the "dark" 12 hr while avoiding appreciable visual stimulation of the subjects. Ambient temperature was maintained at 22±1°C, with relative humidity about 50%. Extraneous auditory stimulation was masked by white noise (82–85 db SPL).

* This research was supported by Grant MH04920 by the U. S. Public Health Service.

PROCEDURE

Each subject of Groups L and D was given a single session in one of the shock boxes, preceded by 24-hr food deprivation in his home cage. Undisturbed exploration was permitted during the first 15 min, followed by 1½ hr of the shock condition, ½ hr of rest, another 1½ hr of the shock condition, ½ hr of rest, and a final 2 hr of the shock condition. While the shock condition was in force, 12-sec periods of inescapable shock (1.6-ma., 0.5-sec pulses, separated by 1.0 sec) were administered on a variable-interval 90-sec schedule (i.e., on the average of once every 90 sec).

Immediately after the 6¼ hr in the shock box the animal was weighed and returned to his home cage, without food and water. One hour later he was sacrificed by sodium pentobarbital overdose. The stomach was removed, inflated with 0.9% saline, ligated, and inspected for ulcers near the line of incision by translumination [3]. After 15 min it was opened along the greater curvature, washed in cold tap water, mounted on glass slides, and reexamined. The criterion for an ulcer was a fixed blood clot on the inner surface, visible to the naked eye. The experimenter's diagnosis was in every instance corroborated by an independent judge who later inspected the preserved tissues without knowledge of the experimental condition.

On each of seven days the same procedure was followed with four subjects, two from Group L in the morning and two from Group D in the evening.

Because of unexpected weight differences developing between the two major groups, three additional groups of four rats each were treated in the same manner as Group D, except that the food deprivation preceding the shock session was 10-, 5- and 0-hr, respectively.

RESULTS AND DISCUSSION

Of the 14 subjects exposed to unpredictable shock during the light, only 3 or 21% developed stomach ulcers, while similar pathologies were found in 13 or 93% of the Group D subjects (upper section of Table 1)*. This difference is highly significant ($p<.01$) according to the Fisher exact probability test. It is apparent that there is a clear circadian rhythmicity in this form of response to stress, of an order of magnitude comparable to the cyclicities reported in susceptibility to audiogenic convulsion [4], to drug injection [5], to infectious agents [6], and to X-irradiation [7]. It may be that the failure to consider interaction with this temporal variable accounts for certain of the inconsistencies found when the effects upon ulceration of sex [8, 9], early experience [10, 11], and prior fear conditioning [12, 13] have been investigated. The implications for adequate experimental design and specification of conditions are obvious.

Unexpectedly, Group L lost slightly more weight during the period of pre-shock food deprivation than did Group D (14.8% vs 12.3%). Since previous research [14, 15] has suggested that the probability of gastric ulceration in response to

* The ulcers were confined to the glandular portion of the stomach. In one instance microscopic examination of the rumen revealed a small pit in the surface, which may have represented an early stage of ulceration.

stress is directly related to degree of food deprivation, it seemed appropriate to examine the relevance of this variable in our experimental situation. Hence the addition of the three small groups shocked in the dark after the lesser degrees of food deprivation. As the lower section of Table 1 indicates, all four of the subjects deprived for 10 hr prior to the shock session developed ulcers, two of the 5-hr-deprived group, and all the 0-hr group. While it must be pointed out that some additional degree of hunger developed during the 6¼ hr in the shock box without food, our procedure—at least when carried out during the dark half of the diurnal cycle—is minimally dependent on food deprivation for its efficacy. Expressed differently, variation in susceptibility appears to be considerably more dependent on phase of the circadian cycle than upon deprivation.

Table 1 Proportion of subjects developing ulcers.

Condition	N	Hours of food deprivation	% Ulcers
L	14	24	21
D	14	24	93
D	4	10	100
D	4	5	50
D	4	0	100

The significant physiological concomitants of the susceptibility cycle have yet to be elucidated. There is evidence [16] that histamine is the chemical mediator of gastric acid secretion and that increased gastric acidity is associated with stomach ulceration [17]. It has been observed [18] that the activity of histidine decarboxylase, the catalyst of histamine found in the acid-secreting cells of the glandular portion of the stomach, increases greatly during stress and is related to both incidence and severity of gastric ulcers; pretreatment with histidine decarboxylase inhibitor reduces the ulceration process. It would be instructive to investigate the circadian variation of the elements in this chemical chain.

SUMMARY

Unrestrained male rats, 24-hr hungry, were subjected to an unpredictable pattern of inescapable shock during one 6-hr session. Gastric ulcers were detected in 93% of the 14 animals shocked during the dark half of the LD 12:12 cycle, and in 21% of the 14 shocked during light. Degree of hunger seemed less important than the circadian variation in susceptibility, for 83% of an additional 12 subjects exposed to the stress of shock in the dark after 0–10 hr of food deprivation developed ulcers.

REFERENCES

1. Ader R. (1964): Gastric erosions in the rat: effects of immobilization at different points in the activity cycle. Science, *145*: 406–407.
2. Ader R. (1967): Behavioral and physiological rhythms and the development of gastric erosions in the rat. Psychosom. Med., *29*: 345–353.
3. Mikhail A. A. and Holland H. C. (1966): Evaluating and photographing experimentally induced stomach ulcers. J. psychosom. Res., *9*: 349–353.

4. Halberg F., Bittner J. J., Gully R. J., Albrecht P. G. and Brackney E. L. (1955): 24-hour periodicity and audiogenic convulsions in I mice of various ages. Proc. Soc. Exp. Biol. Med., *88*: 169–173.
5. Pauly J. E. and Scheving L. E. (1964): Temporal variation in susceptibility of white rats to pentobarbital sodium and tremorine. Int. J. Neuropharmacol., *3*: 651–658.
6. Feigin R. D., San Joaquin V. H., Haymond M. W. and Wyatt R. G. (1969): Daily periodicity of susceptibility of mice to pneumococcal infection. Nature, *224*: 379–380.
7. Pizzarello D. J., Isaak D., Chua K. E. and Rhyne A. A. (1964): Circadian rhythmicity in the sensitivity of two strains of mice to whole-body radiation. Science, *145*: 286–291.
8. Sines J. O. (1959): Selective breeding for development of stomach lesions following stress in the rat. J. comp. physiol. Psychol., *52*: 615–617.
9. Sines J. O. (1962): Strain differences in activity, emotionality, body weight and susceptibility to stress induced stomach lesions. J. genet. Psychol., *101*: 209–216.
10. Ader R. (1965): Effects of early experience and differential housing on behavior and susceptibility to gastric erosions in the rat. J. comp. physiol. Psychol., *60*: 233–238.
11. McMichael R. E. (1961): The effects of preweaning shock and gentling on later resistance to stress. J. comp. physiol. Psychol., *54*: 416–421.
12. Sawrey W. L. and Sawrey J. M. (1964): Conditioned fear and restraint in ulceration. J. comp. physiol. Psychol., *57*: 150–151.
13. Mikhail A. A. (1969): Relationship of conditioned anxiety to stomach ulceration and acidity in rats. J. comp. physiol. Psychol., *68*: 623–626.
14. Brodie D. A. and Hanson H. M. (1960): A study of the factors involved in the production of gastric ulcers by the restraint technique. Gastroenterology, *38*: 353–360.
15. Weisz J. D. (1957): The etiology of experimental gastric ulceration. Psychosom. Med., *19*: 61–73.
16. Levine R. J. (1965): Effect of histidine decarboxylase inhibition on gastric acid secretion in the rat. Fed. Proc., *24*: 1331.
17. Levine R. J. and Senay E. C. (1970): Studies on the role of acid in the pathogenesis of experimental stress ulcers. Psychosom. Med., *32*: 61–65.
18. Levine R. J. and Senay E. C. (1968): Histamine in the pathogenesis of stress ulcers in the rat. Amer. J. Physiol., *214*: 892–896.

TRANSPLANT CHRONOBIOLOGY*

JULIA HALBERG, ERNA HALBERG, WALTER RUNGE, JAMES WICKS, LINDA CADOTTE, EDMOND YUNIS, GEORGE KATINAS, OSIAS STUTMAN** and FRANZ HALBERG

Chronobiology Laboratories, Departments of Pathology and Laboratory Medicine, University of Minnesota Minneapolis, Minnesota, U.S.A.

To examine whether transplant surgery can exploit natural time factors such as rhythms [1], 47 C_{57} Black, subline 6 (C_{57}), 21 Dilute Brown, subline 2 (DBA) and 23 Bagg Albino (C) mice, all males and about 16–21 weeks of age at time of transplantation, were studied. The DBA and C_{57} were received on November 10, 1970, from colonies maintained by the National Institutes of Health, Bethesda, Maryland. Presumably healthy at shipment, some mice appeared ill and died shortly after arrival. The remainder were placed on tetracycline medication, yet 29 out of 50 DBA mice and three out of 50 C_{57} mice died during the several months prior to transplantation from bacterial pneumonia diagnosed at autopsy. During the same months, intraperitoneal temperature of some animals was monitored by special sensors [2].

During the third month after arrival all animals had rectal temperatures within the range delineated by the physiologic rhythm established earlier for similar animals. The transplantation work was then begun. During the ensuing one and a half months of observation three of the 21 Dilute Brown mice and two of the 47 C_{57} Black mice died. Several rectal temperature measurements were made in the course of the study after transplantation. Since these were in the rhythm-determined range, it is assumed that the animals evaluated for the chronobiology of transplantation were healthy. It also seems likely that a sizeable number of surviving mice had been exposed to pneumonia but had apparently fully recovered from this condition by the time of study.

For at least five weeks prior to transplantation all animals were singly housed in light controlled (LD12:12) "chambers" or "hives"; they were given all the food (Purina Rat Chow) and water they could consume. Before transplantation, the mice were disturbed only for the cleaning of cages, for weighing and for temperature measurements at intervals of several days; presumably during these activities there was less disturbance than was associated with daily checking of the transplants later in the study. One third from each of the C_{57} and DBA strains were implanted under pentobarbital sodium (Diabutal®) anaesthesia (0.1 cc/10 gm of body weight of a 1:7 dilution in saline) with Franklin Institute [2] temperature sensors, Type MIVD, another third from each strain received dummy sensors of equal volume and weight [3] and the last third remained unimplanted. One-half of the C-mice were left unimplanted, and the other half received dummy

* Supported by grants from the United States Public Health Service (5-K6-GM-13, 981 1 RO1-CA-14445-01; 1-RO-1-A1 10153-03 and HL-06314-13) NASA and Mr. SAMUEL M. POILEY, Head, Mammalian Genetics and Animal Production Section, National Cancer Institute, NIH, Bethesda, Maryland.

** Research Associate of the American Cancer Society.

sensors. Just prior to transplantation the rectal temperatures of all mice were repeatedly measured around the clock with a thermistor-bridge circuit [4]. Thereby the circadian temperature rhythm phase at the time of transplantation of the separate subgroups operated at different points along the 24-hour scale could be used as reference for any temporal differences in the results of transplantation (Fig. 1). For transplantation, an assembly line was constituted. One person carried the animals to and from their environments; another took the temperature, weighed the mouse and recorded these data. A third injected the anesthetic. Another person shaved the animals and, using a brass stamp especially prepared for this purpose, immersed it in Sudan black and stamped the skin area to be transplanted—in order to draw borders of standardized size. This step should be avoided when skin color changes are to be evaluated. Next, the skin was disinfected, lifted with a forceps and an incision made by means of scissors. The skin was cut along the previously marked rectangle. One person (J.W.) carried out all surgical steps. At 4-hour intervals around the clock, back skin was exchanged either between two mice from the same inbred strain [syngeneically (isograft)] or the back skin of a mouse from one inbred strain was exchanged with that from another inbred strain [allogeneically (homograft)]. After the transplants were completed, each graft was checked and rated at least once daily until those that were allogeneic showed 100% rejection.

Between 1300 and 1500 on January 21, 1971—about 12 days ±12 hours after transplantation—all animals were examined and photographed (results shown in Fig. 2). On the following day subgroups were examined separately at clock hours corresponding to the times of implantation and some grafts were photographed (results shown in Fig. 3) with a set-up appropriate for photogrammetric recording. Such photographs also had been taken when signs of rejection were first noticed. Shrinkage of the graft occurred in most mice and was assessed by serial photogrammetry on each of several animals (e.g., Fig. 4) but for the sake of economy not all mice were photographed at all times.

RESULTS AND DISCUSSION

I. All homografts, irrespective of time of transplantation, eventually were rejected, isografts—although showing shrinkage and/or transient color change—finally took. Initially, certain changes in graft appearance, including size, take place in both allogeneic and syngeneic transplantation and must be distinguished as quantitative rather than qualitative differences.

II. Among homografted mice kept in light from 0600 to 1800 the time-span elapsed to first outward sign of rejection was longest for those transplanted at 1600. Thus, actual rejection appeared to be slowest in those mice transplanted prior to the daily occurrence of the circadian periodic high in body temperature; slowest rejection coincided roughly with transplantation at the time of high blood corticosterone levels established earlier for comparable mice.

III. Serial photogrammetry on the same animal provides an objective way to assess the rejection rate of a skin graft. Percent shrinkage of the graft and changes in color, among other features such as swelling, dessication, and ulceration are thus objectively assessed Fig. 4a–c.

IV. With photogrammetric evaluation least shrinkage (at a fixed postoperative time) was ascertained to occur among those mice transplanted around 1600

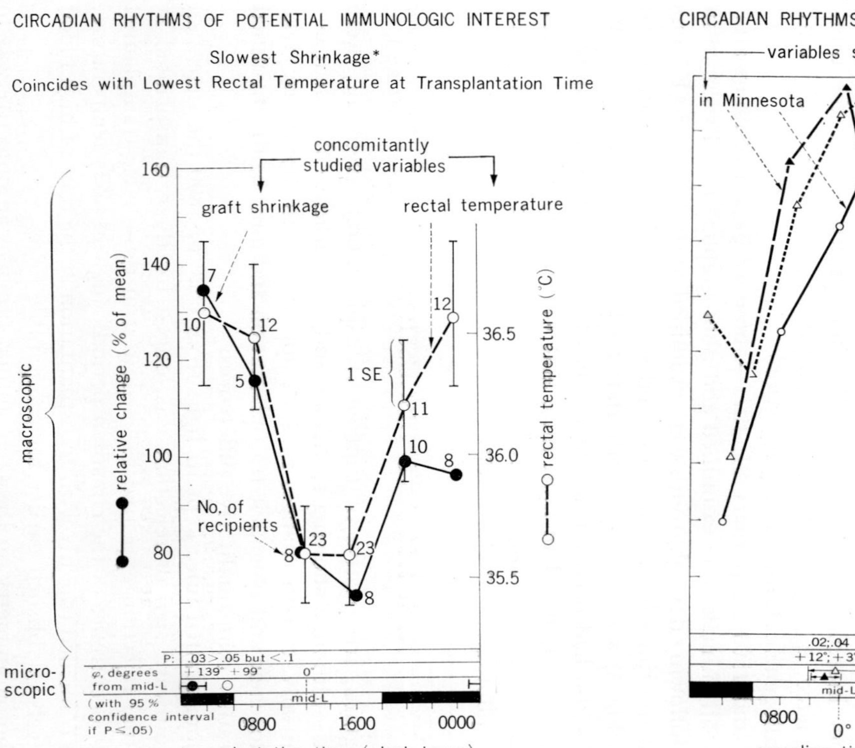

Fig. 1 While graft shrinkage and rectal temperature are at their trough, serum cortiosterone levels, blood lymphocyte counts and epidermal mitosis (extrapolated from an earlier study) are high, but direct causal relationship between the several variables, some expressed as % of mean, cannot be intimated without further evidence. Whether the fate of certain mouse skin homografts depends upon any factors controlling body temperature, serum corticosterone or other rhythms of recipient and/or donor at the time of transplantation and during the processes of healing or rejection is a question for further study.

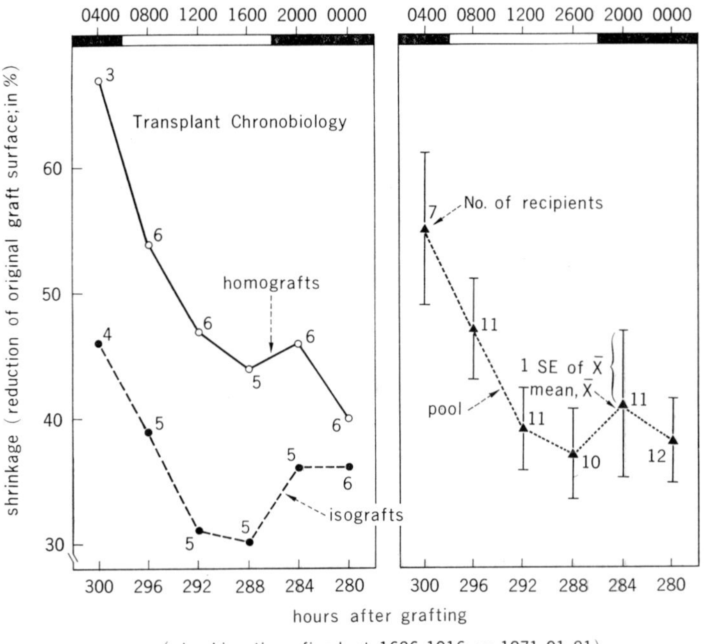

Fig. 2 Percent shrinkage of back skin exchanged between $C_{57}B/_6$ and $DBA/_2$ mice is a function of time elapsed from transplantation but does not seem to be a linear one. Scale at the top indicates the hour at which a given group of animals was grafted, showing also the light and dark span during standardization and checking. Number of animals at each time point given above the plotted points. On the curve showing shrinkage as a function of clock hour of transplantation for groups of mice kept on L0600–1800 D1800–0600, those operated first at 0400 show most shrinkage and those operated last, at 0000, less shrinkage. However, mice grafted around 1600 show least shrinkage although they were operated earlier than the mice grafted at 2000 or 0000. High values of serum corticosterone and high blood lymphocyte counts are anticipated at about 1600.

(Figs. 2 and 3). Mice transplanted at certain other times showed more shrinkage as well as earlier signs of rejection. The trend in the photogrammetric data obtained by photography at a fixed time-of-day—thus at a post-transplantation time that differed among subgroups by up to ±12 hours (Fig. 2)—agreed with results on selected animals photographed at 4-hour intervals at a fixed post-implantation time—Fig. 3.

The strong immunologic barrier between the several strains of mice used herein for exchanging back skin allogeneically accounts for the rejection that occurred in all homografts; work involving a lesser genetic barrier also is indicated. Moreover, the methods used to measure shrinkage and to evaluate graft appearance must be standardized for chronobiologic studies and for any other work involving relatively small differences in the results of skin transplantation. Thereby one may more rigorously assess the processes of rejection and of wound healing.

TRANSPLANT CHRONOBIOLOGY

Homografts at One Circadian System Phase (1600 for mice in L0600–1800)
Shrink More Slowly than Those Made at Other Times

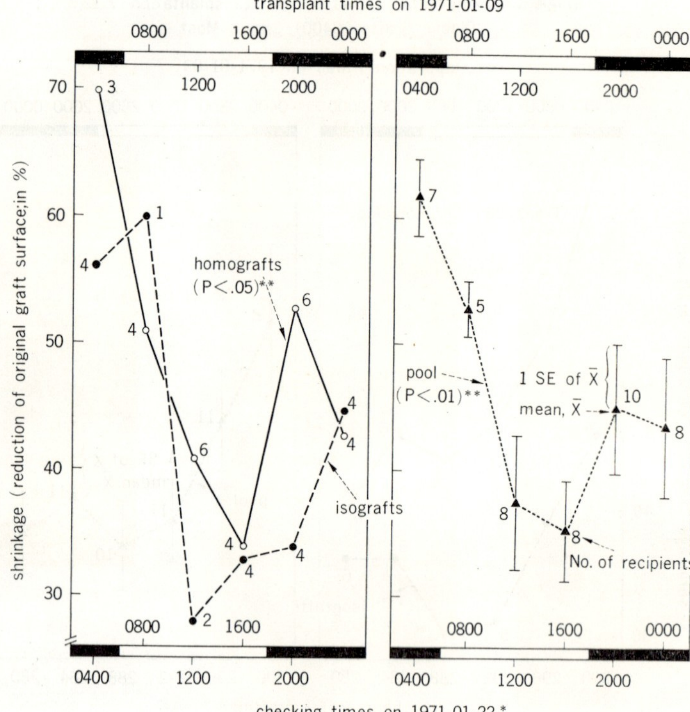

* around the clock; transplants at 0800 on 01-09 checked 13 days later at 0800; those made at 1200 were checked at 1200, etc.

** P from analysis of variance

Fig. 3. Percentage of shrinkage in isografts and homografts of back skin from C57B/6 and DBA/2 mice, made at one of six different circadian phases. Scale on top lists these different transplantation times as clock hours on the 9th of January, 1971, for mice standardized in light from 0600 to 1800 alternating with darkness. The results plotted were obtained by checking different subgroups at 4-hour intervals around the clock, each subgroup at the clock hour on January 22, 1971 corresponding to that chosen for operating those mice 13 days±1.5 hours earlier, as shown on bottom of graph. For homografts and isografts shrinkage is a non-linear function of time after transplantation. In these controlled spot-checks, a lesser shrinkage of those homografts and isografts made at the 1600 or 1200 transplantation times respectively denotes an apparent advantage in keeping with impressions gained also from Fig. 2 and from the pool of all grafts, on the right of this figure. Differences in the circadian system state of donor and/or recipient at transplantation time and circadian rhythms in any one individual graft may all contribute to these statistically significant results.

Fig. 4a Homografting (gradual rejection shown in frames 1–9 taken on the same DBA/2 mouse bearing C57B/6 skin) compared with the take of an isograft [in each, frames 10 (DBA/2) and 11 (Balb/cCr)] Such photographs, taken to spotcheck the time course of transplants, prompted the development of systematic objective methods. Later, the area of the graft on a photograph of the mouse held at a fixed distance from the lens was excised for weighing. Two quantitative indices were obtained, one directly for shrinkage, another for graft appearance involving estimates of the areas showing swelling, desiccation, ulceration and/or color change. A composite index was formed by combining shrinkage and graft-appearance indices. Figs. 4b and 4c (by G. Katinas) stem from these subsequent studies.

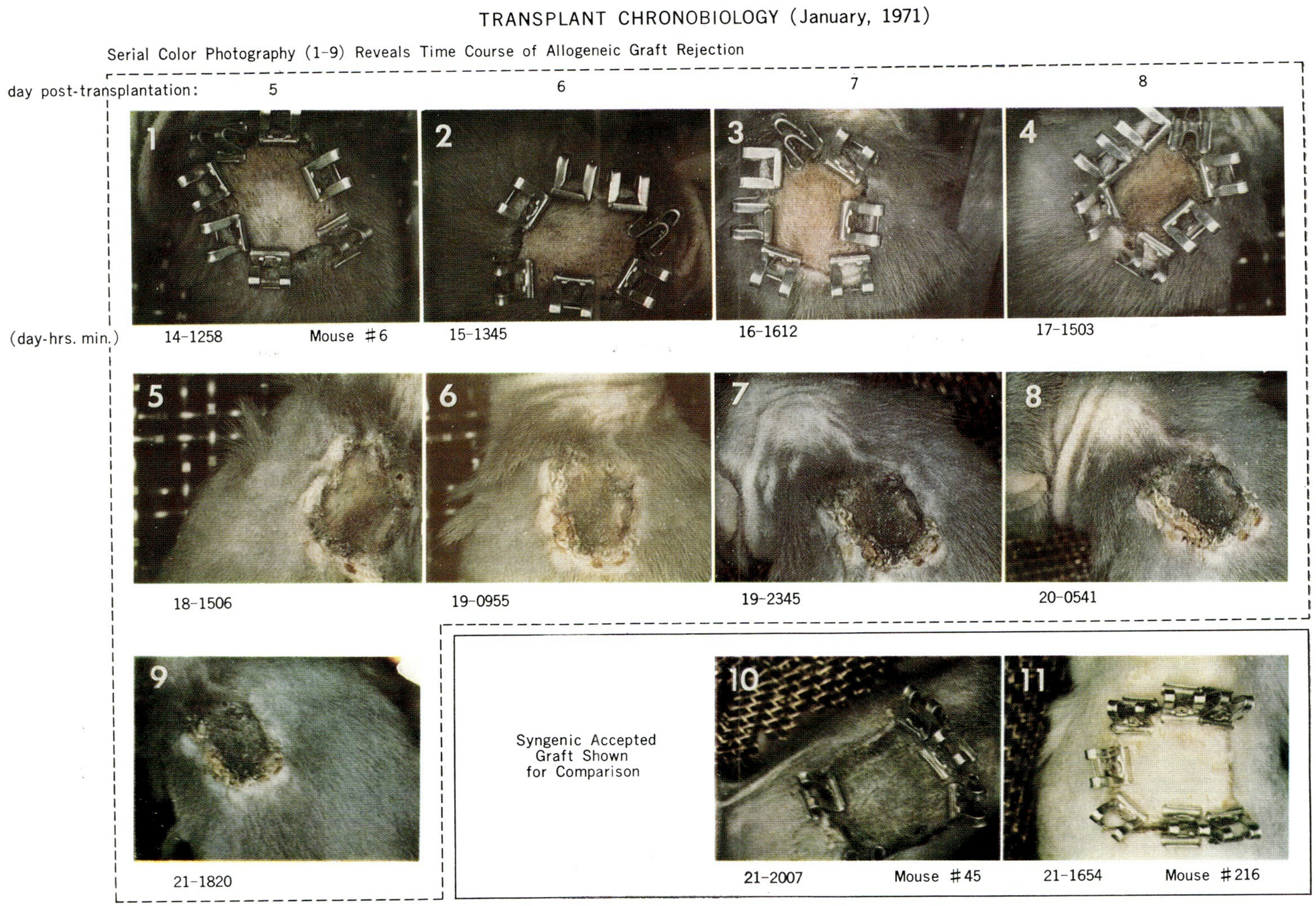

TRANSPLANT CHRONOBIOLOGY (January, 1971)

Serial Color Photography (1-9) Reveals Time Course of Allogeneic Graft Rejection

day post-transplantation: 5 6 7 8

(day-hrs. min.)

1 — 14-1258 Mouse #6

2 — 15-1345

3 — 16-1612

4 — 17-1503

5 — 18-1506

6 — 19-0955

7 — 19-2345

8 — 20-0541

9 — 21-1820

Syngenic Accepted Graft Shown for Comparison

10 — 21-2007 Mouse #45

11 — 21-1654 Mouse #216

Fig. 4 a

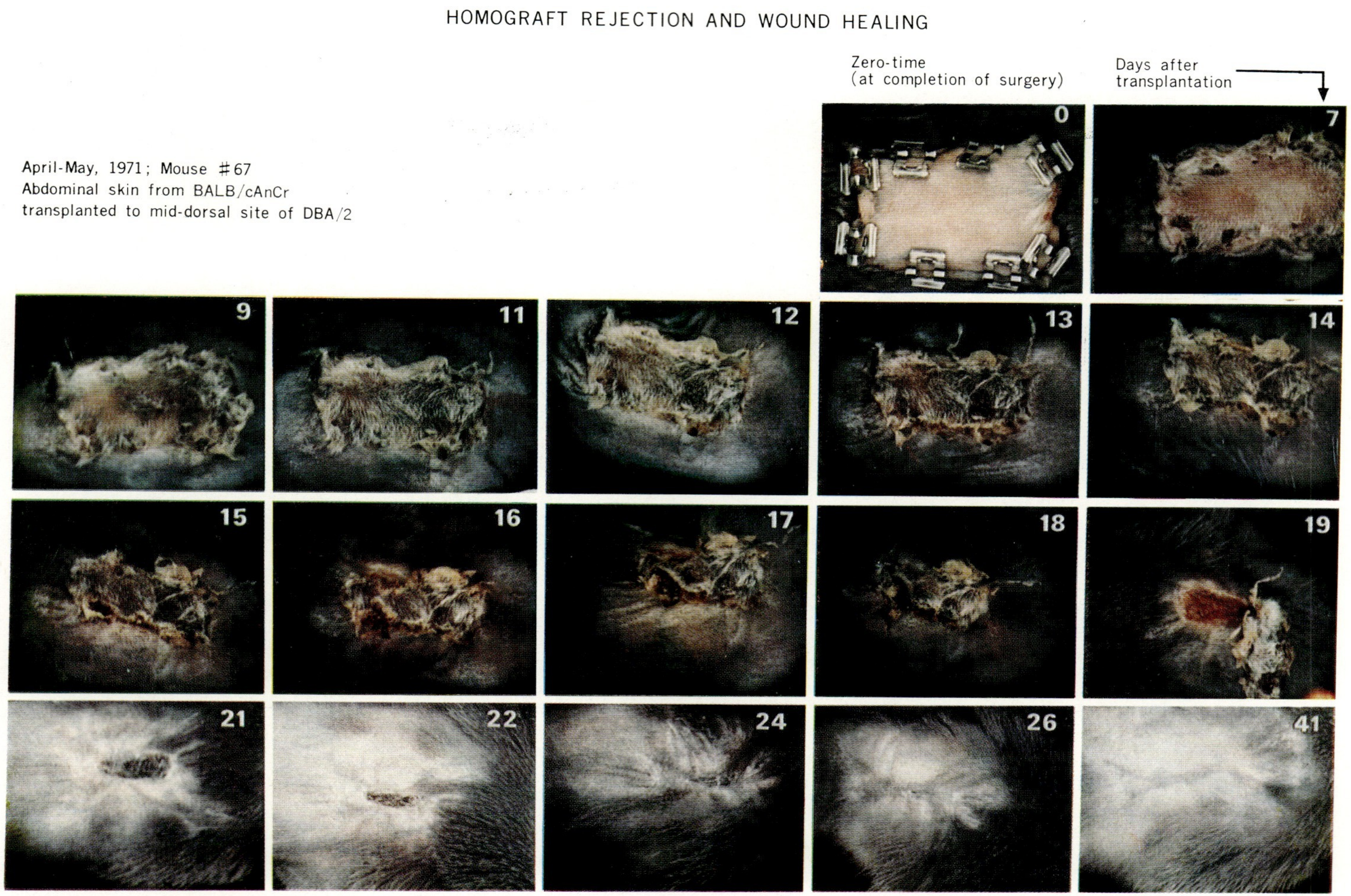

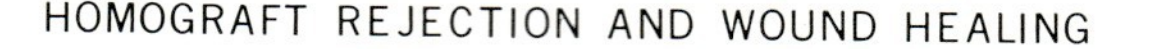

HOMOGRAFT REJECTION AND WOUND HEALING
April-May, 1971; Mouse #67
Abdominal skin from BALB/cAnCr
transplanted to mid-dorsal site of DBA/2
Zero-time
(at completion of surgery)
Days after
transplantation
0
7
9
11
12
13
14
15
16
17
18
19
21
22
24
26
41

Fig. 4 b

TAKE OF ABDOMINAL BALB/cAnCr SKIN ON MID-DORSAL SITE OF RECIPIENT FROM THE SAME INBRED STRAIN (April-June, 1971; Mouse #53)

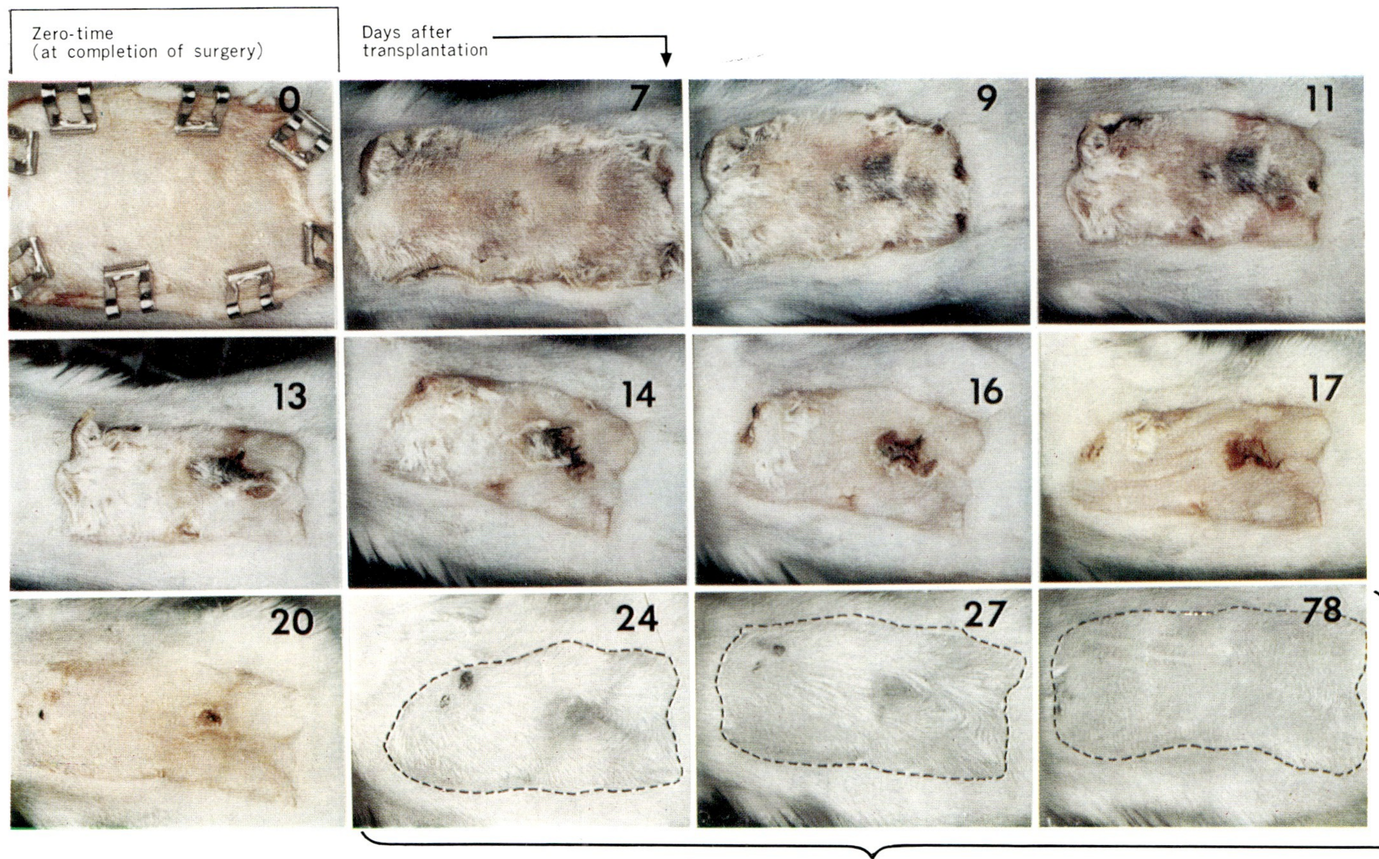

Fig. 4c

The time elapsed till the first signs of change appear in both isografts and homografts as well as the percentage shrinkage of the grafts themselves, is on the average longer or smaller, respectively, in animals transplanted at the time of spontaneously higher adrenal cortical hormone levels, Fig. 1. Many other body functions, some of special interest to immunologists, such as blood lymphocytes in mice [5], Fig. 1 and man [6, 7], change rhythmically. Further transplantation studies and other immunologic experiments are warranted within the framework of chronobiology, on these and other endocrine factors, and also on other graft and recipient rhythms of potential interest to immunologists [8]. Moreover, cellular and humoral immunity may each possess a characteristic and different timing of their rhythms [9].

REFERENCES

1. Halberg F. (1970): Chronobiologic aspects of transplantation (Review). Circulation, *42*: 786.
2. Halberg F., Nelson W., Runge W. J., Schmitt O. H., Pitts G. C., Tremor J. and Reynolds O. E. (1971): Plans for orbital study of rat biorhythms. Results of interest beyond the Biosatellite program. Space Life Sciences, *2*: 437–471.
3. Pitts G. C., Bullard T. R., Tremor J. W., Sebesta P. D., Halberg F. and Nelson W. (1969): Rat body weight and composition: sensor implantation and lighting effects. Aerospace Medicine, *40*: 317–420.
4. Halberg F., Zander H. A., Houglum M. W. and Mühlemann H. R. (1954): Daily variations in tissue mitoses, blood eosinophils and rectal temperatures of rats. Am. J. Physiol., *177*: 361–366.
5. Brown H. E. and Dougherty T. F. (1956): The diurnal variation of blood leucocytes in normal and adrenalectomized mice. Endocrinology, *58*: 365–375.
6. Elmadjian F. and Pincus G. (1946): A study of the diurnal variations in circulating lymphocytes in normal and psychotic subjects. J. Clin. Endocrinol., *6*: 287.
7. Halberg F. (1960): Circadian temporal organization and experimental pathology. Report from VII Conferenza Internazionale della Societa per lo Studio dei Ritmi Biologici inclusa la Basimetria, Siena, Sept. 5–7, 1960. Panminerva Medica, Atti, pp. 1–20.
8. Baldwin W. M. III and Cohen N. (1973): "Weak" histocompatibility antigens generate functionally "strong" humoral immunity as measured by *in vivo* assay. Fed. Proc., *132*: 1006.
9. Fernandes G., Yunis E. J., Nelson W. and Halberg F. (1973): Differences in immune response of mice to sheep red blood cells as a function of circadian phase. *In*: Chronobiology. Proc. Symp. Quant. Chronobiology, Little Rock, 1971, (L. E. Scheving, F. Halberg and J. E. Pauly, eds.), pp. 329–335, Igaku Shoin Ltd., Tokyo.

DIFFERENCES IN IMMUNE RESPONSE OF MICE TO SHEEP RED BLOOD CELLS AS A FUNCTION OF CIRCADIAN PHASE

Gabriel FERNANDES, Edmond J. YUNIS, Walter NELSON and Franz HALBERG

Chronobiology Laboratories, Departments of Pathology and Laboratory Medicine, University of Minnesota Minneapolis, Minnesota, U.S.A.

INTRODUCTION

The importance of physiologic rhythms [1] emerges from their effect upon response to drug administration [2]. Are these same rhythms altered at an early stage of impaired health? If so, attempts to correct these altered rhythms to alleviate certain diseases will be warranted. The former line of abundant evidence [1, 2] and the latter line of thought both prompt medical interest in the field of chronobiology more generally [1], as well as with respect to specific circadian rhythms of the cardiovascular, renal and other systems [3, 4].

Against this same background, it is noteworthy that a number of studies on physiologic rhythms relate to the cellular or humoral immune systems of experimental animals. Indeed, good health depends largely, inter alia, upon the proper function of cellular and humoral immune systems; it is mainly up to these to counteract pathogenic microorganisms and other potentially harmful agents. However, their merits notwithstanding such immune mechanisms present a formidable obstacle to certain tasks related to the alleviation of impaired health; for instance, they underly the rejection of a transplant. A demonstration of rhythmic changes relating to these immune systems accordingly will be of considerable medical interest to all of those who wish to find the time when such responses are minimal so that they can reduce their undesired consequences as well as to those who can benefit from the time of maximal response or "best defense" when immune responses are required.

BACKGROUND

There is considerable evidence that certain blood cell counts exhibit circadian rhythmic changes in experimental animals [5–8] and man [9, 10]. Rhythms with several frequencies much lower than 1 cycle/day also are known to characterize human variables [11]. Hence, lower-than-circadian frequencies are likely to be found as well in the immune functions of the body. Indeed, recent reports deal with relatively low frequency "cyclic" changes in the erythropoiesis following irradiation [12] and chemotherapy [13]. Similar periodicity was also observed in

* Supported by grants from the U. S. Public Health Service (5-K6-GM-13-981 and AI HD 10153-01 1 RO1-CA-14445-01; 1-RO-1-A1 10153-03 and HL-06314-13), NASA and Mr. Samuel M. Poiley, Head, Mammalian Genetics and Animal Production Section, National Cancer Institute, NIH, Bethesda, Maryland.

the immune response of mice against sheep red blood cells (SRBC) [14] in the phenomena of the graft-versus-host-reaction [15], and in the numbers of antibody forming cells following treatment with cyclophosphamide and SRBC [16, 17] or endotoxins [18].

The present experiment was undertaken to study whether the immune response to SRBC depends on the circadian phase at which animals are injected with SRBC and subsequently bled. Since there are reports [19, 20] that hemagglutinins against SRBC occur naturally in mice, probably due to microbial antigens acting as a stimulus to the production of immunoglobulins and so called "auto-antibodies" [21, 22], it was of interest to study not only the SRBC-injected mice but also mice not knowingly exposed earlier to SRBC.

MATERIAL AND METHODS

Male mice of the inbred BALB/c and DBA/2 strains, 10 weeks of age, were obtained from the colonies maintained by the National Institutes of Health, Bethesda, Maryland. On their arrival these mice were housed for 6 weeks in a room for indirect periodicity studies as earlier described [23] with controlled temperature 72°F and 50% humidity. Lights went on automatically at 0900 and off at 2100 each day. The mice were housed 5 per plastic cage. Purina chow and water were available *ad libitum.* Except for cleaning and watering about once weekly, the animals were not disturbed until the experiment was begun. For a rhythm mapping survey, a total of 56 DBA/2 and 100 BALB/c mice were sampled at 4 time periods (0900, 1500, 2100, 0300). 10 BALB/c mice at each time point served as controls and were injected i.p. with 0.2 cc saline; 15 BALB/c mice at each time point served as experimental mice and were injected with 0.2 cc of a 20% suspension of SRBC.

Before each injection, rectal temperature and body weight were measured and the mice bled for hematocrit determinations. Body temperature was measured using a Yellow Springs temperature measuring instrument*, weight was taken with a Mettler balance weighing to the nearest 0.1 g. Mice were bled from retro-orbital capillaries by the insertion of a heparinized capillary micro-hematocrit tube**. The blood was centrifuged in an Adam micro-hematocrit centrifuge***. DBA/2 mice were killed at each time immediately after the bleeding. After 10 days the rectal temperature and weight of each injected BALB/c mouse were again measured and all mice were bled once more, each mouse at the same time point at which it was bled and injected 10 days earlier. Hematocrits were again determined and serum was collected in non-heparinized micro-capillary tubes and stored at −20°C. Immediately after the bleeding, each mouse was killed by decapitation. Spleen and thymus were removed and their weights determined with a torsion balance weighing to the nearest 1 mg.

Hemagglutination titers were determined in disposable "V" plates (conical shaped well bottoms)**** by following the method of Wehman and Smithies [24]. The first well contained undiluted serum; a 1:5 dilution was made at the second well, followed by serial (1:2) dilutions up to 12 wells. The plates were mixed by

* Yellow Springs Instrument Company, Ohio.
** Scientific Products, Illinois.
*** Clay-Adams Inc., New York.
**** Microbiological Associates, Bethesda, Maryland.

gentle rotation, centrifuged and kept overnight at 4°C. Titers were read after keeping the plates at a 45° angle for 20 minutes. The hemagglutination test was repeated for two time point groups with 1:2 dilution at the second well (1500 and 2100); the results were very close to those obtained in the first test. Further, a separate group of 22 uninjected BALB/c mice from the Bittner mouse colony were bled at 2 time points (Fig. 1, Study II) to check on the results on naturally occurring SRBC antibodies.

Conventional statistical summaries were made on an Olivetti desk computer.

RESULTS

The results presented in Fig. 1 indicate statistically significant differences in immune response (hemagglutination titer) as a function of circadian phase in: a) mice undergoing no previous treatment (Study II), b) mice receiving a saline injection 10 days before bleeding (Study I, controls), c) mice injected with SRBC 10 days before bleeding (Study I, experimentals).

CIRCADIAN PHASE
AND IMMUNOLOGICAL RESPONSIVENESS

Natural or Artificially Enhanced

● control mice, injected with 0.2 cc saline
○ experimental mice, injected with 0.2 cc 20% SRBC suspension

Separate groups of BALB/c male mice injected i.p. at time points indicated on abscissa and bled at same time points 10 days later.

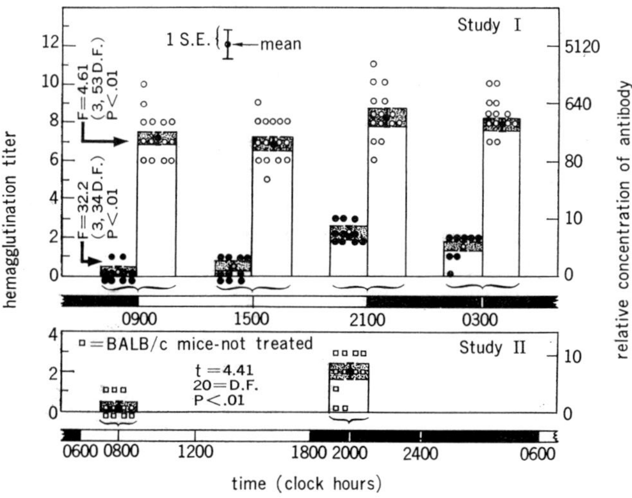

Fig. 1

Although these within-day differences in titer amount to an average change of less than two titer units (i.e., "wells"), it should be noted that large changes in antibody concentration are indicated. For example, in the case of Study I for experimental mice, relative antibody concentration changes from a mean of 171 at 1500 to a mean of 394 at 2100.

In addition to the naturally occurring within-day changes (control animals in Studies I and II), the results from the SRBC-injected mice indicate statistically significant differences (P<.01) in immune response as a function of circadian phase. This statement is based on a consideration of differences in mean titers and relative antibody concentration between experimental and control mice. Values for the latter are very small in relation to the change in antibody concentration for experimental mice between the 1500 and 2100 time points.

Table 1 compares results on other variables studied in intact untreated BALB/c male mice and on animals injected with saline or SRBC (Study I). Body weights were measured before injecting the SRBC or saline. Differences in body weight along the 24-hour scale were not statistically significant for 4 groups of mice on two occasions; i.e., before the injections were given and 10 days post-injection.

Table 1 Circadian variations in male inbred BALB/c mice before injections (C) or after injection with sheep red blood cell antigen (A) or with saline (S).

Variable (unit)	Group*	Mean±S. E. of variables investigated at times given in degrees from mid-L (=1500)				Time factor	
		−270° 0900	0° 1500	−90° 2100	−180° 0300	F**	P
Weight (gm)	C	28.4± 0.7	29.8± 0.5	29.0± 0.5	29.0± 0.4	1.19	>.05
	S	28.7± 0.5	30.5± 0.5	29.0± 0.5	29.5± 0.6	2.53	>.05
	C	28.4± 0.3	29.0± 0.5	29.3± 0.4	30.1± 0.5	2.51	>.05
	A	29.0± 0.3	29.0± 0.6	29.4± 0.4	36.6± 0.2	1.25	>.05
Temperature (C°)	C	35.9± 0.2	35.7± 0.3	36.8± 0.3	36.6± 0.2	4.41	<.01
	S	35.7± 0.2	35.1± 0.3	35.8± 0.3	36.0± 0.1	6.07	<.01
	C	35.9± 0.2	34.8± 0.2	36.2± 0.1	36.6± 0.1	17.87	<.01
	A	35.5± 0.2	34.8± 0.3	35.6± 0.1	35.6± 0.1	0.16	>.05
Hematocrit %	C	50.9± 0.4	52.3± 0.2	50.4± 0.6	50.3± 0.4	5.56	<.01
	S	51.0± 0.4	51.8± 0.4	50.5± 0.3	49.8± 0.4	1.35	>.05
	C	50.8± 0.4	52.7± 0.4	50.8± 0.5	51.9± 0.4	4.48	<.01
	A	51.1± 0.4	52.5± 0.6	51.3± 0.5	51.4± 0.4	2.01	>.05
Relative*** spleen wt.	S	326 ±14.4	368 ±27.2	357 ±21.0	367 ±16.0	0.96	>.05
	A	453 ±28.2	409 ±20.2	446 ±12.0	448 ±30.0	0.72	>.05
Relative*** thymus wt.	S	108.2± 5.6	98.7± 3.0	110.8± 3.2	87.5± 5.8	6.02	<.01
	A	112.1± 4.5	94.3± 4.6	123.5± 6.2	90.0± 3.6	9.26	<.01

* Experimental animals (A) sacrificed 10 days after SRBC injection, controls (S) 10 days after injection of saline, controls (C) before injection.

** F from analyses of variance with 3,36 degrees of freedom for control (10 mice at each of 4 time points,) 3,56 degrees of freedom for experimental group (15 mice at each of 4 time points).

*** In milligrams per 100 gm of body weight.

Statistically significant changes with time are seen for rectal temperature (P<.01) and hematocrit levels (P<.01) in both groups prior to saline and SRBC-injections. However, at 10 days after injection, significant variation was indicated only for rectal temperature of the saline-injected group. Remaining non-significant differences may be due to a possible effect of injection and/or bleeding 10 days earlier.

An overall increase in spleen weight (P<.01) was seen for the groups injected with SRBC but any effect of circadian phase in spleen weight was not detected in any of the injected or control BALB/c animals. Large differences in thymus weight were found among the 4 time points—a highly significant time effect (P<.01) being detected by an analysis of variance for the saline and the antigen (SRBC) injected groups.

Table 2 presents results on circadian variation at 4 time points within 24 hours on intact DBA/2 strain male mice. Besides body weights, rectal temperatures and hematocrits both exhibit statistically significant differences among 4 time points, the hematocrit level being highest when the rectal temperature is lowest, namely at 0° from the middle of the daily light span (Mid-L). Spleen (4 time points) and thymus weight in these mice (measured only at 2 times points) undergo statistically significant changes with time. The thymus is largest in size at −90° from Mid-L at a time when spleens are smallest. Hemagglutination titers against SRBC were not measured in these mice.

Table 2 Circadian variations in intact untreated male DBA/2 mice.

Variable investigated (unit)	variation of means in intact untreated male DBA/2 mice*					
	−270° (16)	0° (10)	−90° (16)	−180° (14)	F**	P
Body weight (gm)	30.6±0.8	28.2±0.4	29.5±0.4	27.9±0.4	5.18	<.01
Rectal temperature (C°)	36.7±0.2	36.0±0.1	36.8±0.2	37.8±0.2	15.39	<.01
Hematocrit %	47.5±0.3	49.4±0.3	48.1±0.3	48.6±0.3	7.40	<.01
Relative spleen weight***	256 ±0.8	289 ±6.0	241 ±5.6	318 ±9.1	7.31	<.01
Relative thymus weight***	43.5±3.3		55.8±0.4		2.47	<.02

* Mid-L=middle of 12-hour light span. No. of mice per time point given in parentheses. 360°≡24 hours, 1 hour=15°. Actual sampling at 1500 and 2100 on both L09–21D21–09 for 0° and −90° values and on L21–09D09–21 for −180° and −270° values.

** F from analyses of variance with 3, 52, DF except for t-test for thymus weight in last row.

*** In milligrams per 100 g of body weight.

DISCUSSION

Circadian rhythms of the pituitary-adrenal system come to mind in reviewing our findings of periodicity in antibody production. Corticosteroid is known to have physiologic effects related to blood volume and lympholysis which can affect the weight of lymphoid organs [25, 26]. Furthermore, the administration of corticosteroids reportedly suppresses delayed hypersensitivity as well as specific immunologic processes and inflammatory responses [27, 28] and reduces the number of circulating lymphocytes and monocytes [29].

Previous studies on BALB/c mice standardized under conditions similar to those here instituted indicate a peak in serum corticosterone at about −48° (−34° to −62°) from mid-L and a peak in temperature at about −171° (−162° to −179°) from mid-L [30]. Since the highest temperature observed in our studies occurred either at the −90° or −180° time points (Table 1) we can safely assume that serum corticosterone had it been determined, would probably have been highest at the time of highest antibody titer. *In vitro* and *in vivo* studies [31, 32] on mice also indicate a circadian rhythm in reactivity of the adrenal gland to ACTH, with least response at the time when corticosterone concentration in serum and adrenals is highest. Further studies with more frequent sampling are required to attempt to describe statistically significant circadian rhythms [33] in the immune systems and also to ascertain whether different inbred strains of mice may differ in their antibody titers against SRBC—natural or immunized—at various time points. Thus, a basis may eventually be found for predicting the phase relation between pituitary-adrenal rhythms and circadian rhythms of the immune systems.

SUMMARY

Statistically significant within-day changes characterize the antibody titer against SRBC, for standardized adult BALB/c male mice, whether or not the animals were injected with SRBC 10 days earlier. For such mice injected with SRBC 10 days earlier, thymus weight exhibited significant time-dependency. In the case of standardized DBA/2 mice, significant time effects were observed in body weight, rectal temperature, hematocrit, relative thymus, and spleen weight.

REFERENCES

1. Halberg F. (1969): Chronobiology. Annual Review of Physiology, *31*: 675.
2. Reinberg A. and Halberg F. (1971): Circadian chronopharmacology. Annual Reviews of Pharmacology, *2*: 455.
3. Halberg F., Good R. A. and Levine H. (1966): Some aspects of the cardiovascular and renal circadian systems. Circulation, *34*: 715.
4. Halberg F., Reinhardt J., Bartter F., Delea C., Gordon R., Reinberg A., Ghata J., Hofmann H., Halhuber M., Gunther R., Knapp E., Pena J. C. and Garcia Sainz M. (1969): Agreement in endpoints from circadian rhythmometry on healthy human beings living on different continents. Experientia, *25*: 107.
5. Halberg F. and Visscher M. B. (1950): Regular diurnal physiological variation in eosinophil levels in five stocks of mice. Proc. Soc. Exp. Biol. and Med. *75*: 846.
6. Panzenhagen H. and Speirs R. (1953): Effect of horse serum, adrenal hormones, and histamine on the number of eosinophils in the blood and peritoneal fluid of mice. Blood, *8*: 536.
7. Halberg F., Visscher M. B. and Bittner J. J. (1953): Eosinophil rhythm in mice, range of occurrence, effects of illumination, feeding and adrenalectomy. Amer. J. Physiology, *174*: 313.
8. Brown H. E. and Dougherty T. F. (1956): The diurnal variation of blood leucocytes in normal and adrenalectomized mice. Endocrinology, *58*: 365.
9. Bartter F. C., Delea C. S. and Halberg F. (1962): A map of blood and urinary changes related to circadian variations in adrenal cortical function in normal subjects. Annals New York Acad. Sci., *98*: 969.
10. Halberg F. (1960): Circadian temporal organization and experimental pathology. VII Conferenza International della societa per lo studio dei ritmi Biologici, Siena, 20.
11. Halberg F., Engeli M., Hamburger C. and Hillman D. (1965): Spectral resolution of low-frequency, small-amplitude rhythms in excreted 17-ketosteroid; probable androgen-induced circaseptan desynchronization. Acta Endocrinologica Supplement, *103*: 54.
12. Morley A. and Stohlman F. Jr. (1969): Periodicity during recovery of erythropoiesis following irradiation. Blood, *34*: 96.
13. Morley A. and Stohlman F. Jr. (1970): Cyclophosphamide-induced cyclical neutropenia. New Eng. J. Med., *282*: 643.
14. Stimpfling J. H. and Richardson A. (1967): Periodic variations of the hemaggluti-

ninin response in mice following immunization against sheep red blood cells and allo antigens. Transplantation, *5*: 1496.
15. Cornelius E. A., Yunis E. J. and Martinez C. (1969): Cyclic phenomena in the graft-versus-host reaction. Proc. Soc. Exptl. Biol. Med., *131*: 684.
16. Many A. and Schwartz R. S. (1971): Periodicity during recovery of immune response after cyclophosphamide treatment. Blood, *37*: 692.
17. Radovich J., Hemingsen H. and Talmage D. W. (1969): The immunologic memory of LAF_1 mice following a single injection of sheep red cells. J. Immun., *102*: 288.
18. Britton S. and Moller G. (1968): Regulation of antibody synthesis against escherichia coli endotoxin. I. Suppressive effect of endogenously produced and passively transferred antibodies. J. Immun., *100*: 1326.
19. Stern K. and Davidsohn I. (1954): Hetero-hemoantibodies in inbred strains of mice 1. Natural agglutination for sheep and chicken red cells. J. Immun., *72*: 209.
20. Baum J. (1969): Naturally occurring hemagglutinins in the New Zealand black mouse. Clin. Exp. Immun., *4*: 453.
21. Weiser R. S., Myrvik A. N. and Pearsall N. N. (eds.) (1970): In: Fundamentals of Immunology. p. 37, Lea & Febiger Publishers.
22. Hammarstrom S., Peremann P., Gustafsson B. E. and Lagercranz R. (1959): Autoantibodies to colon in germ-free rats monocontaminated with clostridium difficile. J. Exp. Med., *124*: 747.
23. Halberg F. (1959): Physiologic 24-hour periodicity; general and procedural considerations with reference to the adrenal cycle. Z. für Vitamin-Hormon- und Fermentforschung, *10*: 225.
24. Wehman T. and Smithies O. (1966): A simple hemagglutination system requiring small amounts of red cells and antibodies. Transfusion, *6*: 67.
25. White A. and Dougherty T. F. (1945): Effect of prolonged stimulation of the adrenal cortex and of adrenalectomy on the numbers of circulating erythrocytes and lymphocytes. Endocrinology, *36*: 16.
26. Smolensky M. H. (1971): Ph.D. Thesis, University of Illinois, Urbana.
27. Casey W. J. and McCall C. E. (1971): Suppression of the cellular interactions of delayed hypersensitivity by corticosteroids. Immunology, *21*: 225.
28. Allison F. (1965): Anti-inflammatory agents. *In*: The Inflammatory Process (B. Zweifach, L. Grant and R. McClusky, eds.), p. 559, Academic Press, New York.
29. Thompson J. and Van Furth R. (1970): The effect of gluco corticosteroids on the kinetics of mononuclear phagocytes. J. Exp. Med., *131*: 429.
30. Nelson W. and Halberg F. (1969): Phase relations of circadian rhythms. *In*: Handbook of Environmental Biology; Fed. Amer. Soc. for Exp. Biol., p. 586.
31. Ungar F. and Halberg F. (1962): Circadian rhythm in the in vitro response of mouse adrenal to adrenocorticotropic hormone. Science, *137*: 1058.
32. Haus E. (1964): Periodicity in response and susceptibility. Anna. N. Y. Acad. Sci., *117*: 292.
33. Halberg F. (1965): Some aspects of biologic data analysis, longitudinal and transverse profiles of rhythms. *In*: Circadian Clocks, (J. Aschoff, ed.), p. 13, North-Holland Publ. Co., Amsterdam.

CHRONOBIOLOGY, PEDIATRICS AND AGING

Chairman: WALTER NELSON

Co-Chairman: THEODOR HELLBRÜGGE

THE DEVELOPMENT OF CIRCADIAN AND ULTRADIAN RHYTHMS OF PREMATURE AND FULL-TERM INFANTS

THEODOR HELLBRÜGGE

Forschungsstelle für Sozialle Pädiatrie und Jugendmedizin der Universität München München, Germany

We know from previous work that:

1. Different physiologic functions develop circadian rhythms independently.
2. Circadian rhythms of the different functions become apparent at different times after birth.
3. During the development of circadian periodicity, an increase in the amplitude occurs in all physiologic functions.
4. The monophasic circadian rhythm seems to originate out of polyphasic cycles.
5. For the development of circadian rhythms, the maturity of the child at birth is important, periodicity developing later in premature than in full-term infants.

This report demonstrates, with power spectra, the development of circadian and ultradian rhythms in body temperature, heart rate, sleep-wake distribution and eye movement of premature and full-term infants. The different studies were carried out by my co-workers BUSSE, HÖGL, ULLNER, WINDORFER and ZINSMEISTER with support from the Deutsche Forschungsgemeinschaft.

Fig. 1 shows the sleep-wake rhythms of full-term and premature infants, in terms of variance spectra computed with a scale linear in period. This analyzes more exactly the circadian rhythm and less exactly the ultradian rhythm.

In the first and second week of life, one cannot find a circadian rhythm either in full-term or in premature infants. With full-term infants, the ultradian rhythm is distinct in the 4-hour range. It appears remarkable that this does not yet show with premature infants. There is a dominance of shorter rhythms with a maximum in the about-two-hour range. It might be conceivable that the ultradian rhythm of about four hours is developed out of these even shorter rhythms.

In the age between the third and the sixth week of life, both groups of infants show a clear maximum of ultradian rhythms around the value of about four hours. Furthermore, one can find for the first time an indication of the circadian rhythm with both groups of infants.

In the age of seven to ten weeks the ultradian rhythm is dominant; on the other hand, the circadian rhythm is more prominent in full-term infants than in the prematurely born.

During the age of ten to thirteen weeks, the circadian rhythm becomes stronger and stronger, again increasing more in the full-term than in premature infants, whereas the ultradian rhythm is lessening in intensity.

Fig. 2 shows variance spectra for eye movement and heart rate of a premature

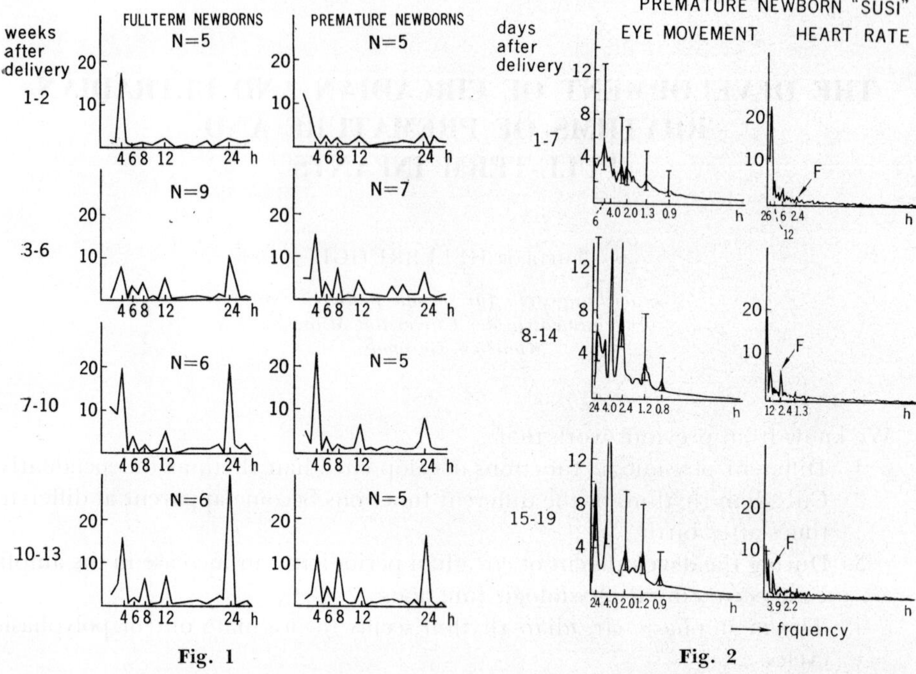

Fig. 1

Fig. 2

girl, computed with a scale linear in frequency. This analysis resolves primarily the range of ultradian frequencies. Whereas Fig. 1 summarized data from groups of infants, Fig. 2 shows the variance spectra of a single infant. As the ultradian rhythm signifies apparently the more primitive rhythm, a longitudinal analysis of a prematurely born infant seemed most significant. The development is shown quite distinctly. With the heart rate there cannot be found an ultradian rhythm between the first and the seventh day of life. With the eye movement a maximum between six and eight hours is distinct. From the eighth day of life on the ultradian rhythm of the eye movement becomes more distinct, specifically about four hours. An equivalent in the heart rate cannot be stated. The maximum (F) caused by the feeding rhythm of 2.4 hours is not an endogenous rhythm, but is due to the feeding schedule. At the age of fifteen to nineteen days this feeding maximum has altered as the feeding was now 4-hourly. The ultradian rhythm has developed into a clear 4-hour rhythm. In addition to that one can find—as with the sleep-wake distribution—secondary maxima that can be explained within the exactness of the analysis as harmonic overtones of a 4-hour rhythm. One can recognize a 24-hour rhythm in eye movement as well as in heart rate in the second and third weeks of life.

To emphasize the circadian rhythm, respective variance spectra were also computed with a scale linear in period (Fig. 3). This reveals a circadian rhythm in eye movement as well as in heart rate during the first week of life. The circadian rhythm of the heart rate decreases distinctly during the second week of life. Both functions show more distinctly again during the third week of life.

During the first week of life the circadian rhythm is apparently still a motherly rhythm that is gradually reduced during the second week of life and replaced by the infant's own circadian rhythm during the third week of life.

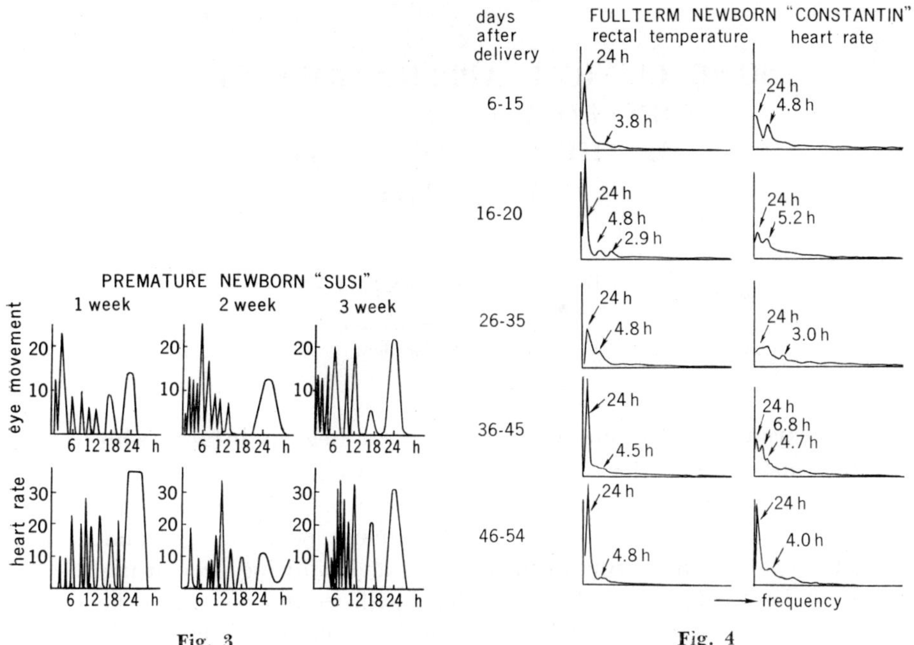

Fig. 3

Fig. 4

Comparison of variance spectra for the two functions indicates possible differences in development between rhythms in heart rate and eye movement. It seems that the circadian rhythms in these two functions during the first two weeks of life do not fully coincide. This is equally true for the secondary maxima and, last but not least, for the ultradian rhythms.

Fig. 4 shows the development of the circadian and ultradian rhythm of the body temperature and the heart rate of a full-term healthy infant, continuously observed from the sixth to the fifty-fourth day of life according to the principles of "self-demand feeding". Variance spectra were computed with scale linear in frequency.

As the frequencies of heart rate and body temperature are largely linked functions it seems remarkable that both functions show already in the second week of life a distinct circadian rhythm, and besides a divergent ultradian rhythm of 4.8 hours with heart rate and of 3.8 hours (indicated) in body temperature. The circadian rhythm in body temperature is more distinct in the third week of life. The ultradian rhythm is divergent in its values.

The findings of the days of life twenty-six to thirty-five seem remarkable. During this time the infant was suffering from an abscess on the head and a light inflammation of the perianal region. The infections are probably the reason for the temporary clear decrease of the circadian rhythm in body temperature and heart rate during this period. With the healing of the infections both rhythms are again distinct between the thirty-sixth and the forty-fifth day of life. Body temperature as well as heart rate show a distinct circadian rhythm, as well as ultradian rhythms of 4.8 and 4 hours, respectively, between the forty-sixth and the fifty-fourth day of life.

SOME CLINICAL APPLICATIONS OF OUR KNOWLEDGE OF THE EVOLUTION OF THE CIRCADIAN RHYTHM IN INFANTS

Raymond MARTIN du PAN
Clinique des Nourrissons
Genève, Switzerland

1. Appearance of the circadian rhythm in healthy infants who live continuously under the same intensity of light

Light plays an important role conditioning an infant. As soon as the optical system has become sufficiently mature, the change between light and dark acts like an indicator; stimulating in the morning and inducing sleep at night.

To our knowledge, no publication deals with studies on infants kept in constant light. On the other hand, a more or less pronounced slowing down of the circadian rhythm has been noted [9, 11] for subjects kept in the dark for fairly long periods.

Two infants were kept in constant light for eight weeks; at the beginning of the test, they were eight days old. These infants lived in a room isolated from the outside world by double windows and insulated walls. They were continuously watched and received care and food at their request. Recording instruments enabled us to follow continuously their EEG, their pulse and rectal temperature.

Fig. 1 summarizes the observations made regarding one of the infants, Constandinos, whose weight at birth was 3,240 g. According to Aldrich's method [1], we have marked on his diary meals by black color, cries by white dots, time awake by black dots and sleep by white color. The vertical lines left and right of the chart represent midnight, the one in the middle noon. Each row represents one day. In order not to make the chart too large, we have indicated only two days per week.

As we see in Fig. 1, Constandinos claimed five meals per day during the second week of his life; they were distributed irregularly over 24 hours. When he was 23 days old, he requested 8, afterwards less. Between 30 and 31 days of age, he slept only 14 to 15 hours out of 24, and still requested six meals. Later, slowly, the system of sleep and being awake becomes organized, but quite differently from what one would have hoped. Indeed, at the age of 53 days, for instance, Constandinos requested 5 meals during 24 hours. He started his day between 1700 and 1800 by requesting a meal and finished at 0500, then he slept very soundly until about 1300, requested a meal and slept again until about 1700. His sleep-wake rhythm was inverted in comparison with his room-mates next door who were stimulated by daylight and slept during the night.

When Constandinos reached the age of 80 days, we switched off the light at night and let him live in the natural rhythm of day and night like the other infants. As we can see in the chart, already after nine days Constandinos had reached a waking-sleeping rhythm related to the periods of day and night. At the

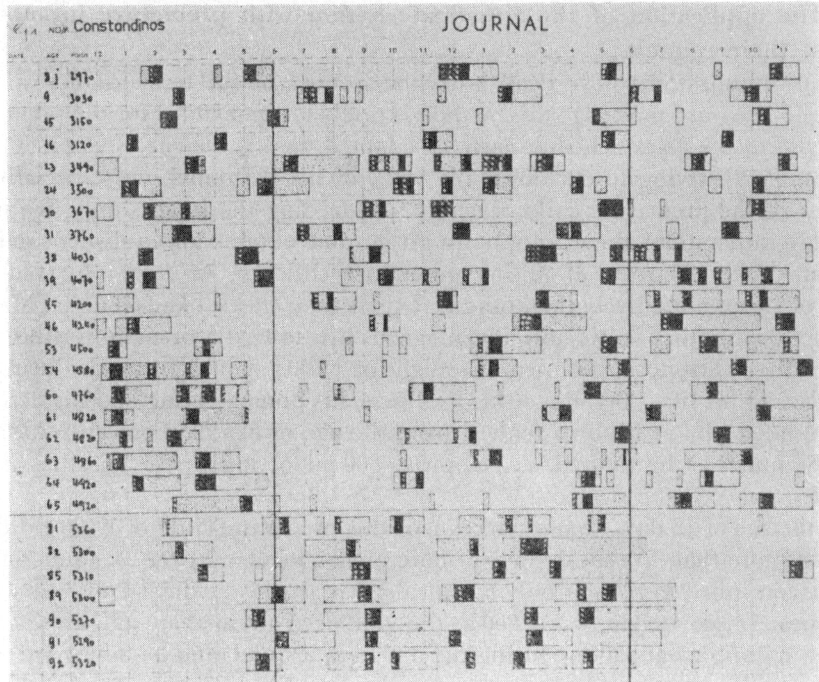

Fig. 1

age of 89 days, he requested no more than four meals per day and slept perfectly well from 1930 to 0500. A test of the excretion of urine and several urinary electrolytes based on different determinations showed a less considerable excretion from 0600 to 1800 than from 1800 to 0600, when the child was in constant light and had a wake-sleep rhythm which was inverted.

However, only a few days after having been exposed to the normal order of light and dark, the diurnal excretion of several elements exceeds the nocturnal excretion (Table 1).

In conclusion, this child as well as another infant also exposed to continuous light, developed a circadian rhythm, but started its activities at about 1800 and finished them at 0600, sleeping especially from 0600 to 1800. When exposed to day light and dark at night, however, the child quickly adapted itself to sleep during the night.

Table 1 Constandinos: Excretion of urine and urinary electrolytes.

Test periods	Test hours	Urine (mg)	Na (mg)	K (mg)	Cl (mg)	PO^4 (mg)
Application of Continuous light						
12.9.–14.9.	0600–1800	280	4.5	3.2	5.5	65
	1800–0600	410	6.1	7.3	7.3	95
Order of day and night						
18.9.–21.9.	0600–1800	260	5.4	7.8	12.8	137
	1800–0600	160	4.5	6.2	10.4	127

2. The application of the circadian rhythm with premature infants fed at their request

Many publications have dealt with the feeding of full-term infants at their request. To our knowledge this method is rarely recommended for the premature child [8] to whom a schedule strictly adapted to his weight is applied [2–4]. GISLAIN [5, 6], trying to cut down the work of his personnel and especially the fatigue of the premature child, suggests tube-feeding the infants only five meals per day rather than the 12 normally given, and obtains an equivalent gain of weight. After 20 years of raising premature children, we have observed that some of them, more lively than others, request their meals like full-term children. The following story shows how advantageous it is to feed a premature child at its own request: Stropoco, born with a weight of 1,800 g, got eight meals during the first days of his life. Because of his liveliness, his hungry crying and his desire to suck, we fed him at his own request from the age of five days onwards and gave him as much as he desired, i.e., approx. 300 ml of milk per day—that is 200 calories.

At the age of 12 days, to our amazement this premature child of 1,700 g did not request more than five meals spread more or less regularly over 24 hours. Later on, he continued to request only five meals; but already at the age of four weeks, Stropoco, whose weight was 2,060 g, did not need more than four meals. He started off his day at 2200. During the day he was quiet, and he only cried when requesting his meals. When we gave him back to his mother at the age of seven weeks, his weight was 2,760 g. Breast-fed, he developed perfectly well later on (Fig. 2).

Summary

In the case of a premature child, feeding at the child's request reduced the number of meals considerably: from eight to five and later on four within four weeks after birth. This strong premature child soon acquired a regular rhythm and developed perfectly well with a reduced number of meals.

This example shows the practical advantage of feeding some premature children at their request. Fed according to its needs, the premature child drinks more rapidly and often less frequently than expected. This method is perfectly adapted to the infant's circadian rhythm; it also has the advantage of considerably cutting down the time a nurse needs to feed premature children.

3. Appraising of circadian rhythm of ill infants

When Joëlle came to the "Clinique des Nourrissons" in Geneva, at the age of 4 months, she was in poor condition with congenital heart malformation and multiple infections.

Born with a weight of 3,500 g, this 4-month-old dystrophic child had a weight of only 4,370 g when measuring 60 cm; suffering from a tetralogy of Fallot, she became cyanotic when making only the smallest effort. She also had a generalized piodermia, complicated by digestive and cutaneous mycosis. Having been treated generously by antibiotics, Joëlle had become allergic to some of them. She also suffered from an obstinate loss of appetite, and generally refused the meals offered to her; she lost 300 g during the month prior to admission. Although she was treated by adequate doses of digitalis, Joëlle became tired quickly when she drank her bottle, and she agitated dangerously when one tried to tube-feed her; for these

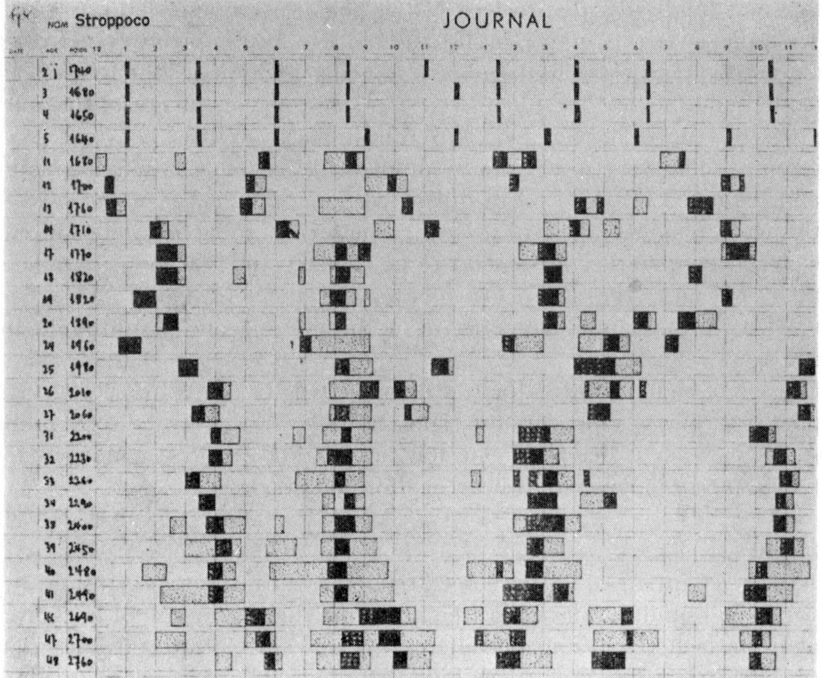

Fig. 2

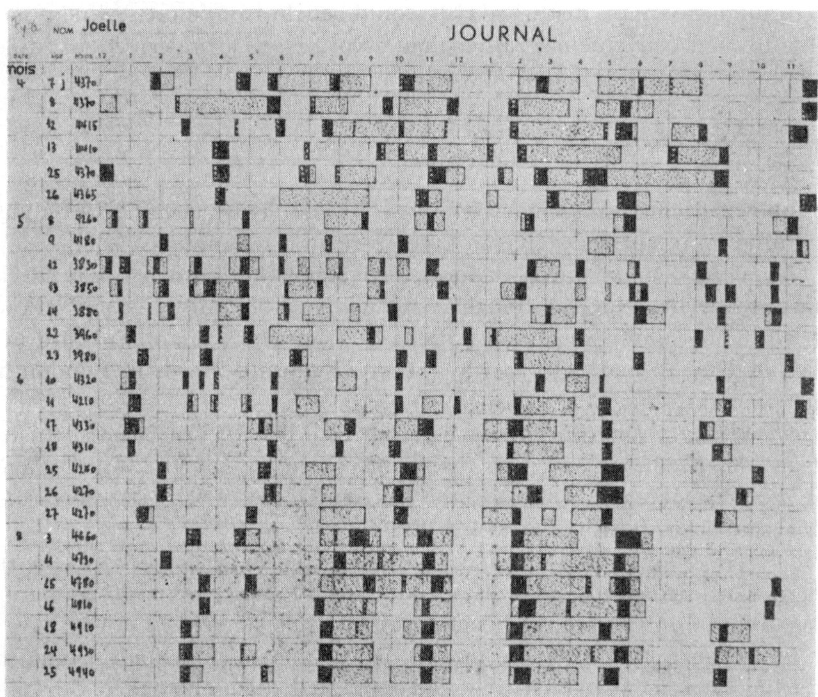

Fig. 3

reasons it was finally decided to feed her at her own request. To our great surprise we found that this infant, accustomed to five meals spread evenly over the day, requested up to eight feeding-bottles during 24 hours. But feeding at her own request and medical treatment were not sufficient: at the age of five months, Joëlle weighed only 3,850 g, her skin condition as well as her mycosis were troublesome. The very weak infant could drink only a very small amount of milk and requested up to 14 meals during 24 hours (Fig. 3).

At the age of 6 months, her circadian rhythm seemed to be inverted: indeed, her highest rectal temperature was found to be at 0300, and her lowest temperature at 1700. Not until the age of 18 months did the circadian rhythm become normal—with a maximum temperature in the evening and a minimum in the morning.

At the age of eight months, Joëlle still requested six meals from 0300 to 2100, her appetite was better, she put on weight regularly; slowly, she adopted a normal rhythm for a child of her age. She left the hospital in good condition for an infant suffering from congenital heart malformation.

Summary

This very sick infant could have been treated by intravenous perfusion; by this treatment, she would not have lost weight during the first month of her stay. But the condition of her skin and her allergy towards antibiotics were such that we were afraid to treat her by intravenous perfusion. Thanks to the care and the continued attention of the personnel in charge of her, and thanks to the adaptation of her circadian rhythm to her needs, Joëlle outgrew a condition which made us really wonder about its final outcome. We suggest that some very sick infants whose circadian rhythm is disturbed [10] would benefit from more frequent meals, according to their convenience, rather than according to a prescribed schedule in a hospital.

CONCLUSION

The following conclusions may be drawn after studying circadian rhythms in healthy full-term infants as well as in premature and sick infants:

1. *The healthy infant* quickly adopts a circadian rhythm of 24 hours. However, during the first days of life this rhythm is rather unstable and allowing feeding on request eliminates the interminable cries which are often heard in the nurseries attached to maternity wards. Several maternity wards in London have put this experience into practice to advantage.

2. *Some strong premature infants* certainly will benefit from meals which are lighter and less frequent, given to them only when requested. This system of feeding would make the work for the personnel easier with fewer meals to distribute and the infants easier to handle.

3. *Some vomiting infants* who have not been able to get adapted to the strict rhythm of scheduled meals, as well as some *mentally defective infants* or infants suffering from *anorexia,* would benefit from feeding at request. We have observed, for instance, that anorexic infants, fed at their own request, drank more milk during three meals which they had requested themselves than during five meals which previously had been scheduled for them.

REFERENCES

1. ALDRICH C. A. and ALDRICH M. M. (1946): Feeding Our Old Fashioned Children. Macmillan Co.
2. CROSS M. (1957): The Premature Baby. J. A. Churchill Ltd., London.
3. DUNHAM E. C. (1948): The Premature Infants Children. Bureau Public., No. 325.
4. FANCONI G. and WALLGREN A. (1967): Lehrbuch der Pediatrie. Schwabe and Co., Basel.
5. GRISLAIN J. R., LEMOINE P., De FERRON C., DELAROCHE V., MAINARD R. and MORIN F. (1961): Arch. Franc. Ped., *18*: 721.
6. GRISLAIN J. R., MAINARD R., De BERRANGER P., De FERRON C. and BRELET G.: XXIIe Congrès de l'Association des pédiatres de langue française, Sept., 1969. Exp. Scient. Franç. Paris.
7. HALBERG F. (1955): Acta med scand. Suppl. 307, 117.
8. HORTON F. H., LUBCHENKO L. and GORDON H. (1952): Yale J. Biol. Med., *24*: 263.
9. JOUVET M. (1967): Phylogénèse et onctogénèse du sommeil paradoxal, son organisation ultradienne. Cycles biologiques et psychiatrie, p. 185, Georg Cie.
10. MENZEL W. (1962): Menschliche Tag-Nacht-Rythmik und Schichtarbeit. Schwabe.
11. MILLS J. N. (1966): J. of Physiol., *189*: 30.

BIOLOGIC RHYTHM OF PLASMA HUMAN GROWTH HORMONE IN NEWBORNS OF LOW BIRTH WEIGHT

THOMAS R. C. SISSON[1], ALLEN W. ROOT[2], LIDA KECHAVARZ-OLAI[1] and ENID SHAW[1]

[1]*Department of Pediatrics, Neonatal Research Laboratory Temple University School of Medicine*
[2]*Department of Pediatrics, Albert Einstein Medical Centre Philadelphia, Pennsylvania, U.S.A.*

The newly delivered infant is not born with a fully synchronized set of biologic rhythms. We know from the work of HELLBRUGGE [1], MARTIN DU PAN [2], PARMELEE [3] and others that gross physical activities, sleep patterns, and even certain endocrine (17-OHCS) levels in plasma do not develop circadian rhythmicity for some weeks after birth.

One might suppose that the immature infant, whose gestational development has been interrupted, would be less capable of demonstrating such a periodicity. To test this hypothesis studies were begun in the nurseries of Temple University Hospital to measure levels of plasma human growth hormone (HGH) in the first days of life and to determine if rhythmic variations exist in this important hormone.

METHODS

Plasma HGH was determined for 48 hours in 37 newly born infants of low birth weight (1500–2500 g) and of less than 38 weeks gestational age. Blood samples were drawn at eight hour intervals commencing at 48–50 hours of age. This time was chosen since it is known that at least during the first 24 hours of life alterations in plasma volume occur by shifts from intra- to extravascular spaces [4]. Such shifts could influence the measured concentrations of plasma HGH. During the first 48 hours of life the immature infant must undergo many other adaptations that might profoundly affect HGH levels.

Infants were placed in four groups, depending on time of birth, so that onset of sampling was rotated to include each two-hour period around the clock, as shown in Table 1. It was not practicable to adjust sampling to hourly periods, though this would have been desirable. Eight-hour intervals of blood sampling were selected to avoid undue trauma to the infants, who are also easily fatigued by handling. Although atraumatic sampling can be accomplished by placement of an in-dwelling umbilical vein catheter, there are hazards to such catheterization, and there was no therapeutic justification for this procedure.

No infant with systemic disease such as sepsis, respiratory distress syndrome, or with hemolytic disease of the newborn was included in the study.

Subjects were kept in a windowless nursery with ordinary lighting (white fluorescent lamps, 100 footcandles luminance) between 0800 and 2200 hours, and

Table 1 Time of blood sampling.

Group No.	No. of infants	Hours of sampling		
1	10	0800	1600	2400
2	9	1000	1800	0200
3	9	1200	2000	0400
4	9	1400	2200	0600

in the dark, except for very dim light (three footcandles) during feedings, between 2200 and 0800 hours. This gave a cycle of 14 l : 10 d.

The infants were fed a standard prepared formula (Similac ®) every three hours, though occasionally feeding was briefly delayed to permit blood sampling no sooner than two hours after feeding. Environmental temperature and humidity were constant, and body temperatures of the infants did not deviate more than 0.3° from normal (99°F). The pattern of feeding did not contribute to the entrainment of a rhythm.

Blood samples were drawn from heel-pricks into heparinized capillary tubes, immediately centrifuged and the plasma separated and frozen until assayed. Plasma human growth hormone was then determined by the method of Root, et al. [5].

RESULTS

Fig. 1 illustrates the mean plasma HGH levels at 2 hour intervals over the 48 hours of study. As may be seen, there is a patterned rise and fall; peak concentrations appearing at 1000, 1600, and 2400–0200 hours. The shortest interval of decrement was two hours between 1000 and 1200 hours; others occupied an interval of four hours. Increments were four hours in duration with one exception —the level between 2400 and 0200 hours was virtually unchanged in both the first and second 24 hour periods. This occurred, in both instances, in the dark cycle.

Fig. 2 illustrates the composite levels of the first and second 24 hour periods of study. Logically this repeats the picture of either or both individual periods and

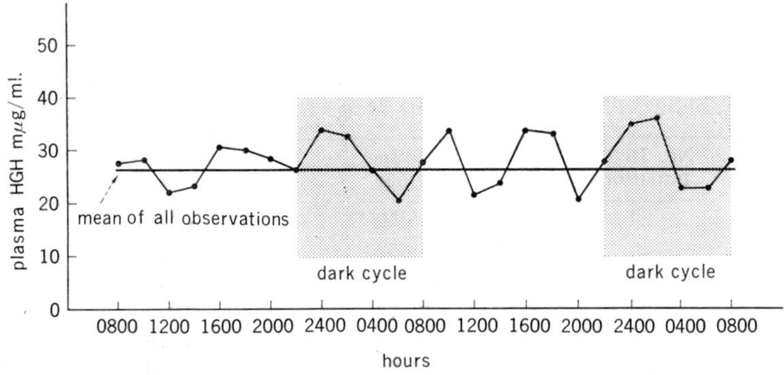

Fig. 1 Mean plasma growth hormone levels of 37 premature in fants from 48 to 96 hours of age. The lighting cycle was 14 1:10 d. The mean of all observations, as a reference point, was 26.2 mμg/ml. Ultradian frequency in the first period was reproduced with greater amplitude in the second period.

indicates the consistency of the rhythms observed. Were there significant differences between the two days of consecutive observation, this graph would reveal the discrepancies.

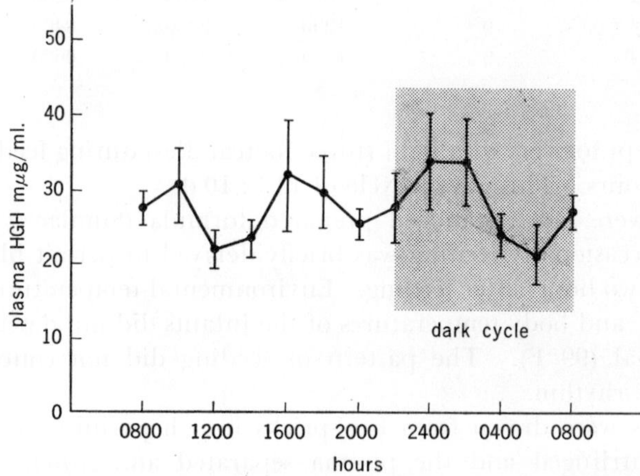

Fig. 2 A composite of the mean plasma HGH in two consecutive 24 hour periods of observation of 29 premature infants, with standard deviation of each mean. The lighting cycle was 14 l: 10 d. Vertical lines represent one S.D.

It is recognized that many enzymatic and metabolic functions in the immature infants achieve a maturity of action and response in the days subsequent to birth. Consideration of the possibility that plasma HGH values found in these subjects might have been influenced by simple maturation and adjustment to extra-uterine life, rather than expressing independent cyclic phenomena, led us to explore this factor. Hence, the data were arranged in relation to hour of age at the time of sampling not in relation to time of day. These mean plasma HGH levels are presented in Table 2. No significant differences were encountered between any of the observed means, which were nearly linear as time after birth progressed. Thus, it is possible to state with some confidence that the rhythm of plasma HGH observed in this study did not depend upon age in hours after birth.

Table 2 Plasma HGH at hour of age in 29 premature infants.

No. Determinations	Hours of age	Mean HGH mμg/ml
29	48—50	25.9
26	56—58	24.6
25	64—66	23.8
26	72—74	23.5
14	80—82	24.9
24	88—90	23.6

DISCUSSION

The plasma HGH concentrations reached peak levels in an uneven rhythm of six then eight hours. This was followed by a two hour plateau or lag between 2400 and 0200 hours, followed then by another eight hour period. Had blood samples been taken at hourly intervals a true peak at 0100 hours might have been seen, but this is conjectural.

Without further investigation under conditions of uncycled lighting and reversed dark-light cycling one cannot assess the influence of a cycled-light environment with strict confidence. However, it is clear from the routine of investigation employed here that the rhythm was present when the infants were first studied—whether in the dark or light cycle. This implies that the rhythm was endogenously derived.

It remains to be seen if endogenous entrainment of plasma HGH rhythm is actually begun *in utero,* and, should this be so, if the entrainment is mediated by maternal rhythms acting upon the fetus. It is doubtful that human growth hormone periodicity could contribute directly, for it has been shown that maternal growth hormone is not transferred to the fetus across the placenta even early in gestation, and that the only source of HGH in the fetal plasma is the fetal pituitary [6].

Sleep-wake cycles were not recorded in the subjects of this investigation. FINKELSTEIN [7] has reported that plasma HGH levels are lowest in periods of quiet wakefulness, similar to basal adult levels, though no relation to sleep stage could be found. The technique of time-rotation in blood sampling used in this study would abolish such sleep—wake relationships.

It is apparent that the rhythm exhibited in the first 24 hour period was repeated in the second period. A rise, lag, fall, and subsequent rise of HGH occurred only in the dark cycles. The amplitude of the rise and fall appears to have increased by the second 24 hour period of observation (Fig. 1). The reinforcement of these rhythmic alterations cannot be explained without further

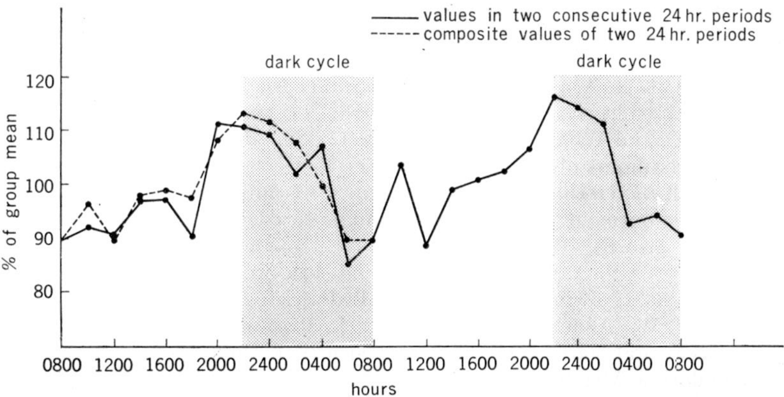

Fig. 3 Human growth hormone values in plasma during two consecutive 24 hour periods of study in 29 newborn premature infants expressed as percentages of the group means. A circadian rhythm is shown.

study. The frequency of the rhythmic variations in HGH are ultradian in character but seemed to be parts of an essentially circadian rhythm, and to demonstrate this possibility more clearly the data were expressed as percentages of the respective daily group means. The results of this treatment of the data are shown graphically in Fig. 3. This representation of the plasma HGH values shows a circadian rhythm characteristic for both 24 hour periods of observation: lower values in early hours of the light cycle, rising in the later hours, reaching peak values in the dark cycle, and declining to lower values in the last third of the dark cycle.

It has not been demonstrated that maintenance of the subjects in the dark had itself initiated the rhythm, but rather that the rhythm mimicked, to a degree, the events observed in adults [8]. We believe that the sequence of the rhythm in these infants, although having ultradian frequencies, is actually circadian.

SUMMARY

Plasma human growth hormone levels were measured under a 14 l : 10 d lighting cycle in 37 immature infants of low birth weight between 48 and 96 hours of age. Blood sampling at eight hour intervals was rotated to include each two hour period around the clock for 48 hours.

A rhythmic rise and fall of mean HGH was recorded. Peak values were found at 1000, 1600, and 2400–0200 hours. This pattern was consistent in both the first and second consecutive 24 hour periods of observation, though the amplitude increased in the second period.

It was concluded that a circadian rhythm of plasma HGH, with ultradian frequencies, is present in the premature infant by 48 hours of age, and that this is independent of feeding or cycled-lighting schedules. These data suggest that the rhythm of plasma HGH is endogenously established.

REFERENCES

1. Hellbrügge T., Lange J. E., Rutenfranz J. and Stehr K. (1964): Circadian periodicity of physiologic functions in different stages of infancy and childhood. Ann. N. Y. Acad. Sci., *117*: 361.
2. Martin du Pan R. (1967): L'apparition du rhythme circadien des 17-hydroxysteroides chez le nourrisson. Sa modification sous l'effet de la consommation de cortico-steroides. Praxis, *56*: 138.
3. Stern E., Parmelee A. H., Akiyama Y., Schultz M. A. and Wenner W. H. (1970): Sleep cycle characteristics in infants. Pediatrics, *43*: 65.
4. Sisson T. R. C. and Whalen L. E. (1960): The blood volume of infants: III. Alterations in the first hours of life. J. Pediat., *56*: 43.
5. Root A. W., Rosenfield R. L., Bongiovanni A. M. and Eberlein W. R. (1967): The plasma growth hormone response to insulin induced hypoglycemia in children with retardation of growth. Pediatrics, *39*: 844.
6. King K. C., Adam P. A. J., Schwartz R. and Teramo K. (1971): Human placental transfer of human growth hormone-I 131. Pediatrics, *48*: 534.
7. Finkelstein J. W., Anders T. F., Sachar E. J., Roffwarg H. P. and Hellman L. D. (1971): Behavioral state, sleep stage and growth hormone levels in human infants. J. Clin. Endocr. & Metab., *32*: 368.
8. Takahashi Y., Kipnis D. and Daughaday W. H. (1968): Growth hormone secretion during sleep. J. Clin. Invest., *47*: 2079.

CIRCADIAN VARIATIONS IN RESIDENTS OF A "SENIOR CITIZENS'" HOME*

LAWRENCE E. SCHEVING[1], CHESTER ROIG, III[3], FRANZ HALBERG[4], JOHN E. PAULY[2] and EDWARD A. HAND[5]

[1,2]*Departments of Anatomy, University of Arkansas Medical Center Little Rock, Arkansas, U.S.A.*
[3]*Louisiana State University School of Medicine*
[4]*The Chronobiology Laboratories, University of Minnesota Minneapolis, Minneso'a, U.S.A.*
[5]*The Coushatta Senior Citizens Home Coushatta, Louisiana, U.S.A.*

INTRODUCTION

A majority of circadian-time-structure studies done on young or middle-aged individuals have documented that most physiological variables undergo partly predictable variation along a 24-hr time scale. Less attention has been given thus far to the analysis of circadian variation in older persons, and it is of interest to report herein results from rhythmometry conducted on five male and four female volunteers residing at a senior citizens home located in a small town within a farm community of Louisiana, USA.

MATERIALS AND METHODS

In this report, the subjects, aged 69, 75, 81, 82, 83, 84, 86 and 86, are frequently referred to as "senior citizens". They reside in what is commonly referred to as a "nursing" home in Coushatta, Louisiana. All had chronic cardiovascular disease and in addition many had other diseases or complications from disease; all were receiving several different medications. [The number of diseases and the different medications involved are too numerous to list, but are available from one of us (LES).] The study was conducted for ten days (total span=231 hours) during June 1970. All sampling or testing took place at 3-hr intervals during the individual's waking span. Sleep was scheduled from 2130 to 0600. All subjects ate only the three scheduled meals, the bland diet probably being typical of diets in most homes for senior citizens. Some of the participants did lie down and rest between sampling times but for the most part they were moderately active and would move around sporadically during their waking hours. All measurements, except those of a biochemical nature, were conducted or supervised by a medical student (CR) under the direction of one of us (EH). The following variables were measured on all subjects 1) oral temperature (using a special electronic thermometer); 2) pulse rate; 3) blood presure; 4) ability to estimate time, and 5) right and left grip strength (determined with a dynamometer). Peak expiratory flow and eye-hand coordination were measured on three males. All tests or measure-

* Supported by USPHS (5-K6-GM-13,981) and (12389-01, Phy); NASA (9-12338); Central Chapter, Connecticut Heart Association.

ments mentioned above have been described elsewhere [1]. In addition, the urine of two males (75 and 85 years of age) was collected over the 10-day span at the same 3-hr intervals. The following constituents of urine were measured by conventional methods: 1) volume; 2) pH; 3) sodium; 4) potassium; 5) chloride; 6) 17-ketogenic steroids; 7) norepinephrine; and 8) epinephrine. The data were analyzed by an inferential statistical method of fitting a 24-hr cosine to each time series, creating vectors from resulting parameters and summarizing the group as a whole by the method commonly referred to as the cosinor: the details of this method have been described elsewhere [2].

RESULTS

The computer-determined parameters mesor, amplitude and acrophase are summarized in Table 1. The acrophase is expressed in degrees, where $360° \equiv$ 24 hr; $15°=1$ hr. The phase reference is local midnight (0000). The mesor and amplitude are expressed in the same units as the variable analyzed. A P value of 0.05 or less indicates that the data of a time series follow a pattern approximated by a 24-hr cosine curve rather than fluctuating randomly.

Oral temperature. The fit of a 24-hr cosine curve to the data for 8 of the 9 subjects showed a statistically significant rhythm at the 1% level (an 81-year-old woman being an exception). When the amplitudes and acrophases of all individuals were summarized by the cosinor technique, the group acrophase occurred at $-253°$ with a .95 confidence interval ($P<0.004$) between $-214°$ and $-283°$, the amplitude being $0.60\pm.35$ (Table 1).

Blood pressure (systolic). Seven of the 9 subjects demonstrated a significant rhythm in systolic blood pressure, with the exception of a 69-year-old woman and a 78-year-old man. When the data were summarized for the group, the acrophase occurred at $-337°$ ($-225°$ to $-12°$) and the amplitude was 6.47 ± 3.06 mmHg (Table 1).

Blood pressure (diastolic). Six of 9 subjects demonstrated a circadian component in diastolic blood pressure. The group-summarized acrophase occurs at $-350°$ ($-242°$ to $-9°$). The amplitude was 4.06 ± 2.69 mmHg (Table 1).

Pulse. When a 24-hr cosine curve was fitted to the data of each individual, 7 of the 9 subjects had a circadian component in their pulse rate, an 86-year-old woman and a 75-year-old man being the exceptions. When the data of the group were summarized by the cosinor technique, no group confidence region could be ascertained, because individual pulse acrophases differed greatly among individuals.

Time estimation. In this measurement the subject estimated the time he or she thought it took for the passage of ten seconds and then one minute. None of the nine individuals demonstrated a circadian rhythm (Table 1) for ten-second time estimation. The passage of a minute data of one individual did show a 24-hr component with the acrophase occurring at $-254°$ ($-202°$ to $-306°$). The group summaries for both time estimations failed to demonstrate a rhythm.

Eye-hand coordination. This was a test of the time it took an individual to place 30 small beads into a special container as well as the number of errors made. Only three males participated in this test, none of them demonstrating a circadian component (Table 1).

Peak expiratory flow. The same three individuals who participated in the eye-hand coordination test were involved in this measurement. Only the data of an 82-year-old man showed a circadian component, with the acrophase occurring at −195° (−173° to −216°).

Grip strength. This was measured with a dynamometer on both right and left hands of all individuals. The cosinor did not detect a rhythm, although the data of one male showed a circadian component (left hand) with acrophase at −87° (−56° to −138°) and the data for this same man and another demonstrated circadian rhythmicity in right grip strength, with acrophases occurring at −98° (−43° to −154°) and −149° (−117° to −181°) respectively (Table 1).

DISCUSSION

It repeatedly has been demonstrated that in younger individuals all parameters measured in these senior citizens fluctuate with a circadian frequency [3, 4]. In this study, rhythms found in temperature, pulse and blood pressure are especially noteworthy because of the age, disease state and medication of the subjects. One major difference between these rhythms in younger individuals [3] and in senior citizens is that the acrophases of the younger group are more closely synchronized, as revealed by smaller confidence intervals when the group data are analyzed. (Hereinafter, when we mention data of a younger group we refer to a study done by us on 13 healthy young (average age 24) soldiers [3].) To illustrate our point, the confidence interval for systolic blood pressure acrophase in a group of young people spanned 90°; for this older group it was slightly larger, i.e., 118°. The confidence arc of the group acrophase for diastolic blood pressure in the young soldiers was 83°, for the older group it extended over 93°. Comparison of the over-all mean pressure for the younger group (119±2 mmHg) with the same mean for the elderly group (146.3±6 mmHg) revealed a highly significant ($P<0.005$) difference. There was no difference in mean diastolic blood pressure of younger (untreated) and older ("treated") groups; both had an over-all mean of 75 mmHg. The confidence arc of the pulse acrophase in the younger subjects was 42°; for the older subjects the group data did not fit a 24-hr cosine curve but most individual series did show a significant fit, with acrophases ranging from as early as −319° to as late as −51°. The over-all mean pulse rate of our senior citizens (73 beats/min) was significantly higher ($P<0.005$) than for the younger group (63 beats/min).

When amplitudes are compared, those for the older group in both oral temperature ($P<0.02$) and pulse ($P<0.002$) are significantly lower than in the younger group. The amplitudes of the blood pressure rhythms did not differ significantly between the two groups.

Only one of three individuals studied revealed a circadian rhythm in peak expiratory flow. From the prior data on younger individuals we would have expected all to demonstrate a circadian rhythm, especially over a 10-day sampling span. It can be concluded that, allowing for exceptions, the elderly have less of a tendency to demonstrate circadian variation in grip strength, eye-hand coordination, ability to estimate time, etc.

There are many problems associated with the collection of urine from this age category. Some of our subjects were incontinent and only two could hold urine for a span of three hours. Because of this problem we had to abandon our original plan to collect urine on all 9 subjects. Again, as for vital signs, most of

Table 1

	N	P*	Mesor (M) ± SE	Amplitude (A) ± SE	Acrophase, ϕ, and (95% Confidence arc) Degrees
Vital signs*					
Oral temperature (°F)	9	.004	97.31 ± 0.3	.60 ± .35	−253 (−214 −284)
Systolic blood pressure (mm Hg)	9	.001	145.84 ± 6.4	6.47 ± 3.06	−337 (−255 − 13)
Diastolic blood pressure (mm Hg)	9	.007	75.28 ± 3.6	4.06 ± 2.69	−350 (−277 − 10)
Pulse (beats/min)	9	.365	73.52 ± 3.4	2.39 ± 4.82	−210
Peak expiratory flow	3	.245	144.8 ± 19.2	5.35 ± 27.09	−215
Performance**					
1–minute time estimation (seconds)	9	.351	50.78 ± 2.81	.61 ± 1.21	−235
10–second time estimation (seconds)	9	.141	8.03 ± .57	.13 ± .18	−352
Dynamometer, right hand	9	.080	29.55 ± 7.1	1.08 ± 1.22	−127
Dynamometer, left hand	9	.481	24.55 ± 6.1	.36 ± .87	−142
Eye-hand coordination***					
Number of seconds elapsed	3	.706	151.9 ± 26.7	5.65 ± 112.9	−50
Number of errors	3	.515	.54 ± .17	.23 ± 2.76	−24
Urinary variables (per hour)***					
Specific gravity	1	.009	5.91 ± .34	2.36 ± .52	−215 (−195 −235)
	1	.002	13.83 ± .57	2.63 ± .92	−200 (−173 −227)
pH	1	.009	5.87 ± .03	.35 ± .04	−70 (− 56 − 84)
	1	.260	5.18 ± .01	.02 ± .01	−125
Volume	1	.009	77.7 ± 7.84	34.72 ± 10.33	−7 (−332 − 42)
	1	.729	73.5 ± 13.9	15.36 ± 19.38	−62
Potassium (mEq/hr)	1	.273	1.79 ± .25	.59 ± .36	−299
	1	.018	2.27 ± .12	.50 ± .17	−214 (−179 −249)
Sodium (mEq/hr)	1	.119	3.29 ± .40	1.12 ± .53	−350
	1	.184	4.73 ± .28	.72 ± .38	−309
Chloride (mEq/hr)	1	.347	3.27 ± .44	.90 ± .61	−326
	1	.952	4.89 ± .30	.12 ± .41	−306
Norepinephrine (μg/hr)	1	.012	.11 ± .01	.04 ± .01	−38 (− 1 − 75)
	1	.059	.25 ± .02	.10 ± .04	−213
Epinephrine (μg/hr)	1	.688	.14 ± .01	.01 ± .02	−353
	1	.068	.21 ± .03	.10 ± .04	−207
17–Ketogenic steroids (mg/hr)	1	.315	.30 ± .03	.07 ± .05	−325
	1	.131	.34 ± .01	.05 ± .02	−206

Documented span in hours is 231 for all variables.

* Rhythm detection. ** Cosinor summary of group data *** Cosinor summary of data of two males

the urine constituents measured are known to be markedly rhythmic for younger individuals; this was not the case for senior citizens. For an 82-year-old man, only potassium, specific gravity and norepinephrine showed a circadian component (Table 1). For the 75-year-old man only urine volume, pH, specific gravity and norepinephrine showed a good fit to a 24-hr cosine curve (Table 1).

Since blood pressure and norepinephrine seem to undergo prominent rhythms in senior citizens, one might consider some causal relationship. Perhaps the medication is serving as the synchronizing agent. Norepinephrine is more predictably rhythmic than epinephrine, based on both published [4] and unpublished work. No explanation can be offered.

It is noteworthy that in spite of age, disease and medication, the 9 subjects selected for this study did represent some of the most alert and active individuals residing at this home and probably would be representative of the more active and alert individuals of most homes of this type. It is the opinion of the senior author that the morale of this group probably was much higher than might be expected in many comparable homes. This was in part due to the fact that the subjects were in many cases lifetime acquaintances who had grown up in the same community; the director of the home had known most of them all his life and they held him in high esteem. We mention this because in an earlier attempt to do similar studies in another home we were unsuccessful because no one volunteered; in fact there was a complete lack of interest, even hostility. The majority of senior citizens in the Louisiana home would like to have volunteered. Playing a role in this study gave each and every one of them special status and the entire population became engrossed in the studies being conducted; it was as though they themselves were involved and this gave them great satisfaction. The success of any study of this type depends in large part on the motivation of the participants.

REFERENCES

1. Halberg F. (1969): Chronobiology. In the Annual Review of Physiology, *31*: 675–725.
2. Halberg F., Tong Y. L. and Johnson E. A. (1967): Circadian system phase—an aspect of temporal morphology; procedures and illustrative examples. *In*: The Cellular Aspects of Biorhythms, (H. von Mayersbach, ed.), pp. 20–48, Springer-Verlag, Berlin.
3. Kanabrocki E. L., Scheving L. E. and Halberg F. (1973): Circadian variation in healthy young men. Space Life Science, *4*: 258–270.
4. Enna C., Scheving L. E., Halberg F., Jacobson R. R. and Mather A.: A study of circadian rhythms of various parameters in patients with leprosy, (in press).

CIRCADIAN TEMPERATURE RHYTHMS AND AGING IN RODENTS*

EDMOND J. YUNIS, GABRIEL FERNANDES, WALTER NELSON
and FRANZ HALBERG
Chronobiology Laboratories, Departments of Pathology and Laboratory Medicine
University of Minnesota
Minneapolis, Minnesota, U.S.A.

The purpose of this report is to present evidence that circadian rhythms are present during aging in several inbred strains of mice and in an inbred strain of rats, and that characteristics of these rhythms may change with age.

In an earlier study [1] it had been shown that both the rhythm-determined mean or mesor (M) and the amplitude (A) of the circadian temperature rhythm decrease with aging in the I strain of mice and that such changes are more pronounced in males, although the mesors were lower in females.

In previous experiments it also was shown that some inbred strains of mice as compared to others undergo relatively early changes associated with immunodeficiency gauged by early thymic involution and waning of cell-mediated immunity resulting in the early occurrence of autoimmune diseases [2, 3]. The present study compares the circadian rhythms of a few young and aging mice from long-lived and short-lived inbred mouse strains. The long-lived ones are autoimmune resistant, the short-lived ones are susceptible. It will be shown that all strains of mice studied during aging show a continuing circadian periodicity and a decrease in temperature mesor. In addition, a decreased circadian amplitude was observed at 12 months of age in Af and NZB mice but not in CBA or C3Hf mice.

MATERIALS AND METHODS

Male CBA/H, C3Hf, Af and NZB mice of various ages were used. These mouse strains were derived from the colonies of Drs. J. J. BITTNER and C. MARTINEZ. A detailed description of the strains has been reported [4] and designated as University of Minnesota colony sublines (Umc). A pertinent summary of these strains is given in Table 1. The mice were not kept singly housed but were raised in cages in groups of 2 to 5 in a room illuminated by artificial light only. A clock controlled switch turned the lights on at 0600 and off at 1800. Purina lab chow and tap water were available to the mice from the time of weaning throughout the experiment. The temperature of the room was kept at 75°±1°F and humidity at 50 to 55%.

Rectal temperatures were obtained from mice with a Yellow Springs temperature-measuring device [5]. In the first experiment temperatures were determined on each mouse every four hours during a 48-hour span. The resulting time series from each animal was first grossly inspected and thereafter analyzed microscopical-

* Supported by grants from the U.S. Public Health Service (5-K6-GM-13-981, AI HD 10153-01 1-RO1-CA-14445-01; 1-RO-1-AI 10153-03 and HL-06314-13) and NASA.

Table 1 Some characteristics of strains investigated.

Symbols in text	Strain symbols	No. of mice	Life span 50%, days, $\bar{X}\pm SE$	Comment
CBA/H	CBA/H/ Umc	15 15	females 760±8 males 740±5	Autoimmune resistant. Relatively little decrease of cell-mediated immunity during aging.
C3Hf	C3HfC57–BL/Umc	20 20	females 480±9 males 530±7	Autoimmune resistant. Develop amyloidosis during aging.
Af	AfCBA/ Umc	15 15	females 575±6 males 550±7	Develop renal lesions, antinuclear antibodies, amyloidosis and decline of cell-mediated immunity during aging.
NZB	NZB/Umc	25 25	females 370±11 males 415±10	Develop autoimmune hemolytic anemia, antinuclear antibodies and kidney lesions before one year of age.

ly [6]. For the latter purpose, it was fitted with a 24-hour cosine curve by a least-squares procedure [7]. The average values of circadian rhythm parameter estimates thus obtained were then calculated for each strain and age. In an attempt to reduce effects of disturbance, a second experiment was performed with temperature measurements on each mouse at intervals of 28 hours during a 6-day span to again yield data at the same six clock hours as in the first study. For this second study, mean temperature values were first computed for each strain and age at each clock hour. The resulting series of mean values representing a given strain and age was then fitted with a 24-hour cosine curve by least-squares to yield circadian rhythm parameter point-and-interval estimates.

In a study of possible age-related changes in body temperature rhythms of the rat, data were obtained by telemetry from intraperitoneally-implanted temperature transensors [8]. Nine female rats from the inbred Minnesota Sprague-Dawley (MSD) strain were used, the same individuals being studied at different ages. The animals were singly-housed throughout the studies in a regimen consisting of light from 1200 to 0000 alternating with darkness from 0000 to 1200, with food and water freely available. Temperature-calibrated transensors were surgically implanted when the rats were about three months old and were replaced with fresh units ten months later. Although transensor calibrations were not re-checked after each span of data collection, experience with similar units indicated no appreciable change with time in characteristics of the transensor response to temperature change. Data obtained at ten minute intervals throughout 3-week spans when the rats were about four months old and again when they were about 19 months old were analyzed by the cosinor method [7].

RESULTS

Fig. 1 shows, as an example, the chronograms from the studies on CBA/H mice at different ages. The rhythmic change in temperature stands out clearly.

Fig. 2 summarizes the two experiments; it shows the results obtained in fitting 24-hour cosine curves to the data from mice of different ages and strains. The mesor undergoes a lowering with age in animals sampled repeatedly under the same condition. The amplitude also exhibits a decline with age for all strains in the second study (Fig. 2, right half) but only for the CBA strain in the first

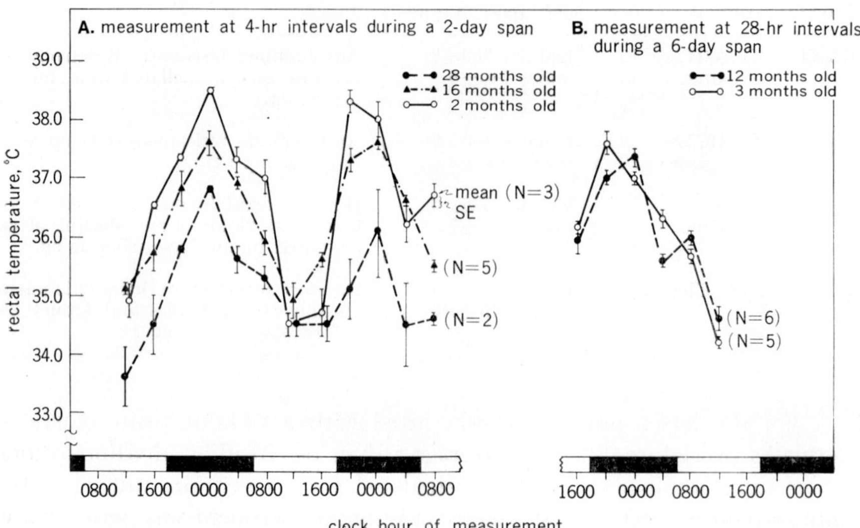

Fig. 1 Circadian rhythm in rectal temperature of inbred CBA/H male mice at different ages.

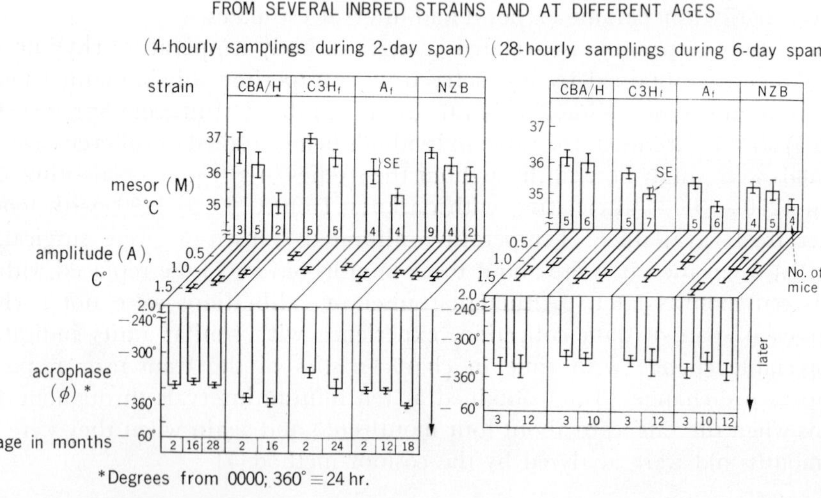

Fig. 2 Rhythmomeric summary of rectal temperature in male mice from several inbred strains and different ages.

study (Fig. 2, left half). This difference between studies conceivably could be related to the frequency of disturbance. There was no consistent change in acrophase with age, at least between 3 and 12 months.

Results of cosinor analysis for the group of 9 female MSD rats at the two ages studied are summarized in a cosinor plot, Fig. 3. In this polar plot a vector [a line directed outward from the pole (center)] represents the amplitude of the

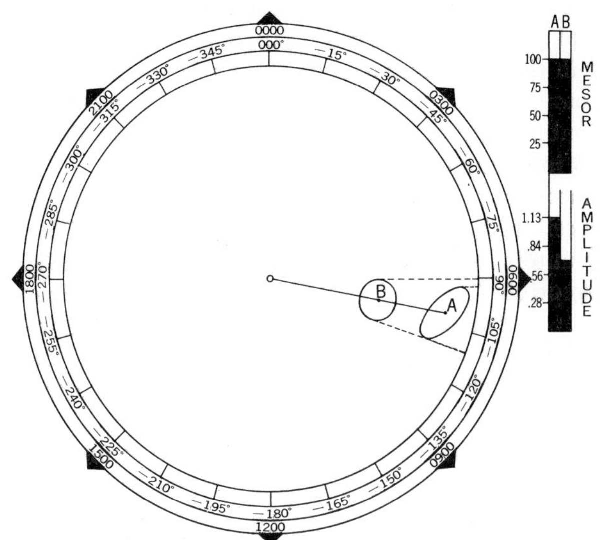

Fig. 3 Cosinor summary of the circadian rhythm in intraperitoneal temperatures of nine female rats when they were 4 months old (A) and again when they were 19 months old (B). Rats singly housed with lights from 1200 to 0000 alternating with darkness from 0000 to 1200. Bar plots at upper right present circadian mesor (above), and amplitude estimates for the two ages, with scale units in °F.

circadian rhythm. The angular position of the vector relative to the reference point midnight (with 360°≡24 hours) indicates the rhythm's acrophase. The ellipse at the head of the vector depicts its 95% confidence region. A significant reduction in amplitude of the circadian temperature rhythm is indicated between the ages of 4 months (vector A) and 19 months (vector B), while the acrophase and mesor apparently do not change. The fact that data at the two ages were obtained from the same individuals argues against the possibility that a reduction in amplitude with age is due to a difference between young animals and a select group of old survivors. By summarizing sets of rhythm parameter estimates from individuals, the cosinor method also assures that a reduction in the group circadian amplitude with age is not a consequence of circadian desynchronization among individuals [1].

DISCUSSION

Several parameters used to measure aging in a population of animals have been reviewed by Walford [9]. The most common way to compare the rate of aging of an experimental and control population is by recording their respective 50% survival points, provided that the slope of the curve on each side of the 50% point is the same [10]. The time at which the incidence of a particular disease reaches its maximum may in some case be taken as an estimate of the physiologic age of the population [11]. However, several diseases of aging can be altered by outside factors such as nutrition and x-irradiation. For instance, irradiation shifts the incidence and time of onset of amyloidosis [12] and nephrosclerosis [12] and dietary experiments show marked prolongation of life span in rats on reduced caloric intake due to delayed onset of tumors and other diseases [14]. Morphologic and chemical changes have also been used as a measurement of aging [9].

There is a lag in the beginning of cell division for tissue explants derived from old animals as compared to younger animals [15]. However, tissue culture scores of number of doubling times of a cell in culture cannot distinguish between a young and an old individual [9].

Another parameter of aging is immunological. There is a decline of immune capacity, mainly cell mediated immunity, during aging [9]. This decline may be accompanied by autoimmune manifestations that serve to divide the mice into autoimmune susceptible (NZB, Af, NZW, etc.) and autoimmune resistant (DBA/1, CBA/H, etc.). Autoimmune-susceptible strains show an earlier decline of immune activity because of an earlier involution of thymic function [2, 3]. In other words, immune deficiency is associated with aging although it may not necessarily explain primary aging or senescence.

To our knowledge there is not a single parameter that measures primary aging independently of the diseases of aging. This study was an attempt to survey several strains of mice and a strain of rats with respect to circadian periodicity during aging. Our results are preliminary but corroborate previous findings of Halberg et al. [1] that there is a decrease in circadian amplitude and mesor of body temperature during aging of mice. The first experiment presented here suggests that there is periodicity of temperature in young and old mice from autoimmune susceptible and resistant strains studied and that the temperature mesor may drop with increasing age. However, in the first study, a decrease of amplitude with age was observed only in the CBA mice. It is quite possible that mice were greatly disturbed by sampling every four hours for 24 hours or longer. The second experiment done by sampling every 28 hours suggests a decrease of circadian amplitude during aging in CBA/H, C3Hf, Af and NZB mice and that this decrease in circadian amplitude may be larger in autoimmune-susceptible strains (NZB, Af) than in autoimmune-resistant strains. More studies are required to establish whether or not an age-associated decrease in circadian amplitude is a parameter of primary aging or is due to a disease of aging. One may ask, for instance, whether or not a given percentage decrease in amplitude of the circadian temperature rhythm occurs at a comparable percentage of the total life span in long-lived and short-lived mice. The limited samples used thus far cannot answer such questions but they illustrate methods for objectively scrutinizing aging.

The extent to which a decrease in circadian amplitude for intraperitoneal temperature of the rat and rectal temperature of the mouse may be related to a change in mesor also awaits further study. Larger and more systematically and repeatedly sampled groups of rats or mice from short-lived and long-lived strains will serve to be studied with the foregoing considerations in mind.

REFERENCES

1. Halberg F., Bittner J. J., Gully R. J., Albrecht P. G. and Brackney E. L. (1955): Proc. Soc. Exp. Biol. Med., *88*: 169.
2. Yunis E. J., Stutman O. and Good R. A. (1971): Annals of the New York Academy of Sciences, *183*: 205.
3. Yunis E. J., Fernandes G. and Stutman O. (1971): Am. J. Clin. Path., *56*: 280.

4. Staats J. (1969): Inbred Strains of Mice (Editor). Jackson Laboratories, Bar Harbor, Maine. N-6-60, July.
Maine. N-6-60, July.
5. Halberg F., Zander H. A., Houglum M. W. and Mühlemann H. R. (1954): Am. J. Physiol., *177*: 361.
6. Halberg F. (1969): Annual Review of Physiology, *31*: 675.
7. Halberg F., Tong Y. L. and Johnson E. A. (1967): *In*: The Cellular Aspects of Biorhythms. Springer-Verlag, Berlin.
8. Halberg F., Nelson W., Runge W. J., Schmitt O. H., Pitts G. C., Tremor J. and Reynolds O. E.: Space Life Sciences *2*: 431, 1971.
9. Walford R. L. (1969): The Immunologic Theory of Aging. Munksgaard, Copenhagen.
10. Storer J. B. (1962): Rad. Res., *17*: 878.
11. Simms H. S., Berg B. N. and Davies D. F. (1959): In: Ciba Found. Colloq. on Aging, Vol. 5, p. 72, Little, Brown and Co., Boston.
12. Leaker S., Grahn D. and Sallese A. (1957): J. Nat. Cancer Inst., *19*: 1119.
13. Lamson B. G., Billings M. S. and Bennet L. R. (1959): Arch. Path., *67*: 471.
14. McCay C. M. (1952): *In*: Cowdry's Problems of Aging, (A. I. Lansing, ed.), p. 139, Williams & Wilkins, Baltimore.
15. Comfort A. (1964): "Aging: The Biology of Senescence", London: Routledge and Kegan Paul.

ONTOGENY OF NOCTURNAL FEEDING RHYTHM IN RATS AND EFFECT OF PROLONGED DIURNAL FEEDING EXPERIENCE*

MARY SCHILD**

University of Virginia
Charlottesville, Virginia, U.S.A.

The ontogeny of the nocturnal feeding rhythm in the rat has received attention intermittently during 50 years of study of rat behavioral rhythms. RICHTER's early observation [1] of its preweaning appearance at 16–17 days of age has, for example, long been part of the standard literature on feeding rhythm development. More recently, however, BOLLES and WOODS [2] saw no feeding cyclicity during their observation of the first three weeks of the preweaning period. Small wonder, perhaps, that ZUCKER [3] (p. 120) should recently state: "Little is known concerning developmental aspects of rhythms in eating. . . ."

ZUCKER himself recently reported the observation that nocturnal rhythms of food and water intake are "present in essentially adult form in rats 23 days old. . . ." [3] (p. 115) which have been recently weaned. He also concluded that the "early perinatal period is not critical for entrainment of eating and drinking rhythms. . . ." to the nocturnal portion of a light/dark cycle ([3], p. 115).

In my data, however, rats even at 32 days of age did not all show the "adult form" of the eating cycle which later appeared, and the cycles of some 32-day-old rats were even diurnal rather than nocturnal. My data also demonstrate that, whether or not the *early* perinatal period is considered critical for entrainment, what happens for an extended period afterward apparently does influence the ease and degree of entrainment. The remainder of this presentation will be a preliminary report of these data.

Three groups of Sprague-Dawley albino rats were reared for 102 days with either little or no opportunity to eat in the dark part of an artificially-maintained light/dark cycle (LD 14:10). Before weaning, all three groups lived in continuous light (LL) with mothers which had been placed there at least a week before casting their litters. After weaning, an *LL-ad-lib* group remained in LL on *ad-libitum* feeding; the other two groups were changed to the LD cycle, with no eating allowed in the dark for a *light-fed-only* group and only .10 per cent quinine-adulterated food geing available in the dark for a *dark-food-quinine* group. Both of the latter groups had unadulterated food available in the light, and the aversiveness of the nighttime diet of the dark-food-quinine animals resulted in their self-imposing a primarily diurnal feeding cycle.

An LD-*ad-lib* group was born into the LD cycle and lived there for the entire experiment with *ad-libitum* feeding. After 102 days of age, all three of the ex-

* This research was supported primarily by Grant No. MH104290 from the Public Health Service to Dr. F. W. Finger and was done in the Department of Psychology at the University of Virginia.

** Now at Columbus College, Columbus, Georgia.

perimental groups also lived in LD with *ad-libitum* feeding for the final four weeks of the experiment.

Liquid diet was introduced during the second postnatal week and constituted the sole food supply thereafter; a water supply was not considered necessary.* Weaning occurred at 31 days of age. Animals lived for one day in individual cages in the same preweaning illumination condition with food available while groups were being matched for weight. They lived thereafter in activity wheels.

White noise which rather effectively masked both intra- and extra-room sounds was always present, and all external light was completely excluded. Temperature was reasonably constant at 77°F in LD and 79°F in LL.

Forty animals were used, 10 per feeding condition. The experiment was divided into two replications, with five animals per group in each replication. Measurement of amount of intake began at wheel entry and was recorded as number of milliliters drunk from calibrated Richter tubes. Light and dark intakes were recorded separately in order to derive the proportion of total daily intake occurring in the light, which is presented for each animal in the accompanying Figs. 1 and 2.

Of the 10 LD-*ad-lib* animals, four had *diurnal* eating cycles (greater than 50 per cent in light) the first week, and three of these did not become nocturnal for the first three weeks or longer. About half of the LD-*ad-lib* animals clearly did *not* exhibit initially the "adult form" which developed gradually over several weeks. As adults, some of them ate as little as about 5 per cent in the light, while others consistently ate 30–40 per cent in the light.

By the final week, however, all 10 control animals and most of the 30 experimental animals did have nocturnal feeding cycles. However, nocturnal cycles were *not* manifest at that time by five light-fed-only animals, two dark-food-quinine animals, and one LL-*ad-lib* animal, despite their having had four weeks of *ad-libitum* feeding in LD after the 10-week experimental phase.

The patterns of feeding exhibited by experimental animals during the final four weeks of *ad-libitum* feeding lead to two conclusions: (1) The tendency (of whatever origin) to eat mainly in the dark is very strong, since none of the experimental animals maintained the exclusively or primarily daytime feeding pattern imposed for so long from birth. (2) The existence of a nocturnal feeding cycle and the proportion of nocturnality do not derive entirely and automatically from an "innate" nature. (In this experiment, neither the existence nor the proportion of nocturnality was entirely independent of previous feeding experience.)

The differences in proportions and differences in the temporal changes in those proportions exhiibted by some control animals (i.e., LD-*ad-lib* animals) during the experiment, as well as the differences among some experimental animals during the four *ad-libitum* weeks at the end, suggests that individuals may differ to a greater extent than usual descriptions of light-vs.-dark feeding cycles (see, for example, Le Magnen [4], pp. 12–13) would lead us to surmise. Such differences suggest a further significant implication also: individuals may differ greatly also in their capacity to be influenced by the same experimental manipulation. Attempts to delineate either the general developmental characteristics or general adult attributes of feeding in relation to light/dark cycles must therefore include

* The formula for the diet was that used by Epstein and Teitelbaum [5]. Previous investigators [6, 7] have reported that little or no water is drunk by rats when maintained on this liquid diet, and preliminary observations in the present study were consistent with those reports.

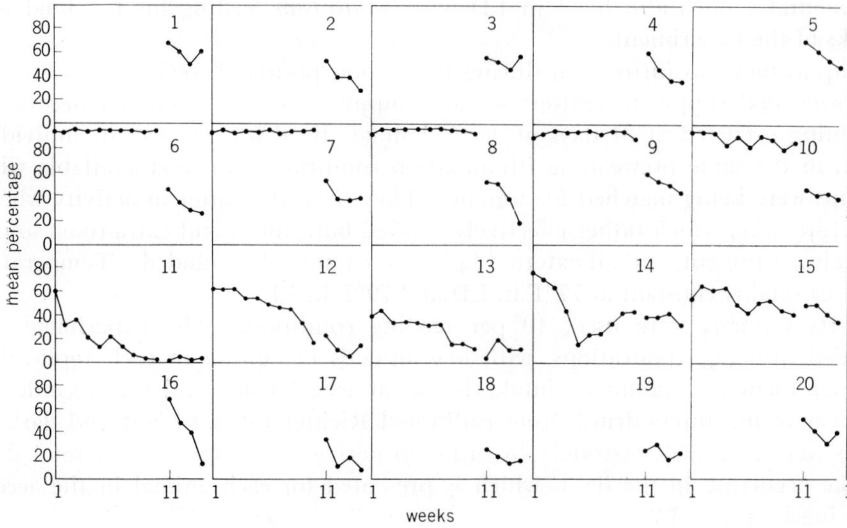

Fig. 1

Note: Nos. 1– 5 and 21–25 are light-fed-only;
Nos. 6–10 and 26–30 are dark-fed-quinine;
Nos. 11–15 and 31–35 are LD-ad-lib;
Nos. 16–20 and 36–40 are LL-ad-lib.

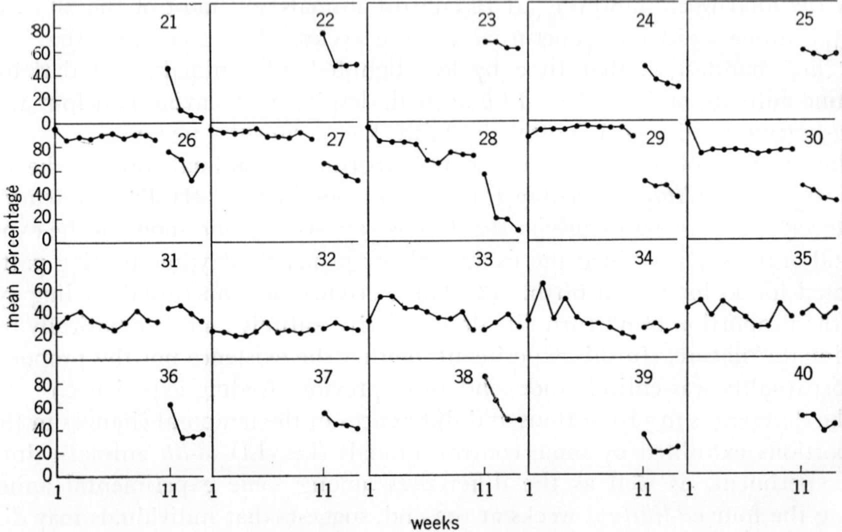

Fig. 2 (see Note of Fig. 1)

attention to individual differences in order to provide a more informative and more accurate "general" description.

Though Zucker [3] alludes to the question of latency needed for initially establishing "distinct rhythms" and points out that a slight delay exhibited by one group of his animals may have been due to an immediately prior two-week experience of light-feeding-only, he does not directly raise the more subtle question. Once a distinct rhythm had actually stabilized, *does* the extent or proportion of its nocturnality evidence continuing influence from the animal's earlier experience with a light-feeding-only regimen?

That such continuing influence does occur is suggested even within his own data. Whereas several apparently typical control groups were reported to distribute 20 per cent, 28 per cent, and 25–30 per cent of their eating in the light portion, the proportions for LL-reared animals first placed into LD at 59 days of age were slightly higher than those figures for the second three days in LD (30–35 per cent). For LL-reared animals placed into LD at 70 days, but with food available only in the light until 84 days, proportions stabilized after a week at even somewhat higher levels (35–40 per cent). Whether the apparent stabilization at the higher levels would have been permanent is not known.

My data speak specifically also to this question of long-term influence, as can be seen by comparing values in the figure for the proportions of intake occurring in the light during the last week. Though ranges overlap, the higher proportions are indeed exhibited by the experimental animals, reflecting continued influence of the earlier feeding experience.

Regardless of whether the differences in Zucker's recent observations [3] of postweaning feeding and mine are attributable to differences in experimental situation (e.g., homecage vs. wheel and pellets vs. liquid diet), it is still of interest that at least under some conditions—the present ones—the feeding schedule during early development continues to influence the feeding rhythms of at least some animals even when normal *ad-libitum* feeding has prevailed for several weeks.

In summary, two points concerning ontogeny of feeding rhythms may be emphasized: (1) Individual differences in the emergence of the feeding rhythm and its nature covered a rather *wide* range. (2) As adults, some rats exhibited *diurnal* feeding rhythms rather than the more typical nocturnal ones, and the diurnal deviations were related to previous light/dark feeding experiences.

REFERENCES

1. Richter C. P. (1927): Animal behavior and internal drives. Quart. Rev. Biol., *2*: 307–343.
2. Bolles R. C. and Woods P. J. (1964): The ontogeny of behaviour in the albino rat. Animal Behaviour, *12*: 427–441.
3. Zucker I. (1971): Light-dark rhythms in rat eating and drinking behavior. Physiol. Behavior, *6*: 115–126.
4. Le Magnen J. (1967): Habits and Food Intake. *In*: Handbook of Physiology. Section 6: Alimentary Canal. Volume I. Food and Water Intake, (C. F. Code and Werner Heidel, ed.), Chapter 2, pp. 11–30, Williams & Wilkins, Baltimore.
5. Epstein A. N. and Teitelbaum P. (1962): Regulation of food intake in the absence of taste, smell, and other oropharyngeal sensations. J. Comp. Physiol. Psychol. *55*: 753–759.
6. Teitelbaum P. and Campbell B. A. (1958): Ingestion patterns in hyperphagic and and normal rats. J. Comp. Physiol. Psychol., *51*: 135–141.
7. Wright J. H. (1965): Modifications in the rat's diurnal activity pattern as a function of opportunity for reinforcement by ingestion. J. Comp. Physiol. Psychol., *59*: 463–465.

CHRONOBIOLOGIC EDUCATION AND HEALTH CARE DELIVERY

Chairman: ANDREW AHLGREN

Co-Chairman: HARRY S. LIPSCOMB

ROLE OF CHRONOBIOLOGICAL STUDIES IN HUMAN PHYSIOLOGY

Harry S. LIPSCOMB

Xerox Center for Health Care Research
Baylor College of Medicine
Houston, Texas, U.S.A.

Temporal integration of body function in humans constitutes one of the fundamental dimensions of organization lending itself to rigorous quantitative validation.

From the earliest studies of circadian rhythmicity in plant physiology through cosinor techniques measuring modifications of time-series endpoints, we now possess tools with which to examine in infinite detail the functional capacity of biological systems.

It is no longer sufficient to measure single body systems when it is now possible to measure multiple organ systems with appropriate tools. Furthermore, it is no longer reasonable to measure single isolated events of physiological or biochemical function when dynamic, kinetic measurements are possible, analyzable, and fruitful. It is no longer sufficient to classify health as the absence of disease. We now possess tools and the rigorous statistical methodology which would allow us to ask of any functional organism, more specifically man, "what is your maximal and minimal functional capacity to perform a variety of tasks?" or of a system "what is the maximal and minimal functional capacity of this system at any one time?" Tacit in this question is the assumption that capacity and functional capabilities vary throughout the day and night, and the earliest reflectors of derangement of functional capacity may, in fact, lie within an analysis of rhythmic, temporally oriented events. In our studies, we have found it economically unreasonable to study single-organ systems, and rather have gone to systems which examine the entire organism. Starting from very fundamental assumptions regarding function, we have made discreet measurements, without recognition, however, of the time base of variability in these various functional systems. It seems now reasonable to examine these various systems throughout a period of time sufficient to give us a picture of the temporal organization of the system. Documentation of this sort, under conditions of "normal" function may give us the earliest base line upon which to assess ultimate dysfunction, as reflected in temporal disorganization.

Clearly, chronobiological studies and recognition of the temporal organization of functional man may provide us the first valid glimpse of what constitutes "health" as distinct from disease.

Moreover, it is inconceivable that in 1971 we would continue to adhere to tables of "normal" values for a physiological parameter without full recognition that such "normal" values constitute single points in time, with no recognition of the dynamic nature of all living systems. Put another way, perhaps the only circumstance in the future where measurement of single non-temporally oriented parameters may be justifiable will be in non-living systems.

BLOOD PRESSURE SELF-MEASUREMENT FOR COMPUTER-MONITORED HEALTH ASSESSMENT AND THE TEACHING OF CHRONOBIOLOGY IN HIGH SCHOOLS*

Franz HALBERG, Erhard HAUS, Andrew AHLGREN, Erna HALBERG, Harold STROBEL, Anthony ANGELLAR, Jürgen F. W. KÜHL, Russell LUCAS, Eugene GEDGAUDAS and Joseph LEONG

Chronobiology Laboratories, Departments of Pathology, Radiology and Curriculum Studies, University of Minnesota Minneapolis, and the Department of Pathology, St. Paul-Ramsey Hospital, St. Paul, Minnesota, U.S.A.

Fifty high school students learned the techniques of self-measurement and computer analysis of rhythms in their data, in relation to concepts of biologic rhythmicity and its implications for health care. Both the feasibility and importance of autorhythmometry are demonstrated by individualized "usual" ranges for such variables as blood pressure, as well as by the consistent description of rhythms in physiologic and psychologic performance variables. (The potential importance of individualized usual ranges is illustrated by a separate example of a child's disease-dependent blood pressure elevation that occurred entirely within the usual rhythm-unqualified range for the corresponding sex and age group.)

High blood pressure is the major debilitator and killer in developed countries; often the condition can be corrected efficiently. The earlier high pressure or the risk of its occurrence is detected and treated, the less irreversible damage is done. Given sufficient properly scheduled measurements, computer methods can characterize pressure variation over time—to detect "afternoon" or "odd-hour" hypertension and other forms of labile or fixed pressure elevation. However, measurement series by health-care personnel are impractical when the majority of the public lives without even a single measurement of pressure per decade. Hence, there is great utility in: a) training public school teachers and their students in self-measurement techniques; b) providing computer analysis of data as "health summary" charts; c) teaching the rationale for interpreting results, and d) motivating as well as facilitating subject-physician contact and communication.

In two studies on school children 13–18 years of age, carried out at Minneapolis Southwest High School and the Twin Cities Summer Institute, 1971, the teaching of chronobiology was combined with autorhythmometry. Students measured body temperature, heart rate, systolic and diastolic blood presure, short-term memory, 1-minute estimation (or tempo) and eye-hand skill (by a so-called nut-threading loop and by finger counting). At the same sampling sessions the students also rated their vigor and mood. The spans of observation ranged

* Supported by U.S. Public Health Service (5-K6-GM-13,961), NSF, Manned Spacecraft Center of NASA, Houston (BB-321-79-1-343P) NSF-GW 7163, Central Chapter, Connecticut Heart Association and St. Paul Ramsey Hospital Medical Research and Education Foundation.

between 16 and 64 days, the number of measurements per day varying from one subject to another. Here we wish to summarize physiological and psychological information for all students from both studies.

VARIABLES

The techniques used to collect and analyze the data are described elsewhere [1], but some general comments concerning the choice of variables seem appropriate. Oral temperature is of interest from several points of view. From a physiologic viewpoint, body core temperature has a very stable circadian rhythm [2]: it may be used as a physiologic reference standard to which one relates information on other rhythms, and as an indicator of overall metabolism. For the psychologist, temperature is of interest because it has been reported to correlate rather well with certain types of performance [3]. From a medical viewpoint, the finding of an elevated body temperature can be complemented by scrutiny of possible changes in rhythm characteristics and the eventual association of such changes with impending or established disease. Moreover, body temperature is readily measured. To the extent that temperature correlates with certain performance measures, it may eventually be substituted for tests that cannot be repeated at frequent intervals during the 24-hour span because they are too time-consuming.

Much work remains to be done in the admittedly complex field of gauging performance. Some mental aspects of performance (such as short-term memory and time estimation) as well as measurements of physical performance (such as eye-hand skill evaluated by finger counting or by a nut-threading ring) also can be rapidly assessed repeatedly during the day. The tests should be non-boring, preferably motivating and, of course, non-painful. The self-rating of physical vigor and mood along a 7-point scale repeatedly during the day may compare favorably with a single qualitative reply to the question usually asked by the physician—"How have you felt since I last saw you?" A standardized, semi-quantitive series of self-ratings is more valuable both in that it is obtained as one goes rather than by recall after days (or even years!), and in that it reveals variation—some portion of which may by rhythmic and thus predictable.

Three measures of the circulation—systolic and diastolic arterial blood pressure and heart rate—were included in the self-measurement sessions, because individualized ranges for these particular variables may be especially useful for the early detection of a circulatory risk-state such as a blood-pressure elevation. A few training sessions with relatively inexpensive equipment led to student blood pressure self-measurements of satisfactory accuracy (about ±3 mmHg), as judged by simultaneous stethoscope monitoring with a Y-tube [4].

RESULTS

Rhythms approximated by least-squares fitting of single cosine curves to the data can be displayed on a polar "cosinor" plot, where length of the vector indicates the amplitude of the rhythmic variation, and the angle of the vector indicates the time of the peak (acrophase). The uncertainty of estimation is represented by the 95% "error ellipse" centered on the end of the vector; thus the statistical significance of the fit is indicated by an error ellipse that does not cover the plot pole [5].

Table 1 Dividing lines among "Hypotension", "Normotension" and "Hypertension" amplified on PICKERING's Table 3.1 in "Hypertension." *Causes, Consequences and Management* by George PICKERING. London: J. & A. Churchill, 1970, p. 28.

Reference no.	Criterion Syst/Diast	Author, year (Reference, title)
		"Hypotension" and Normotension"
1	110/–	*British Medical Dictionary*. (ed. ARTHUR S. MACNULTY, 1963. Philadelphia, J.B. Lippincott, Co., p. 721).
2	110/66 (male) 105/62 (female)	DALLY, J.F. HALLS, 1928, (*Low Blood Pressure. Its Causes and Significance*. New York, William Wood and Co., pp. 27, 31).
3	90/–	AMATUZIO, D.S., Department of Medicine, University of Minnesota Medical School, personal communication, 1971.*
		"Normotension" and "Hypertension"
4	– /90	AMATUZIO, D.S., Department of Medicine, University of Minnesota Medical School, personal communication, 1971.*
5	120/80	ROBINSON, S.C. and M. BRUCER, 1939. (Arch. intern. Med. *64*: 409–444; Range of normal blood pressure. A statistical and clinical study of 11,383 persons).
6	130/70	BROWNE, F.J., 1947. (British medical Journal *ii*: 283–287; Chronic hypertension and pregnancy).
7	140/80	AYMAN, DAVID, 1934. (Arch. intern. Med. *53*: 792–802; Heredity in arteriolar (essential) hypertension: A clinical study of the blood pressure of 1,524 members of 277 families).
8	140/90	REID D.E. and TEEL H.M. 1939. (Amer. J. Obstet. Gyn. *37*: 886–896; Nonconvulsive pregnancy toxemias).
9	140/90	CHESLEY L. and ANNITTO J. 1947. (Amer. J. Obstet. Gyn. *53*: 372–381; Pregnancy in the patient with hypertensive disease).
10	140/90	PERERA G.A., 1948. (Am. J. Med. *4*: 416–422; Diagnosis and natural history of hypertensive vascular disease).
11	140/90	SCHROEDER H.A., 1953. (*Hypertensive Diseases. Causes and Control*. Philadelphia, Lea and Febiger, p. 29).
12	140/90	MAYER J.H. and FLYNN J. 1971. (The role of arterial hypertension in coronary atherosclerosis; in *Coronary Heart Diseases*, H.I. RUSSEK and B.L. ZOHMAN, eds. Philadelphia, J.B. Lippincott, pp. 138).
13	150/90	THOMAS, C.B., 1952. (Am. J. Med. Sci. *224*: 367–376; The heritage of hypertension).
14	150/90	HOOBLER, S.W., 1959. (*Hypertensive Disease. Diagnosis and Treatments*. New York, Paul B. Hoeber, p. 70).
15	150/95	MOSER, M. and GOLDMAN, A.G. 1967. (*Hypertensive Vascular Disease. Diagnosis and Treatment*. Philadelphia, J.B. Lippincott Co., p. 11).
16	150/100	ASK-UPMARK, E., 1967. (*A Primer of High Blood Pressure*. Bonniers, Svenska Bokförlget, p. 14).
17	160/100	BECHGAARD, P., 1946. (Acta med. scand. Suppl. 172; Arterial hypertension: a follow-up study of one thousand hypertonics).
18	180/100	BURGESS, A.M., 1948. (New Eng.J. Med. *239*: 75–79; Excessive hypertension of long duration).
19	180/100	EVANS, W., 1956. (*Cardiology*. London, Butterworth, 2nd edition, p. 386).
20	90/63 (3 yrs) 115/75 (16 yrs) 140/90 (40 yrs) 150/90 (over 40)	GOLDRING, W. and CHASIS, H. 1944. (*Hypertension and Hypertensive Disease*. New York; The CommonwealthFund, p. 9).
21	140/90 to 150/100	SMIRK, F.H., 1957. (*High Arterial Pressure*. Oxford: Blackwell Scientific Publication, p. 18).

* Clinical medicine standards as practiced at the University of Minnesota Hospitals, Minneapolis, Minnesota.

By this method, 24-hour-synchronized rhythms for the total group were found for all variables except systolic blood pressure. Cosinor plots for these nine variables appear in Fig. 1. Temperature, vigor, mood, and pulse are characterized by an acrophase occurring late in the afternoon, whereas short-term memory peaks in late evening and diastolic blood pressure has a morning acrophase. Cosinors for the performance variables finger-counting, nut-threading, and 1-minute estimation represent the longest test duration; hence the peak performances occur twelve hours away from the night-acrophase on the graph, i.e. in midafternoon.

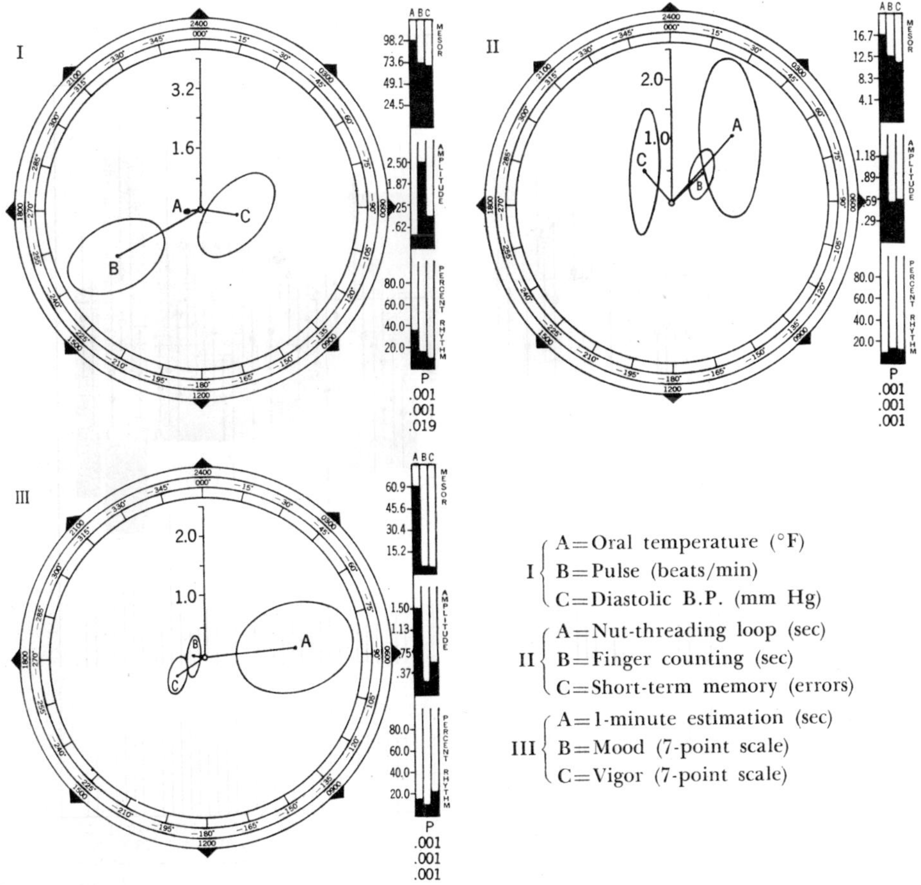

Fig. 1 Self-measurements for one or several weeks (according to the individual's preference) analyzed by cosinor yield confidence regions that do not overlap pole. Thereby a circadian rhythm is detected for several variables of high school students found to be "healthy" in a physical examination, including chest x-ray, ECG and urine screen for certain drugs.

Fig. 2 displays usual ranges as well as coefficients of variation for the systolic blood pressures of each student. Such individualized usual ranges are desirable for the quantitative assessment of health and for detection of disease. As may be seen from the section on males in Fig. 2, a given systolic blood pressure value—say, around 115 mmHg—can be completely below the range of values for one in-

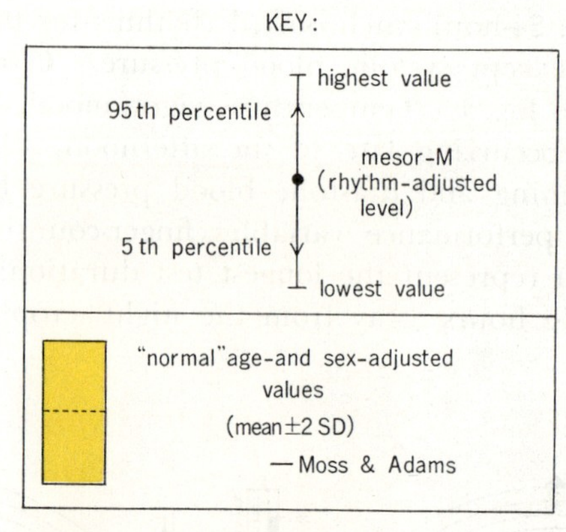

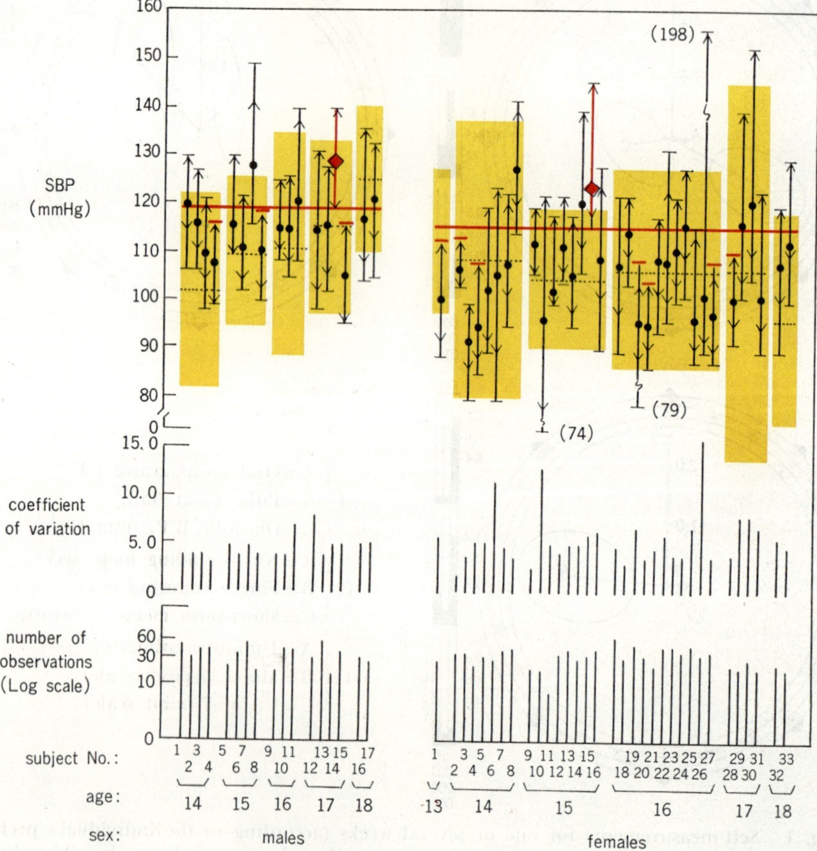

Note: Lowest value of one subject ( ), is higher than highest values ( ) of other subjects. Such findings are obscured by exclusive reliance upon a group range.

Fig. 2 Even when data do not suffice for rhythm description, the provision of usual ranges for a given individual (rather than for a group) constitutes an invaluable dividend of autorhythmometry. It is important to know indeed that the systolic blood pressure of say, 115 mm Hg can be "too high" for one child but "too low" for another of the same sex and age (see text).

HIGHER BLOOD PRESSURE DURING NEPHRITIS (left), THAN AFTER RECOVERY (right), —YET ALL VALUES ARE WITHIN RANGE GENERALLY CONSIDERED NORMAL

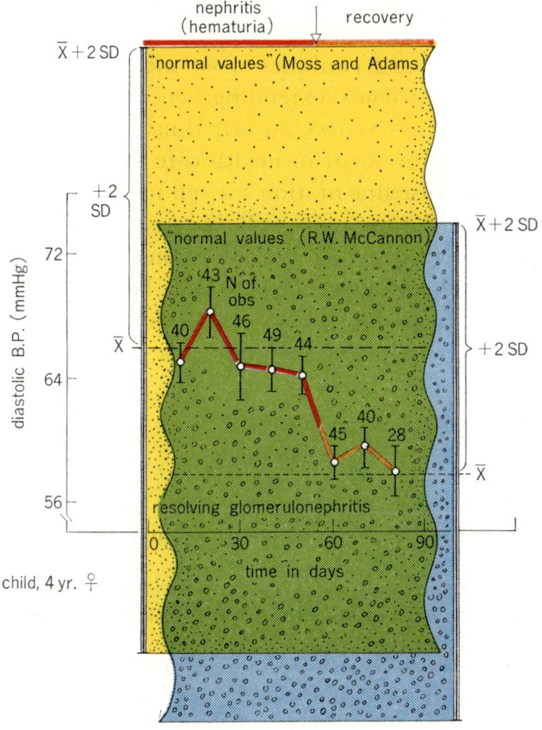

Fig. 3. Exclusive reliance upon "usual values" for a given age and sex-cohort draws a curtain of ignorance over the range of physiological variation in health, states of risk or overt diseases. This curtain can be lifted by rhythmometry; thus one detects (within the "normal range" of a peer cohort) "individualized" mesorhypertension [a statistically significant elevation in the rhythm-adjusted mean of blood pressure assessed in relation to an individualized mesor (Fig. 2)].

dividual (Subject 13) and above the range of values for another (Subject 14). Obviously, the range of "usual" values has to be defined for each individual if hypertension is to be optimally recognized and treated.

An example from another study [to be discussed elsewhere by one of us (E. HAUS)] is given in Fig. 3, which shows that the diastolic pressure in a case of glomerulonephritis is within the "usual range" defined by two standard references, yet the same pressure is seen to be elevated when compared to the given child's own usual range. Hence, reliance upon the range of a cohort, as is currently the practice, may be complemented by interpretation of rhythm characteristics.

It should be reemphasized that, apart from rhythms, the 100% ranges of two 17 year-old boys, graphed next to each other in Fig. 2, do not overlap. Lack of overlap also can be seen for the 100% ranges of several girls in the same Fig. 2. Exclusive reliance upon the individualized range is not advocated, yet it is clear that a diagnosis of blood pressure elevation is likely to be greatly facilitated if one uses as reference standard the individual's usual ranges (as well as his rhythm characteristics) in health. Rhythmometric standards will have to become available, unless one indeed refrains from using any dividing line between blood pressure in health and disease [6] or arbitrarily chooses such a line from the diversity of suggestions available in the literature—for example, 120/80 [7], 140/90 [8] or 180/100 [9].

Eventually the rationale and techniques of self-measurement can be introduced by students to relatives and friends, extending benefits broadly in the adult community. In terms of cost-effectiveness, autorhythmometry in school constitutes a critical step toward an improvement in health care delivery achieved at minimal cost. The chief cost is in instrumentation for self-measurement and analysis but, insofar as such instrumentation may well become a component of materials regularly used in science teaching, it does not constitute a health cost as such. The computer used by students also will eventually be part and parcel of standard high school education; the analysis of self-measurements merely provides an immediately rewarding and interesting computer use in place of other, more contrived instruction.

REFERENCES

1. HALBERG F., JOHNSON E. A., NELSON W., RUNGE W. and SOTHERN R. (1972): Autorhythmometry—procedures for physiologic self-measurements and their analysis. Physiology Teacher, *1*: 1–11.
2. HALBERG F. (1970): Frequency spectra and cosinor for evaluating circadian rhythms in rodent data and in man during Gemini and Vostok flights. COSPAR Life Sciences and Space Research,*8*: 188–214.
3. STEPHENS G. and HALBERG F. (1965): Methodologic aspects of human time estimations with special reference to 24-hour synchronized circadian rhythms. Nursing Research, *14*: 310–317.
4. AHLGREN A. (1971): Experimental comparison of blood pressure instruments for making self-measurements. Multilith. Center for Educational Development, University of Minnesota.
5. HALBERG F., TONG Y. L. and JOHNSON E. A. (1967): Circadian system phase—an aspect of temporal morphology; procedures and illustrative examples. *In*: The Cellular Aspects of Biorhythms, (H. von MAYERSBACH, ed.), pp. 20–48, Springer-Verlag, Berlin.
6. PICKERING G. (1968): High blood pressure, 2nd ed., Grune & Stratton, New York.
7. ROBINSON S. C. and BRUCE M. (1939): A statistical and clinical study of 11,383 persons. Arch. Intern. Med., *64*: 409–444.
8. MAYER J. H. and FLYNN J. (1971): The role of arterial hypertension in coronary atherosclerosis. *In*: Coronary Heart Diseases, (H. I. RUSSEK and B. L. ZOHMAN, eds.), p. 138, J. B. Lippincott, Philadelphia.
9. BURGESS. A. M. (1948): Excessive hypertension of long duration. New Eng. J. Med., *239*: 75–79.

AUTORHYTHMOMETRY METHODS FOR LONGITUDINAL EVALUATION OF DAILY LIFE EVENTS AND MOOD: PSYCHOPHYSIOLOGIC CHRONOTOGRAPHY*

CHARLES F. STROEBEL

Laboratories for Experimental Psychophysiology
Institute of Living Hospital
Hartford, Connecticut, U.S.A.

Chronobiologists and psychologists classically have assumed constancy in one anothers' domains in order to simplify experiments designed to examine variables of particular interest to each [11]. A more clinically realistic, combined approach, namely, the comparatively recent discipline of psychophysiology [14], has challenged the constancy assumption with experimental evidence demonstrating the interdependence of behavioral and biologic functioning in both humans and animals [6]. However, the psychophysiologists' strategy of making fewer assumptions of constancy yields multivariate experiments which are more costly and complicated than the classical precedents; i.e., more actual biologic and behavioral variables are studied simultaneously as *measured* variables.

Ideally, the psychophysiological approach should be sensitive to the entire spectrum of human experience (mental, behavioral, physiologic, pathologic) as viewed longitudinally. This would resolve periodic changes over time and the effects of past life stresses on subsequent functioning. Unfortunately, practical considerations (cost, limited laboratory facilities and the endurance of subjects and experimenters) have tended to limit psychophysiologic investigations to several occasions per subject. For example, the discipline has experienced its greatest success in the study of correlates and sequences of sleep stages, a situation where subjects are easily confined to a fixed location, and are relatively unaware of the considerable instrumentation which must be worn (this often includes EEG, EMG, EOG, EKG, and GSR electrodes; penile and respiration strain gages, etc.).

Radio telemetry and/or portable recorders permit extension of the physiologic aspects of these investigations to the waking state, but incur new complications: 1) increased recording artifact due to physical motion; 2) difficulty in simultaneous, objective recording of changing behavioral and mental states which may have produced physiologic changes (and vice-versa), and 3) the *Uncertainty Principle;* (Re: quantum mechanics, where the process of measurement itself changes the basic state one is attempting to measure) i.e., conscious or unconscious (Freud) awareness by a subject of monitoring and its related apparatus changes the natural condition of the subject which one is trying to measure. These complications will continue as obstacles in the years ahead.

I recognize that the longitudinal coupling of volunteer human chronobiology

* This research was supported in part by grants MH-08870 and MH-08552 from the National Institute of Mental Health, and by the Gengras Foundation. The author is indebted to Prof. Franz HALBERG and Dr. B. C. GLUECK, Jr., for encouragement in the pursuit of those studies.

subjects to a biochemical autoanalyzer, polygraph, or computer (even via telemetry)—akin to 24-hour occupancy on an intensive care unit—*will* provide significant chronopsychophysiologic information, but will be a costly research effort not fully sensitive to the richness and variety of human experience, and, in the foreseeable future, will have limited applicability to the larger population needing preventive medical care. The resulting longitudinal chronologic psychophysiologic "maps" (see footnote)* will provide a substantive scientific foundation, but will be too tedious to apply on a daily basis to large numbers of individual, idiosyncratic, semi-cooperative, and possibly desynchronized (potentially diseased) persons.

This dilemma could hinder the application of the exciting potential of psychophysiologic chronotography to the healthcare system, especially in the areas of preventive medicine and psychiatry. What we need is the development of inexpensive, practicable techniques for evaluating the temporal relations of behavioral and biologic changes in single individuals which will be potentially correlatable with the more elegant and complete reference standard data collected in the chronobiology laboratory.

The present research was designed with this goal in mind: how to obtain meaningful serial (chronologic) psycho- and physiologic information reliably, inexpensively, easily, and regularly from *large* populations of human beings. Four premises were adopted at the outset as follows:

1) Because of the large number of anticipated subjects (conceivably the entire world population), the procedure should be amenable to computer scoring and analysis.

2) Human beings, unlike animals, can be trained to make reliable and valid observations about themselves (self report) and others (observer report) at both psychologic and physiologic levels, if provided a practical means for doing so.

3) Daily changes in mood, life stress, habits, and bodily functioning—semi-covert information about periodic functioning which most of us quickly forget—could be documented on a daily or more frequent basis as a data source for psychophysiologic chronotography. In computer form, these patterns of daily living which ordinarily are considered trivial (because of our own poor memory of them), become available for quantitative analysis.

4) The measurement system must be flexibly sensitive to time, to the degree the observer is capable, willing, and interested. For example, a dedicated chronobiologist may be willing to drop all other activity to measure or evaluate himself every three hours on the hour; few other human beings would do so consistently on a voluntary basis. Flexibility implies that the measuring procedure could be used by a dedicated, compulsive reporter (self or observer) making observations four or five times daily to detect ultradian or circadian changes, as well as by the more casual reporter making observations only twice daily (conceivably limiting detection to infradian changes).

* Defined here as *Psychophysiologic Chronotography,* where indices or factors underlying psychologic and physiologic functioning (ordinate) are quantitatively plotted or "mapped" on a correlative basis over time (abscissa) based on data acquired under controlled laboratory conditions with defined, synchronized population as a reference standard. Like the revelation of the paper chromatogram which separates an organic compound into its components and makes them visible by colors, a psychophysiologic chronotogram will identify those specific components of behavior and physiology which progressively evolve over time in health and in disease (see Fig. 2).

Conceptually, our psychophysiological chronotography studies have evolved in identifiable stages as three rating instruments for evaluating daily life events, body changes and mood over time:

1. Computerized Psychiatric Nursing Notes (nurse or trained aide as observer)

The procedure for routine charting of psychiatric nursing notes was computerized at the Institute of Living in 1967 [3–5]. Nursing reports are made on each patient twice daily (reports could be made as often as desired, were extra staff available) by routine nursing personnel on a special IBM 1232 mark sense form that is designed for computer scoring. Eleven areas of noninferential patient behavior are rated on the front side of the form which is shown in Fig. 1. Temperature, pulse, blood pressure, the nature of daily activities, and physical complaints are reported in a similar fashion on the reverse side of the form. The computer produces two types of output from each report: the first is a narrative summary to be filed in the patient record as a legal document; the second is a set of twenty daily factor scores describing the patient's behavior numerically as compared to his hospital residence unit (normalized in T-score units where the mean for each factor=50, and the standard deviation=10). The original factor loadings and normalization were obtained from a factor analysis-varimax rotation of 2,325 reports. Each factor possesses independence since no single item loads on more than one factor. Further, nursing personnel are not required to make complicated inferences about behavior such as "depression" since they are unaware of which items load on which factor. Among the 20 factors are measures of 1) Acceptable behavior; 2) Disorganization; 3) Depression; and 4) Anxiety. Fig. 2 illustrates a longitudinal display of these four factors over a sixty day period in a manic-depressive patient undergoing psychiatric treatment. Since 1967, nursing note data in this form on 2400 patients with runs of data from three weeks to four years have been accumulated, and are being analyzed.

2. Behavior Profile Chart (a self report form)

This rating instrument was developed in 1968 at the Institute of Living to identify time changes in physiologic functions and life events within a 24-hour time domain using a single form [12]. The front side of a scored form is shown in Fig. 3. The Behavior Profile Chart form is not computer scorable and requires tedious keypunching of data for computer plotting and analysis. Extensive runs of data have been collected from six control subjects, two psychiatric patients and 21 psychophysiology sleep study subjects. This form was regarded as developmental, since all subjects noted considerable difficulty in reliably and quantitatively rating their mood changes on a day to day basis.

3. Psychophysiological Diary (a self report form)

This new rating instrument designed for use by normals, inpatients, and outpatients, has been published as a step-down booklet into which a new computer scored mark sense form is inserted each day [13]. The Psychophysiological Diary incorporates the best features of automated nursing notes and the Behavioral Profile Chart, but also includes entirely new sections. For example: 1) 16 specific mood indices are rated quantitatively on ten step transitive scales which are anchored by descriptive phrases; 2) Seventy somatic and physical complaints are

PATIENT NAME — CASE NUMBER — UNIT — DATE

OBSERVER STAFF IDENT NUMBER: 0 1 2 3 4 5 6 7 8 9

PATIENT CASE NUMBER: 0 1 2 3 4 5 6 7 8 9

DAY OF MONTH: 0 1 2 3 4 5 6 7 8 9

REPORTING AREA: 0 1 2 3 4 5 6 7 8 9

SHIFT 1 DAY 2 NIGHT

INSTITUTE OF LIVING

PATIENT BEHAVIOR INDEX • NURSING

PERSONAL HABITS
- WITHDRAWN
- HAS TO BE REMINDED WHAT TO DO
- ANNOYS PERSONNEL BY TOUCHING THEM
- SMOKES INCESSANTLY
- SLOW TO FOLLOW ROUTINE
- RESENTS UNIT ROUTINE
- FOLLOWS ROUTINE ACCEPTABLY
- NEEDS HELP WITH PERSONAL HYGIENE
- REFUSES TO DO ROUTINE THINGS EXPECTED OF HIM
- DOES ODD, STRANGE THINGS
- SEE NARRATIVE
- UNABLE COMMENT

APPEARANCE
- FUSSY, FASTIDIOUS
- LOOKS TIRED, WORN OUT
- INAPPROPRIATELY, INFORMALLY DRESSED
- CAREFULLY DISORDERED
- CLEAN, NEAT, APPROPRIATELY DRESSED
- SLOPPY, UNKEMPT
- OVERDRESSED FOR THE OCCASION
- DRAMATIC, THEATRICAL
- BIZARRELY DRESSED
- LOOKS YOUNGER THAN IS
- SEE NARRATIVE
- UNABLE COMMENT

SLEEPING AND EATING HABITS
- SLEEPS DURING DAY
- COMPLAINED OF NOT BEING ABLE TO SLEEP
- SKIPPED MEAL
- RETIRED EARLY
- NOT UP FOR BREAKFAST
- SLEPT WELL
- EATS WELL
- SLEEPS RESTLESSLY DURING NIGHT
- FOOD INTAKE INADEQUATE
- WAKES EARLY
- SEE NARRATIVE
- UNABLE COMMENT

UNIT RELATIONSHIPS
- PREFERS COMPANY OF PERSONNEL
- ENJOYS SADISTIC HUMOR
- COMPLAINS ABOUT BEING IN HOSPITAL
- SUSPICIOUS OF ACTIONS OR MOTIVES OF PERSONNEL OR OTHER PATIENTS
- SPENDS GREAT DEAL OF TIME IN ROOM
- SATISFACTORY ADJUSTMENT TO UNIT
- RARELY GOES OFF UNIT ON HIS OWN INITIATIVE
- MUST BE REMINDED TO ATTEND CLASSES
- PRANKISH
- HAS TO BE TOLD TO COME OUT OF HIS ROOM
- SEE NARRATIVE
- UNABLE COMMENT

SOCIAL BEHAVIOR
- SECLUSIVE
- CANNOT TOLERATE DELAYS OR DENIAL OF HIS WISHES
- MEMBER OF "CLIQUE"
- FORMAL, RESERVED
- QUIET
- BOISTEROUS
- MAINTAINS A CLOSE RELATIONSHIP WITH ONE OTHER PATIENT
- FRIENDLY AND COOPERATIVE
- RELAXED, AT EASE
- RESTLESS, FIDGETY
- SEE NARRATIVE
- UNABLE COMMENT

SOCIAL INTERACTION
- EXCESSIVELY FAMILIAR WITH MEMBER OR MEMBERS OF OPPOSITE SEX
- IS SEDUCTIVE
- CONVERSES ONLY ON APPROACH
- OVERLY FAMILIAR WITH SAME SEX
- UNPOPULAR
- IMPULSIVE
- AVOIDS OPPOSITE SEX
- TEASES
- IMPOLITE
- PARTICIPATES IN GROUP ACTIVITY
- SEE NARRATIVE
- UNABLE COMMENT

MOOD
- SAD
- IRRITABLE
- MOODY, CHANGEABLE
- SMOOTH, EVEN DISPOSITION
- SHOWS LITTLE FEELING
- SEEMS AFRAID OF SOMETHING
- IS PLEASED WITH HIMSELF
- ANGRY
- TEARFUL
- PREOCCUPIED, OFTEN SEEMS TO BE DAYDREAMING
- SEE NARRATIVE
- UNABLE COMMENT

ATTITUDE
- SEEMS PLEASANT, YET IS OBSTRUCTIVE
- MAKES EXCUSES FOR HIS ACTIONS
- PLEASANT
- HOSTILE TO ONE PERSON IN PARTICULAR
- OFTEN DEMANDS ATTENTION OR PRAISE
- DEMONSTRATES FEELINGS OF INADEQUACY
- ARGUMENTATIVE OR UNCOOPERATIVE
- CAN'T MAKE UP MIND, INDECISIVE
- MANIPULATIVE
- FEELS REJECTED
- SEE NARRATIVE
- UNABLE COMMENT

VERBALIZATION
- USES STRANGE WORDS, PHRASES
- LOGICAL, CLEAR
- RAMBLES
- VOICE FLAT, MONOTONOUS
- VULGAR LANGUAGE
- SPEAKS SLOWLY, HESITANTLY
- SAYS THINGS ARE HOPELESS; HE IS NO GOOD
- SARCASTIC
- TALKATIVE
- REPEATS THOUGHTS, WORDS OR PHRASES OVER AND OVER
- SEE NARRATIVE
- UNABLE COMMENT

INTELLECTUAL BEHAVIOR
- EXPRESSES FEW THOUGHTS
- INAPPROPRIATE LAUGHTER
- CONFUSED
- STATEMENTS OR THOUGHTS INAPPROPRIATE TO MOOD OR SITUATION
- MOSTLY SELF-CENTERED
- FORGETFUL
- ALERT AND RESPONSIVE; CONCENTRATES WELL
- STATES PEOPLE ARE UNFAIR OR MEAN TO HIM
- GIDDY, CHILDISH
- DOESN'T PROFIT FROM MISTAKES
- SEE NARRATIVE
- UNABLE COMMENT

MISCELLANEOUS
- STATES NEED FOR LEAVING HOSPITAL
- HELPFUL
- OVERACTIVE
- WELL-MANNERED
- TENSE
- SLUGGISH OR DROWSY
- PACING
- UNREALISTIC IDEAS ABOUT HIMSELF, OTHERS OR HIS SURROUNDINGS
- NEGLECTS RESPONSIBILITIES
- UNUSUAL FACIAL EXPRESSIONS, GRIMACES
- BECOMES UPSET EASILY
- ENGAGES IN SOLITARY ACTIVITIES ON UNIT

INSTRUCTIONS

1. MAKE YOUR MARKS WITH A NO. 2 BLACK LEAD PENCIL.
2. FILL EACH MARK POSITION COMPLETELY.
3. ERASE COMPLETELY ANY MARKS YOU WISH TO CHANGE.
4. DO NOT STAPLE OR FOLD THIS SHEET.
5. PRINT NARRATIVE STATEMENT(S) ON REVERSE WHEN "SEE NARRATIVE" IS MARKED IN ANY CATEGORY.

IOL FORM 1232.1 P.C 7-66

Fig. 1 Front side of the automated patient behavior index from for rating by psychiatric nursing personnel on each shift.

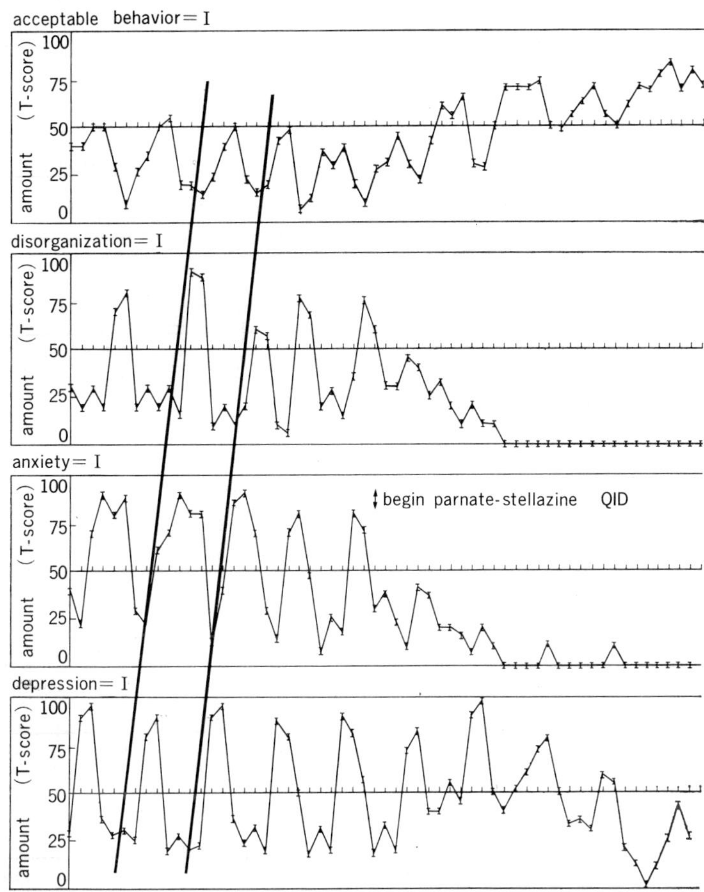

Fig. 2 Computer produced longitudinal graphs of four automated nursing note factors over a sixty day period from a manic-depressive patient receiving psychiatric treatment. Note the gradual damping of the 6–7 day periodicity in each of the factors shortly after medication was initiated (arrow).

listed and scored for time of onset, duration, and intensity; 3) The HOLMES and RAHE [2] 42 item scale of life stress events is presented for rating once each day. The latter provides cumulative predictions of future illness.

The Psychophysiologic Diary is available in two forms:

FORM A: Ratings are made on arising (AM temperature, pulse, and sleep/dream data) and retiring (mood during activity period, somatic complaints, PM temperature and pulse, etc.) which require time of day information for many items. Total rating and measurement time is 13 minutes.

FORM B: Ratings may be made as frequently as specified within each 24 hour time period. Total rating and measurement time per rating is 7 minutes.

The Psychophysiological Diary is being used to evaluate all Institute of Living laboratory subjects taking part in sleep and other studies in the months preceding and following actual laboratory evaluations with polygraph and biochemical serial measures. We are hopeful that many psychophysiology and chronobiology laboratories using human volunteer subjects will also use the Psychophysiological Diary as a baseline instrument for correlation of its indices with objectively

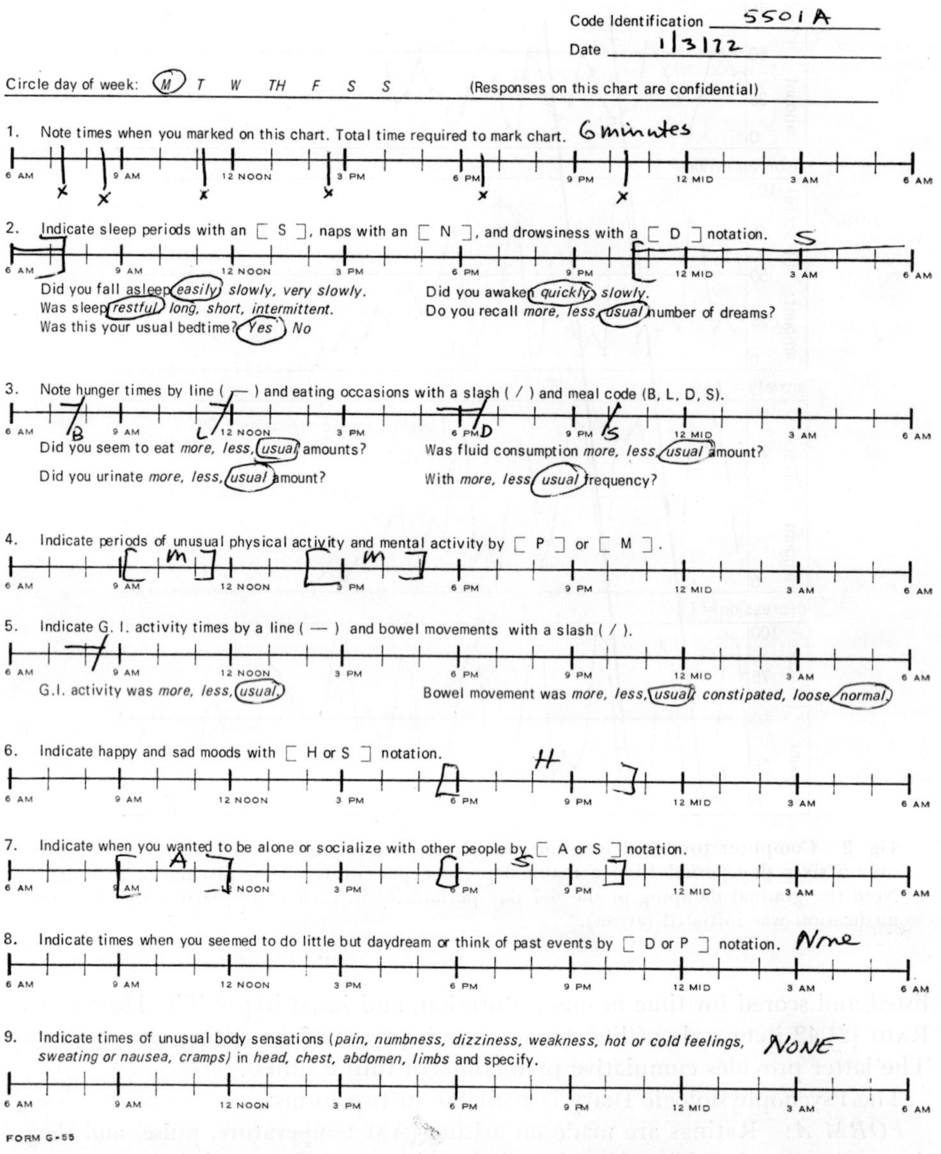
THE INSTITUTE OF LIVING

BEHAVIOR PROFILE CHART

Code Identification 5501 A

Date 1/3/72

Circle day of week: M T W TH F S S (Responses on this chart are confidential)

1. Note times when you marked on this chart. Total time required to mark chart. 6 minutes

6 AM 9 AM 12 NOON 3 PM 6 PM 9 PM 12 MID 3 AM 6 AM

2. Indicate sleep periods with an [S], naps with an [N], and drowsiness with a [D] notation. S

6 AM 9 AM 12 NOON 3 PM 6 PM 9 PM 12 MID 3 AM 6 AM

Did you fall asleep *easily, slowly, very slowly.* Did you awaken *quickly, slowly.*

Was sleep *restful, long, short, intermittent.* Do you recall *more, less, usual* number of dreams?

Was this your usual bedtime? *Yes No*

3. Note hunger times by line (—) and eating occasions with a slash (/) and meal code (B, L, D, S).

6 AM 9 AM 12 NOON 3 PM 6 PM 9 PM 12 MID 3 AM 6 AM

Did you seem to eat *more, less, usual* amounts? Was fluid consumption *more, less, usual* amount?

Did you urinate *more, less, usual* amount? With *more, less, usual* frequency?

4. Indicate periods of unusual physical activity and mental activity by [P] or [M].

6 AM 9 AM 12 NOON 3 PM 6 PM 9 PM 12 MID 3 AM 6 AM

5. Indicate G. I. activity times by a line (—) and bowel movements with a slash (/).

6 AM 9 AM 12 NOON 3 PM 6 PM 9 PM 12 MID 3 AM 6 AM

G.I. activity was *more, less, usual.* Bowel movement was *more, less, usual; constipated, loose, normal.*

6. Indicate happy and sad moods with [H or S] notation.

6 AM 9 AM 12 NOON 3 PM 6 PM 9 PM 12 MID 3 AM 6 AM

7. Indicate when you wanted to be alone or socialize with other people by [A or S] notation.

6 AM 9 AM 12 NOON 3 PM 6 PM 9 PM 12 MID 3 AM 6 AM

8. Indicate times when you seemed to do little but daydream or think of past events by [D or P] notation. None

6 AM 9 AM 12 NOON 3 PM 6 PM 9 PM 12 MID 3 AM 6 AM

9. Indicate times of unusual body sensations (*pain, numbness, dizziness, weakness, hot or cold feelings, sweating or nausea, cramps)* in head, chest, abdomen, limbs and specify. NONE

6 AM 9 AM 12 NOON 3 PM 6 PM 9 PM 12 MID 3 AM 6 AM

FORM G-55

Fig. 3 Front side of the behavior profile chart form which is marked either prn or a minimum of six times daily by each subject.

measured variables acquired with laboratory precision. Extensive field tests with psychiatric and medical patients, and a large scale control population are planned, supplemented by a computer scored Review of Medical Systems (ROS) [1] and the Minnesota Multiphasic Personality Inventory (MMPI).

Statistical measures of significant change have already been developed for some of the Psychophysiological Diary indices [9]. Computer detection of such changes in presumably normal individuals will be used as an "alert" or "flag" indicating the need for laboratory or clinical scrutiny. Hence, it is conceivable that the Psychophysiological Diary will find application in the area of preventive medicine in future healthcare delivery systems [10].

REFERENCES

1. Donnelly J., Rosenberg M. and Fleeson W. P. (1970): The evolution of the mental status—past and future. Amer. J. Psychiat., *126*: 121–126.
2. Holmes TH and Rahe RH (1967): The social readjustment rating scale. J. Psychosom. Res. *11*: 213–218.
3. Glueck B. C. Jr. (1965): A psychiatric observation system. *In*: 7th IBM Med. Symp. pp. 317–322, IBM Corp., Yorktown Heights, New York.
4. Glueck B. C. Jr. and Stroebel C. F. (1969): The computer and the clinical decision process: II. Amer. J. Psychiat., *125*: 2–7.
5. Rosenberg M., Glueck B. C. Jr. and Stroebel C. F. (1967): The computer and the clinical decision process. Amer. J. Psychiat., *124*: 595–599.
6. Sternbach R. A. (1966): Principles of Psychophysiology. Academic Pres, New York.
7. Stroebel C. F. (1965): Biologic rhythm approach to psychiatric treatment: an application of process control computers. *In*: 7th IBM Med. Symp., pp. 215–242, IBM Corp., Yorktown Heights, New York.
8. Stroebel C. F., Bennett W., Ericson P. and Glueck B. C. Jr. (1967): Designing computer information systems: problems and strategy. Comp. Psychiat., *8*: 491–508.
9. Stroebel C. F. and Glueck B. C. Jr. (1970): Computer derived global judgements in psychiatry. Amer. J. Psychiat., *126*: 41–50.
10. Stroebel C. F. and Glueck B. C. Jr. (1972): Computers in medicine. *In*: The Practice of Medicine, Vol. II, Chap. 61, Harper and Row, New York.
11. Stroebel C. F. (1972): Psychophysiologic pharmacology. *In*: Handbook of Psychophysiology, (N. S. Greenfield and R. Sternbach, eds.), pp. 787–838, Holt, Rinehart and Winston, New York.
12. Stroebel C. F. (1969): Circadian Behavioral Profile Chart. Institute of Living, Hartford, Conn.
13. Stroebel C. F., Luce Gay and Glueck B. C. Jr. (1971): The psychophysiological diary: a computer scored daily record of moods, body changes, and life events. Institute of Living Hartford, Conn.
14. NOTE: The Society for Psychophysiological Research was founded in 1961. Its journal, Psychophysiology, first appeared in 1964.

ESTABLISHING REFERENCE STANDARDS BY AUTORHYTHMOMETRY IN HIGH SCHOOL FOR SUBSEQUENT EVALUATION OF HEALTH STATUS*

LAWRENCE A. SCHEVING, LAWRENCE E. SCHEVING and FRANZ HALBERG

Chronobiology Laboratories, University of Minnesota
Minneapolis, Minnesota, U.S.A.
Department of Anatomy, Medical Center, University of Arkansas
Little Rock, Arkansas, U.S.A.
Brown University, Providence, Rhode Island

While presumably in satisfactory physical and mental condition one of us (LAS), 18 years of age, carried out self-measurements as he attended Ben Franklin High School in New Orleans, Louisiana—a school for high achievers. His general state of physical and mental health may be deemed satisfactory; he was a member of the school's football team, had an A-average, and suffered from no apparent illness during the measurement span and for the year thereafter.

Despite pressures attending the several weeks before graduation from High School, LAS self-measured the variables listed in the first column of Table 1 and collected his urine at unequal intervals approximately 6 times during each wakefulness span, as well as for the usually uninterrupted sleep span. The total number of observations analyzed is shown at the bottom in Table 1. The procedures and their background have been described elsewhere [1, 2].

Following graduation, LAS continued self-measurement while moving to the Chronobiology Laboratories at the University of Minnesota for several profile studies including, *inter alia,* the automatic monitoring of blood pressure under various conditions imposed upon himself.

Chemical determinations on the urines collected by LAS were carried out in the laboratory of LES and involved conventional methods.

All of the time series thus obtained were fitted by least squares with one or several cosine curves and the results of these analyses were summarized in tables and displays prepared directly on microfilm, as part of an eventual readily portable and conveniently readable health summary (eventually visualized in card form).

RESULTS FROM AUTORHYTHMOMETRY BY SELF-MEASUREMENT

For each variable investigated, the second and third columns of Table 1 provide first the lowest and highest values and then a 90% range of "usual values". The total range is desirable to identify any suspect extreme values— whatever their origin—whether they result from excess loads impinging upon the individual or from recording punching errors. The 90% range, obtained by omitting the

* Supported by grants from the U.S. Public Health Service (5-K6-GM-13, 981) Central Chapter, Connecticut Heart Association, and NASA (NGR 24-005-006).

Table 1 Rhythmometric summary.

Larry Scheving (0570). Age 18. ten days (70/05/16–25) — Chronobiologic window from 28.0 to 20.0 hours

	Data summary					Analysis summary using least squares fitted single cosine at														
						24-hour period											Best fit			
	Variable	Lowest and highest values		90% range		PR	P	Mesor ±SE		Amplitude ±SE		Acrophase ϕ					Period	PR	P	A
												DEG	±SE	CK HR	(.95 CL)					
1	Oral temp.	96.6	99.2	97.0	99.0	28	.009	97.86	.06	.46	.10	−261	10	1724	1559	1850	24.0			
2	Urine temp.	97.0	99.5	97.5	99.5	40	.009	98.44	.07	.66	.11	−264	8	1737	1630	1844	24.0			
3	Syst BP	100.	148.	106	142	08	.072	119	1.42	4.24	1.92	−277	26	1831			24.5	.13	.019	5.0
4	Diast BP	50.	76.	54	72	02	.449	62	.98	1.97	1.55	−227	31	1509			25.5	.08	.082	2.7
5	Pulse	50.	84.	52	82	25	.009	61	.97	5.43	1.30	−279	14	1837	1643	2031	24.1	.25	.009	5.4
6	Peak flow	413.	553.	438	532	20	.009	486	3.96	23.60	6.19	−243	11	1614	1447	1740	24.1	.22	.009	24.4
7	E-H skill	29.	46	29	41	40	.009	35.98	.42	4.09	.66	− 34	6	217	122	311	24.1	.41	.009	4.1
8	Urine volume	25.	176.6	32.0	116.6	02	.445	67	4.11	7.15	5.59	− 8	45	34			20.7	.09	.068	12.3
9	Urine 17KGS	.1	[illegible]4.0	.6	[illegible]4.0	32	.009	2.14	.12	.87	.17	−209	11	1359	1227	1530	24.5	.36	.009	.9
10	Urine Na	22.6	887.9	75.2	784.3	21	.009	338.61	30.66	146.52	41.60	−194	16	1259	1050	1507	24.1	.21	.009	148.4
11	Urine K	4.2	616.6	16.6	574.0	33	.009	139.68	17.38	112.91	23.39	−190	12	1243	1107	1418	24.3	.35	.009	118.6
12	Urine Cl	23.5	1164.3	61.0	1005.9	39	.009	379.66	35.67	255.29	47.13	−181	11	1204	1036	1332	24.1	.39	.009	257.9
13	Urine Ne	.0	[illegible]6.4	.0	5.7	01	.604	1.03	.23	.32	.32	−263	55	1733			21.0	.09	.072	.6
14	Urine E	.1	[illegible]3.7	.0	[illegible]2.7	03	.385	.83	.09	.18	.13	−266	40	1744			22.0	.13	.016	.3

(Series No. .N) 1 59 2 59 3 59 4 59 5 59 6 59 7 59 8 56 9 53 10 49 11 49 12 49
13 57 14 57

Start time for series 1970 516 9; documented span in hours is 225.0

PR=Percent rhythm=(sum of squares due to fitted 24-h cosine/total SS)×100.
P=Probability of zero amplitude (null hypothesis)
DEG=degrees; CK=clock; CL=confidence limits

upper and lower 5% of the original values, serves, in its turn, as an individualized reference standard, desirable if not indispensable for any subsequent evaluation of status in terms of health or disease. For instance, if at some later date clinical reasons prompt scrutiny of the subjects' blood pressure and an elevation of the latter is a matter for concern, one can for diagnostic purposes compare the two 90% ranges (as well as other parameters, see below) obtained at different points in time. As a criterion added to the index "90% range" one may wish also to compare the mesors (or means) to see whether—for a given person—the variability of blood pressure has changed in terms of its location along the conventional scale as well as in terms of its dispersion. Quite apart from rhythms, an individualized usual range and (mesor or) mean seem to be highly desirable for the positive assessment of health and as a basis for drawing objective conclusions concerning a given individual's condition.

The need for such an individualized approach has been eloquently discussed by Smirk [3] who points out the dangers of (the alternative of) ignoring individualized reference standards. Two individuals may have identical blood pressures, say 130 mmHg. However, one who earlier may have had an average of 100 mmHg may now have a nephritis, whereas another, previously (and now) at the same 130 mmHg value, is presumably in perfect health. Unless the lower earlier range of the nephritic subject is known, the individual will be labeled normotensive and may suffer the consequences of this error.

As noted, the rhythmometric summary in Table 1 provides ranges of usual values not only for systolic and diastolic blood pressure but also for pulse, peak expiratory flow, eye-hand skill (measured by the time needed to insert thirty beads into a tube), oral temperature and a number of urinary variables. The foregoing considerations apply not only to blood pressures but to these other variables as well, in that an individualized range of usual values promises to be diagnostically useful. Indeed, if physiologic function is variable rather than constant, the extent of this variability must be known before using the function for assessing health.

By the same token it is important to quantify any predictable part of this variability, such as any prominent rhythms. In other words, if some of the variability is rhythmic, then a range of values does not appropriately describe the functional dynamics; one then searches (in addition to defining an overall range) for characteristics that describe variability in health and how it may change when a risk develops or when a disease process sets in. Rhythm parameters, also given in Table 1, are such characteristics. Although those here provided are not to be regarded as definitive measures, they represent fledgling approximations of characteristics that will be more meaningful when more ample data allow a more rigorous parameter estimation, with an assessment of the waveform as well.

In the table, we can first note a P value, which indicates the extent to which a statistically significant rhythm can be described by the fit of a 24-hour cosine curve. With these considerations in mind, we find that 28 or even 40% of the overall variability in oral and urinary temperature can be accounted for by the fit of a 24-hour cosine curve, a point seen from the values found in the column PR (or percent rhythm) under the results from a 24-hour cosine fit. From a technical viewpoint, it is pertinent that the urinary temperature is more rapidly measured than the oral one; it has the added advantage of exhibiting the more promient circadian rhythm, as demonstrated earlier by Simpson [4].

A statistically highly significant rhythm also was detected by the fit of a 24-hour cosine curve for heart rate, peak expiratory flow, eye-hand skill, urinary 17-ketogenic steroids, sodium, potassium and chloride. Blood pressures, however, were better approximated by a 24.5-hour cosine curve, as will be discussed below. A circadian rhythm could be described with the fit of a 22-hour cosine curve for urinary epinephrine below the 2% level and for urinary volume and norepinephrine only at the 10% level of statistical significance.

The acrophases of the more prominent electrolyte rhythms were narrowly synchronized, differing only by a few degrees (Table 1). The 17-ketogenic steroid and other urinary electrolyte rhythms led in acrophase the rhythms in peak expiratory flow, oral and urinary temperature and pulse (Table 1).

Scrutiny of Table 1 thus shows that as a rule a good fit was found for many variables at a period rather close to 24.0 hours, if not precisely at 24.0 hours. It is re-emphasized that systolic blood pressure, with a period of 24.5 hours (P=.019), could not be described by a precise 24-hour cosine curve. In the case of systolic pressure the 24.5-hour cosine curve accounted for about 30% of the total variability, as did a best-fitting period of 22 hours for the urinary epinephrine. All of the foregoing results were obtained by autorhythmometry in New Orleans while the subject followed his regular high school routine.

Fig. 3 compares the vital sign acrophases described in Table 1 (T=225 hrs) with those determined in a previous study done over an 833-hr span. The 833-hr sampling was compatible with the rigorous schedule of a high school student. It consisted of sampling once daily upon arising (0600) on Sunday through Friday; on Saturday sampling was done more often, yet during waking hours only (0800, 1100, 1700, and 2000). The study began 3/3/70 and ended 4/28/70. Fig. 3 shows the reproducibility of the timing from one study to the next and also suggests that the less frequent sampling was acceptable in this case for both rhythm detection and acrophase estimation. In the first (longer) study the systolic blood pressures were well fitted by a 24-hour cosine curve whereas in the later (225-hour) study systolic blood pressure data were best approximated by a 24.5-hour cosine curve.

The failure to describe changes in norepinephrine and epinephrine by the fit of a 24-hour cosine curve may be related to the fact that sampling intervals for these variables were infrequent, by virtue of an uninterrupted sleep span of ~8 hours.

RESULTS FROM AUTOMATIC MEASUREMENT

Figs. 1 and 2 present some of the data obtained in Minnesota in a follow-up study of systolic and diastolic blood pressure carried out with an automatic blood pressure monitor; a best-fitting cosine curve also is shown at the top of these figures.

The remarkable variability of systolic and diastolic blood pressures in one and the same healthy individual stands out clearly. It seems important to emphasize this variability, since it is usually ignored in statements regarding the time course of *averages* of blood pressure in a population. To cite but one example, a Survey of the National Center for Health Statistics [5] reports that *on the average*, human blood pressure varies during the day by no more than a very few—3 or 4—milli-

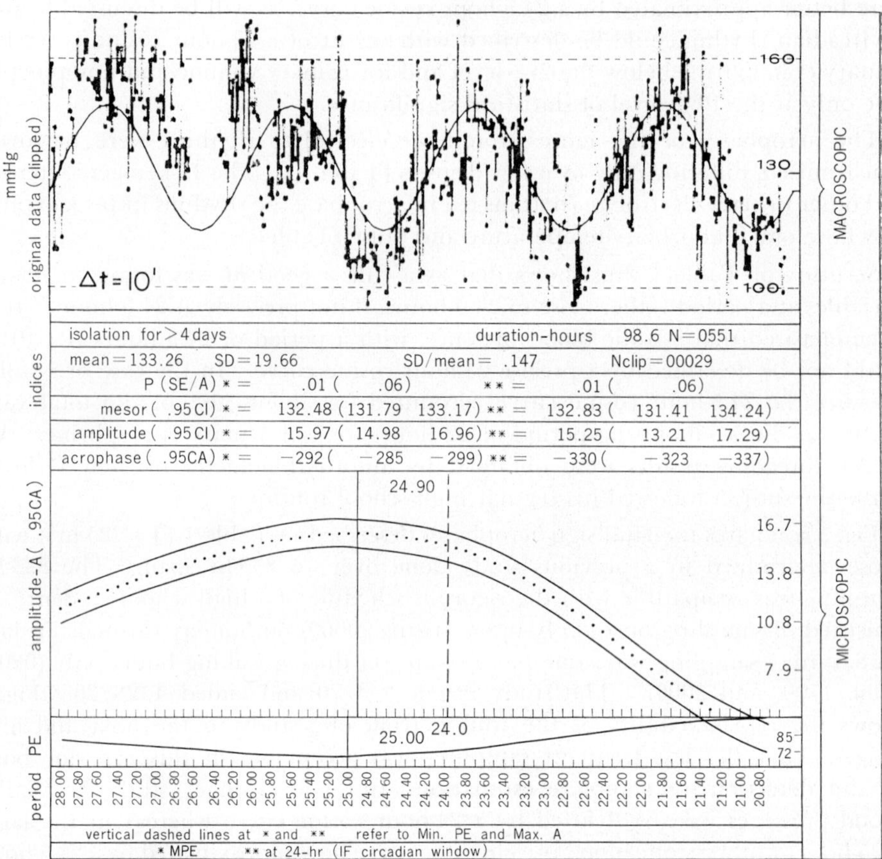

Fig. 1 Systolic blood pressure of healthy young man (LAS) from non-invasive monitor (Type 15100–A kindly provided by Mr. W. Gruen, Ambulatory Monitoring, Ordsley, N.Y., USA).

meters Hg. Indeed this is true on the average, yet for any given subject the behavior of averages not only is out of context but is grossly misleading. The *individual's* blood pressure varies incomparably more than those who rely on averages care to discuss, with but few notable exceptions [6]. Fig. 1 shows occasional systolic readings considerably below 100 mmHg and well above 160 mmHg. Fig. 2 shows diastolic measurements below 60 mmHg as well as above 95 mmHg—in a physically and mentally fit subject during a vacation.

The lower halves of Figs. 1 and 2 show a so-called chronobiologic window prepared in the spectral region between 28.0 and 20.6 hours—with largest amplitude at 24.9 hours for systolic and at 25.1 hours for diastolic pressures. However, as compared to the corresponding values at a trial period of 24.0 hours, the amplitude at a trial period of 25.0 hours is not much higher and the unaccounted (residual) variance or error is not greatly reduced. The span covered by the data is too limited to discuss the question of synchronization or desynchronization of the circadian rhythms in diastolic and systolic blood pressure. However, the

DIASTOLIC BLOOD PRESSURE OF
HEALTHY YOUNG MAN (LAS)
FROM NON-INVASIVE MONITOR (TYPE 15100-A)

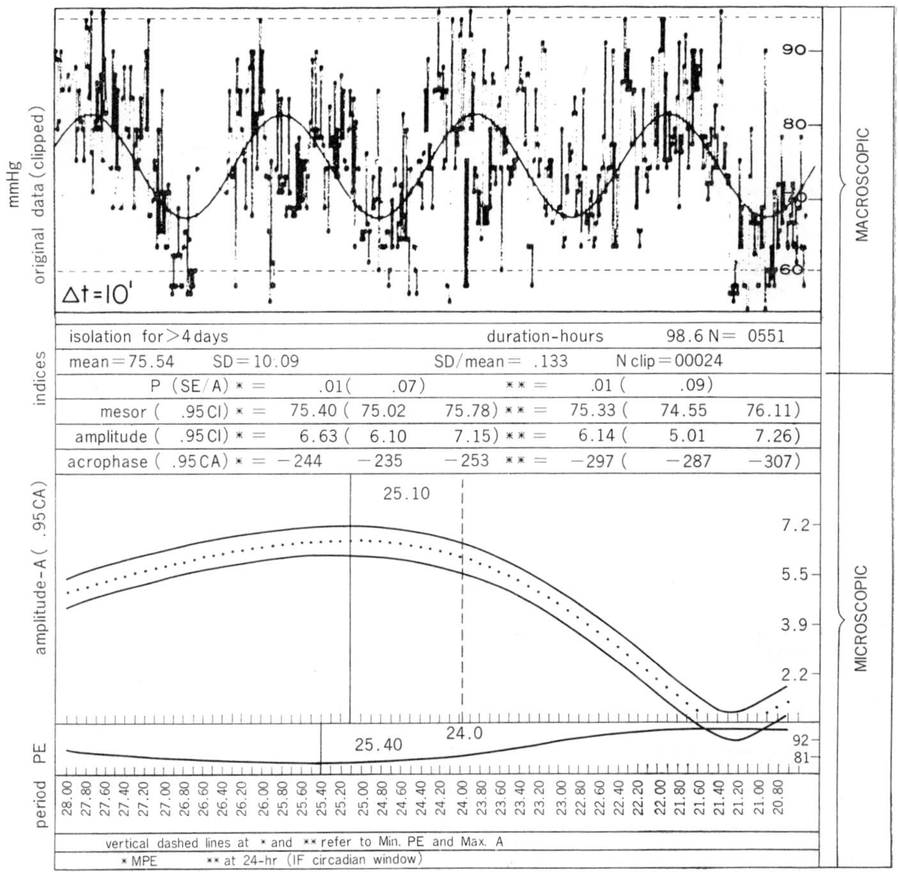

Fig. 2 Diastolic blood pressure of healthy young man (LAS) from non-invasive monitor (Type 15100–A kindly provided by Mr. W. Gruen, Ambulatory Monitoring, Orsley, N.Y., USA.).

analyses certainly suffice to describe a circadian rhythm in the face of very great variability and to estimate its parameters, amplitude, acrophase and mesor. From a physiologic viewpoint, the analyses at a trial period of 24 hours reveal a small but statistically significant acrophase difference between human systolic and diastolic blood pressures, which has already been noted in another case of a presumably healthy, self-measuring woman [7].

The results here presented, as suggested earlier [1, 2], will be valuable as reference standards for any subsequent evaluation of health status.

SUMMARY AND CONCLUSION

Autorhythmometry represents a tool *par excellence* not only for the physiologist interested in human variability but also for the family physician. With self-measured or automatically collected data available, he need not be overburdened by endeavors to improve his information base and thus the quality of care.

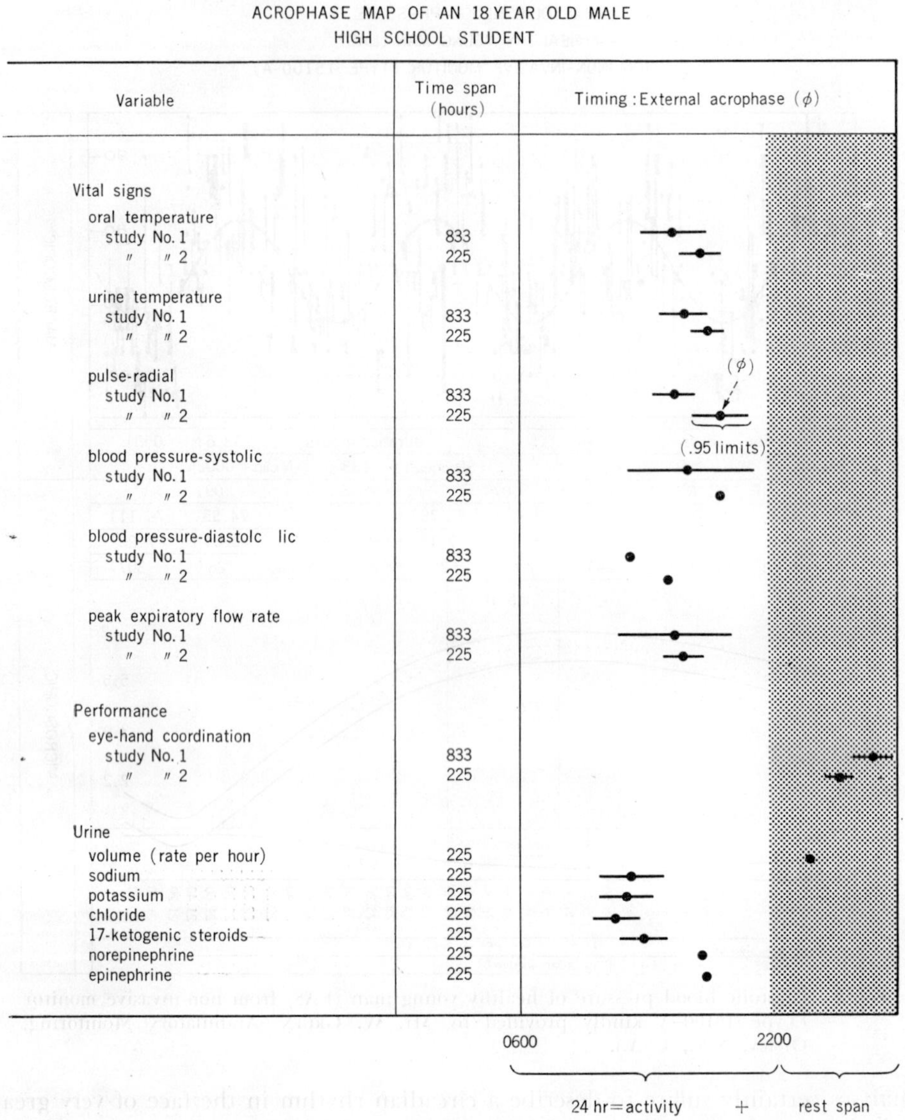

ACROPHASE MAP OF AN 18 YEAR OLD MALE
HIGH SCHOOL STUDENT

Variable	Time span (hours)	Timing : External acrophase (ϕ)
Vital signs		
oral temperature		
study No. 1	833	
〃 〃 2	225	
urine temperature		
study No. 1	833	
〃 〃 2	225	
pulse-radial		(ϕ)
study No. 1	833	
〃 〃 2	225	(.95 limits)
blood pressure-systolic		
study No. 1	833	
〃 〃 2	225	
blood pressure-diastolc lic		
study No. 1	833	
〃 〃 2	225	
peak expiratory flow rate		
study No. 1	833	
〃 〃 2	225	
Performance		
eye-hand coordination		
study No. 1	833	
〃 〃 2	225	
Urine		
volume (rate per hour)	225	
sodium	225	
potassium	225	
chloride	225	
17-ketogenic steroids	225	
norepinephrine	225	
epinephrine	225	

0600 2200

24 hr = activity + rest span

Fig. 3

By suggesting self-measurement he can cease to depend upon a single time-unqualified and hence unrepresentative measurement of variables such as blood pressure, heart rate or body temperature and other functions, as documented herein. When autorhythmometry is too time-consuming, or when work cannot be carried out sufficiently frequently by self-measurement, automatic measurements are in order and can provide information relatively rapidly and with a frequency that is not readily obtainable by self-measurement.

In private and hospital practice alike the interpretation of single casually or "basally" recorded vital signs in relation to a crude population "normal range" can be complemented by the scrutiny of more pertinent kinds of individualized information. Thus, the abnormality or normalty of new time series or even of a single value may be assessed in relation to temporally-qualified reference standards appropriate to an individual as well as to his cohort.

REFERENCES

1. Halberg F.: Education, biologic rhythms and the computer. *In*: Engineering, Computers and the Future of Man. Proc. Conf. on Science and the International Man: The Computer. Chania, Crete, June 29–July 3, 1970. International Science Foundation, Paris, (in press).
2. Halberg F., Johnson E. A., Nelson W., Runge W. and Sothern R. (1972): Autorhythmometry—procedures for chronophysiologic self-measurements. Physiology Teacher, *1*: 1–11.
3. Smirk F. H. (1944): Casual and basal blood pressures. IV. Their relationship to the supplemental pressure with a note on statistical implications. Brit. Heart J., *6*: 176–182.
4. Simpson H. W. (1970): Urine temperature measurement in human circadian rhythm studies. J. Physiol., *212*: 29–30P.
5. National Center for Health Statistics, Division of Health Examination Statistics. "Blood pressure of adults by race and area, United States 1960–1962." U. S. Dept. of Health, Education and Welfare. U. S. Govt. Printing Office, Washington, D. C. 1964.
6. Pickering G. (1970): Hypertension. Causes, Consequences and Management. J. & A. Churchill, London.
7. Halberg F., Halberg E. and Montalbetti N. (1969): Premesse e sviluppi della cronofarmacologia. Quad. med. quant. sperimentazione clin. controllata, *7*: 5–34.

AUTORHYTHMOMETRY, ANXIETY AND THE DELABELING PROCESS*

Fritz REEKER
Chronobiology Laboratories, Department of Pathology
University of Minnesota
Minneapolis, Minnesota, U.S.A.

INTRODUCTION

While it is common for any individual to worry about his health, the alleviation of such fears—whether they be reasonable or not—is generally difficult and therefore far less common. When the self-measurement of body functions analyzed by computer methods (autorhythmometry), briefly AR, is being introduced at an early age [1, 2], the dangers of creating hypochondriacs and, in particular, those of aggravating concern about an imagined as well as real stigma, loom large. Does one compound an already complex situation with AR by a child stamped with any mark of inferiority early in life; will it become difficult if not impossible to instill a sense of genuine self-worth into that individual later in life? More specifically, one who is for one reason or another labeled as having a particular health defect or at least to be at risk is likely to live with his anxiety until such fear is recognized and dealt with. In this paper it will be shown how AR first aggravated, then relieved a pre-existing anxiety concerning blood pressure.

CASE REPORT

The subject (FR) first appeared upon the scientific scene in 1963 when a Behavior Day record kept by his parents was analyzed and published, Fig. 1 [3–6] and the development of a circadian sleep-wakefulness rhythm subsequently gauged by computation of the Circadian Quotient (CQ), Table 1 [6]. Now 20 years of age and a student at Carleton College, Minnesota, he is maintaining slightly less than a B+ average scholastically and is a member of the school tennis team. Among his major interests are chess, athletics and writing (prose and poetry).

When FR was first offered the opportunity to study circadian rhythms in blood pressure through self-measurement, he was hardly enthusiastic about the task. Initially, he declined to take part in that facet of the study despite his knowledge that the measurement of blood pressure was neither complex nor painful. After 24 hours FR decided to take a blood pressure reading despite his continuing discomfort and was perturbed when the systolic reading exceeded 140 mmHg. He then hesitated a few moments and pumped the cuff up again—only to find the pressure surpassing the 150 mmHg point on the instrument. At that moment, FR immediately removed the cuff and spent several agonizing days in silence

* Supported by the U. S. Public Health Service (5-K6-GM-13,981) and NASA (NGR-24-005-006).

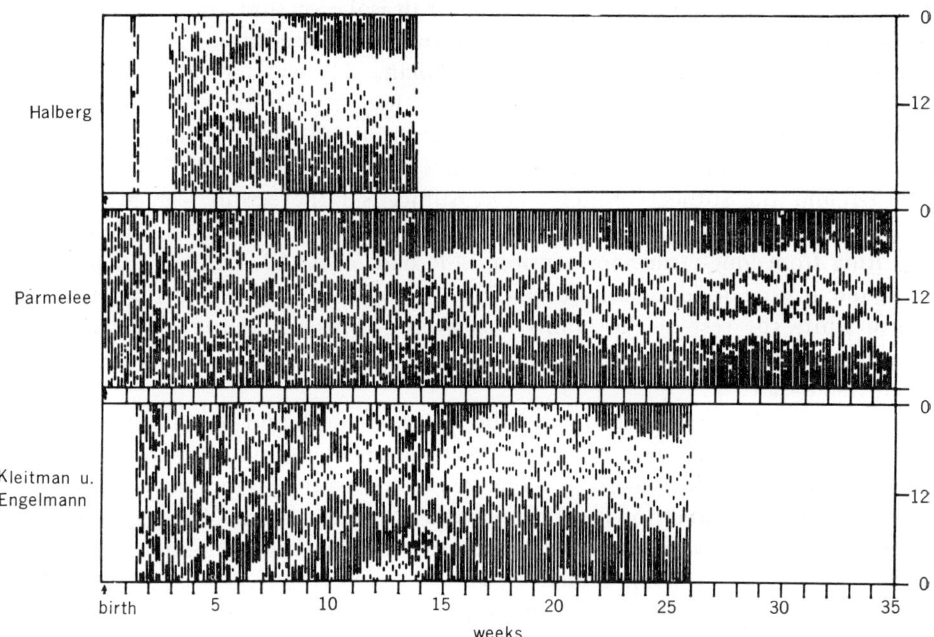

Fig. 1 Behavior-Day Charts of healthy infants raised on "self-demand", visualizing problems encountered in quantifying certain rhythms (3) and near-rhythms and leading to the development of a circadian quotient (4–6), cf. Table 1.

Table 1 Circadian quotient (CQ) gauging development of circadian sleep-wakefulness rhythm in a healthy boy.*

Week of life	Observation period** 1951 (Mo./Day)	Variance	CQ (%)	Interval of CQ confidence	
				.05 limit	.95 limit
4th	3/21 to 3/28	3.75	9	6	13
5th	3/28 to 4/4	3.75	9	6	12.6
6th	4/4 to 4/11	3.77	13	9	18
7th	4/11 to 4/18	3.73	21	14	30
8th	4/18 to 4/25	3.72	27	18	38
9th	4/25 to 5/2	3.74	45	30	62
10th	5/2 to 5/9	3.78	48	32	66
11th	5/9 to 5/16	3.88	58	38	80
12th	5/16 to 5/23	3.88	50	33	70
13th	5/23 to 5/30	3.88	58	38	80
14th	5/30 to 6/6	3.80	58	38	82

* Obtained from spectal analysis of variance according to Halberg, F. and Panofsky, H.: I.: Thermo-variance spectra; method and clinical illustrations. Exp. Med. & Surg. *19*:284–309, 1961; in such spectral analysis m determines the resolution. m=96; Δt (interval between observations)=.25 hours.

** Male infant born February 28, 1951, at 01:55; N=672/observation period, except for 14th week, with N=624.

before reporting his results to an adviser (FH) at the University of Minnesota, who suggested that the subject place himself in isolation and allow a non-invasive automatic blood pressure monitor to determine the readings. For several days the subject was unwilling to agree to this plan. He was virtually certain there was something wrong with him but was in no way eager to have his fears verified.

During these initial stages of non-cooperation, the reason for the subject's anxiety was becoming more clear to several people, including FH, the subject's mother and the subject himself. The subject's father, who seemed a strong, nearly indestructible individual, had died five years earlier of complications resulting from long-term hypertension. FR had found his father, who was forty years old at time of death, on the sofa in the family living room during a thunderstorm while the rest of the family was away. This traumatic experience now manifested itself in sub-consious fear that the subject himself might have developed the disease.

FR was gently persuaded to enter the isolation room and to subject himself to automatic blood pressure recordings; the readings of the monitor were inconclusive and somewhat disconcerting. The systolic pressure was extremely labile with some pressures around 155 mmHg and others much lower. The diastolic pressure, while more stable, averaged around 90 mmHg—higher than would be expected for the subject's age group.

These readings warranted a complete conventional physical and laboratory examination. This work, carried out by Dr. Erhard Haus, Chairman, Division of Laboratories, Ramsey County Hospital, St. Paul, Minnesota, yielded negative results, in all respects. The examination thus suggested that the high readings were probably "psychosomatic". FR then went again into an isolation room and subjected himself to further automatic monitoring of his pressure. As before, readings were taken every ten minutes. Initial readings were as high as those which had entailed the physical examination in the first place. But this time, a gradual reduction in pressure was recorded as the subject slowly relaxed with the knowledge that he had no demonstrable reasons for worry about his health (Table 2).

Table 2 Statistical analysis. Comparison of initial fifty arterial blood pressure self-measurements by a subject suffering anxiety with fifty directly subsequent self-measurements, as subject "delabeled" himself.

Blood pressure	Anxiety		Delabelling	
	Mean	Standard deviation	Mean	Standard deviation
Systolic	135.5	7.96	127.06	5.99
Diastolic	85	4.44	79.5	4.25

"t" test comparing these two sets of blood pressure indicates a difference at the .01 level of significance.

When the subject resumed AR, his readings began near the original high values; they continued in the vicinity of 130/85 mmHg for several weeks, before the subject was completely adjusted to AR. He thereafter recorded pressures consistently in the neighborhood of 120/80 mmHg and below—Table 3.

Table 3

Chronobiology Laboratories – University of Minnesota — Minneapolis Minnesota 55455 USA 612–373–2920

Rhythmometric Summary

F Reeker self measurement study/701031–710408 — Chronobiologic window from 28.0 to 20.0 hours

	Data summary						Analysis summary using least squares fit of single cosine at											
							24–hour period							Best fit				
	Variable	Lowest and highest values		90% range		Coef. var.	P	PR	Mesor	±SE	Amplitude	±SE	Acrophase ϕ Degrees	Period	P	PR	A	ϕ
1	Sys BP	104	150	111	136	7.4	.810	0	122.19	.55	.48	.83	−23	24.1	.009	7	2.9	.210
2	Dias BP	65	103	75	94	7.0	.128	1	83.29	.41	1.38	.68	−50	23.9	.009	8	2.5	.204

Series No. 1 to 2 has 314 time PTS

Start time for series 1970 10 31 8; documented span in hours is 3818

PR=Percent rhythm=Percent of total variability contributed by fitted curve =(Sum of squared deviations from mean (SS) of values derived from fitted cosine curve at sampling times/SS of data themselves×100.

P=Result of testing zero amplitude=Probability of obtaining estimated parameters if sinusoidal rhythm with period were absent.

The delabeling of this subject by AR, while unpleasant for him at the given time, was well worth the trouble, apart from any merit to the results of AR in themselves. His subconscious fear of hypertension eliminated, the subject began to take an interest in factors underlying the variability of blood pressure and is now seriously considering and working toward a career related to medicine.

In this instance, one physician successfully dealt with a problem which could have caused anxiety for the subject throughout his life, had self-measurement not been gradually implemented to calm the fears brought about by earlier trauma-related experience. It is important to note that if a consistent blood pressure elevation of any degree had been detected in this subject, current conventional medicine is prepared to control this condition and can do so with relatively safe medication.

Just as a hospital room with its doctors, nurses and batteries of tests may throw a patient into depression, the very act of measuring such a physiological function as blood pressure is capable of bringing about anxiety far beyond the nuisance associated with the measuring procedure. A hospital stay is sometimes useful and necessary for the swift diagnosis of a disease. By the same token, autorhythmometry may be vital to the quantification of health, as well as the recognition of risk or disease. AR is the far cheaper approach and it is on occasion more pertinent, since it covers by more extensive data a number of real life situations.

REFERENCES

1. Halberg F.: Education, biologic rhythms and the computer. *In*: Engineering, Computers and the Future of Man. Proceedings of Conference on Science and the International Man: The Computer, Chania, Grete, June 29–July 3, 1970. International Science Foundation, Paris, (in press).
2. Halberg F., Johnson E. A., Nelson W., Runge W. and Sothern R. (1972): Autorhythmometry—procedures for physiologic self-measurements and their analysis. Physiology Teacher, *1*: 1–11.
3. Halberg F. (1969): Chronobiology. Annual Review of Physiology, *31*: 675–725.
4. Halberg F. (1963): Periodicity analysis—a potential tool for biometerologists. Int. J. Biometerology, *7*: 167–191.
5. Halberg F. (1964): Stein M., Diffley M., Panofsky H. and Adkins G. (1964): Computer techniques in the study of biologic rhythms. Annals N.Y. Acad. Sci., *115*: 695–720, July.
6. Reeker E. and Diffley M. (1963): Development of circadian sleep-wakefulness rhythm in a healthy boy gauged by circadian quotients. Minnesota Academy of Sciences Proceedings, *31*: 63–64.

AUTORHYTHMOMETRY WITH SERIAL SECTIONS IN HEALTH AND IN THE SEARCH FOR CHRONOPATHLOGY*

RONALD SHIOTSUKA and FRANZ HALBERG

Chronobiology Laboratories, University of Minnesota Minneapolis, Minnesota, U.S.A.

INTRODUCTION

Histologists and pathologists started to focus upon tissues in health and in disease as soon as light microscopic techniques became available. Attempts to define and standardize a time structure in man depended in turn upon the development of chronobiologic techniques, greatly facilitated by the availability of the electronic computer. Human variables have thus been explored for rhythms by MENZEL [1, 2] and HALBERG [3] among others, quite often on single cases studied longitudinally in time as well as in transverse group approaches. Without extrapolating beyond the scope of three subjects, rhythmometric analysis [4] will here be applied to data from a healthy subject and two patients. In the healthy subject, 14 variables will be described by the least squares fit of a cosine curve having a period of 24 hours. The time course of some of these variables will be further examined by chronobiologic serial sections, a technique also applied to results on blood pressure self-measured by a patient with so-called "labile hypertension", as well as on rectal temperature automatically recorded from a patient with hypernephroma.

PRESUMABLY HEALTHY MALE, 24 YEARS OF AGE (RS)

A rhythmometric summary [5] of self measurements by RS in Table 1 shows most functions to be rhythmic and 24-h synchronized. The variability accounted for by the fit of a 24-h cosine curve ranged from 60% in the case of urine temperature to as little as 4% in the case of pulse. Complementary chronobiologic serial sections (SS) for systolic blood pressure, diastolic blood pressure and pressure amplitude (the difference between systolic and diastolic pressures at a given measurement) are given in Fig. 1, and for oral temperature in Fig. 2.

A serial section analysis summarizes data by fitting a 24-h cosine curve to 480 hours of data (interval) displaced longitudinally in increments of 24 hours along the entire time series. In these serial sections the top row (Raw Data) represents the actual data points collected in time. In Fig. 1, such values for systolic blood pressure range from about 107 to 155 mmHg. The second row shows the good-

* Supported by U.S. Public Health Service (5-K6-GM-13,981) Central Chapter, Connecticut Heart Association, and NASA (NGR-34-005-006). Dr. Werner MENZEL, Professor of Medicine, University of Hamburg, Hamburg, West Germany, provided the original clinical observations and collected the data on patients here analyzed, among many other invaluable contributions to medical chronobiology [8].

Table 1 Autorhythmometric summary of RS, presumably healthy male, 24 years of age who followed his usual daily activity pattern with about 7 hours of sleep per day during this period; tests are described in detail in the Physiology Teacher [5].

Data summary							Analysis summary—Least squares fit of a single cosine with a period of 24 hours*							
Variables	N	Lowest and highest values		90% range		Coef var	PR	Mesor M±SE		Amplitude A±SE		Acrophase [.95CA] (clock hr)		
1 Oral Temperature	391	96.1	98.8	97.0	98.5	.5	48	97.7	.01	.5	.02	1727	[1705	1749]
2 Urine Temperature	275	96.5	99.5	97.0	99.0	.6	60	98.0	.02	.7	.03	1714	[1654	1735]
3 Pulse per Minute	400	52	86	56	80	10.5	04	66.2	.04	2.5	.6	1652	[1526	1818]
4 Systolic Blood Pressure	400	108	152	118	146	6.1	21	129.6	.4	6.1	.6	1644	[1508	1719]
5 Diastolic Blood Pressure	400	60	94	68	86	7.4	05	75.8	.3	2.3	.5	1552	[1443	1701]
6 Blood Pressure Amplitude	400	28	80	42	65	13.9	11	52.9	.4	3.9	.6	1717	[1519	1815]
7 Peak Expiratory Flow	275	4.5	8.1	5.1	7.6	11.0	08	6.6	.04	.4	.07	1521	[1419	1522]
8 Dynamometer-Right Hand	395	73	131	97	124	7.7	27	111.0	.4	6.4	.6	1830	[1749	1911]
9 Dynamometer-Left Hand	395	58	133	93	121	8.6	19	106.5	.5	6.3	.7	1733	[1649	1817]
10 Bead Intubation Speed	396	33.3	97.2	43.6	73.0	15.6	21	56.6	.4	6.6	.7	1640	[1624	1656]
11 Finger Counting Speed	396	9.7	16.1	10.5	14.0	9.5	21	12.2	.1	1.0	.1	1536	[1524	1548]
12 Adding Speed	392	31.5	58·8	35.5	49.5	10.5	24	42.2	.2	3.8	.3	1556	[1544	1608]
13 One-minute Estimation	400	44	88	51	71	10.3	08	61.0	.3	3.1	.5	1608	[1540	1636]
14 Vigor	400	1	5	2	5	22.5	16	4.0	.04	.6	.1	1658	[1615	1741]

Start time for series: 1970 09 22 2000

Documented span in hours: 2076.5

PR=Percent rhythm=Percent of total variability contributed by fitted curve=(sum of squared deviations from mean, SS, of values derived from fitted cosine curve at sampling times/SS of data themselves)×100.

P=Result of testing zero amplitude=Probability of obtaining estimated parameters if sinusoidal rhythm with stated period were absent.

*P values for all variables≤.009

ness of fit of the cosine curve to the data, with the dashed line representing a P value of 0.05. The third row shows changes in time for two additional parameters of serial section analysis. The mesor (M) and M minus one standard error (SE), is represented by the lower line and its corresponding set of dots. The difference between M and the next higher line is the amplitude (A), and the dots above the upper line represent A plus 1SE. The computative acrophase (ϕ) shown in the fourth row uses local 0000 as zero reference for the timing of the fitted curve, with the span between the two dotted lines equal to 360° or 24 hours. The last row indicates the number of data points recorded in the interval (480 hours) studied.

Thus the ϕ in the SS of systolic and diastolic blood pressure, pressure amplitude and oral temperature in Figs. 1 and 2 for the healthy male subject (RS) reveal that circadian rhythms in these variables are well synchronized to the 24-h social routine. Changes in mesor, more or less mirrored by similar changes in the amplitude may be seen in blood pressure serial sections. A circadian rhythm is not demonstrated at the 5% level in parts of the data section for this subject.

HANNAH C., PATIENT WITH "LABILE HYPERTENSION", 72 YEARS OF AGE

Personal history: Born 27/6/1899, subject had many of the childhood diseases and did not learn to walk before 2 years of age. At 15 years of age, she had gastritis with jaundice which bothered her for two years. She remembers her menarche at 15 years of age. With her first pregnancy in 1920, she was hospitalized three months for kidney disease. With her second pregnancy in 1927 she was hospitalized with a kidney disease for five months and gained weight. With the third pregnancy in 1934 she had thrombosis, lung embolism, uremia, cerebral hemorrhage, paralysis of the tongue, disturbances of vision and from this time on had a record of high blood pressure. Her pregnancy in 1934 was interrupted as were pregnancies in 1935 and 1937; a spontaneous abortion occurred in 1936. Yet another pregnancy was interrupted in 1938; she was then sterilized. Since that time she has been treated for hypertension and a host of other diseases. In 1967 she was hospitalized for hypertension, obesity and pancreatopathy. Her hypertension remains complicated by adiposity, varicosis, arthrosis and recurring pancreatitis after cholecystectomy. Against this background, notably because of her labile hypertension, it seemed of great interest to analyze her self-measured blood pressure by serial sections. Again intervals of 480 hours and increments of 24 hours were used in the analysis.

A circadian rhythm is well described during most but not all of the observation span for systolic pressure (Fig. 1). Lack of statistical significance occurs for short durations rather well documented by data. All in all, one can see that the mesor and amplitude undergo marked changes. The circadian amplitude of the systolic pressure more or less reflect the changes in mesor. The acrophase occurs during the late evening hours consistently before midnight. Evidence for desynchronization is not found for systolic pressure but is suggested for diastolic pressure as can be seen in the adjacent serial section.

Conclusions for diastolic pressure, however, are qualified since a statistically significant rhythm description is feasible only during the last fourth of the record. As noted in discussing systolic pressure, changes of mesor are also seen for diastolic

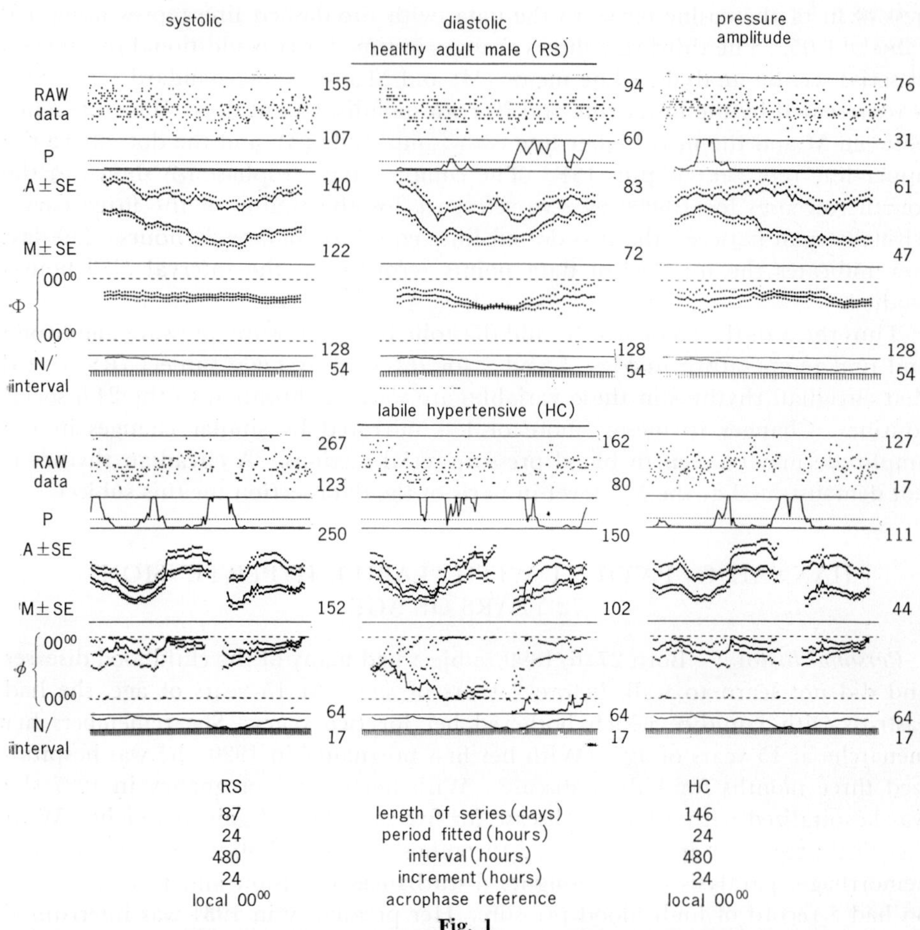

Fig. 1

Top row: serial sections of blood pressure from a healthy subject (RS).
Bottom row: serial sections of blood pressure from a patient with labile mesorhypertension (HC).
Disregard sections with P values above dashed horizontal 5% line in P row.

pressure. These changes in mesor provide the label "labile" for such patients. In contrast to the 24-h synchronization shown by the relatively stable acrophase for systolic blood pressure, diastolic pressure shows a gradual acrophase drift during the first third of the time series where the acrophase occurs earlier each day which is indeed compatible with a 23.6-h period found in an analysis of the total time series. If indeed a desynchronization occurred as suggested along the lines discussed above, the question may be raised as to whether subjects with labile mesorhypertension should be resynchronized. If resynchronization were possible, it may be of further interest to see whether it would be to the subject's advantage. Quite clearly the patient had decreasing mesors while the blood pressure rhythm apparently desynchronized from the 24-h social schedule with a 23.6-h period. Her blood pressure in turn gradually rose while she was resynchronized thereafter for reasons as yet unknown. This situation is the more remarkable since similarly in an unpublished case, a gradual decrease in mesors occurred while the blood pressure rhythm was apparently free-running, a stage followed by spontaneous resynchronization and an increase in mesor. The implication of a free-running rhythm in diastolic blood pressure is that antihypertensive therapy of such patients timed by rhythms cannot be done according to a fixed hour of the day [6, 7] since the rhythm's placement differs each day.

The blood presure amplitude is significant through most of the observation span analyzed. Mesor and amplitude undergo changes reminiscent of the changes found in systolic and diastolic blood pressure. The acrophase seems to vary within a very small time span during the few hours preceding midnight.

In treating the disease these rhythms should be taken into account for two reasons. First, one should not attribute to a drug what happens naturally as a consequence of characteristic rhythmic variations. Secondly, the changes here recorded in Fig. 1, if predictable may serve to prevent an adverse situation for the patient by properly timed additional therapy.

ELLEN R., A PATIENT WITH HYPERNEPHROMA (Fig. 2)

Born 17/2/1907; no pertinent early history; nephrectomy in June, 1970 for a hypernephroma. Thereafter subject received "telecobalt irradiation"; since that time she is depressed and weak with occasional fever and tachycardia. On x-ray, multiple round foci are found in the lung. A continuous record of her rectal temperature was analyzed by fitting cosine curves by the least squares technique to data points read off the continuous record at 10-minute intervals. For this data transfer, the temperature values were estimated to the nearest tenth of a

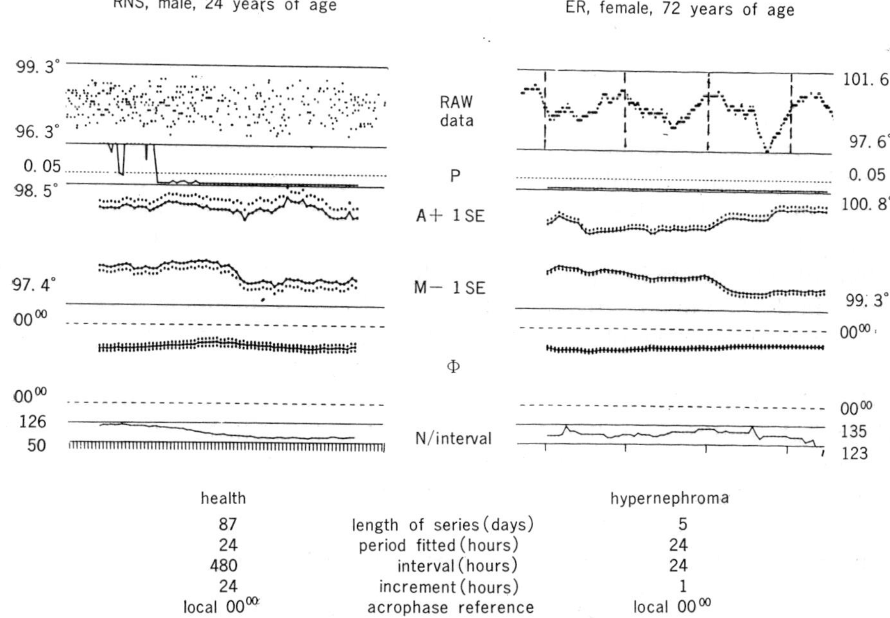

Fig. 2. Serial sections of oral temperature from a healthy subject RNS (left) and rectal temperature from a continuous recording by ER, a patient with hypernephroma, taken off at 10-minute intervals for analysis. Note different length of the two series. Vertical shortmarks indicate days of recording, many on the left, few on the right.

degree Celsius and later converted to Fahrenheit for purposes of comparison. The patient was not ambulatory during the entire measurement period.

A circadian rhythm was grossly evident by looking at the original data recorded on graph paper. Its presence was confirmed by the computer fitting of a 24-h period yielding a mesor of 99.6±.01°F, an amplitude of .77±.03°F and an acrophase of −281° (−278°, −285°) [since 360°≡24 hours, 15°=1 hour]. Reappraisal of the original data with the time lines drawn as shown in Fig. 2 reveals a clear right-ward shift of the temperature peak from day to day. By fitting cosine curves with periods ranging from 28.0 to 20.0 hours, the best fitting period was found to be at 24.7 hours, with a mesor of 99.6±.01°F, an amplitude of .18±.02°F and a computative acrophase of −253° (−250°, −256°), thus confirming a small but consistent drift in the acrophase. The fit accounts for approximately 67.4% of the overall variability.

Werner MENZEL (from inspection by the naked eye) believed to detect the occurrence of periods about 2 hours in length. Thus in addition to the circadian rhythm, it was of interest to search for any ultradian rhythms. The microscopic technique used did not corroborate this impression. When periods of 4 hours or less, decreasing by a tenth of an hour or lesser increments were fitted, the only period with a "p" value less than .05 (.03) was found to be 3.43 hours, its amplitude being .06±.02 accounting for approximately 1.7% of the variance.

Thus, the patient's temperature data show a very prominent circadian and minor ultradian rhythm. The extent to which the non-circadian components are separate entities of the circadian waveform cannot be discussed without further study. The acrophase drift may possibly be a consequence of the non-ambulatory condition of the patient or a result of her hypernephroma.

SUMMARY

Serial sections allow one to scrutinize a time course of rhythm characteristics during health as well as in certain diseases. The extreme variability of diastolic pressures from 80 to 160 mmHg and of systolic pressures from 125 to 260 mmHg as seen in a patient with labile hypertension may thus be scrutinized for changes not only in mesor but also in other rhythm characteristics as a function of time. Evidence suggestive of a desynchronization (from a 24-h social schedule) of diastolic blood pressure in a patient with labile hypertension can here be reported to occur in the face of a relatively synchronized circadian rhythm in systolic blood pressure and pressure amplitude. A small but consistent daily shift can be observed in the temperature data from the patient with hypernephroma. A causal relationship cannot be drawn between such rhythm alterations and a given pathological condition until a larger population is studied. Nevertheless, it seems reasonable that therapy should take these changes in rhythm characteristics into account. Moreover, from an etiologic viewpoint, the tools now available allow us to search for parallel changes in possible factors underlying such variability and thus lead to new and more pertinent modes of treatment.

REFERENCES

1. MENZEL W. (1952): Über den heutigen Stand der Rhythmenlehre in Bezug auf die Medizin. Z. f. Altersforschung, *6*: 26, 104.

2. MENZEL, W. (1962): Periodicity in urinary excretion in healthy and nephropathic persons. Ann. N. Y. Aca. Sci., *98*: 1107–1117.
3. HALBERG F. (1969): Chronobiology. Ann. Rev. Physiol., *31*: 675–725.
4. HALBERG F., TONG Y. L. and JOHNSON E. A. (1967): Circadian system phase—an aspect of temporal morphology; procedures and illustrative examples. *In*: The Cellular Aspects of Biorhythms, (H. von MAYERSBACH, ed.), pp. 20–48, Springer-Verlag, Berlin.
5. HALBERG F., JOHNSON A., NELSON W. RUNGE W. and SOTHERN R. (1972): Autorhythmometry—procedures for physiologic self-measurements and their analysis. Physiology Teacher, *1*: 1–11.
6. REINBERG A. and HALBERG F. (1971): Circadian Chronopharmachology. Ann. Rev. Pharm., *11*: 455–492.
7. HALBERG F., HALBFRG E. and MONTALBETTI N. (1969): Premesse e sviluppi della cronofarmacologia. Quad. med. quant. sperimentazione clin. controllata, *7*: 5–34.
8. MENZEL W. (1962): Menschliche Tag-Nacht-Rhythmik und Schichtarbeit, (Benno SCHWABE, ed.), pp. 189, Basel, Stuttgart.

CLINICAL ASPECTS OF BLOOD PRESSURE AUTORHYTHMOMETRY

Howard LEVINE*[2] and Franz HALBERG**[1]

[1]Chronobiology Laboratories, Department of Laboratory Medicine and Pathology
Health Sciences Center, University of Minnesota
Minneapolis, Minnesota, U.S.A.
[2]Department of Medicine, University of Connecticut Health Center
Hartford, Connecticut, U.S.A.
New Britain General Hospital
New Britain, Connecticut, U.S.A.

GENERAL CONSIDERATIONS

Whenever health is to be quantified or illness defined and treated, a few selected variables should be evaluated. One readily quantifies physiologic variability in body temperature, heart rate, blood pressure, peak expiratory flow and psychomotor performance; grip strength, eye-hand skill, addition speed, short-term memory and time estimation, as well as ratings of mood and vigor, already can be assessed in easy tests [1, 2]; contentment and productivity need evaluation as important sociologic as well as psychologic facets of an individual's health.

Rhythmic variables should be evaluated not only for health assessment but also in order to adjust the kind and timing of treatment according to their characteristics. It is important to measure performance (productivity) as well as contentment, since changes in such thus far ignored parameters may well be considered when therapeutic schedules are assessed. Within certain limits an individual may be prepared to reduce his productive output in exchange for a longer life; however, this certainly will not always be the case. Such heretofore philosophical considerations can now be weighed in the light of results from fast yet reliable self-measurement. Physical as well as intellectual performance will then have to be compared against not only selfmeasurements but perhaps automatic blood pressure monitoring.

An individualized data base for cardiovascular as well as psychomotor performance thus should become available. For this purpose, chronobiology introduces the individualized rhythm-adjusted total ranges, waking (or sleeping) ranges, 90% (or other) percentiles, waking (or sleeping) mesors, M, amplitudes, A, and acrophases, ϕ(Fig. 1A). These parameters and/or others are needed not only for health assessment but also for identifying and treating illness. Without rhythmometry one must ignore most if not all changes predictably recurring with time.

In clinical medicine the need to introduce a challenge or load to diagnose illness has been emphasized by the occasionally "normal" electrocardiogram in some patients with coronary artery disease and the "normal" fasting blood sugar in some diabetic patients. To the concept of "loading" there must be added the criterion of timing the challenge. Patients with Cushing's disease are more readi-

* Connecticut Regional Medical Program; Central Chapter, Connecticut Heart Association.
** USPHS (5-K6-GM-13,981), NSF-GW 7613, and NASA.

ly recognized when the withdrawal of a single blood sample for the determination of 17-hydroxycorticosteroids is timed in relation to circadian rhythms. By the same token, the chances of picking up in a casual urine sample an abnormal constituent can be improved by timing. F. C. BARTTER recognized the patient who has an elevated blood pressure at one clock hour but not at another; a similar situation for a case of mild mesorhypertension under treatment is shown in Fig. 1B. The so-called afternoon diabetes is a result of performing glucose tolerance tests at unconventional hours. Thus properly timed single samples can reveal disease that otherwise may not be apparent, as was demonstrated for Wuchereria bancrofti microfilariae about a century ago.

As a minimal requirement, loading or other tests should be specified not only in terms of time of day but also in relation to the subject's sleep-wakefulness schedule. As an optimum, one should specify the stage of the individual's self-measured rhythms in tests carried out at a single specified circadian time or on several such occasions. Already, the responses of heart rate, blood pressure, electrocardiographic response and other aspects of performance to graded exercise loads, e.g., on the treadmill, are being studied as a circadian system function in several laboratories.

The high incidence of conventional "hypertension" in the population and the frequency of complications involving strokes, cardiac infarction, and cardiac and renal failure make it a desirable subject for further study. However, "hypertension" has not been defined precisely nor characterized quantitatively, nor is this likely to be done unless the remarkable intra-individual temporal variability of blood pressure and of related variables is assessed. Since a substantial share of the individual variability in many physiologic functions, including human blood pressure, is contributed by circadian (about-24-hour), circannual (about-1-year) and other rhythms (Fig. 2B), the need for assessing any sizable changes is great. Ultradian-to-infradian rhythmometry, as opposed to focus upon a single cardiac cycle, is likely to provide exact knowledge with respect to predictable changes in blood pressure and related variables. Circadian and other rhythms account for the drastic differences in response to a variety of agents including drugs. Hence, rhythmometry would further provide guidance to the optimal treatment of hypertension timed according to variations such as those shown in Fig. 2. For instance, at 1200, an individual's diastolic pressure is, on the average, 82 mmHg, under chlorothiazide treatment (Fig. 2A). By this criterion one could say that he is well controlled. However, at 1600, the corresponding mean is 93 mmHg and the 95% confidence interval extends from 88 mm to 98 mmHg. By the same token, an average systolic pressure of 164 mmHg (152, 176) at 1600 certainly differs from 141 mmHg as a basis for therapeutic action (Fig. 2A).

Moreover, blood pressure behavior may show several kinds of dyschronism, i.e., an alteration of one or several rhythm characteristics at one or several periods, including infradian [2] ones (Fig. 2B): thus, there may be a change in amplitude (A), acrophase (ϕ) or wave-form [pairs of amplitude and acrophase (A, ϕ) at statistically validated harmonics] as well as in rhythm-adjusted mean or mesor.

Before there be a more or less persistent statistically significant increase in overall mesor—fixed mesorhypertension—blood pressures may rise only occasionally and unpredictably—labile mesorhypertension—or there may be a consistent change in one or several other rhythm characteristics (τ and/or ϕ, A as well as wave-form).

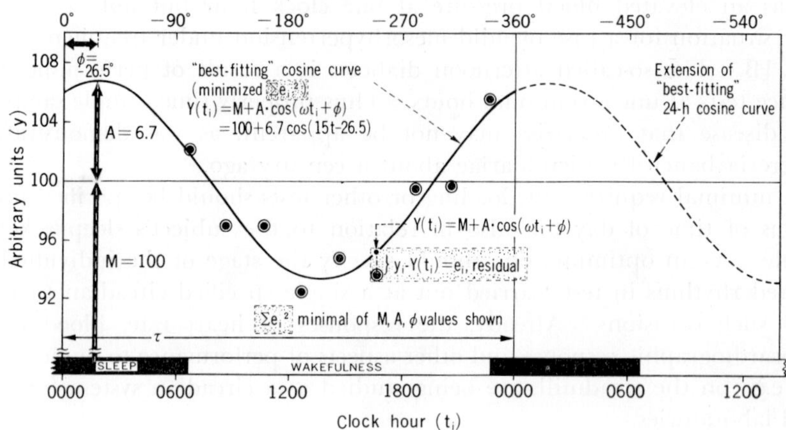

Fig. 1A

Fig. 1 Demonstration of curve fitting used, in abstract form (A, above) and in relation to real data (B, next page). Least squares spectral windows (B) show how a single cosine curve can locate the rhythm in noisy (B, top) or nonsinusoidal (B, bottom) data on systolic blood pressure and physical vigor (step function), respectively. The acrophase will be an appropriate measure of timing in these instances; other indices such as the mesor for physical vigor will be compromised by the fit of an inappropriate wave from as is the case for the approximation of a step function by a single cosine curve.

For monitoring the treatment of mesorhypertension, labile or fixed, mild or severe, one may advocate repeated self-administered tests of blood pressure, performance and psychologic rating. Results of such tests may guide the dosage of now conventional drugs to achieve, first, a lowering of blood pressure and, second, in the case of other rhythm alteration the reconstitution of a physiologic (e.g., circadian) rhythm. Moreover, if a dyschronism of certain rhythms be found before therapy is initiated or if it be brought about by treatment, does its correction (e.g., the correction of acrophase alteration or the synchronizaiton of desynchronized rhythms including circadian ones) result in desirable effects upon mental and physical performance as well as upon mood? Indeed, rhythmometry can reveal effects that are not grossly apparent by conventional methods of study, as shown also in a case study [2].

CASE REPORT

Self-measurements reported here were made by a 55-year-old physician with mild to moderate hypertension of ten years' duration. He had been in excellent health until age 45 when sustained elevation of blood pressure up to 180/100 mmHg and a slight aortic diastolic murmur were noted. Conventional medical evaluation during hospitalization initially revealed no correctable condition. More recently the diagnosis of primary aldosteronism was made and hypertension corrected by 100 mg. Spironolactone taken daily at 0800. Both parents had mild

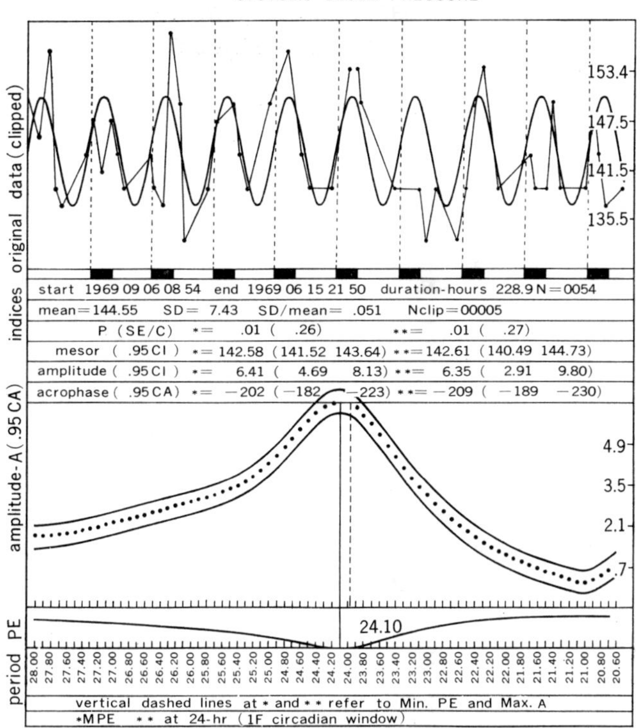

PHYSICAL VIGOR

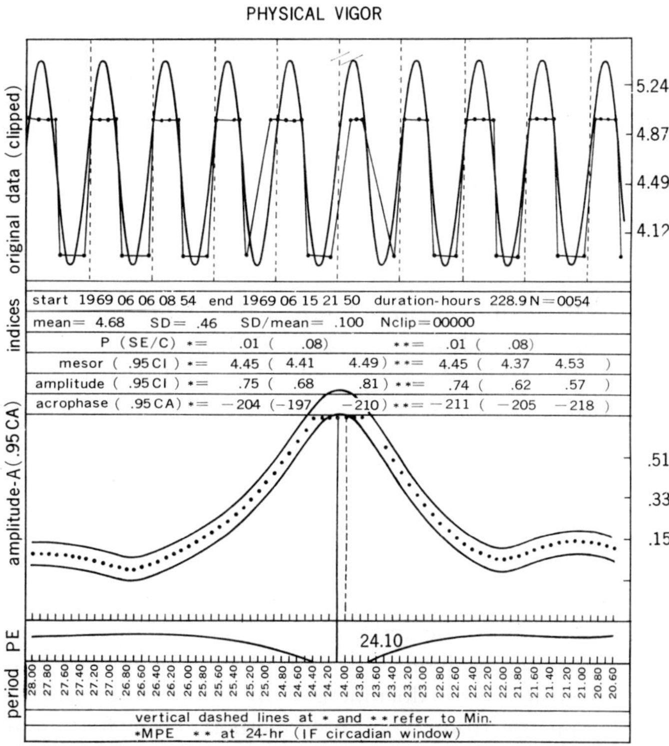

Fig. 1B

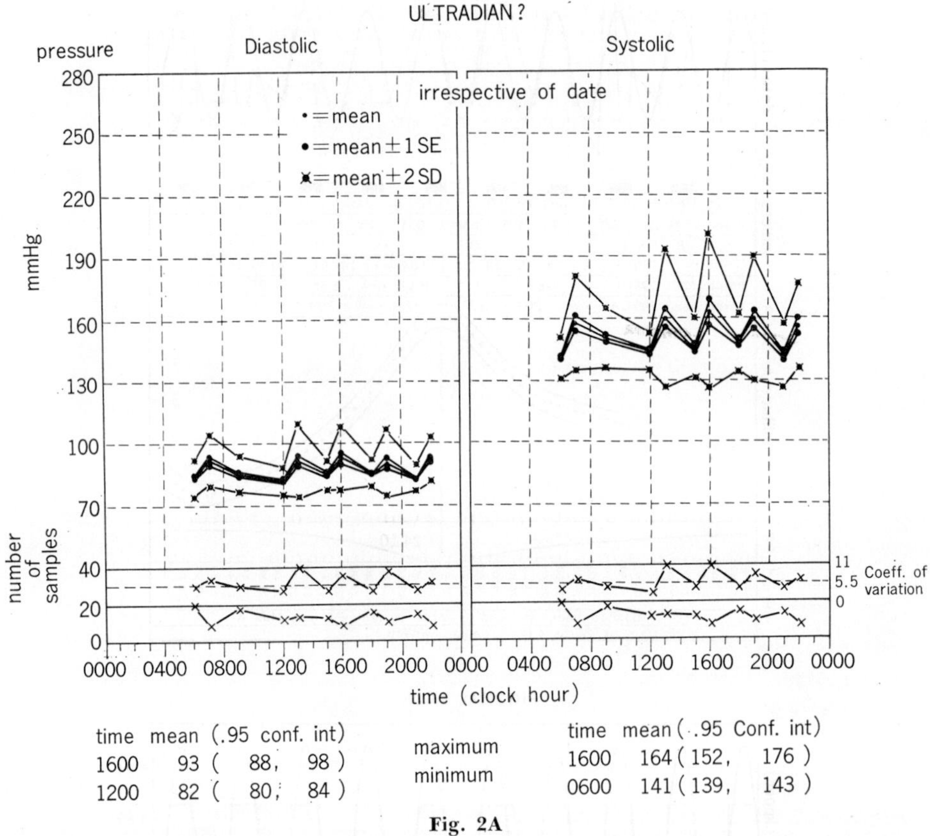

Fig. 2A Self measurement data obtained by a mildly mesorhypertensive physician. Plexograms for diastolic and systolic blood pressure suggest the occurrence of ultradian changes and/or effects of aliasing.

hypertension. Three siblings were normotensive; one of these recently died suddenly from aortic stenosis due to rheumatic heart disease.

Ten years ago chlorothiazide, 0.5 gm, with 15 mEq KCl twice daily at 0700 and 1900, induced "casual normotensive" values. One year later the patient developed minimal pulmonary tuberculosis. Treatment consisted of sanatorium care for three months with 100 mg isoniazid and 4 gm para-amino salicyclic acid three times a day after each major meal for eighteen months. The subject discontinued, then resumed antihypertensive treatment during the ensuing six years while well and active in professional, medical and educational work. Some detailed data and charts are available elsewhere [2].

Circadian rhythms in systolic and diastolic blood pressure, pulse, oral temperature, hand-grip strength, eye-hand skill and other variables in this physician with mild mesorhypertension were documented by self-measurements analyzed with so-called microscopic electronic computer methods (autorhythmometry, or AR). With the methods used, rhythms in certain physical aspects of performance

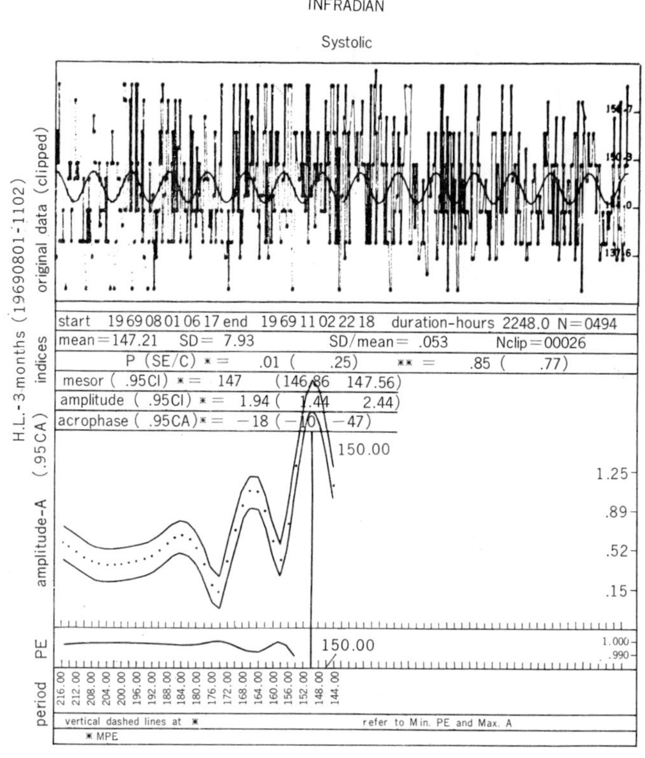

Fig. 2B

Fig. 2B Chronologic window of systolic blood pressure demonstrates circaseptan component with small amplitude.

(Fig. 3) were demonstrated as being statistically significant throughout the entire study span when a 24-hour period was fitted to intervals of 20 days displaced in 1-day increments (Figs. 4 and 5). Detection and quantification of rhythms also was feasible part of the time for systolic blood pressure, but not invariably—e.g., not during spans when the subject carried an obvious psycho-social burden or was readjusting after a transmeridian flight.

In the mildly mesorhypertensive subject here investigated, the mesors of both systolic and diastolic blood pressure were reduced and certain performance measures improved after the institution of treatment with chlorothiazide. With changes in the timing of treatment a further very small decrease was recorded in the mesor of systolic as well as diastolic blood pressure and in certain performance measures: this very slight, yet statistically significant decrease in blood pressure again coincided with a small increase in grip strength and an improvement in eye-hand skill. A "learning" and/or "training" effect contributed substantially to these performance changes and without placebos no conclusions are here drawn concerning treatment effects. Performance actually may decrease with a therapeutically intended lowering of blood pressure and may serve as a gauge of the optimal blood-pressure mesor.

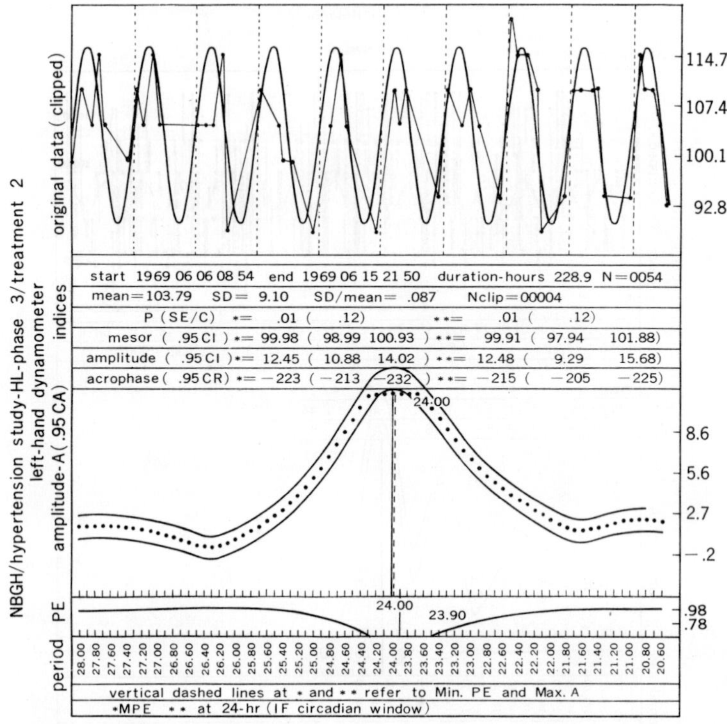

Fig. 3 Detection and quantification of a circadian rhythm in grip strength data covering ~9 days.

INFERENCES

We can suggest, with data, that physical and mental performance measures provide an objective basis for assessing the desirability of a given physiologic change, large or small, spontaneous or induced—say, a blood pressure decrease following institution of treatment or any added lowering of pressure after changing the timing of therapy. Such studies will have to be complemented on groups rather than individuals by a search for long-term effects, notably for complications from unduly low or high blood pressures—at certain circadian system acrophases.

A performance decrement was seen as a marked decrease in rhythm-adjusted mesor of eye-hand skill and grip-strength following an intercontinental flight from west-to-east but not following the return east-to-west flight (Figs. 4 and 5). This performance change in grip-strength demonstrated for a mildly hypertensive physician was relatively small [2]. A loss of eye-hand skill, apparent also in 8 other presumably healthy subjects studied concurrently as a function of similar flights, was remarkable. In this subject the post-flight decrement in performance differed among variables. This decrement constitutes an important practical problem amenable to 1) resolution by the development of special computer programs discussed elsewhere [1–3] and, hopefully also to 2) correction, if not 3) prevention, by the development of chronobiotic medication.

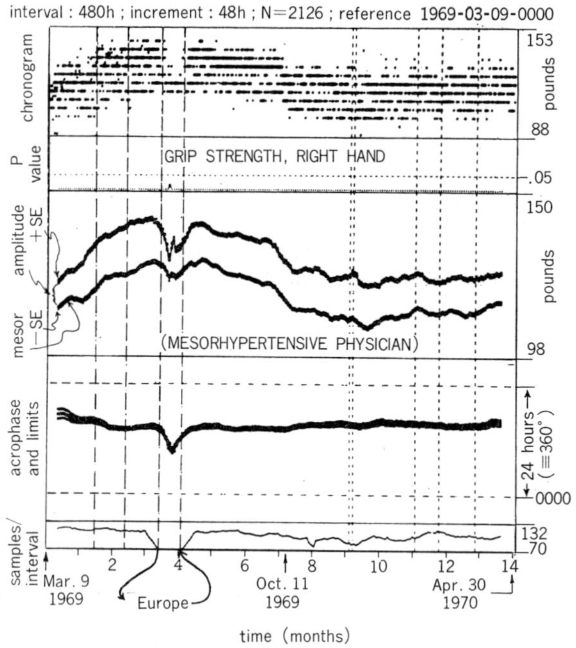
RELATIVE PERFORMANCE DECREMENT WITHIN PRIOR RANGE
interval : 480h ; increment : 48h ; N=2126 ; reference 1969-03-09-0000
chronogram
153
pounds
88
P value
GRIP STRENGTH, RIGHT HAND
.05
amplitude +SE
mesor −SE
150
pounds
98
(MESORHYPERTENSIVE PHYSICIAN)
acrophase and limits
24 hours (≡360°)
0000
samples/ interval
132
70
2
4
6
8
10
12
14
Mar. 9 1969
Europe
Oct. 11 1969
Apr. 30 1970
time (months)

Fig. 4

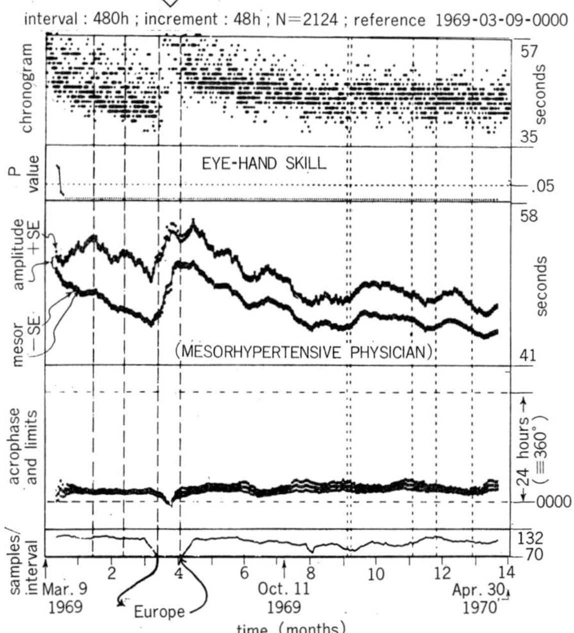
ABSOLUTE PERFORMANCE DECREMENT OUTSIDE PRIOR RANGE.
interval : 480h ; increment : 48h ; N=2124 ; reference 1969-03-09-0000
chronogram
57
seconds
35
P value
EYE-HAND SKILL
.05
amplitude +SE
mesor −SE
58
seconds
41
(MESORHYPERTENSIVE PHYSICIAN)
acrophase and limits
24 hours (≡360°)
0000
samples/ interval
132
70
2
4
6
8
10
12
14
Mar. 9 1969
Europe
Oct. 11 1969
Apr. 30 1970
time (months)

Fig. 5

REFERENCES

1. HALBERG F., JOHNSON E. A., NELSON W., RUNGE W. and SOTHERN R. (1972): Autorhythmometry—procedures for physiologic self-measurements and their analysis. Physiology Teacher, *1*: 1–11.
2. LEVINE H. and HALBERG F. (1972): Circadian rhythms of circadian system. Lit. Rev.; Computerized case study of transmeridian flight and medication effects on a mildly hypertensive. U.A. Air Force Report, April, 64 pp.
3. HALBERG F., ENGELI M., HAMBURGER C. and HILLMAN D. (1965): Spectral resolution of low-frequency, small-amplitude rhythms in excreted 17-ketosteroid; probable androgen-induced circaseptan desynchronization. Acta Endocrinologica Supplement, *103*: 54.

SPACE BIOLOGY AND MEDICINE

Chairman: Franz HALBERG

Co-Chairman: Hubertus STRUGHOLD

CYCLOECOLOGY IN SPACE ON THE MOON AND BEYOND

HUBERTUS STRUGHOLD

Honorary Consultant to the Aerospace Medical Division
Brooks Air Force Base
Texas, U.S.A.

Ecology—(Greek "oikos"=habitat, E. HAECKEL, 1869), studies the relations between environment and living organisms. With regard to the various environmental components we differentiate between *chemo-ecology, thermo-ecology, photo-ecology, ionizing radiation ecology, gravi-ecology,* and *magneto-ecology*.

All of these basic ecological conditions show local or topographic differences (topoecology) and chronographic variations (*chrono-ecology* or *cyclo-ecology*).

All processes in the universe are cyclic as manifested in the rotation of the galaxies, the existence of pulsating stars or pulsars, the 11-year activity cycle of the Sun and its rotation of about 25 terrestrial days, rotation of the planets and their moons, revolution of both of them around their primaries, periodic gravitational tidal effects associated with these movements; regular reappearance of certain comets, periodic occurrence of meteor showers, and—on a miniature scale—vibration of the molecules and atoms, and the orbit-like movement of the electrons around the atomic nucleus. Cyclicity or rhythmicity is the rule in the physical universe, in the macrocosmos and the microcosmos (*chronocosmology*).

Rhythmicity is also a dominant characteristic in the living world on Earth, as manifested in the *exogenous* biological rhythms imposed by and dependent on external environmental cycles; and in the *endogenous* rhythms, inherent and independent of—but synchronized with—external physical cycles. Only the physical cycles within our solar system (caused directly or indirectly by the Sun) have an effect upon both of these types of biorhythms on Earth. This is obvious in biological effects of the 11-year cycle of maximal solar activity (tree rings), of the circannual cycle seasons (growth rate, color, blossom-time, mating, reproduction, hibernation, aestivation, migration etc.), of the New Moon-Full Moon cycle of 27 days and of the lunar gravitational tidal effects. The best known biorhythm is the one associated with the day-night cycle caused by the Earth's rotation in the electro-magnetic radiation field of the Sun—the circadian rhythm found in man, and in the animal and plant kingdom. Since ancient times the rotating Earth has been so-to-speak the external photo- and thermoecological "Zeitgeber" for its whole biosphere within its home habitat.

But in our era of modern technology man, by means of airplanes, crosses numerous time zones in a matter of hours; furthermore rockets carry him completely out of the terrestrial sphere of time zones into the extraterrestrial space environment and exposes him to artificial astronautically produced ecological cycles within the natural astronomical cycloecological environment of space and on other celestial bodies.

Since two decades the time zone effect in air travel has been a frequent topic in chronobiology and aviation medicine and with the beginning of manned space flight, the sleep and wakefulness regime of the astronauts and cosmonauts has become increasingly the concern of the space medical physicians in charge of the missions. I had the privilege to observe, as a guest of NASA, most of the missions of the Gemini and Apollo projects in the Control Center of the Manned Spacecraft Center in Houston and was particularly interested in the astronaut's biological clock. But in this paper I shall confine myself primarily to the physical ecological cycles encountered in space with some remarks about the physiological side of the problem. More details about the latter is found in reports of NASA, in the Journal of Aerospace Medicine, in the Russian publication, Space Biology and Medicine and in the Astronautica Acta of the International Academy of Astronautics.

Beginning with the first phase of manned space flight—*orbital flight in near earth space,* which requires the first astronautical or cosmic velocity i.e. 8 km/sec, we face a completely novel photo-ecological situation: the customary geographic day-night cycle is replaced by a short sunlight—earth-shadow cycle.

Within the radiation-safe altitude range from 200 to 800 km below the van Allen Radiation Belt, the orbital periods last from 90 to 130 minutes, Table 1a. About 30% of this time—depending upon the orbit's inclination—the spacecraft is in the Earth's shadow cone. This orbital external light-shadow or photoscopic cycle is not longer than one-tenth of the day-night cycle on Earth. Furthermore, it is rhythmically modified by earthshine and moonshine with a permanent velvet black sky in the background.

Faced with this short orbital photoperiodicity in near earth space, the astronauts as terrestrial creatures, in the arrangement of their sleep and activity regime, have to follow the "tick-tock" of their internal clock. It has to be more or less isochronus with their inborn circadian pattern and preferably also synchronous with the time zone of the space flight Control Center or Launching Center.

In addition to the absence of a suitable external light-dark cycle, the absence of weight enters the life of the astronauts. Fortunately, weightlessness, as such, seems to be not a disturbing factor for sleep. One of the reasons—no gravitational pressure points; furthermore, under weightlessness the parasympatheticus is dominant (parasympathicotonia), just the same as during sleep. All of our astronauts and the Russian Cosmonauts had a relatively good sleep when noise and uncomfortable conditions were kept at a minimum. In case that two or three astronauts are in a space ship, it has been suggested in 1965 by Ch. BERRY, Chief of Medical Space Operations and Research (NASA), in order to avoid any disturbance for the sleeping partner, that they sleep at the same time, synchronized with the Control Center's time zone.

Confining myself to the presently two long record flights—during the longest American orbital flight of Gemini 7 (Dec. 1965, 14 days, 206 revolutions) the two astronauts, Frank BORMAN and James A. LOVELL, had no significant sleep difficulties. The inside of the spacecraft was artificially darkened by covering the windows. In this way, they had a microenvironmental day and night of their own, synchronized with the time zone at Cape Kennedy.

Whereas, the sleeping time in the earlier Russian Soyuz spaceships was kept more or less in tune with the night time at the launching center near Baikonur in Kazakhstan, the two cosmonauts in Soyuz 9, (Jan. 1970, 17 days, 17 hours) had to shift the sleeping cycle by 12 hours because they passed always over Russian at night and they had to land in the early morning. They slept at Russian daytime and worked at night. The Commander Cosmonaut, Andrian NICOLAYEV reported, "Our sleep in flight was normal. After the sleep we felt refreshed and quite efficient."

All in all, the recorded and reported sleep and wakefulness time patterns in orbital space flight reflect by and large the inherited rhythmostatic nature of the astronauts in terms of the circadian time scale. But it has to be fixed into the flight schedule as required by "orbital mechanics."

In future *space stations* with a larger team of up to 12, a properly arranged rotating shift in the sleep and duty regime of the flight operational crew will be possible due to more comfortable special sleep compartments.

At this point I like to include a personal communication from Dr. W. R. HAWKINS, Chief of Flight Operations, NASA, Houston, which summarizes the factors influencing the Astronauts' clock. "The primary factors that contributed to the fact that inflight sleep was less than that obtained on Earth were (1) cyclic noise disturbances resulting from such events as thruster firings, communication, or movement within the spacecraft; (2) staggered sleep periods; (3) significant displacements of the astronauts' normal diurnal cycle; (4) the so-called command-pilot syndrome; (5) the unfamiliar sleep environment; and (6) excitement.

During the Apollo Program, no new sleep problems have been encountered. Apollo missions are necessarily tailored around an operational trajectory which, by nature, is highly inflexible and constraining. The astronaut must be integrated into this fixed mission plan in the best possible way. That is, man is required to accommodate to the mission and not the converse."

A mission to the *Moon* requires a fixed time schedule of several more astronautical actions than in orbital flight namely after insertion into an Earth orbit, injection into the transmoon trajectory, insertion into a lunar orbit, separation of the lunar module from the command module etc.; this makes the programming of the sleep and duty cycle more complicated. The light/dark cycle in circumlunar orbits are shown in Table 1b.

During a longer stay on the Moon as in a future *lunar research laboratory,* which has been since 1965 the objective of the Lunar International Laboratory (LIL) Committee of the International Academy of Astronautics, the sleep-activity cycle will be completely independent of the 27 terrestrial days long light/dark cycle of the Moon. During the light part of the cycle, solar illuminance on the Moon amounts to 140,000 lux, the same as above the Earth's atmosphere. In addition, there is also periodically earthshine, which is seventy-five times stronger at "full earth" than is moonshine on Earth at full moon. Such is the general photorama on the Moon.

Sunrise and sunset on the Moon do not provide a Zeitgeber comparable to the one on Earth. Therefore, the astronauts inside a lunar laboratory would have to schedule their sleep-and-activity cycles in terms of the terrestrial circadian pattern. They can also be arranged in shifts among the members of the operational team, in comfortable sleep facilities.

With regard to gravity, sleep on the Moon might be better than on Earth due to its lower gravitational force, 1/6 of 1 G. A hypnogram or actogram of a sleeping astronaut will probably show fewer of the body movements which often interrupt our sleep on Earth.

For comparison, the Russian moon robot Lunakhod 1 had to go into a state of hibernation during the two weeks lunar night. With sunrise it became active again for the same period telecontrolled by engineers on Earth. Thus an instrumented robot shows in fact a real lunar circadian cycle in contrast to astronauts who stick to their terrestrial circadian rhythm.

The first postlunar planetary target for a manned mission will be the *planet Mars,* as envisioned for the mid-1980s by Wernher von BRAUN, Deputy Director of NASA. If this interplanetary journey were based on a minimum energy trajectory, it would last about seven months, which is from a space medical point of view too long and must be shortened considerably. Nuclear propulsion systems are expected to achieve this.

The Mars ship, with a crew of six to eight is at a distance of 1,300 million km beyond the earth's shadow cone and will be in constant sunshine against a velvet-black sky. In this non-periodic light environment along the trans-Mars trajectory, the occupants of the spaceship must arrange their sleep, rest, and activity regime corresponding to their circadian rhythm on Earth. Furthermore, exercise, which must be carried on to prevent certain types of physiological deterioration, will automatically contribute to a good sleep.

Within half a million kilometers from the Martian surface, the Mars ship will be in Mars' gravisphere and can be inserted into a circum-Martian orbit in preparation for the landing maneuver. If an altitude of 100 km is chosen, the occupants will observe a cycle of sunshine and Mars shadow of about the same duration as in the departure orbit in near-earth space, Table 1c.

On Mars itself, the day/night cycle is only 37 minutes longer than that on Earth. Solar illuminance on the Martian surface at noon may reach one-third of that on Earth or about 30 lux. Thus, the dark/light cycle on Mars offers a Zeitgeber sequence familiar to terrestrial visitors.

If there should be indigenous life in the form of vegetation in the dark, blue-green surface regions of Mars, thriving on some kind of photosynthesis, it would be active only during about five daylight hours. At night it would pass into a dormant state, due to the extremely low temperature. Nocturnal biology would be cryobiology. Experiments in so-called Mars jars, have shown that certain lower plants survive a simulated Martian circadian freeze-thaw cycle.

According to data provided by ground-based and space-bound astronomy, Mars will probably be the only postlunar astronautical target for a manned landing mission. All the other planets have extremely hostile cyclic thermoecological environments. Venus' rotation takes 243 terrestrial days in retrograde direction —i.e., opposite to the direction of its revolution around the Sun. The slowly rotating planet Mercury, 58½ days, closest to the Sun, is too hot on the sunlight side and too cold on the opposite side. The outer planets, Jupiter, Saturn, Uranus, Neptune and Pluto rotate within periods of 10 to 14 hours.

Several billion kilometers beyond Pluto, in interstellar space, solar illuminance with no periodicity drops below the light minimum for reading and color vision. This is a realm of eternal night, with the Sun of a stellar magnitude not much different from that of other stars. In interstellar flight, which requires the third astronautical velocity i.e. the escape velocity from the Sun's gravisphere (40 km per sec) and even higher fractions of the speed of light the behavior of the body clock of interstellar space travelers must be imagined in the perspective of the phenomenon of time dilation. It's physiological rhythm might be many times

Table 1 Velocities and periods of revolution of vehicles in circular orbits at various altitudes.

Altitude above surface (km)	Orbital velocity (km/sec)	Period of revolution (hr	min)
	a) in near earth space		
200	7.88	1	28
300	7.729	1	30
400	7.672	1	32
500	7.616	1	34
600	7.561	1	37
700	7.507	1	39
800	7.455	1	41
	b) in circumlunar space		
100	1.633	1	58
200	1.590	2	08
500	1.480	2	38
1,000	1.338	3	34
	c) in circummartian space		
100	3.51	1	45
200	3.46	1	49
300	3.41	1	54
500	3.32	2	01
1,000	3.13	2	27

After O. L. Ritter and H. Strughold

longer than 24 hours, but the interstellar space travelers would not become aware of it. The reason: at speeds approaching that of light, molecular movement slows down. This would prolong the action of all body organs, including the function of the biological clock. Whereas interstellar flight is at present a matter of fantasy and science fiction, manned interplanetry flight as far as Mars is in the realm of reasonable, realistic science vision.

So far I have discussed the energies in the vacuum of space. But space is not completely empty of matter; it contains material such as the already mentioned subatomic particles found especially in the solar plasma wind and molecular masses in the form of dust, meteoroids and asteroids. In the following, I would like to make a few remarks about one type of *meteoroids,* which appears in cycles. This happens, when the Earth moves through a stream of those meteoroids, which are the debris of a disintegrated comet. Astronomy knows of about a dozen regular permanent meteor stream-crossing dates of the Earth sometimes manifested in a spectacular meteor shower. Table 2. The best for the sky watchers are the Perseid shower (Aug. 12), the Orionids (Oct. 20), the Leonids (Nov. 6) and the Germinids (Dec. 13).

But how good or bad are these cyclic events for the astronauts? So far during the total space flight time, no meteoroid incident has occurred and four extravehicular excursions in space and eight on the Moon have been made without micrometeoritic interference. Moreover, what is not so well-known, two Russian and six American spaceships have been in space at the time when the Earth

Table 2

Earth's Crossing of Meteor Streams		Space Flight	
Name	Date	Name	Dates
Quadrantids	January 3		
Aurigids	February 9	Apollo XIV,	Jan. 21–Feb. 9, 1971
Lyrids	April 21		
Eta Aquarids	May 4		
Draconids	June 28		
Delta Aquarids	July 29	Apollo XV,	July 26–Aug. 7, 1971
Perseids	August 12	Vostok III,	Aug. 11–25, 1962
		Vostok IV,	Aug. 12–15, 1962
Orionids	October 20	Apollo VII,	Nov. 11–22, 1968
Taurids	October 31		
Arietids	November 12	Gemini XII,	Nov. 11–15, 1966
Leonids	November 16	Apollo XII,	Nov. 14–24, 1969
Geminids	December 13	Gemini VII,	Dec. 4–18, 1965
Ursids	December 22	Apollo VIII,	Dec. 21–27, 1968

crossed a meteor stream and even during the most conspicuous showers just mentioned and underlined in Table 2. This is certainly a good omen for the future of sky labs and permanent space stations. All this indicates that the picture of meteoritic hazards in the Earth-Moon region looks brighter than had been expected before manned space flight became reality. The reasons—the meteor streams become less dense and the meteoroid material of cometary origin is soft "fluffy stuff" consisting of dirt and ice. This is different concerning the meteoroids of asteroidal origin which are stony, iron-stony, and irons. But they are sporadic, not cyclic and not so frequent; however their frequency increases beyond Mars in the belt of the asteroids, their birthplace. Be that as it may, protective devices called meteor bumpers suggested as early as 1951 by Fred Whipple, might protect the spacecraft against meteoroids of puncture capabilities, and even the space suits have a protective layer concerning micrometeoroids.

In conclusion in our chronobiological studies of the present ecological cycles in the cosmos we must also look into the past—into the paleontological development of astronomical cycles. For example, was some 100,000 years ago the Earth's rotation slower? This is conceivable, if P. A. Dirac's hypothesis is correct, according to which the gravitational constant slowly decreases and consequently the duration of the Earth's rotation could have been somewhat different in ancient times. If this was the case, then certain deviations of the human circadian rhythm from the exact 24 hours observed on subjects kept under constant photic conditions could be considered as an evolutionary remnant of the fossil man. Or was the 11-year cycle of solar maximal activity always the same? A study of the rings of trees in the petrified forest in Arizona could give the answer. I tried to look into this, but it is difficult to get a clear picture of the rings of this stony material. There may be some more examples for comparison between present day chronobiology and paleochronobiology which might be also of interest for astronomy

REFERENCES

1. ASCHOFF J. (1964): Significance of circadian rhythms for space flight. *In*: Third International Symposium on Bioastronautics and the Exploration of Space, (T. C. BEDWELL and H. STRUGHOLD, eds.), Aerospace Medical Division, Brooks, AFB, Texas.
2. BERRY C. A., COONS D. D., CATTERSON A. D. and KELLY G. F. (1966): Gemini Midprogram Conference: Part 1. NASA Manned Spacecraft Center, Houston, Texas.
3. BERRY C. A. (1970): Medical experience in the Apollo manned spaceflights, Aerospace Medicine, *41*(5): 500.
4. CONROY R. T. W. L. and MILLS I. N. (1970): Human Circadian Rhythms. I & A Churchill, London.
5. de RUDDER B. (1941): Über Sogenannte Kosmische Rhythmen Beim Menschen (About So-called Cosmic Rhythms in Man). George Thieme Verlag, Leipzig.
6. GAZENKO O. G. (1965): Medical investigations on Space ships, Vostok and Voskhod. Chapter XVIII, Bioastronautics and the Exploration of Space, (T. C. BEDWELL, Jr. and H. STRUGHOLD, eds.), Aerospace Medical Division, Brooks AFB, Texas.
7. GOLTRA E. R. (1959): Time dilation and astronaut. *In*: Man in Space, (KENNETH F. GANTZ, ed.), Duell, Sloan and Pearce, New York.
8. HALBERG F. (1964): Physiologic rhythms. *In*: Physiological Problems in Space Exploration, (J. D. HARDY, ed.), Charles C. Thomas, Springfield.
9. KLEIN K. E. et al. (1970): Circadian rhythms of pilots' efficiencies and effects of multiple time zone travel. Aerospace Medicine, 41.
10. LUCE G. G. (1971): Body Time, Random House, New York.
11. MANDROVSKY B. V. (1971): Soyuz 9 Flight, a manned biomedical mission. Aerospace Medicine, Vol. 42, No. 2.
12. MENZEL D. H. (1971): Astronomy, Random House, New York.
13. NININGER H. H. (1952): Out of the Sky, Denver Press.
14. PARIN V. V., VOLYKIN Y. M. and VASSILYEV P. V. (1964): Manned Space Flight. COSPAR Symposium, Florence, Italy.
15. PICKERING T. S. (1958): Astronomy, 1001 Questions and Answers. Grosset and Dunlap Publ., New York.
16. SIFFRE M. (1964): Beyond Time. McGraw Hill, New York.
17. STRUGHOLD H. (1967): Lunar medicine. *In*: Proceedings of 2nd Lunar International Laboratory (LIL) Symposium, International Academy of Astronautics, pp. 112–21, Madrid, 1966, Pergamon Press, Inc., New York.
18. STRUGHOLD H. (1971): Your Body Clock, Scribners and Son, New York.
19. STUHLINGER E. et al. (1966): Study of nerva-electric manned mars vehicle. AIAA/AAS Stepping Stones to Mars Meeting, Baltimore, Md., American Institute of Aeronautics and Astronautics, New York.
20. WARD R. R. (1971): The Living Clocks. Alfred A. Knoph, New York.
21. WATSON F. G. (1956): Between the Planets. Harvard University Press, Cambridge, Mass.
22. WHIPPLE F. L. (1969): Earth, Moon and Planets. Blakiston Co., Philadelphia.

BIORHYTHMS OF A NONHUMAN PRIMATE IN SPACE

TAKASHI HOSHIZAKI
Space Biology Laboratory, Brain Research Institute
University of California
Los Angeles, California, U.S.A.

The effects of terrestrial environments on the various biorhythms have been well documented. The roles of daylight, temperature and humidity in initiating and controlling these rhythms are established. What has not been firmly established, however, is whether magnetic and gravitational fields influence the circadian rhythms exhibited by various organisms. An opportunity to study the effects of magnetic and gravitational fields came when Biosatellite III program was initiated by NASA. As one of the principal participants, the Space Biology Laboratory of the University of California, Los Angeles executed a series of experiments to study the effect of a space environment on the physiological and neurological systems of a nonhuman primate [1, 2, 4]. Among the many areas studied was the biorhythmicity of various physiological and metabolic rhythms such as temperature, heart rate, blood pressure, respiratory cycles and the sleep/wake activity cycle. The environmental conditions experienced by the subject during the Biosatellite III flight were monitored as to the magnetic field through which the satellite traversed, pressure and humidity within the capsule, capsule temperature and also the partial pressure of the carbon dioxide within the space capsule [3]. Pressure, humidity and capsule temperature did not show cyclic variations. The slight changes that did occur were well within the operational range and specification of the space craft. The magnetic flux density encountered by Biosatellite III was obtained from the geomagnetic model which the Goddard Space Flight Center uses for all experimenters requiring such information. Deviation from this model during the flight of Biosatellite III was probably no less than 0.002 gauss. The predominant periods of these data obtained through spectral analysis relate mainly to the orbit parameters and are 1.6, 23.5 and 70.4 hours. The partial pressure of carbon dioxide (pCO_2) reflects the metabolism of the animals. The gas management system which circulates the air (one change every 40 sec. assuming perfect mixing) in the space craft removes only 10% of the CO_2 during each air change and one may, using the constants of the gas management system, calculate the CO_2 production of the animal. This measure is the result of the subject's respiration, and its periodicity estimate was found to be greater than 25 hours. The subject's data, consisting of brain and body temperatures, heart rate, blood pressure and sleep/wake cycles were analyzed for circadian periods [3, 5]. The heart rate and the brain and body temperatures revealed a periodicity greater than 25 hours, similar to the pCO_2 rhythm. On the other hand, blood pressure data indicated rhythms that were not longer than 24 hours. The sleep/wake cycle of the subject was 24 hours in length but had a phase angle difference of two hours from the imposed night and day modes. These findings of a 24 hour rhythm in blood pressure and sleep/wake activity and a 26 hour rhythm in the heart rate, body and brain temperature and pCO_2 may indicate that internal desynchronosis occurred in the flight subject and may have contributed to the physiological pathology that prompted the termination of the flight. The desynchronosis found in the flight animal was not observed in the control subjects. Precise 24 hour rhythms were observed in all parameters measured in the 9 control animals tested under 1 G conditions.

Circadian rhythms were discernible in the EEG spectra data from the flight subject [10]. However, their brief duration and variable morphology, most likely due to rapid shifts in the sleep/wake states, made a fine specification of their periods difficult (Fig. 1). Circadian rhythmicity was noted best in the EEG spectra obtained from the left amygdala and especially through the 3 to 15 Hz band, with a general increase in intensity noted early in the night period. The strongly circadian fluctuation of brain temperature may have been closely related to the intensity peaks noted, since the peaks in the daily values of brain temperature were observed at this time. Other daily changes were recorded. However these changes, resulting from eye movement and chewing, were associated with the behavioral task presented twice during the light period.

Although the second Biosatellite experiment concerned with nonhuman primate was cancelled, baseline experiments were started for a follow-up experiment using a chimpanzee instead of a monkey. The planned flight duration was 180 days. Results of one of these baseline experiments were recently presented [6, 8, 9]. A chimpanzee was tested for 30 days in a high performance isolation chamber. The subject was entrained during the first ten days of isolation to a 12 hour light: 12 hour dark regimen. Next came a 10 day period of continuous light, followed by 10 days of a 12 hour light: 12 hour dark regimen. As with the experiments with the Macaque monkey, physiological, central nervous and sleep/wake activity patterns were measured. Under the entrainment conditions of the first and last 10 days, the chimpanzee's sleep/wake cycle was 24 hours. During the continuous light period the mean duration was 24.8 hours. The significantly longer free running sleep/wake rhythm was thought to be due to the animal's spending more time in the wake and the REM sleep stages. Urine volumes and voiding times of the chimpanzee during the 30 days of isolation were found to exhibit clear circadian micturition rhythms, with the voiding peak occurring immediately after the subject awoke and the urine flow rhythm reaching a maximum volume in the morning hours (Fig. 2). When the subject was entrained to a 12 hour light: 12 hour dark regimen a 24 hour micturition rhythm was seen, and when he was exposed to continuous light the rhythm was 24.8 hours. A possible underlying 24 hour micturition rhythm was also seen during the continuous light period. During the last three days of continuous light the time of awakening and the micturition rhythm appeared to have locked on to around 1700 hours and did not progressively shift as did the beginning of sleep (compare slopes of beginning sleep to those of awakening during the last three days in Fig. 2c). It is curious to note that 1700 hours is also the time of the afternoon micturition during the first ten days of isolation. It appears that there was a transition of the afternoon micturition to the morning micturition which gives credence to the existence of an underlying 24 hour rhythm even during the continuous light period. Similar findings were reported for human subjects exposed to continuous daylight in the Arctic [7].

Baseline data for a space experiment have indicated that a second species, a

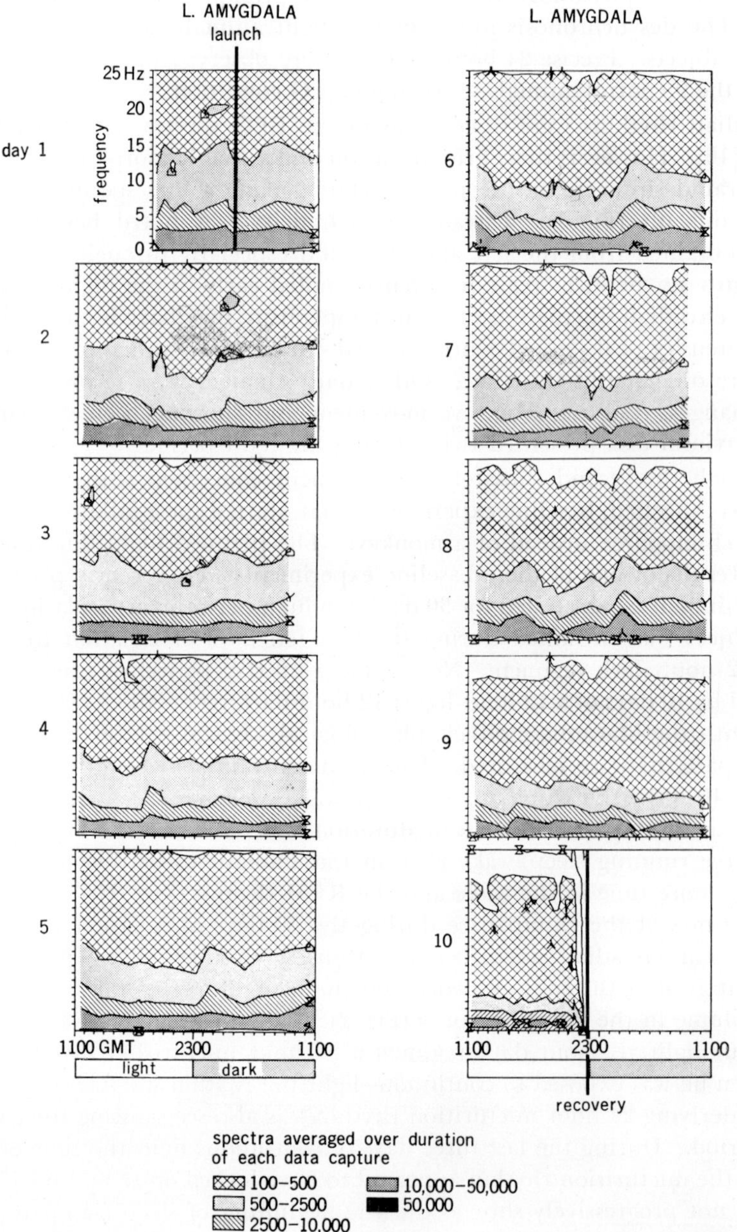

Fig. 1 Summary flight spectra. Contours of spectral intensity are integrated across the data captures available for each separate capsule day. Contours are calibrated in counts/Hz. (10^4 counts/Hz=64×10^{-8} volts^2/Hz.) [10].

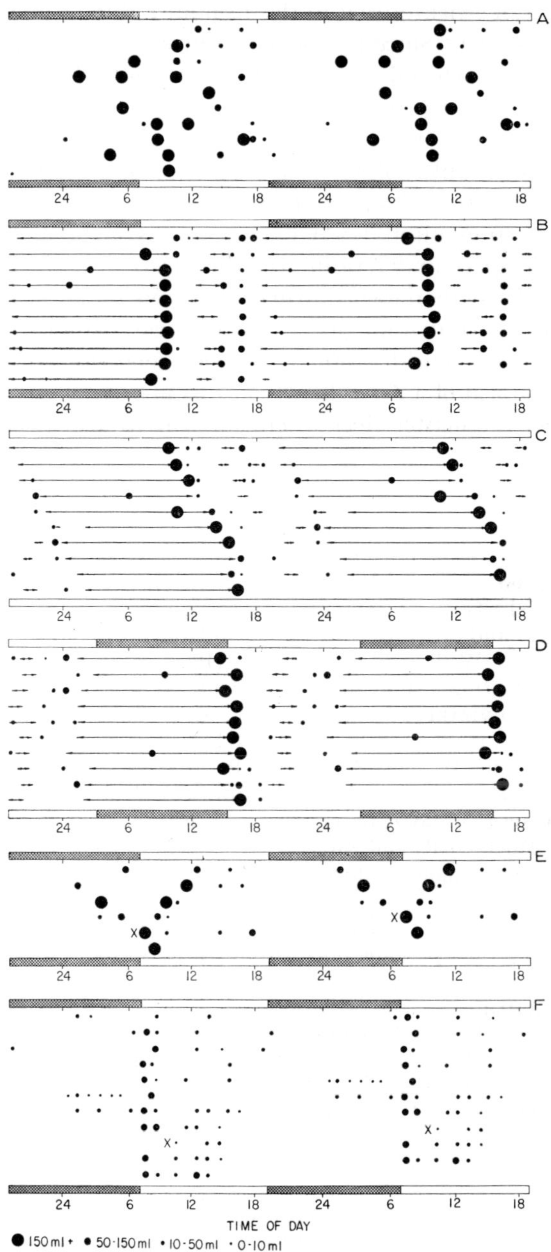

Fig. 2 Micturition and urine volume of an unrestrained male chimpanzee. A. Pre-isolation adaptation period—10 days 12L:12D B. Isolation—10 days of 12L:12D. C. Isolation—10 days of continuous light. D. Isolation—10 days of 12L:12D. E. Post Isolation—6 days of 12L:12D. F. Control—10 days of 12L:12D in home cage. Sleep is indicated by arrows for the isolation period only. Sleep data was not available for other periods. The graph has been double-plotted on a 48-hour basis. Days run from top to bottom. X—collector malfunction, actual volumes collected are presented but time is uncertain. Open bar—light at 10 ft. c.; hatched bar—light at 0.3 ft. c. [6.]

chimpanzee, was well synchronized to the 12 hour light: 12 hour dark regimen in all parameters measured. Since the ground controls of the Biosatellite III experiment were also well synchronized to the 12 hour light: 12 hour dark regimen in their physiological and neurological responses [4], one wonders whether the response of the chimpanzee under a space environment might again be similar to that of the Macaque monkey Bonny. It would be of great interest to seek these data if and when opportunities arise in such programs as Skylab. Also, it would be most interesting to monitor closely the astronauts' metabolic and neural physiological rhythms in space to determine whether desynchronization might occur in the astronauts, and whether such desynchronization might lead to a decrease in performance level or contribute to a physiological pathology. An interesting point arises from these considerations. If space conditions did invoke a permanent change by removal of a Zeitgeber, then this might give us an opportunity to determine whether the master clock of circadian rhythms is possibly paced by an exogenous factor.

REFERENCES

1. Adey W. R., Kado R. T. and Walter D. O. (1967): Analysis of brain wave records from Gemini flight GT-7 by computations to be used in a thirty day primate flight. *In*: Life Sciences and Space Research. North-Holland Pub. Co., Amsterdam.
2. Berkhout J., Adey W. R. and Campeau E. (1969): Simian EEG activity related to problem solving during a simulated space flight. Brain Res., *13*: 140–145.
3. Hahn P. M., Hoshizaki T. and Adey W. R. (1971): Circadian rhythms of the *Macaca nemestrina* monkey in Biosatellite III. Aerospace Med., *42*: 295–304.
4. Hoshizaki T., Adey W. R., Meehan J. P., Walter D. O., Berkhout J. I. and Campeau E. (1969): Central nervous, cardiovascular and metabolic data of a *Macaca nemestrina* during a 30-day experiment. Bibl. Primat., *9*: 8–38.
5. Hoshizaki T., Durham R. and Adey W. R. (1971): Sleep/wake activity patterns of a *Macaca nemestrina* monkey during nine days of weightlessness. Aerospace Med., *42*: 288–295.
6. Hoshizaki T., McNew J. J., Sabbot I. M. and Adey W. R. (1971): Effects of 30 days of isolation on the periodic micturition patterns of an unrestrained chimpanzee. *In*: Proceedings, Ann. Sci. Meeting of Aerospace Med. Assoc., Houston, Texas, 1971, pp. 244–245 (Abstr.).
7. Lobban M. C. (1965): Dissociation in human rhythmic functoin. *In*: Circadian Clocks. Proc. of the Feldafing Summer School, (J. Aschoff, ed.), pp. 219–227, North-Holland Pub. Co., Amsterdam.
8. McNew J. J., Burson R. L., Hoshizaki T. and Adey W. R. (1971): Effect of continuous light on the sleep-wake cycle of an unrestrained isolated chimpanzee. *In*: Proceedings, Ann. Sci. Meeting of Aerospace Med. Assoc., Houston, Texas, 1971, pp. 242–243 (Abstr.).
9. Sabbot I. M., McNew J. J., Hoshizaki T. and Adey W. R. (1971): Calcium and phosphorus excretion during short term stress and prolonged stress in the unrestrained chimpanzee. *In*: Proceedings, Ann. Sci. Meeting of Aerospace Med. Assoc., Houston, Texas, 1971, pp. 17–18. (Abstr.).
10. Walter D. O., Berkhout J. I., Buchness R., Kram E., Rovner L. and Adey W. R. (1971): Digital computer analysis of neurophysiological data from Biosatellite III. Aerospace Med., *42*: 314–321.

PHASE RELATIONSHIPS BETWEEN CIRCADIAN RHYTHMS AND THE ENVIRONMENT IN HUMANS DURING HYPOKINESIS

CHARLES M. WINGET, JOAN VERNIKOS-DANELLIS, CAROLYN S. LEACH, and PAUL C. RAMBAUT

Environmental Biology Division, Ames Research Center, NASA Moffett Field, California, U.S.A.
Preventive Medicine Division, Manned Spacecraft Center, NASA Houston, Texas, U.S.A.

ABSTRACT

The effect of restricted activity (bed-rest) on certain physiologic rhythms in eight healthy male subjects maintained in a defined environment was investigated. A photoperiod of 16L:8D was maintained for a six-day ambulatory, pre-bed-rest equilibration period, 56 days of bed-rest and a 10-day post-bed-rest recovery period. Four of the subjects exercised during bed-rest. Body temperature (BT) data were obtained using ear thermistors and heart rate (HR) was measured from pulse rates and by EKG sensors connected to a cardiotachometer. Circulating cortisol, triiodothyronine (T_3) and thyroxine (T_4) concentrations were determined in the blood samples drawn at four hourly intervals for 48-hour periods before, 10, 20, 30, 42, and 54 days during, and 10 days post-bed-rest. During bed-rest mean HR increased while BT and steroid outputs decreased. Neither exercise nor the 10-day post-bed-rest ambulatory period prevented or corrected this effect. HR remained more stable throughout bed-rest than the other rhythms studied. In contrast, the amplitude of the T_4 rhythm appeared to increase as bed-rest progressed and the total serum T_3 concentrations increased during the latter part of the bed-rest. The data indicate the daily change in phase and amplitude of BT, T_3, and T_4 rhythms are due to the position of the body. The observed low grade hypothermia and minor tachycardia are characteristic of the hypokinetic syndrome in man.

INTRODUCTION

The daily rhythm of human physiological functions is a complex reaction, produced at least in part by adaptation of the organisms to daily changes in the earth environment. The complex, dynamic interrelationships produced by social and physical environmental factors underlie the formation of diurnal rhythms in human physiological functions. The biorhythm literature is ample indicating the influence of certain environmental factors, *Zeitgebers,* on circadian rhythms. However, little is known about the influence of hypokinesis on changes in diurnal periodicity of certain human metabolic and endocrine rhythms. The purpose of this study was to quantitatively assess the influence of decreased muscular activity on these physiological functions.

METHODS

Data for quantitative description of dynamic physiologic systems were collected from eight healthy, male, human subjects confined to bed for 56 days, following a six-day ambulatory period. A constant 16L:8D photoperiod was maintained. Heart rates (HR) and body temperatures (BT) were obtained six times daily (0330, 0730, 1130, 1530, 1930, and 2330 hours). Circulating cortisol, triiodothyronine (T_3) and thyroxine (T_4) concentrations were determined in the blood samples drawn at four hourly intervals for 48-hour periods before, 10, 20, 30, 42, and 54 days during, and 10 days post-bed-rest. The subjects were confined two per room, affording comparisons of rhythms between roommates.

A circadian, or 24-hour period, was assumed in the biologic data for all subsequent analyses and on *a priori basis,* the data was assumed to be non-stationary in time. Furthermore, the concept of the "train of vectors" or the summation-dial, forms the basis for the methods of data analyses [1].

RESULTS

Fig. 1 shows the relationship of the rhythm in circulating levels of cortisol, thyroxine (T_4) and triiodothyronine (T_3), and body temperature (BT) in the eight subjects during their ambulatory pre-bed-rest control period. All three hormones showed a significant diurnal fluctuation with maximal levels occurring at 0730 hours, anticipating lights-on. In addition, thyroxine showed a secondary peak at 3:30 P.M. The amplitude of both thyroid hormone rhythms was much smaller than that of the corticosteroid. The BT rhythm showed a minimum value four hours after the lights were turned off.

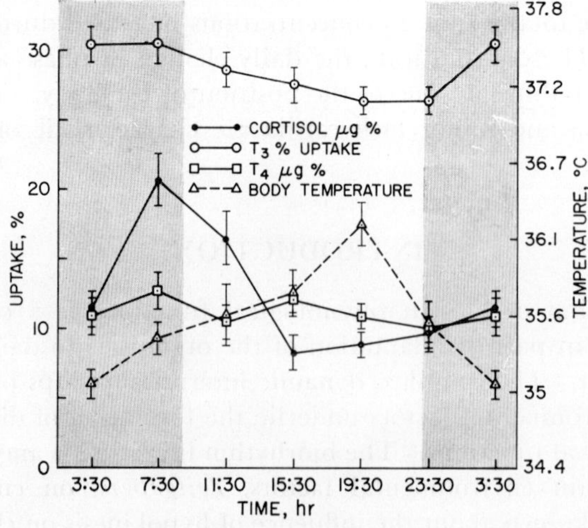

Fig. 1 Diurnal rhythms in mean circulating cortisol (●), thyroxine (□), and triiodothyronine (○) and in body temperature in eight normal ambulatory subjects. Vertical lines represent standard error of the mean. Stippled area represents lights-off period.

Fig. 2 shows the BT and HR rhythms during and post-bed-rest. Peak HR during bed-rest was noted beginning about three hours after lights were turned on while post-bed-rest it occurred at the end of the light period. During bed-rest, minimum HR occurred four hours after lights-off but three hours after lights-on in the post-bed-rest period.

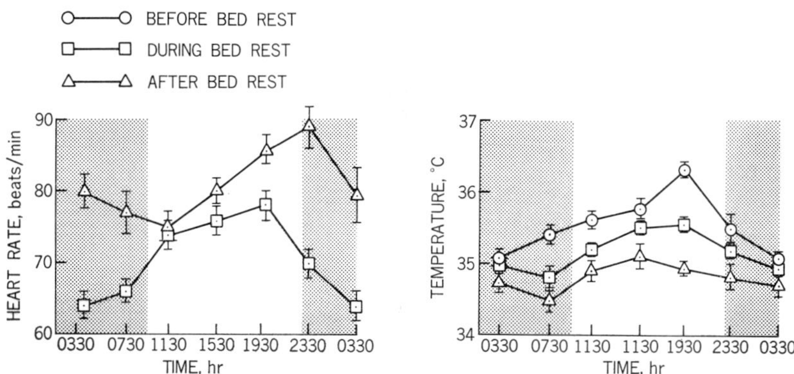

Fig. 2 Mean heart rate and body temperature values (± standard error) for eight subjects before (○—○) during (□—□) and after (◇—◇) 56 days of bed-rest. The lights were turned on at 0900 hours and off at 2300 hours.

It is apparent that bed-rest did not affect the BT wave forms. However, bed-rest did produce a depression of mean BT which did not return to pre-bed-rest values 10 days later. Bed-rest presumably had no effect on HR wave form, however, a change in amplitude and time of peak are noted during the post-bed-rest period.

Fig. 3 shows the plasma cortisol rhythm at various time intervals during the study in the four subjects who exercised and the four who did not. Bed-rest had little effect on the circadian rhythmicity of this hormone. A significant fluctuation ($p<0.001$) in plasma cortisol was apparent with peak levels occurring around 7:30 A.M. throughout the experiment. However, progressive bed-rest reduced the amplitude of the steroid rhythm and exercise did not prevent nor did the 10-day post-bed-rest ambulatory period correct this reduction in amplitude. The occurrence of the peak in plasma cortisol was more variable in the exercised subjects as shown in Fig. 4. This illustrates the use of the summation dial to depict these results and shows that the exercised group peaked about one hour before the non-exercised subjects. In contrast to the remarkable stability of the plasma cortisol rhythm, both thyroid hormones showed more variable rhythms during the bed-rest period with a return to original rhythmicity at the post-bed-rest collection period (Fig. 5). However, exercise did not appear to increase the stability of these rhythms.

The observed phase relationships between the measured metabolic (HR and BT) and endocrine indicators (CS, T_3, and T_4) are presented in Fig. 6. These data leave no doubt about the stable phase relationship between BT and CS, between HR and CS, and between BT and HR. Whereas, there are changes in the phase relationships between T_3, T_4, and the other circadian parameters with perhaps the exception of T_3 and HR rhythms. Approximately the same rela-

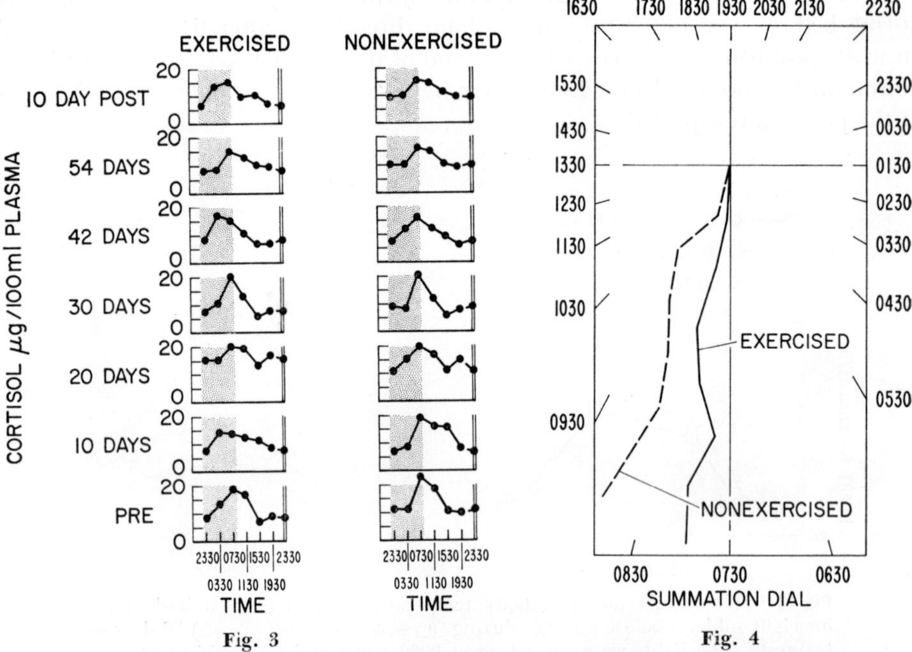

Fig. 3 Fig. 4

Fig. 3 Plasma cortisol rhythm in four exercised and four non-exercised subjects before, during, and following 56 days of bed-rest. Stippled area represents lights-off period.
Fig. 4 Summation dial of plasma cortisol rhythm in four exercised and four non-exercised subjects during 56 days of bed-rest.

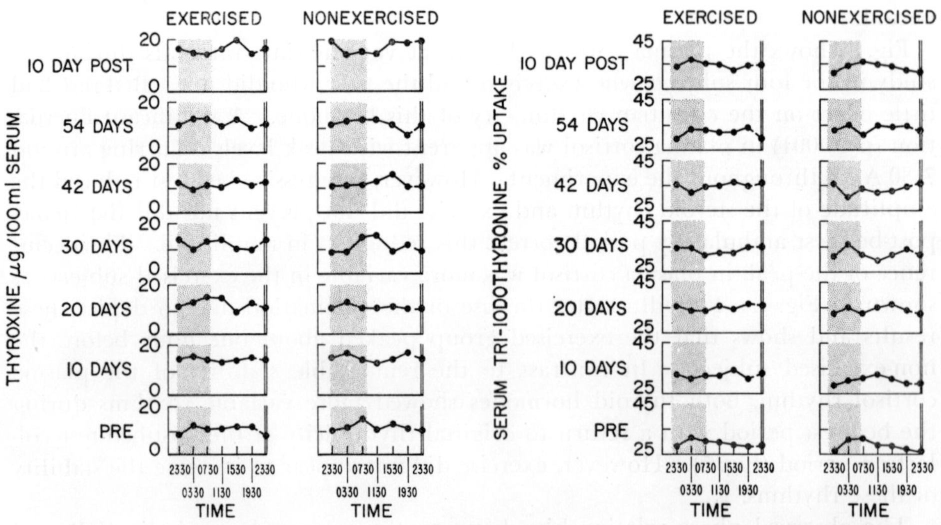

Fig. 5 Serum thyroxine and triiodothyronine in four exercised and four non-exercised subjects before, during, and following 56 days of bed-rest. Stippled area represents lights-off period.

tionships held true between the exercised and non-exercised groups in these parameters.

There is some degree of individual variation observed throughout the study. A subject such as 2A for whom the BT was not well defined, exhibited one of the better defined HR rhythms. On the other hand, Subject 8A who had one of the better BT rhythms, was the only subject to show a random change in phase of the HR rhythms. For some subjects, the two parameters were out of phase during bed-rest. This magnitude of individual variation has been reported previously in the literature [2]. This is not true in the subhuman primate which shows very stable and well defined daily oscillations of HR and BT [3, 4].

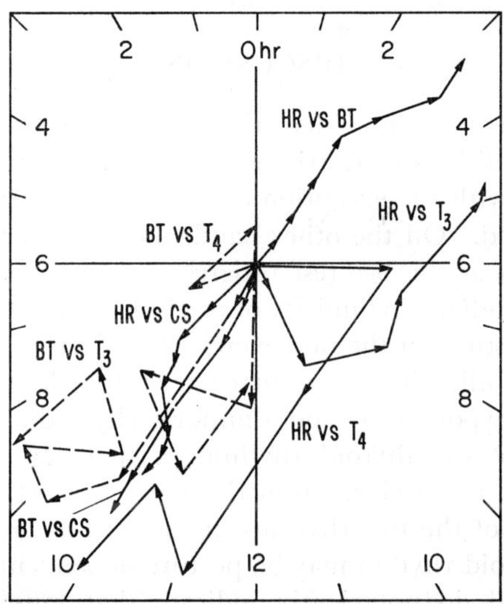

Fig. 6 The phase relationships between the various metabolic (HR and BT) and endocrine (CS, T_3 and T_4) rhythms.

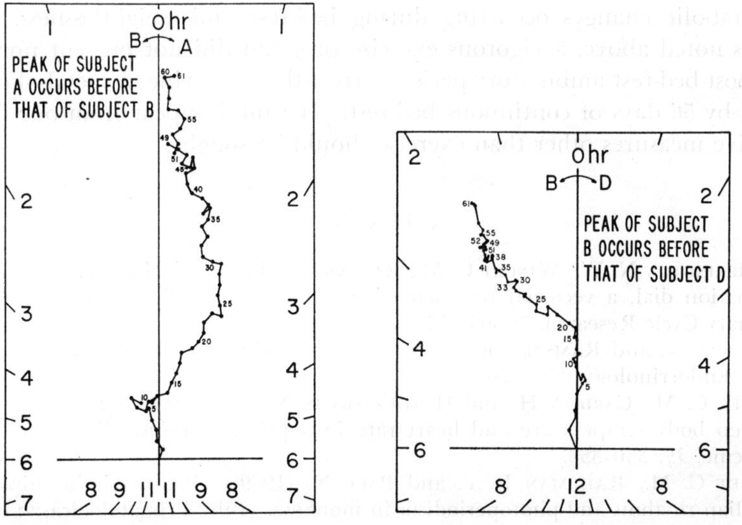

Fig. 7 The vector difference summation dial indicating the influence of social interaction between Subjects A and B and on the right, between Subjects B and D.

The influence of social interaction on circadian rhythms is depicted by the vector difference summation dial of BT rhythms of Subjects A and B (Fig. 7a). During pre-bed-rest, Subject A's BT peaks led those of Subject B by 1.5 hours. The roommates were out of phase for Days 6–14. For the next 10 days (15–24), Subject B led Subject A by about 3.5 hours. The subjects were synchronized toward the end of the study, i.e., the peak times were the same for both subjects.

The social interaction of Subjects B and D, non-roommates, are compared as to time of BT peaks in Fig. 7b. These individuals were synchronized for the first 19 days of the study. Thereafter, to Day 40, peak times of Subject B occurred about four hours earlier than those of Subject D, and about two hours earlier after the 40th day.

DISCUSSION

From the data presented it is evident that diurnal fluctuations of BT, HR, and T_3 change under conditions of restricted mobility. Bed-rest had little effect on the overall T_4 levels with the exception of the sharp increase observed when the subjects got out of bed. On the other hand, there was a prompt and sustained elevation of T_3 levels as the bed-rest progressed. Fifty-six days of bed-rest had little effect on the circadian rhythmicity of cortisol, the peak occurring at the same time throughout the study in the non-exercised subjects. The exercised subjects showed greater variability in the occurrence of the daily peak. Bed-rest also induced a low grade hypothermia and a minor tachycardia.

The dissociation of the thyroid rhythm, particularly T_3, from the cortisol rhythm during bed-rest as well as from the light and activity schedule and the prompt reassociation of the two rhythms in the post-bed-rest ambulatory period suggests that the thyroid rhythm may be posture dependent.

The data presented do not clearly indicate that muscular activity is more important than other external or internal factors of the organism's environment for modulating the dynamics of circadian oscillations. Exercise has been used or at least suggested as the prophylactic or preventive measure for cardiovascular and metabolic changes occurring during bed-rest and weightlessness. In this study, as noted above, a vigorous exercise program did not prevent nor did the 10-day post-bed-rest ambulatory period correct the endocrine or metabolic changes induced by 56 days of continuous bed-rest. It would appear from our data that preventive measures other than exercise should be sought.

REFERENCES

1. Hetherington N. W., Winget C. M., Rosenblatt L. S. and Mack P. B. (1971): The summation dial, a vectorial representation of time series data. Journal of Interdisciplinary Cycle Research, *2*: 365–377.
2. Kleitman N. and Ramsaroop A. (1948): Periodicity in body temperature and heart rate. Endocrinology, *43*: 1–20.
3. Winget C. M., Card D. H. and Hetherington N. W. (1968): Circadian oscillations of deep body temperature and heart rate in a primate (*cebus albafrons*). Aerospace Medicine, *39*: 350–353.
4. Winget C. M., Rahlman D. F. and Pace N. (1969): Phase relationships between circadian rhythms and photoperiodism in monkeys. Bibl. Primatologica, *9*: 64–74.

RHYTHMIC VARIATION IN HEART RATE AND RESPIRATION RATE DURING SPACE FLIGHT—APPOLLO 15

JONN A. RUMMEL

Environmental Physiology Laboratory
NASA, Manned Spacecraft Center
Houston, Texas, U.S.A.

INTRODUCTION

As man travels into space and beyond the confines of his evolutionary environment, he must be provided with a "capsule" of his terrestrial surroundings. Although we have prescribed the physiological limits for many of the required environmental parameters, we have not defined what part of man's temporal environment must be included [1].

There are two basic questions in relation to circadian rhythms and space travel: (1) Are synchronized 24-hour rhythms of physiological processes, which have set phase-relationships, necessary for the physiological and psychological performance required during space flight? and (2) What effect, if any, does an altered input from subtle geophysical factors (i.e., magnetic field, gravity, radiation) have on basic circadian mechanisms?

Up to and including the present time, operational constraints have prevented a detailed scientific evaluation of manned space flight and circadian rhythms. Not only are crewmembers unavailable for longitudinal evaluation prior to and after space flight, but it has not been possible to measure physiological variables of choice. Additionally, those variables that have been measured have been sampled irregularly.

As part of the operational biomedical monitoring for Apollo manned missions, ECG and respiration rate are telemetered at selected intervals to mission control. The data presented in this paper were collected as part of this monitoring program. These data were evaluated for circadian and ultradian rhythmicity because of the uniqueness of this data, but no attempt can be made under the present constraints to evaluate specific scientific hypotheses.

METHODS

The electrocardiogram (ECG) and impedance pneumograph (ZPN) were measured on the three crewmembers of the Apollo 15 mission (launched on July 26, 1971, at 08:34 CDT). The equipment for these measurements was part of the standard operational bioinstrumentation. These signals were telemetered as one minute values of heart rate and respiration rate.

The initial work/rest schedule for this mission was maintained on Cape Kennedy time and approximate 8-hour sleep periods were initiated at 2':34 CDT of launch day. On day three of the mission and again on day five the beginning of the sleep period was advanced 1 hour. Additional advances in the beginning of the sleep period occurred on day six (2 hours) and day seven (2 hours) for a total shift of approximately 6 hours. The rest period was then rescheduled on day eight to agree more closely with the original (launch) schedule.

For circadian analysis one minute averages of heart rate and respiration rate were sampled at 30-minute intervals. Since data transmission was not continuous, any available data points within ±5 minutes of each 30 minutes were selected. The exact mission times of data sampling were maintained. There was limited preprocessing of the raw data to eliminate any spurious noise spikes on the telemetered data stream.

Heart rate and respiration rate were analyzed for circadian periodicity utilizing a least-squares technique [8]. Analysis included a least-squares spectrum computed over the entire flight as well as an incremental trend analysis. The latter was accomplished by forcing a 24-hour period through the data over small and successive time intervals and observing phase relationships. The data window utilized in the trend procedure was 72 hours, incremented in 24-hour periods. This process smoothed individual evaluation points since each estimate contained 48 data points the estimate on either side.

For ultradian analysis one minute heart rate data values were utilized with each sequential anlysis comprising an 8-hour block of time. The same least-squares program was utilized with the analysis window set between 60 and 160 minutes in one-minute increments.

An additional procedure (iterative multiple regression) has been developed for minimizing the inter-effects of multiple periods within the same data. Multiple periods can occur as a result of true biological rhythms in different time periods [3] or can be erroneously introduced as a result of blocks of missing data.

The following steps are utilized in this procedure:

1. A least-squares spectrum is computed.
2. Significant periods are selected on the basis of amplitude/standard error ratios.
3. A multiple time series regression analysis is computed utilizing all significant periods found in Step 2.
4. Periods originally selected which are no longer significant (as a result of being a side lobe of a major period) are deleted from the analysis.
5. Remaining significant periods are iterated to either side of their original values to remove any interaction in the initial analysis.
6. A least-squares spectrum is repeated using significant periods in a multiple regression model.
7. Steps 2–7 are repeated if any new periods are detected.

RESULTS

Fig. 1 shows the least-squares spectrum between 20 and 30 hours for respiration rate and heart rate for the three crewmen during the entire flight. The abscissa shows the period in hours and the ordinate the amplitude/standard error ratio. All three crewmen exhibited significant variability in the circadian range in both heart rate and respiration rate. However, only in the case of respiration rate for Astronaut 1 and heart rate for Astronaut 3 were the 24-hour estimates the most significant.

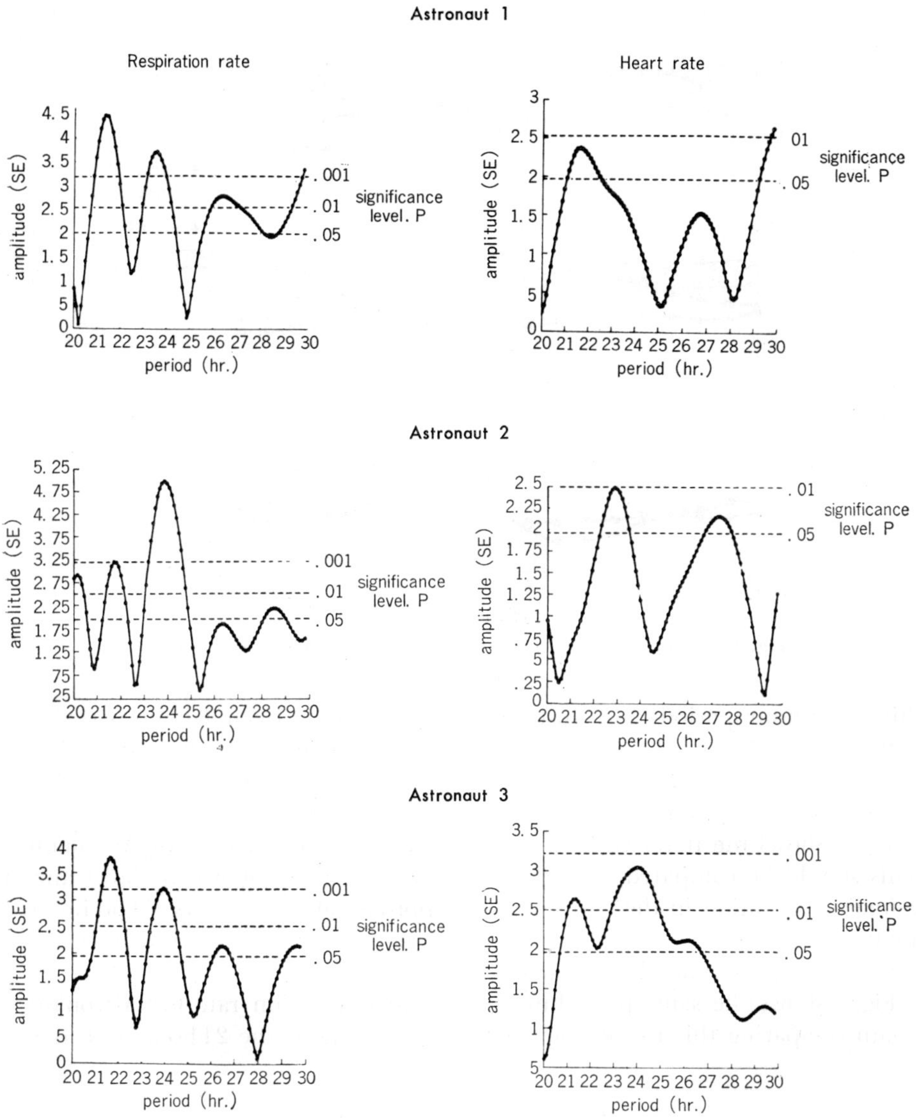

Fig. 1 (NASA-S-71-54315).

The baseline (±S.E.)/amplitude (±S.E.) values for the most significant circadian variability in respiration rate (breaths/minute) for the three astronauts were 6.9±0.3/3.0±0.8, 8.5±0.2/2.8±0.5, and 10.2±0.3/2.7±0.7, respectively. For heart rate (beats/minute) these values were 72±1.7/9.6±3.5, 67±1.2/6.0±2.4, and 76±1.4/8.9±2.9.

The stability of the near 24-hour periods is demonstrated in Fig. 2 which shows the acrophase of a forced 24-hour period in 72-hour segments of the data. All three crewmen exhibited shifts which came in the same time period as a shift in the scheduled work rest cycle.

The analysis for ultradian heart rate variability in the range of 60–160 minutes produced significant results in three 8-hour time periods for Astronaut 1, in

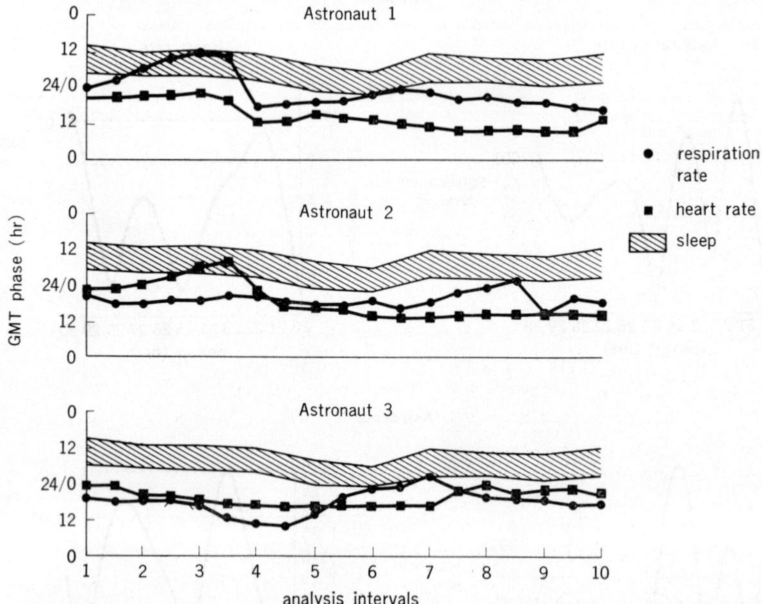

Fig. 2 Acrophase respiration rate—heart rate (NASA-S-71-3443-S).

thirteen 8-hour periods for Austronaut 2, and in fourteen 8-hour periods for Astronaut 3. The total number of 8-hour periods analyzed was thirty-three. The average amplitude (±S.D.) in beats/minute for this variability was 6.3±4.1, 4.2±1.3, and 5.6±3.8, respectively, for the three astronauts.

Fig. 3 shows the iterative least-squares analysis for heart rate for Astronaut 1. This should be compared with the comparable Fig. 1 plot which shows an undefinable alteration in the major peak at approximately 24 hours. The iterative analysis procedure permitted separation of the two periods with a distinct period detected at approximately 24 hours.

Fig. 4 shows the same procedure applied to respiration rate for Astronaut 2. Again comparing this to the comparable plot of Fig. 1, the 24-hour peak is now

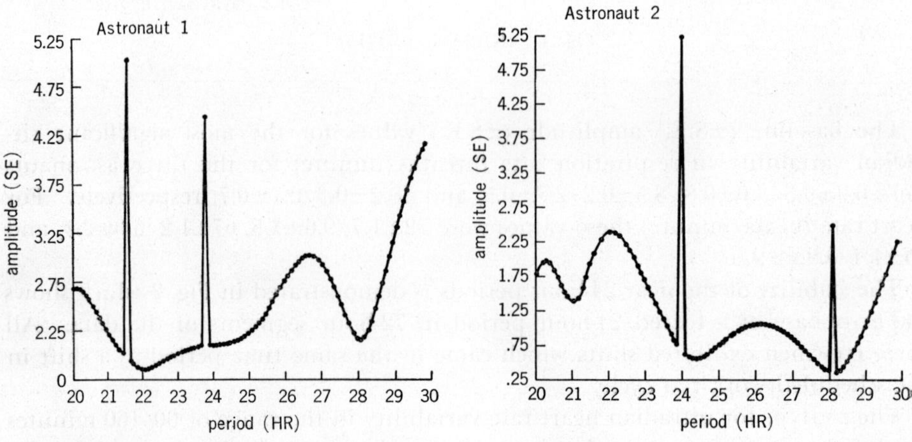

Fig. 3 Heart rate (NASA-S-71-3442-S). **Fig. 4** Respiration rate (NASA-S-71-3440-S).

well defined instead of the original 23–25 hour spread in the original least-squares analysis. A smaller amplitude 28-hour period is also visible after the 24-hour variability has been removed from the overall spectrum.

DISCUSSION

The ability to detect and quantitate biorhythms in living systems during space flight is an important aspect of evaluating hypotheses concerning the underlying mechanisms of these phenomena. Although circadian variation in heart rate during space flight is demonstrated in this paper and has been reported previously [4, 8], the close relationship between this variable and the imposed activity cycle prevent any detailed discussion regarding basic mechanisms. Circadian variation has also been observed in ECG waveform parameters (unpublished data) but the lack of comparable groundbased data makes the evaluation of these findings impossible at the present time.

The amplitudes of the circadian periods in heart rate for the three astronauts are of the same order of magnitude as previously reported [5]. The amplitudes of the respiration rate circadian rhythms are larger than those reported in previous studies [2, 7]. Whether this is the result of environmental factors associated with this flight or a result of data sampling cannot be determined.

The detection of ultradian periodicity of approximately 90 minutes [6] has become a new area of investigation. Significant variability in this range of biorhythms was detected in the present data but it cannot be compared to any pre- or postflight data. One of the crewmembers had much less variability in these periods but again the reason for this is unknown. There was no particular portion of the day in which the rhythmicity occurred and there were several days when it occurred throughout the day.

The collection of biological time series data during space flight may be influenced by irregular collection intervals as well as periods of complete data loss. The analysis of these data cannot be accomplished utilizing techniques which require even data spacing. The least-squares technique championed by Dr. Franz Halberg appears to be the only feasible approach on data of this type and has been quite successful in providing a quantitative "microscopic" description of biorhythmic phenomena. However, this author feels that a remaining unknown in this approach is the inter-dependence and influence of multiple significant periods in the same data set.

The approach described in the present paper attempts to eliminate this problem by simultaneously accounting for the significant periods in a multiple regression analysis. In analyzing generated time series data it has been found that period discrimination is much better than the theoretical limit. The reason for this is that the technique is iterative and essentially subtracts detected periods and their resultant variability from the spectrum. The statistical properties of this technique are currently being investigated utilizing generated time-series data with known properties.

The detection and analysis of biorhythms during space flight offer a unique opportunity to evaluate basic unknown properties of these phenomena. However, operational constraints of present manned vehicles have prevented a detailed evaluation. During the interim period until such limitations are removed we should develop techniques and rationale which will maximize the amount of information resulting from data that are available. One such area of investigation which may provide additional insight is the study of circadian variation in ECG waveform parameters. Also data analysis techniques should provide accurate "microscopic" quantitation in order to detect subtle changes.

REFERENCES

1. FRAZIER T. W., RUMMEL J. A. and LIPSCOMB H. S. (1968): Circadian variability in vigilance performance. Aerospace Medicine, *39*: 383–395.
2. GÜNTHER R., KNAPP E. and HALBERG F. (1969): Referenznormen der Rhythmometrie: circadiane acrophasen von zwanzig Körperfunktionen (II). Z.f. Ang. Bäder und Klimaheilkunde, *16*: 123–153.
3. HALBERG F., ENGELI M., HAMBURGER C. and HILLMAN D. (1965): Spectral resolution of low-frequency, small amplitude rhythms in excreted ketosteroid; probable androgen-induced circaseptan desynchronization. Acta Endo. Suppl., *103*: 5–54.
4. HALBERG F., VALLBONA C., DIETLEIN L., RUMMEL J., BERRY C., PITTS G. and NUNNELEY S. (1968): Human circadian circulatory rhythms during weightlessness in extraterrestrial flight or bedrest with and without exercise. Space Life Sciences, *1*: 14–28.
5. HALBERG F. (1969): Chronobiology. Ann. Rev. of Physiol. *31*: 675–725.
6. HARTMAN E. (1968): The 90-minute sleep-dream cycle. Arch. Gen. Psy., *18*: 28–286.
7. HAUTY G. T. (1962): Periodic desynchronization in humans under outer space conditions. Ann. N. Y. Acad. Sci., *98*: 1116–1125.
8. RUMMEL J., SALLIN E. and LIPSCOMB H. (1967): Circadian rhythms in simulated and manned orbital space flight. Rassegna di Neurologia Vegetativa, *21*(1–2): 41–56.

HYDROCORTISONE AND ACTH LEVELS IN MANNED SPACEFLIGHT

Carolyn S. LEACH[1] and Bonnalie O. CAMPBELL[2]

[1]*Head, Endocrine Laboratory, NASA/MSC*
[2]*Department of Physiology, Baylor College of Medicine*
Houston, Texas, U.S.A.

Numerous studies have described the circadian periodicity of pituitary-adrenal function. Circadian variation of plasma 17-hydroxycorticosteroid levels has been well documented in the human [2, 3]. The constancy of this pattern during total bedrest [5], night work [7], illness [20, 21], as well as the difficulty in altering this pattern with changes in work-rest routine [23] and the problems in assessing the effect of stress-induced changes on this documented pituitary-adrenal periodicity have been discussed [11].

It is the purpose of our laboratory to assess the hormonal-metabolic response of man to the spaceflight environment. The conditions which can be considered stresses that man must face in spaceflight include weightlessness, acceleration, confinement, restraint, unique atmospheres, long-term maintenance of high levels of performance and possible desynchronosis. It must be considered that subjects escaping from the earth's gravity are exposed to changing day and night schedules. The results of this transition could produce compromising circumstances for the crewmen of extended space voyage.

We have chosen to consider the reactions of the pituitary-adrenal axis as one indicator of the effect of spaceflight on the Apollo crewmen. The spacecraft size and the operational complexities which had to be mastered during the Apollo program made it necessary to defer attempts to conduct more elaborate in-flight biomedical studies. However, detailed pre- and postflight observations were conducted in order to better comprehend the effects of spaceflight.

METHODS

Blood samples were drawn 30, 14 and 5 days before the flight for analysis to ascertain the health status of the crew and to establish baseline values for postmission comparison. Similar samples were drawn immediately after spaceflight recovery, usually about 2 hours, and on future days until the return to baseline status had been attained for most of the variables. Except for the sample drawn immediately after splashdown, all samples were obtained in the early a.m. immediately after arising with the men in a fasting condition.

Twenty-four-hour urine samples were collected from each man (usually) starting on the same day as the blood collection. The crew consumed a diet of their own choice during the pre- and postflight periods and ate the provided Apollo diet throughout the mission.

Analysis of the blood plasma samples included, among others, hydrocortisone, utilizing the binding globulin technique of Murphy [17], and adrenocorticotrophic hormone (ACTH) using the method of radioimmunoassay [4]. The 24-

hour urine samples were analyzed for hydrocortisone by the binding globulin technique [17]. The immediate postflight results were compared to a mean of the preflight values by the Student's T test [25].

RESULTS

Table 1 shows the plasma hydrocortisone values for each Apollo crewman. The immediate postflight samples are significantly lower than a mean of the preflight samples ($p<.001$) when the samples for the entire group are compared.

Table 2 gives the plasma ACTH values for the two missions on which this test was conducted. When the postflight sample is compared to a mean of the preflight samples, no significant difference is noted ($p>.05$).

Table 3 shows the values for the 24-hour excretion of hydrocortisone. The first postflight samples are significantly higher than a mean of the preflight samples ($p<.01$).

Table 1 Plasma hydrocortisone values (μg/100 ml).

Apollo flight	Crewman	Preflight F−30	F−14	F−5	$\bar{X}\pm$SE	Postflight ASAP*	R+1	R+4	R+6	R+16
8	1	14.1	9.1	12.7	12 ±1.5	6.5	11.8		10.8	
	2	12.8	14.5	19.3	15.5±1.9	5.0	15.4		11.5	
	3	16.6	12.2	18.8	15.9±1.9	6.9	10.9		14.5	
9	1		6.0	12.0	9.0±3	14.2	18.8			9.8
	2		9.2	16.5	12.9±3.7	5.6	14.6			8.2
	3		13.5	17.0	15.3±1.8	10.4	9.4			12.5
10	1		10.2	9.0	9.6±0.6	5.2	13.9			
	2		15.1	24.8	20.0±4.9	5.4	13.1			
	3		12.1	10.1	11.1±1.0	6.1	11.2			
11	1	11.0	14.3	18.4	14.6±2.1	20.2			24.4	14.5
	2	13.8	11.1	20.6	15.2±2.8	10.7			17.2	15.6
	3	13.5	9.9	13.3	12.2±1.2	5.6			15.4	15.0
12	1	13.0	15.0	17.2	15.1±1.2	9.4			11.6	
	2	13.2	24.0	21.4	19.5±3.3	14.5			14.4	
	3	10.4	16.8	13.6	13.6±1.8	12.2			18.8	
13**	1	29.0	21.5	25.5	25.3±2.2	44.0		10.6	23.9	18.9
	2	30.4	17.2	27.6	25.1±4.0	40.6		25.4	29.8	12.8
	3	18.2		4.6	11.4±6.8	15.3		33.4	31.8	13.8
14	1	12.0	16.8	17.8	15.5±1.8	9.2	12.8		10.8	14.0
	2	17.6	14.8	22.4	18.3±2.2	12.2	19.6		22.4	20.2
	3	17.6	18.4	23.2	19.7±1.7	8.0	19.2		10.4	22.0
15	1	12.0	18.4	29.0	19.8±5.0	7.2	23.6		20.8	16.0
	2	15.0	14.4	20.8	16.7±2.0	7.6	14.8		34.4	12.0
	3	12.4	19.0	15.8	15.7±1.9	15.2	28.0		23.2	15.0

* Differs from preflight mean $p<.001$

** Not included in analysis

DISCUSSION

The significant decrease in plasma hydrocortisone immediately following spaceflight is an interesting finding. A search for the possible explanation for this initial observation has been one of the main interests of our laboratory. In con-

Table 2 Plasma ACTH values (μg/ml).

Apollo flight	Crewmah	Preflight			Mean $\bar{X}\pm SE$	Postflight			
		F−30	F−14	F−5		ASAP*	R+1	R+6	R+16
8	1		38	35	37±1.5	27		22	
	2		30	27	29±1.5	18		23	
	3		22	40	31±9.0	30		25	
15	1	41	45	20	35±7.8	54	22	33	47.4
	2	48	52	59	53±3.2	20		50	39.3
	3	100	90	90	93±3.3	32	68	72	52.3

* Not significantly different from pre-flight mean p>.05

Table 3 Urinary hydrocortisone values (μg/24 hour volume).

Apollo flight	Crewman	Preflight			Mean $\bar{X}\pm SE$	Postflight				
		F−30	F−14	F−5		ASAP*	R+1	R+2	R+6	R+16
3	1		40.0	62.0	51.0±11.0	131.0			63.0	
	2		49.0	100.0	74.5±25.5	34.0			86.0	
	3		136.0	52.0	94 ±42.0	182.0			50.0	
9	1		73.8	59.4	66.6± 7.2	102.0				
	2		76.9	104.7	90.8±13.9	80.4				
	3		103.6	125.0	114.3±10.7	106.0				
11	1	46.8	50.8		48.8± 2.0	51.1			56.5	26.4
	2	51.8	68.7		60.3± 8.4	56.9			187.7	162.8
	3	23.3	53.1		38.2±14.9	51.6			18.6	46.1
13**	1	118.5	99.5	54.7	90.9±18.9	94.7	119.0			
	2	55.0	87.5	43.0	61.9±13.2	42.3	106.7			
	3			57.5		77.3	49.6			
14	1	17.5	105.5	75.2	66.1±25.8	103.9	109.1	62.4	74.6	
	2	16.2	36.7	71.3	41.4±16.1	74.3	64.4	77.4	49.0	
	3	39.9	100.0	109.2	83.0±21.7	121.0	86.0	43.5	39.2	
15	1	95.5	52.5	74.7	74.2±12.4	136.8	85.3	101.4	53.3	80.8
	2	36.4	46.9	94.5	59.3±17.9	86.0	84.4	110.4	115.6	71.2
	3	19.3	42.2	22.7	28.1± 9.1	62.0	136.0	34.8	185.0	44.3

* Differs from pre-flight mean p<.01
** Not included in analysis.

sideration of these findings, certain areas have been examined and the following possibilities have evolved.

There is evidence that adrenal-cortical function is suppressed during spaceflight [15]. The decreased 17-hydroxycorticosteroid excretion measured in urine from the inflight Gemini 7 metabolic-endocrine experiment was noted but not explained. Therefore, our results could reflect an insensitive adrenal gland due to decreased inflight function. However, this does not seem likely since there is a general increase in the urinary excretion of hydrocortisone in the first 24 hours postflight, indicating a responsive gland. Furthermore, the normal values of the plasma ACTH indicate that the adrenal gland was functioning properly; ACTH would have been elevated if the adrenals were not secreting adequate hydrocortisone.

Another possible explanation might be in the review of the adrenal-cortical hormonal response in plasma to exercise. Since there is no way to quantitate the physical stress of spaceflight, and more particularly to splashdown and recovery, it is difficult to compare the response to any known test. However, the literature does contain references noting a decrease in plasma adrenocortical hormones following stressful exercise [9, 22, 26]. While these results are not fully explained, the authors do suggest several possibilities. CORNIL et al. and STAEHELIN et al. propose an increased utilization of cortisol occurring with stress which could produce a stimulation of adrenocortical activity and also bring about an increase in the metabolism of the steroids demonstrated by a fall in the blood level. It is possible that the same process occurs postflight in the Apollo crewmen. This hypothesis could be verified by using the isotope dilution method to determine the production rate of cortisol during the stress event.

The circadian variability is a third possibility for the observed adrenal-cortical response postflight. The importance of work-rest cycles in the adrenal-cortical rhythm is evident and has been studied in relation to shift work, time-zone shifts, and light/dark cycles [8, 10, 18, 24]. However, the effect of altered work-rest cycles on the adrenal-cortical rhythm is not clear. MIGEON et al. [16] reported that night workers and blind persons have the same diurnal variation as normal persons. However, there is a question as to these subjects' complete adaptation to the phase shift. In contrast, CONROY [7] found that plasma corticosteroid rhythm does adapt in persons habitually on nightwork. His subjects demonstrated peak values about the time of awakening. Furthermore, PERKOFF et al. [20] in a study of longer duration found that a reversal of sleep periods resulted in inverted plasma 17 OHCS rhythm. Another study which imposed 12-, 19- or 33-hour work/rest cycles resulted in a disturbed corticosteroid rhythm. Several days after these schedules were enforced, maximal values were seen about the time of awakening [19]. A recent study conducted by A. N. LITSOV [13] studying frequently alternating work and rest cycles concluded that schedules involving significant deviations from normal rhythms can be tolerated by subjects for only short periods, to be followed by more favorable cycles; he did not consider the adrenal cycle.

Recognizing this potential problem, our laboratory took advantage of a 3-day chamber simulation of an Apollo lunar mission to study the effect of an abnormal work-rest cycle on one subject. This simulation resulted in a fatigued subject who, in the face of clinical and biochemical evidence of stress, still retained a 23.9-hour hydrocortisone rhythm. These results agree with CURTIS and FOGEL who conducted a study with six subjects on "scrambled" sleep schedule. After a week, they found that despite the random schedule, adrenal steroid levels seemed to reach a peak in the early morning, with secondary peaks after sleep periods [14]. Our findings also support those of HALBERG et al. who found the persistence of the adrenal-cortical rhythm in subjects continuously active for 2 days [10].

Although rhythmic influence is an acknowledged area of concern in manned spaceflight, the operational constraints of the missions to date have not allowed the inclusion of these considerations in the mission scheduling. This has resulted in imposed phase shifts in work-rest cycles of as much as 12 hours in 5 days on some mission plans, as exemplified in Fig. 1. It has been established that the peak of adrenocortical secretion usually occurs just before wakefulness [12], and that persons continually shifted in sleep-wake schedule alter the adrenal-cortical rhythm [20]. The conditions imposed on the Apollo crewmen resemble those ground base studies involving the impositions of abnormal work-rest time schedules. However, data comparing the actual work-rest to the mission plan is not available.

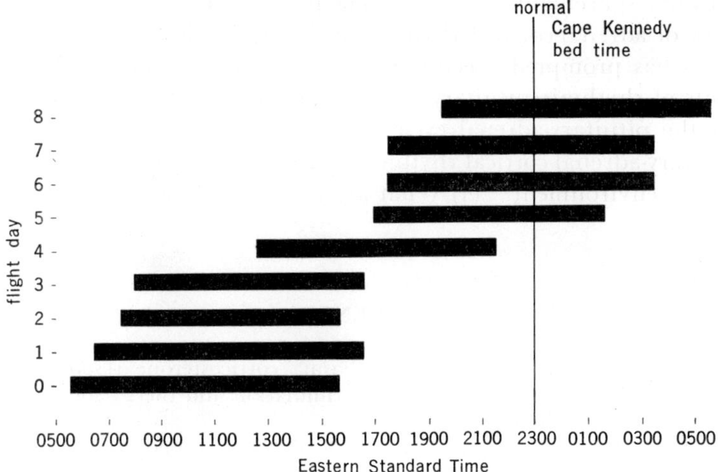

Fig. 1 Apollo 12 crew work/rest periods.

The Apollo flights are launched from Kennedy Spaceflight Center, Florida on Eastern Standard Time and the entire missions are maintained on EST for their durations. Table 4 shows the time schedule for our blood withdrawals pre- and post-spaceflight. It is of interest to note that the low plasma hydrocortisone values generally found post-flight are explained if one considers only the circadian variation. However, we believe the stress imposed by the events of recovery are sufficient to significantly increase plasma hydrocortisone values at any point in the cycle. This is suggested because there is evidence of an increased response by the adrenal cortex to stress when the stimulation is superimposed upon the trough of the cycle in animals [1].

Table 4 Apollo program blood collection times (NASA-S-71-3386-S).

Flight number	Preflight Cape Kennedy Est	Immediately postflight recovery ship		Mission durations Days
		Local time	Est	
8	7:30—8:30 AM	6:00 AM	12 noon	8
9	7:30—8:30 AM	7:30 AM	1:30 PM	10
10	7:30—8:30 AM	9:00 AM	2:00 PM*	8
11	7:30—8:30 AM	9:00 AM	2:00 PM*	8
12	7:30—8:30 AM	11:00 AM	5:00 PM	10
13	7:30—8:30 AM	10:30 AM	3:30 PM*	6
14	7:30—8:30 AM	10:00 AM	4:00 PM	10
15	7:30—8:30 AM	12:30 PM	5:30 PM*	13

* Daylight saving time

Even if there is a decrease in responsiveness of the adrenal gland to ACTH during the trough, as has been proposed, we are of the opinion that the central nervous system control of the pituitary-adrenal axis [6] would override the normal circadian cycle under conditions of stress.

The information presented here describes data collected on the Apollo crewmen before and after spaceflight. Limited data preclude firm conclusions about the maintenance of adrenal-cortical rhythms during spaceflight. However, a review of these data has prompted specific questions; these include: (1) What are the determinants of rhythmic pituitary-adrenal functions in a stressful environment? (2) How is the pituitary-adrenal cycle affected by variable work/rest schedules? (3) Are pituitary-adrenal-cortical rhythms necessary for man to function adequately in the space environment? (4) What are the best methods to use for studying these changes?

REFERENCES

1. ADER R. and FRIEDMAN S. B. (1968): Plasma corticosterone response to environmental stimulation: effects of duration of stimulation and the 24-hour adrenocortical rhythm. Neuroendocrinology, *3*: 378–386.
2. BARTTER F. C., DELEA C. S. and HALBERG F. (1962): A map of blood and urinary changes related to circadian variations in adrenal cortical function in normal subjects. Ann. N. Y. Acad. Sci., *98*: 969–983.
3. BLISS E. L., SANDBERG A. A., NELSON D. H. and EIK-NES K. (1953): The normal levels of 17-hydroxycorticosteroids in the pheripheral blood in man. J. Clin. Invest., *32*: 818–823.
4. CAMPBELL B. O., LEACH C. and LIPSICOMB H. S. (1971): Radioimmunoassay of plasma corticotrophin in normal males and females. The Physiologist, *14*: 118.
5. CARDUS D., VALLBONN C., VOGT F. B., SPENCER W. A., LIPSCOMB H. S. and EIKE-NES K. B. (1965): Influence of bed rest on plasma levels of 17-hydroxycorticosteroids. Aerospace Medicine, *36*: 524–528.
6. CLAYTON G. W., LIBRIK L., GARDNER R. L. and GUILLEMIN R. (1963): Studies on the circadian rhythm of pituitary adrenocorticotropic release in man. J. Clin. Endocr., *23*: 975–980.
7. CONROY R. T. W. L. (1967): Circadian rhythm of 11-hydroxycorticosteroids in night workers. J. Physiol., *191*: 21–22.
8. CONROY R. T. W. L., ELLIOTT A. L. and MILLS J. N. (1970): Circadian rhythms in plasma concentration of 11-hydroxycorticosteroids in men working on night shift and in permanent night workers. Brit. J. Industr. Med., *27*: 170–174.
9. CORNIL A., DECOPINSCHI G. and FRANCKSON J. R. M. (1965): Effect of muscular exercise on the plasma level of cortison in man. Acta Endocrinol., *48*: 163–168.
10. HALBERG F., FRANK G., HARNER R., MATTHEWS J., AAKER H., GRAVEM H. and MELBY J. (1961): The adrenal cycle in men on different schedules of motor and mental activity. Experientia, *17*: 282–284.
11. KLEIN K. E., WEGMANN H. M. and BRUNER H. (1968): Circadian rhythm in indices of human performance, physical fitness and stress resistance. Aerospace Medicine, *39*: 512–518.
12. KRIEGER D. T., ALLEN W. RIZRZO F. and KRIEGER H. P. (1971): Characterization of the normal temporal pattern of plasma corticosteroid levels. J. Clin. Endocr. Metabol., *32*: 266–284.
13. LITSOV A. H. (1971): Diurnal rhythm of human physiological functions and performance in a schedule with frequent alteration of sleep and wakefulness. Kosmicheskaya Biologiya Meditsina, *5*: 44–52.
14. LUCE G. G. (1970): Biological rhythms in psychiatry and medicine. Public Health Service Publication, *2088*: 137.
15. LUTWARK L., WHEDON G. D., LACHANCE P. A., REID J. M. and LIPSCOMB H. S. (1969): Mineral, electrolyte and nitrogen balance studies of the Gemini-VII space flight. J. Clin. Endocr., Metabol., *29*: 1140–1156.
16. MIGEON G. J., TYLER F. H., MAHONEY J. P., FLORENTIN A. A., CASTLE H., BLISS E. L. and SAMUELS L. T. (1956): The diurnal variation of plasma levels and urinary excretion of 17-hydroxycorticosteroids in normal subjects, night workers and blind subjects. J. Clin. Endocr., *16*: 622–633.

17. MURPHY B. E. (1967): Some studies of the protein-binding of steroids and their application to the routine micro and ultramicro measurement of various steroids in body fluids by competitive protein-binding radioassay. J. Clin. Endocr., *27*: 973–990.
18. ORTH D. N. and ISLAND D. P. (1969): Light synchronization of the circadian rhythm in plasma cortisol (17-OHCS) concentration in man. J. Clin. Endocr., *29*: 479–486.
19. ORTH O. N., ISLAND D. P. and LIDDLE G. W. (1967): Experimental alteration of the circadian rhythm in plasma cortisol concentration in man. J. Clin. Endocr., *27*: 549–555.
20. PERKOFF G. T., EIK-NES, K., NUGENT C. A., FRED H. L., NIMER R. A., RUSH L., SAMUELS L. T. and TYLER F. H. (1959): Studies of the diurnal variation of plasma 17-hydroxycorticosteroids in man. J. Clin. Endocr., *19*: 432–443.
21. RAIMONDO V. C. Di and FORSHAM P. H. (1956): Some clinical implications of the spontaneous diurnal variation in adrenal cortical secretory activity. Am. J. Med., *21*: 321–323.
22. ROSE I., FRIEDMAN H. S., BERRING S. C. and COOPER K. H. (1970): Plasma cortisol changes following a mile run in conditioned subjects. J. Clin. Endocr., *31*: 339.
23. SHARP B. W., SLORACH S. A. and VIGOND H. J. (1961): Diurnal rhythm of keto- and ketogenic steroid excretion and the adaptation to changes of the activity-sleep routine. J. Endocr., *22*: 377–385.
24. SIMPSON H. W. and LOBBAN M. C. (1967): Effect of a 21-hour day on the human circadian excretory rhythms of 17-hydroxycorticosteroids and electrolytes. Aerospace Medicine, *38*: 1205–1243.
25. SNEDECOR G. W. (1956): Statistical Methods Applied to Experiments in Agriculture and Biology. pp. 45–47, The Iowa State University Press, Ames. Iowa.
26. STAEHELIN D., LABHART A., FROESCH R. and KAGI N. R. (1955): The effect of muscular exercise and hypoglycemia on the plasma level of 17-hydroxysteroids in normal adults and in patients with adrenogenital syndrome. Acta Endocrinol., *18*: 521–529.
27. ACTH studies were conducted at Baylor College of Medicine, Houston, Texas. NIH AM-04122 and NAS 9-10547.

THE HUMAN CIRCADIAN SYSTEM AND AEROSPACE TRAVEL*

HUGH W. SIMPSON

Department of Pathology, University of Glasgow
Glasgow, Scotland
Chronobiology Laboratories, University of Minnesota
Minneapolis, Minnesota, U.S.A.

The fact of aerospace travel and the thought of life on other celestial bodies, provokes one to ask what effect the unusual day lengths will have on human rhythms; this is particularly pertinent in the case of our circadian system which appears to be an evolutionary adaptation suitable only for this earth and any other bodies with a similar period.

Other day/night cycles may be simulated in various ways such as in bunkers, caves, etc. [1], but the method I and others have used extensively is that pioneered by KLEITMAN namely the study of individuals in the Arctic when summer daylight is continuous and 'night' may be created in huts. Such considerations, amongst others, led me to organize two expeditions (Spitsbergen 1960; Devon Island 1969) to study the effect of a 21-h day/night cycle on eight adults for up to

Table 1 Cosinor evidence for a presumably intrinsic period, ($24.0\ h<\tau<24.8\ h$) in human subjects living on a 21-h 'day' for up to 5 weeks: Acrophase estimates based on fitting a 24-h cosine curve. ϕ reference=00^{00} (local midnight) with $360°\equiv$ 24 h.

		Urinary excretory variables				
		17–OHCS	K^+	H_2O	Na^+	Cl^-
ϕ during pre-control 24 h		−190° (−154 to −235)	−204° (−163 to −307)	−217°	−213°	−206° (−176 to −239)
Difference in ϕ between successive spans	21-h 'day' wks I	− 3°	− 16°	− 15°	− 36°	− 22°
	III	− 53°	− 30°	+ 3°	− 4°	− 29°
	V	− 53°	− 50°	− 40°	− 43°	− 7°
Cumulative ϕ drift		−109°	− 96°	− 52°	− 83°	− 58°
ϕ after return to "24-h day"		−202° (−159 to −236)	−197° (−177 to −219)	−207°	−212°	−206° (−173 to −257)

All available subjects. Note that in successive spans on 21-h time there is a drift to a later acrophase on 14 out of 15 occasions and that the cumulated drift usually exceeds the 95% confidence limits of phase of the control 24-h days before and after the study. In view of this it seems likely that the data are characterized by a >24-h period. Week I, 7 subjects; Week III, 5 subjects, Week V, 3 subjects.

* Supported by grant from the Medical Research Council.

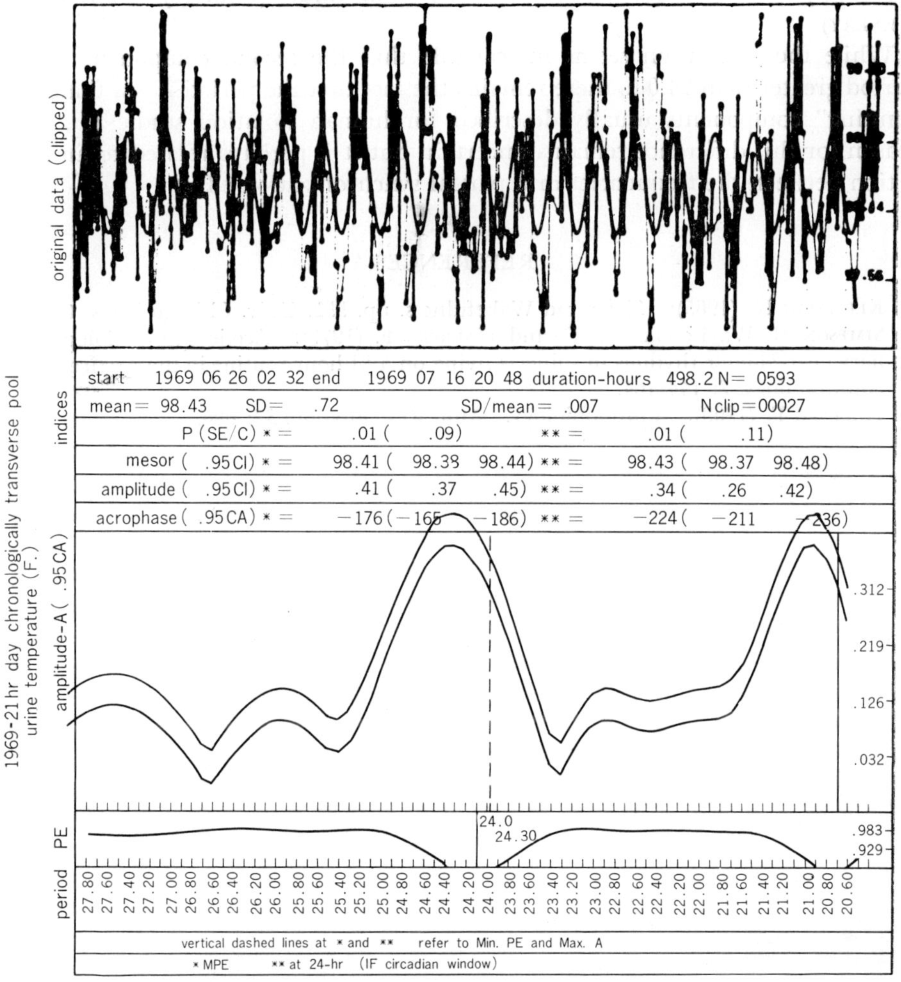

Fig. 1 Urine temperature. Three calendar weeks data with five human subjects on a 21-h routine. Top, the raw data. Bottom, the serial amplitudes of the best-fitting cosine waves in spectral range 20.6 h to 28.0 h. Note the two prominent frequences.

7 weeks. The endpoints were the circadian rhythms in body temperature and the excretion of electrolytes and 17-hydroxycorticosteroids. An extensive previous publication summarizing earlier statistical analyses has already appeared [2]. The conclusions of this work were that a 21-h routine in man resulted in a split of the normally single 24-h circadian component into an environmentally timed 21-h component and an about 24-h component (see Fig. 1).

The subject of this communication is to re-emphasize the non-24-h nature of this intrinsic component and to give evidence that it is at a period significantly greater than 24.0 h.

The evidence is derived from the fact that when a precisely 24-h cosine wave approximating function is serially fitted by least squares to each calendar week of data, the acrophase estimates are progressively later and later (see Table 1). Furthermore spot checks of the same phase estimates with the 21-h component previously removed from the data reveal similar values (Courtesy of Dr. Robert HENERY).

While the conclusion seems inescapable that the second component is at a period greater than 24.0 h, there remains the fact that the period is less than "free running" components usually described for human isolation studies [3]. The reason for this difference remains uncertain but it is perhaps related to an interaction between the frequencies or a partial synchronization effect.

REFERENCES

1. KLEITMAN N. (1965): Sleep and Wakefulness. pp. 552, Univ. Chicago Press, Chicago.
2. SIMPSON H. W., LOBBAN M. C. and HALBERG F. (1970): Arctic chronobiology. Urinary near-24-hour rhythms in subjects living on a 21-hour routine in the Arctice. Arctic Anthropology, *7*: 144–164.
3. HALBERG F. (1969): Chronobiology. Ann. Rev. Physiol., *31*: 696.

CHANGES IN INTERNAL PHASE RELATIONSHIPS DURING ISOLATION

JÜRGEN KRIEBEL

Max-Planck-Institut für Verhaltensphysiologie
Erling-Andechs, Germany

INTRODUCTION

Circadian rhythms are endogenous and persist under constant conditions with periods which deviate to some extent from 24 hours [1–3]. Under constant conditions the rhythmic functions tend towards a new phase-relationship different from that under conditions of entrainment with normal social environment [4].

MATERIALS AND METHODS

In order to compare the phase-relationship of various functions in synchronization and isolation an experiment with a young male subject was performed. The experiment was divided into three parts:

1. Seven days of synchronization to 24 hours of normal social life routine with the subject doing laboratory work during the day time.
2. A 17-day isolation period during which the subject lived alone in a special isolation room of about 20 m^2 relatively removed from Zeitgebers. Outside contacts were possible only by letters; the exact timing of measurements or urine sample collections etc. was marked on an event recorder.
3. A 13-day postisolation period again in normal social life routine as described in 1. For most considerations the transients (first four days in isolation) and the resynchronization time (first four days in postisolation) are excluded [5].

The following physiological and psychometric functions were observed:

1. Sleep-wakefulness cycle, recorded when the subject switched the lights on and off
2. Physical vigor (Phys. vig.) and mental state (M. st.): Self rating by using two 8-point scales ranging from 'inactive' to 'very active' and from 'depressed' to 'good' respectively.
3. Time estimation (Time est.): Counting from 1 to 120, as an estimate of 2 minutes.
4. Body temperature (Temp.) measured sublingually in degrees Centigrade.
5. Blood pressure (BP) taken with an ercameter of the Riva-Rocchi-type.
6. Pulse rate (P) counted during 60 seconds.
7. Respiration rate (Resp.): Number of in- and expirations during 60 seconds.
8. Pulse-respiration-quotient (PRK): Computed from 6. and 7.

9. Expiratory peakflow (EPF): Recording of maximal expiration force in 1/sec with a peakflow meter [6].

Furthermore the following urinary variables:

10. Urinary volume (Vol.).
11. Adrenaline (A) and Noradrenaline (NA) indicating activity of the sympathetic nervous system and the adrenal medulla [7].
12. 17-Hydroxycorticosteroids (17-OHCS) and the 17-Ketosteroids (17-KS) indicating the activity of the adrenal cortex [8, 9].

The missing time cues in isolation excluded the possibility of equidistant measurements. In order to have comparable conditions throughout the whole experiment the urine collections (in average 4.3/day) and estimations of the other physiological and psychometric variables (in average 7.3/day) took place in spontaneous, i.e. non equidistant intervals in isolation as well as during synchronization. For evaluation the data were grouped in comparable fractions of each period, i.e. 24 hours during synchronization and about 26 hours during isolation. A full period, measured from one onset of activity to the following is taken as 360°. With this graduation all measurements can be normalized on one scale, irrespective of the actual value of the period. As intervals for grouping the data 30° or 60° respectively were chosen (60°-intervals for urine estimations, 30°-intervals for the other variables). For a given sample of urine the time plotted was not its collection time, but rather the midpoint between this and the foregoing collection time. For every 30°- or 60°-interval all data were averaged and drawn with their standard deviations.

Of course in a longitudinal experiment such as this the statistical prerequisite for serial independence of data is not fulfilled. All given statistical criteria are therefore of descriptive value only.

RESULTS

The four diagrams in Figs. 1 and 2 include a summary of the entire experiment. In each diagram the horizontal bars represent activity time. Subsequent activity times are drawn beneath each other. The horizontal grid lines mark the beginning and end of isolation. On the average the period between the onset of activity lasts 24.0±7 h during preisolation, 24.0±.5 h during postisolation and 26.1±.3 h during the steady state part of isolation, resulting in a steady phase shift of about 2 h/day against local time. The total phase shift is 420°, that is 60° more than one whole period [10, 11]. By definition we are therefore dealing with a really free running rhythm.

This also is shown by the continuous roughly synchronized shifts of the maxima of the various functions [12]. In Fig. 1 the maxima of the urinary variables and body temperature are illustrated with different symbols. Fig. 2 demonstrates the same result for the psychometric variables and vital signs. There like in Fig. 1 the roughly synchronous shifts of the maxima during isolation demonstrate that all variables remained internally synchronized. This is confirmed by the power spectra of Fig. 3 with crests around 26.0 h corresponding to the free running period of the activity rhythm [13].

Whereas during social synchronization (in post- as well as in preisolation) subsequent maxima occur rather late in the second half of activity time the maxima

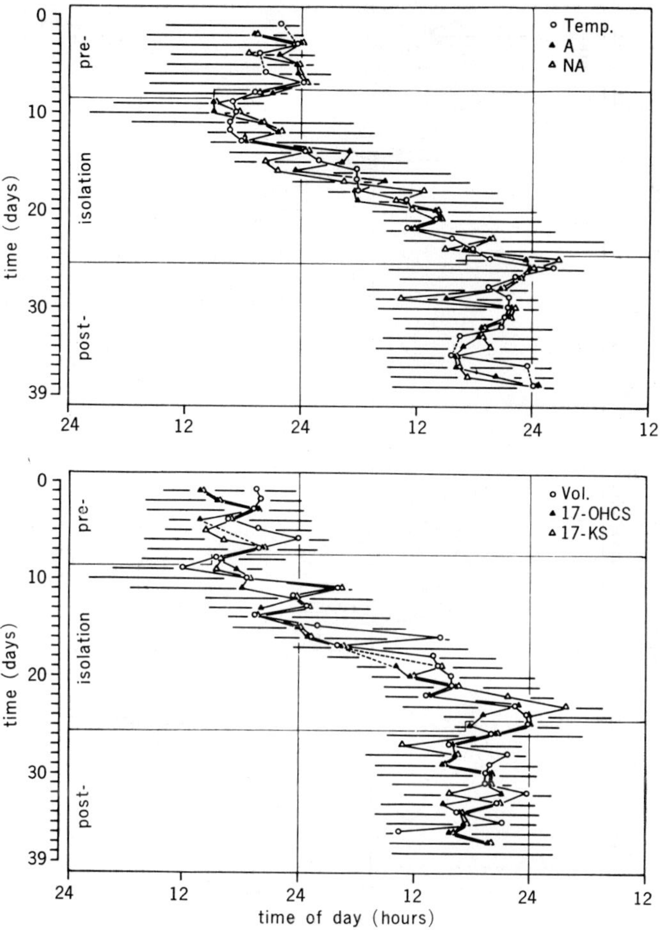

Fig. 1 Timing of the maxima of urinary variables and body temperature. Subsequent activity times (horizontal bars) are drawn beneath each other.

during isolation move toward the onset of activity (Figs. 1 and 2). This shift of the maxima in the autonomous (free running) system results in changes in the shapes of the curves, since the minima are advanced to a lesser degree. Figs. 4 and 5 demonstrate the different wave forms for synchronization and isolation and especially the different phase relationships. Each abscissa is graduated in degrees up to one full 360° period. The horizontal broken lines show the overall average of the functions during isolation and synchronization. This is arbitrarily taken as 0 and the ordinate scales then show the percentage deviation of the 60°- and 30°-interval values from this reference. The data obtained during synchronization are drawn in dashed lines, those obtained in isolation in solid lines. With exception of urine volume, blood pressure amplitude and pulse-respiration-quotient, the maxima of all functions move toward the onset of activity during isolation. Despite the unchanged timing of the maxima of the urinary volume and blood pressure amplitude the tendency of the average curve to be displaced in the same direction is obvious, as it is for some minima as well. This information obtained by visual inspection of the new data would be lost by fitting

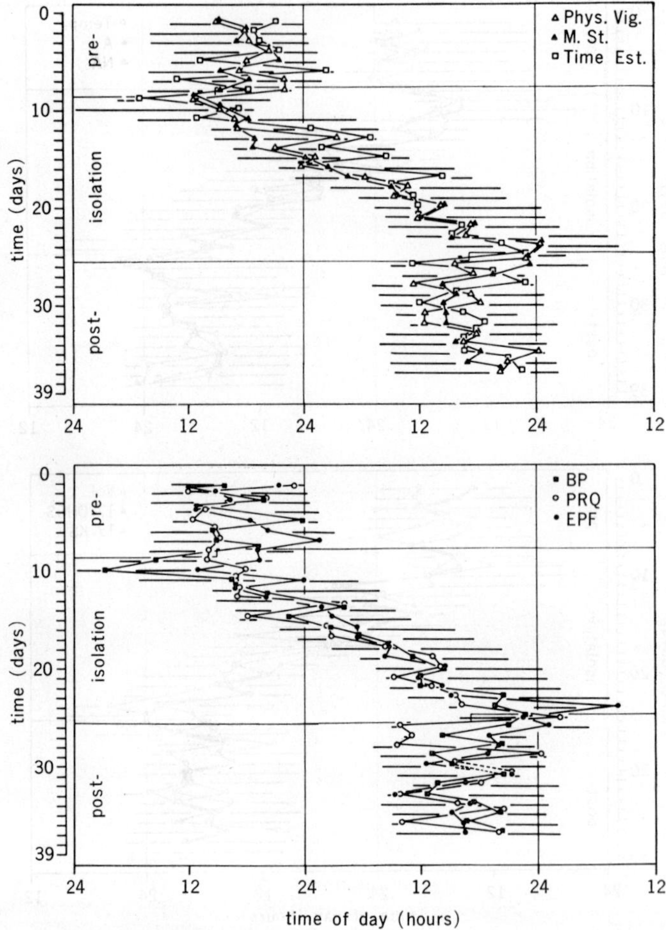

Fig. 2 Timing of the maxima of psychometric, cardiovascular and respiratory variables. Subsequent activity times (horizontal bars) are drawn beneath each other.

of harmonic functions and comparison of the resulting acrophases. Similarly bimodal patterns (Temp., NA, BP) would disappear.

If one takes the maxima of body temperature as phase references we can summarize for the urinary variables a change of the phase angle differences from negative to positive values for the catecholamines and from higher positive to lower positive values to the steroids [4, 14]. Fig. 6 combines the exact amount of the phase shifts and changing internal phase relationships with respect to the average onset of activity, left ordinate, and the average midpoint of activity, right ordinate, on the left diagram for the urinary variables and body temperature on the right diagrams for the psychometric, cardiovascular and respiratory variables.

According to the multi- and group oscillatory theory [15] the amount of the displacements is identical for both the catecholamines as it is for both the steroids in this experiment, and it differs for the other variables. As mentioned already the pulse-respiration-quotient only moves toward the opposite direction. The mean phase advances of the urinary variables are 96.0°±80.5°, for the psycho-

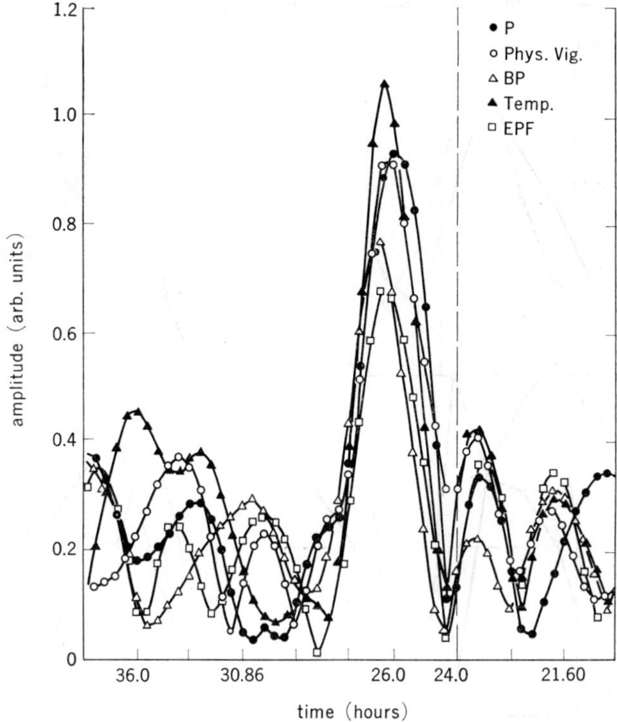

Fig. 3 Chronobiologic windows of psychometric variables and vitals signs indicating internal synchronization during isolation.

metric variables and vital signs 51.8°±46.7°. The larger phase shift of the urinary variables may be caused to some extent by the 60°-data-intervals.

In order to compare the resulting changes of the circadian wave forms Fig. 7 combines the phase maps of all variables; the upper ones always indicate synchronization, the lower ones isolation. The ordinates show again the percentage deviation from the mean value, with different scales for the urinary- and psychometric, cardiovascular and respiratory functions [16, 17].

According to the different position of the maxima the ascent to and the descent from the maxima differs widely for the various functions. All in all the curves appear to be more harmonic in isolation. This should be taken into account if harmonic functions are fitted to data obtained during synchronization and during isolation. The range of oscillation is higher for the urinary variables, than for the psychometric variables, on the other hand the range of oscillation is smaller during isolation than during synchronization. The difference between these variables is statistically significant on the 2.5% level with the t-test. No statistically significant differences could be validated for the urinary and other cardiovascular and respiratory data except the expiratory peak flow.

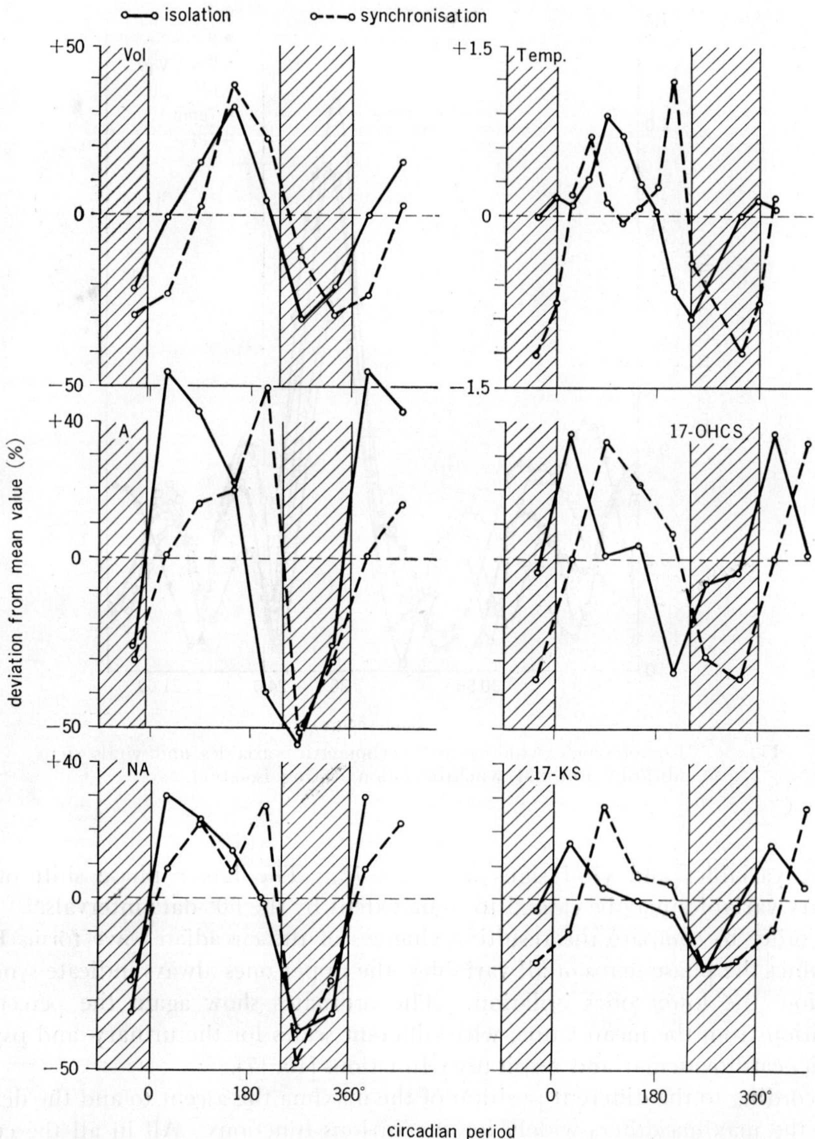

Fig. 4 Changes of circadian rhythms of urinary variables and body temperature in isolation and synchronization. Hatched areas: dark time (for further explanations, see text).

SUMMARY

Circadian rhythms of 15 functions are described for a young male subject over a period of 37 days, including 17 days of isolation.

All these variables remained internally synchronized and showed free running periods of about 26-h during the steady state part of the isolation time.

With the onset, or midpoint of activity, or the timing of the maxima of body temperature as phase reference the phase relationships between activity or tem-

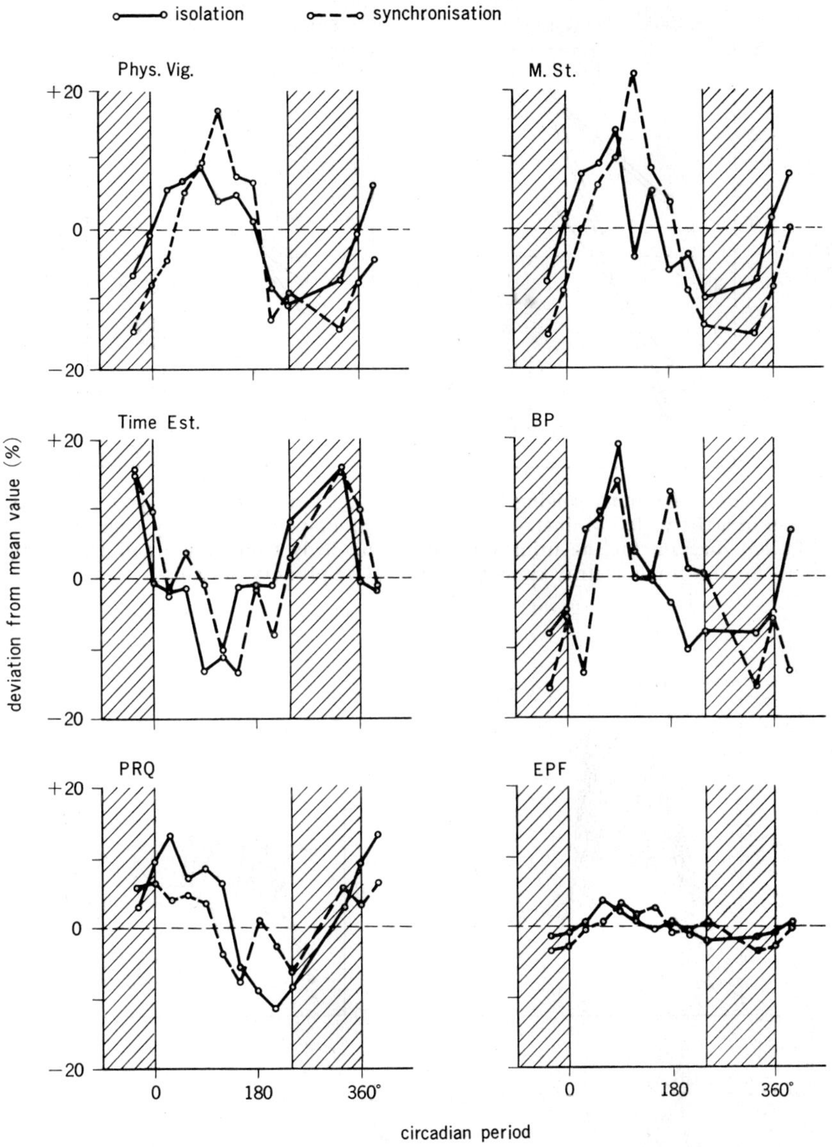

Fig. 5 Changes of circadian rhythms of psychometric, cardiovascular and respiratory variables in isolation and synchronization. Hatched areas: dark times (for further explanations, see text).

perature and the other variables under study were different in social synchronization and isolation. In isolation all maxima but one were moving toward the onset of activity resulting in changes of the wave forms.

According to the multioscillatory theory the amount of the displacements of the maxima was different for the various functions.

The ranges of oscillations were greater for the urinary variables and showed no marked differences during social synchronization and isolation, whereas the ranges of oscillation of the psychometric variables and expiratory peakflow were statistically smaller during isolation.

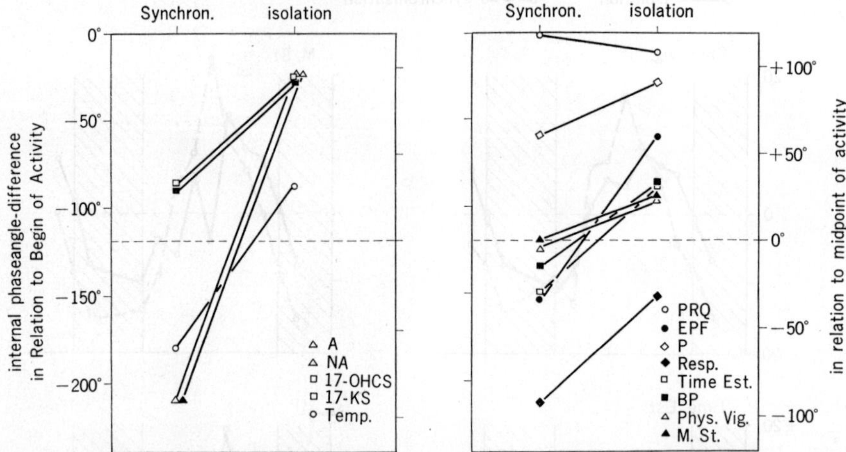

Fig. 6 Phase shifts and changing internal phase relationships with respect to the average onset of activity (left ordinates) and the average midpoint of activity (right ordinates). On the left diagram for the urinary variables and body temperature and on the right diagram for the psychometric, cardiovascular and respiratory variables.

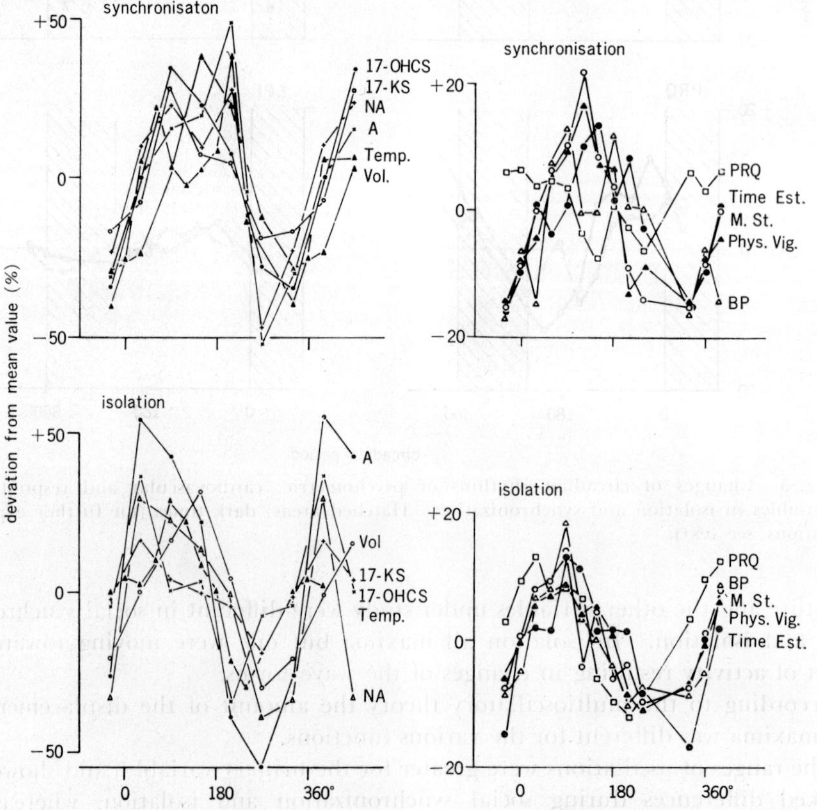

Fig. 7 Phase maps for synchronization and isolation with different scales for the urinary and psychometric, cardiovascular and respiratory variables (For further explations, see text).

ACKNOWLEDGEMENTS

This experiment was performed at the Chronobiology Laboratories of the University of Minnesota, Minneapolis, with the generous help of Prof. Dr. F. HALBERG, Dr. W. NELSON and R. B. SOTHERN.

REFERENCES

1. ASCHOFF J. and WEVER R. (1962): Spontanperiodik des Menschen bei Ausschluss aller Zeitgeber. Naturwisenschaften, *49*: 337–342.
2. HALBERG F. (1959): Physiologic 24-hr periodicity: general and procedural considerations with reference to the adrenal cycle. Z. Vitamin-, Hormon- und Fermentforsch, *10*: 225–296.
3. SIFFRE M. et al. (1966): L'isolement souterrain prolongé. Etude de deux subjets adultes sains avant, pendant et après cet isolement. Presse Medicale, 915–919.
4. ASCHOFF J. (1965): The phase-angle difference in circadian periodicity. *In*: Circadian Clocks, (J. ASCHOFF, ed.), North-Holland Publ. Comp., Amsterdam.
5. ASCHOFF J., KLOTTER K. and WEVER R. (1965): Circadian vocabulary. *In*: Circadian Clocks. (J. ASCHOFF, ed.), North-Holland Publ. Comp. Amsterdam.
6. HILDEBRANDT G. (1963): Die Bedeutung der Atemstossmessung (Pneumometrie) für die Atemfunktionsdiagnostik in der Praxis. Ärzt. Forsch., *17*: 571–578.
7. ROBINSON R. L. and WATTS D. T. (1965): Automated trihydroxyindole procedure for the differential analysis of catecholamines. Clin. Chem., *11*: 986.
8. PETERSON et al. (1965): The physiological disposition and metabolic fate of hydrocortisone in man. J. Clin. Invest., *34*: 1779–1794.
9. PETERSON R. and PIERCE C. E. (1960): Methodology of urinary 17-ketosteroids lipids and the steroid hormones in clinical medicine, (F. W. SUNDERMAN, ed.), p. 158, Lippincott, Philadelphia.
10. WEVER R. (1969): Untersuchungen zur circadianen Periodik des Menschen mit besonderer Berücksichtigung des Einflusses schwacher elektrischer Wechselfelder. Forschungsber. W 69–31, Bundesmin. f. wiss. Forsch.
11. KRIEBEL J. (1970): Circadiane Periodik der Catecholamine beim Menschen. Naturwiss., *57*: 500–501.
12. HAUS E., HALBERG F., NELSON W. and HILLMAN D. (1968): Shifts and drifts in phase of human circadian system following intercontinental flights and in isolation. Federation Proc., 224.
13. HALBERG F. (1969): Chronobiology. Ann. Rev. Physiol., *31*: 675–725.
14. HALBERG F. and SIMPSON H. (1967): Circadian acrophases of human 17-hydroxycorticosteroid excretion referred to mid-sleep rather than midnight. Human Biol., *39*: 405–413.
15. ASCHOFF J. and WEVER R. (1961): Biologische Rhythmen und Regelung. Bad Oeynhausener Gespräche, 1–15.
16. HALBERG F. (1965): Some aspects of biological data analysis: Longitudinal and transverse profiles of rhythms. *In*: Circadian Clocks, (J. ASCHOFF, ed.), pp. 14–22, North-Holland Publ. Comp., Amsterdam.
17. HALBERG F. (1960): Temporal co-ordination of physiologic function. Cold Spring Harbor Symp. Quant. Biol., *25*: 289–310.

GRAVITATIONAL CONSIDERATIONS WITH ANIMAL RHYTHMS*

CHARLES C. WUNDER

Department of Physiology and Biophysics
University of Iowa
Iowa City, Iowa, U.S.A.

Access to orbital flight (zero G) and to the lunar surface (0.17 G) offers opportunities unavailable on Earth for studying the two most biologically specific features of man's environment on Earth: the planet's gravitational field intensity and the planet's period of rotation. Unfortunately, most environmentalists have neglected gravity's biomedical significance. Moreover, these new opportunities to investigate the significance of these agents to life have generally been relegated as secondary in priority to that of space or lunar exploration.

In the absence of space-based, one-G, control centrifuges it will be difficult to resolve which of these two attributes of space flight (i.e. altered period or altered gravity) may cause biologically observable effects. Gravitational biologists are likely to attribute to mechanical considerations what will actually result from time-setting cues arising from the gravitational time pattern. Likewise, rhythm biologists are likely to attribute to time-setting cues what actually results from mechanical considerations. As a better exchange of information is, therefore, obviously necessary between gravitational and rhythm biologists, the author (a gravitational biologist) is most happy to have been invited to participate in this session.

Effects of Gravity Upon Animals. Galileo in his *Discourses* of 1638 was among the first to recognize gravity's role as a biological determinant. In 1954, our own laboratory was the first to establish, with a continuing program, chronic animal centrifugation as a method for experimentally investigating the "high-G side" of this determinant's role [1].

As established in our laboratory and largely confirmed by others, these simulated high-G environments (as reviewed elsewhere, 1–7):

1. influence growth and development of animals as small or smaller than baby turtles, sometimes accelerating and sometimes decelerating these processes,
2. result in many functional changes or adjustments in feeding, metabolism, circulation, fluid balance, and structures for support, and
3. influence life expectancy.

Of the many animals we studied at high G, baby turtles promise to be the closest relative of man ideally suited for developmental investigations at low G because of:

1. low payload and life support requirements in space,

* This work was supported in part with NASA funding from Grant No. NGR-16-001-031 and Contract No. NAS2-6064.

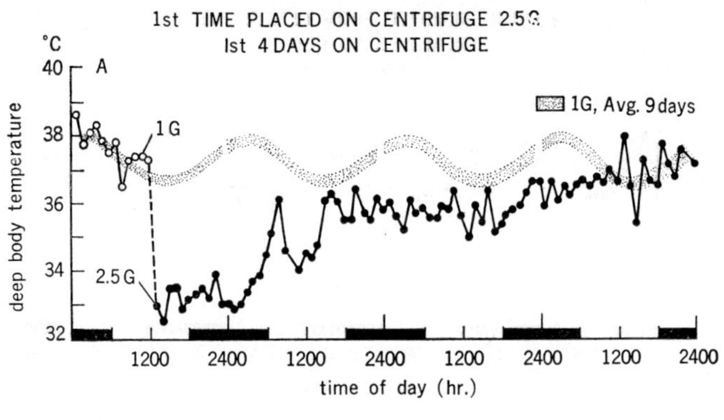

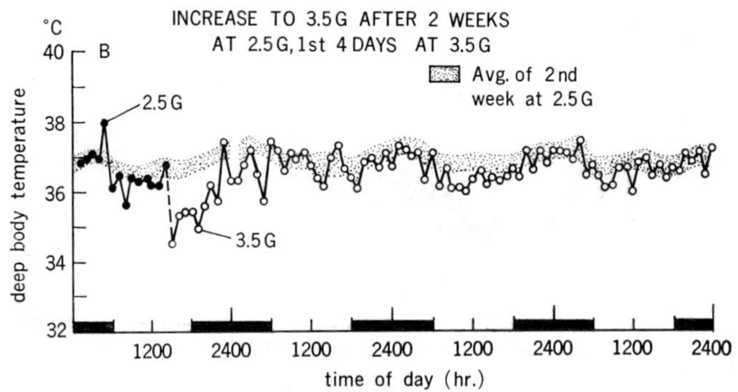

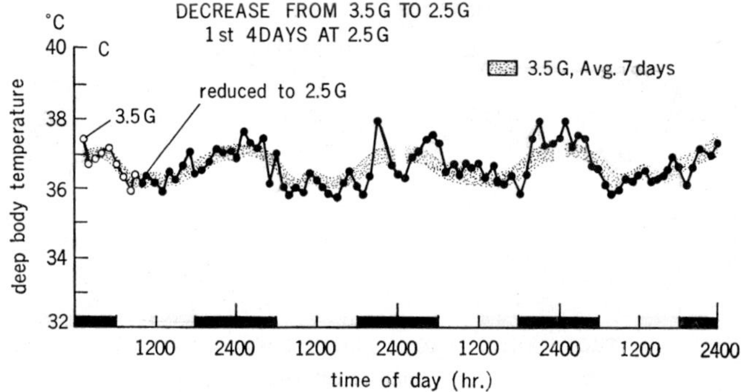

Fig. 1-1 (A-C)

Fig. 1 (A–F) Typical response in hourly body temperature values of a female, white rat exposed to different inertial environments. Courtesy of Dr. Jiro OYAMA [8].

2. rapid developmental responses to small gravitational change,
3. high tolerance to environmental situations, and
4. a shell for attachment of remote measuring and restraining devices.

Effects of Simulated High Gravity Upon Biological Rhythms. I know of no rhythm laboratories which have investigated the influence of gravity. OYAMA, et al. [8] by use of radiotelemeters implanted in rats were the first to examine rhythms at high G. They found that the body temperature, although initially

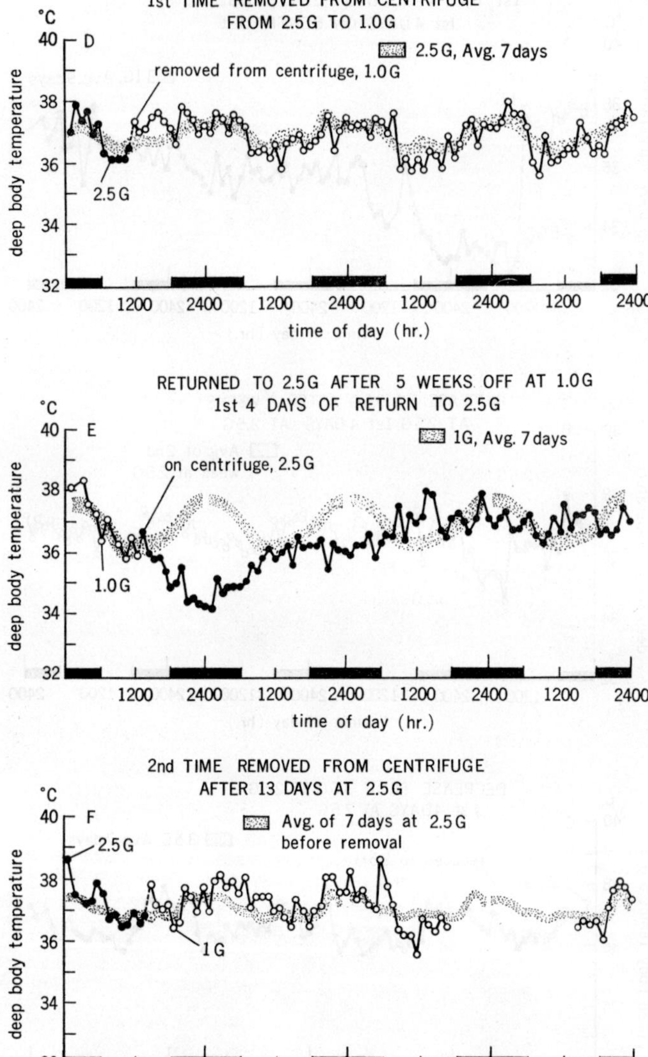

Fig. 1-2 (D-F)

reduced, retained its day-night rhythm for which no change with fields as high as 3.5 G's (Fig. 1) could be detected.

Our own laboratory has recently developed an assembly of equipment [9] (Fig. 2) suitable for measuring oxygen consumption of small mammals as influenced by chronic centrifugation and/or by day-night rhythms. For male, white mice studied two at a time in the same chamber with artificial light daily from 6:30 A.M. to 6:30 P.M. our preliminary findings suggest:

1. confirmation of the previously reported [10] day-night rhythm with 6 base-line experiments and 6 control experiments (solid line of Fig. 3),

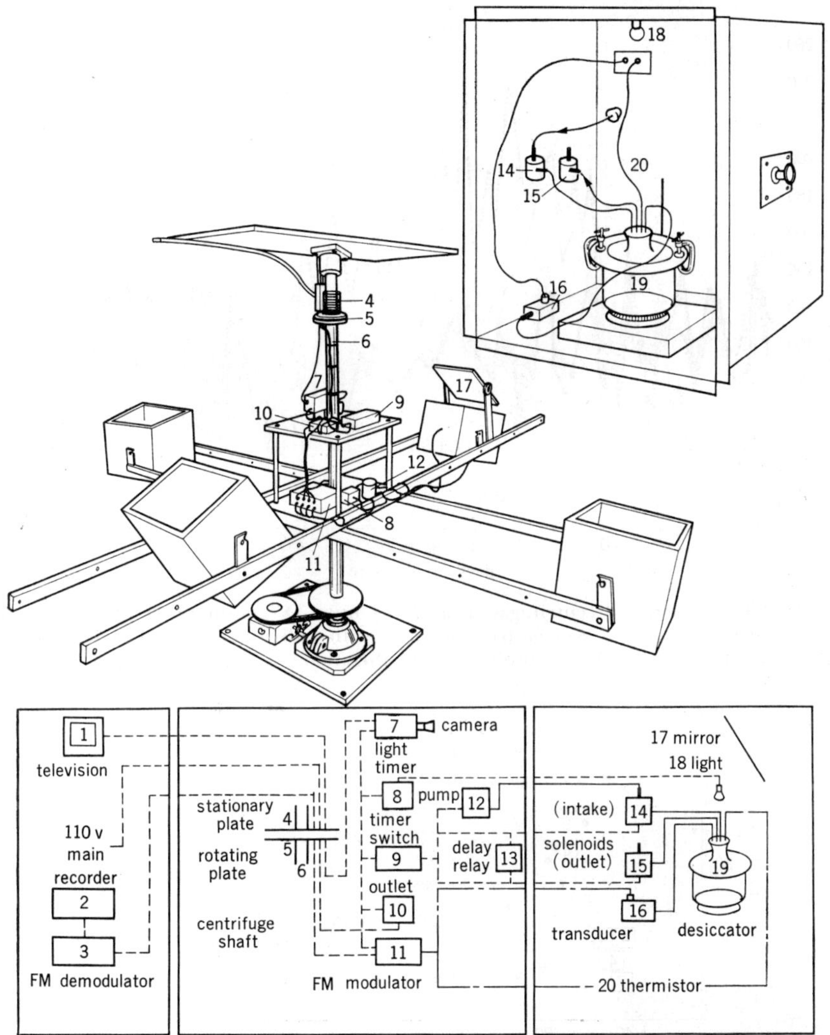

Fig. 2 Assembly for O_2 consumption measurements in Second Iowa Centrifuge [9].

2. some persistence of the rhythm with one experiment at 2 G's, 3 at 4 G's, and 2 at 7 G's (broken lines of Fig. 3) during and after one to seven days of chronic centrifugation,
3. no effect at 2 G's,
4. increased metabolism at 7 G's,
5. a consistent "dampening" of the rhythm at 4 and 7 G's, throughout the first two days of exposure (Figs. 4 and 5) and possibly lasting for as long as seven days, and
6. concern that previous measurements of decreased metabolism [11, 12] after chronic exposure should be reexamined over a 24 hour period and/or in light of rhythmic phenomena.

Increases in metabolism are probably attributable to increased adrenalin levels. "Dampening" probably results from restraint of the animal's activity.

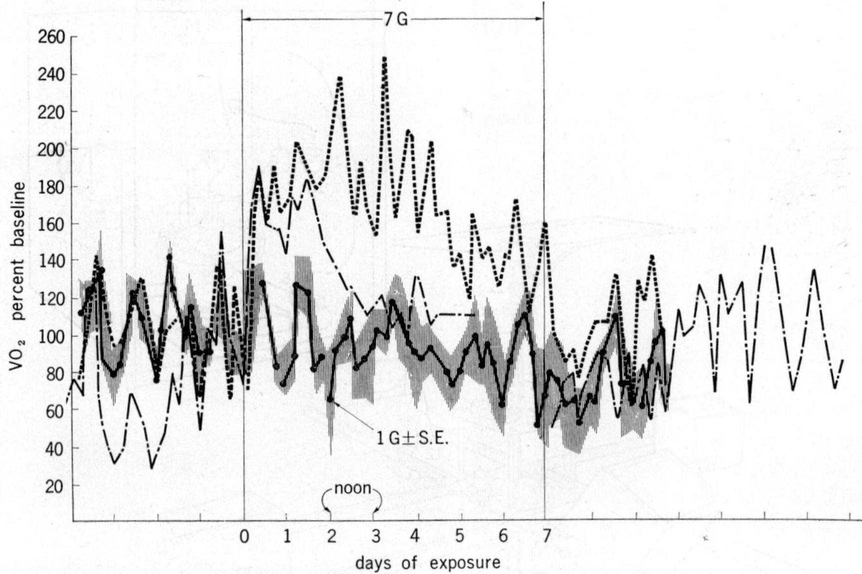

Fig. 3 Effect of chronic centrifugation at 7 G's upon O_2 consumption [9]. Values are normalized relative to the average baseline value with individual pairs of male mice over the 24 hour period immediately preceding centrifugation.

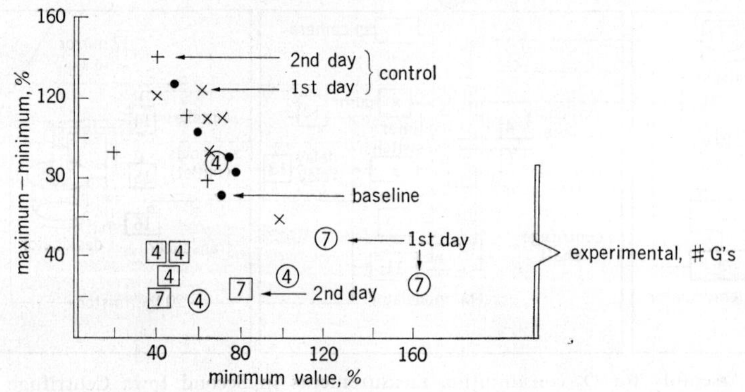

Fig. 4 Daily excursion of metabolic rate relative to 24 hour baseline average for chronically centrifuged mice.

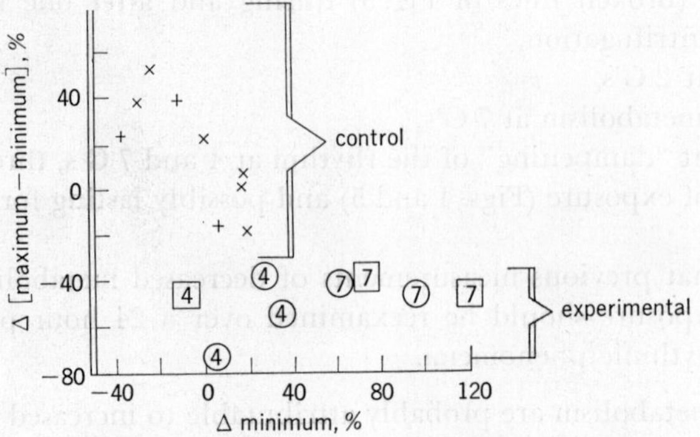

Fig. 5 Shift from baseline values for measurements shown in Fig. 4.

REFERENCES

1. WUNDER C. C. (1955): Gravitational aspects of growth as demonstrated by continual centrifugation of the common fruit fly larvae. Proc. Soc. Exptl. Biol. Med., *89*: 544–546.
2. WUNDER C. C. and LUTHERER L. O. (1964): Influence of chronic exposure to increased gravity upon growth and form of animals. *In*: International Reviews of Gen. and Exp. Zool. Vol. 1 (W. J. FELTS and R. J. HARRISON, eds.), pp. 333–416, Academic Press, New York.
3. WUNDER C. C. (1965): Care and growth of animals during chronic centrifugation. *In*: Methods of Animal Experimentation, Vol. 2, (W. I. GAY, ed.), pp. 371–449, Academic Press, New York.
4. WUNDER C. C. (1966): Life into space. *In*: An Introduction to Space Biology, pp. 121–181, F. A. Davis Co.
5. WUNDER C. C., DULING B. and BENGELE H. (1968): Gravity as a biological determinant. *In*: Hypodynamics and Hypogravics: Physiology of Inactivity and Weightlessness, (Michael MCCALLY, ed.), pp. 1–69, Academic Press, New York.
6. SMITH A. H. and BURTON R. R. (1971): Effects of chronic acceleration on animals. *In*: Gravity and the Organism, (S. A. GORDON and M. J. COHEN, eds.), pp. 371–388, University of Chicago Press, Chicago and London.
7. WUNDER C. C. (1971): The effects of chronic acceleration of animals. *In*: Gravity and the Organism, (S. A. GORDON and M.J. COHEN, eds.), pp. 389–411, University of Chicago Press, Chicago and London.
8. OYAMA J., PLATT W. T. and HOLLAND V. B. (1971): Deep body temperature changes in rats exposed to chronic centrifugation, Am. J. Physiol., *221*: 1271–1277.
9. FETHKE W. G., COOK K. M., PORTER S. M. and WUNDER C. C. (1973): Oxygen consumption measurements during continual centrifugation of mice. J. Appl. Physiol. (in press).
10. PEARSON O. P. (1947): The rate of metabolism of some small mammals. Ecology, *28*: 127–145.
11. WUNDER C. C., CRAWFORD C. R. and HERRIN F. W. (1960): Decreased oxygen consumption of fruit fly larvae after continual centrifugation. Proc. Soc. Exptl. Biol. Med., *104*: 749–751.
12. WUNDER C. C. (1963): Altered respiration of growing animals following continuous centrifugation. Abstracts of 7th Annual Meeting of the Biophys. Soc., paper W. E. 6.

27-HOUR-DAY EFFECTS ON REPRODUCTION AND CIRCADIAN ACTIVITY PERIOD IN RATS*

FREDERICK M. BROWN**

University of Virginia
Charlottesville, Virginia, U.S.A.

A basic problem remaining in the study of biologic rhythms is the determination of long-term effects of cyclic adiurnal illumination schedules on the circadian rhythms of organisms [1]. This is becoming more pressing with the advent of extra-terrestrial human travel which will require more than a few days or weeks in transit. Long-term reproduction studies using rats [2–4], eggs of chickens [5] and lizards [6, 7], which simply exposed animals to adiurnal illumination regimens or constant-illumination (LL) conditions, indicated that the circadian activity period of progeny closely approximated that of the parent.

However, it is also known that simple exposure to a 28-hr day resulted in an increased mortality of mice offspring prior to weaning [8]. Similarly, although it originally was a secondary observation, an apparent decrease in numbers of litters and offspring occurred in nine female albino rats which were synchronized to a 26-hr day of LD 13:13 (13 hr of illumination followed by 13 hr of darkness) prior to breeding [9]. They were maintained on this regimen until the four litters of young which were produced were weaned at about 25 days of age. Of primary significance was the fact that upon subsequent testing in constant darkness (DD) the mean circadian wheel-running activity period in the surviving 14 offspring was longer than that of a control group raised from conception in LD 12:12 until DD testing after weaning [10]. At the end of the first week of testing the average difference in period between groups was 57 minutes; for the second week it was 22 minutes. The mean period for the 26-hr group was 24 hr 45 min and 24 hr 27 min, respectively. After an additional three months in DD, the mean period for another test week was 24 hr 31 min, but with no control data for comparison of lengthened period.

A recent study [11] has replicated the observations of interference in reproduction and lengthening of the circadian wheel-running activity period in rats synchronized to artificial days of 27-hr duration. Portions have been reported elsewhere [12].

Female Sprague-Dawley Derived albino rats were constituted into two groups according to total wheel-running in DD, clarity of visual activity records, and ease of synchrony to LD 14:13. The gradual-transition method [13] was used for synchronization to the 27-hr day. Eighteen rats remained synchronized to the experimental 27-hr day while 13 returned to a control 24-hr (LD 12:12) day. All were bred for 13 days, returned to the wheel for the next five days, and then

* This investigation was supported by the Public Health Service, grant MH04920, awarded to professor Frank W. FINGER, University of Virginia, Charlottesville, Virginia 22901.

** Present address: Psychology and Psychobiology Programs, Division of Science and Mathematics, Centre College of Kentucky, Danville, Kentucky, 40422.

placed in individual maternity cages. Wheel-running data obtained prior to parturition closely demonstrated continued synchrony to their respective lighting regimens. The mean activity period for each group as measured from visual event records is shown in Fig. 1. A Pearson product-moment correlation for the activity period of the two maternal groups in DD was $r \geqslant .94$; for LD 14:13 it was $r \geqslant .95$.

Among the 13 control females, the members of 10 out of 12 viable litters survived until weaning at 25±3 days of age. In contrast, 13 of the 18 27-hr-day females produced known litters of which members of 10 survived until weaning. The comparative percentages of litters at weaning are 77 to 56, respectively. The latter reproduction statistic is similar to the 45 percent mortality of litters found by KAISER [8] in mice. Analysis of litter sizes and sex distribution indicated no significant differences.

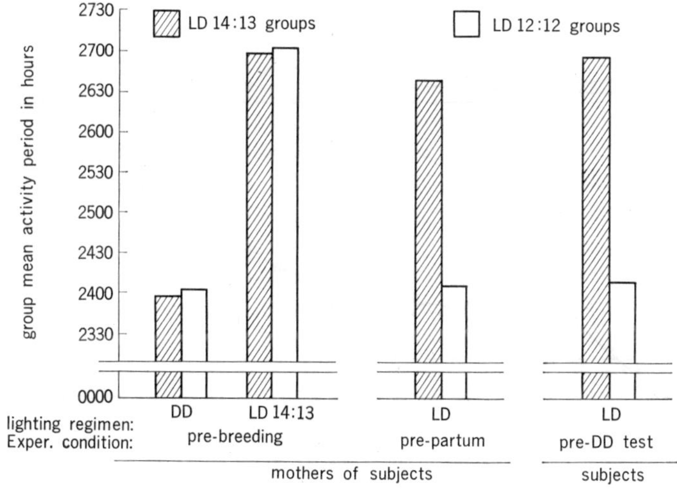

Fig. 1 Comparison of the activity period in hours under different experimental and illumination conditions for the mothers of the subjects and the subjects themselves. The left set of histograms (pre-breeding) indicates the precision of maternal matching along the DD and LD 14:13 illumination dimensions before breeding. The next histogram (pre-partum) indicates the relative LD synchrony of the two maternal groups after breeding and before birth of the subjects. The right histogram (pre-DD test) demonstrates the synchrony of the subjects to their respective LD regimen prior to DD testing.

At weaning 12 littermate trios were selected from the 27-hr-day mothers. They were maintained, a trio per cage, on LD 14:13 until testing. Then each trio member was placed in a Wahmann LC-34 activity wheel, the first at 26, the second at 61, and the third at 94 days of age. Each was habituated to wheel-running for five LD cycles, after which activity during the next five was recorded. Thereafter DD was begun and the free-running activity period was measured for four test blocks, each of which was equivalent in duration to five 27-hr LD cycles. Control offspring were treated simultaneously, according to their LD 12:12 regimen.

Activity-wheel revolutions were cumulatively counted at 10-min intervals by remote electrical recording on print-out counters. Data from individual animals were analyzed using a 24-hr spectral window least-squares fit [14, 15]. Average activity level (C_0), amplitude at the crest (C), and activity crest, or acrophase (ϕ'), were determined for each 24-hr test cycle. A computer-derived chronogram

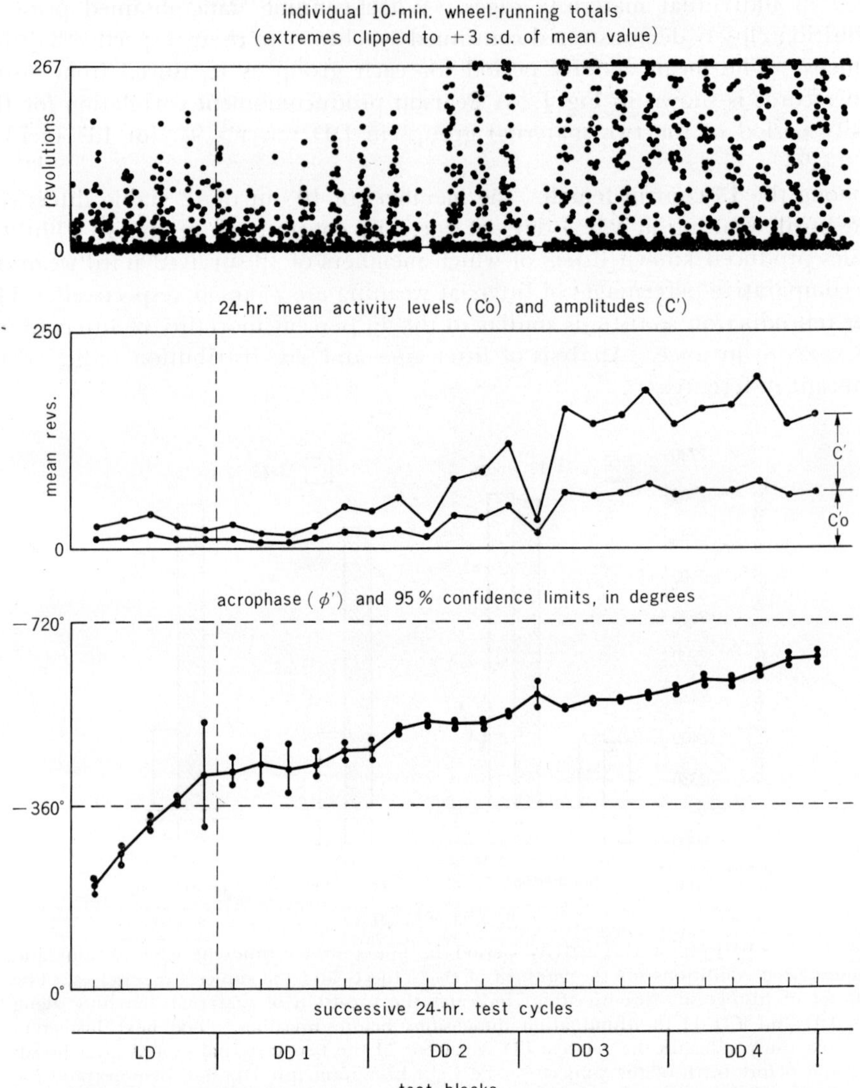

Fig. 2 Serial section for a 27-hr whole-life subject after using a 24-hr spectral window least-squares fit of 10-min. wheel-running activity data. In the top graph each dot is the total wheel-revolutions per 10-min. interval. In the middle graph each vertical pair of dots is a mean 24 hr level of activity (below) and amplitude (above). The two-cycle graph at the bottom shows the progressive lag of the acrophase across the test cycles. The 95 percent confidence limits are shown by the vertical lines connecting two dots. It should be noted that each test block is slightly longer than five 24-hr. cycle lengths. This is because a test block for 27-hr. subjects is equivalent to five 27-hr. cycle lengths.

displaying the analysis for one animal is shown in Fig. 2. Following individual data analysis, the group mean activity period for each experimental condition then was derived by cosinor analysis [16].

Group computative acrophases (ϕ) obtained from the cosinor analysis for each test block were rotated to external acrophase (φ), after BATCHELET [17]. These are shown for all conditions in Fig. 3. Unweighted acrophases were chosen for statistical analysis because of ease of adaptability to parametric analysis and also because relative, rather than absolute, amplitude measurements were used in the second step of the cosinor analysis. The latter was made necessary because of the peculiarities in the type of data collected, and after consultation with Professor Franz HALBERG. Pearson product-moment correlations for weighted and unweighted acrophases within illumination conditions and overall were $r \geqslant .99$. According to HALBERG et al. [16], under similar conditions the use of unweighted instead of weighted acrophases in cosinor analysis would appear as an acceptable approximation. All group acrophases (φ) among the experimental conditions were converted to period measurements in hours and minutes following a procedure described in BROWN [11].

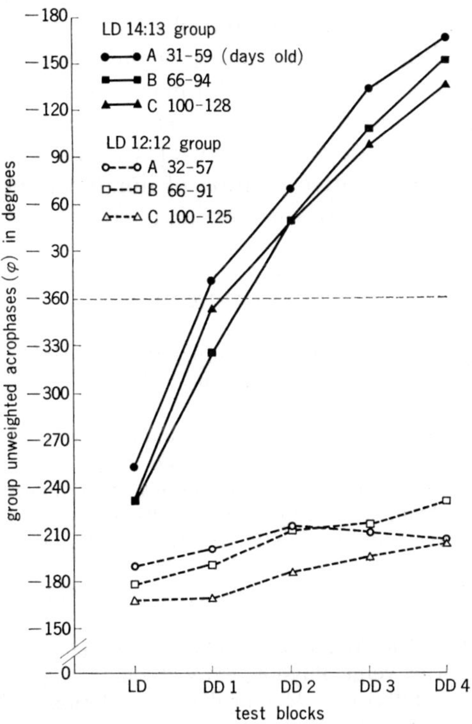

Fig. 3 Mean external acrophases (unweighted) for matched age groups of rats under different post-weaning exposure conditions to days of LD 12:12 or LD 14:13. Each test block is composed of a time span equal to five respective 24-hr. or 27-hr. LD cycles.

A 2×3×5 factorial design with repeated measures (Case I) [18] was used in the statistical analysis. For the 27-hr-day subjects the circadian wheel-running activity period in each of the first and third main effects, lighting regimens and DD test blocks, was found significantly the longer ($p<.01$). Post-hoc analysis within DD test blocks between the two illumination groups indicated that all means were significantly the greater ($p<.05$) at 25 hr 35 min, 24 hr 49 min, 24 hr 47 min, and 24 hr 35 min, respectively. An apparent significant difference ($p<.05$) for the second main effect, age of testing, did not persist ($p>.10$) with further analysis among the three groups. The only significant interaction was lighting regimen × test blocks ($p<.01$).

The 27-hr day reproduction results strongly support the suggestion of earlier studies that synchrony, if not just simple exposure, to artificial day lengths, at some time after conception and prior to weaning, adversely affects the reproductive process in rats and mice. Whether the effect is predominantly a result of changes in maternal behavior attending and following birth, or a decrease in viability of the offspring is not well established. A tentative comment is that these results would seem to suggest, by extrapolation, that the possible consequences of adverse alterations in human reproduction might occur during prolonged exposure to artificial days. This possibility must be explored.

The circadian wheel-running activity period in rats is lengthened nearly three weeks by synchrony to a 27-hr illumination regimen of LD 14:13 from conception until weaning. During which early developmental stage, pre-natal or post-natal until weaning, the adiurnal illumination effect is prepotent has not yet been determined. The implications of the course of the lengthened circadian activity period substantially support the alleged innateness of a circadian timer and reject the notion of rhythmicity as solely a conditioning phenomenon. These conclusions are discussed in detail in BROWN [11]. There in data are also presented in support of an interactive position of innate timing and environmental conditioning for the expression of the circadian activity rhythm.

REFERENCES

1. HALBERG F. (1969): Chronobiology. Annual Rev. Physiol. *31*: 675–725.
2. BROWMAN L. G. (1952): Artificial sixteen-hour day activity rhythms in the white rat. Amer. J. Physiol., *168*: 694–697.
3. ASCHOFF J. (1955): Tagesperiodik bei Mäusestämmen unter konstanten Umgebungsbedinungen. Pflügers Arch. Physiol., *262*: 51–59.
4. ASCHOFF J. (1960): Exogenous and endogenous components in circadian rhythms. Cold Spring Harbor Symposia on Quantitative Biology, *25*: 11–28.
5. ASCHOFF J. and MEYER-LOHMAN J. (1954): Die Schubfolge der lokomotorischen Activität bei Nagern. Pflügers Arch. Physiol., *260*: 81–86.
6. HOFFMAN K. (1957): Angeborene Tagesperiodik bei Eidechsen. Naturwissenschaften, *44*: 359–360.
7. HOFFMAN K. (1959): Die Activitätsperiodik von im 18- und 36-Stunden-Tag erbrüteten Eidechsen. Zschr. vergl. Physiol., *42*: 422–432.
8. KAISER I. H. (1967): Effect of a 28-hour day on ovulation and reproduction in mice. Amer. J. Obst. Gyn., *99*: 772–784.
9. BROWN F. M. (1969): Synchronization of non-24-hour light-dark cycles affecting reproduction in animals. Paper presented at the meeting of the Kentucky Academy of Science, Murray, October 1969, (unpublished).
10. BROWN F. M. (1968): Lengthened circadian activity period in the rat. Paper presented at the meeting of the Virginia Academy of Science, Roanoke, May, 1968. Virginia J. Sci., *19*: 210 (Abstr.).
11. BROWN F. M. (1971): Circadian rhythm modification in rats by whole-life synchrony to a 27-hour day. Unpublished doctoral dissertation, University of Virginia.
12. FINGER F. W. and BROWN F. M. (1971): The effect of lengthened illumination cycles on littering and upon offsprings' freerunning activity. Gegenbaurs Morph. Jb., *117*: 103–104.
13. TRIBUKAIT B. (1956): Die Aktivitätsperiodik der weissen Maus im Kunsttag von 16–19 Stunden Länge. Zschr. vergl. Physiol., *38*: 479–490.
14. HALBERG F. and PANOFSKY H. (1961): Thermo-variance spectra: Method and clinical illustrations. Exp. Med. Surg., *19*: 284–309.
15. PANOFSKY H. and HALBERG F. (1961): Thermo-variance spectra: Simplified computational example and other methodology. Exper. Med. Surg., *19*: 323–338.

16. HALBERG F., TONG Y. L. and JOHNSON E. A. (1965): Circadian system phase—an aspect of temporal morphology; procedures and illustrative examples. *In*: The Cellular Aspects of Biorhythms, (H. von MAYERSBACH, ed.), pp. 20–48, Springer-Verlag, Berlin.
17. BATSCHELET E. (1965): Statistical methods for the analysis of problems in animal orientation and certain biological rhythms. American Institute of Biological Sciences, Washington.
18. WINER B. J. (1962): Statistical Principles in Experimental Design. McGraw-Hill, New York.

ULTRADIAN RHYTHMS AND SLEEP

Chairman: DANIEL F. KRIPKE

ULTRADIAN RHYTHMS AND SLEEP: INTRODUCTION

DANIEL F. KRIPKE

Department of Psychiatry, Universty of California San Diego, California, U.S.A.

The papers assembled here represent an effort by investigators involved with sleep physiology to re-explore their findings with chronobiologic methods. The studies span several frequency ranges. Common experience tells us that there are circadian (about 24-hr) sleep-wakefulness rhythms in adult humans and many animals. We also know that human infants have ultradian (more than one cycle per 24 hr) sleep-wakefulness cycles related to feeding, superimposed on circadian fluctuations, so that the sleep-wakefulness continuum is being modulated by more than one frequency. In the papers which follow, WEITZMAN et al. describe a 180 min. sleep-wakefulness cycle in adult humans, and STERMAN reports a 100 min. sleep-wakefulness cycle in adult cats, as well as circadian sleep wakefulness cycles.

It is confusing to consider so-called sleep-wakefulness cycles of more than one frequency, even in a single experimental model, but sleep studies have forced us to consider another ultradian frequency range modulating the same data. The authors represented below have awarded these higher frequencies an extraordinary range of appellations:

BRAC (Basic Rest-Activity Cycle)	short-term sleep cycle
rest-activity cycle	encephalic cycle
performance cycle	"on-demand" feeding cycle
REM non-REM cycle	ultradian rhythm
REM cycle	the ultradian cycle
paradoxical sleep cycle	ultradian REM periodicity
ultradian oscillation	

These names refer to cycles with mean periods of about 90–100 min. in adult man, 50–60 min. in the neonatal human or juvenile monkey, and 20 min. in adult cats. The many names for these rhythmic phenomena indicate our confusion in describing them. We are like the blind men of the fable touching an elephant—or perhaps it is a herd of animals. One of us touches the eye; another the mouth; another the limb; and so forth, and each is calling the beast by a different name. We sleep researchers are accustomed to working in the dark, but we have to find out if we are all hanging on to the same elephant and how to grasp it.

The recent history of sleep studies provides the background for our thinking about these ultradian rhythms. The 90–100 min. periodicity during adult human sleep drew consistent attention only with the advent of the "sleep staging" method [1]. During sleep, rapid eye movements, a hypotonic electromyogram, and a low voltage, mixed frequency electroencephalogram often appear in association, and this state has been called "the Rapid Eye Movement sleep stage," or Stage REM. Stage REM soon attracted great interest because it was found to be correlated with the most hallucinatory dream imagery, genital engorgement, high frequency motor neuron discharges, gastric acid secretion, circulatory changes, etc. [2]. Psychiatrists and psychologists hoped that Stage REM was a physiological correlate of emotions and drives, and thus began an intensive study of sleep as an approach to the mind/body interface. It was immediately recognized that Stage REM may occur periodically, and discussion of "REM cycles" crept rapidly into our vocabulary, however, a biorhythm perspective was neglected in the excitement surrounding the reification of Stage REM. Quantitative measurement of Stage REM attracted great enthusiasm. Stage REM was discussed as a substance, or at least the epiphenomenon of a substance, e.g., "REM juice." The wave theory was forgotten for quantum mechanics. We now know the sleep stages are not nearly so discrete as had originally been assumed, and "sleep stages" as we measure them do not always correlate well with psychologic, behavioral, or pharmacologic measures. Some of the same things seen in Stage REM sleep also sometimes appear in interrupted sleep or wakefulness. As disenchantment grew with the study of nocturnal sleep stages, as far as revealing mind/body correlations was concerned, attention returned to the suggestion by KLEITMAN [3] that a cycle like that of Stage REM in sleep might be present also during waking states. GLOBUS began a series of experiments [4, 5] in which he collected evidence that the "rapid eye movement cycle" can be identified during wakefulness. Several studies using a variety of methods have now appeared confirming this view [6–8], particularly when eating or "oral" behavior is measured. Thus we have renewed our hope that a "wave" model may illuminate the neurophysiology of drive states and explain phenomena a "quantum" model cannot. For one thing, we cannot measure "rapid eye movement *sleep*" as such during wakefulness —the concept is self contradictory. Definitions of Stage REM describe it as a sleep state only. Also, since we have so far failed to recognize any discrete qualitative state in wakefulness analogous to Stage REM in sleep, it is likely that any waking analog is a quantitative variation rather than a qualitative change in waking function. Thus we are led to conceptualize an ultradian oscillatory system which may express itself during sleep by the periodic occurrence of Stage REM, but which can also be revealed during intermittent sleep or steady wakefulness by the quantitative modulation of physiologic or behavioral variables. All of the papers in this chapter are concerned with examining some aspect of this model.

The papers reveal a number of the difficulties in studying ultradian rhythms. In sleep recordings, the Stage REM cycle is regularly manifest, and meets criteria defining a biorhythm [9], yet everyone agrees the Stage REM periodicity is most irregular, even within a single night's recordings, and stable frequency estimates are only achieved by extensive averaging. Some authors assert that the Stage REM cycle increases consistently in frequency over a single night's recordings [10], as it plainly increases in amplitude. We have also found suggestions that this oscillator may be subject to circadian frequency modulation over the whole 24 hr [11]. Stage REM cycle frequencies certainly decrease during development and differ from species to species. It is difficult to designate this oscillator by its frequency, since the frequency is so changeable. We have yet to agree on a commonly acceptable name for it, if it is a discrete oscillatory system. The many sources of variability make it difficult to establish if an ultradian oscillations in a variable, once recognized, arises from the same oscillator as does Stage REM, unless Stage REM itself can be measured simultaneously. Even if the oscillations in some variables are at the same frequency, a multioscillator model cannot be easily excluded. Many of us have been inclined to name an ultradian oscillation by the variable in which it was measured, thus dignifying any number of dependent measures with teleological significance. An important goal for future research is to find means of demonstrating when we are dealing with multiple oscillators, and when we are measuring a single, multi-faced oscillatory system. We must distinguish peripheral manifestations from central clocks.

Beyond this, we must explore the interrelations among oscillators. For example, the infant feeding rhythm is apparently harmonically coupled to Stage REM periodicity [3]. GLOBUS [4] has proposed that Stage REM is phase coupled to circadian synchronizers. A great deal of work will be required before all of these oscillatory phenomena can be sorted out.

Nevertheless, the "wave" approach fostered by chronobiology points toward a worthwhile goal. The evidence is encouraging that by studying oscillations during wakefulness, we will learn more about the drive systems with which these oscillators are correlated. Modern models of instinct, comprising a waxing appetency and consummatory release, imply the existence of relaxation oscillators. We might expect such oscillators to be rhythmic but unstable, as the Stage REM periodicity is. The study of ultradian rhythms, and the attempt to demonstrate a generalized ultradian oscillatory system, may indicate whether many instincts and drives are parts of a biologically unified system.

REFERENCES

1. DEMENT W. and KLEITMAN N. (1957): Cyclic variations in EEG during sleep and their relation to eye movements, bodily motility, and dreaming. Electroenceph. Clin. Neurophysiol., *9*: 673–690.
2. KALES A. (1969): Sleep: Physiology & Pathology. J. B. Lippincott, Philadelphia.
3. KLEITMAN N. (1963): Sleep and Wakefulness (2nd Ed.). Chicago, University of Chicago Press.
4. GLOBUS G. G. (1966): Rapid eye movement cycle in real time. Arch. Gen. Psychiat., *15*: 654–659.
5. GLOBUS G. G., PHOEBUS E. and MOORE C. (1970): REM "sleep" manifestations during waking. Psychophysiology, *7*: 308.
6. KRIPKE D. F. (1972): An ultradian biological rhythm associated with perceptual deprivation and REM sleep. Psychosomatic Med., *34*: 221–234.
7. FRIEDMAN S. and FISHER C. (1967): On the presence of a rhythmic, diurnal, oral instinctual drive cycle in man: a preliminary report. J. Amer. Psychoanol. Ass., *15*: 317–343.
8. OSWALD I., MERRINGTON J. and LEWIS H. (1970): Cyclical "on-demand" oral intake by adults. Nature, *225*: 959–960.
9. HALBERG F. (1968): Physiological considerations underlying rhythmometry, with special reference to emotional illness. *In*: Cycles Biologiques et Psychiatrie, (J. de AJURIAGUERRA, ed.), pp. 73–126, Masson & Cie, Paris.
10. KRIPKE D. F., REITE M. L., PEGRAM G. V., STEPHENS L. M. and LEWIS O. F. (1968): Nocturnal sleep in rhesus monkeys. Electroenceph. Clin. Neurophysiol., *24*: 582–586.
11. KRIPKE D. F., HALBERG F., CROWLEY T. J. and PEGRAM G. V. (1970): Ultradian rhythms in rhesus monkeys. Psychophysiology, *7*: 307–308.

ON THE DEVELOPMENT OF ULTRADIAN RHYTHMS: THE RAPID EYE MOVEMENT ACTIVITY IN PREMATURE CHILDREN

Rolf E. Ullner

Institut für Frühpädagogik
München, Germany

It was interesting to study the chronophysiology of infants to examine developmental aspects of periodicity.

METHOD

Two premature children, born at gestational ages of 33 wk., were measured continually day and night—Susi for the first 30 days of life and Anton for 10 days starting at a gestational age of 38 wk. The infants were fed at regular intervals under constant light. Feeding synchronizer effects were removed in later analyses. The *rapid eye movement activity* (REMA) was measured by watching the EOG and judging each 12 sec as active if it contained at least one rapid eye movement. The mean of each 25 measurements was taken as one point of a time series. Thus time series were created consisting of equal 5 min. intervals. EKG, breathing, motility, and general behavior were measured at the same time.

Statistical analyses of these data from premature children were difficult, as random noise generated by the clinical routine was relatively great. Another biological difficulty arose from the fast alterations of the frequency patterns during early childhood: within a few days the frequency patterns changed, thus obscuring findings of variance spectra computed from time series of sufficient length. Therefore, it was necessary to compromise by cutting the entire time series into intervals of 4 or 5 days, so that the frequency might appear stable within the interval. In addition, the lag had to be relatively long in order to obtain spectral estimates of the interesting circadian periods. This caused greater statistical instability in the computed spectra. In order to get better spectral estimates in the more unstable high frequency region, spectra with shorter lags were also computed.

To check the statistical methods and to verify the results, a hybrid study was also performed on 28 premature children, each observed for at least 4 days. Spectra and circadian and ultradian quotients were computed for these data with confidence interval estimates.

RESULTS

Polygraphic chronograms did not show any distinct patterns, as functions are not yet internally synchronized. Well-defined behavioral patterns were not visible, but each observed function was revealed as a multi-oscillatory system composed of different frequency patterns, changing with age.

REMA is a gross indicator of brain activity, as shown already in polygraphic sleep investigations. This parameter showed distinct frequency patterns when evaluated by variance spectra. The spectra showed gradual changes in frequency patterns and amplitudes depending on age. *Outstanding spectral components* were detectable during the whole span of observation, changing more in amplitude than in frequency. These *outstanding spectral* components are shown in Fig. 1, in relation to age. The highest frequency detected was about 1 cycle per 0.9 hr. It is well known as the frequency of sleep cycles in the newborn. Also, a frequency of about 90 min. per cycle was evaluated, similar to that of sleep cycles in adults. The other spectral components do not seem to be harmonics, as they changed their values independent of one another. The biological correlates of the 3.5, 6.0, 8.2, 12.0, and 16.0 hr per cycle components is still questionable.

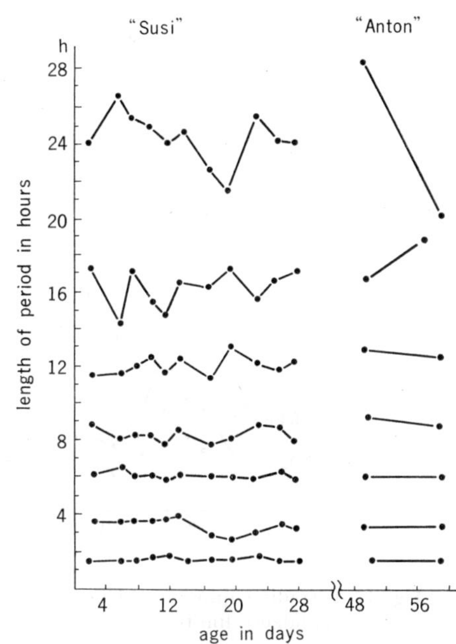

Fig. 1 Relation of several spectral maxima to age.

Fig. 2 shows the period length of the *dominating ultradian period* of each spectrum at different ages. This period shifted to the low end of the frequency spectrum with increasing age. At ages of 1 to 4 days, the ultradian period of 90 min. was dominating. At the end of the observations of both children, the dominant period had shifted to a length of 12–17 hr.

Fig. 3 shows the ultradian quotients of the dominating ultradian component of each interval in relation to age. Fig. 4 shows the circadian quotients in relation to age. During the first days of observation, the circadian quotient was higher than the ultradian quotient. At an age of about 10 days, the ultradian quotient was dominant. At a gestational age of 39 wk. or an age of 59 days, the circadian component was dominant again.

These results were substantially confirmed by the results of the hybrid study.

DISCUSSION

The results of these two studies are consistent with findings of Hellbrügge,

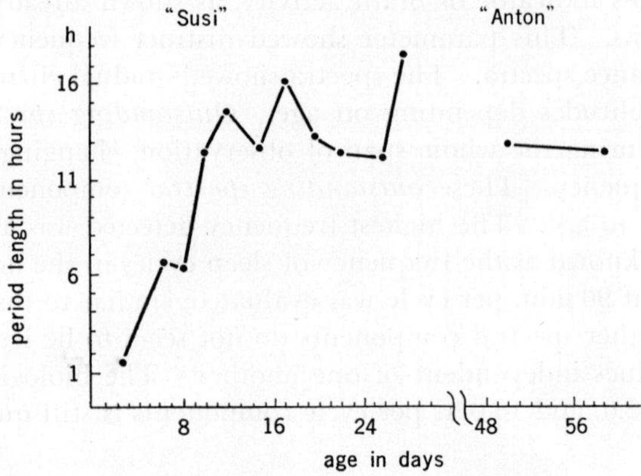

Fig. 2 Relation of *dominating ultradian period* to age.

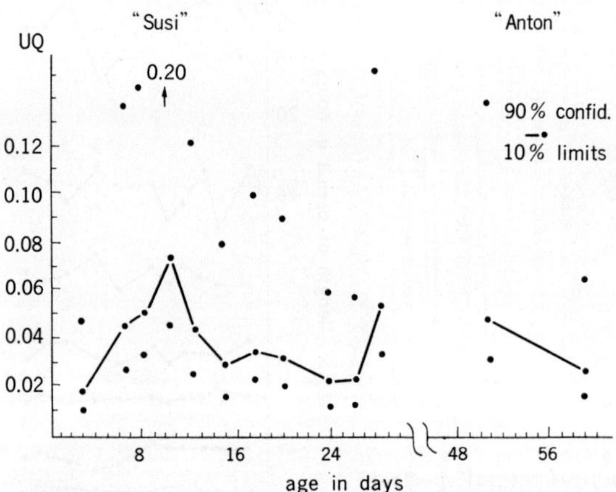

Fig. 3 Ultradian quotients (UQ) in relation to age (dots show 80 percent confidence limits).

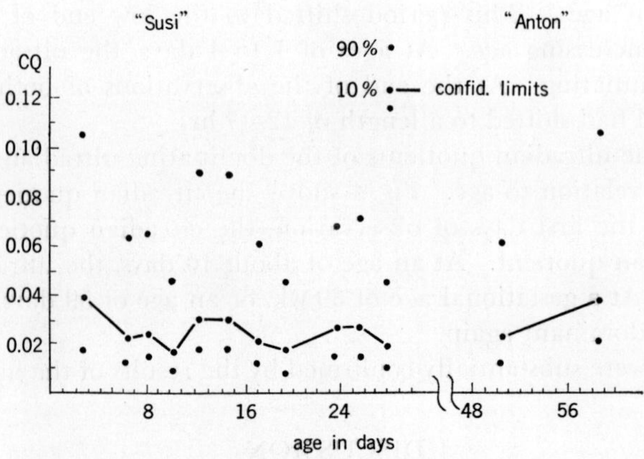

Fig. 4 Circadian quotients (CQ) in relation to age (dots show 80 percent confidence limits).

EHRENGUT-LANGE, and ULLNER (1969) (shown in Fig. 5). In the latter study, the circadian spectral components were dominant at an age of 2 to 5 days. The maximum of the ultradian components was at ages of about 2 mo. The circadian quotient reached its final height at an age of about 1.5 yr. The surprising result is that in both studies, the circadian components were dominant first.

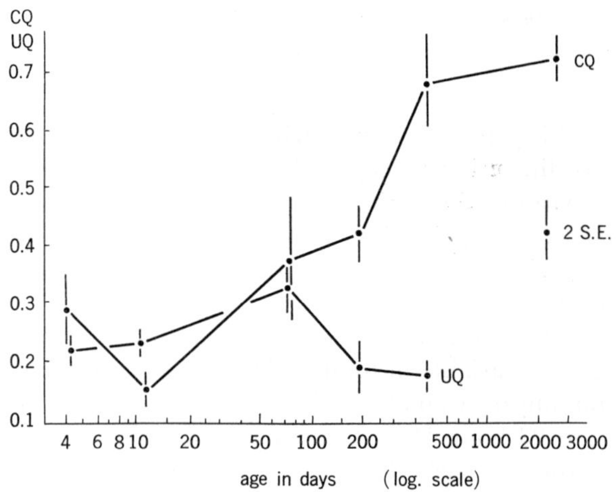

Fig. 5 Ultradian quotient (UQ) and circadian quotient (CQ) in relation to age (mean and standard error). From a study of heart rate in full term children.

This may be due to some maternal synchronization. In the study of REMA in premature infants, the minimum of both ultradian and circadian components was reached at an age of about 5 days, and in the heart rate study of full term children, it was reached at an age of about 6 days. Until this age, newborns are clinically in a state of disordered adaptation to extrauterine life. If we regard the minimum of outstanding spectral components, i.e., a nearly flat variance spectrum, as the expression of a chronophysiologic disorder, we may find a correspondence between the clinical and physiologic findings.

ACKNOWLEDGEMENT

The subject of this field study was suggested by Professor HELLBRÜGGE. These studies, together with those of WINDORFER, Jr. analysing EKG in the same children, were supported by Deutsche Forschungsgemeinschaft at the University Children's Hospital, Erlangen.

REFERENCES

1. HELLBRÜGGE TH., EHRENGUT-LANGE J., ULLNER R. (1969): On the Development of the Circadian Periodicity of Heart Rate in Childhood. 2nd Internat. Symposium on Experimental and Clinical Chronobiology, Florenz, Manuscript.
2. ULLNER R. (1971): Über die Entwicklung periodischer Vorgänge der Augenbewegungsaktivität bei zu früh geborenen. Dissertation, München.

THE RHYTHMS OF SLEEP AND WAKING

WILSE B. WEBB

Department of Psychology, University of Florida
Gainesville, Florida, U.S.A.

The purpose of this paper is to summarize the cyclical components of sleep and waking along two dimensions: sleep and waking relations within the circadian period and cycles within sleep periods.

Sleep and waking: Considered as a two phase process (sleep-awake) within a twenty-four hour period three variations of sleep are possible: total amount of sleep; number of sleep events; temporal placement of sleep. All three of these show a wide range as a function of age, species, and individual differences. I shall focus on age as a variable in human sleep-wake patterning.

The average amount of sleep of the human neonate is about 16½ hours with a range from 10½ to 23 hours [1]. At three weeks the average is 14½ hours [2]. From one year to the late teens sleep length is reduced by about five hours (12½ to 7½ hours) in an essentially linear decline. In short, about one-fourth of the total reduction in sleep amount occurs in the first three weeks and somewhat less than half in the first year. Beyond adolescence the average sleep length is remarkably stable across a wide variety of cultures and conditions. In the aged subject, there appears to be less of a change in average length than an increase in between subject variability and in a redistribution of sleep [3].

Changes in number of episodes are less marked. The neonate has about five to eight short bursts of awakenings between relatively long sleep periods. By about six months the pattern which will hold true until the third or fourth year has been formed; there is a long nocturnal sleep period (with occasional awakenings) and typically two nap periods interspersed in the day. By five years of age the basic byphasic sleep-wake pattern has been developed in most subjects.

The early developmental aspects of the temporal placement of sleep are clearly outlined by KLEITMAN [4] based on the Kleitman and Englemann data cited above. The increase in the night/day ratio is, up to the 12th week, a function of increasing night sleep and decreasing day sleep. Beyond the 12th week, the ratio is a function of the decreasing day sleep. By the age of one, sleep during the day constitutes only about 10–15% of the total sleep time and by the age of four about 5%. From the age of five into the sixties, with the exception of the "siesta" pattern of the tropical climates, this nocturnal pattern of sleep continues. In the sixties, naps again become a prominent feature of sleep and the night period shows intrusions of awakenings although the basic nocturnal placement is maintained [3].

Infrarhythms of sleep: The infrastructure of sleep, as defined by the electroencephalogram, also may vary along three major dimensions: the total amount of each of the five identifiable stages of sleep; the number of sleep stage events, and the temporal ordering of the stages within sleep. These factors also show a wide range heavily influenced by age, species, and individual differences. For this paper, we shall concentrate on human sleep and the factors which influence two of the major stages of sleep—stage 4 (or deep sleep) and stage REM (or dream sleep)—with particular emphasis on the temporal aspects.

Both stage 4 and stage REM are clearly responsive to age changes. REM sleep constitutes about 50% of the sleep of neonates. By the age of two REM sleep is about 30% of total sleep time indicating the marked influence of the early developmental process on this stage of sleep. There is a stabilization of REM amount in the early teens at about 20–25% and this percent remains essentially constant into the sixties. If there is a decline of REM in the elderly, it is quite slight. The influence of age on stage 4 sleep is in sharp contrast. The amount of stage 4 peaks in the teens at about 20% and begins a gradual but clear decline in the twenties. By the fifties, the average amount of stage 4 is less than 10% and it is not unusual to find little or no stage 4 in one-fourth of the subjects measured. With the decline in stage 4, there is corollary increase in the amount of light sleep and awakenings.

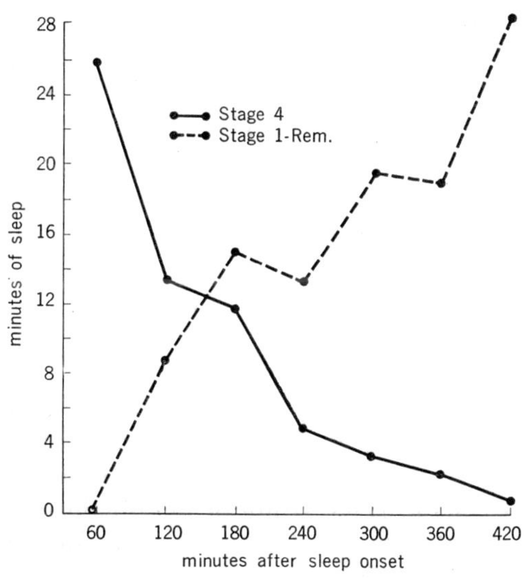

Fig. 1

Stage 4 and REM sleep show contrasting distributions within sleep. Fig. 1 presents the relationship between time asleep and average amount of stage 4 and REM. Minutes of sleep per hour are on the ordinate and hours of sleep are on the abscissa. The data are drawn from 82 nights of laboratory sleep of 29 young adults (three nights per subject).

The relationship between amount of prior wakefulness and the amount of stage 4 obtained in the first three hours of sleep appears to be essentially linear up to about 24 hours where the amount obtained begins to asymptote [5]. In contrast REM amounts appears to be essentially unrelated to time of prior wakefulness. Again the data was drawn from young adults and sleep onset time is held constant at 11 P.M. (circa).

The time of sleep onset does not appear to have a marked effect on stage 4 but has a marked distributional effect on REM sleep. Fig. 2 presents the characteristics of stage 4 and REM sleep with a sleep onset time of 8 A.M. with time of wakefulness of 17 hours [6]. When compared with Fig. 1, one can see only a limited distortion of stage 4 but a clear modification of the REM distribution.

REM sleep has particularly interesting features for those interested in biorhythms. It displays a strong rhythmic character within sleep and it is quite sensitive to circadian variations. This analysis will focus on these aspects. We should note in passing, however, that REM appears remarkably insensitive to extra sleep variables such as time or prior wakefulness, ago (beyond early adolescence), psychopathology, or presleep experiences. In contrast, this stage of sleep is very sensitive to a broad spectrum of drugs (c.f., KALES [7]). To us these facts point to a strong endogenous control.

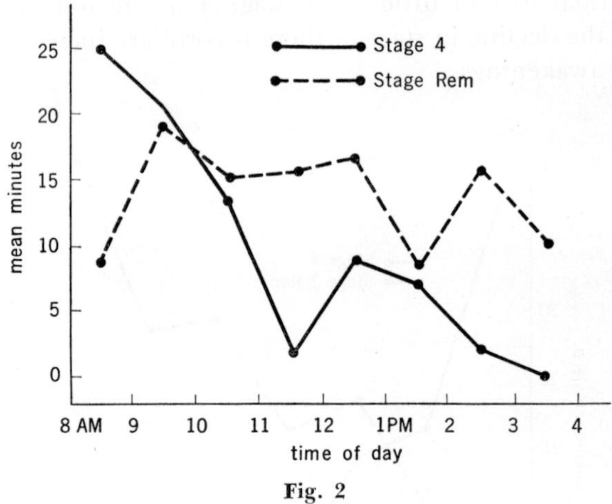

Fig. 2

Let us consider the rhythmic quality of REM. We all know that there is a 90 minute (circa) REM cycle. What we don't all seem to know is that this 90 minute cycle is a *mean* around an enormous variability of onset times and that each episode is different in length. The nature of the obtained curve of REM sleep in Fig. 1 results from a wide range of onset times around 90 minute intervals which overlap and from each successive interval having a tendency to be somewhat longer than the preceding one.

Let us examine in detail the REM cycling of 22 young adults during 3 nights each. A REM episode was defined as one in which sleep was preceded or succeeded by at least 30 minutes of non-REM and an interval is defined from onset times.

First, it is true that REM occurs in "bursts." Two percent of the subject has six episodes, 28% had five episodes, 58% had four episodes, and 12% had three episodes. Table 1 displays the range of onset times. The first row gives the onset times of 66 first REM episodes and the second row the interval between the first and second episode. The range of both sets of figures are considerable. Approximately 10% of the first REM episodes occurred before 60 minutes or after 130 minutes, an interval of more than an hour. The mean time of onset was 93 minutes—close enough to 90 minutes to be impressive—but only 17% of the onset times fell within a twenty minute interval around 90 minutes. The time interval between the first episode and the second episode was slightly narrower in range.

Table 1 Latency of onset of first REM period and interval between REM periods I and II. (N=66)

	Minutes				
	51–70	71–90	91–110	111–130	131–150
Onset I	19	24	12	5	6
Interval I–II	3	27	21	12	3

Although no interval was shorter than 60 minutes, 14% exceeded 120 minutes and 41% of the times fell within 20 minutes of the ninety minute time period.

In summary, then, although it does appear that there is an approximately 90 minute cycle associated with REM, we must recognize this as a group characteristic which contains a considerable variance.

REM sleep shows an interesting circadian responsivity. Earlier studies, in which time of prior wakefulness was not controlled, had demonstrated that when sleep onset was displaced from its "normal" period (11 P.M.–7 A.M.) REM amount in the first few hours of sleep was sharply enhanced by sleep onsets near normal sleep termination time where REM tendencies are maximum (see Fig. 1). The potentiality of REM in early sleep was a decreasing function from that time period to normal sleep onset and about 90 minutes thereafter [6, 8]. The study from which Fig. 2 was taken, in addition to controlling prior wakefulness, adds two provocative pieces of information. As in the earlier reports, the amount of REM obtained during the sleep period was not different from that obtained during baseline sleep. In short, not only did the clock continue to function, but an averaging process seemed to be present as well.

From this array of facts, let me select a few features to emphasize. The patterning of sleep, both in its circadian and ultradian character, is clearly developmental. Similar documentation could be developed for species determinants and for stable individual differences. This suggests a heavily endowed biorhythm system. The infrastructure of sleep is apparently comprised of at least two differentially responsive systems—stage 4 and REM. Stage 4 appears to be highly responsive to variations in wakefulness. The REM system, on the other hand, shows evidence of being a strongly intrinsic rhythm (albeit variable) which is quite resistive to imposed conditions.

REFERENCES

1. Parmelee A. Shultz H., Disbrow M. (1961): Sleep patterns of the newborn. J. Pediat., *58*: 241.
2. Kleitman N. and Englemann T. (1953): Sleep characteristics of infants. J. Appl. Physiol., *6*: 269.
3. Webb W. and Swinbure H. (1971): An observational study of sleep of the aged. Percept. & Mot. Skills, *32*: 895.
4. Kleitman N. (1963): Sleep and Wakefulness. 2nd ed., University of Chicago Press, Chicago.
5. Webb W. and Agnew H.: The influence of time course variables on stage 4 sleep, (in press).
6. Webb W. and Agnew H. (1971): Effect on sleep of a sleep period time displacement. Aerospace Med., *42*: 152.
7. Kales A. (1969): Sleep: Physiology and Pathology. J. B. Lippincott Co., Philadelphia.

8. Webb W., Agnew H. and Sternthal H. (1966): Sleep during early morning hours. Psycho. Sci., *6*: 277.
9. Webb W. and Agnew H. (1967): Sleep cycling within twenty-four hour periods. J. Exp. Psychol., *74*: 158.
10. Webb W. and Agnew H. (1969): Measurement and characteristics of nocturnal sleep. *In*: Progress in Clinical Psychology, (L. E. Abt and B. Reiss, eds.), Grune & Stratton, New York.

THE PARADOXICAL SLEEP CYCLE REVISITED

STUART A. LEWIS*

Department of Psychiatry, University of Edinburgh
Edinburgh, Scotland

Sleep abounds in mythology. The purpose of this paper is to examine two recently avowed articles of faith. While I do not set myself up as an iconoclast, it is perhaps appropriate that someone hailing from the city of John Knox should attempt to pick holes in our pontifications.

Paradoxical sleep, it is said, is a periodic phenomenon having a cycle length of 90 min. While I will accept that paradoxical sleep appears, disappears and re-appears throughout the night, what I question is whether this alternation of orthodox sleep—paradoxical sleep is rhythmic. In other words, how stable is the 90 min. cycle?

If the appearance of paradoxical sleep was periodic then the standard deviation should be small. Furthermore, the most stable measure of the rhythm may not be from the start of one paradoxical phase to the start of the next; it may be end to end or mid-point to mid-point. The data** of Table 1 demonstrate that the *mean* cycle length is indeed 90 min., however a cycle is defined. However, the standard deviations not only show individual differences, they are large. It seems

Table 1 Mean cycle length and standard deviation.

Subject	N.†	Start—Start	N.	End—End	N.	Mid-pt—Mid-pt
1	16	80.4±14.4	16	84.1±10.1	16	84.7±17.6
2	13	93.9±19.1	11	104.5±22.7	11	100.8±23.4
3	27	98.3±24.9	25	104.1±27.5	25	111.3±66.2
4	33	83.9±21.6	30	89.0±18.0	30	86.0±15.4
5	11	85.0±11.3	11	91.3±14.7	11	90.7±20.5
6	10	103.1±18.1	9	120.0±25.6	9	110.7±18.2
7	11	114.7±29.7	8	131.1±24.2	8	123.1±28.6
8	13	92.0±22.9	13	94.3±30.3	13	95.1±23.5

† N=number of periods of paradoxical sleep summed across nights. One period is defined as continuous provided no more than 15 minutes of orthodox sleep or waking intervenes. There also had to be at least 20 minutes of orthodox sleep between the last paradoxical sleep and the end of the record for this paradoxical sleep to be included.

to be unreasonable to imply rhythmicity when, to account for 95 percent of the distribution, the range has to be of the order of 50 min. to 130 min. (these values are in line with those of HARTMAN, 1968). While this could be considered a

* Present address: University of Nottingham Medical School, Nottingham, England.

** All the data presented were obtained from the baseline nights for subjects who subsequently took part in already published drug studies. It should be noted that all had two laboratory adaptation nights prior to sleep recordings and that there were at least 4 nonconsecutive nights available for analysis.

"rhythm" it would seem to be of little predictive value. On the other hand, the fact that there is some "order" to the appearance of paradoxical sleep should tell us something about the internal organization of sleep. It must also pose problems for those seeking daytime equivalents of paradoxical sleep.

Assuming paradoxical sleep to have a rhythm, albeit a weak one, then it is a reasonable assumption that it is not "switched" on and off by sleep. The more parsimonious hypothesis is that our current means of detecting paradoxical sleep make it invisible during waking hours. However, there are three areas of study claiming a 90-min. daytime rhythm: The "on demand" feeding cycle [1, 2], spontaneous seizure discharges [3], and psychomotor performance [4]. The first of these is especially interesting in that the "on demand" feeding cycle in babies is related to the hourly cycle and their paradoxical sleep has a mean periodicity of about 60 min.

Following from the hypothesis that paradoxical "sleep" is endogenous and present throughout the 24 hr., then its time of appearance after falling asleep would have no particular distribution. This is so since our time of going to sleep varies over a considerable time period. Of course, the drive for going to bed may be stronger when we are approaching the time for a block of paradoxical sleep. If this were the case, then would it not be a logical conclusion that paradoxical sleep should occur very rapidly after the onset of sleep? This we know is not the situation. It is well known that there is normally an obligatory 45 min. of orthodox sleep before the first paradoxical sleep period. The distribution of the delay to the first paradoxical sleep period shows that a bimodal distribution with peaks at 60 min. and 130 min. This implies that there is something about the onset of sleep that controls the onset of paradoxical sleep. Globus [5] has suggested that it is "clock on the wall" that is the determining factor. Let me enumerate some predictions from his hypothesis.

If paradoxical sleep is unrelated to sleep per se but is linked to a "real time" clock then:

i. The variance of the time of the first paradoxical period should be less than the variance of either the time of going to bed or the onset of sleep.

ii. The alternative formulation of this is that the variance of the delay to sleep (from sleep onset) should be greater than the delay to the first paradoxical sleep (from the time of going to bed).

iii. There should ge a high positive correlation between the sleep onset time.

In other words, the correlation between delay to sleep onset and delay to first paradoxical sleep should be high and positive.

The data presented in Tables 2, 3, and 4 indicate that these deductions from the Globus hypothesis are not upheld either for the group or for individuals. This would certainly strengthen the affirmation that the switching mechanisms for the internal structure of sleep are endogenous rather than in the external environment.

The implication of these findings is that paradoxical sleep has nothing to do with sleep but it is a basic period phenomenon of the body. In turn this highlights the problem of the definition of this physiological state.

Customarily, as is implied by the name rapid eye movement sleep, we define the state in terms of its peripheral manifestations. Under certain circumstances this can lead to difficulties. For example, the effect of the monamine oxidase

Table 2 Group values.

	Mean	S.D.	Median	S.I.Q.R.*
Time in bed	23.42	21.04	23.48	10.6
Time of sleep onset	00.04	22.07	00.03	8.35
Time of 1st REM	01.25	39.04	01.21	21.4
Delay to sleep onset (d)			20.2	7.8
Delay to 1st REM from sleep onset (D)			65.3	18.3
d+D			92.5	19.5

(Obtained from 50 subjects, each subject contributing 1 set of data)
* The semi-interquartile range (S.I.Q.R.) is used since delays are non-normal

Table 3 Means and standard deviations for individuals.

Subject	N.*	Time in bed		Time asleep		Time of 1st REM		Delay to sleep (d)		Delay to 1st REM (D)		d+D	
		Mean	S.D.	Mean	S.D.	Mean	S.D.	Mean	S.D.	Mean	S.D.	Mean	S.D.
1	12	2342	18.1	2349	19.1	0127	41.2	8.4	2.6	87.5	31.9	95.9	33.5
2	11	2340	14.5	2353	12.2	0102	37.1	10.8	7.0	74.6	40.9	85.8	42.9
3	12	2348	13.1	0018	19.1	0127	25.4	29.6	15.6	69.3	22.9	98.8	27.8
4	11	2352	20.1	0013	19.4	0114	26.6	20.6	16.4	60.6	10.6	81.3	21.5
5	9	2354	24.8	0014	29.3	0131	34.1	19.3	13.8	76.6	21.1	95.9	22.5
6	12	2353	10.2	0004	14.3	0128	23.0	32.5	13.6	64.4	17.0	96.9	16.6
7	10	2338	12.4	2354	9.8	0140	52.9	14.9	7.0	106.3	52.5	123.2	50.4
8	10	2340	19.3	0000	17.8	0056	18.5	20.0	18.0	56.3	13.7	76.3	21.3

* *N*=number of nights

Table 4 Correlation d* v D* (Spearman rank).

Subject	N.*	P
1	12	0.5997
2	11	0.3160
3	12	0.1871
4	11	−0.1454
5	9	−0.0708
6	12	−0.0454
7	10	−0.0818
8	10	−0.0484

* d, D, and N. as defined in Tables 1 and 2

inhibitor, phenalzine, is ultimately to abolish paradoxical sleep [6, 7]. After this drug has been given for some time, the muscle tone is not periodically reduced and there are no eye movements, but there do appear to be changes in the EEG at times when paradoxical sleep might have been expected (Dunleavy, personal communication). Perhaps if PGO spikes or some hitherto unknown parameter were more amenable to recording in man, it would be found that the MAOI's do not abolish paradoxical sleep.

This paper has concentrated on one or two aspects of paradoxical sleep. It has not considered the problem of the lengthening of the paradoxical sleep phases across the night. Nor have I discussed stages 3+4 sleep. These orthodox stages of sleep appear usually twice nightly with the bulk appearing in the early part of the night. This, together with preferential recovery of stages 3+4 over paradoxical sleep following total sleep deprivation must raise problems for OSWALD's [8] hypothesis of the function of sleep.

Briefly, OSWALD suggests that stage 3+4 sleep reflect body repair processes while paradoxical sleep affects brain repair processes. It follows then from the deprivation results, that body repair takes precedence over brain repair. This does not seem a wise system to operate. Stage 3+4 sleep tends to be neglected; it may turn out to be more important than paradoxical sleep.

REFERENCES

1. FRIEDMAN S. and FISHER C. (1967): On the presence of a rhythmic, diurnal, oral instinctual drive cycle in man. J. Amer. Psychoanal. Ass., *15*: 317.
2. OSWALD I., MENINGTON J. and LEWIS H. (1970): Syclic "on demand" oral intake by adults. Nature, *225*: 959.
3. STEVENS J. R., KODAMA H., LONSBURY B. and MILLS L. (1971): Ultradian characteristics and spontaneous seizure discharges recorded by radio telemetry in man. Electroenceph. Clin. Neurophysiol, *31*: 313.
4. GLOBUS G. G., DRURY E. P. and BOYD R. (1971): Ultradian rhythms in performance. Paper read to A. P. S. S., Bruges, June, 1971.
5. GLOBUS G. G. (1966): Rapid eye movement cycle in real time. Arch. Gen. Psychiat., *15*: 654.
6. WYATT R., KUPFER D. J., SCOTT J., ROBINSON D. S. and SNYDER F. (1969): Longitudinal studies of the effects of monoamine oxidase inhibitors on sleep in man. Psychopharmacol., *15*: 236.
7. AKINDELE M. O., EVANS J. I. and OSWALD I. (1970): Mono-amine oxidase inhibitors, sleep and mood. Electroenceph. Clin. Neurophysiol., *29*: 47.
8. OSWALD I. (1969): Human brain protein, drugs and dreams. Nature, *223*: 893.

RAPID EYE MOVEMENTS DURING SLEEP AND WAKEFULNESS

EKKEHARD OTHMER and MARY HAYDEN

Department of Psychiatry, Renard Hospital
St. Louis, Missouri, U.S.A.

KLEITMAN has hypothesized that rapid eye movement (REM) sleep is a manifestation of a basic rest and activity cycle which is present during sleep and wakefulness [1]. Several authors have presented some evidence to support this hypothesis [2–13]. The results of recent polygraphic 24-hour studies on three female college students are compatible with this hypothesis.

METHOD

The subjects were kept in bed for 24 hours on each of four non-consecutive days. The studies started at 10 p.m. After a morning breakfast, subjects were asked to try to stay awake but to keep their eyes closed. Subjects fell asleep shortly after the beginning of the study. Standard electroencephalic (EEG), electromyographic (EMG) and electrooculographic (EOG) recordings were obtained during the entire 24 hour period. Whenever sleep occurred during this period it was scored in accordance with a standardized manual [14]. Special emphasis was given to the scoring of rapid eye movements (REM). The criteria for REM-scoring during sleep are broadly accepted (stage 1-REM). Its occurrence is believed by many sleep researchers to be associated with dreaming [15]. There exist, however, no criteria for the scoring of different polygraphic stages of wakefulness. In the present study a new score for rapid eye movement during wakefulness (stage W-REM) was introduced using the following criteria:

1) Eye movements had to be similar in amplitude, duration and shape to those seen during REM sleep.

2) The muscle tonus (EMG) had to be lower than the one observed during wakefulness prior to sleep onset at night or those periods of wakefulness that were accompanied by eyeblinks and frequent body movements.

3) The EEG had to show a predominance of alpha-waves.

4) Eye movements differing in shape and amplitude from those present in REM sleep or eye movements mixed with eyeblinks or accompanied by high muscle tonus or frequent body movements were not scored as stage W-REM.

RESULTS

The results of one part of the experiment are shown in Fig. 1. This figure shows the plots obtained from 24-hour polygraphic records of each of three subjects. The ordinates represent different stages of sleep. Stage W represents wakefulness and stages 1–4 are different stages of sleep which are roughly determined

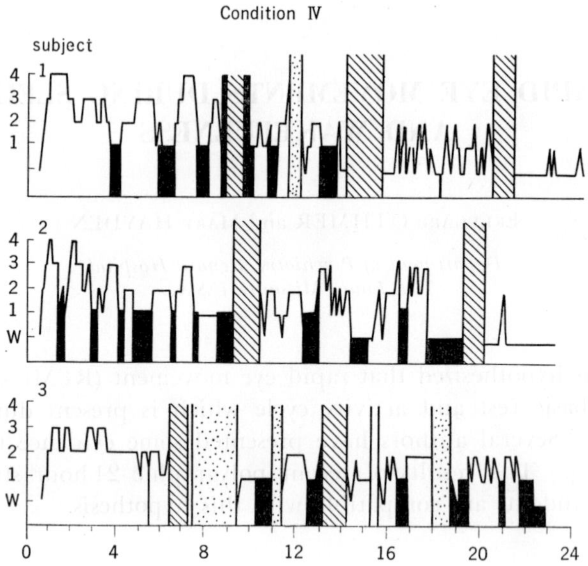

Fig. 1 Results of 24-hour polygraphic studies on 3 subjects. For description, see text.

by the amount of slow waves (delta waves) present in a 30-second epoch. The different sleep stages correlate significantly with increasing auditory threshold levels and, therefore, have often been interpreted as different levels of deepness of sleep. The abscisas represent the time in hours. Time 0 is approximately 10 p.m. The striped columns indicate meal breaks and the dotted columns, interruptions due to technical problems. The solid areas represent the occurrence of rapid eye movements. If rapid eye movements occur during sleep (stage 1-REM), the column height is drawn up to stage 1; if rapid eye movements occur during wakefulness (stage W-REM), the column height is drawn up to stage W. The figure shows that stage W-REM occurred only once in the recording of subject 1, for a relatively short time during the 19th hour of recording. In subject 2, W-REM was observed during the 15th, 17th and 18th hours of recording and in subject 3, at the 19th and 23rd hours of recording. In subject 3, W-REM appeared first intermittently during a stage 1-REM period (19 hour) and then subsequent to a stage 1-REM period (23rd hour). Interviews that were obtained during W-REM in all three subjects revealed mental activity characterized by visualizations of a scene, emotional involvements, and a dramatic story-like flow of events, while during stage W without rapid eye movements, abstract and non-visual thoughts were reported.

Fig. 1 shows also that the occurrence of rapid eye movements is not dispersed randomly throughout the whole time of subjects' wakefulness, but occurs similarly to stage 1-REM in an episodic fashion with a scorable onset and end of the episode. It is tempting to speculate that stage W-REM episodes occur with a similar episode length and degree of regularity as stage 1-REM.

After breakfast (first striped column in each of the three plots), extended periods of sleep were prevented by awakening the subject whenever stage 2 sleep exceeded 3 minutes. Subjects returned to sleep relatively rapidly and had to be awakened again. Surprisingly, this interference with sleep did not prevent the onset of stage 1-REM periods. Indeed, stage 1-REM periods occurred in subject 1 immediately after breakfast at the onset of sleep ("sleep onset REM"). Sleep onset REMs are not observed during the beginning of night sleep in normal adult subjects. These results are typical of those obtained during the 3 other 24-hour studies made with each subject.

DISCUSSION

The occurrence of stage W-REM led to the conclusion that the appearance of rapid eye movements is not limited to sleep. The phenomenologically similar appearance of stage W-REM to stage 1-REM leads us to suggest that both phenomena should be analyzed statistically to determine if their interval properties are identical. The roughly similar distribution of stage W-REM and stage 1-REM, the occurrence of visualizations of story-like events during both states and the occurrence of sleep onset stage 1-REM lead us to speculate that both phenomena might have the same central origin. However, only the detection and measurement of common underlying neurophysiological processes can assure us that peripheral phenomena that show some similarity with each other are indeed centrally connected and can therefore be considered as manifestations of an encephalic cycle during sleep and wakefulness.

REFERENCES

1. Kleitman N. (1963): Sleep and Wakefulness. Univ. of Chicago Press, Chicago.
2. Globus G. C. (1966): Rapid eye movement cycle in real time. Arch. Gen. Psychiat., *15*: 654–659.
3. Friedman S. (1968): Oral activity cycles in mild chronic schizophrenics. Amer. J. Psychiat., *125*: 743–751.
4. Friedman S. and Fisher C. (1967): On the presence of a rhythmic, diurnal, oral instinctual drive cycle in man: a preliminary report. J. Amer. Psychoanal. Assoc., *15*: 317–343.
5. Oswald I., Merrington J. and Lewis H. (1970): Cyclical "on demand" oral intake by adults. Nature, 225: 959–960.
6. Kripke D. F.: An ultradian biological rhythm associated with perceptual deprivation and REM sleep. Psychosomat. Med., (in press).
7. Othmer E. (1967): Schlaf-und wachverhalten. Forum der Psychiatrie, H.8, Enke Verlag Stuttgart.
8. Othmer E., Hayden M. P. and Segelbaum R. (1969): Encephalic cycles during sleep and wakefulness. Science, *164*: 447–449.
9. Dement W., Kelley J., Laughlin E., Carpenter S., Simmons J., Sidoric K. and Lentz R. (1971): Life on "the basic rest-activity cycle": Sleep studies of a ninety minute day. Presented at the First International Congress of the Association for the Psychophysiological Study of Sleep, Bruges, Belgium.
10. Sterman M. D., Lucas E. and MacDonald L. (1971): The basic rest-activity cycle in the cat. Presented at the First International Congress of the Association for the Psychophysiological Study of Sleep, Bruges, Belgium.
11. Lucas E. A. and Sterman M. B. (1971): Effects of performance, forebrain lesions and drugs on sleep-wake and rest-activity cycles in the cat. Presented at the First International Congress of the Association for the Psychophysiological Study of Sleep, Bruges, Belgium.
12. Passouant P., Popoviciu L., Velok G. and Baldy-Moulinier M. (1968): (Polygraphic study of narcolepsy during a 24-hour period.) Etude polygraphique des narcolepsies au cours du nycthemere. Rev. neurol. (Paris), *118*: 431–441.

13. HAYDEN M. P. (1969): Sleep-onset REM in normals. Electroenceph. Clin. Neurophysiol., *27*: 685, (Abstr.).
14. RECHTSCHAFFEN A. and KALES A. (1968): A Manual of Standardized Terminology, Techniques and Scoring System for Sleep Stages of Human Subjects. Public Health Service, U.S. Government Printing Office, Washington D.C.
15. HARTMANN E. (1967): The biology of dreaming. Boston State Hospital Monograph Series No. 2, (I. NEWTON KUGELMASS, ed.), Charles C. Thomas, Springfield.

REM'S ULTRADIAN RHYTHM DURING 24 HOURS IN NARCOLEPSY

P. PASSOUANT

University of Montpellier, Faculty of Medicine
Montpellier, France

Narcolepsy is a sleep anomaly which permits a significant approach to the 24-hr periodicity of REM.

Polygraphic studies have shown that in a great number of cases, sleep onsets during the day as well as sleep onsets at night, begin with REM. The appearance of numerous sleep attacks during the day underlines the problem of periodicity in these attacks.

It is important to note that in narcolepsy, sleep may start with slow sleep and not with REM, and in some cases only slow sleep appears without rapid sleep. This latter case is characterized clinically by the absence of the accompanying signs of narcolepsy–hallucinations, memories of rich and sometimes disagreeable dreams, and cataplexy–and may be called hypersomnia. There are, indeed, numerous intermediate forms between the type with immediate appearance of REM and the type without REM during the day and with slow sleep at nocturnal sleep onset. The name dysomnia may be preferable to group these diverse forms.

This study includes three parts: (1) The periodicity of REM during the 24 hrs, (2) The influence of imipramine on the periodicity of REM, and (3) The relation of cataplexy to the periodicity of REM.

Periodicity of REM throughout the 24 Hours. Periodicity was studied in 25 narcoleptic patients, 23 men and 2 women, from 16 to 63 years old. Often several observations were made of the subject at intervals of several months or years to study the spontaneous evolution of the disease or the effect of treatment.

The 24 hr polygraphic records were divided into 10 min. periods. In every period, the proportion of each of three states of vigilance was measured: wakefulness, slow sleep, and REM. These intervals were used to develop 24 hr plots.

Spontaneous Periodicity of REM during the Day. This was observed in certain narcoleptics and corresponded to a period of 2 hr or longer.

(1) A 2 hr periodicity, comparable to that of REM at night, was observed in 3 subjects. In one of them, five narcoleptic attacks, commencing with REM and followed by a short period of slow sleep, appeared regularly every 2 hr, and this periodicity was more regular than that of REM in the night (Fig. 1). In another subject, six sleep attacks were produced with a brief period of slow sleep followed by REM, with the exception of the sixth, consisting only of slow sleep.

(2) A periodicity of 4 hr was recognized in another narcoleptic during the day, while at night the cycles varied from 1 hr 30 min. to 3 hr 40 min. It should be noted in this subject that periods of nocturnal wakefulness, common in narcoleptics, preceded or followed the occurrence of REM.

The Periodicity of Daytime Induced REM. Every 2 hr, daytime sleep was facilitated by a brief recumbent rest in a darkened chamber. In one narcoleptic, REM appeared with the regularity of a clock and was often followed by slow sleep. At night the duration of REM was very impressive, about 50 percent of total sleep (Fig. 2). In a second subject, each sleep attack began with slow sleep, and REM followed somewhat later. The mean periodicity of REM during the day was 2 hr 15 min. During the night it was shorter: 1 hr 45 min.

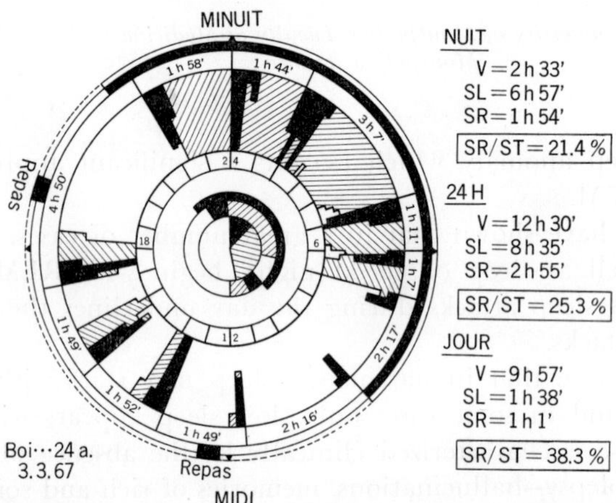

Fig. 1 Spontaneous periodicity of REM about every 2 hr. during the day in a narcoleptic (black: REM; hatched: slow sleep; white: wakefulness; black circumference on clock: lights out.)

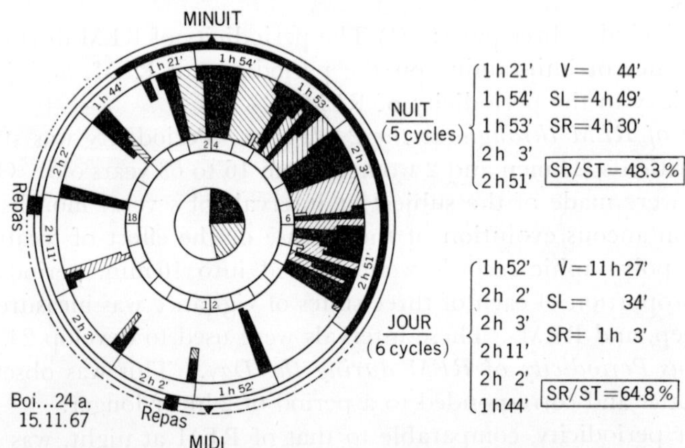

Fig. 2 Induced periodicity in REM of 2 hr. during the day in a narcoleptic. There was an important increase in REM at night.

The Periodicity of REM during 24 Hour Sleep Facilitation. To encourage sleep, one recumbent subject was placed in the dark for a whole day. Under these conditions, 17 sleep cycles were identified, varying from approximately 45 min. to 1 hr 45 min., without any notable difference in period from day to night. The total duration of REM was about 6 hr, that of slow sleep about 11 hr 45 min., and that of wakefulness 7 hr. (Fig. 3).

Comment. These results on the 24 hr periodicity of REM permit two types of conclusion: (1) A spontaneous periodicity of 2 hr can be identified in certain narcoleptics. It is augmented by facilitating sleep. (2) The periodicity of REM varies according to the type of narcolepsy. Precise in the evolving forms, it becomes difficult to identify in stabilized forms, either spontaneously or as a result of treatment. Indeed, in certain cases only slow sleep survives during the day, and REM appears only at night as in the normal subject.

Variations in the periodic production of REM are functionally related to two other states of vigilance: wakefulness and slow sleep.

In all cases of narcolepsy, a decrease in the level of vigilance causing sleep attacks may be considered the common factor. The types translated into immediate production of REM seem to be facilitated by slow sleep insufficiency, partially indicated in the characteristics of the sleep EEG (rarity of spindles and Stage 4). The transformation of this type of narcolepsy into the type with preponderance of slow sleep may be understood in terms of competition between the two sleeps. This competition appears particularly significant when sleep is facilitated over 24 hr. The duration of the cycles, those of REM, of slow sleep, and of wakefulness, are profoundly modified, and the sleep wakefulness organization over the 24 hr is comparable to that of the newborn. Following this analogy, the narcoleptic anomaly may be compared to an involution in states of wakefulness (Fig. 3).

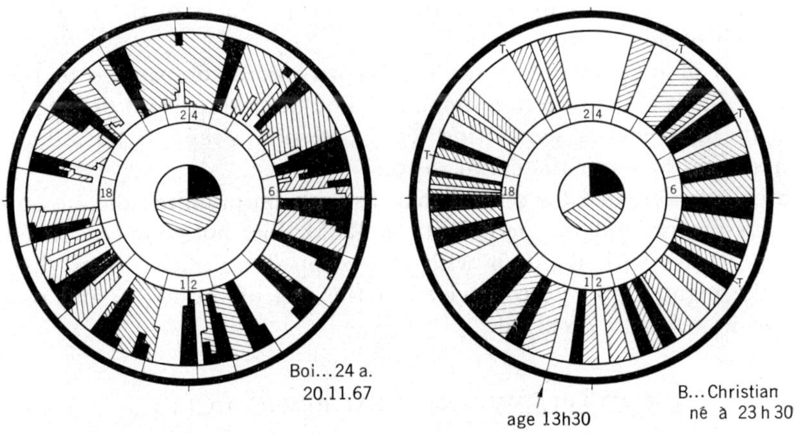

Fig. 3 A comparative study of organization of the 24 hr. states of vigilance. On the left, a narcoleptic patient is shown under sensory deprivation. On the right is a recording of a newborn commenced 13 hr. 30 min. after birth (T=feeding.)

Ontogenetic studies in animals and man have demonstrated the primitive organization of REM and the later development of wakefulness and slow sleep. Studies of cerebral monoamines in the kitten from birth to 2 months have demonstrated development of neuronal catecholamine terminals in the first days, while serotonergic neurons develop terminals only after the second month.

The ultradian rhythm in REM only appears slowly, and our studies of newborns demonstrate a very great variability over the 24 hours in the duration of these cycles from 7 min. to 2 hr. On this basis, one can assert that wakefulness and slow sleep are the synchronizers of the ultradian rhythm of REM.

In the normal subject, wakefulness delays the periodic development of REM. Its insufficiency in narcoleptics permits REM to develop.

In the normal subject, slow sleep permits the periodic development of REM. Its insufficiency facilitates the norcoleptic anomaly with the immediate appearance of REM at sleep onset, a pattern difficult to interpret in terms of the metabolic data on sleep cycles.

IMIPRAMINE AND THE ULTRADIAN RHYTHM OF REM

Imipramine is a retarder or supressor of rapid sleep and has been suggested in the treatment of narcolepsy.

The effect of imipramine on the nocturnal sleep of narcoleptics has been studied over 25–30 nights and complemented by telemetric recordings during the day.

For the first three placebo nights, the characteristic narcoleptic's sleep (sleep onset REM, augmentation of total sleep duration) was not modified.

Imipramine given for 4 days in progressive steps from 25–100 mgm produced specific changes. From the first night, after only 25 mgm of imipramine, REM was supressed. Periods of Intermediate REM, in which slow sleep and eye movements were associated, survived at the end of the night. Cessation of imipramine was associated with persistance of Intermediate REM, REM only appeared gradually, and the narcoleptic anomaly did not recover until the 25th night. There was no REM rebound. During the day, the sleep attacks of these subjects consisted entirely of slow sleep.

The effects of imipramine in the narcoleptics were comparable to those obtained in animals and in normal humans. The discrete periods of Intermediate REM in the narcoleptics, associating REM and slow sleep, are an indication of the competition between these two sleeps.

In narcoleptics treated for several months with imipramine, the action of the drug is less precise, but the periodicity of REM does not recover. In one subject, sleep attacks with REM or with slow sleep diminished. Slow sleep commenced the night, rapid sleep survived at the end of the night, and numerous periods of wakefulness occurred.

CATAPLEXY AND REM PERIODICITY

The cataplectic symptom of narcolepsy, interpreted as a disassociated expression of REM, has no periodicity.

A narcoleptic suffering 80 attacks a day permitted study of this point of view. Cataplexy, associated with a waking EEG, is commonly triggered by an emotion. It can occur, as in the case of this patient, during the most varied states: dining, conversation, etc. This symptom has no periodicity, although it is more frequent at the beginning than at the end of the day. Imipramine has a definite influence on cataplexy.

It thus appears that cataplexy, which is probably of reflex origin, should be differentiated from narcoleptic attacks, which appear to be related to the REM periodicity in typical forms of narcolepcy.

STUDIES OF THE TEMPORAL ORGANIZATION OF FUNCTIONAL STATES

M. B. STERMAN

Sepulveda V. A. Hospital
Department of Anatomy, University of California
Los Angeles, California, U.S.A.

Surprisingly, the pontine cat, the human fetus and the human neonate have in common a basic aspect of their respective behavioral repertoires. Each shows a continuous cycling of somatic quiescence and activity throughout the day and night. In the decerebrate cat and human neonate, the periodic recurrence of activity is accompanied by rapid eye movements and other physiological criteria which define the state known as active or REM sleep. Research in our laboratory has traced the development of the human REM sleep cycle into the prenatal period [1]. Fetal activity was recorded in eight normal pregnant women 21–28 years of age. These data were obtained during all-night recordings taken in the last two trimesters of pregnancy. After an initial night of adaptation to the sleep laboratory and recording procedures, each subject returned on 3–5 subsequent nights for experimental data collection. Recordings were obtained at approximately 30-day intervals until term. Fetal activity was monitored by the application of three pressure-sensitive electrodes to the abdominal surface. A fourth electrode was fixed to the dorsal surface of the thigh to register maternal activity. Pressure changes produced by fetal movemenst were amplified and displayed together with the mother's physiological measures on a 10-channel polygraph. This technique for recording fetal activity proved to be quite sensitive and disclosed several different types of movement. Thirty all-night recordings of fetal activity were collected in this manner from eight subjects. Quantitative analysis of these data was achieved by power-spectral analysis. It was found that a 30–50 minute cycle, as indicated by periods of fetal somatic activity and quiescence, was present at the earliest ages recorded (21 weeks gestation age) and continued unaltered throughout gestation. This cycle was found to be of a similar period to that observed directly in the newborn and showed developmental changes independent from the postnatal emergence of a sleep-waking rhythm (Fig. 1).

The pontine cat, with midbrain and forebrain removed by aspiration, shows an alternation of tonic and atonic phases, the latter associated with rapid eye movements and other physiological changes. The alternation of these patterns is often identical in its periodicity to the approximately 20 minute REM cycle in the intact cat [2]. Yet it is difficult for a number of reasons to attribute sleep and waking states to this animal. These and the above findings suggested that the REM cycle is more primitive than the sleep-waking rhythm and, in fact, that it transcends sleep, as suggested many years ago by Kleitman [3].

In recent studies we have attempted to measure carefully the temporal organization of behavior in the cat in order to pursue this question further. Periodicity in physiological patterns during sleep and in rate of performance during wake-

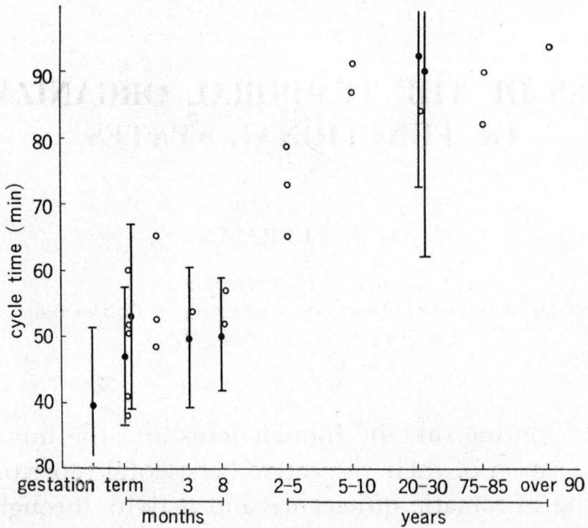

Fig. 1 Survey of the development of the human rest-activity or REM state cycle during ontogeny. Solid dots indicate reliable sample means for which adequate sample size and statistical description were available in the articles reviewed. Bars indicate the standard deviation of these means. Open circles represent mean values derived from other data presented or merely stated without further statistical information. Collectively, these data indicate a slight trend toward lengthening of the cycle from the prenatal sample to 8 months of age; however, it is impossible to determine the statistical validity of this trend from these data and the overlapping variabilities suggest no reliable developmental change during this period. A clearly significant increase in cycle duration is evident by 20–30 years of age, which appears to be sustained into senility. The precise period during which this shift occurs cannot be determined here, but the data suggest a transition between 2–10 years of age.

fulness was measured and compared in four separate experiments [4]. Twenty animals were prepared surgically with indwelling electrodes for the chronic recording of appropriate physiological parameters and some, additionally, for positive brain stimulation. Measurements of the periodicity within sleep (the REM cycle) and waking (performance cycles) were based upon specification of strict criteria for baseline states. Data obtained from these experiments provided an overall mean REM cycle of 19.20±2.20 minutes. Periodicity in performance was measured through statistical and computer power-spectral analyses. Food or self-stimulation reward was utilized to establish either EEG or bar press conditioned responses in four different studies. Periodicities in performance were manifest as either the rate of response or the episodic occurrence of performance. A dominant performance cycle was demonstrated with an overall mean value of 21.41±1.97 minutes. The REM and performance cycles did not differ statistically from one another or across studies (Table 1). Evidence was obtained, also, which indicated that the waking cat spent the same proportion of time in performance as the sleeping cat did in REM.

Measurement of the sleep-waking rhythm in the cat was achieved also through the specification of strict physiological and behavioral criteria. A reliable rhythm with a period of approximately 100 minutes was thus obtained. Moderate-sized lesions placed bilaterally at the basal and rostral preoptic level of the hypothalamus (basal forebrain area) produced a marked decrease in the period of the sleep-waking rhythm (by decreasing sleep) but had no effect on REM or performance cyclicity. Comparable lesions placed in the mesencephalic tegmentum and

Table 1 Comparison of mean performance and REM cycle values obtained from 20 cats in four independent experiments. Analysis consisted of midpoint-to-midpoint measurement of intervals between successive episodes or peaks during strictly defined baseline waking and sleep states.

Experimental treatment	Number of cats	Mean performance cycle (min)	Number of cycles	Mean REM cycle (min)	Number of cycles
A. Continuous performance followed by sleep (3–5 hrs.). SMR for food.	6	19.85±7.71	19	19.12±7.41	12
B. Continuous and episodic performance (12–24 hrs.). SMR for positive brain stimulation.	4	23.89±3.74	75	—	—
C. Episodic performance and sleep (24 hrs.). SMR for food.	6	19.80±3.85	40	17.03±3.72	86
D. Episodic performance and sleep (18–24 hrs.). Bar press for food.	4	22.13±6.96	47	21.45±5.25	66
Grand Means		21.41±1.97		19.20±2.20	
Anova (Between Treatments)		F=0.2066 (N.S.)		F=0.9460 (N.S.)	

From STERMAN et al. [4]

posterior thalamus had no effect on either of these periodicities. Low dose administration of amphetamine (0.4 mg/kg) prolonged the sleep-waking rhythm (by extending wakefulness) without altering the performance cycle, and Nembutal (6 mg/kg) had the opposite effect, namely no influence on the sleep-waking rhythm, but a slowing of the ultradian cycle.

We conclude that the general temporal organization of behavior in the cat can be expressed in terms of at least two independent periodicities, a sleep-waking rhythm and a separate rest-activity cycle upon which this is superimposed. Forebrain influences appear to organize the sleep-waking rhythm but do not influence the rest-activity cycle. Barbiturates can influence the latter, presumably by virtue of their action on its brainstem substrate. Evidence was also obtained which suggested that temporal mechanisms in the brain can transcend specific state patterns resulting from other aspects of neural integration.

The physiological basis of the ultradian basic rest-activity cycle is as yet undetermined. Its manifestation in the pontine cat and fetal human indicates a neural substrate in the caudal brain stem. It has been suggested that the frequency of this cycle is determined by the rate of metabolic activity. While our knowledge of the rest-activity cycle is limited to the observation of epiphenomena during sleep and waking, the characteristic variability of cycle values so obtained suggests that it is a pacemaker or feedback mechanism rather than a true biological clock. This would be consistent with the previous suggestion, since a periodicity in neural excitability sensitive to the conditions of tissue metabolism could bring behavior into closer harmony with physiological need.

REFERENCES

1. STERMAN M. B. and HOPPENBROUWERS T. (1971): The development of sleep-waking and rest-activity patterns from fetus to adult in man. *In*: Brain Development and

Behavior, (M. B. STERMAN, D. J. MCGINTY and A. M. ADINOLFI, eds.), pp. 203–227, Academic Press, New York.

2. STERMAN M. B. (1972): The basic rest-activity cycle and sleep: developmental considerations in man and cats. *In*: Sleep and the Maturing Nervous System, (C. CLEMENTE, D. PURPURA and F. MAYER, eds.), Academic Press, New York.
3. KLEITMAN N. (1963): Sleep and Wakefulness. 2nd Edition. University of Chicago Press, Chicago, Illinois.
4. STERMAN M. B., LUCAS E. A. and MACDONALD L. R. (1972): Periodicity within sleep and operant performance in the cat. Brain Res., *38*: 327–341.

STUDIES ON ULTRADIAN RHYTHMICITY IN HUMAN SLEEP AND ASSOCIATED NEURO-ENDOCRINE RHYTHMS

E. D. WEITZMAN, D. FUKUSHIMA, C. NOGEIRE, L. HELLMAN, J. SASSIN, M. PERLOW and T. F. GALLAGHER

Departments of Neurology, Oncology and the Institute of Steroid Research Montefiore Hospital and Medical Center Bronx, New York, U.S.A.

It is now well established that all mammals have a recurrent ultradian cyclic pattern of sequenital sleep stages during the sleep portion of the circadian sleep-waking rhythm. Although species differ in regard to specific measured physiological variables as well as in the duration of the cycle length, the evidence amassed during the past 15 years supports the concept that the CNS of all mammals alternates between two distinct states of physiologic activity. This alternating sleep pattern has been called the short term REM-Non-REM sleep cycle [1].

There is general agreement that during a night's sleep in man, a rhythmic alteration of REM-Non-REM sleep cycles takes place with a period of circa 90–100 minutes [2–4]. This cycle has been shown to be quite stable under typical laboratory conditions in normal young adults. In addition to this short-term sleep cycle, there is a distribution of sleep stage pattern as a function of time within the nocturnal sleep period [5]. Stage 3–4, the high voltage slow wave pattern, is present predominantly during the first 2–3 hours, whereas the REM stage increases progressively throughout the night, occupying approximately 40–50% of total sleep during the last 2 hours of the sleep period. It is important to emphasize that age, habituation, psychiatric health, daytime waking activity, and stability of the circadian sleep-wake cycle are all important factors which have been shown to affect the temporal organization of sleep stage patterns [1].

In a series of studies in man, our laboratory has demonstrated that during the latter half of the nocturnal sleep period, cortisol is secreted in an episodic manner [6, 7]. Utilizing the technique of frequent sampling (every 20 minutes) with an indwelling I.V. catheter, we have extended these observations for the 24 hour period. We have recently reported that under a [8] defined and stabilized sleep-wake 24 hour schedule, normal young adult subjects had an average of 9 secretory episodes (range 7–13), with the subjects spending an average of 24% of the time in active secretion. The temporal pattern of episodic secretion of the 24 hour sleep-wake cycle could be divided into 4 unequal temporal phases: Phase 1. A 6-hour period of "minimal secretory activity" (4 hours before and 2 hours after lights out); Phase 2. A 3-hour period called "preliminary nocturnal secretory episode" (3rd to 5th hour of sleep); Phase 3. A 4-hour period, the "main secretory phase" (6, 7, 8 hours of sleep and 1st hour after awakening); and Phase 4. The 11 hours of "intermittent waking secretory activity." No evidence for a "basal level" or "steady state" of cortisol concentration was found. Changes in cortisol output during the 24-hour day appear to be due to differences in frequency and duration of secretory episodes and not to major changes in secretory rate. This consistent ultradian temporal pattern of cortisol secretion within the 24 hour sleep-waking cycle was found to have a mean periodicity of 99 minutes for the period from sleep onset to 8 hours later, 117 minutes for the next 8 hours and 247 minutes for the subsequent 8 hours.

As part of our study of the effect of sleep-wake cycle shifts on sleep and neuroendocrine patterns we carried out a study of the effect of a prolonged 3 hour sleep-wake cycle on a group of seven subjects. We determined the effect of such a drastic cycle change on sleep stages, cortisol and growth hormone secretion patterns. The subjects underwent the following 3½ week (24 day) scheduled sleep-wake cycle changes, while living on a hospital (Clinical Research Center) unit. Following one week of baseline nocturnal sleep polygraphic recording (11 P.M. to 7 A.M.), the subjects assumed a schedule for 10 days of 2 hours waking followed by 1 hour sleep for each 24 hour period. This was then followed by 1 recovery week of nocturnal sleep (11 P.M. to 7 A.M.). Rectal temperature and urine samples were obtained every 3 hours for the entire 24 day period except for the nocturnal sleep periods. On days 5–6 and 13–14 an intravenous catheter was used to obtain plasma samples every 20 minutes for a 24 hour period. Cortisol was determined by the competitive protein binding method and Human Growth Hormone by the radio-immunoassay technique.

When average sleep time was obtained for six subjects, a partial ultradian sleep-wake cycle was established. However, a prominent superimposed circadian sleep-wake cycle persisted throughout the entire 10 day experimental period. During this period, total sleep time was consistently reduced by approximately one-third compared to baseline. Most sleep was obtained during the 4 hours of 3 A.M.–4 A.M., 6 A.M.–7 A.M., 9 A.M.–10 A.M. and noon to 1 P.M. Therefore, a phase shift of peak sleep time of approximately 5 hours (from approximately 3 A.M. to 8 A.M.) was established during the first three days, and then maintained throughout the experimental period. The amount of time utilized for sleep was dramatically reduced from approximately 90% at 9 to 10 A.M. to approximately 25% at 9 to 10 P.M. Therefore, in spite of significant chronic sleep deprivation, the subjects were generally less able to utilize the 3 hourly periods between 6 P.M. to 1 A.M., consistently sleeping less than one-third of each hour. The 24 hour distribution of REM sleep closely paralleled that of total sleep with the maximum occurrence between 3 A.M. and 1 P.M., and virtually no REM sleep present between 9 P.M. and 1 A.M. However, stage 3–4 sleep was more evenly distributed throughout the 24 hour period with a slight tendency for circadian periodicity.

Measurement of body temperature (rectal, every 3 hours) also demonstrated a persistence of the circadian rhythm throughout the ten day 3 hour sleep-wake period with a partial shift, delay of approximately 4–6 hours.

Analysis of the pattern of cortisol secretory activity for both baseline and ultradian 24 hour periods revealed no significant differences in average 24 hour cortisol output, number of secretory episodes and total secretory time. However, in the ultradian experimental condition the secretory episodes appeared to be entrained to the 3 hour sleep-wake cycle. In this 3 hour cycle cortisol output and secretory time were maximal for the first hour after awakening, was less for the second hour of awakening and was minimal for the third hour, i.e., the hour of the next available sleep period. Despite the establishment of this 3 hour pattern, a clear circadian rhythm was also present with maximal secretory activity occurring between 4 A.M. and 4 P.M. coinciding with that of the baseline rhythm and with the time of maximal sleep. However, the usual maximal secretory phase seen in baseline, between 4 A.M. and 8 A.M., was not prominent on the ultradian 24 hour curves.

Analysis of human growth hormone in five subjects during the ultradian period revealed that GH was released in 19 of 28 periods of sleep. There were an additional 17 episodes of hormonal release not associated with sleep. Several of these were clearly associated with venipuncture or other stressful events. The temporal organization of the 24 hour GH release pattern was altered by the establishment of a 3 hour sleep-wake cycle although the previously demonstrated relationship of GH release to sleep persisted under these experimental conditions [9].

Therefore, we were unable to eliminate the well established circadian rhythm of sleep-waking activity, body temperature and cortisol secretory rhythm by the 3 hour sleep-wake schedule. However, the timing of the acrophase was shifted (delayed by approximately 4–6 hours) and a partial entrainment of cortisol secretion was established to the 3 hour sleep-wake experimental schedule.

REFERENCES

1. Kales A. (1969): Sleep: Physiology and Pathology, A Symposium. J. B. Lippincott Co., Phila.
2. Globus G. G. (1970): Quantification of the REM sleep cycle as a rhythm. Psychophysiology, *7*: 248–253.
3. Globus G. G. (1970): Rhythmic function during sleep. Int. Psych. Clinics, *7*: 15–21.
4. Webb W. B. (1970): Length and distribution of sleep and intra-sleep process. Int. Psych. Clinics, *7*: 29–31.
5. Webb W. B. and Agnew H. W., Jr. (1969): Measurement and characteristics of nocturnal sleep. *In*: Progress in Clinical Psychology, Vol. 8, Dreams and Dreaming, (L. E. Abt and B. F. Reiss, eds.), pp. 2–27, Grune and Stratton, New York.
6. Weitzman E. D., Schaumburg H. and Fishbein W. (1966): Plasma 17-Hydroxycorticosteroid levels during sleep in man. J. Clin. Endocr. and Metab., *26*: 121–127.
7. Hellman L., Nakada F., Curti J., Weitzman E. D., Kream J., Roffwarg H., Ellman S., Fukushima D. K. and Gallagher T. F. (1970): Cortisol is secreted episodically by normal man. J. Clin. Endocr., *30*: 411–422.
8. Weitzman E. D., Fukushima D., Nogeire C., Roffwarg H., Gallagher T. F. and Hellman (1971): The twenty-four hour pattern of the episodic secretion of cortisol in normal subjects. J. Clin. Endocr. and Metab., *33*: 14–22.
9. Sassin J. F., Parker D. C., Mace J. W., Gotlin R. W., Johnson L. C. and Rossman L. G. (1969): Human growth hormone release: relation to slow-wave sleep and sleep-waking cycles. Science, *165*: 513–515.

RHYTHMS OF THE BIOGENIC AMINES IN THE BRAIN AND SLEEP*

P. J. MORGANE and W. C. STERN

Laboratory of Neurophysiology
Worcester Foundation for Experimental Biology
Shrewsbury, Massachusetts, U.S.A.

This paper discusses some problems involved in studying how different regions of the brain known to be involved in sleep are phase-locked with respect to rhythms of brain chemistry. Since the chemical theories of sleep have centered around the role of serotonin in slow-wave sleep and norepinephrine in rapid eye movement (REM) sleep, it is important to determine if brain serotonin or norepinephrine rhythms lead in phase with reference to the electroencephalographic (EEG) output for each state. It is well known that the commonly used EEG frequency bands, such as the delta, theta, alpha, and beta bands also undergo circadian periodic changes in EEG output and it follows that the underlying chemical fluxes must be exceedingly complex in these fast alternating changes in brain wave activity.

NEUROCHEMISTRY AND NEUROANATOMY OF SLEEP

JOUVET's cross-disciplinary studies [1, 2] were the beginnings of attempts to attempts to bring together data bearing on the relation of the biogenic amines to sleep-waking cycles. He first put forth that there was a three-way correlation between the extent of destruction of the serotonin-containing nuclei of the raphe complex, the percent of sleep, and the amount of serotonin in the brain (whole forebrain anterior to lesion). The key word here is "correlation," but it goes without saying that correlation is not causation. However, with present methods we have not been able to go far beyond these types of correlations in defining the chemical bases of the sleep states. JOUVET showed that the most significant alterations in sleep following large anterior raphé lesions (Fig. 1) were a lowering of the percentage of slow-wave sleep with less effect on REM sleep. Similarly, we have found with lesions limited strictly to the nucleus raphé dorsalis and/or medialis that the dominant effect is a decrease in slow-wave sleep. JOUVET showed that the most significant alterations in sleep following large anterior raphé lesions (Fig. 1) were a lowering of the percentage of slow-wave sleep. Similarly, we have found with lesions limited strictly to the nucleus raphé dorsalis and/or medialis that the dominant effect is a decrease in slow-wave sleep. JOUVET showed that the most critical chemical change following raphé lesions was a lowering of serotonin, but not norepinephrine, in the whole brain anterior to the raphé (whole diencephalon and telencephalon) whereas we, using a more regional approach, have shown this lowering of serotonin to be limited to the hypothalamus, basal forebrain area and discrete limbic forebrain regions, i.e., areas which con-

* Supported by Grants MH-02211 and MH-10625, NIMH.

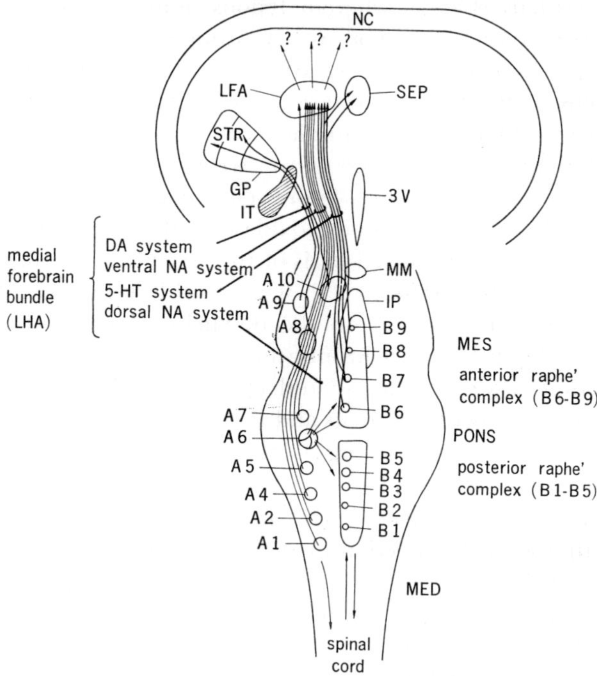

Fig. 1 Scheme of chemically coded neural circuits in the brain of the cat. The anterior raphé complex (B6 to B9) contains serotonergic (5-HT) neurons and projects, via the mid-lateral hypothalamic area, to the basal fore-brain area in the medial component of the medial forebrain bundle (5-HT system). Lesions in this system decrease slow-wave sleep and increase waking and are associated with a decrease of serotonin in the limbic forebrain area (LFA). The mesencephalic, pontine and reticular nor-adrenergic (NA) system (A1 to A5 and A7) also projects, via the far-lateral hypothalamic area, to the limbic forebrain area but does not exactly overlap the projection field of the medial 5-HT system. It is more ventrally disposed in the medial forebrain bundle and is termed the ventral NA system. A special part of this lateral nor-adrenergic system, the locus coeruleus (A6), has been related to REM sleep (lesions in this nucleus abolish the muscle atonia seen in the REM state). It has elaborate projections dorsally to the cerebellum, medially to the anterior and posterior raphe complex, and caudally to the spinal cord. Most importantly, it projects rostrally in the dorsal aspect of the medial forebrain bundle (dorsal NA system) to the limbic forebrain area and histofluorescent mapping has traced it into the cerebral neocortical formations. Another special part of the lateral catecholaminergic system are nuclei A8 to A10. A8 forms the meso-limbic system while A9 and A10 project in the more lateral aspects of the medial forebrain bundle, penetrate the internal capsule, pass through the globi pallidi and into the caudate and putamen (dopamine (DA) system). It appears that rhythms of brain chemistry regionally are in large part regulated by activity in these chemo-specific circuits. Understanding of these trajectories and their inter-relations may shed light on how amine-specific pathways in the brain interact to modulate distant brain chemo-architecture and behavior. *Abbrev.* NC—neocortex; SEP—septal area, LFA—limbic forebrain area; STR—striatum; GP—globus pallidus;; IT—internal capsule; 3V—third ventricle, MM—mammillary body; IP—interpeduncular nucleus; MES—mesencephalon; MED—medulla, LHA—lateral hypothalamic area. NA grouping A3 is not a discrete nucleus and is not shown in this figure.

tain the most terminals of the medial forebrain bundle projections from the raphé nuclei [3]. It is significant that the lowering of serotonin in these specific areas is "associated" with marked depressions of slow-sleep and increases in waking, especially since these areas are the sites of origin of the descending inhibitory circuits where electrical and chemical stimulation evoked synchronized sleep [4, 5] and where lesions produce increased wakefulness [6].

The locus coeruleus (Fig. 1), wherein lesions obliterate the muscle atonia of REM sleep [7], can be looked upon as a possible generator of aspects of REM sleep by nature of its many neural relations with the raphé, cerebellum, spinal cord, and nucleus of the tractus solitarius. This nucleus also sends contingents forward in the medial forebrain bundle to the hypothalamus, basal forebrain area and cerebral neocortex [8] and is the source of origin of the dorsal noradrenergic system identified by histofluorescence mapping. The locus coeruleus is high in norepinephrine, monoamine oxidase, and acetylcholinesterase, and, in the light of theories of the chemical basis of the ascending "arousal" system being cholinergic [9] or noradrenergic [10, 11], it is possible that one chemically coded pathway in this system is related to the desynchronization of the REM state while another is related to desynchronization of waking. These lines of investigation suggest that future studies should examine rhythms of chemical agents in such key areas as the locus coeruleus and individual raphé nuclei, which studies might then come closest to providing a "chemical neuroanatomy" of sleep.

Other indirect supporting evidence for the chemical theory of sleep has come from pharmacological experiments. The most convincing of these have been the demonstration that a single injection of p-chlorophenylalanine (PCPA), an agent that blocks the synthesis of serotonin, decreases slow-wave sleep predominantly. Elevating serotonin levels by giving its precursor, 5-hydroxytryptophan, restores normal sleep profiles in close temporal relation to restoration of brain serotonin levels. However, when PCPA continues to be administered chronically, associated with chronic depression of serotonin, the sleep profiles tend to return to normal values while serotonin remains low. Our data on the effects of reserpine, which lowers brain serotonin and norepinephrine, and decreases REM sleep, shows a similar non-correspondance between "levels" of regional biogenic amines and sleep profiles. The sleep profiles approach normal values two days after a single 0.15 mg/kg dose of reserpine given intraperitoneally, whereas regional brain amine levels are approximately one-half of normal values at this time. These latter two studies are difficult to put in direct context of a chemical basis of each sleep state based on absolute amount of a specific chemical agent. It must be remembered, however, that to date absolute levels of amines have been the main measures, usually in whole brain, and sleep-waking patterns have not been related to *regional* changes and *regional* chemo-dynamics (such as turnover studies *in vivo* as an index of utilization). Although the more widely accepted view is that REM is triggered by a noradrenergic mechanism, we and others have shown that alpha-methyl-para-tyrosine, which blocks tyrosine hydroxylase and thus decreases dopamine and norepinephrine synthesis, increases REM sleep by about 50% with little effect on slow-wave sleep. This evidence appears contradictory to any theory that norepinephrine "produces" REM sleep. Thus, presently it appears that changes in biogenic amines by lesioning and drug manipulation are associated with drastic alterations of the sleep states but direct relations between the amount of a specific biogenic amine and a given sleep state are not yet clear. Interestingly, with respect to the importance of the regional approach, we have found that certain brain areas may show twice the change in amine levels in response to amine-altering drugs as compared to other areas. This cross-correlation of chemo-architecture regionally is under investigation presently in our laboratory in relation to the sleep states.

NEUROCHEMICAL RHYTHMS AND SLEEP

Despite the previously discussed problems, the weight of evidence from a variety of types of data do implicate the biogenic amines in sleep-waking behavior and suggest that each sleep state, slow-wave and REM sleep, may have its own chemical transmitter and neural substrate. A group of workers not studying sleep have so far given us the best lines of approach and the best understanding of rhythms of chemical profiles in the brain. Some of these studies, such as those of SCHEVING et al. [12] in the rat, have pointed up certain relationships between rhythms of brain chemistry and general activity levels. They carried out hourly sacrifices of animals and measured whole brain amine levels and found a circadian rhythm for serotonin and an ultradian rhythm for dopamine and norepinephrine. Comparison of the phasing of the serotonin rhythm with the typical motor activity rhythm of the colony showed a correlation of maximal serotonin levels with rest and minimal levels with activity. With respect to sleep, it would have been valuable had these types of studies been carried out in chronic animals wired for recording and on which sleep profile data had been collected. Also, an additional consideration is the effect of time-lags in these types of studies, i.e., the animals are awakened in order to sacrifice them for brain chemistry. It is hard to determine the value of chemical data obtained from normal animals on which chronic sleep profiles have been run since the chemistry measurements are taken from the animal after a recording session with the intervening effects of waking and stress producing results which might differ from those obtained from animals sacrificed actually *during* sleep. A further consideration is that, in the context of sleep-chemical rhythms, we cannot say whether the observed phase-locking between chemistry and sleep results from the neurochemical agent producing sleep or from the sleep states themselves inducing neurochemical changes.

Evidence from our laboratory and that of others, especially REIS et al. [13, 14] and QUAY [15], has shown that many brain areas have their own unique rhythm of chemical flux, some circadian, some ultradian, and that the excitability of these regions varies according to their changing chemo-architecture which, in turn, is regulated by key fiber systems passing from raphé, locus coeruleus, and other areas of the brain stem (Fig. 1). The studies of REIS et al. and QUAY, though not specifically carried out with regard to sleep, hence no chronic recording, are still by far the best attempts to describe regional brain chemistry in regard to their fundamental rhythms. Most importantly, brain dissections are easily done in cats and can be carried out as we do in our laboratory with measures in over 20 samples per brain. Specific brain regions, and even nuclear groups, known to be involved in sleep are then studied for chemical profiles at various times of the day. REIS et al. and QUAY have shown diurnal fluctuations of amines in specific loci in the brain for both serotonin and norepinephrine. In particular, REIS et al. reported that the diurnal variations of norepinephrine in the lateral hypothalamic area were masked when a larger block of brain tissue was removed for analysis. This illustrates the disadvantages of any such measure as large brain samples, whole brain, or pooled brain studies. REIS and WURTMAN [16] postulated that cyclic changes in brain norepinephrine metabolism are probably involved in generating other biologic rhythms but, as pertains to sleep, these links still remain to be forged. They also reported data showing that there are regional diurnal variations in serotonin concentrations which were not usually in the same regions as those showing diurnal norepinephrine rhythms. Also, the regional rhythms of norepinephrine and serotonin are not synchronized but have independent cycles with peak values at different times in the day. The regions with serotonin rhythms are located preponderantly in the upper brain stem, particularly the mesencephalon which, of course, correlates with the locality of the anterior raphé complex. Also, the basal forebrain area, which contains the raphé neuron terminals, have serotonin rhythms. Even more interesting is the finding that the nature of the serotonin rhythm in the cat differs from region to region, being circadian in some areas and ultradian in others. So far, no one has sampled raphé, locus coeruleus, basal forebrain area, and certain other critical areas known to be related to sleep but these experiments are in progress in our laboratory.

In the past there have been methodological problems in the study of the relationship between rhythms of biogenic amines and sleep, e.g., the tendency to study chemical profiles in either whole brain or even pooled brains from many animals, thus masking regional effects. There has been, in addition, a noticeable lack of 24-hour sleep profile studies in association with chemo-architectural analyses. It has been impossible, so far, to get truly simultaneous direct measures of chemistry and sleep on the same subject. In this regard, it is possible to sample cerebrospinal fluid (CSF) in animals wired for long-term recording but we do not really know the relation of CSF chemical contents to that in the brain. It still remains to be determined how CSF substances reflect dynamics in specific regions of the brain. Push-pull cannulae are valuable in some types of studies, such as Myers' [17] work on the monoamine theory of temperature control, but sampling brain tissue perfusates or "washings" in this manner may not give the same chemical profile as a brain homogenate. Soulairac et al. [18] have recently described a method for biopsy of small subcortical brain samples in the chronic animal. This sort of procedure is now being instituted in our laboratory to obtain samples for brain chemistry following which the animal can still be studied and brain recordings taken in the chronic period. This *in vivo* sampling may prove to be a valuable tool for correlating brain chemistry with behavior. *In vivo* sampling methods represent, perhaps, the most promising approach to understanding the dynamic relationships between on-going regional neurochemical events and rapidly alternating physiological states.

In conclusion, at least some generalities can be derived for approaching the relation of rhythms of brain chemicals with respect to the sleep states. The complexity of the distribution and of the timing of the monoamine cycles in the brain makes it hazardous to draw simple correlations between rhythmic physiological functions or behavioral states such as sleep, psychomotor activity, endocrine rhythms, etc. and regional fluctuations in the levels of a biogenic amine. It seems probable that the amine rhythms in a region reflect fluctuations in activity either in cell discharge or in metabolism of the cells of origin of amine-containing terminals which, as noted, are localized in specific regions of the brain stem. The regional specificity of serotonin and norepinephrine rhythms, and their asynchrony, indicates that the regional rhythms of each of these amines are independnetly regulated in the cat brain. This would fit well with a theory of an independent anatomical and chemical substrate for the two states of sleep.

REFERENCES

1. JOUVET M. (1967): Mechanisms of the states of sleep: A neuropharmacological approach. *In*: Sleep and Altered States of Consciousness, (S. KETY, E. EVARTS and H. WILLIAMS, eds.), pp. 86–126, The Williams & Wilkins Co., Baltimore.
2. JOUVET M. (1969): Biogenic amines and the states of sleep. Science, *163*: 32–41.
3. MORGANE P. J. and STERN W. C. (1972): Relationship of sleep to neuroanatomical circuits, biochemistry, and behavior. Annals of the New York Academy of Sciences, *193*: 95–111.
4. STERMAN N. B. and CLEMNTE C. D. (1962): Forebrain inhibitory mechanisms: Cortical synchronization induced by basal forebrain stimulation. Exp. Neurol., *6*: 91–102.
5. HERNÁNDEZ-PEÓN R., CHAVEZ-IBARRA, G., MORGANE P. J. and TIMO-IARIA C. (1963): Limbic cholinergic pathways involved in sleep and emotional behavior. Exp. Neurol., *8*: 93–111.
6. MCGINTY D. J. and STERMAN M. B. (1968): Sleep suppression after basal forebrain lesions in the cat. Science, *160*: 1253–1255.
7. JOUVET M. and DELORME F. (1965): Locus coeruleus et sommeil paradoxal. C.R. Soc. Biol., *159*: 895–899.
8. FUXE K., HÖKFELT T. and UNGERSTEDT U. (1970): Morphological and functional aspects of central monoamine neurons. Int. Rev. Neurobiol., *13*: 93–126.
9. SHUTE C. C. D. (1969): The distribution of cholinergic and monoaminergic neurones in the brain. J. Anat., *104*: 579.
10. FUXE K., HAMBERGER B. and HÖKFELT T. (1968): Distribution of noradrenaline nerve terminals in cortical areas of the rat. Brain Res., *8*: 125–131.
11. UNGERSTEDT U. (1971): Stereotaxic mapping of the monoamine pathways in the rat brain. Acta Physiol. Scand., Suppl. 367, pp. 1–48.
12. SCHEVING L. E., HARRISON W. H., GORDON P. and PAULY J. E. (1968): Daily fluctuation (circadian and ultradian) in biogenic amines of the rat brain. Am. J. Physiol., *214*: 166–173.
13. REIS D. J., CORVELLI A. and CONNERS J. (1969): Circadian and ultradian rhythms of serotonin regionally in cat brain. J. Pharmacol. Exp. Therap., *167*: 328–333.
14. REIS D. J., WEINBREN M. and CORVELLI A. (1968): A circadian rhythm of norepinephrine regionally in cat brain: Its relationship to environmental lighting and to regional diurnal variations in brain serotonin. J. Pharmacol. Exp. Therap., *164*: 135–145.
15. QUAY W. B. (1968): Differences in circadian rhythms in 5-hydroxytryptamine according to brain region. Am. J. Physiol., *215*: 1448–1453.
16. REIS D. J. and WURTMAN R. J. (1968): Diurnal changes in brain noradrenalin. Life Sci., *7*: 91–98.
17. MYERS R. D. (1971): Hypothalamic mechanisms of pyrogen action in the cat and monkey. *In*: Ciba Foundation Symposium on Pyrogens and Fever, (G. E. W. WOLSTENHOLME and J. BIRCH, eds.), pp. 131–153, J. & A. Churchill, London.
18. SOULAIRAC A. TANGAPREGASSOM A. M. and TANGAPREGASSOM M. J. (1970): Prélevèment du noyau supra-optique chez le Rat vivant, intérêt dans les recherches comportementales. Ann. Endocrinol., *31*: 939–946.

CHRONOBIOLOGY AND RHYTHMS IN PSYCHIATRY AND PSYCHOLOGY

Chairman: Arne SOLLBERGER

Co-Chairman: Hugh W. SIMPSON

CHRONOBIOLOGY AND RHYTHMS IN PSYCHIATRY AND PSYCHOLOGY

ARNE SOLLBERGER

Departments of Physiology and Information Processing, and the Medical School, Southern Illinois University Carbondale, Illinois, U.S.A.

Psychiatry and psychology are relative newcomers to biological rhythm research, though notable exceptions exist. Among the latter are GJESSING's careful records of periodic catatonia, which have now been reanalyzed (see SIMPSON et al. in this panel), or GEORGI's observations of inverse circadian rhythms in psychiatric patients, or the early studies of periodicities in mental performance (e.g. time estimation, addition of numbers, clearness of memory images; GLASS, LANGE, PHILPOTT, VOSS).

Several factors may have contributed to slow down development in this field. For a biological rhythm to be decently discernible (and therefore accepted) or to be meaningfully analyzed, any measured variables should be quantifiable and have a low noise level, and any recorded events should be well defined.

In psychiatry, however, mental instability, and therefore variability, is pronounced. Though periodic catatonia was described already by KRAEPELIN, pure cases are rare. The so-called "periodic psychoses" usually present such varied pictures that the consensus at a recent conference on rhythms in psychiatry was a certain doubt as to the existence of true periodic disease entities (AJURIAGUERRA). Furthermore, the psychiatric state of a patient, as well as animal or human mood, may be hard to quantify; such tools as the psychological test batteries or the neurophysiological analogues of psychological states are recent developments.

Many other processes are event variables rather than continuous in character. The present problem is here to develop suitable methods of analysis. Sine curve fitting can, however, still be used successfully (cf. POCHOBRADSKY).

The psychiatric patient is also hard to motivate. He may refuse to cooperate or misinterpret instructions. It may be difficult to demand of him the strict regimens necessary in biological rhythm research, or to isolate him in a single room, suspiciously similar to a cell. Such procedures might even be harmful. This should force us to develop discrete and unobtrusive telemetry systems before we can start studying our patients efficiently (cf. the panel on chronobiological techniques). The contribution on prediction of psychiatric states (REYNOLDS, this panel) nevertheless demonstrates that one can successfully use sophisticated mathematical-statistical techniques in the analysis of psychiatric periodic data.

There is, today, a rapidly growing interest in biological rhythms among the neurosciences (cf. SOLLBERGER). In psychology proper, more and more variables turn out to have rhythmic components (DAVIS; in this panel ANTHONEY et al., POIREL and TERNES), e.g. the startle response, or hiccoughing.

Perhaps as a reflection of the difficulties mentioned, some panel papers impinge heavily upon the neighbouring neurosciences, e.g. neurophysiology (with the concomitant problem of central nervous processing of stimuli and modification of response, FRAISSE) and neuroendocrinology (CURTIS et al.; cf. also the panel on chronobiology, endocrines, neurohumors and reproduction).

Judging from the wide range of subjects treated in this panel, the emphasis of interest is perhaps less on the circadian rhythms (POIREL, TERNES) than on other frequency regions, such as reflex activity (tapping, FRAISSE), 14–21 day cycles (REYNOLDS, SIMPSON et al.), or monthly and seasonal rhythms (CURTIS et al.).

REFERENCES

1. AJURIAGUERRA J. (ed.) (1968): Cycle biologiques et Phychiatrie, Masson, Paris.
2. DAVIS M. and SOLLBERGER A. (1971): Twenty-four-hour periodicity of the starle response in rats. Psychon. Sci., *25*/1, 37–39.
3. GEORGI F. (1944, 1947): Psychophysische Korrelationen. Schweiz. Med. Wschrft., *74*: 539; *77*: 1276.
4. GJESSING R. (1932, 1936, 1939): Beiträge zur Kenntnis der Pathophysiologie periodischer Katatoner Zustände I–IV. Arch. Psychiatr., *96*: 319, 393; *104*: 355; *109*: 525.
5. GLASS R. (1888): Kritisches und Experimentelles über den Zeitsinn. Phil. Stud., *4*: 423.
6. KRAEPELIN E. (1913): Klinische Psychiatrie, Barth, Leipzig.
7. LANGE N. (1888): Beiträge zur Theorie der sinnlichen Aufmerksamkeit und der aktiven Apperception. Phil. Stud., *4*: 390.
8. PHILPOTT S. J. (1932): Fluctuations in human output. Brit. J. Psychol., *6*/Monogr. #XVII.
9. POCHOBRADSKY J. (1971): Periodogram analysis of menstrual cycles. J. Interdisciplin. Cycle Res., *1*/4: 315.
10. SOLLBERGER A. (1971): Biological rhythms and their control in neurobehavioural perspective, Neurosciences Research, Vol. 4, Academic Pres, New York.
11. VOSS G. V. (1899): Über die Schwankungen der geistigen Arbeitsleistung. Psychol. Arb., *2*: 399.

CUES IN SENSORI-MOTOR SYNCHRONIZATION

PAUL FRAISSE

Université de Paris V
Laboratoire de Psychologie Expérimentale et Comparée
Paris, France

The synchronization of two rhythms raises many problems in biology. Some of them are related to the synchronization between physical (e.g. light vs. dark) and biological rhythms.

Among all the possible synchronizations, we have taken a particular interest in an aspect of human behavior which apparently has no equivalent in animals, namely the synchronization of movements to periodic stimuli, usually auditory ones.

Synchronization may be understood in two ways. In a weak sense, it means that two series of events have the same period (Fig. 1A), the lag between the two series being constant. In a strong, restrictive, sense not only do the two series of events have the same period, but also the events are simultaneous (Fig. 1B).

In the A sense, if stimuli (S) are one series of events and responses (R) are the other, the series of R may be a series of reactions to stimuli, which is the usual case in behavior. But in the B sense, there cannot be synchronization without some anticipatory mechanism which enables R to be simultaneous with S.

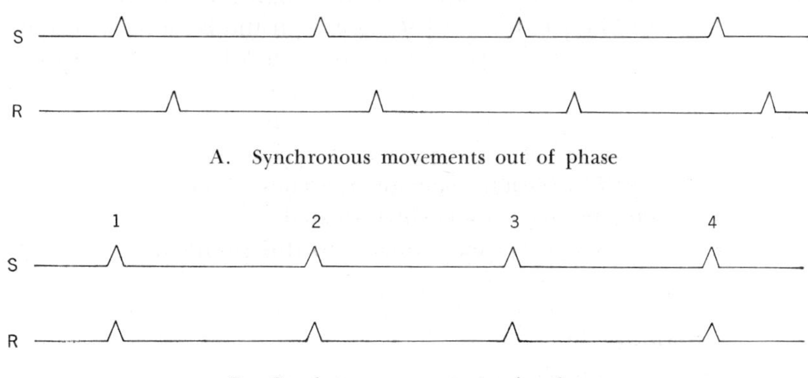

A. Synchronous movements out of phase

B. Synchronous movements in phase

Fig. 1

Anticipatory processes appear only in periodic phenomena. Several have been investigated in biology but sensori-motor synchronization of type B has some special characteristics:

1. *Synchronization is perceived,* i.e. S and R are not perceived as successive but as simultaneous. This perception is possible only because synchronization is achieved with quite good precision. The lag between S and R is on the average less than 50 ms and not greater than 100 ms. Then variability (σ) ranges from 20 to 40 ms [1].

It has to be underlined at this point that, on the average, the response slightly anticipates the stimulus.

However, the synchronization is accurate only when the intervals between S_1-S_2-S_3 . . . lie within the range 300 and 1500 ms. With faster cadences, there are both sensory (lack of sensory definition of stimuli) and motor (difficulties of controlling reactions) difficulties. With slower cadences the difficulties have rather a mnemonic origin [2]. In fact it has been established that when the intervals are greater than 1500 ms simultaneity is no longer perceived, and the subject either anticipates his response or reacts to the stimulus after he has heard it, as in a reaction time task.

The synchronization phenomena which we are studying range within a rather limited band of frequencies.

Even within this band different levels may be distinguished: optimum synchronization occurs with 700 to 800 ms intervals; within the limits of 500 to 1000 ms synchronization is very easy. Below and beyond these values, it is more difficult and then becomes impossible.

2. From a developmental point of view this *synchronization it early*. Spontaneous synchronization phenomena are observed in one year old children (rocking the head or the trunk while hearing music). Voluntary synchronization is also early and in the 7 year old child it is equivalent to that observed in the adult [3].

3. *Its rapidity of response*. If the subject listens to a cadence before responding, his first response may already be synchronised to the stimulus (for instance a dancer going onto a dance floor).

If the subject begins to respond from the very beginning of a series of stimulations, synchronization has already appeared by the third stimulation [4].

4. *Its irresistible character*. With intervals ranging from 500 to 1000 ms, it is very difficult for the subject not to synchronize (in the B sense) with the stimulus. Whether he wants to try to conform to situation A by attempting to respond to the stimulus as if it was a reaction time task, or whether he wants to perform a syncopated movement, by interpolating a response between two stimuli, he cannot succeed, or he gives very irregular and variable responses [5].

Having defined synchronization behavior, the cues which guide the subject in this anticipatory behavior have to be investigated.

We have devoted two series of experiments to this problem.

First study: *the nature of the cues coming from the response made to the stimulus.*

If, for instance, the subject responds to an auditory cadence by pressing a key or tapping a board, three indices provided by the response can be distinguished : kinesthetic cues due to movement (setting up or restraining), tactile cues at the moment of the tapping (direct or through the intermediary of an instrument), auditory cues from the noise of the tapping.

We have analyzed these three components in a study where the subject was asked to beat with his arm while holding a rigid flag. Kinesthetic cues were inferred from recordings of the mechanical movements involved. There were tactile cues when the beat was executed on a sound-insulated surface, and auditory cues were produced by the passage of the flag before a photo electric cell releasing a sound which could be delayed (Fig. 2).

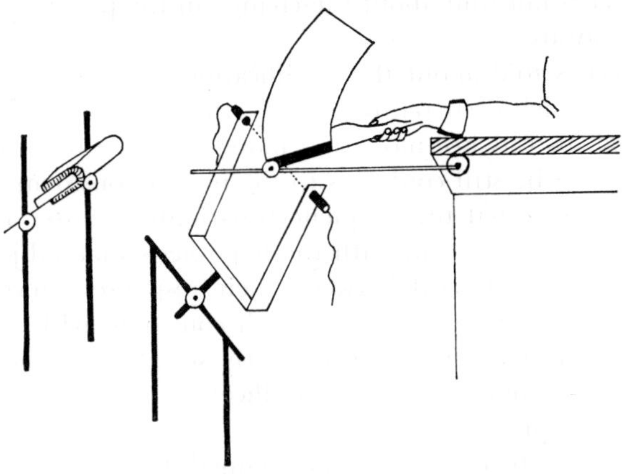

Fig. 2

Kinesthetic cues were obviously always present. They are sufficient to make synchronization possible, but the S-R lag is twice as variable as when there are also tactile or auditory cues. As the sound could be delayed, it was possible to create a conflict between auditory and kinesthetic or (and) tactile cues. Two observations have been made: (a) tactile cues are the most important, kinesthetic and auditory cues coming in the second and third position, respectively. This predominance of tactile cues is revealed by the spontaneous behavior of subjects who are receiving rhythmical stimulations: they generally beat to the rhythm by taping with their finger, hand or foot on a surface. (b) The subjects seek a compromise between the cues by trying to make them temporally close to the sound-stimulus, i.e. by minimizing the lag between the cues available and the stimulus [6, 7].

This means that subjects do not try to produce a response simultaneous with one of the cues but rather a stable pattern including both S and the different aspects of R.

Second study: *the specific role of stimulus-response lag in synchronization.*

Under the sense B, synchronization implies the maintenance of a fixed interval between responses equal to the interval between stimuli as well as the simultaneity of each stimulus and each response. What is the relative role of these variables?

To elucidate that question, an apparatus enabling the suppression of the lag between S and R was devised, such that the subject produced S by giving R. The experimental situation is simple and had two phases: first the subject was asked to synchronize his responses with an auditory cadence, in the usual way; later on, at a given time *t*, the experimenter suddenly changed the experimental situation, and the independent series of stimuli was replaced by stimuli which were produced by the responses of the subject. In that case, there was no lag between S and R or R and S. This condition will be referred to as "pseudo-synchronization".

The behavior of the subject differed, depending on whether he had or had not been given information concerning the situation:

(a) The subject is not told about the change in the production of sounds and does not become aware of it.

(b) The subject is told about the modification.

In the first case, which occurred frequently, there was a systematic effect. The subject accelerated his cadence more and more, reaching often the speed of tapping. He thinks that he still conforms to the instructions demanding synchronization and believes also that the frequency of stimuli is faster and faster.

This behavior is not consistent with the hypothesis that subjects try primarily to maintain a constant interval between their responses, since this hypothesis cannot explain the acceleration. Neither is it consistent with the view that subjects try primarily to give a response perfectly simultaneous with the stimulus. This assumption does not explain a modification in the behavior and consequently the acceleration.

On the contrary, this behavior can be explained if it is remembered that, when synchronizing, the subject spontaneously and without being aware of it, gives responses which, on the average, are slightly anticipating stimuli. What the subject does in reality is to couple R with S, R preceding S by some ms or cs. When S and R are coincident, the same set elicits a permanent attempt to give a response slighlty prior to the stimulus, the result of which, in the experimental situation, is a constant acceleration until physiological limit is nearly reached.

When the subject is informed about the experimental situation, there is no acceleration. His cadence keeps nearly stable. With 700 or 800 ms cadences, the variability of the differences in duration between successive intervals is of the same order in the phase of pseudo-synchronization and in the first phase of strict synchronization. With 1600 and 3200 ms cadences, the subject is much less regular. Why? The computation of auto-correlations between successive intervals gives the answer. During synchronization with slow cadences, auto-correlations become negative. This means that too short an interval is followed by too long an interval, and the theoretical value of auto-correlations tends toward $r=-0.5$. In practical terms, it means that the subject rectifies the successive intervals that he produces by referring to the time-lag between his response and the stimulus, and he has a tendency to give too long an interval after one which is too short, and so on. This strategy has no sense when there cannot be an interval between S and R. Auto-correlations become positive, which indicates that the successive values of the intervals are about of the same magnitude in relation to the mean and when there are changes, they are progressive. Which means that the successive intervals produced by the subject are in conformity with the pattern they have memorized (Table 1).

Thus, the two types of behavior are very different with slow cadences. With faster cadences, auto-correlations within the limits of strict synchronization are approaching zero. The subject produces a series of intervals which conform to a memorized pattern. Furthermore, if his responses deviate too much from the optimum interval of anticipation (and not of simultaneity) he corrects them from time to time, but very rarely. For what reason? Because the variability of successive intervals does not entail a variability in the lags between R and S which would exceed the limits of synchronization.

To conclude these observations, it may be seen that the synchronization of a series of tapping-responses to a series of isochronous sounds is based on the production of a series of intervals fairly equal to the interval of the cadence. The

Table 1 Results (in ms) for the informed subjects

Cadence	1	2		3	
	mx	$\sigma_{xi-xi-1}$		$r_{xi.\,xi-1}$	
	PS	S	PS	S	PS
400	395	33	30	−.15	−.15
800	758	54	46	−.12	+.03
1600	1435	197	109	−.23	+.30
3200	2920	599	224	−.38	+.42

1: Mean of the intervals with post-synchronization (PS)
2: Mean of the differences between successive intervals in synchronization (S) and (PS)
3: Auto-correlations of successive intervals in S and PS.

subject has direct control of the interval between S and R. This interval which is null perceptually corresponds to a slightly anticipatory R (from 30 to 50 ms). The correction mechanism is the same whatever the cadence but two zones may be distinguished.

First the zone where the subject can achieve a synchronization which he perceives. It corresponds to cadences the intervals of which range from 300 to 1500 ms. Within these limits, the variability, i.e. differential sensitivity, is such that the variations between successive intervals are about the same as for intervals below the threshold of temporal distinction of S and R. If too large an interval appears, which sometimes happens, the subject makes a correction. These few corrections slightly increase the variability of the intervals. EHRLICH (1958) has shown that when the tempo is 600 ms, intervals of synchronization are slightly more variable than intervals of spontaneous tempo.

Beyond that zone of good synchronization, the variability of successive intervals increases in absolute value (though it does not increase very much in relative value). The result is that lags between S and R are larger, their duration being often too long to enable the perception of simultaneity. Consequently, the subject makes continuous corrections in order to try to maintain the periods of R and S synchronous.

These corrections explain that beyond 1500 ms the variability of the intervals produced in the course of synchronization is higher than when there is no synchronization. In the latter case, the mean duration of the intervals is slightly different from the initial pattern.

Consequently, the fundamental basis of sensori-motor synchronization is mnemonic. The successive intervals which are produced by the subject follow the pattern stemming from the perceived cadence. The approximate coincidence between R and S is useful only in maintaining equal the periods and does not play the same role with slow and fast cadences. This role is defined by the relation between differential sensitivity to S-R lag and the differential sensitivity which controls the regularity of the production of successive intervals.

REFERENCES

1. Ehrlich S. (1958): Le mécanisme de la synchronisation sensorimotrice. Année psychol., *58*: 7–24.
2. Bartlet N. R. and Bartlet S. C. (1959): Synchronization of a motor response with an anticipated sensory event. Psychol. Rev., *66*: 203–218.
3. Fraisse P., Pichot P. and Clairoin G. (1949): Les aptitudes rythmiques. Etude comparée des oligophrènes et des enfants normaux. J. Psychol. norm. et path., 309–330.
4. Fraisse P. (1966): L'anticipation de stimulus rythmiques. Vitesse d'établissement et précision de la synchronisation. Année psychol., *66*: 15–36.
5. Fraisse P. and Ehrlich S. (1955): Note sur la possibilité de syncoper en fonction d'une cadence. Année psychol., *55*: 61–65.
6. Fraisse P. Oléron G. and Paillard J. (1958): Sur les repères sensoriels qui permettent de contrôler les mouvements d'accompagnement de stimuli périodiques. Année psychol., *58*: 321–338.
7. Oléron G. (1961): Attitude et utilisation des repères dans la synchronisation à des stimuli périodiques. Année psychol., *61*: 59–78.
8. Fraisse P. and Voillaume C. (1971): Les repères du sujet dans la synchronisation et dans la pseudo-synchronisation. Année psychol., *71*: 359–369.
9. Voillaume C. (1971): Modèles pour l'étude de la régulation des mouvements cadencés. Année psychol., *71*: 347–358.

LOW AMPLITUDE INFRADIAN CYCLES OF URINARY 17-HYDROXYCORTICOSTEROID EXCRETION IN A HEALTHY MALE SUBJECT*

GEORGE C. CURTIS and DONALD MC EVOY

Eastern Pennsylvania Psychiatric Institute
and Departments of Psychiatry and Biochemistry
University of Pennsylvania
Philalphia, Pennsylvania, U.S.A.

In data collected from himself by Dr. Christian HAMBURGER over 16 years, HALBERG et al. [1] detected low amplitude but statistically significant cycles of 17-ketosteroid excretion, with periods of about a week (circaseptan), about 20 days (circavigintan), about 30 days (circatrigintan), and about a year (circannual). In data from another healthy adult male subject, cycles of 17-hydroxycorticosteroid excretion with similar amplitudes, and with similar and different periods, are now reported.

METHODS

One of the investigators (GCC) was the subject. The study extended from August 22, 1961, when he was 34 years old, through September 8, 1965, when he was 38. During this interval a total of 795 urine specimens were collected, mostly of 24 hours duration, but occasionally of 48 or 72 hours. Sampling was interrupted at various points and for various time intervals during the 4 years.

During the study the subject's only illnesses were occasional head colds, and the only intake of drugs was aspirin. On the rare occasions when aspirin was taken, the urine for that day was discarded. A fairly regular scientific and academic routine, with daytime work and nighttime sleep, was maintained. Usually the subject worked from 0830 till about 1730 each weekday, until about 2200 on Monday and Thursday evenings, and until about noon Saturdays.

Urinary 17-OHCS were measured by the method of GLENN and NELSON [2]. Quality control data on the analytical procedure indicated a measurement error of 3–4% when replicate aliquots of the same urine sample were assayed simultaneously in the same batch of unknown samples. When replicate aliquots were run on different days in separate batches of unknowns, the error of measurement was then around 7%.

The data obtained were analyzed by the Chronobiology Laboratories, University of Minnesota, by the fitting of cosine curves as described by HALBERG et al. [3]. Spectra linear in frequency were done with several intervals among consecutive trial periods. While several statistically significant components were thus detected in the low frequency domain of biological rhythms, only a few of these are singled out, either because of their statistical prominence or because of earlier findings of biological rhythms with similar periods in the same or related variables.

* Supported in part by PHS Grant MH-08806.

RESULTS AND DISCUSSION

Table 1 presents a summary of some of the more statistically prominent cycles detected. Their periods included 8.5, 11, 18, 30, 45.6, 52, 182, and 348 days. The periods of 8.5, 18, 30, and 348 days correspond approximately to the circaseptan, circavigintan, circatrigentan, and circannual cycles of 17-ketosteroid excretion previously reported by HALBERG et al. [1].

Despite the high levels of statistical significance of rhythm detection, the percent of variance accounted for by these trial periods (variance ratio) was always less than 6%.

Table 2 presents the amplitudes of the various tentatively detected cycles of urinary 17-OHCS, expressed as percentages of the mean (mesor) 17-OHCS excretion and as percentages of the amplitude of the circadian cycle of urinary 17-OHCS in the same subject determined from three brief series of measurements in 1970 and 1971. All infradian amplitudes were less than 10% of the mesor and less than 15% of the circadian amplitude. Corresponding percentages were sometimes as high as 30% in the earlier report on 17-KS excretion.[1].

Table 1 Summary of cycles of 17-OHCS excretion by a healthy male subject between August 22, 1961, and September 7, 1955. Rhythm detection at all indicated periods was significant at $p<.01$. Variance Ratio (VR)=% of total variance accounted for by the cycle detected. Urinary 17–OHCS expressed as μg/hr.

Period (τ) Hours	days	Mesor M±S.E.	Amplitude A±S E.	SE/A	Variance ratio	Phase (ϕ) with 95% C.I. (0°=Aug. 22, 1961)
(8760)	(365)	(288.36±2.50)	(12.50±3.33)	(.282)	(1.6)	[−330 (−301, −358)]
8358	348	287.90±2.35	12.73±3.42	.269	1.7	−346 (−318, −374)
4380	182		21.25±3.33	.154	5.1	−117 (−101, −134)
1248	52		15.85±3.20	.202	3.0	
1095	45.6		19.53±3.25	.166	4.4	
720	30		10.66±3.26	.306	1.2	
432	18		10.86±3.24	.278	1.4	
264	11		13.75±3.33	.235	2.2	
203.7	8.5		11.75±3.26	.278	1.6	

Table 2 Amplitudes of detected infradian and circadian cycles expressed as percentages of mesor and of circadian amplitude.

Period days	Amplitude as a % of: Mesor	Circadian amplitude
348	4.4	7.8
182	7.4	13.2
52	5.5	9.8
45.6	6.8	12.1
30	3.7	6.6
18	3.8	6.7
11	4.8	8.5
8.5	4.1	7.3
1	45.0	100.0

With only four years of data, the apparently free running circannual cycle of 348 days cannot be reliably differentiated from a 365 days synchronized annual cycle. The results with the latter trial period have been included in Table 1, and also indicate a statistically significant component, even though it did not yield the best fit in this spectral region. The acrophase of the annual component occurred in late July, in contrast with the November peak of the 17-ketosteroid cycle reported by HALBERG et al. [1]. Variable acrophases would be expected if circannual free running biological cycles existed. On the other hand it is noteworthy that 17-OHCS and 17-ketosteroids do not reflect identical secretory processes even though there is some overlap of their constituent compounds. 17-OHCS are derived exclusively from the adrenal crotex, while 17-KS reflect both adrenal cortical and testicular secretions.

The 182 day component represents a period of almost exactly ½ year and was the most prominent component detected, both statistically and in terms of amplitude. Its crests occurred approximately in late October and late April. The late October crest is in better agreement with the November crest of 17-KS [1]. The possibility that there may be free running or multi-frequency cycles of 17-OHCS excretion indicates caution in interpreting findings such as Uete's, which was not evaluated statistically, that 17-OHCS excretion was higher in winter than in summer [4], of OKAMOTO et al. [5] of significantly higher 17-OHCS excretion in January than in September, or of HALE et al. [6] that 17-OHCS excretion progressively decreased as summer heat subsided.

The periods of the cycles of 17-OHCS excretion tentatively detected here do not correspond precisely to known environmental or societal cycles, and if confirmed, would presumably be of endogenous origin. In this respect an endocrine cycle in men, corresponding in period to the female menstrual (circatrigintan) cycle, and now reported for the second time, would be of special interest. A free running cycle corresponding approximately in period to the societal week would also be of considerable interest, both to biology and to sociology and anthropology.

SUMMARY

In measurements made on a healthy male subject over a 4 year period, low amplitude cycles of urinary 17-OHCS excretion with periods of about 8.5, 11, 18, 30, 45.6, 52, 182, and 348 days have been tentatively detected.

ACKNOWLEDGMENTS

The authors are indebted to Stephen BRUMBERG, R. Leon JOSHLIN, and Sarah PERRY for technical assistance, to Dr. Max L. FOGEL for statistical consultation, and to Dr. Franz HALBERG for the data analyses.

REFERENCES

1. HALBERG F., ENGELI M., HAMBURGER C. and MILLMAN D. (1965): Spectral resolution of low-frequency, small-amplitude rhythms in excreted 17-ketosteroids; probable androgen-induced circaseptan desynchronization. Acta. Endocr., 50, *103* (Suppl.): 5–54.
2. GLENN E. M. and NELSON D. H. (1953): Chemical method for the determination of 17-hydroxycorticosteroids and 17-ketosteroids in urine following hydrolysis with β-glucuronidase. J. Clin. Endocr., *13*: 911–921.

3. HALBERG F., TONG Y. L. and JOHNSON E. A. (1967): Circadian system phase–an aspect of temporal morphology; procedures and illustrative examples. *In*: The Cellular Aspects of Biorhythms, (H. von MAYERSBACH, ed.), pp. 20–48, Springer-Verlag, Berlin.
4. UETE T. (1961): Excretion of unconjugated and conjugated corticosteroids in healthy subjects. Metabolism, *10*: 1045–1051.
5. OKAMOTO M., KOHZUMA K. and HORIUCHI Y. (1964): Seasonal variation of cortisol metabolites in normal man. J. Clin. Endocr., *24*: 470–471.
6. HALE H. B., ELLIS J. P., Jr., BALKE B. and MCNEE R. C. (1962): Excretion trends in men undergoing deacclimatization to heat. J. Appl. Physiol., *17*: 456.

HOMEOSTASIS OF BEHAVIOR

Thomas D. REYNOLDS

Laboratory of Human Behavior
Division of Special Mental Health Research
National Institute of Mental Health
Saint Elizabeths Hospital
Washington, D.C., U.S.A.

In this talk we shall raise some basic questions about behavior, whose definitive answers will require many years of investigation. We wish to ask in what sense is behavior regulated? We also wish to ask whether or not the study of behavior can contribute to a genuine understanding of central nervous processes.

These general questions suggest more specific ones. Do behavioral time-series exhibit any of the characteristics of homeostatic systems, such as negative feedback, damped oscillations, etc? From a rigid environmentalist point of view, we might expect that behavioral time-series will simply be a record of moment-to-moment organism-milieu interactions, and therefore mathematically random, that is, empty of deterministic signals except those imposed by environmental periodicities. However we are probably all familiar with the periodic behavioral disturbances described in human beings by Gjessing, and in both animals and humans by Richter. In these disturbances deterministic signals clearly predominate over noise. In addition to pathological phenomena, we have the familiar experience that our own tensions and emotions do not subside immediately, but must decay at rates which seem subjectively natural and inevitable. Therefore it seems worthwhile to look at more or less unselected behavioral time-series for evidence of regulation.

An examination of behavioral output, with some attention to environmental conditions, may suggest hypotheses above the over-all design of its regulatory agencies. It has seemed to me necessary to approach the general mathematical properties of behavioral time-series, before detailed examination of input-output relations. Indeed it is difficult and often impossible to identify significant inputs in human beings except through perturbation of expected output.

To orient ourselves we require some simple models, provided that we do not take them seriously except as aids to thought. We might form a simple pair of linear differential equations as follows, where E is the level of some excitatory process proximal to the final common pathway, and I is the level of some responsive inhibitory process.

$$\frac{dE}{dt}=C_0-C_1E-C_2I$$

$$\frac{dI}{dt}=K_0-K_1I+K_2E$$

It is easy to show, by eliminating I that E may be described by a second-order linear differential equation. This is already a fertile model, as processes with such a system-equation can exhibit steady oscillation, damped oscillation, or exponential decay, and even a linear trend, depending on the state of the parameters. The empirical significance of such a model is that such a system will possess a second-order linear difference equation, that is, an equation of the form:

$$Y_k = A_0 + A_1 Y_{k-1} + A_2 Y_{k-2}$$

where the y's are any three successive equally-spaced values of the time-series, and the A's are constant coefficients. A second-order difference equation is nothing more or less than a second-order autoregressive scheme, and its coefficients may be determined empirically. That is, we can approach a time-series and hope to suggest a theoretical model from the data itself. That is precisely what I have attempted to do, and I shall use one particular non-periodic time-series to illustrate the methodology.

The data are extremely simple: our particular patient is observed, along with 14 other psychiatric patients, roughly 150 times per week during rest-periods when the patients are free to do as they please except leave the ward. Observers work in pairs so that reliability can be assessed at any time. This particular patient is an episodically restless deteriorated schizophrenic man. The only behavioral item present in sufficient amplitude to work with, is simple wandering, versus sitting or lying down. The more clinically tense and agitated he is, the more he wanders, so that a simple percentage of times-observed that he is either standing or pacing, forms a good record of behavioral arousal generally, and is less subjective than clinical estimation.

We have 175 weeks of data at present available for examination, during which no drugs were administered and during which the patient has been on a fairly uniform ward setting. In Fig. 1, we see the first 50 weeks of wandering rates plotted against time.

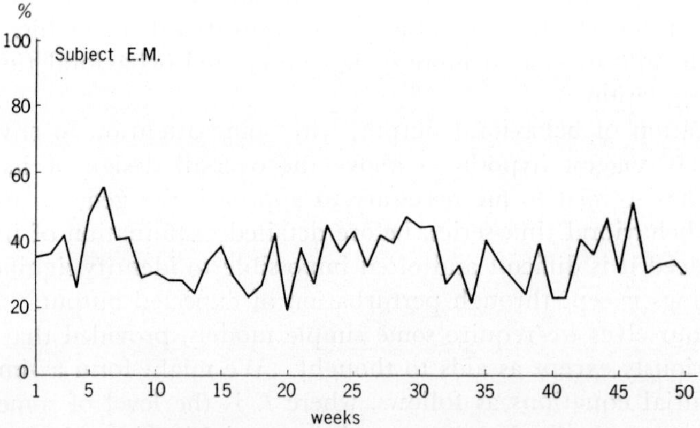

Fig. 1 Per cent positive check times for "wandering."

For his first 50 weeks autoregressive coefficients for schemes up to ninth order were determined by least-squares. Polynomial terms were added to first, second, and third order schemes to see if non-linear modification would improve predictive fit. The results of prediction into weeks 51 through 75 were compared with moving averages up to nine terms, and of course the mean of all previous observations, as a basic simple random model for sake of comparison. Sums of absolute errors of prediction were used as measures of fit. After each set of 25 predictions, models were recalculated, and a variety of non-linear models devised as we went along. Varying filter-lengths were employed in calculating the coefficients. The conclusions are fairly easy to state: the best predictor is a second-order autoregressive scheme computed from all previous data. For 125 predictions, the sum of absolute errors for a second-order scheme (AR-2) was *991*, followed by *1003* and *1006* for third and forth order schemes respectively. Use of a simple mean (of all past data) yielded *1041*, and a moving average of five terms 1034. No other predictor could compete with the AR schemes, including non-linear models, exponentially weighted moving averages, and seasonal models of various kinds. These are not very impressive differences, but we do have reason to accept a second-order linear AR scheme as the most parsimonious account of the time-series. If we follow the criteria described by Box, we satisfy the criteria for at least tentative identification of our system as autoregressive of second order. The reason for rather meager success of the AR-2 scheme as a predictor, however, is easily discovered by plotting the AR coefficients themselves against time. We see fairly systematic fluctuation that might permit prediction by extrapolation. Fig. 2 exhibits the time-course of the coefficients of our autoregressive scheme each set computed from fifty values at intervals of five weeks.

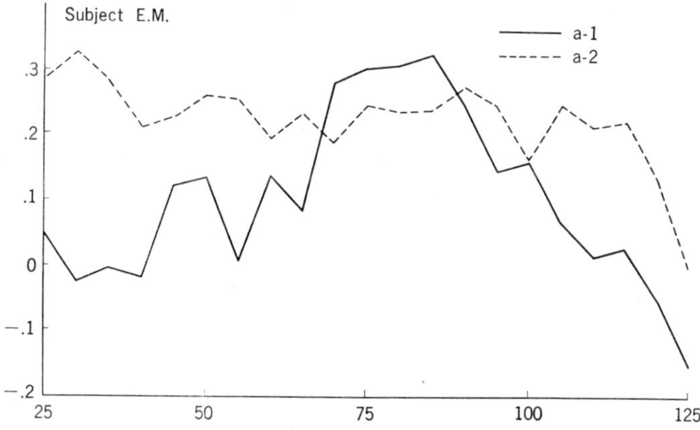

Fig. 2 Auto-regressive coefficients as a function of time.

In any case, we will have to continue our trials to make a stronger case, and presumably the final answer will depend upon whether indeed with sufficient data we can begin to extropolate the actual trends in the coefficients. The cause of the curious dip in value of both coefficients near the end of the observation period is unexplained.

To illustrate the use to which we might put such an analysis, let us compute our AR-scheme for all 175 weeks. We obtain

$$Y_k = 22.468 + .134Y_{k-1} + .260Y_{k-2}$$

This is a second-order difference equation whose solution in conventional notation is

$$Y = 37.07 + C_0 e^{-.54t} + C_1 e^{-.80t} \cos \Pi t$$

that is, it consists of an exponential decay term and a damped oscillatory term with roughly a two-week period length. The corresponding differential equation is not second, but third order, namely, in operator notation

$$(D^3+2.51D^2+11.39D+5.71)Y=211.54$$

We must therefore reject our simple E-I linear model, and inquire after the next simplest model which will yield a third-order differential equation.

We note that our E-I model was an entirely closed system: that is, only endogenous processes were related to outcome. This is surely unreasonable. If we break up E into compartments, the first endogenous and the second responsive to the milieu, both effected in different fashion by the inhibitory process, and allowing also that the two excitatory compartments mutually influence one another, we can form a simple linear system of differential equations. These generate a third-order differential equation for each variable considered independently. This would be the simplest system compatible with our data.

In conclusion it would appear that a simple linear model with excitatory-inhibitory feed-back, making a distinction between endogenous and exogenous sources of stimulation is the most parsimonious theoretical model consistent with the time-series analysis we have performed. To strengthen such conclusions, we must be able to present the coefficients as functions of time, and be able to extrapolate their values in making our predictions. The handling of only one case can be merely suggestive, but the general approach seems fairly clear.

REFERENCES

1. Box and Jenkins (1970): Time Series Analysis. Holden-Day, San Francisco.
2. Gjessing and Gjessing (1961): Some main trends in the clinical aspects of periodic catatonia. Acta Psychiat. Scand., *37*: 7.
3. Morrison (1969): Introduction to Sequential Smoothing and Prediction. McGraw-Hill, New York.
4. Reynolds (1968): Short-term prediction of human behavior. J. Psychiat. Res., *6*: 237.
5. Richter (1960): Biological clocks in medicine and psychiatry. Proc. Nat. Acad. Sci., *46*: 1506.

THE HUMAN HICCUP: TIME RELATIONSHIPS AND ETHOLOGICAL SIGNIFICANCE*

TERENCE R. ANTHONEY, SHARON L. ANTHONEY
and DEANNA J. ANTHONEY

School of Medicine and Department of Zoology
Southern Illinois University at Carbondale
Carbondale, Illinois, U.S.A.
School of Medicine
University of California at Irvine
Irvine, California, U.S.A.

ABSTRACT

The determination of time relationships has been an integral part of an ongoing study on the phylogeny and ontogeny of human hiccup, underway since the summer of 1965. Most data to date have been obtained on humans, including data from questionnaires and interviews of more than 50 individuals, as well as records on essentially all hiccups to occur over continuous periods of 3 mo.–5½ yr. for more than 20 additional people. No attempts were made to stop hiccups, except during specified periods (hiccup "deprivation", *v.i.*).

Though few obvious rhythms were noted, several of the preliminary findings involved non-random occurrence of hiccups in time ("temporal structure"):

1) Hiccups became less frequent as subjects matured.

2) Hiccups were usually much more frequent in young adult females than in males of the same age.

3) If more than a few hiccups occurred in a series, the hiccups seemed to become "established" and a "hiccup bout" ensued and did not stop until a certain "minimum" number of hiccups had occurred.

4) For many subjects, two or more bouts of hiccups frequently occurred upon the same "hiccup day", even though days on which bouts occurred were weeks apart.

5) Hiccup-hiccup intervals within each bout often became gradually longer as the bout progressed. A critical range of values for hiccup-hiccup intervals seemed to exist, above which a bout would likely terminate and below which a short series of hiccups would likely become a hiccup bout.

6) For some subjects, the interval between two bouts occurring on the same day showed a positive linear correlation with the ratio–"no. of hiccups in succeeding bout/no. of hiccups in preceding bout".

7) For some subjects, the number of hiccups on a hiccup day was directly correlated with the number of days since the preceding hiccup day.

8) Incidence of hiccup bouts was correlated with sexual cycles. During menstrual cycles, most bouts occurred in the follicular stages, the peak incidence oc-

* The research of T.R.A. was partially supported by U.S.P.H.S. Grant 1-SOL-FR-5367-04-5, administered by the Univ. of Chicago, and by N.I.M.H. Post-Doctoral Research Fellowship F2-MH-39876. The indispensable services of many colleagues and friends in providing data are gratefully acknowledged.

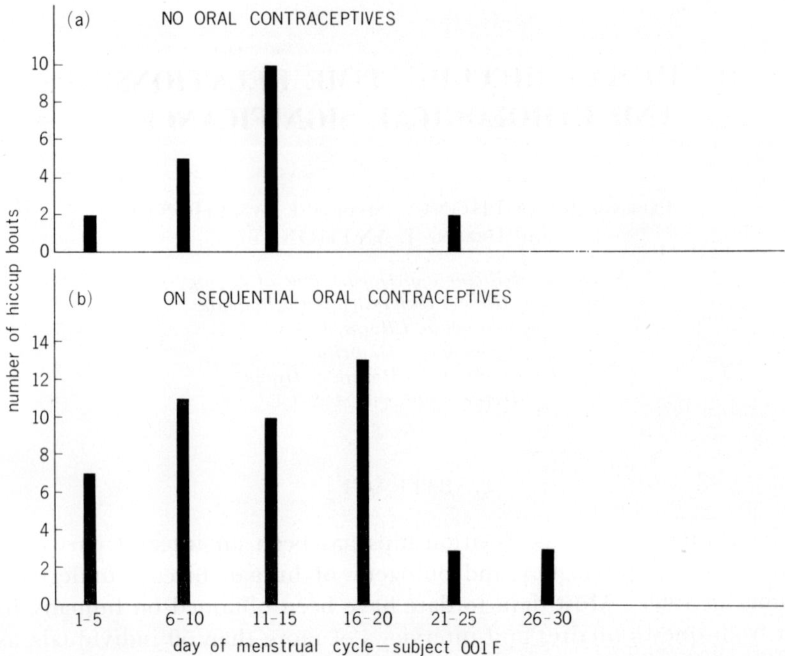

Fig. 1 Example of distribution of hiccup bouts within menstrual cycles of a subject-001F (21–25 years old). (a) Distribution during 13 consecutive cycles in which no oral contraceptives were taken. (b) Distribution during 28 consecutive cycles while on sequential oral contraceptives (an estrogen preparation usually taken on days 6–20, an estrogen-gestagen preparation on days 21–25).

curing in the few days just preceding ovulation (Fig. 1a). Subjects on sequential oral contraceptives hiccupped most while taking only the estrogen, least while taking the gestagen (Fig. 1b). During pregnancy, the incidence of hiccups was decreased.

9) For many subjects, peak incidence of hiccups occurred during the evening hours (Fig. 2).

10) When subjects were "deprived" of hiccups by stopping each bout prematurely, the frequency of bouts increased and the bouts became harder to stop.

Several characteristics of human hiccup, some of which have been better defined by study of time relationships, suggest that it is not simply a senseless behavior or pathologic reflex of the diaphragm, but rather an example of a "fixed action pattern" (FAP), the concept of which is extremely important theoretically in *ethology*—the study of behavior within an evolutionary and ecological context, as well as with regard to physiologic mechanisms within the given species [1–2]. The human hiccup has the following in common with other FAP's (e.g., patterns of courtship or nest-building in many avian species): It is highly stereotyped and reflects the coordinated activity of several muscles [3]; it is of species-wide distribution and present before birth [4]; its performance frequently results from quite specific types of stimuli ("releasers"); and it possesses "action-specific energy" (i.e., a gradual build-up of "pressure" to hiccup, decreased by performance, seems to occur, as evidenced, for example, by increasing length of hiccup-hiccup intervals within a bout, by positive correlations between the number of hiccups in a

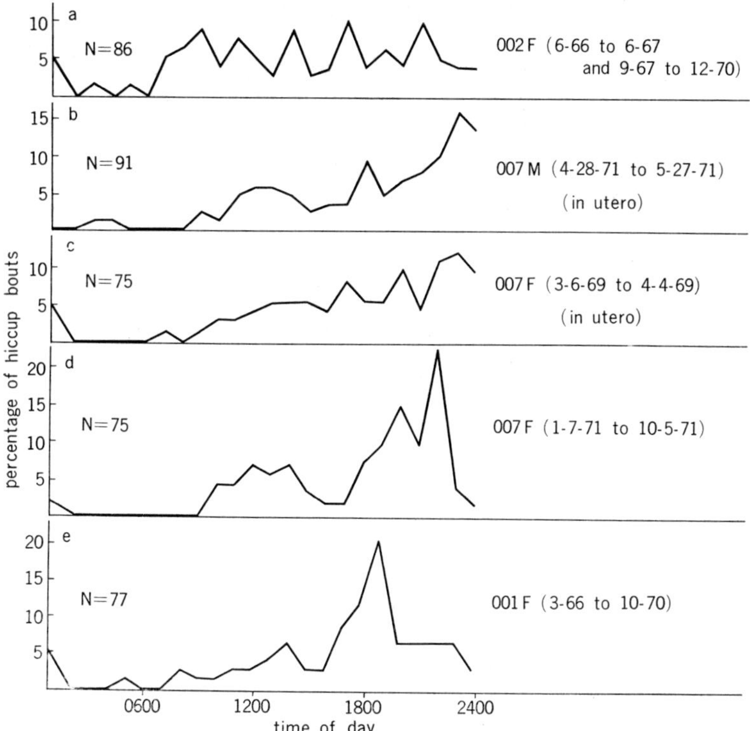

Fig. 2 Examples of diurnal distribution of hiccup bouts, expressed as percentage of daily totals for each hour. Subjects 001F (birthdate: 9-26-44), 002F (birthdate: 1-23-47), 007F (birthdate: 4-5-69), and 007M (birthdate: 5-28-71); separate distributions for 007F before and after birth.

bout or on a hiccup day and the length of time since the previous bout or day, respectively, and by the increased frequency of bouts and increased difficulty of stopping hiccups in individuals whose bouts are being terminated prematurely).

Since FAP's can be treated as structures of a species, much like a bone, muscle, or gland, possible functions of human hiccup may be explored by looking for homologous behaviors in other species and examining the functions they subserve. Some mammals, including both non-primates [5–6] and non-human primates, exhibit behaviors which are very similar to human hiccup with regard to form, probable stimuli, results, and relations to other motor patterns. One evolutionary interpretation of particular interest to the authors at present is that the human hiccup is a behavioral remnant, homologous to motor patterns utilized by certain other mammals to regurgitate food for conspecifics. Though the human hiccup may no longer have any regurgitative function, its presence and several of its characteristics seem quite reasonable when viewed in the context of ethological findings and theory. Other possible functions of human hiccup are also being investigated.

REFERENCES

1. Hinde R. A. (1966): Animal Behaviour: A Synthesis of Ethology and Comparative Psychology. McGraw-Hill, New York.

2. Marler P. R. and Hamilton W. J. III (1966): Mechanisms of Animal Behavior. John Wiley & Sons, Inc., New York.
3. Newsom Davis J. (1970): An experimental study of hiccup. Brain, *93*: 851–872.
4. Norman H. N. (1942): Fetal hiccups. J. comp. Psychol., *34*: 65–73.
5. Martins T. (1949): Disgorging of food to the puppies by the lactating dog. Physiol. Zool., *22*: 169–172.
6. Van Lawick-Goodall H. and Van Lawick-Goodall J. (1971): Innocent Killers. Houghton Mifflin Co., Boston.

PHASE ANALYSIS OF THE SOMATIC AND MENTAL VARIABLES IN GJESSING'S CASE 2484 OF INTERMITTENT CATATONIA*

H. W. SIMPSON, L. GJESSING, A. FLECK, J. KÜHL and F. HALBERG

University Department of Pathology, Royal Infirmary Glasgow, Scotland
Chronobiology Laboratories, Department of Pathology University of Minnesota Minneapolis, Minnesota, U.S.A.

GJESSING's extensive metabolic studies in periodic catatonia, published in Archiv für Psychiatrie* [1–3]**, await a long overdue reappraisal in view of developments in endocrinology and chronobiology. Possibly they contain a clue to the pathogenesis of this disease.

As a preliminary exercise, we have reinvestigated GJESSING's case designated 2484, a male of 22 years diagnosed clinically as a "periodic catatonia". There was a family history of schizophrenia. He was a bright student at primary and secondary schools and there was no hint of mental disease but as his education was nearing completion, character changes were noted. He became selfish, hypersensitive and lazy, though, interestingly enough, he was an exceptional sprinter (100 metres in 11.3 secs).

At the age of 21, there was an unexpected explosive onset of mental symptoms including violent changes of mood, hyperkinesia and delusions. After one month in hospital his illness was quiescent for 6 months. Then after another outburst he was transferred to Dikemark hospital. On admission his weight was 68 kg and height 1.73 metres. It was noted he had severe acne. Between this time and the fall of the same year, a 3-week periodicity was noticed in his symptoms. One extreme was characterized by a stage of depression and then increasing loss of affect, until he became completely withdrawn. While in this condition—usually at night—there would be a sudden change preceded by whistling and singing in bed when he should have been asleep. The restlessness became more marked. He talked, laughed or cried incessantly. He shouted erotic anecdotes and was hyperkinetic. Over the next 9 days or so this stage subsided slowly, to be replaced again by a stupor.

GJESSING set up a comprehensive metabolic balance study to correlate any cyclic somatic and mental changes. The patient was housed in a single room and generally confined to bed. The latter rested on a 'seismograph' to record body movement. The patient was given a wholly fluid diet (so called 'H' diet) of eggs, milk, cream, vitamins, sugar and salts. Uneaten residues were measured. Daily feces were collected for nitrogen assay and also all urines for nitrogen (total and NH_3), sulphate, titratable and phosphoric acid, pH, and chloride, over a 106 day span.

* Grant support: Medical Research Council: U.S. Public Health Service (5-KM-GM-13, 981) and NASA (NGR-24-005-006).

** An English translation is in preparation (GJESSING L. personal communication).

His mental, physical and intellectual states were rated by the nursing staff on 3 scales. Mental concentration was scaled from 0–3 where 'O' was relative normality. In '2' he could talk responsibly about his future but with his mind tending to wander. The low point on this scale [3], indicated a stage of incoherency.

After a 160-day observation span the patient became well enough to be discharged home on dessicated thyroid therapy and some six years later he was living as a normal individual with an apparently complete remission. This is our last record.

METHODS

Our assessment of this case involved cosine wave fitting on the data available at 24-hour intervals, the so-called least squares spectral analysis. It was done linearly in period in the region from 72 to 2160 hours, with 1-day increments between consecutive trial periods.

RESULTS

A best-fit (defined as the trial period with the minimal percent error) at 21 days was found for nearly all variables (See Fig. 1). The four exceptions–namely: O_2 consumption, diastolic blood pressure, white blood count, nitrogen–had minimal percent errors at 20 or 22 days.

The acrophases for the circavigintan rhythms are summarised in a circular diagram where 360° ≡ 21 days (Fig. 1).

The stupor stage of the "mental state" has been used as an acrophase reference (0°) for all other variables. The nitrogen retention peak leads it slightly while the acrophases of temperature and chloride excretion are synphasic or slightly later. These are followed on day 3 by the acrophases of body weight and urine volume. This is a remarkable and apparently paradoxical synphasia since some would have expected body weight to move in the opposite direction to urine volume.

The excitement stage is heralded by acrophases in phosphoric acid (day 8) nitrogen, sulphate, hemoglobin (day 9) white blood count, blood pressure, oxygen consumption, carbon dioxide output (day 10–12) and urine pH. These are synphasic with excitement while the low point of the respiratory quotient follows a day or so later (increased fat metabolism?) along with the high of urine specific gravity and nitrogen (NH_3) excretion.

DISCUSSION

Gjessing emphasized that the pathology of intermittent catatonia was related to intermittent nitrogen retention and this idea bring to mind the fact that there was *no* detectable rhythm in nitrogen input, or in fecal nitrogen. Thus the fecal nitrogen in the stupor stage (0.44 g/day) was not significantly different from that in the excitement stage (0.48 g/day).

On the other hand the daily urine nitrogen excretion in the stupor state was significantly lower (8.8 g) than in the excitement stage (10.8 g; $P < .001$) and hence the changes in nitrogen balance.

The metabolic periodicity might be due to a periodic variation in anabolic

CIRCAVIGINTAN SYSTEM OF R.GJESSING'S CASE ♯2484
♂ 22yrs., CLINICAL DIAGNOSIS:'PERIODIC CATATONIA'

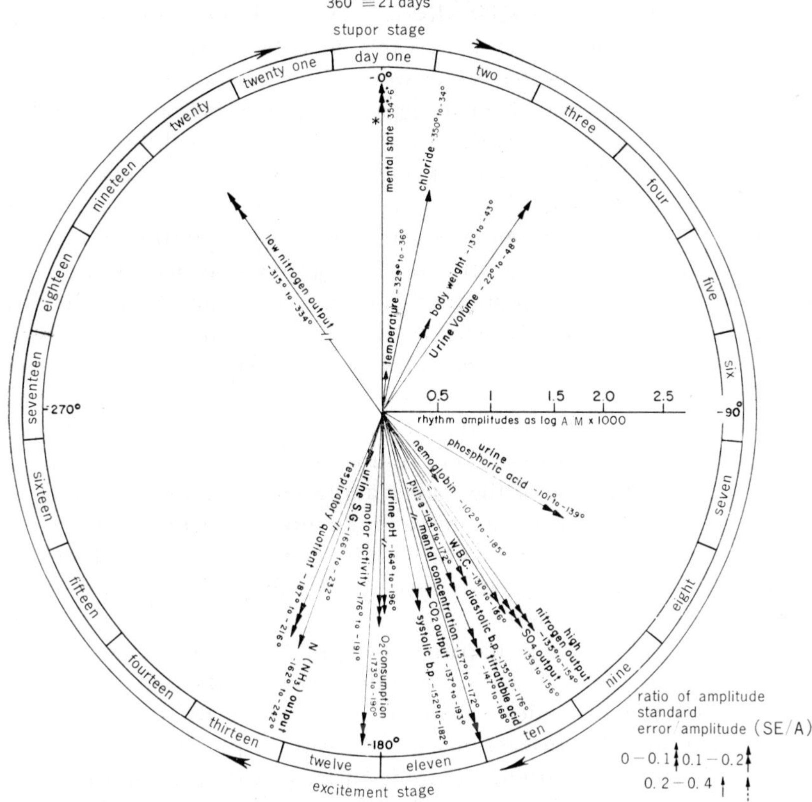

Fig. 1 Internal timing along the 21-day scale in intermittent catatonia. "State" assessed clinically on points scale in which catatonic stupor is at the maximum. All other acrophases referred to this maximum, except pH, RQ and Mental Concentration, which are referred to their low points. Nitrogen (total) in Urine referred to low and high points. Broken vectors used for 4 variables referred to low point.

* Figures at vector heads refer to 95% limits of acrophase and are only quoted when "F" test (of zero amplitude) describes a rhythm at a probability of $<.05$.

hormones. Two main possibilities come to mind, namely, growth hormone or androgens. The patient did not exhibit the outward physical manifestations of acromegaly moreover, blood sugar and glucose tolerance tests were unremarkable. Androgens are also a reasonable suggestion in view of the patient's build, athletic performance and the presence of severe acne. Further support comes from the demonstration of a temperature peak during the catatonia since some androgens are known to be thermogenic [4]. The temperature peak seems particularly relevant since it unexpectedly occurs when the patient was least physically active.

The possibility of a 21-day androgen rhythm as a mediator of some or all of the metabolic events is also strengthened by increasing direct and indirect information of such a period in normal man. Thus, a least squares analysis of ketosteroid excretion in one man's urine collected daily over 16 years, revealed an about-3-week period in the 4×4 yr subspans [5]. Also a synphasic 16-day cycle of beard growth and body weight has been shown in longitudinal studies on healthy man [6]. Furthermore, a circavigintan eosinophil cycle was found in the case of periodic manic depressive disease described originally by BRYSON and MARTIN [7]. Presumably this cycle was mediated by adrenal corticosteroids.

If androgens were implicated what would be their origin? There might be a possibility of a compensated virilising adrenal hyperplasia with a relative enzyme lack; in this respect the family history suggests the possibility of an inherited disease and at the IX International Congress of Paediatrics CARA and GARDNER [8] described a form of this disease with periodic fever. Puberty here, might, perhaps be the final precipitating event in view of the inhibition of corticosteroid hydroxylation by androgens [9]. It is interesting too, that in a published autopsy report of a case of periodic catatonia mention is made that both adrenal glands contained "adenomas" [10]. ROWNTREE and KAY [11] studied two cases of recurrent schizophrenia and noted acne, pyrexia and raised urinary ketosteroids during the "attack". Also GUNNE & GEMZELL [12] studying another two cases of intermittent catatonia, concluded that there was a 'poor' response to ACTH.

To return to the present case, there is also the fact that the maximum of body weight occurred at the time of the stupor and a few days after the peak nitrogen retention. At first sight this fact fits a build up of the nitrogen into protein; but one has also to reconcile the coincident maxima of urine and chloride output KRUSKEMPER [13]; if the nitrogen is used for protein build up, then there must have been relative dehydration.

This diuresis and probable natriuresis might be explained as a "salt-losing" type adrenogenital syndrome with low aldosterone production [9].

Further interest in the idea of an adrenal enzyme defect is gained by the knowledge that some corticosteroids are narcotic–especially progesterone compounds [14, 15] and intermittent catatonia in women is often associated with the later, progestational stage of the cycle [16]. Progesterone compounds are, of course, also implicated in the adreno-genital syndrome. This idea, naturally, remains speculative in the present case.

CONCLUSION

Phase analysis of the metabolic events in a case of intermittent catatonia revealed a sequence of nitrogen retention, catatonic stupor, pyrexia, weight gain and a salt and water diuresis in that order. A possible pathogenesis for this case has been discussed implicating an adrenal enzyme lack and an anaesthetic effect of abnormal corticosteroid metabolites. The object of presenting the discussion is to provide a working hypothesis for prospective research.

REFERENCES

1. GJESSING R. (1932): Beiträge zur Kenntnis der Pathophysiologie des katatonen Stupors. Mitteilung I: Über periodisch rezidivierenden katatonen Stupor, mit kritischem Beginn und Abschluss. Arch. Psychiatr., *96*: 319–392.
2. GJESSING R. (1932): Beiträge zur Kenntnis der Pathophysiologie des katatonen Stupors. Mitteilung II: Über aperiodisch rezidivierend verlaufenden katatonen Stupor mit lytischem Beginn und Abschluss. Arch. Psychiatr., *96*: 393–413.
3. GJESSING R. (1936): Beiträge zur Kenntnis der Pathophysiologie des katatonen Stu-

pors. Mitteilung III: Über periodisch rezidivierende katatone Erregung, mit kritischem Beginn and Abschluss. Arch. Psychiatr. *104*: 355–416.

4. Segaloff A., Bowers C. Y., Gordon D. L., Schlosser J. V. and Murison P. J. (1957): Hormonal therapy in cancer of the breast. The effect of etiocholanolone therapy on clinical course and hormonal excretion. Cancer, *6*: 1116–1118.
5. Halberg F., Engeli M., Hamburger C. and Hillman D. (1965): Spectral resolution of low frequency, small amplitude rhythms in excreted 17-ketosteroids; probable androgen induced circaseptan desynchronisation. Acta Endocr., (suppl.), *103*: 54.
6. Sothern R. B. (1974): Low-frequency rhythms in beard growth of a man. *In*: Chronobiology. Proc. Symp. Quant. Chronobiol., Little Rock, 1971, (L. E. Scheving, J. E. Pauly and F. Halberg, eds.), pp. 241–244, Igaku Shoin Ltd., Tokyo.
7. Halberg F. (1968): Physiologic considerations underlying rhythmometry, with special reference to emotional illness. *In*: Symposium Bel-Air III, Cycles Biologiques et Psychiatriques. (J. de Ajuriaguerra, ed.), pp. 73–126, Georg et Cie, Geneve.
8. Cara J. and Gardner L. I. (1960): Two new sub-variants of virilising adrenal hyperplasia. Paediatrics, *57*: 461–470.
9. Visser, H. K. A. (1966): Congenital disorders of adrenocortical function. Triangle (Sandoz), *7*: 220–233.
10. Cammermeyer J. and Gjessing R. (1951): Fatal myocardial fat embolism in periodic catatonia with fatty liver. Acta Med. Scand., *139*: 358–367.
11. Rowntree D. W. and Kay W. W. (1952): Clinical biochemical and physiological studies in cases of recurrent schizophrenia. J. Ment. Sci., *98*: 100–121.
12. Gunne L. M. and Gemzell C. A. (1956): Adrenocortical and thyroid function in periodic catatonia. Acta Psychiatr. neurol. Scand., *31*: 356–378.
13. Kruskemper H. L. (1968): Anabolic Steroids, Translated by Doering. Academic Press, New York.
14. Raisinghani K. H., Dorfman R. I., Forchielli E., Gyermek L., Genther G. (1968): Uptake of intravenously administered progesterone, pregnanedione and pregnanolone by the rat brain. Acta Endocr., *57*: 395–404.
15. Gyermek L., Genther G. and Fleming N. (1967): Some effects of progesterone and related steroids on the central nervous system. Internat. J. Neuropharm., *6*: 191–198.
16. Hatotani N., Ishida C., Yura R., Maeda M., Kato Y., Nomura J., Wakao T., Takekoshi A., Yoshimoto S., Yoshimoto K. and Hiramoto K. (1962): Psychophysiological studies of atypical psychoses—endocrinological aspect of periodic psychoses. Fol. Psychiatr. Neurol. Jap., *16*: 3.

SOME CIRCADIAN RHYTHMS IN EXPERIMENTAL ETHOLOGY AND COMPARATIVE PSYCHOPATHOLOGY*

CHRISTIAN POIREL

Laboratoire de Psychophysiologie et Psychopathologie
Université du Québec, Chicoutimi, Canada
Laboratoire de Psychophysiologie, U.E.R. de
Biologie Expérimentale
Université de Toulouse, France

The present study is a preliminary attempt to interpret behavior circadian variations in the field of neuropsychology. These observations give some information about the different behavioral circadian rhythms in Swiss/Albino (Rb) Mice. Particularly, these experiments have required ethological methods and non-parametric statistical procedures such as the Friedman analysis of variance [1–3, 12]. According to chronobiological methodology, we have found and described experimentally several psychophysiological and psychological circadian patterns.

We have found circadian variations in generalized epilepsy (Fig. 1, SE) which are statistically significant ($P<0.05$). The highest incidence of seizures is between 1600 and 2000 hours, the lowest incidence is between 2400 and 0400 hours. Circadian periodicity of seizures would be caused by exogenous and endogenous factors (photoperiodicity, sleep-wakefulness rhythm; internal regulation of motor system mechanisms). There are functional relations between temporal fluctuations of central activation (Fig. 1, CA) and circadian occurrence of tonico-clonic seizures. Circadian rhythms of epileptic convulsions would suggest that temporal seizure occurrence would be controlled by several internal clocks. These findings are of interest in understanding the neuro-biological mechanisms of epilepsy [2, 4] and may also supply ethological data about Rodents [5, 6].

Several circadian rhythms of emotional behavior have been found and described. Particularly, these observations suggest theoretically the existence of two "temporal systems" in the neurophysiological mechanisms of Emotion [7, 11]. Experimental analysis shows that the basic rhythm of Emotion (Fig. 1, BE) and the temporal fluctuations of reticular central activation (Fig. 1, CA) present the same "circadian periodicity". But the temporal occurrence of neurotic disorders (Fig. 1, Nb) would be caused by an endogenous regulation under the influence of environmental factors (photoperiodicity, socio-ecological synchronizers). The two rhythms of Emotion are characterized by chronobiological variations which are statistically significant ($P<0.01$). Particularly, we can observe a very marked peak of neurotic manifestations at 2000 hours. The recorded regular oscillations suggest circadian programmes of diurnal and nocturnal modifications of vegetative functions and behavioral functions by several "Psychophysiological Clocks". Experimental studies on the circadian organization of locomotor activity would confirm these interpretations [8–10].

* Work supported by the "Centre National de la Recherche Scientifique, France" and the National Research Council of Canada (Grant No. A 7893).

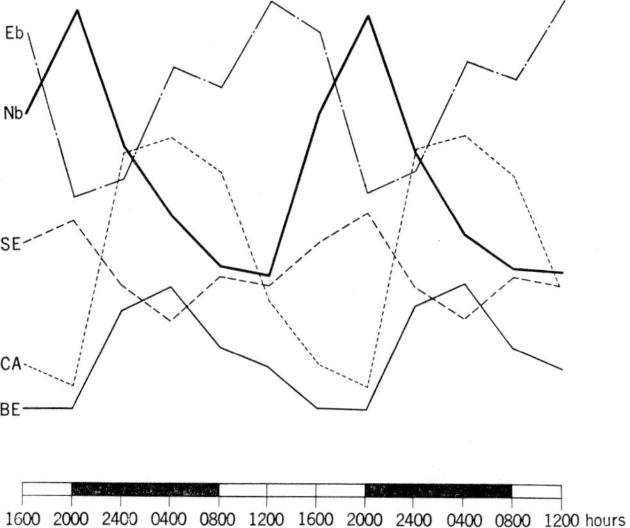

Fig. 1 Circadian rhythms in psychophysiology and in comparative psychopathology. Abscissae represent diurnal and nocturnal hours on which observations were conducted. Abbreviations on the left indicate the variations of different psychophysiological and psychological functions (CA: Central Activation; BE: Basic Emotion; SE: Susceptibility to generalized Epilepsy; Nb: Neurotic behavior; Eb: Exploratory behavior).
Photoperiodicity (L/D: 12/12; Light between 0800 and 2000 hours; Darkness between 2000 and 0800 hours). Constant Temperature of 24° C$\pm$1. Relative humidity in the range 40 to 50%.
These different variations are statistically significant ($P<0.05$).

In the field of experimental psychology, the diagram (Fig. 1, Eb) shows also circadian variations of exploratory behavior. These variations are statistically significant ($P<0.02$). The highest peak of exploratory behavior is observed at 1200 hours and the lowest peak is observed at 0400 hours. Consequently, bimodal oscillations of exploration indicate the biological or psychological complexity of this motivation [5]. Thus, for example, primary and complex motivations change with diurnal and nocturnal hours. Experimental studies on variations of central activation are of great interest in understanding the biology of psychomotor syndroms and psychological drives.

Moreover, in these experiments on circadian rhythms, different psychophysiological parameters are positively correlated (for example: basic emotion and central activation: $P<0.01$ with the test T of Wilcoxon), others are negatively correlated (for example: susceptibility to tonic-clonic seizures and central activation; $P<0.01$ with the test H of Kruskall-Wallis).

In order to verify the circadian form of different variation curves obtained, we have experimentally modified parameters of photoperiodic entrainment. The new re-entrained rhythm curves obtained (Fig. 2) confirm the amplitude, form and phase relations between the circadian rhythms shown in the diagram of Fig. 1. They are also statistically significant ($P<0.05$) and coefficients of correlation calculated between central activation level and behavioral functions present also significant tendencies. This work exhibits, consequently, characteristic chronograms (Fig. 1 and Fig. 2) in the field of neuropsychology.

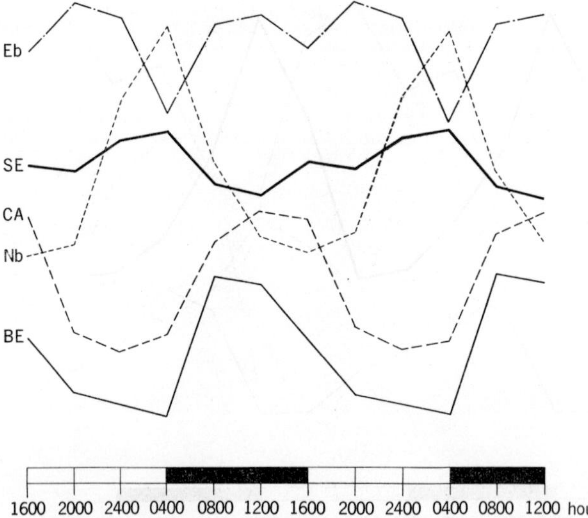

Fig. 2 Photoperiod-shift of the circadian variations in psychophysiology and in comparative psychopathology. In these experiments circadian variations were entrained by a new photoperiodic program (L/D: 12/12; Light between 1600 and 0400 hours; Darkness between 0400 and 1600 hours). These different variations are also statistically significant ($P<0.05$).

The macroscopic analysis of these different circadian patterns lead to several comments:

1) Desynchronization experiments confirm the circadian periodicity of psychophysiological and behavioral functions studied.

2) These observations confirm the important role of the photoperiodicity as entrainment agent of psychophysiological and behavioral rhythms.

3) These chronobiological investigations also confirm experimentally psychological hypotheses concerning functional relations between the central activation level and the organization of drives.

Thus, ethological investigations allow to draw several interpretations concerning the problem of circadian rhythms in the field of general psychology and comparative psychiatry.

REFERENCES

1. Siegel S. S. (1956): Non-parametric statistics for the behavioral sciences. Vol. 1, p. 312, McGraw-Hill, New York.
2. Poirel C. (1967): Mise en évidence de variations nycthémérales de la *susceptibilité* à la crise audiogène chez la Souris Swiss/Albinos. C.R. Soc. Biol., *161*: 1461–1465.
3. Poirel C. (1971): Mise en évidence de relations fonctionnelles entre le niveau d'activation centrale et le degré de réactivité émotionnelle primaire chez la Souris. Can. J. Zool., N.R.C.C. Publ., Ottawa, *49*: 877–878.
4. Poirel C. and Dantzer R.: Psychophysiological Research on Circadian Rhythms in Epilepsy. Presse Med., in press.
5. Poirel C. (1968): Variations temporelles du comportement d'exploration chez la Souris. C.R. Soc. Biol., *162*: 2312–2316.
6. Dantzer R. et Poirel C. (1969): Action ménagée du phénobarbital sur l'épilepsie acoustique de la Souris. C.R. Soc. Biol., *163*: 1435–1439.

7. Poirel C. (1970): Recherches expérimentales sur les variations nycthémérales de la réactivité émotionnelle chez la Souris. Psychol. Fr., C.N.R.S. Ed., Paris, *15*: 3–14.
8. Richter C. P. (1960): Biological clocks in medicine and psychiatry. Proceedings of the National Academy of Sciences, *46*: 1506–1530.
9. Richter C. P. (1971): Inborn nature of the rat's 24-hour clock. J. Comp. Physiol. Psychol., *75*: 1–4.
10. Dechambre R. P. et Gosse C.: Rhythmes nycthéméraux chez les Rongeurs de laboratoire. Rev. Med. Exp. Pathol. Comp., in press.
11. Poirel C. (1968): Les rythmes biologiques en psychopathologie. Ann. Med. Psychol., *1*: 345–362.
12. Poirel C.: Etude comparative de la distribution circadienne de phénomènes psychopathologiques chez l'Homme et al Souris. Rev. Med. Exp. Pathol. Comp., in press.

CIRCADIAN CYCLIC SENSITIVITY TO GAMMA RADIATION AS AN UNCONDITIONED STIMULUS IN TASTE AVERSION CONDITIONING*

JOSEPH W. TERNES**

Department of Psychology, Florida State University Florida, U.S.A.

The first clear demonstration of the motivating effects of radiation was provided by GARCIA, KIMELDORF and KOELLING [1]. They added saccharin to water in order to make it discriminative to rats and then made this solution available to the animals during irradiation. The taste stimulus was then paired with water in order to show that a conditioned avoidance to saccharin solution could be elicited in the absence of the radiation stimulus in a post-irradiation preference test. The results of the study showed that the sham-irradiated rats maintained their pre-irradiation preference for saccharin while the animals in the 30-r exposure group decreased their saccharin consumption and the animals in the 57-r exposure group almost completely avoided the saccharin solution. The study also demonstrated that the strength of the conditioning, measured by both degree of the initial aversion as well as the resistance to extinction of the aversion, was dose-dependent. The authors went on to posit gastrointestinal disturbances, which are known to be disturbed during irradiation, as the physiological events which motivated the animal in this learning situation. Since the appearance of this study, the general effect of radiation as an unconditioned stimulus (US) for conditioning of a taste aversion has been extensively documented in a wide variety of conditioning experiments designed to study the parameters of the phenomenon [2–4].

According to KIMELDORF and HUNT [3], the incidence or the degree of the conditioned aversion (indicating the strength of the radiation-induced motivation) to a gustatory conditioned stimulus (CS) depends primarily on the total accumulated dose. However, no taste aversion studies have considered the possibility of circadian system phase differences in conditionability. The present experiment was designed to investigate this possibility. It studied the effects of total accumulated dose and the time of exposure. Total doses of 30-r and 60-r were used to condition an aversion to saccharin at three different times of day.

METHOD

Subjects

The subjects (*Ss*) were 90 Charles River CD strain male albino rats which were approximately 80 days old at the start of the experiment.

* This research was supported by United States Atomic Energy Commission Contracts AT-(40-1)-2903 and AT-(40-1)-2903 and AT-(40-1)-2690 (Division of Biology and Medicine).

** Present address: Department of Psychology, University of Puerto Rico, Rio Piedras, P. R.

Apparatus

The *Ss* were individually housed in Hoeltge HB-11A cages and allowed free access to Purina rat chow throughout the entire experiment. A General Electric time switch automatically controlled the light-dark cycle. The experimental isolation room was illuminated from 8:00 AM to 8:00 PM daily. The gamma irradiations were produced by a Model 150 C Gamma-beam Irradiator (Atomic Energy of Canada Limited) containing 1553 curies of CO^{60}. All Ss were irradiated in plexiglass boxes mounted on an electric motor driven carousel cycling in front of the cobalt source. The carousel was placed at a distance from the CO^{60} source such that the mean dose rate (air) measured inside a revolving chamber at the perimeter of the wheel was 2-r/min.

Procedure

All *Ss* were maintained for 2 weeks prior to irradiation under strict conditions of isolation, cycling illumination and limited maintenance disturbances.

The *Ss* were divided, according to their time of exposure, into six groups of 15 animals each. The time of irradiation was designated according to an arbitrary time scale which used the moment that the light came on each day as its reference point. Thus, the arbitrary synchronizer time (AST) 0000 corresponded to the exact time that the light came on. AST 0300 indicated that the light had been on for 3 hours, etc. The three times of radiation exposure were 0100, 0900, and 1700 hours AST. Two radiation doses, 30-r and 60-r, were given to two separate groups of rats during each exposure time.

In an attempt to minimize variability of results, four days of habituation to water deprivation and one sham exposure trial were given prior to irradiation. Ninety-six hours prior to irradiation the water bottles were removed from *Ss*' home cages at the predetermined time of eventual irradiation. Twenty-three hours and 30 minutes later the *Ss* were given access to a bottle of water for 30 minutes. The following days the same procedure was followed. The day before irradiation the *Ss* were watered at the appropriate time and then placed in plexiglas boxes, transported to the radiation source, and sham exposed. On the day that the *Ss* were irradiated, saccharin solution (0.1% by weight) was provided for 30 minutes instead of water. Immediately following the 30 minutes of access to the saccharin the *Ss* were transported to the radiation source where they were exposed to either 30-r or to 60-r. All *Ss* were then returned to their home cages and given *ad lib* food and water until the start of the preference test. Twenty-four hours after exposure to radiation the two bottle preference test between tap water and 0.1% saccharin solution was initiated. The difference in the weight of each drinking bottle and of its contents before and after a 24 hour period was used as an estimate of fluid consumption. In order to minimize the effect of bottle position on the consumption scores, the placement of the saccharin bottle was alternated each day after it had been refilled. Thus each consecutive left and right measurement pair were combined for each animal and only the 48 hour values were reported. The dependent variable was the "saccharin score" which is defined as the percent of the total fluid intake during the 48 hour period which was saccharin flavored water, i.e., saccharin intake in ml divided by total of saccharin and water intake in ml. The design for the experiment is summarized in Table 1.

Table 1 Experimental design.

		Time of exposure			
		0100	0900	1700	N
Dose in	30	15	15	15	45
Roentgen	60	15	15	15	45
	Total	30	30	30	90

Note: The entry in each cell is the number of *S*s.

RESULTS AND DISCUSSION

Fig. 1 shows the group mean saccharin scores plotted as a function of the time of their respective irradiations. These data demonstrate the tendency for the groups which were irradiated in the afternoon, 0900 hours AST, to be more sensitive to the effects of irradiation regardless of whether the dose was 30-r or 60-r.

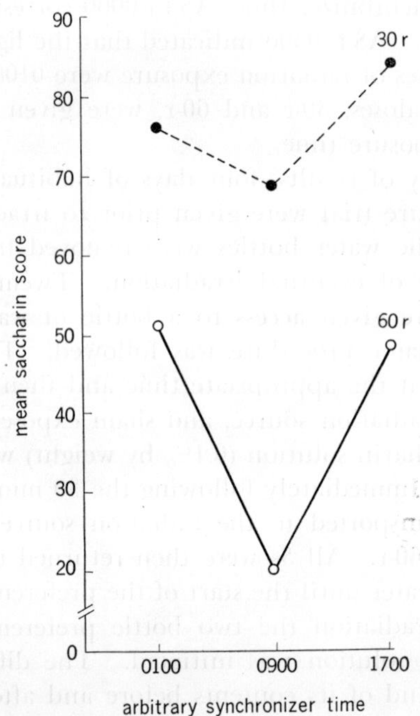

Fig. 1 Conditioned aversion to saccharin as a function of time of exposure to radiation.

This figure also shows the differential effectiveness of the two doses. The 60-r groups patently exhibited better conditioning. It should be noted, however, that only one group, the 60-r group irradiated at 0900 hours AST, exhibited a strong aversion to saccharin. A 2×3 analysis of variance [5] was performed on these data. Table 2 is an analysis of variance summary table of the results of the analysis. The main effects, (A) time of exposure (F=5.60 df=2/84, p<.01) and (B) radiation dose (F=36.43 df=1/84, p<.01) were both significant. The interaction of time of exposure by dose (A×B) was not significant.

Table 2 Analysis of variance.

Source	*df*	*MS*	*F*
Time of exposure (A)	2	22280	5.60**
Dose (B)	1	144881	36.43**
A×B	2	5738	
Error	84	3976	

** $p<.01$

These results indicate that the effectiveness of CO^{60} as an US varies as a function of the time of presentation if the organism to be conditioned is synchronized to a 12 hour light 12 hour dark illumination schedule. Such findings are consistent with recent studies documenting the existence of circadian rhythmic sensitivity both to whole body irradiation [6–8] and to the learning of emotional responses [9]. The present study demonstrates differential learning of a gustatory aversion as a function of the circadian system phase of the learner.

REFERENCES

1. Garcia J., Kimeldorf D. J. and Koelling R. A. (1955): Conditioned aversion to saccharin resulting from exposure to gamma radiation. Science, *122*: 157–158.
2. Garcia J., Kimeldolf D. J. and Hunt E. L. (1961): The use of ionizing radiation as a motivating stimulus. Psychol. Rev., *68*: 383–395.
3. Kimeldorf D. J. and Hunt E. L. (1965): Ionizing Radiation: Neural Function and Behavior. Academic Press, New York.
4. Tucker D. and Smith J. C. (1969): The chemical senses. Annual Rev. Psychol., *20*: 129–158.
5. Winer B. J. (1962): Statistical Principles in Experimental Design. McGraw-Hill, New York.
6. Pizzarello D. J., Witcofski R. L. and Lyons E. A. (1963): Variations in survival time after whole-body radiation at two times of day. Science, *139*: 349.
7. Pizzarello D. J., Isaak D. and Chus K. E. (1964): Circadian rhythmicity in the sensitivity of two strains of mice to whole-body radiation. Science, *145*: 286–291.
8. Darenskaya N. G. and Kuznetsova S. S. (1967): Diurnal fluctuations of the radiosensitivity of mice, AEC-TR-6891.
9. Stroebel C. F. (1967): Behavioral aspects of circadian rhythms. *In*: Comparative Psychopathology, (J. Zubin and H. F. Hunt, eds.), Grune and Stratton, New York.

RHYTHMS AND CHANGES IN ENVIRONMENTAL SCHEDULES

Chairman: RICHARD N. HARNER

Co-Chairman: JOHN N. MILLS

NEURAL CONTROL OF CIRCADIAN RHYTHMS

RICHARD N. HARNER

Department of Neurology, The Graduate Hospital
University of Pennsylvania
Philadelphia, Pennsylvania, U.S.A.

An important aspect of the study of biological rhythms is the investigation of responses produced in an organism by an alteration in its environmental schedule. Practical application of such investigations has been emphasized by the need to optimize human performance after travel across several time zones and during journeys into the polar night, under the sea, or into outer space where social, visual, atmospheric, and gravitational cues are reduced.

In this chapter, GERRITZEN demonstrates the utility of urinary measurements for studies of phase shifts during intercontinental flights and MILLS applies appropriate statistical techniques to separate individual rhythms in electrolyte excretion during an induced phase shift.

Many of the factors which might be implicated in an organism's response to an altered environmental schedule (Table 1) are considered by KLEIN in his studies of the effects of transmeridian flight on human performance. The importance of circadian placement of sleep in man is discussed by TAUB and BERGER. ANDREWS suggests a role for internal desynchronization in population control of the arctic lemming.

Table 1 Some factors to be considered in studies of human response to environmental alterations.

1.	Fatigue, related to prolonged activity
2.	Anxiety related to travel or study protocol
3.	Direct effects of the mode of phase displacement
4.	Sleep loss, qualitative or quantitative
5.	Number, variety, and quality of environmental phase signals
6.	Integrity of human phase receptor-transducer systems
7.	Reduction in total level of response
8.	Displacement of circadian acrophase of response
9.	Distorted internal relations among rhythms related to response

Studies of pineal ablation in sparrows provide BINKLEY with a model of disturbed receptor-transducer function in the central nervous system and QUAY has developed a mathematical means of describing the phase response in such systems.

Throughout, the potential role of the central nervous system in phasing circadian rhythms under different conditions cannot be overestimated, even though details and localization of receptors, transducers, transmitters, and metabolic controls remain obscure.

For example, localized infarction of the brain stem in man can produce marked reduction in the amplitude and quality of circadian rhythms of sleep and wakefulness. Of more general importance is the abolition of synchronized circadian rhythmic activity *outside* the central nervous system produced by infarction of the pons in one such patient whom we observed for several months [1]. The free-running period estimate for rectal temperature obtained over a period of 41 days (Table 2) is similar to that observed by SIFFRE in humans during prolonged subterranean isolation.

Table 2 Effect of chronic pontine infarction on circadian rhythms in a 72 year old woman.

Patterns of Sleep-Wakefulness	not present
Variance Reduction by Fitting 24-hr Cosine:	
Rectal Temperature (every 4 hr)	15%
EEG Frequency (every 10 min × 3 d)	5%
Serum Cortisol (every 4 hr × 3 d)	2%
Free-Running Period Estimate (41 d)	
Rectal Temperature (every 4 hr)	25.36 ± .72 hr

This study documents the critical role of the brain in phase-setting. Further studies of the localization of this function are badly needed.

In these days of relevant research it should be mentioned that there is a clinical disorder characterized by impaired thermal regulation in the elderly (spontaneous hypothermia) which is responsible for thousands of deaths annually in a country the size of Great Britain. A patient with this disorder who survived was studied for three weeks in our Clinical Research Center. Results in Table 3 were calculated from data taken at 4 hr intervals (EEG, 1 hr) 7 days after stabilization on a controlled lighting (L 0730–2330) and dietary schedule and again 7 days following a sudden and persistent 12 hr delay in the environmental schedule, beginning with the onset of the daily dark period. In this case, phase relations among rhythms are initially maintained, but the rate of phase response (P, P vs P) and the circadian contribution (A, VR) are less than expected for a younger, healthier individual (compare with Table 4). These findings suggest a central regulatory defect which is not limited to control of temperature.

Table 3 Effect of a 12-hour phase delay on circadian rhythms in a patient with spontaneous hypothermia.

	Level	Amplitude	Phase (hrs)	Variance reduction
		Temperature, Rectal		
Pre-shift	98.21	0.44	−18.39	46.7%
Post-shift	98.70	0.19	− 8.27	13.7%
		Plasma Cortisol		
Pre-shift	9.30	1.58	− 6.60	21.6%
Post-shift	9.59	1.25	−16.38	10.9%
		EEG Freq (R)		
Pre-shift	8.87	1.24	−15.41	28.0%
Post-shift	9.53	1.73	−23.69	32.7%
		EEG Freq (L)		
Pre-shift	8.32	1.45	−15.14	39.8%
Post-shift	9.12	1.52	− 0.76	37.1%

Finally, studies of phase shifting in man now suggest an additional parameter of physiological phase response which is likely to be controlled by the central nervous system. Review of the data in Table 3 reveals a potential ambiguity with respect to the *direction* of phase response because only two time points are presented, pre-shift and post-shift, without intervening data. Apparent phase

Table 4 Effect of a 12-hour phase delay on circadian rhythms in a patient with right temporal lobe seizures (M.B., 18 yr).

	Level	Amplitude	Phase (hrs)	Variance reduction
		Temperature		
Pre-shift	99.03	0.56	−16.37	88.5%
Post-shift	98.69	0.57	− 4.14	85.0%
		Paroxysmal Discharge/30 min		
Pre-shift	25.65	8.41	− 2.57	12.7%
Post-shift	32.73	10.16	− 2.95	68.2%
		Plasma Cortisol		
Pre-shift	13.08	4.01	− 8.38	64.7%
Post-shift	9.90	2.59	−19.08	28.0%

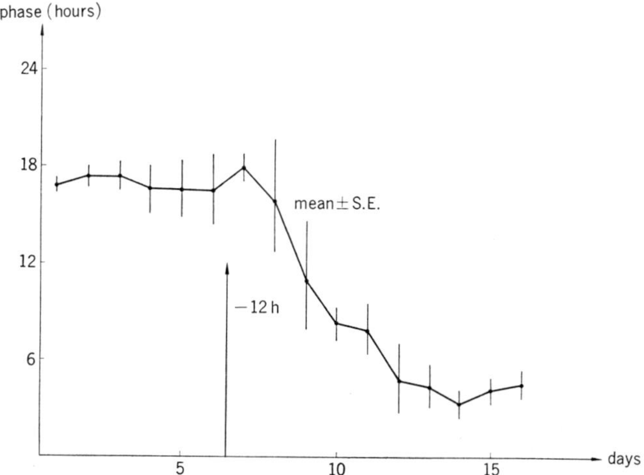

Fig. 1 Phase advance in circadian acrophase for rectal temperature in response to a 12 hr delay in evironmental timing. Patient M. B., male, age 18 yr, temporal lobe epilepsy.

advance from −18.39 hr to −08.27 hr in the case of rectal temperature is interesting but not statistically different from the expected but incomplete phase delays for plasma cortisol and EEG frequency.

We were stimulated to re-examine earlier data from an 18 years old male with temporal lobe epilepsy (and no apparent defect in thermal regulation) in whom a 12 hr phase delay was produced under similar experimental conditions and a full 12 hr phase shift in rectal temperature was found (Table 4). Fig. 1 shows the data obtained daily and analyzed by a least-squares curve-fitting technique over intervals of 3 days with increments of 1 day. There is a clear phase *advance* of rectal temperature in response to a phase delay in the environment.

The extensive data of H. Levine (this volume) and personal communications with F. Halberg and J. Mills suggest that this "antidromic phase response" may have been present but not emphasized in previous instances and should be looked for in future studies of phase shifting.

It is now abundantly clear that environmentalists and students of the nervous system alike must join with general biologists in the application of sensitive chronobiological methods in order to learn more about the neural control of body functions in a changing environment.

REFERENCE

1. Ostergren K., Runk L., McEvoy D. and Harner R. N. (1972): Reduction of circadian EEG, cortisol, and temperature rhythms in a patient with chronic pontine infarction. Psychophysiol., *2*: 134.

ADAPTATION OF CIRCADIAN RHYTHMS IN URINARY EXCRETIONS TO LOCAL TIME, AFTER RAPID AIR TRAVEL

F. GERRITZEN and TH. STRENGERS

Wassenaar, Holland

In order to study the problems that arise after global flights in East-West direction or *vice versa*, we flew with a different number of test persons:

	Number of test persons
from Amsterdam to New York and return	2
from London to Johannesburg and return	4
from Amsterdam to Anchorage and Tokyo	5
from Amsterdam to Anchorage and Tokyo	7

In all of these flights the test persons, healthy students, arrived at their destination with circadian rhythms determined by an environmental schedule synchronized initially by local time at the point of departure. This was different from local time at arrival in all cases except the North-South flight from London to Johannesburg and the return flight to London (Fig. 2).

The rhythm in urinary excretion was determined with the following method: the test persons were in the same recumbent position during the entire experiment. They got each hour the same amount of food—2 biscuits of known composition—and they drank 30 ml of water every hour during the first day, 40 ml of water during the second day and 50 ml of water until the end of the experiment. The circadian rhythm was recorded during night and day, before, during and after the flight.

Every hour they urinated, taking care to empty their bladder completely. The urine was measured individually and electrolytes (Cl, Na, K), creatinine, and corticosteroids were determined afterwards.

This procedure results in a very smooth and regular diuresis, where maximum and minimum of the circadian rhythm can easily be located by visual inspection of the average (Fig. 1). This was very valuable at a time when statistical treatment of rhythmic phenomena was not yet developed, as it is now by the efforts of HALBERG and SOLLBERGER. Still it remains to determine to what extent mathematical treatment can compensate for a lack of experimental conditions, a limited number of test persons or a restricted duration of the experiment.

In our North-South flights, without crossing of time zones, no discongruence was observed between the peak of urinary excretion and local time (Fig. 2).

In all of our Westbound flights the test persons arrived at their destination with circadian phase determined at their point of departure. Thus there existed a phase difference between their endogenous rhythms and the local time, a difference that is believed to be the cause of the symptoms of malaise or fatigue experienced after rapid flights in East-West direction or *vice versa* (Fig. 3).

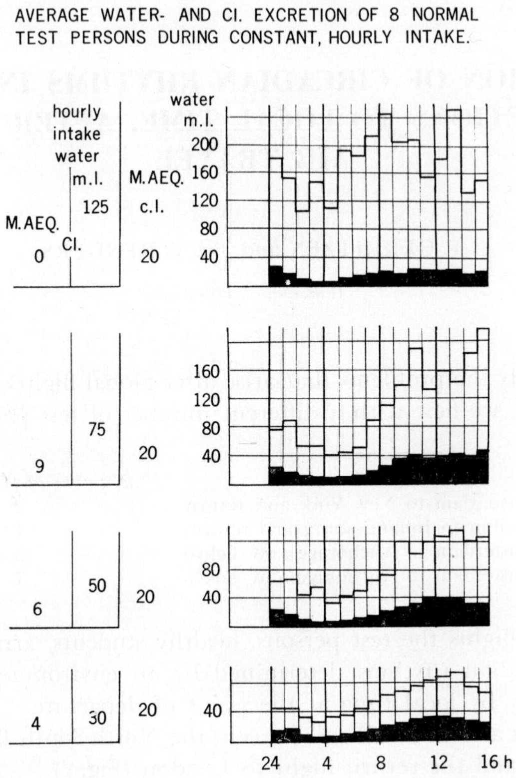

AVERAGE WATER- AND Cl. EXCRETION OF 8 NORMAL TEST PERSONS DURING CONSTANT, HOURLY INTAKE.

Fig. 1 The importance of restricting the hourly fluid intake when studying circadian rhythms in urinary excretion. Note the ill-defined maxima when the water intake is greater than 50 ml/hour.

I have done a few experiments on people who travelled by boat, and I found that this slow movement did not give rise to an obvious phase difference between the endogeneous rhythm and the local time. But when the displacement is faster, the difference becomes manifest and this problem will gain in importance when airplanes fly still faster in the future.

Having stated that there exists a discongruence at the point of arrival between the endogenous circadian rhythm of the test persons and local time, two questions arise:

1. How long does the adaptation to local time take?
2. How can we shorten the period of adaptation, thereby diminishing the symptoms of fatigue?

The first question is difficult to answer. We found quite substantial differences in the duration of the period of adaptation. After a flight from Amsterdam to New York 4 days were not enough and in our first flight to Anchorage it took a week, whereas after our second flight to Anchorage and Tokyo adaptation was complete in 3 days. There existed also individual differences.

We think there are three potential ways of shortening the period of adaptation: (a) prior phase alteration of the circadian rhythm, for example by an altered lighting schedule, (b) induction of natural or artificial sleep, and (c) use of pharmacological agents to speed adaptation of neuroendocrine systems.

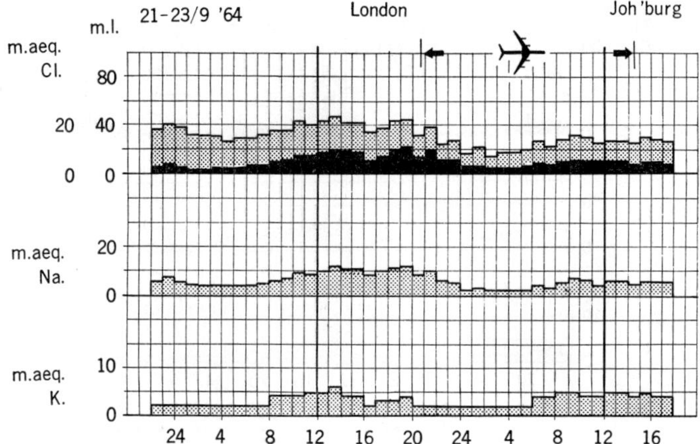

Fig. 2 Absence of the discongruence between the endogenous rhythm and local time, after a North-South flight and *vice versa*.

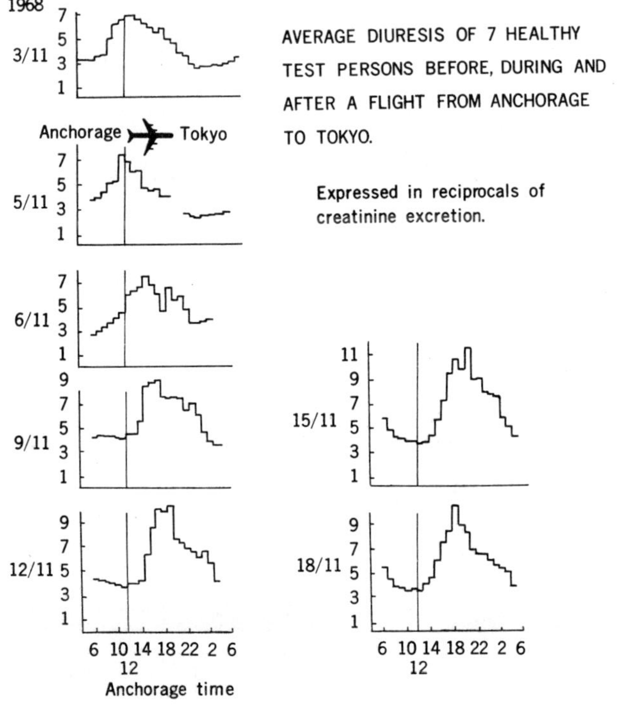

Fig. 3 Location of the maximum by visual inspection of the average diuresis of 7 test persons.

In the case of (a), we were able to invert the circadian rhythm of urinary excretion in one study (Figs. 4 and 5) but were unable to reproduce this finding a second time. Moreover, four other test persons exposed to inverse light did not adapt faster than three controls following intercontinental flight across several time zones.

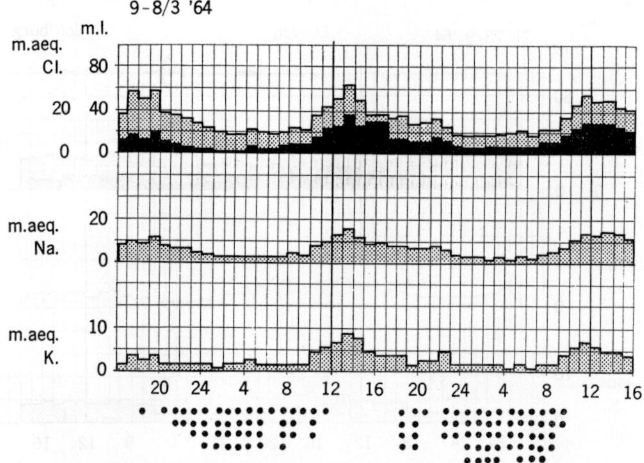

Fig. 4 Average diuresis, Cl^-, Na^+, and K^+ excretion of 5 healthy students under normal illumination: light from 0600–1800 and dark from 1800–0600. • = sleep.

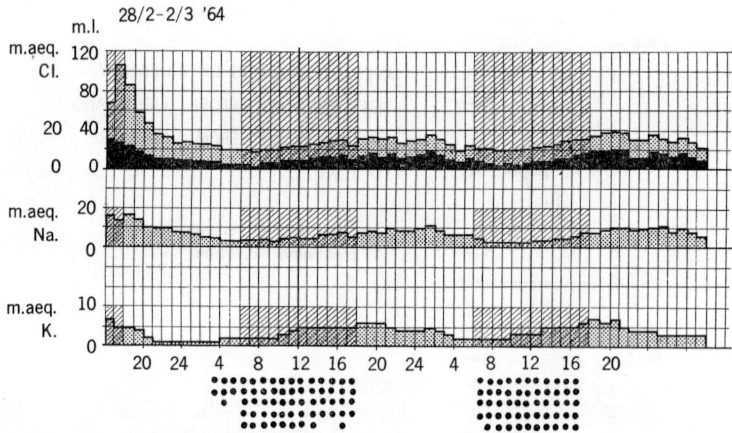

Fig. 5 Average diuresis, Cl^-, Na^+ and K^+ excretion of 5 healthy students under inverse illumination: light from 1800–0600 and dark from 0600–1800. • = sleep.

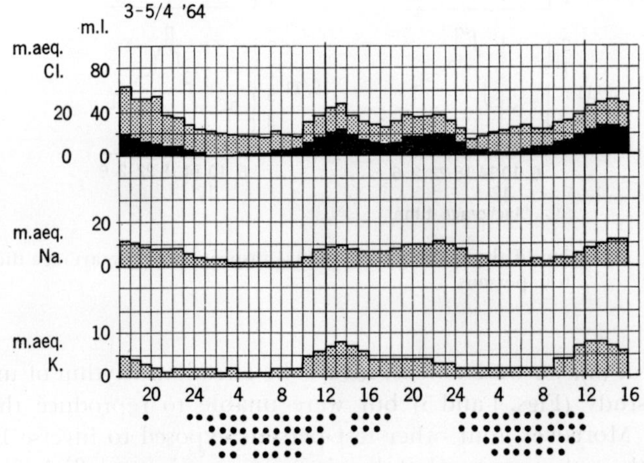

Fig. 6 Average diuresis, Cl^-, Na^+ and K^+ excretion of 4 healthy students under continuous illumination. • = sleep.

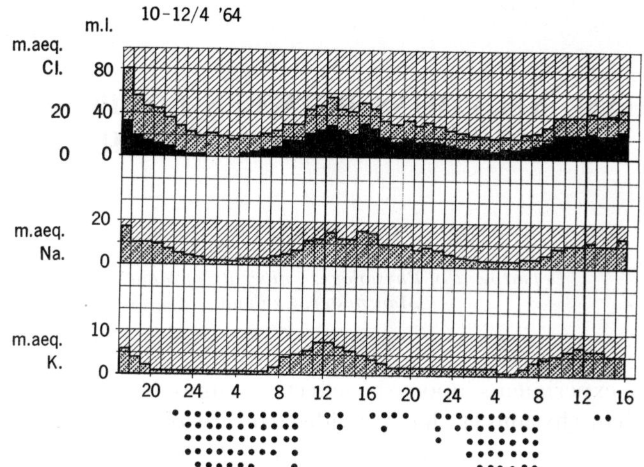

Fig. 7 Average diuresis, Cl^-, Na^+ and K^+ excretion of 5 healthy students in continuous dark. • = sleep.

With respect to (b), sleep was not an independent variable in our experiments but was observed to coincide regularly with the time of minimum diuresis even during constant light or constant darkness (Figs. 6 and 7). We do not know whether an enforced sleep schedule would have speeded adaptation in our subjects.

The pharmacological possibilities mentioned in (c) are many but remain speculative at this time.

At the end of these short remarks, I would like the reader to recall that the basic causes of the urinary excretory rhythm, peaking at the same physiological time (about 1300–1400) in persons all around the world, remain unknown.

PHASE RELATIONS BETWEEN COMPONENTS OF HUMAN CIRCADIAN RHYTHMS

JOHN N. MILLS

Department of Physiology, University of Manchester
Manchester, England

A series of experiments upon the effect of simulated time zone shifts upon human circadian rhythms [1] has provided control observations from which circadian maps may be drawn. Groups of three to five subjects spent around 10 days in an isolation unit from which obvious Zeitgebers such as light and external noise were excluded. They spent two or three control days on a regular routine of sleep, meals and activity, timed by a clock which was then advanced or retarded by 6 or 8 hours to simulate a time zone shift. Among other physiological variables, oral temperature and urinary excretion rate of electrolytes were measured. Sine curves were fitted to each successive nychthemeron, using a conventional least-squares method for the temperature data, and the method of FORT and MILLS [2] for the urinary excretion. Significance was assessed from the ratio of variance due to the fitted sine to residual variance, and 161 cycles whose fit fell within 95% confidence limits were analysed; 97 fell within the 99% confidence limit.

Table 1 Acrophase of oral temperature and renal excretion of potassium, sodium and chloride in subjects living on normal time in isolation unit.

	Oral temp.	Renal excretion		
		K^+	Na^+	Cl^-
No. of subjects	14	21	21	22
No. of nychthemera	27	44	43	47
Mean time of acrophase [clock hours]	1815	1518	1756	1620
Variance between subjects [h^2]	0.240	3.850	6.537	6.125
Variance within subjects [h^2]	1.915	1.307	1.903	2.188
S. D. Within subjects [h]	1.38	1.14	1.38	1.48

The procedure of computing the acrophase separately for each nychthemeron permits a comparison of the variability in any one subject with that between subjects; analysis of variance (Table 1) shows that for all four variables studied the between-subject variance was significantly greater ($P<0.01$) than the variance within subjects; thus each subject has his individual acrophase, even when they live together in a closed community. Table 1 also presents the overall means for the acrophases, with the standard deviations of the values for individual subjects. The means were closely enough grouped to obviate any difficulty over calculating the mean of a circular measure. The standard deviations arise from random uncontrolled influences, analytical error, and statistical error in computing the acrophase.

The earliest sodium acrophase was 1352 hr and the latest 2225 hr, and the acrophases of other components were also widely dispersed; this scatter permits consideration of how far they are associated, and whether an acrophase may be better defined by reference to another component than to clock time. There is a very strong relationship between sodium and chloride, although the acrophase for sodium is over an hour later than that for chloride. The mean difference is 1.10–0.12 h (S.E. of mean of 38 values); and the variance of the external chloride acrophase is seven times that of its internal acrophase with reference to sodium. This relationship can also be shown graphically (Fig. 1), and indicated by the correlation coefficient of 0.924 between sodium and chloride acrophases. This close association is not merely a consequence of the existence of 'larks' and 'owls', with all their rhythms early or late; the 'within subject' variance of the external chloride acrophase is still 3.7 times that of the internal acrophase referred to sodium.

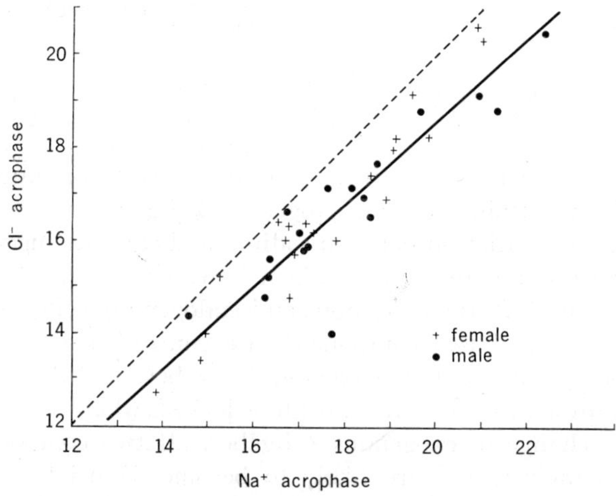

Fig. 1. Chloride and sodium excretory acrophases in subjects living on normal time in isolation unit; each point represents a separate nychthemeron. Continuous line; regression of chloride upon sodium. Broken line: identical acrophases.

There is a similar but much weaker relationship between the sodium and potassium acrophases (Fig. 2). The potassium acrophase was earlier, with a mean difference of 2.59±0.34 h (34), and the correlation is much less (r=0.483); the total variance of the external sodium acrophase is much the same as the variance of its internal acrophase referred to potassium, and the 'within subject' variance is considerably less. Temperature acrophase was later than those of electrolytes: the mean time difference between temperature and potassium acrophases was 4.17±0.35 h, but the temperature acrophase was not correlated with those of urinary electrolytes.

Similar analyses have been performed upon temperature and urinary thythms of a man who lived alone for a year, recording temperature rectally and collecting urine over a series of separate 32-hour periods. Though the external acrophases varied widely over the course of the year, the internal acrophases were similar to those of subjects in the isolation unit. Chloride led sodium by a mean of 1.47±0.33 h (9), potassium led sodium by a mean of 2.05±0.55 h (8). Temperature followed potassium by a mean of 0.85±0.65 h (7).

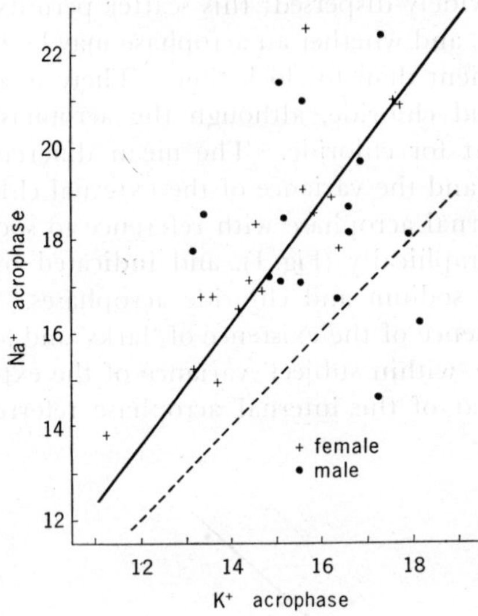

Fig. 2 Sodium and potassium excretory acrophases in subjects living on normal time in isolation unit; each point represents a separate nychthemeron. Continuous line: regression of sodium upon potassium. Broken line: identical acrophases.

On a usual nychthemeral routine, even in the isolation unit, physiological rhythms may be determined by an exogenous Zeitgeber or by an endogenous clock, and the close association between sodium and chloride suggests merely that the same influence dominates both. When however the subjects experienced a simulated time-zone shift, the exogenous and endogenous Zeitgebers were in conflict. If the rhythms of two components are controlled entirely by the same endogenous clock, or by a shifted exogenous Zeitgeber, or if they share a common final effector pathway, they will remain phase-locked, whatever the magnitude or direction of the change in exogenous Zeitgeber. If two components are differently controlled their phases are likely to become dissociated at first, though eventually the previous phase relation may be restored. The variance of the phase differences has therefore been calculated for all nychthemera where two components oscillated sinusoidally after simulated time-zone shift (Table 2). All the phase differences became more variable; but although the increase in variance was significant, the variance of the sodium–chloride difference—the internal acrophase of chloride in reference to sodium—was still small, indicating that their control is closely linked, whilst the linkage between the other components is much looser.

Fig. 3 shows the acrophases of potassium and sodium excretion in a subject after a simulated eastward flight; the sodium rhythm adapted almost immediately, while the potassium rhythm adapted slowly over 4 days. Fig. 4 shows the temperature and potassium acrophases in a subject after a simulated westward

Table 2 Variance of internal acrophase [phase difference] between pairs of constituents in subjects in isolation unit, before and after simulated time zone shift. All values in h^2.

	Control	After time shift
Na:Cl	.586	1.347
Na:K	3.808	7.346
Temperature:K	5.244	16.828

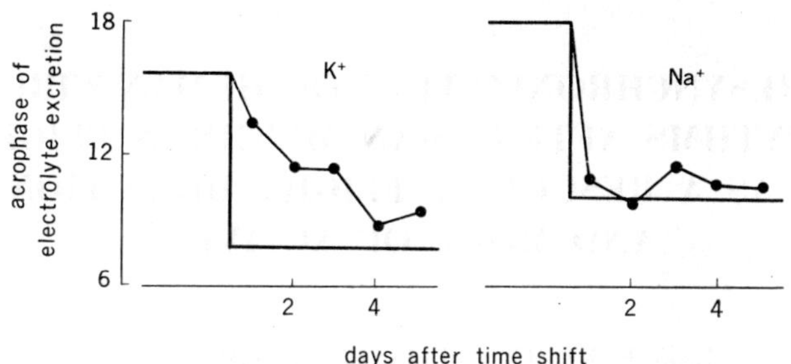

Fig. 3 Acrophases of urinary potassium and sodium in subject in isolation, before and after simulated 8-hour eastward time shift. Horizontal lines indicate the acrophase of a single sine curve fitted to three control days, and the expected shift if the rhythms adapted immediately to experimental time. ●——● acrophases of single nychthemera after time shift.

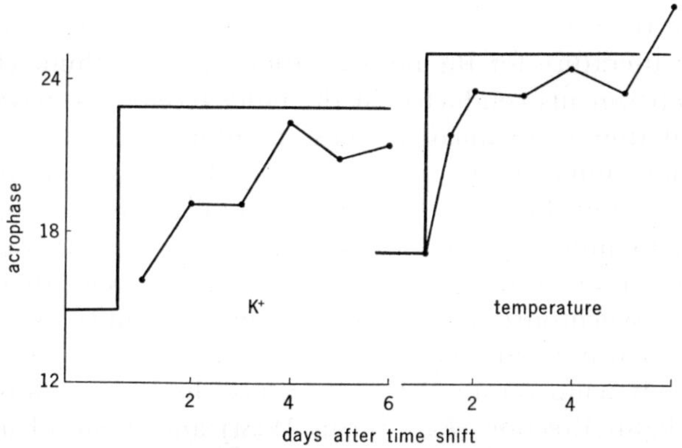

Fig. 4 Acrophases of urinary potassium and body temperature in subject in isolation, before and after simulated 8-hour westward time shift. Conventions as in Fig. 3.

flight; here also potassium took about 4 days to adapt, and temperature adapted immediately. The lack of any phase locking between urinary potassium and etiher urinary sodium or body temperature is clearly shown.

ACKNOWLEDGMENT

I am indebted to my collaborators Ann Fort, D. S. Minors and J. M. Waterhouse, to Beryl Wilson and A. Onuegbu for technical assistance, to the Nuffield Foundation for the construction of the isolation unit, and to the Medical Research Council for an equipment and expenses grant.

REFERENCES

1. Elliott Ann L., Mills J. N., Minors D. S. and Waterhouse J. M. (1971): The effect of real and simulated time zone shifts upon the circadian rhythms of body temperature, plasma 11-hydroxycorticosteroids, and renal excretion in human subjects. J. Physiol., *221*: 227–257.
2. Fort Ann and Mills J. N. (1970): Fitting sine curves to 24-h urinary data. Nature, *226*: 657–658.

THE RESYNCHRONIZATION OF HUMAN CIRCADIAN RHYTHMS AFTER TRANSMERIDIAN FLIGHTS AS A RESULT OF FLIGHT DIRECTION AND MODE OF ACTIVITY*

Karl E. Klein and Hans-Martin Wegmann

Institut für Flugmedizin (DFVLR)
Bonn-Bad Godesberg, Germany

Translongitudinal air travel is followed by an instant desynchronization of human circadian rhythm and periodic time cues in the environment [1–10]. The abnormal phase relationship involves physiological and psychological processes. Times given in literature for the normalization of dysrhythmic changes are discordant, and experiments conducted in the more recent past have also failed to reveal consistent trends in human response pattern.

Flight direction, number of time zones crossed, specific nature of biological functions, stress of travelling and the mode of post-flight activity are factors which have been said to influence intensity and duration of dysrhythmic symptoms. However, so far we are still lacking data showing the exact time course of circadian rhythms resynchronization under the different conditions. Therefor, from the data of translantic flight studies we have computed the postflight course of adaptation of body temperature and performance rhythm as it was observed with differences in flight direction (East versus West) and mode of activity (indoor versus outdoor).

METHOD

The data used were obtained from 2 groups of 8 students which were flown independently of each other between Germany and the U.S.A.; the flights resulting in a shift in local time of 6 hours. Both groups were tested on three days before the flights and on day 1, 3, 5, 8, and 13 (only in one group) after the outgoing and the return flight. The tests were repeated every three hours over a period of 24 hours from 0900 to 0900; between the 2400 and the 0900 test session lights were turned down and subjects were encouraged to sleep. Tests included urine collection, body temperature measurements (in one group only) and evaluation of performance; one test session fasted 40 minutes. Air travel was by civil airlines and followed their regular schedule. The sojourn abroad was 18 days. The data were analysed with the periodic regression technique and subsequent analysis of variance developed by Bliss [11]. To the derived set of phase angles for one post-flight period, a quadratic equation was fitted. For the expression of phase angle radians in clock hours, the reference is 0900, i.e., the mean of time of the first test session in the morning.

* Part of the research reported in this paper was sponsored by the Aerospace Medical Research Laboratory, Aerospace Medical Division, Air Force Systems Command, Wright Patterson AFB, Ohio, under Contract No F 33615-70-C-1598 with Wright State University, Dayton, Ohio.

RESULTS

For rectal temperature Fig. 1 presents the postflight changes observed in the 24-hour mean value. The 24-hours mean drops slightly on each of the first postflight days; it has returned to the level of the preflight oscillation by the third day. In opposition to the very transitory response of the 24-hour mean value it takes about 11–12 days after the westbound flight and approximately 15 days after travelling in the opposite direction for the phase of the temperature rhythm to resynchronize with local time again (Fig. 2).

Similar to temperature is the response pattern of some performance phenomena. Again there is an immediate slight depression in the 24-hour mean values after the flight and a rapid normalization in the first days (Fig. 3). Though there seems to be a slight preponderance in the changes following the eastbound travel the differences in connection with flight direction are not very clear in the 24-hour mean. However, a difference is obvious again with the phase of circadian performance rhythm (Fig. 4). If compared with the preflight norm, postflight changes in the phase of the rhythm were generally more marked after the advance shift (West-East travel); this was most pronounced in psychomotor performance.

In addition to the difference in phase adjustment due to the travel direction there is a difference demonstrable in connection with different postflight activities of the subjects. Under otherwise very similar conditions, in one experiment the subjects were kept inside of the test facility for 7 days after the flights in relative isolation from the environment ("indoor activity"); in another study the subjects were allowed to leave the test facility for outdoor activity on those postflight days—2, 4, 6, and 7—when tests were not performed ("in- and outdoor activity").

For the averaged raw data of psychomotor performance, in Fig. 5 the difference in the rhythms postflight response pattern is demonstrated.—In Fig. 6 the phase difference between pre- and postflight performance rhythm is presented; it is clearly discernible that:

(1) The phase angle difference between biological rhythm and local time is almost identical in the two groups on the first postflight day, and
(2) Later on, the speed of phase adjustment is increasingly higher in the group with the intermittent outdoor activity.

Figures for the speed of phase adjustment of the psychomotor performance rhythm are presented in Table 1; they are listed in h/day as "average speed" of phase shift up to the particular day they are shown for. On the first postflight day phase shift was faster by 70–77% following westbound travel than after the flights in the opposite direction. This marked difference decreases as resynchronization progresses, and by the time the phase shift was completed in both directions, there is only 9.1% or 8.6% left in favour of the westbound travel direction.

The influence of the type of "activity" points in the other direction. From practically no difference on the first postflight day there is an increasing difference in phase shift speed in favour of the group which intermittently was allowed outdoor activity. By the time the biological rhythm in both groups has completely synchronized with local time, the group with outdoor activity has accomplished synchronization in half the time (53% higher average speed).

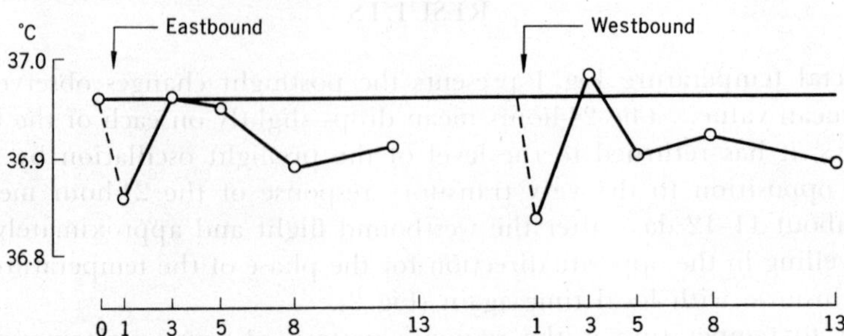

Fig. 1 The 24-hour mean values of rectal temperature circadian rhythm after transmeridian flights. (Time Difference: 6 hours; N = 8)

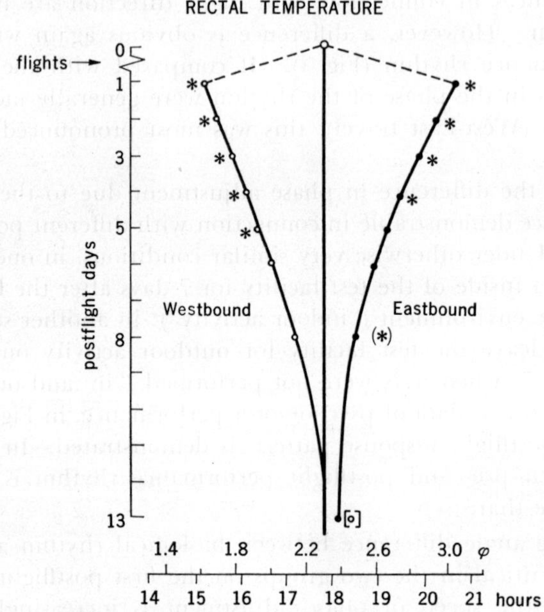

Fig. 2 The phase of rectal temperature circadian rhythm after transmeridian flights. (*), 0.1>P>0.05; *, 0.05>P>0.01; **, P<0.01. (For further explanation: see Fig. 1)

DISCUSSION AND CONCLUSIONS

There is a pronounced difference in the time-course of normalization of phase and 24-hour mean of human circadian rhythms after transmeridian flights. Though the reason is not clear, there are indications that the mechanism, which provokes a general depression of the 24-hour mean is different from that which induces the changes of phase. The depression of the mean seems related to the time in transit rather than to the number of time zones crossed [9], it is often more pronounced after the eastbound night flight than after the day flight in westbound direction, and it normalizes much faster than phase [8]. Finally, is there no decrement in the 24-hour mean when the phase relationship of physio-

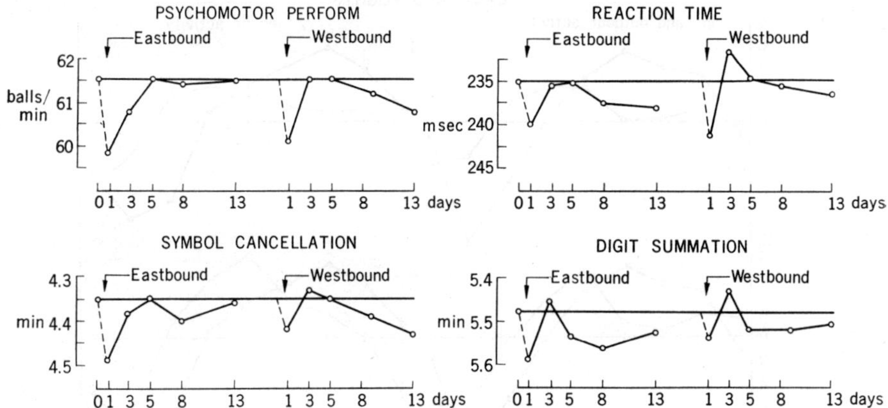

Fig. 3 The 24-hour mean value of performance circadian rhythm after transmeridian flights. (For further explation: see Fig. 2)

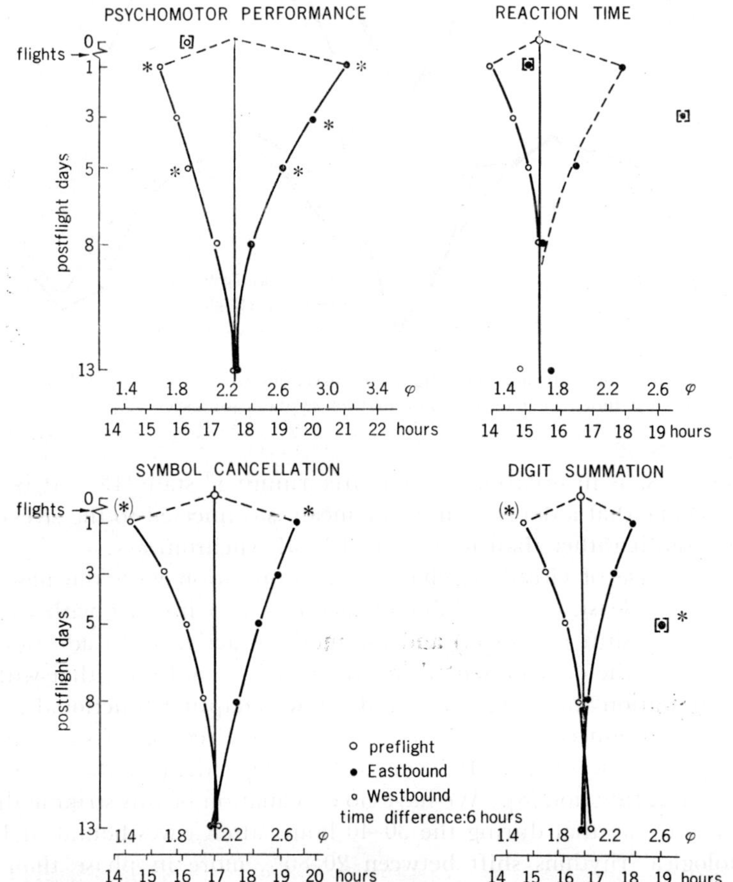

Fig. 4 The phase of human performance circadian rhythm after transmeridian flights. (For further explanation: see Fig. 2)

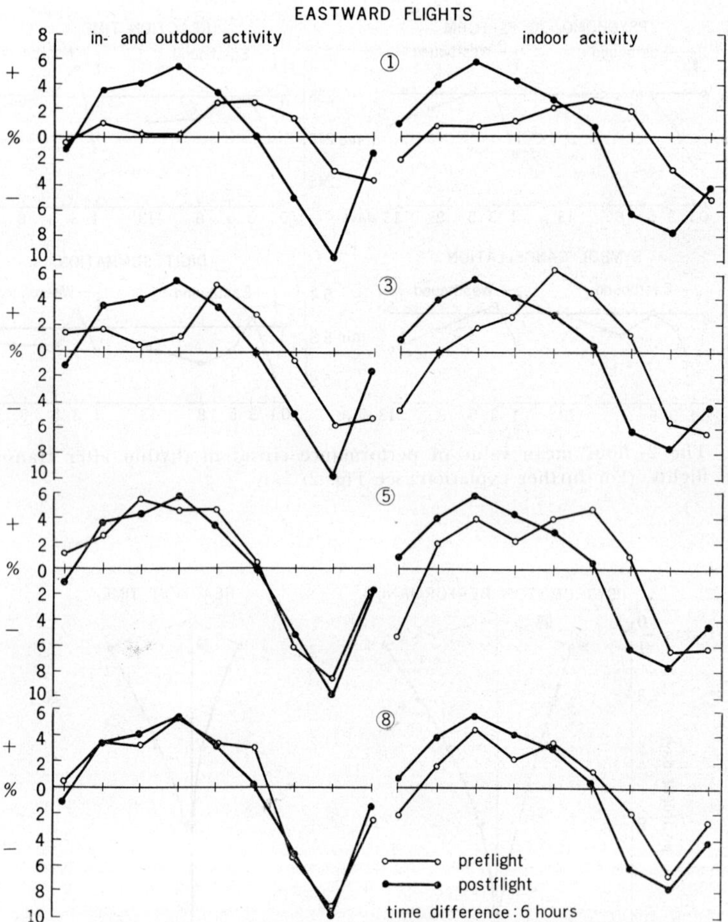

Fig. 5 The effect of transmeridian flights on psychomotor performance rhythm: Difference with mode of activity. (For further explanation: see Fig. 1)

logical functions, as in isolation, is in a "free running" state [13]. It is therefore logical to assume that a depression of the mean is connected to the stress and loss of sleep in transit rather than to a shift in local synchronizers.

The time course of circadian phase resynchronization with the new environment depends at least, on two different factors which interact with each other: The direction of shift (or travel) and the mode of activity. (In addition it must be assumed from the data presented in literature [2, 7, 8, 13, 14] that with respect to resynchronization there are "fast" and "slow" adapting functional systems.)

The most pronounced difference in the effect of direction of shift on phase is seen shortly after the flights. It seems to be independent of the specific nature of the biological function [8]. We have no explanation of this striking difference, which means to say that during the 30–40 hours after a westbound flight (phase delay) biological rhythms shift between 20–80% more in phase than they do when the flight is eastbound (phase advance).

Later on the daily shift often becomes more rapid after an advance shift. This is in concordance with the idea that the periodic agents in the environment which

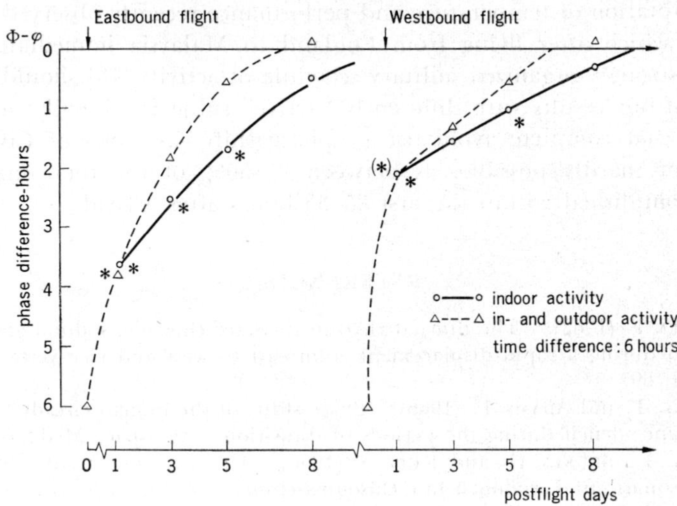

Fig. 6 Phase shift of psychomotor performance rhythm after transmeridian flights: Difference with mode of activity. (For further explanation: see Fig. 2)

Table 1 The influence of flight direction and mode of activity on the shift in phase of psychomotor performance rhythm after transmeridian flights (Time difference: 6 hours).

Postflight Days	Indoor activity West h/day	Indoor activity East h/day	$\left(\frac{W-E}{E}\right)\times 100$	In- and outdoor activity West h/day	In- and outdoor activity East h/day	$\left(\frac{W-F}{E}\right)\times 100$	West and East Out-door h/day	West and East In-door h/day	$\left(\frac{O-I}{I}\right)\times 100$
1	3.90	2.30	+70	3.90	2.20	+77	3.05	3.10	− 1.6
3	1.47	1.17	+26	1.57	1.37	+15	1.47	1.32	+11.4
5	0.96	0.84	+14	1.08	1.10	− 1.8	1.09	1.90	+21.1
8	0.70	0.67	+ 4.5	—	—	—	—	—	—
At Completion of Phase Shift	0.06	0.55	+ 9.1	0.93	0.85	+ 8.6	0.89	0.58	+53.4

make the biological rhythm shift are the more potent and effective, the greater the difference in phase between rhythm and external synchronizer [8]. In the mathematical description of the phase shift (i.e. after fitting a quadratic equation to the derived set of phase angles) the exponential term is consequently more pronounced when adaptation is being accomplished to an advance shift, or a West-East flight [9].

From all that is known about the influence of time cues on biological phase shifting, a difference in the speed of adaptation between the groups with different activities could be expected. Of course, it may not be the type of activity which causes the difference, but the fact that the group which was allowed to intermittently leave the relatively isolated atmosphere of the experimental facility was more intensively exposed to a greater number of time cues from the new environment. Since both groups were aware of the local time as well as the light-dark shift, the difference of 50% in the resynchronization time emphasizes again the role of social synchronizers in man [15]. We are convinced that by making the difference in the experimental conditions more extreme, the difference in the phase adjustment will become even more pronounced. The obviously "instant" diurnal adaptation of temperature and performance recently observed in a group of soldiers which after flying from England to Malaysia immediately was exposed to a strongly organized military schedule of activity [16] should be seen in the lights of our results with differently "active" subjects. Finally, it should be mentioned that complete avoidance of phase shift—as crews of Civil Airlines prefer—seems hardly possible, as between 25–50% of the total shift is often already accomplished within the first 25–35 hours after arrival.

REFERENCES

1. GERRITZEN F. (1962): The diurnal rhythm in water chloride, sodium and potassium excretion during a rapid displacement from east to west and vice versa. Aerospace Med., *33*: 697.
2. HAUTY G. T. and ADAMS T. (1966): Phase shifts of the human circadian system and performance deficit during the periods of transition. Aerospace Med., *37*: 668, 1027.
3. KLEIN K. E., BRÜNER H. and RUFF S. (1966): Untersuchungen zur Belastung des Bordpersonals auf Fernflügen mit Düsenmaschinen. Z. Flugwissenschaften, *14*: 109.
4. HAUS E., HALBERG F., NELSON W. and HILLMAN D. W. (1968): Shifts and drifts in phase of human circadian system following intercontinental flights and isolation. Fed. Proc., *27*: 224.
5. CHAPEK A. U. (1969): Circadian rhythm of physiological functions in flight personnel. Space Biol. Med., *3*: 30; Translation (1969): IPRS *48416*: 45.
6. KLEIN K. E., BRÜNER H., HOLTMANN H., REHME H., STOLZE J., STEINHOFF W. D. and WEGMANN H. M. (1970): Circadian rhythm of pilots' efficiency and effects of multiple time zone travel. Aerospace Med., *41*: 125.
7. WEGMANN H. M., BRÜNER H., JOVY D., KLEIN K. E., MARBARGER J. P. and RIMPLER A. (1970): Effects of transmeridian flights on the diurnal excretion pattern of 17-hydroxycorticosteroids. Aerospace Med., *41*: 1003.
8. KLEIN K. E., BRÜNER H., GÜNTHER E. JOVY D., MERTENS J., RIMPLER A. and WEGMANN H. M. (1971): Psychological and physiological changes caused by desynchronization following transzonal air travel. *In*: Aspect of Human Efficiency: The diurnal rhythm and loss of sleep, (W. P. COLQUHOUN, ed.), English Universities Press, London.
9. KLEIN K. E., WEGMANN H. M. and HUNT B. J.: Desynchronization of body temperature and performance circadian rhythm as a result of outgoing and homegoing transmeridian flights. Aerospace Med., (in press).
10. CHRISTIE G. A., MOORE-ROBINSON M., GULLET C. C. GULLET and BERGIN A.: Project Pegasus. Proc. Symp.: Circadian rhythm and new aspects of corticosteroids. Clinical Trials Journal 1971, (in press).
11. BLISS C. J. (1958): Periodic regression in biology and climatology. Agric. Exp. Station Bull. 615, New Haven.
12. SCHAEFER K. E., CLEGG B. R., CAREY C. R., DOUGHERTY J. H. and WEYBREW B. B. (1967): Effects of isolation in a constant environment on periodicity of physiological functions and performance levels. Aerospace Med., *38*: 1002.
13. ASCHOFF J. (1969): Desynchronization and resynchronization of human circadian rhythms. Aerospace Med., *40*: 844.
14. GERRITZEN F., STRENGERS T. and ESSER S. (1969) : Studies on the influence of fast transportation on the circadian excretion pattern of the kidney in humans. Aerospace Med., *40*: 264.
15. ASCHOFF J., FANTRANSKA M., GIEDKE H., DOERR P., STAMM D. and WISSER H. (1971): Human circadian rhythms in continuous darkness: Extrainment by social cues. Science, *171*: 213.
16. COLQUHOUN W. P. and ADAMS J. M.: Paper read at NATO-Symposiums on effects of diurnal rhythm and loss of sleep on human performance. Strasbourg, July 1970.

EFFECTS OF ACUTE SHIFTS IN CIRCADIAN RHYTHMS OF SLEEP AND WAKEFULNESS ON PERFORMANCE AND MOOD*

JOHN M. TAUB and RALPH J. BERGER

Department of Psychology, University of California
Santa Cruz, California, U.S.A.

The extent to which impaired performance resulting from changes in sleep length might be due to the disruption of an established biological rhythm has not been systematically investigated.

TAUB, GLOBUS, PHOEBUS and DRURY [1] found significant decrements in performance following nights of extended sleep compared with when subjects had slept for their usual 8 hr. Through the application of sufficiently sensitive behavioral measures it was only recently that definitive performance decrements were reported for partial sleep deprivation [2, 3]. These findings indicate that optimal sleep length falls within narrow limits, with lowered efficiency resulting from either excess sleep or sleep loss.

There is also evidence of performance deficits associated with phase-shifts in the sleep-wakefulness cycle resulting from such conditions as shift-work and multiple time zone travel [4–6]. Thus, the possibility must be considered that the detrimental effects of lengthening or shortening sleep might be partially or even wholly due to the displacement of habitual circadian sleep time, since when sleep is lengthened or shortened, the subject is going to bed or getting up earlier or later than usual. The purpose of the present investigation was to study the effects on performance and mood of shifting the time when persons were allotted their habitual 7–8 hr of sleep.

METHOD

Subjects

Ten male college students were selected on the basis of showing no indication of sleep disturbance, medical or psychiatric disorder or of being frequent users of alcohol or drugs and, of habitually retiring at 0000–0030 and obtaining 7–8 hr of sleep nightly.

Procedure

During the experiment the subjects were instructed not to nap, or to drink coffee or tea after midday, but to maintain their usual level of physical activity. They were told the the purpose of the experiment was to study the relationship between sleep and personality.

* Supported by PHS Research Grant No. 1R01 MH 18928-01A1 from the National Institute of Mental Health and a National Institutes of Health Fellowship MH 47945-01 from the Behavioral Sciences Training Branch.

An adaptation night from 0000–0800 preceded a 0000–0800 condition of habitual sleep and four conditions of shifted sleep. These consisted of 2000–0400 and 2200–0600 (early shift) conditions, and 0200–1000 and 0400–1200 (late shift) conditions spaced one week apart and systematically counter-balanced such that each of five pairs of subjects received the five experimental treatments in a different sequence.

The subjects sat in bed engaged in quiet, leisurely activities with all external time cues eliminated until they were asked to sleep. Upon awakening the subjects were served 6 oz of fruit juice as a partial nutritional control for blood sugar level.

A 5 min version of the WILLIAMS and LUBIN [7] experimenter-paced calculation task was given 30 min after the subjects awakened. This was followed by a 30 min vigilance task [2]; the measure of performance used was the total number of signals missed (misses) and incorrectly detected (false reports).

Subjects then completed a 57-item forced-choice adjective check list (ACL) which was used to assess mood. It was scored for scales of activity, anxiety, cheerfulness, concentration-confusion, depression, fatigue and friendliness each consisting of 5–8 adjectives. A group of 5 miscellaneous items served as a response set measure. The vigilance task and mood measure were administered again at 1230 and 1730.

EEG, electromyographic, and electrooculographic activity were recorded according to standard procedures [8]. Differences in performance due to sleep conditions was examined for each task by two-factor analyses of variance with repeated measures on one factor, and the Tukey test was used to compare specific means [9]. The Wilcoxon test [10] was used to compare mood from the ACL data between treatments and time of day.

RESULTS

Combined number of misses and false reports on the vigilance task compiled over the three testing sessions were significantly less after the 0000–0800 condition compared to the conditions of shifted sleep ($p<.01$). There were significantly fewer errors in the final minute of the calculation task in the 0000–0800 condition compared to the late-shift conditions ($p<.01$) and early-shift conditions ($p<.05$) (Fig. 1). Scores on the friendliness mood factor were significantly lower in the morning ($p<.025$) for the early-shift conditions compared to the 0000–0800 condition; and for the late-shift conditions compared to the 0000–0800 condition morning scores for anxiety were significantly higher ($p<.05$), activity mood scores were significantly lower ($p<.05$), and there were significantly higher fatigue scores at 1730 ($p<.05$) (Fig. 2). Differences between conditions for the response set measure were not significant indicating the absence of artifacts in subjects' responses on the ACL.

An approximation of total hours spent asleep indicated no systematic differences between conditions (2000–0400 $\overline{M}=7.59$ hr, S.D.$=26$; 2200–0600 $\overline{M}=7.55$ hr, S.D.$=.24$; 0900–0800 $\overline{M}=7.47$ hr, S.D.$=.24$; 0200–1000$\overline{M}=7.7$ hr, S.D.$=.43$; 0400–1200 $\overline{M}=7.63$ hr, S.D.$=.30$)

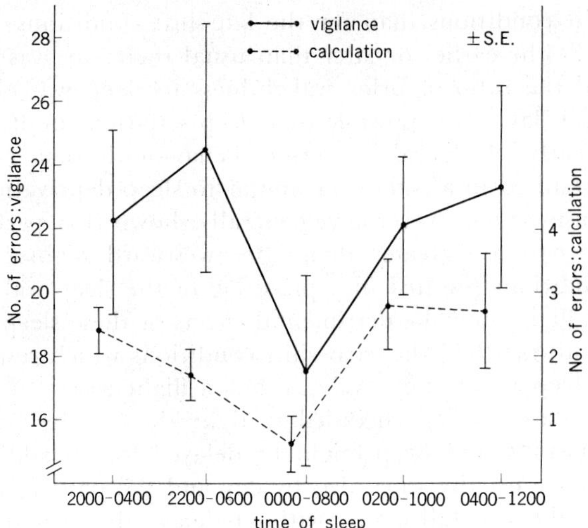

Fig. 1 Errors on the calculation and vigilance tasks after habitual sleep, from 0000–0800, and after conditions of shifted sleep.

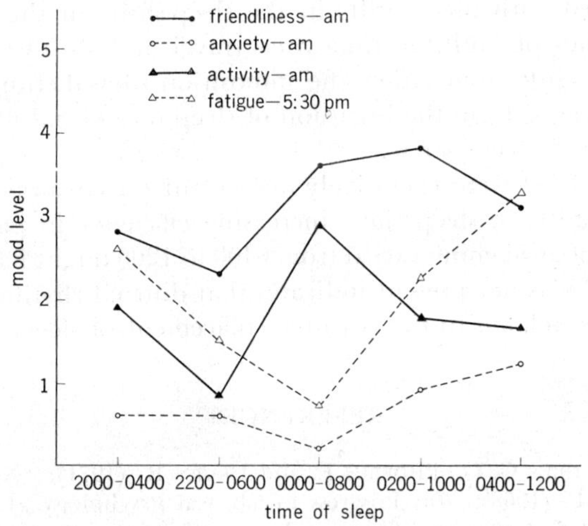

Fig. 2 Self-ratings of mood on the adjective check list after habitual sleep, from 0000–0800, and after conditions of shifted sleep.

DISCUSSION

These results indicate that altering the time in the circadian rhythm when habitual sleep is taken impairs performance on signal detection and calculation tasks, and is deterimental to subjectively assessed mood. Previous studies of sleep deprivation [2, 3, 11, 12] and recent experiments on extended sleep [1, 13, 14, 15] have shown that deleterious behavioral and psychological effects result from shortening and lengthening sleep.

In this study the statistical significance of the impairment on the calculation task, and in mood (but not vigilance), was less relative to the control condition for the early-shift conditions than for the late-shift conditions. When the subjects fell asleep 2–4 hr earlier or later than usual their day was either shortened or extended and the ratio of prior wakefulness to sleep was either reduced or increased the next day. It is possible that sleep satiation resulted when the subjects fell asleep early which was less serious behaviorally than falling asleep late which possibly resulted in a condition similar to sleep deprivation.

Studies of transmeridian travel have generally shown that performance deficits following flights east are greater than after westward flights [5, 6]. However, the late-shift conditions resulted in a phase-lag of the sleep wakefulness cycle as does a westward flight, and the detrimental effects of these sleep conditions were of a greater magnitude than the early-shift conditions which resulted in a phase-advance of the sleep-wakefulness cycle as does a flight east.

Eastward flights are usually scheduled at night, the traveler arrives at his destination in the morning and sleep might be delayed for an additional 12 hours. Westward flights are usually scheduled by day and the passenger, arriving at his destination by night, can fall asleep without delay. In view of such timing of transmeridian flights, differences in the amount of wakefulness prior to sleep might temporarily counteract effects of the resulting phase-advance or phase-lag.

Recent experiments have shown that the normal pattern of sleep is not radically altered when the sleep-wakefulness cycle is reversed [16–18]. Therefore, lowered efficiency associated with phase-shifts in the sleep-shifts in the sleep-wakefulness cycle and in studies of multiple time zone travel and shift-work [4–6] indicates that such effects result from either the imposition of wakefulness upon normal circadian sleep time or from the intrusion of sleep on normal waking time rather than from disrupted sleep *per se*.

If performance level were exclusively dependent on circadian factors preexisting prior to the shifts in sleep time, increasing efficiency in postawakening performance as time of awakening varied from 0400 to 1200 might have been expected. That such a trend was not present indicates that diurnal rhythms of performance and mood must be related to the circadian placement of sleep.

REFERENCES

1. Taub J. M., Globus G. G., Phoebus E. and Drury R. (1971): Nature, (in press).
2. Wilkinson R. T. (1968): *In*: Progress in Clinical Psychology, (L. A. Abt and B. F. Reiss, eds.), Vol. 7, Grune and Stratton, New York.
3. Wilkinson R. T. (1970): *In*: Int. Psychiat. Clin., (E. Hartmann, ed.), Little Brown, Boston.
4. Coloquhoun W. P., Blake M. J. and Edwards R. S. (1968): Ergonomics, *11*: 527–546.
5. Hauty G. T. and Adams T. (1966): Aerospace Med., *37*: 1027–1033.
6. Klein K. E., Brüner H., Holtmann H., Rehme H., Stolze J., Steinhoff W. D. and Wegmann H. M. (1970): Aerospace Med., *41*: 125–132.
7. Williams H. L. and Lubin A. (1967) : J. Exp. Psychol., *73*: 313–317.
8. Rechtschaffen A., and Kales A. (1968): A Manual of Standardized Terminology, Techniques, and Scoring Systems for Sleep Stages of Human Subjects. U.S. Government Printing Office, Washington, D.C.
9. Winer B. J. (1962): Statistical Principles in Experimental Design. McGraw-Hill, New York.
10. Siegel S. (1956): Noparametric Statistics. McGraw-Hill, New York.
11. Johnson L. C. (1969) : *In*: Sleep: Physiology and Pathology, (A. Kales, ed.), J. B. Lippincott, Philadelphia.

12. WILKINSON R. T. (1965): *In*: The Physiology of Human Survival, (O. G. EDHOLM and A. L. BACHARACH, eds.), Academic, New York.
13. GLOBUS G. G. (1969): Psychosom. Med., *31*: 528–535.
14. GLOBUS G. G. (1970): *In*: Int. Psychiat. Clin., (E. HARTMANN, ed.), Little Brown, Boston.
15. TAUB J. M. and BERGER R. J. (1969): Psychon. Sci., *16*: 204–205.
16. BERGER R. J., WALKER J. M., SCOTT T. D., MAGNUSON L. J. and POLLACK S. L. (1971): Psychon. Sci., *23*: 273–275.
17. WEBB W. B., AGNEW H. W. and WILLIAMS R. L. (1971): Aerospace Med., *42*: 152–155.
18. WEITZMANN E. D., KRIPKE D. F., GOLDMACHER D., MCGREGOR P. and NOGEIRE C. (1970): Arch. Neurol., *22*: 483–489.

EXPERIMENTAL MODELS OF BEHAVIORAL-PHYSIOLOGICAL INTERACTIONS IN SOCIAL CHRONOBIOLOGY*

Richard V. ANDREWS

Department of Biology, Creighton University
Omaha, Nebraska, U.S.A.

Interactions of animals in a societal setting are subject to temporal schedules of circadian and circannual frequencies in behavioral and physiological activities. In nature, these rhythms are synchronized by changing schedules of photoperiod and temperature cycles which depend upon the earth's daily rotation and annual orbit. Both climatic changes and the intensity of social interaction are anticipated by individuals within animal populations, so that differences in onset and peak of their daily behavioral schedules become apparent. Moreover, the impact of social status on these individuals affects their physiologic adjustment so that differential survival is a feature of the several small mammal populations we have studied. Our field observations indicate that desynchronization of behavioral and endocrine rhythms may contribute to lower fertility and a higher probability of disease and predation susceptibility among subordinate animals than is apparent in their dominant cohorts.

In the arctic, drastic environmental changes are imposed upon rodents indigenous to the tundra. I was interested in determining whether Arctic cricetids displayed circadian patterns in secretion (documented for hamsters and mice in temperate regions). Indeed, several species of arctic rodent displayed daily rhythms of adrenal secretion, measured by serial sacrifice or organ culture techniques [1, 3] (Fig. 1). Fortuitously, our measurements extended over several years, so that we were able to document a surge in lemming numbers, and an ensuing crash. Adrenal secretory rates fluctuated according to seasonal climatic conditions and population density [2, 5], Table 1. Moreover, when animals were serially sacrificed throughout the summer, the secretory rhythm appeared to free-run during the time when photoperiodic signals were erratic [4] (Fig. 2). Although lemming secretory rhythms tend to free-run around the summer solstice, individual animals were synchronized with each other.

Similar free-running rhythms in locomotor activity of Point Barrow lemmings kept in a greenhouse have been observed. Swade [9, 10] has reported similar free-running locomotor rhythms for a wide variety of individual arctic and subarctic rodents which were housed under conditions where either natural or artificial lighting regimens showed little regular variation. However, when the daily photoperiod reaches a critical maximum : minimum ratio in intensity (value varies

* Work supported by the Arctic Institute of North America, under contractual agreement with the Office of Naval Research (AINA-ONR 416), the United States Public Health Service EC00340), the National Science Foundation (GB8307) and the Creighton Faculty Development Fund.

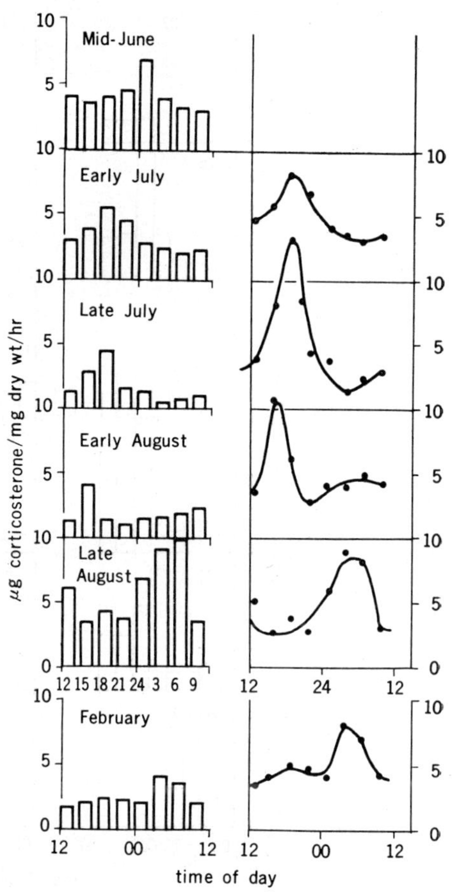

Fig. 1 Daily variations in adrenal secretory rate as evidenced by serial sacrifice of 5 or 6 animals (left series of curves). The mean secretory rates are plotted against time of day at various portions of the summer.

with individual animals and species) with the approach of the arctic sunset, locomotor rhythms become synchronized to a "fixed" local time.

Through the course of several summers we collected lemmings by both "hand-catching" and live-trapping techniques. Field excursions during the evening hours yielded more success than those conducted through morning and afternoon hours. Our live trapping grids yielded more quantitative data, since we checked trap lines every 2–4 hours so that animals would not be lost to marking-recapture estimates of density (Fig. 3). More free-ranging lemmings were caught between 1800 and 0200 hours on our grids. Similar capture frequencies of both *Lemmus trimucronatus* and *Dicrostonyx groenlandicus* were reported by other investigators running snap-trap and live-trap lines in the Barrow vicinity (Melchior, Pitelka, pers. comm.). In years of lemming abundance most ermine and least weasel sightings occurred between 1800 and 2400.* Closer examination of our trapping records indicated that capture of mature, healthy adults are more probable during the evening while juveniles, subadults and scarred (subordinate) adults are more likely to be trapped at other times of the day during mid-summer, particularly from high density populations. In short, it appears that, extended

* These species together with arctic foxes and wolves are the major mammalian predators of lemmings. Intensive hunting activities of avian predators, jagers and snowy owls, were extended from morning hours until about 0200 local time.

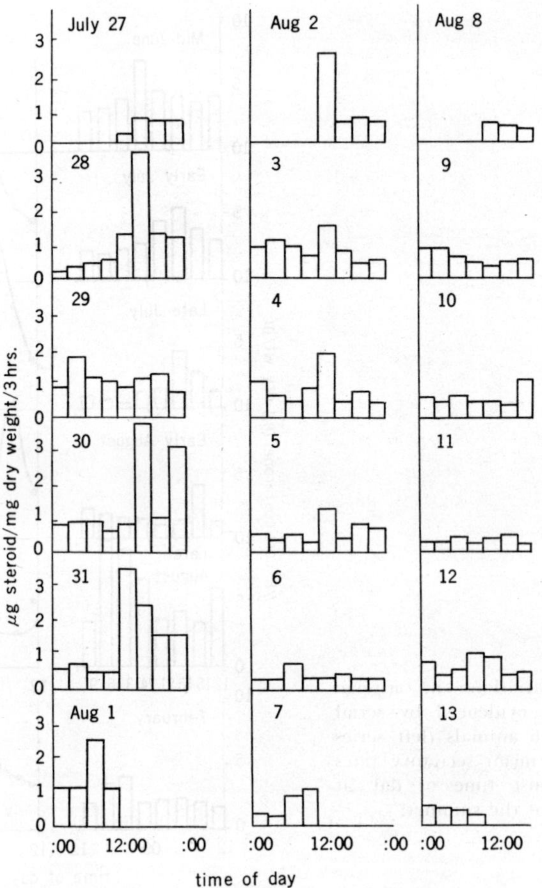

Fig. 2 Quantity of adreno-cortical steroids (per milligram dry weight of tissue) secreted into the nutrient medium at three-hour intervals over daily sequences. The three vertical allignments (represented by designations July 27, August 2 and August 8 at the top of the figure) display daily variations in steroid secreted by 14 adrenal cultures in each experiment. Note that the beginning and end of each experiment is at 1200 hours (AST).

foraging trips by lemmings are more likely when the sun angle is lower (temperature cycles are also at daily minimum). Dispersal of daily activity occurs at all times, but is increased when population densities are higher.**

The development of asynchrony between locomotor rhythms and endocrine rhythms in mid-summer lemming populations may be related not only to absence of clear-cut environmental cues, but also to intensity of social contact among these very aggressive mice. Increased activity and resulting hyperadrenalism manifest by high density lemming populations contribute to the summer decline in numbers caused by predation and metabolic disease. Although we should monitor behavioral and physiological rhythms of individually marked animals (whose social rank can be established) in free ranging populations to verify our hypothesis, social dispersal and desynchronization seem plausible contributors to increased mortality elicited during this population crisis.

** Spatial dispersal is also more prominent during lemming highs, as judged from finding animals in less favorable habitats.

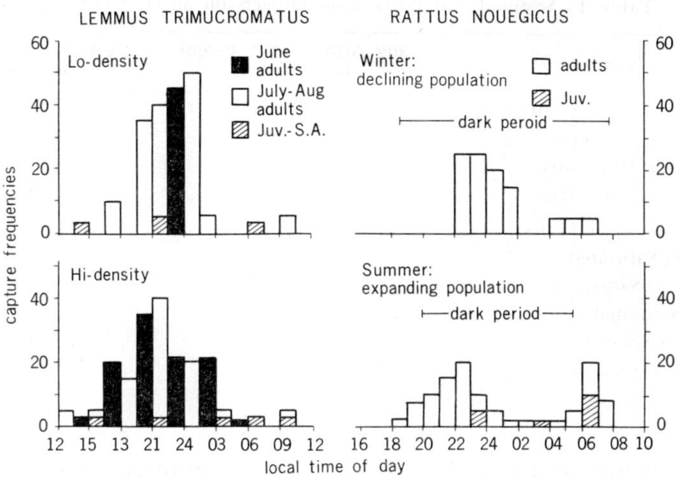

Fig. 3 Capture frequencies of lemmings and rats from differering seasons and population structures.

A second population displays different dynamics and social order. Wild Norway rats, freely ranging in an open dump or in structured open (50′×50′) enclosures, adjust recruitment rates to nearly match mortality rates, and hence display "stable" numbers. Our early observations complemented those of CALHOUN [7], in that the rats are organized into colonial, highly structured social groups which concentrate most of their foraging activities from dusk until 0200 hours (Fig. 3). During spring and summer months a highly ordered schedule of activities is displayed by various social cohorts. Aggressive behavior is much more pronounced among adult males and litter bearing females at this time, but is not so apparent during the winter when little or no breeding occurs in either saturated or expanding populations [6]. Breeding season activities are divided so that early evening (before light intensity has diminished to sunset intensity) and early morning hours are occupied by foraging of juvenile, subadult and subordinate older adults. Dominant, more aggressive adult males and females utilize later evening to midnight hours (2100–0100 CDST) for their activities. Hence, the greatest trapping success for dominant rats occurred between discreet times, while subordinate rats were caught throughout the night into early morning hours.

Since all classes of animals evaluated in preliminary experiments showed lower sustained corticosteroid secretion rates between 0600 and 1400 (but higher values between 2000–0200) we sacrificed most of the animals at 0800 hours on the morning following capture for population analysis. Adrenal secretion was seasonally higher in the fall and early winter and following emergence of the young in the saturated population (Table 1), and was distinctly seasonal in expanding populations where juveniles and subadults were not competing for limited harborage. During the winter, however, foraging activities of all classes of animals were restricted to between 1800–2400 hours with little evidence of the fighting or avoidance behavior seen during the summer. Winter numbers had been reduced by increased mortality which followed the spring recruitment period and continued to drop during the deep winter because of the ravages of disease aided by chronic cold stress. Moreover, reproductive successs was reduced to a minimum during

Table 1 Seasonal and population impacts on adrenal function.

Sample designation	mg Adrenal Wt/100gm body weight	μg Corticosteroid/100mg adrenal/hour
Brown Lemming—Adult Males		
Summer Lo-density	26.9±2.1	8.2± 1.0
Summer Hi-density	44.0±4.1	158.6±15.3
Winter Lo-density	30.1±2.6	15.6± 2.8
Norway Rat-Adult Males		
Spring-Saturated	31.5±2.1	6.8± 0.5
Summer-Saturated	23.2±1.2	4.7± 0.5
Fall-Saturated	32.0±1.3	5.6± 0.6
Winter-Saturated	36.8±1.9	10.4± 1.1
Summer-Social Distruption	38.1±1.9	8.1± 0.6

the fall and winter as a consequence of sex hormone regressions among the majority of animals.

Although the population dynamics of our brown rats are quite different from the microtines, largely because of reproductive control of recruitment, these animals probably also display desynchronization of locomotor and endocrine rhythms. Diminished fertility among females following a recruitment surge might be attributed to behavioral and endocrine desynchronization mediated by adrenal secretion as documented by Swartz for domestic lab rats [11]. Synchronization of social schedules to times of peak physiological efficiency would indeed have adaptive value for the animals we have studied by placing the stronger animals in the most favorable time-space circumstance for survival. Desynchronization of subordinates favors their survival while transient adjustments in density take place by reducing social conflict, but places them at a further disadvantage should densities remain high.

Social synchronization in other animal societies is probable; primates and wolves make good models of smaller social groups. We plan experiments in our local zoo and in the arctic to test for behavioral-physiological synchronies in a social setting. Certainly, a better appreciation of an individual's rhythmic profile would be achieved if we were able to place him into the natural context of social groups or populations. Such studies have already yielded valuable information and turned up challenging questions when human subjects have been used (See Luce, 1970 for bibliography) [8].

REFERENCES

1. Andrews R. V. (1970): Effects of seasonal variations in climate and photoperiod on adrenal rhythms of lemmings, voles and mice. Acta Endocr., *65*: 645.
2. Andrews R. V. (1970): Effects of climate and social pressure on the adrenal response of lemmings, voles and mice. Acta Endocr., *65*: 639.
3. Andrews R. V., Folk G. E. and Hedge R. (1965): Metabolic periodicity in adrenal glands cultured from arctic rodents. Fed. Proc., *24*: 508.
4. Andrews R. V., Keil L. C. and Keil N. N. (1968): Further observations on the adrenal secretory rhythm of the brown lemming. Acta Endocr., *59*: 36.
5. Andrews R. V. and Strohbehn R. (1971): Endocrine adjustments in Comp. Biochem. Physiol., *38*: 183.
6. Andrews R. V. Belknap R. W., Southard J., Lorincz M. and Hess S. (1971): Physi-

ological, demographic, and pathological changes in wild Norway rat populations over an annual cycle. Comp. Biochem. Physiol., (in press).
7. Calhoun J. B. (1962): The ecology and sociology of the Norway rat. PHS Publ. 1008, Washington, D.C.
8. Luce G. G. (1970): Biological rhythms in psychiatry and medicine. PHS Publ. 2088, Washington, D.C.
9. Swade R. H. (1963): Circadian rhythms in the arctic. Ph.D. thesis, Princeton Univ.
10. Swade R. H. and Pittendrigh C. S. (1967): Circadian locomotor rhythms of rodents in the arctic. Amer. Nat., *101*: 431.
11. Nequin L. G. and Schwartz N. B. (1971): Adrenal participation in the timing of mating and LH release in the cyclic rat. Endocrinology, *88*: 325.

PINEAL AND MELATONIN: CIRCADIAN RHYTHMS AND BODY TEMPERATURES OF SPARROWS

SUE A. BINKLEY

Department of Biology, University of Notre Dame
Notre Dame, Indiana, U.S.A.

In 1968, GASTON and MENAKER [1] reported that the circadian rhythm of perching activity was lost after pinealectomy in the house sparrow, *Passer domesticus*. BINKLEY, KLUTH, and MENAKER [2] subsequently subjected the apparently aperiodic activity of such pinealectomized sparrows in constant conditions (DD) to autocorrelation, power spectral analysis, and daily totaling. These analyses strengthened the conclusion that the activity was indeed arrhythmic and added the information that the level of activity was not altered by pinealectomy. Since then, I have tested the sparrow arrhythmia using the periodogram method of ENRIGHT [3] and have obtained the same result (Figs. 1 and 2).

BINKLEY, KLUTH, and MENAKER [4] also showed that another circadian rhythm, that of body temperature, was lost in pinealectomized sparrows in DD (Fig. 3). Moreover, the amplitude of the temperature oscillation was reduced in the pinealectomized birds when they were entrained to a light cycle (LD12:12) as well as when they were in constant conditions. The body temperatures of the pinealectomized birds did not drop to the normal daily minimum (Table 1).

The results of these experiments are significant for two reasons: (1) we interfered anatomically with the circadian time-keeping machinery which very few investigators have done [5], and (2) we discerned an effect of pinealectomy on body temperature. The pineal was first linked to thermoregulation in lizards by STEBBIS [6] who found that pinealectomy lowered cloacal temperatures; and the pineal has since been tied to thermoregulation in rats by MILINE [7] who reported chemical changes in mammalian pineals in response to cold.

Now I have found that the pineal constituent, melatonin, affects sparrows in ways which suggest that it may mediate the above-mentioned actions of pinealectomy on rhythms and temperature lowering. I injected melatonin into unoperated sparrows (IM, during the light portion of their daily cycles) and observed: (1) a greater lowering of cloacal temperature in birds injected with melatonin than in the controls only injected with the vehicle, EtOH ($P \leqq .02$, Table 2, Fig. 4), (2) a condition lasting several hours in the melatonin birds which was similar to "sleep" (or to "anesthesia" produced by Equithesin injections), and (3) more deaths in the experimentals than the controls within a few days after the injection. The temperature drop in the melatonin injected sparrows was of the same magnitude whether the birds received 1.2, 2.0, or 2.5 mg (0.1 mg was ineffective). This temperature drop, 4.7°C, is comparable to the normal nightly temperature drop in sparrows in light cycles, 4.4°C.

Melatonin has also been shown to produce temperature lowering in mice [8] though not in rabbits given smaller doses than those given to the mice and the sparrows [9]. BARCHAS et al. [9] also reported that melatonin potentiates bar-

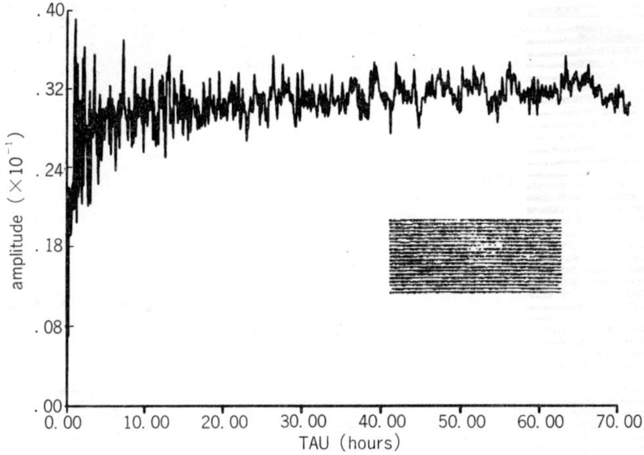

Fig. 1 Enright periodogram and the corresponding event record from a pinealectomized bird in DD. Tach line of the event record contains 24 hours of perching activity data and the lines have been arranged vertically in chronological order. The periodogram was calculated using a Univac 1107 computer and the curve was plotted with a Calcomp Plotter. There are no obvious rhythms. Only one of six pinealectomized birds tested with this mathematical technique had a discernible circadian activity rhythm while all of four control birds had definite circadian rhythms.

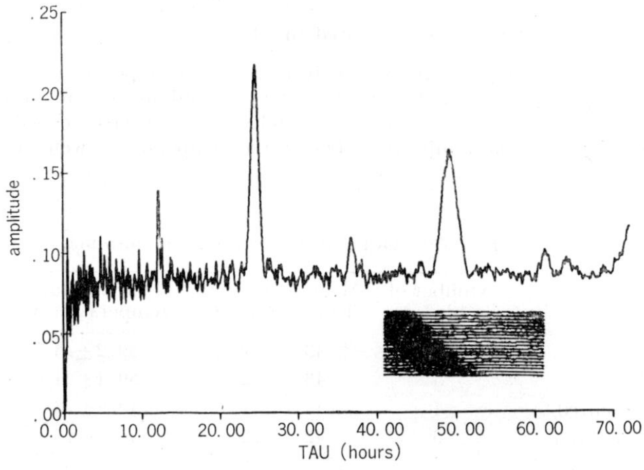

Fig. 2 Enright periogram and the corresponding event record from a control bird exhibiting a circadian rhythm in constant darkness. The event record and periodogram are as for Fig. 1 except that the scale of the y-axis of the periodogram is different. The largest peak in the periodogram occurs at the circadian period. A difficulty associated with the analysis, multiple peaks, can be seen.

biturate "sleep" in mice (as tested by the righting reaction) and that it induced "sleep" (roosting posture) in 4-day old chicks. Two other studies in birds are meaningful for interpretation of our experiments in sparrows. HISHIKAWA, KRAMER, and KUHLO [10] found that melatonin produced sleep as measured by not only behavioral, but also EEG, criteria. LYNCH [11] and RALPH, HEDLUND, and MURPHY [12] using a bioassay have found that there is a circadian rhythm of melatonin in several species of birds, and, further, that the peak of this rhythm

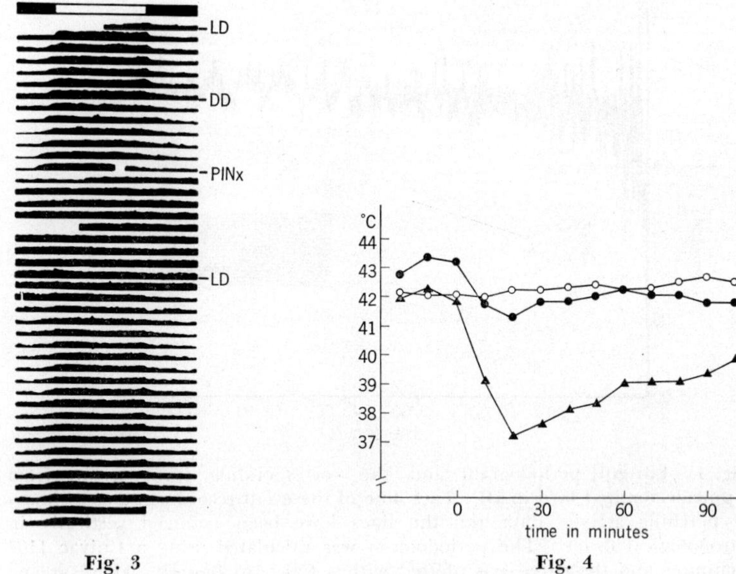

Fig. 3 **Fig. 4**

Fig. 3 Temperature record from a sparrow. Each line contains 24 hours of data and the lines have been arranged vertically in chronological order. The bird's temperature was rhythmic in a light cycle (LD12:12 indicated by the bar at the top of the record) and the rhythm freeran when the bird was placed in constant darkness (DD). When the bird was pinealectomized (PINx) the temperature became aperiodic. The rhythm re-established gradually when the bird was replaced in LD.

Fig. 4 Time course of typical sparrow body temperature responses in injection experiments. Injections were made at 0 minutes (after a 20 minute control interval). Three individual birds" responses are shown: open circles saline injected; closed circles, EtOH injected; triangles, melatonin injected. The initial temperatures were not identical in the birds.

Table 1 Daily temperature maxima and minima.

Class	Number of birds	Mean maximum Temperature°C	Mean minimum temperature°C	Amplitude of variation°C
LD normal	5	43.6±.2	39.2±.3	4.4±.5
DD normal	5	43.1±.2	39.4±.4	3.8±.4
LD pinealectomized	4	43.7±.2	40.6±.6	3.0±.3
DD pinealectomized	6	42.4±.2	40.5±.3	1.9±.1

Table 2 Injection experiements.

Class	Number of birds	Temperature change °C
Saline injected	4	0 ±.4
EtOH injected	5	−1.4±.5
Melatonin injected	11	−4.7±.3

Temperatures are given ± the standard error in both Table 1 and Table 2. The temperature change in Table 2 was found by taking the difference between the minimum body temperatures recorded before and after the injections.

occurs during the dark portion of a light cycle or during inactivity in birds free-running in constant conditions. These facts, taken together with my experiments in sparrows, permit me to conclude that it is possible that melatonin is the pineal agent for the functions revealed by our earlier pinealectomy studies, namely: (1) the daily lowering of body temperature, and (2) the persistence of the circadian activity and temperature rhythms in constant darkness. Since, however, pinealectomized sparrows are inactive in the dark portion of a light cycle, something more than pineal melatonin must be involved in the sleep-wake cycle of sparrows.

Thus the pineal has some role in the regulation of body temperature in sparrows, and the pineal is necessary for the normal persistence of the freerunning circadian rhythms of both activity and temperature. Our experiments with injected melatonin and the results of other workers are consistent with the hypothesis that the pineal functions we have observed in sparrows are mediated by melatonin. However, the effects of pinealectomy on sparrows have not directly been shown to be due to a loss of melatonin, nor is melatonin the only possible means for pineal function. Melatonin has, as mentioned above, effects on mammals similar to those obtained in birds but the requirement for pineal tissue for the persistence of the endogenous circadian rhythm has not been found in other vertebrate groups (reptiles, UNDERWOOD and MENAKER [13]; mammals, QUAY [14] though a pineal role in phase shifting rhythms has been reported in rats [15, 16].

ACKNOWLEDGEMENT

I thank D. TAYLOR, R. ABRAHAM, T. POULSON, J. McGRATH, R. GOODFELLOW, Mrs. MILLER, and Mr. MAXWELL. The work was supported by NIHFR 07033-05 (to K. ADLER, University of Notre Dame). The data re-analyzed by the Enright method, and the temperature data in Table 1 and Fig. 3 were supported by a NIH traineeship (5T01 GM-00836-08) to S. BINKLEY and NIH program project grant (HD-03803-02) at the University of Texas.

REFERENCES

1. GASTON S. and MENAKER M. (1968): Science, *160*: 1125.
2. BINKLEY S., KLUTH E. and MENAKER M. (1971): in preparation.
3. ENRIGHT J. T. (1965): J. Theoretical Biology, *8*: 426–428.
4. BINKLEY S., KLUTH E. and MENAKER M. (1971): Science, *174*: 311–304.
5. NISHIITSUTSUJI-UWO J. and PITTENDRIGH C. S. (1968): Z. für Vergleichende Physiologie, *58*: 1.
6. STEBBINS R. C. (1960): Copeia, 276.
7. MILINE R. (1971): CIBA Foundation Symposium on the Pineal Gland. 29–31, 373, J. & B. Churchill, London.
8. ARUTYUNYAN M. D., MASHKOVSKY M. D. and ROSCHINA L. F. (1964): Fed. Proc., *23*: T1330–T1332.
9. BARCHAS J., DACOSTA F. and SGECTOR S. (1967): Nature, *214*: 919–920.
10. HISHIKAWA Y., CRAMER H. and KUHLO W. (1969): Exp. Brain Research, *7*: 84–94.
11. LYNCH H. J. (1971): Life Sciences, *10*: 791–795.
12. RALPH C. L., HEDLUND L. and MURPHY W. A. (1967): Comp. Biochem. Physiol., *22*: 591–599.
13. UNDERWOOD H. and MENAKER M. (1970): Science, *170*: 190–193.
14. QUAY W. B. (1968): Physiol. Behav., *3*: 109–118.
15. QUAY W. B. (1970): Physiol. Behav., *5*: 353.
16. KINCL F., CHANG C. C. and ZBUZKOVA V. (1970): Endocrinology, *87*: 38.

PHASE-SHIFTS OF CIRCADIAN RHYTHMS: DEFINITIVE REPRESENTATION AND QUANTITATIVE ANALYSIS FROM COMPUTER APPLICATION OF THE BETA-DISTRIBUTION AS A MODEL*

WILBUR B. QUAY**

Department of Zoology, University of California
Berkeley, California, U.S.A.

Circadian rhythms of man and other animals are composed of phases which have, when entrained and fully stabilized, more or less precise temporal relations with environmental synchronizers or Zeitgebers. Often a particular feature or event within or at the beginning of a phase has greater temporal consistency through a series of cycles, and may, at least superficially, appear to be most closely associated with the presumed synchronizer. Such a feature or event may serve as a landmark in the tracing and analysis of temporal changes during phase-shifting of an animal's physiological or behavioral circadian rhythm ($\Delta\phi$) in response to a shifting in the timing of the environmental rhythm ($\Delta\Phi$). However, other features or arbitrary points, such as acrophase or phase midpoint, may be utilized similarly and have advantages in some cases [1, 2].

Following a reversal in the timing of the environmental rhythm or synchronizer, approximately a week is required for complete phase-shift (=entrainment) to occur in the dependent physiological and behavioral circadian rhythms. However, recent experiments have shown that the rate of phase-shift of the rat's circadian rhythm in running activity can be modified by surgical removal of the pineal organ [3–5]. Furthermore, changes in the phase-shift response curve can be caused, also, by other intracranial operations [6, 7], and the direction and magnitude of the effect of pinealectomy can be modified by characteristics of the primary Zeitgeber (light) and the age of the subject [8, 9]. Since circadian phase-shifts are usually non-linear responses with time, and since they can differ greatly in form and variability, accurate representation and satisfactory quantitative analyses are difficult, if not impossible, by classical methods. This report summarizes the nature and applicability of a model for this non-linear response system based upon the Beta-distribution [10]. A more comprehensive description of the model and methods of computation for its use are to be presented elsewhere [11].

METHODS

The Model. In phase-shift curves we are concerned with Y ($=\Delta\phi$ or % shift

* This work was supported by a research grant (NS-06296) from the National Institute of Health, U.S. Public Health Service, and aid from Biomedical Sciences Support grant funds, University of California, Berkeley. The author is especially grateful to Dr. Nora SMIRIGA, Department of Mathematics, San Francisco State College, and to Willie Sue HAUGELAND, Department of Computer Sciences, University of California, Berkeley for major collaborative efforts in the development and application of the primary method summarized here.

** Present address: Waisman Center on Mental Retardation and Human Development, University of Wisconsin, Madison, Wisconsin, U.S.A.

of some feature in the rhythm) as a function of X (=days post $\Delta\Phi$). The relationship $Y=f(X)$ is non-decreasing $(0 \leqq Y \leqq 1)$. Therefore, a cumulative distribution function should be a good model for the data describing phase-shift curves. Moreover, the cumulative distribution should be very flexible so that observed extremes of $\Delta\phi$ would have a good fit. The Beta-distribution offers the flexibility and other characteristics required under these circumstances, and is by definition.

$$I_x(\alpha, \beta)=B_x(\alpha, \beta)/B(\alpha, \beta) \tag{1}$$

where $B_x(\alpha, \beta)$ is the incomplete Beta-function, i.e.,

$$B_x(\alpha, \beta)=\int_0^x u^{\alpha-1}(1-u)^{\beta-1}du \tag{2}$$

and $B(\alpha, \beta)$ is the Beta-function, i.e.,

$$B(\alpha, \beta)=\Gamma(\alpha)\Gamma(\beta)/\Gamma(\alpha+\beta) \tag{3}$$

where Γ is the Gamma-function, i.e.,

$$\Gamma(\alpha)=\int_0^\infty u^{\alpha-1}e^{-u}du. \tag{4}$$

In this definition the domain of X is $0 \leqq X \leqq 1$. Where in the example experimental data $\Delta\phi$ is studied over 12 days after $\Delta\Phi$, the domain of X is $0 \leqq X \leqq 12$, and the model becomes,

$$I_{\frac{x}{12}}=\frac{\int_0^x u^{\alpha-1}(12-u)^{\alpha-1}du}{12^{\alpha+\beta-1}B(\alpha, \beta)}. \tag{5}$$

Computations. A non-linear regression program is used to fit the data to the distribution $I_{\frac{x}{12}}(\alpha, \beta)$. Essentially, the program attempts to find those values of α and β which will will minimize the relative sum of squares,

$$\phi(\alpha, \beta)=\sum_{j=0}^{12}\left(\frac{Y_j-I_{\frac{xj}{12}}(\alpha, \beta)}{Y_j}\right)^2 \tag{6}$$

where $X_j=j=$ the jth day, and Y_j is the % shift ($\Delta\phi$) observed at time X_j, with $1.000=100\%$ shift. A previously described [12] non-linear regression program is modified for the obtaining of the best fit. For the published program there must be supplied,

$$I_{\frac{xj}{12}}(\alpha, \beta)=B_{\frac{xj}{12}}(\alpha, \beta)/B(\alpha, \beta). \tag{7}$$

The integral $B_{\frac{xj}{12}}(\alpha, \beta)$ is numerically evaluated using WYNN's [13] algorithm, and the Beta-function, $B(\alpha, \beta)=\Gamma(\alpha)\Gamma(\beta)/\Gamma(\alpha+\beta)$, is evaluated by using WRENCH's [14] algorithm. The numerical evaluation of $I_{\frac{xj}{12}}(\alpha, \beta)$ thus obtained has been compared for several values of X_j, α and β with that obtained through the use of K. PEARSON's [10] tables. These tables provide only seven digits after the decimal point. The computations yield thirteen digits (when done on a CDC 6400, as in the present study). Rounding the results to seven digits shows a perfect agreement with the tables.

The above least squares program requires as input an initial guess for the parameters α and β. Although this can be accomplished by use of The Biometrika Tables [15], the choice of the initial values of α and β is not critical because the program is not particularly sensitive to the initial choice of parameters; convergence is achieved rapidly and usually before twenty iterations.

Comments. Arrival at appropriate observed values of Y for a particular set of data from a phase-shift may require selection or transformation of the Y scale, if there are peculiarities in values (or entrainment) either at the beginning or the end of the phase-shift. Assignment of Y values depends on the exact determination of points of origin ($Y=0$) and saturation or entrainment ($Y=1.0$) for a particular set of data. Means from replicate observations (entrained cycles) can be used to increase the accuracy of these points.

Observed data for Y are not uniform in statistical variance, but have coefficients of variation which follow a non-linear relationship over successive days (X) of a phase-shift. It is simple to calculate coefficients of variance or other measures of variability from pooled data points or replications, and then as input to the computer program, weight the data points for Y inversely in relation to their variance.

RESULTS

Fig. 1 demonstrates some of the results obtained by this method when it is applied to phase-shifts in circadian wheel-running activity by pinealectomized and sham-operated laboratory rats. Five pinealectomized [16] and five sham-operated female S_1 strain (agouti) rats were studied continuously for over a year in adjacent cages and through nineteen reversals in the timing of the artificial photoperiod (LD 12:12; $\Delta\Phi_1 \rightarrow \Delta\Phi_{19}$). The computed phase-shift curves are shown for selected examples of phase-shifts. More rapid than normal $\Delta\phi$ is shown by the pinealectomized animals when immature and young adults, and slower than normal $\Delta\phi$ when they are older [9]. The most obvious advantage obtained by these curves is the definitive (best answer) representation of particular phase shifts, which would otherwise be shown by scattered data points. However, the curves can also serve to provide precise theoretical values for the time required

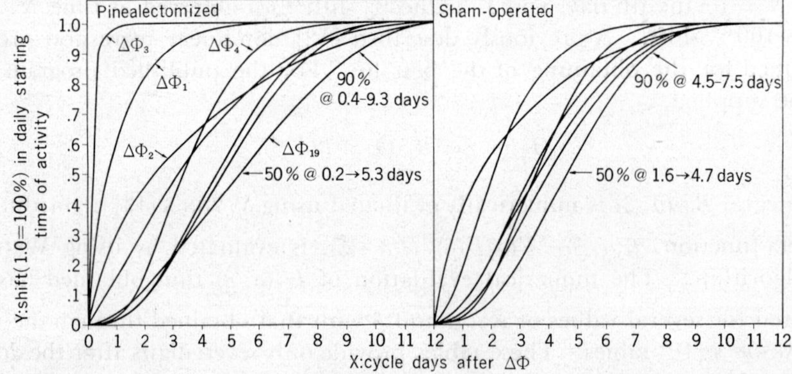

Fig. 1 Example computed phase-shift curves. These are derived from an experiment in which sibling pinealectomized (n=5) and sham-operated (n=5) female rats were studied continuously through 19 reversals of the environmental photoperiod ($\Delta\Phi_1 \rightarrow \Delta\Phi_{19}$) over the span of a year. The starting time of the daily period of running activity is taken as the parameter for tracing the phase-shift. Phase-shift ($\Delta\phi$ S') is seen to be more rapid than normal in pinealectomized animals when immature to young adult, and is slower than normal in the same animals when adult and older.

for particular phase-shifts to occur. In this regard the times for 50% and 90% shifts are noted in the figure.

CONCLUSIONS

It is proposed that both in laboratory and in clinical studies of circadian phase shifts, definitive computed phase-shift curves provide a suitable basis for exact quantitative comparisons to be made between the effects of different treatments, between different individuals or cases under study, and between different environmental factors significant in phase synchronization or in modulation of phase-shift responses to imposed shifts in timing of environmental synchronizer(s).

REFERENCES

1. HALBERG F., TONG Y. L. and JOHNSON E. A. (1965): Circadian system phase–an aspect of temporal morphology; procedures and illustrative examples. *In*: The Cellular Aspects of Biorhythms, (H. von MAYERSBACH, ed.), pp. 20–48, Springer-Verlag, Berlin.
2. ASCHOFF J. (1965): The phase-angle difference in circadian periodicity. *In*: Circadian Clocks, (J. ASCHOFF, ed.), pp. 262–276, North-Holland Publ. Co., Amsterdam.
3. QUAY W. B. (1970): Physiological significance of the pineal during adaptation to shifts in photoperiod. Physiol. Behav., *5*: 353–360.
4. QUAY W. B. (1970): Precocious entrainment and associated characteristics of activity patterns following pinealectomy and reversal of photoperiod. Physiol. Behav., *5*: 1281–1290.
5. KINCL F. A., CHANG C. C. and ZBUZKOVA V. (1970): Observations on the influence of changing photoperiod on spontaneous wheel-running activity of neonatally pinealectomized rats. Endocrinol., *87*: 38–42.
6. QUAY W. B. (1971): Effects of cutting nervi conarii and tentorium cerebelli on pineal composition and activity shifting following reversal of photoperiod. Physiol. Behav., *6*: 681–688.
7. QUAY W. B. (1971): Dissimilar functional effects of pineal stalk and cerebral meningeal interruptions on phase shifts of circadian activity rhythms. Physiol. Behav., *7*: 557–567.
8. QUAY W. B. (1972): Pineal homeostatic regulation of shifts in the circadian activity rhythm during maturation and aging. Trans. N.Y. Acad. Sci., *34*: 239–254.
9. QUAY W. B. (in press): Behavioral effects of the mammalian pineal gland: Quantitative analysis and elicitation by environmental and intracranial factors. *In*: Pineal Workshop, (D. C. KLEIN, ed.), Raven Press, New York.
10. PEARSON K. (1968): Tables of the Incomplete Beta-function, 2nd ed., Cambridge Univ. Press, Cambridge.
11. QUAY W. B., SMIRIGA N. G. and HAUGELAND W. S. (1972): Mathematical model and computer representation for nonlinear response systems. Computers and Biomedical Research, *5*: 239–246.
12. BAER R. M. (1962): Nonlinear regression and the solution of simultaneous equations. Communs. Ass. Comput. Machi., *5*: 397–398.
13. WYNN P. (1965): An arsenal of algol procedures for the evaluation of continuous fractions and for effecting the epsilon algorithm. Mathematics Research Center, United States Army, The University of Wisconsin, MRC Technical Summary Report, No. *537*: 36–37.
14. WRENCH J. W., Jr. (1968): Concerning two series for the Gamma Function. Mathematics of Computation, *22*: 617–626.
15. PEARSON E. S. and HARTLEY H. O. (1954): Biometrika Tables for Statisticians. Cambridge Univ. Press, Cambridge.
16. QUAY W. B. (1965): Experimental evidence for pineal participation in homeostasis of brain composition. Prog. Brain Res., *10*: 646–653.

CHRONOBIOLOGY AND INSECT RHYTHMS

Chairman: Laurence K. CUTKOMP

THE EFFECT ON DIAPAUSE OF PHOTOPERIOD MANIPULATION AT DIFFERENT TEMPERATURES

DORA K. HAYES, ALVIN N. HEWING, DALE B. ODESSER, WILLIAM N. SULLIVAN and MILTON S. SCHECHTER

Entomology Research Division, Agricultural Research Service, USDA Beltsville, Maryland, U.S.A.

Our group has previously made studies [1–3] which showed that artificial lengthening of the photophase in the fall prevents overwintering of pest species exposed out of doors. The studies reported in this paper were made in the laboratory to determine what light regimens would prevent diapause. We hoped to find an effective regimen that would require less total light energy than the 17-hr photophase that was effective [1]. Short light breaks introduced during the night were not effective outdoors in the late summer and early fall [2].

The larvae of the European corn borer, *Ostrinia nubilalis* Hübner, used in the study were furnished by the European Corn Borer Laboratory, at Ankeny, Iowa on the diet of LEWIS and LYNCH [4] and this same medium or the medium of VANDERZANT-ADKISSON [5] was used throughout the tests. Two trials were made with the larvae held at 24°C±1.5°C under 40 lux illumination from a 4-watt fluorescent light in the lard can holder of DUTKY et al. [6]. Table 1 shows the results obtained when a one-hour light break was introduced into the dark period at various times counted in hours after the beginning of the light period. Interruption of the dark period with the hour of light was obviously successful in preventing diapause to a large extent at almost any time of night.

In a second test, developing larvae were held at temperatures ranging from 10–30°C, and light breaks were introduced either early or late in the dark period. The two regimens were therefore LDLD 10:11:1:2 and LDLD 10:4:1:9. Fig. 1

Table 1 Effect of photoperiod manipulation at 24°C on incidence of diapause in European corn borer larvae.*

Photoperiodic regimen	% Pupation based on surviving insects
LD 12:12	29
LD 15:9	93
LDLD 12:2:1:9	100
LDLD 12:3:1:8	90
LDLD 12:4:1:7	94
LDLD 12:5:1:6	96
LDLD 12:6:1:5	88**
LDLD 12:7:1:4	100**
LDLD 12:8:1:3	89
LDLD 12:9:1:2	100**

* Two trials with 20 insects per trial under each regimen except where noted.

** Only one trial under this regimen.

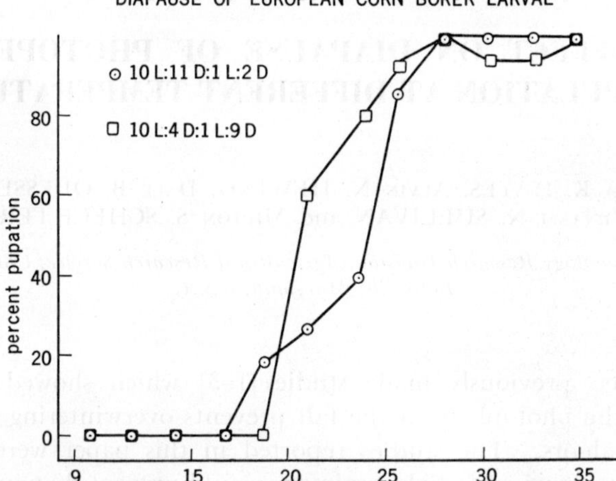

Fig. 1 The effect of light breaks introduced early and late in the dark phase on the incidence of diapause in larvae of the European corn borer held at different temperatures.

summarizes the results. At or above 25°C temperatures, a one-hour light break prevented diapause of the corn borer larvae; thus a one-hour break was adequate to prevent diapause. However, as the temperature was decreased, a greater percentage apparently entered diapause, i.e., the rate of pupation was lower, so that a one-hour break was no longer adequate.

In a third experiment, we tested light regimens which were broken several times during the first portion of the dark period at a temperature of 20°C since this lower temperature more nearly simulates fall temperatures, while still permitting the test to be completed in a reasonable time. The regimens used and the results are outlined in Table 2. When one or two 5-minute light breaks were

Table 2 Effect of photoperiod manipulation at 20°C on incidence of diapause in European corn borer larvae*.

Photoperiodic regimen	Trial	% Pupation based on living insects	95% Confidence limits
LD 10:14	1	17**	(9–28)
LD 12:12	2	0	0–6
LD 16:8	1	92	83–97
	2	85	75–92
LD 12:12 with two 5-min breaks during first 15-min of each of first 6 hours	1	82	72–90
LD 12:12 with one 5-min break every 2 hr of first 6 hr of darkness	1	70	58–82
LD 12:12 with one 5-min break every hr of darkness until 18 hr after beginning of light period	2	76	64–86

* Seventy larvae were used in each trial in each regimen tested.

** This control run at 24°C.

introduced every hour after the beginning of the dark period until 18 hours after the beginning of the light period, a greater percentage pupated than when one 5-minute break was introduced every 2 hr.

The combination of decreasing temperature and decreasing length of the photophase in the fall may cause shifts in the acrophase of sensitivity to light. In larvae of the European corn borer reared at the two photoperiodic regimens, LD 10:14 and LD 16:8, we believed that we saw differences in the acrophase of monoamine oxidase activity [7] and there may also be a difference in the phosphodiesterase activity (Fig. 2). We have also seen a lengthening of the period, τ, of oxygen consumption of diapausing larvae of the codling moth, *Laspeyresia pomonella* (L.), as temperature decreases [8].

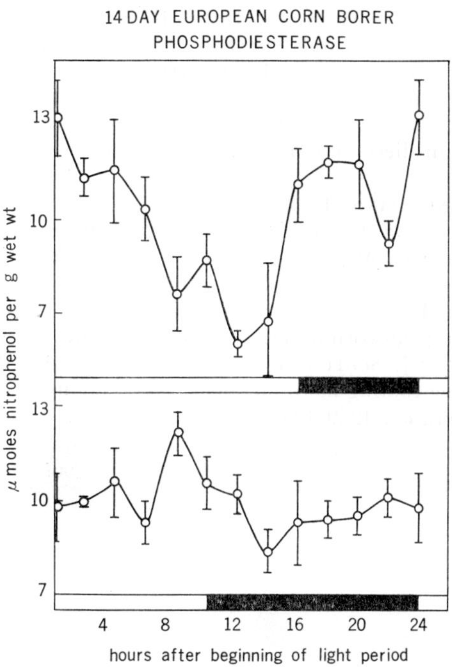

Fig. 2 Levels of phosphodiesterase in homogenates of 14-day-old larvae of the European corn borer reared at 24°C±1.5°C with two photoperiodic regimens.

The initial photochemical event in the photoperiodic prevention or induction of diapause may thus be temperature-independent, but the series of steps that follows need not be. For example, the Q_{10} of the reactions which are triggered by light may dictate that at lower temperatures the trigger must act repeatedly to cause the build-up of a diapause-preventing compound. Then either longer periods of light or multiple light breaks during the dark phase may be necessary at lower temperatures if diapause is to be prevented.

In summary, at a temperature of 20°C or below, diapause is not prevented by a one-hour light break introduced either early or late in the dark period following a photophase of 10 hr. However, at 20°C short light breaks (5 minutes every hour) successfully prevent diapause in 76% or more of a population of larvae of the European corn borer on medium.

ACKNOWLEDGMENT

We thank Gary L. REED, Wilbur D. GUTHRIE, and Stanley CARTER, ENT, for providing larvae of the European corn borer.

REFERENCES

1. HAYES D. K., SULLIVAN W. N., OLIVER M. Z. and SCHECHTER M. S. (1970): Photoperiod manipulation of insect diapause: A method of pest control? Science, *169*: 382–3.
2. SCHECHTER M. S., HAYES D. K. and SULLIVAN W. N. (1972): Manipulation of photoperiod to control insects. Israel J. Entomol., *VI*: 143–166.
3. SCHECHTER M. S., SULLIVAN W. N. and HAYES D. K. (1974): The use of artificial light outdoors to control agricultural insects. *In*: Chronobiology. Proc. Symp. Quant. Chronobiol., Little Rock, 1971, (L. E. SCHEVING, F. HALBERG and J. E. PAULY, eds.), pp. 617–621, Igaku Shoin Ltd., Tokyo.
4. LEWIS L. C. and LYNCH R. E. (1969): Rearing the European corn borer, *Ostrinia nubilalis* Hübner, on diets containing corn leaf and wheat germ. Iowa State J. Sci., *44*: 9–14.
5. ADKISSON P. L., VANDERZANDT E. S., BULL D. L. and ALLISON W. E. (1960): A wheat germ medium for rearing the pink bollworm. J. Econ. Entomol., *53*: 759–762.
6. DUTKY S. R., SCHECHTER M. S. and SULLIVAN W. N. (1962): A lard can device for experiments in photoperiodism. J. Econ. Entomol., *55*: 575.
7. HAYES D. K., WASH D. B. and SCHECHTER M. S. (1972): Monoamine oxidase activity in larvae of the European corn borer. J. Econ. Entomol., *65*: 1229.
8. HAYES D. K., HORTON J., SCHECHTER M. S., and HALBERG F. (1972): Rhythmic oxygen uptake in diapausing codling moth larvae at several environmental temperatures. Ann. Entomol. Soc. Amer., *65*: 1229–1232.

AESTIVATION IN RELATION TO OVIPOSITION INITIATION IN THE CEREAL LEAF BEETLE*, **

STANLEY G. WELLSO

Entomology Research Division, Agricultural Research Service, USDA
East Lansing, Michigan, U.S.A.

"Diapause" and "aestivation" are periods of arrested growth and development that are endocrine-controlled and allow an insect to avoid exposure to an unfavorable environment. I use the term diapause to indicate that this suspension of development occurs during seasonally low temperatures; aestivation is used when the period occurs during exposure to high temperatures or to dry conditions.

The cereal leaf beetle, *Oulema melanopus* (L.), has a univoltive life cycle. HILTERHAUS [1] reported that the female does not usually mate and seldom, if ever, oviposits in the field the same year it emerges. CASTRO et al. [2] found that overwintered adults usually mated and laid eggs from April to June; the adult progeny emerged from these eggs in late June or early July in Michigan and then fed for 2–3 weeks before entering diapause sites. GRISON et al. [3] and CONNIN and HOOPINGARNER [4] established that only females diapause, and CONNIN and HOOPINGARNER also found that males of all ages mated with females after the period but not before; thus, male activity is apparently dependent upon the physiological state of the female. CONNIN et al. [5] also found that if diapause female cereal leaf beetles were treated topically with *trans trans* 10, 11-epoxy-farnesenic acid methyl ester, and placed with males, the treated females would mate and oviposit within 4–6 days, or as soon as untreated control postdiapause beetles. TEOFILOVIC [6] found that diapause in the cereal leaf beetle could be averted by rearing the beetle under high temperature and humidity, 30°C and 90% RH, on young small grain seedlings. However, CONNIN and JANTZ [7] found that the duration of diapause of caged beetles in the laboratory was affected by the length of time they spent in cold storage and the length of the constant photophases to which they were exposed after storage: eggs were laid after beetles were stored 4 weeks at 3.3°C and then exposed to a 16-hr photophase at 23.9°C, however, beetles exposed to 8- or 24-hr photophases laid eggs only after 16 weeks of storage. Moreover, the average number of eggs laid/female increased when the female was exposed to a 16-hr photophase as the length of storage increased from 4–20 weeks. In contrast, HILTERHAUS [1] concluded that the duration of diapause was little influenced by long days or cold. Furthermore, in Germany, diapause was terminated at its earliest in mid-December; thereafter, the beetles become quiescent.

* Coleoptera: Chrysomelidae.

** Michigan State Agricultural Experiment Station Journal no. 5480. Part of a cooperative project between the Entomology Research Division and the Plant Sciences Research Division, Agr. Res. Serv., USDA, and the Department of Entomology, Michigan State University, East Lansing, Mich., 48823.

This paper will present additional information to determine whether the cereal leaf beetle aestivates and/or diapauses in the field, the relationship between beetle age and oviposition, and the status of the beetle prior to and during winter. The termination of aestivation and the initiation of oviposition with controlled photophases were also studied.

MATERIALS AND METHODS

Newly emerged cereal leaf beetles collected in early July (1969 and 1970) from late-planted oats at the Michigan State University Kellogg Gull Lake Biological Research Station, Hickory Corners, Mich. were held in a screened cage (6×6×12 ft) outdoors at East Lansing, Mich. and provided with barley seedlings and corn leaves for 2–3 weeks after capture. Beginning in early July and subsequently about every 15 days (for 3 months in 1969 and for 6 months in 1970), a sample of these beetles was brought into the laboratory and sexed by visual inspection [8]. Equal numbers of these males and females (from 10–17 pairs) were then confined in lantern globes (1–4 on each date) on barley seedlings growing in 9-cm-diam plastic pots. New seedling were provided every 3–4 days or more often, if needed. The globes were held in bioclimatic cabinets set at 26.7°C, 70±10%RH under constant photophases of 8-, 12- or 16-hr.

The number of days of exposure to each photoperiod required to induce oviposition in each age group was noted to determine the physiological activity of the female beetles relative to their diapause and/or aestival development [9]. Also egg counts were made when the food was changed. The sex of each dead beetle was verified by dissection.

RESULTS AND DISCUSSION

Prevailing Conditions During Aestivation and Oviposition. The environmental conditions that normally prevail during aestivation of the cereal leaf beetle and during oviposition are discussed below. As in other insects, aestivation intervenes prior to the environment becoming totally unfavorable to continued development of the species. Thus, aestivation occurs in the cereal leaf beetle when the availability of succulent grasses is diminishing; oviposition occurs when most grasses are succulent and rapidly growing. The temperatures encountered during aestivation are not markedly different from those to which the insects are exposed in June while they are ovipositing. Also, the photophases during oviposition and aestivation are quite similar in duration except that they are increasing during oviposition and decreasing during aestivation.

Eggs/Female and Oviposition Period. The duration of oviposition and the number of eggs laid/ovipositing female are shown in Table 1. The number of females living at the time oviposition was initiated was used to determine the average number of eggs/female. Then, since the actual number of females ovipositing was not verified by dissection, the eggs/female is at most a relative value. The average number of eggs laid/female increased from early July until Nov. 16 when the average was 107.7 eggs, but some of this difference in production probably occurred because fewer females were laying eggs in July. The average duration of the oviposition period was quite variable, from 11.8 to 28 days.

Response to Different Photophases. The number of days of exposure to 8-,

Table 1 Number of eggs/female and duration of oviposition of cereal leaf beetles brought to the laboratory ca. every 15 days. 1970.

Date initiated*	No. of ♀**	Avg No. of eggs/♀	Avg days of oviposition
July 1	39	10.5	12.5
July 15	27	23.4	21.6
July 28	37	13.7	11.8
Aug. 14	44	21.1	22.3
Aug. 31	32	20.3	21.3
Sep. 14	19	32.3	17.0
Sep. 28	13	30.7	15.0
Oct. 15	14	29.0	14.0
Nov. 2	15	40.2	19.0
Nov. 16	15	107.7	16.0
Nov. 30	10	10.0	21.0
Dec. 14	22	38.1	20.0
Dec. 30	11	46.9	28.0

* Date beetles were confined with barley seedings in lantern globes at 26.7°C, 70±10%RH with a 16-hr photophase.

** Females alive at the initiation of oviposition.

Table 2 Number of days of exposure to 8-, 12- and 16-hr photophases before oviposition was initiated in cereal leaf beetles of known age.

Date initiated*	Age of beetles when tested**	Duration of test (days)	Days prior to oviposition Hours of light/day 8	12	16
July 11	11	68	—	—	58
July 28	28	118	87		69
Sep. 15	77	79	54		23
Nov. 16	139	46	14		14
Nov. 30	153	68	14	11	11
Dec. 14	167	53	10	10	10
Dec. 28	181	39	11	11	11

* Date cereal leaf beetles were transferred from a cage in the field to barley seedlings in the laboratory at 26.7°C and 70±10%RH.

** From July 1 as newly emerged (day 1).

12- or 16-hr photophases to induce oviposition in cereal leaf beetles of known ages is shown in Table 2. Only long days (16 hr) terminated aestivation during the 68-day test begun July 11. However after July, the number of days of exposure in the labroatory to constant long photophases necessary to induce oviposition diminished. Then after Nov. 16 a 16-hr exposure was not more effective in inducing mating and subsequent oviposition than a short photophase, that is, after the beetles were ca. 20 weeks old, they oviposited equally well whether they were exposed to short or long days. The cereal leaf beetle therefore completes aestivation by mid-November, does not diapause or require a cold period for diapause development, and overwinters in a state of quiescence. When spring temperatures become proper for activity, the beetles become active and are, thus, synchronized with their hosts.

Comparison Between 1969 and 1970. Fig. 1 shows the sequential change in the date when oviposition was initiated by beetles held in the outdoor cage for a known period. The trend was similar for both years in that aestivation termination and subsequent oviposition did not occur until after 50 days of exposure to a 16 hr photophase for beetles brought into the laboratory in July. However, as the beetles aged, repetitive stimuli were needed to break aestivation. There was a general trend to require fewer days of exposure to the 16 hr photophase to initiate oviposition from late July until late October. From Aug. 15 to Sept. 30 little change in aestival development had occurred, because the number of long days needed to induce oviposition remained about the same; i.e., 21–22 days.

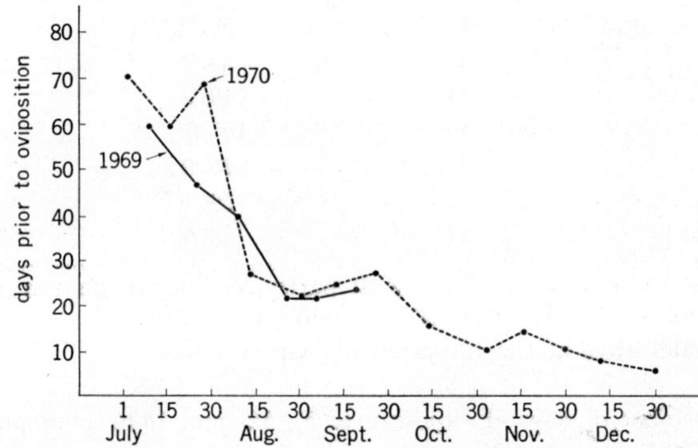

Fig. 1 Number of days of exposure to a 16-hr photophase and a temperature of 26.7°C required to terminate diapause and initiate oviposition in cereal leaf beetles of known age.

After Sept. 30, there was a general decrease in the number of days required to induce oviposition. Under optimum laboratory conditions the average time at 26.7°C for oviposition of stored beetles was 5 days. Because these experiments were checked every 3–4 days, there may be an error of 3–4 days for the date of initiation of oviposition.

The data presented in this paper agree well with those of HILTERHAUS [1] and of CONNIN and JANTZ [7]. Whether the beetles were ca. 20 weeks old and held in the laboratory in the dark at 3.3°C (7) or held in the field exposed to the prevailing environment, aestivation was completed at a similar rate based on the initiation of oviposition. Only after these beetles are ca. 20 weeks old do they respond equally well to long and short days. Thus, aestival development in the cereal leaf beetle is not solely temperature nor photoperiodic dependent. Perhaps some age-dependent metabolic factor(s) govern the duration of aestivation.

SUMMARY

Females of the cereal leaf beetle, *Oulema melanopus* (L.), aestivate from mid-November in the field and then apparently enter quiescence; mating and oviposition do not occur until the following spring. However, beetles brought into the laboratory in mid-November and held at optimum conditions (26.7°C and 70± 10%RH) mated and oviposited whether they were exposed to a short (8 hr) or a long (16 hr) photophase. Thus, the duration of aestivation did not appear to be governed solely by prevailing temperature and photoperiods but seemed to be age dependent, and may be related to the deposition or loss of some metabolic product(s). Activity in spring appeared to be temperature-dependent.

REFERENCES

1. Hilterhaus V. (1965): Biologisch-ökologische Untersuchungen an Blattkäfern der Gattungen *Lema* und *Gastroidea* (Chrysomelidae, Col.) Ein Beitrag zur Agrarökologie). Z. Angew. Zool., *52*: 257–95.
2. Castro T. R., Ruppel R. F. and Gomulinski M. S. (1965): Natural history of the cereal leaf beetle in Michigan. Mich. Agr. Exp. Sta. Quart. Bull., *47*: 623–53.
3. Grison P., Laveyie V., Jourdhevil P., Remaudiere G. and Balachowsky A. S. (1963): Famille des Chrysomelidae. *In*: Entomologie Applique a l'Agriculture, (A. Balachowsky, ed.), pp. 567–873 Part 1, Vol. 2. Masson et Cie, Paris.
4. Connin R. V. and Hoopingarner R. A. (1971): Sexual behavior and diapause of the cereal leaf beetle, *Oulema melanopus*. Ann. Entomol. Soc. Amer., *64*: 655–60.
5. Connin R. V., Jantz O. K. and Bowers W. S. (1967): Termination of diapause in the cereal leaf beetle by hormones. J. Econ. Entomol., *60*: 1752–3.
6. Teofilovic Z. M. (1969): Contribution to the study on morphology and development of cereal leaf beetle (*Lema melanopus* L.) and influence of ecological factors on its life activity. Rev. Res. Work Inst. Small Grains in Kragujevac Edit. Year IV pp. 29–124.
7. Connin R. V. and Jantz O. K. (1969): Some effects of photoperiod and cold storage on oviposition of the cereal leaf beetle, *Oulema melanopus* (Coleoptera: Chrysomelidae). Mich. Entomol., *1*: 363–6.
8. Myser W. C. and Schultz W. B. (1967): Sexing the adult cereal leaf beetle, *Oulema melanopus* (Coleoptera: Chrysomelidae). Ann. Entomol. Soc. Amer., *60*: 1329.
9. Andrewartha H. G. and Birch L. C. (1954): The Distribution and Abundance of Animals. p. 782, U. of Chicago Press, Chicago.

CIRCADIAN OXYGEN CONSUMPTION RHYTHM OF THE FLOUR BEETLE, *TRIBOLIUM CONFUSUM* WITH SPECIAL REFERENCE TO PHASE-SHIFTING

YOSHIHIKO CHIBA, LAURENCE K. CUTKOMP and FRANZ HALBERG

Entomology, Fisheries and Wildlife and Department of Pathology
Chronobiology Laboratories, University of Minnesota
St. Paul and Minneapolis, Minnesota, U.S.A.

Rhythmic events, described macroscopically, denote a reliably periodic aspect of data displayed as a function of time. In "microscopic" terms an objectively quantified *period*, an *amplitude*, and an *acrophase* (crest time of the function used to approximate a rhythm) are used, among other features of a rhythm. The acrophase is very sensitive with respect to manipulation of a synchronizer or other agents, as may be seen from work in many laboratories utilizing computative acrophases–summarized in recent Annual Review articles in Physiology and Pharmacology [1, 2].

Other definitions of "phase" are available; for example, "the instantaneous state of an oscillation within a period, represented by the value of the variable and all its timed derivatives" [3]. Mathematically, phase has been defined as the angle $(\phi+\omega t)$ in a cosine equation $Y=A\cos(\phi+\omega t)$ in our definition, where ϕ corresponds to the acrophase [4].

The present study involves the use of phase-shifting (as here studied, a single abrupt–rather than gradual–displacement in time of a periodicity) the lighting regimen to determine rules governing the (here gradual) phase-shifting of an LD12:12 synchronized circadian rhythm in oxygen consumption of the flour beetle, *Tribolium confusum* Duval.

Adults of both sexes of *Tribolium confusum* were taken at random from stock cultures in the laboratory. They were transferred to the center well of manometer flasks for respirometry determinations. Stone-ground whole wheat flour placed in the center well served as food and medium for the insects.

A recording Gilson differential respirometer was used for continuous determinations of oxygen consumption, read out at hourly intervals. A single insect was confined in the micro-reaction vessel of the flask for studies lasting as long as 35 days. Resetting of the equipment and introduction of ambient air was done once a day with no evidence of microbial contamination even though the beetles were not sterilized. The environmental temperature was kept at 25°±1°C; fluorescent light intensity at manometer flask level was 430 lux.

The design of the experiment was to carry out oxygen consumption measurements continuously on insects exposed first to LD12:12 (a 12-hr light, 12-hr dark regimen) for several days, then to advance or to delay the lighting regimen by 90° (6 hours) and determine the rapidity and completeness of change, while oxygen consumption determinations continued to be made. Other conditions were held as constant as possible.

Statistical treatment applied to the information included chronograms, the least squares fit of a 24-hour cosine function, a determination of the waveform by the statistically significant harmonics and a serial section.

RESULTS

A chronobiologic window (Fig. 1) shows that a circadian oxygen consumption rhythm in the flour beetle can be synchronized by the lighting regimen. The original data seen at the top reveal fluctuations that are prominently circadian. This becomes apparent from the fact that during 24-hour spans, demarcated by vertical dashed lines, high and low levels of O_2 consumption alternate. The row below the original data gives the start and end times, the number of observations here summarized and shows that the record covers only 232 hours. Information

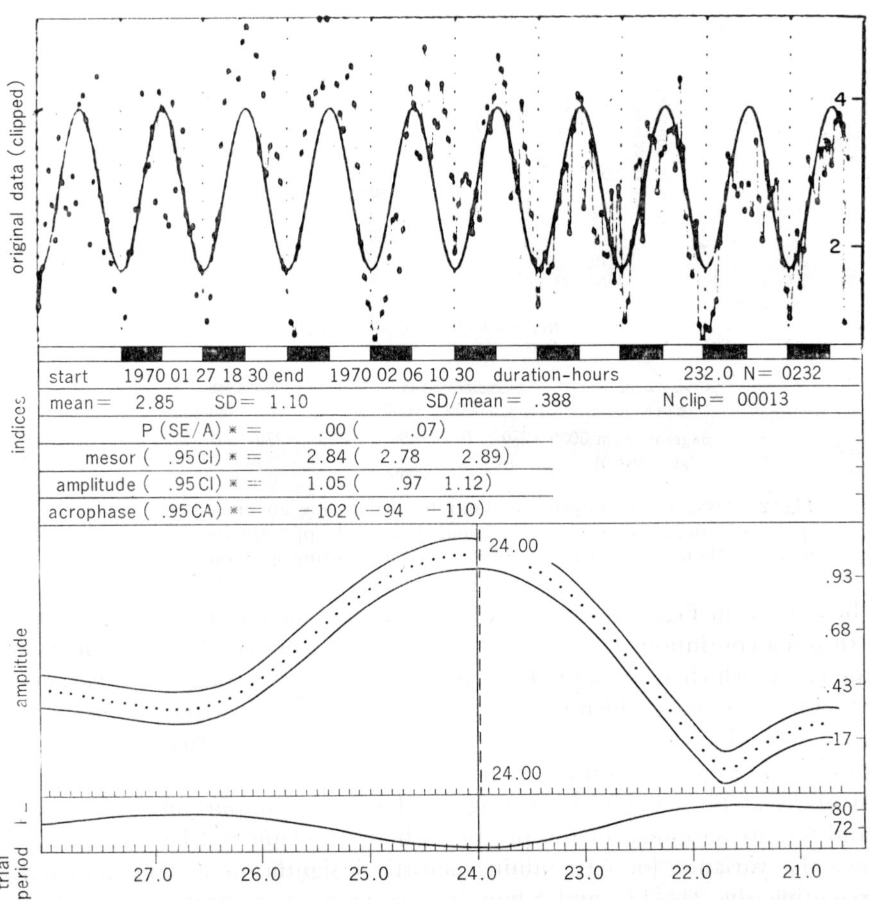

Fig. 1 Oxygen consumption of flour beetles showing daily fluctuations over period of 9 days. Lower portion presents amplitude determinations at 1/10 hour trial periods indicating a maximum amplitude at 24.00 hours.

from the analyses demonstrated in Fig. 1 reveals that the circadian amplitude is highest at the 24-hour trial period as compared to other trial periods within the spectral region from 28 to 20.6 hours. Amplitudes for shorter or longer trial periods, differing in steps of 0.1 hour, are given to the right and to the left, respectively, of the precise 24-hour trial period. Apparent from the gross display of original data (but not analyzed) are rather drastic super-imposed changes with higher-than-circadian frequencies, the so-called ultradian variations. Without further analyses these variations can not be qualified as rhythms or pararhythms. The same may be said for changes with a frequency lower than circadian that cannot be rigorously analyzed from the relatively short 232 hour record.

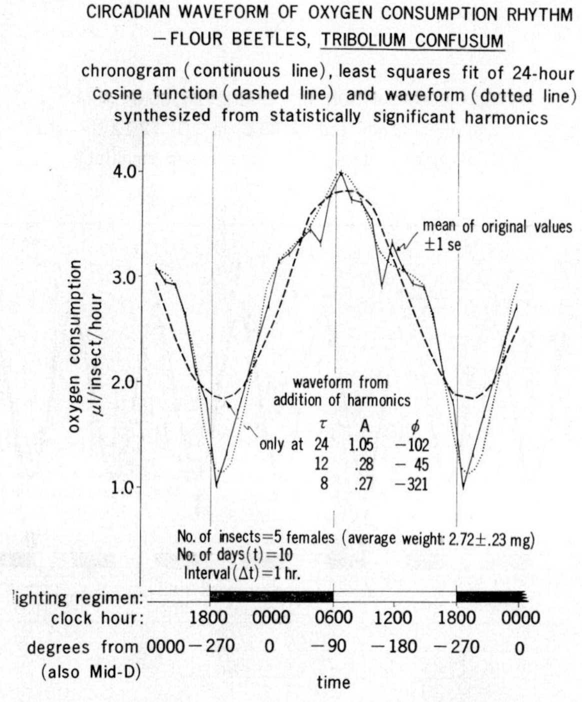

Fig. 2 Oxygen consumption of flour beetles during an "idealized" 24-hour span (utilizing mean values of original data) and approximating actual data with a 24-hour cosine function and a waveform using harmonics.

The data from Fig. 1 is presented as an average for an "idealized" day which constitutes a continuous line in Fig. 2. This line is compared to a single 24-hour cosine curve, which is a reasonable approximation of the original data, and calculation by harmonics which more nearly approximates the data. Thus, in Fig. 2 it can be seen that the 24-hour cosine curve (dashed line) approximates the major features of the continuous line which averages the original data, but is imperfect, particularly at the trough of the original data. It is important, then, that the 12 and 8 hour harmonic of the precise 24-hour fundamental be used to further reduce the variance for determining statistical significance. The dotted line, representing the 24-, 12-, and 8-hour curves with appropriate amplitudes and acrophases, is more clearly presented by a fundamental and two harmonics.

The chronogram (Fig. 3) shows the original data on the phase-shifting of the

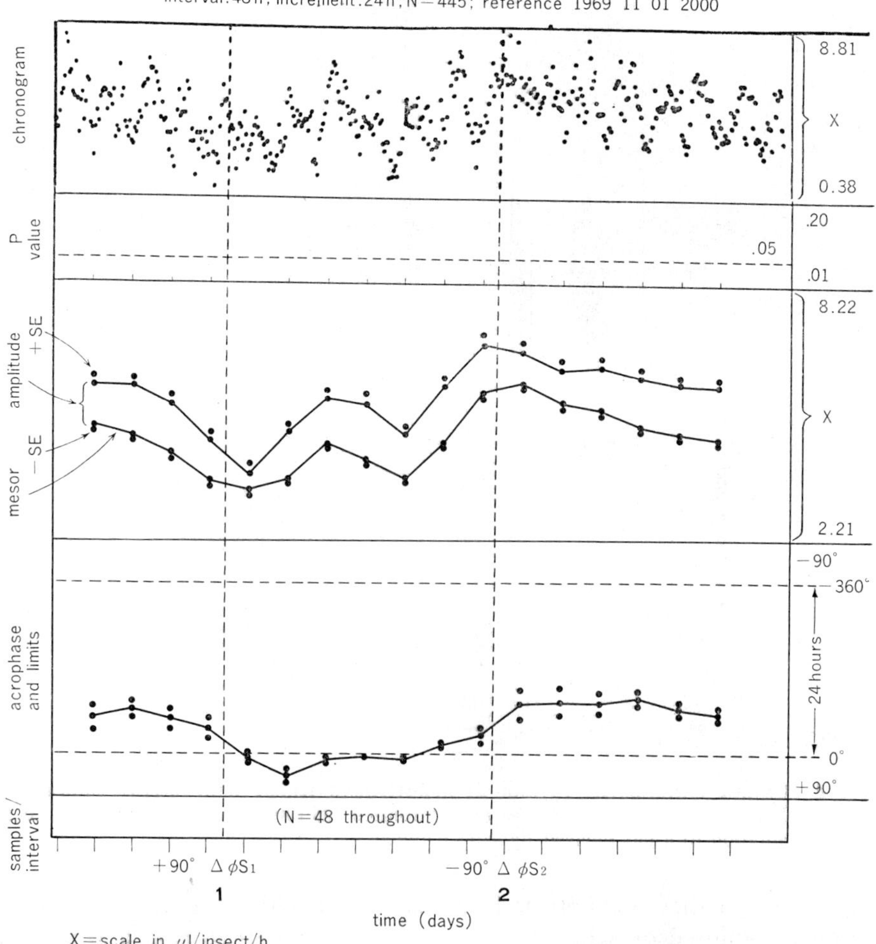

Fig. 3 Chronogram giving original data of oxygen consumption determinations of flour beetles (top) between 0.38 and 8.81 μl per insect per hour under LD12:12. Amplitude and mesor (middle). Change at shifts of lighting shown by acrophase (bottom). At mark 1 on abscissa light span was shortened from 12 to 6 hours; at mark 2 light prolonged from 12 to 18 hours. Acrophase adjustments may be noted.

beetles (top) and reveals a statistically significant circadian rhythm before and after the beetles were subjected to manipulations of the lighting regimen. At mark 1 along the scale of days, shown at the bottom of Fig. 3, a single light span was shortened from 12 to 6 hours; at mark 2 a light span was prolonged from 12 to 18 hours to return, then, to the original regimen. In the bottom row the acrophase indicates that the adjustment to an advance (after mark 1) was slower than the almost immediate response to a delay.

The results shown for a group of flour beetles as a whole presented in Fig. 4 show a slower advance as compared to a faster delay. The comparative physio-

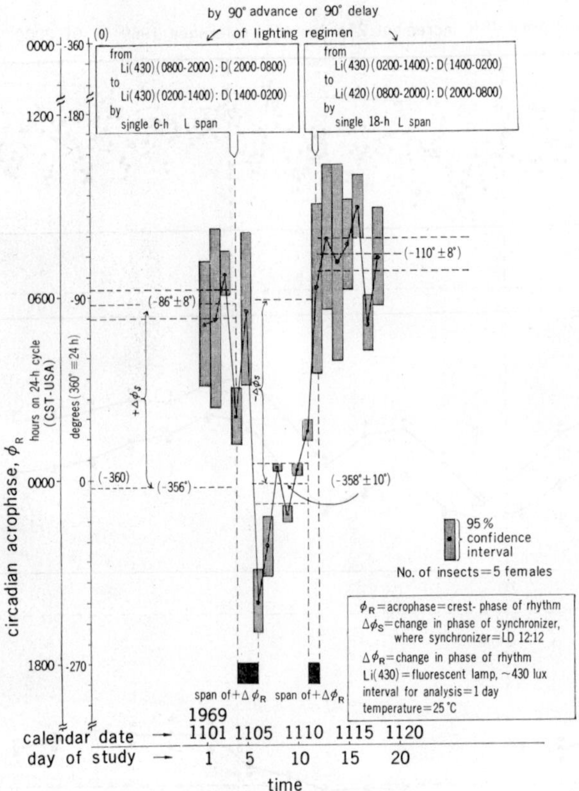

Fig. 4 Oxygen consumption rhythm in non-sterilized female flour beetles showing a slower adaptation to a 90° advance of a synchronized lighting regimen as compared to a faster adaptation to a 90° delayed lighting regimen.

logic interest of these data stems from the fact that the advance resembles the adjustment required of man following a flight from west to east, and the delay that occurs after fast travel from east to west. Most organisms seem to adjust faster in one direction as compared to another. Data from most men and experimental rats indicate relationships which are similar to those found in flour beetles.

REFERENCES

1. Halberg F. (1969): Chronobiology. Ann. Rev. Physiol., *31*: 675–724.
2. Reinberg A. and Halberg F. (1971): Circadian Chronopharmacology. Ann. Rev. Pharmacology, *11*: 455–492.
3. Aschoff J. (1965): Circadian Clocks. North-Holland Pub. Co., Amsterdam.
4. James G. and James R. C. (1968): Mathematics Dictionary, Third Edition. Van Nostrand Reinhold Company, New York.

CIRCADIAN RHYTHMS IN FLORIDA MOSQUITOES*

JAI K. NAYAR and DONALD M. SAUERMAN, Jr.

Entomological Research Center, Florida Division of Health
Vero Beach, Florida, U.S.A.

Circadian rhythms in mosquitoes have been found in pre-adult developmental stages and in adults, e.g., the pupation and emergence rhythms [1–5], flight activity [6–9], oviposition [10, 11] and feeding rhythms [12]. These rhythms are either unimodal or bimodal and they seem to be adaptations to fluctuations of environmental factors. Internal oscillators are assumed to control these rhythms and they are synchronized by external timing elements called synchronizers or entraining agents, such as photoperiod and temperature.

In the course of studying developmental and aging phenomena in 17 species of Florida mosquitoes, we have found circadian rhythms in pupation and in the flight activity of adults. Pupation happening but once to an individual, the rhythm is a population phenomenon. Flying, on the other hand, is a recurrent activity and therefore displayed as a circadian rhythm by both individual and population.

We have studied both of these rhythms extensively in the black salt marsh mosquito *Aedes taeniorhynchus*, which we adopted as a model on which to base a study of circadian rhythms of 16 other species of mosquitoes. These circadian rhythms were found to be always species-specific. One advantage of working with mosquitoes is that it is very easy to synchronize hatching of larvae from eggs in most species, either by artificial stimulation or by synchronous egg laying [2, 3], thus making it easy to have a common starting point for a study population. Larval ecdyses are also easily synchronized by providing the larvae with proper amounts of food.

We will first consider these biorhythms in detail in *A. taeniorhynchus* and then consider the species-specific nature of these biorhythms in other Florida mosquitoes.

ENDOGENOUS DIURNAL RHYTHMS OF PUPATION AND EMERGENCE

a. In *A. taeniorhynchus* [2, 4]

A. taeniorhynchus has, in continuous darkness, an endogenous circadian rhythm of pupation with a period 21.5 hours [2]. This period is lengthened to 23.5 hours under LD 12:12 on standard rearing conditions, but can be entrained to a 24-hour day only under duress, such as dense larval crowding and high salinities,—conditions often occurring in nature. The entrained pupation rhythm persists in continuous DD, but becomes arrhythmic under LL. A nonrepeated

* This work was supported in part by Grant no. AI-06587, National Institutes of Health, U. S. Public Health Service.

light pulse will synchronize each peak of pupation and establish a new phase for the rhythm [4]. A nonrepeated light pulse of at least 4 hours duration is effective in phase-establishment as early as immediately after simultaneous hatching of the eggs and as late as 72 hours after hatch. The amount of phase shift is proportional to the duration of the light stimulus [4]. A light pulse, given at different phases of the subjective light-dark (LD) cycle, showed that sensitivity to light stimuli was greater during the first 12 hours than during the second 12 hours of the subjective LD cycle, and phase response curve obtained for the pupation rhythm under standard conditions was very similar to those reported for different single animals and populations [13]. LD cycles of less than 24 hours, varying from 12 to 22 hours, imposed during the larval stages of development do entrain the pupation rhythm, although these LD cycles were not learned.

Phase-shifts, transients, and phase-resetting by temperature perturbations differ from those generated by light perturbations. Temperature stimuli produce spectacular transients with relatively little phase-shift in the ultimate steady-state, whereas light pulses have spectacular effects on the steady-state phase-shift. Temperature does not affect the circadian period, i.e., at any temperature between 22° and 34°C, the periodicity of pupation peaks varying from 22.5 to 23.5 hours, thus showing a high degree of temperature compensation.

Since only temperature determines the duration of the pupal period, a population exposed to the same temperature regime exhibits an emergence rhythm almost identical to the antecedent pupation rhythm.

b. In other species [5]

Of the 17 species studied, 6 (*A. taeniorhynchus, A. sollicitans, Psorophora confinnis, Culex bahamensis, C. nigripalpus* and *Anopheles crucians bradleyi*) show distinct circadian rhythmicity in pupation under all combinations of factors in the rearing environment. *In Culex pipiens quinquefasciatus* a diurnal rhythm becomes clear only when larvae are reared on the minimum quantity of food and it is bimodal. This bimodal rhythm was modified to unimodal when larvae were *temporarily* crowded and fed during the duration of crowding. A bimodal diurnal rhythm first appears in *P. ferox* when larvae are reared crowded, but it becomes distinct when larvae are either fed or starved during the period of temporary crowding. In 9 other species no indications of diurnal rhythmicity of pupation are evident.

The emergence pattern in all the species follows the pupation pattern which is separated by an interval affected by temperature only.

A growth variability index calculated from the effects of different environmental factors on growth showed that all species possessing endogenous diurnal rhythms of pupation have higher values of this index [5].

CIRCADIAN RHYTHMS OF FLIGHT ACTIVITY

a. In *A. taeniorhynchus* [9]

In an LD 12:12 regime, flight activity occurs both at light-off and light-on, forming a bimodal 'alternans' pattern [9]. This basic pattern of flight activity persists with a periodicity of 23.5 hours under continuous DD, but under continuous LL it is masked over by irregular excessive outbursts of activity. This flight activity rhythm originates strictly with the adult; it is not carried over from

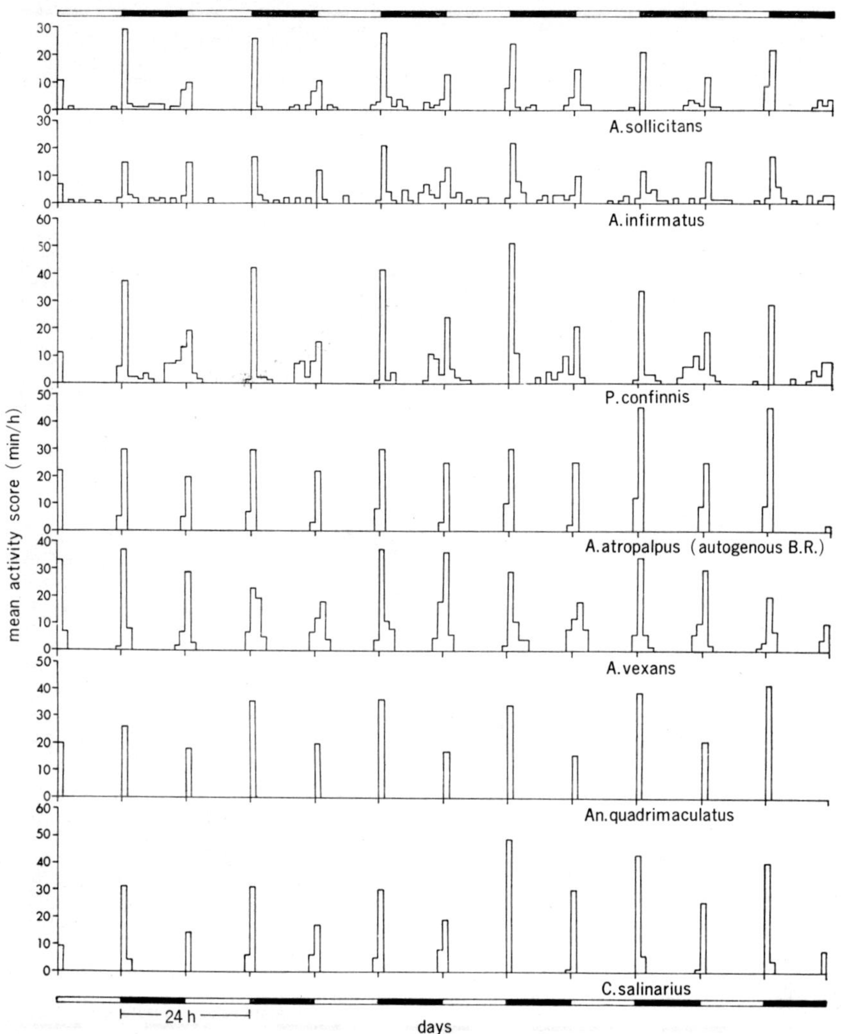

Fig. 1 Mean flight activity patterns in different species of Florida mosquitoes under LD 12:12 at 27°C.

pupation or emergence rhythms. It can be entrained to a new light regime within 24–36 hours, which is rather fast. An early light-off does not reset the phase of the rhythm, but a delayed light-off does. It can be entrained to 24-hour light regimes other than LD 12:12. But a stimulus of less than 12 hours is not effective in initiating the bimodal circadian rhythm. Frequency demultiplication within certain limits can entrain the flight activity rhythm to 24 hr.

b. In other species

In all the species investigated under an LD 12:12 regime, flight activity usually exhibited a bimodal pattern (Figs. 1 and 2). *A. sollicitans, A. infirmatus, P. confinnis, A. atropalpus* (an autogenous strain BR), *A. vexans, An. quadrimaculatus, C. salinarius,* and *C. p. quinquefasciatus* exhibited an alternans pattern, with peaks at light-on and light-off times only, with very little flight activity during

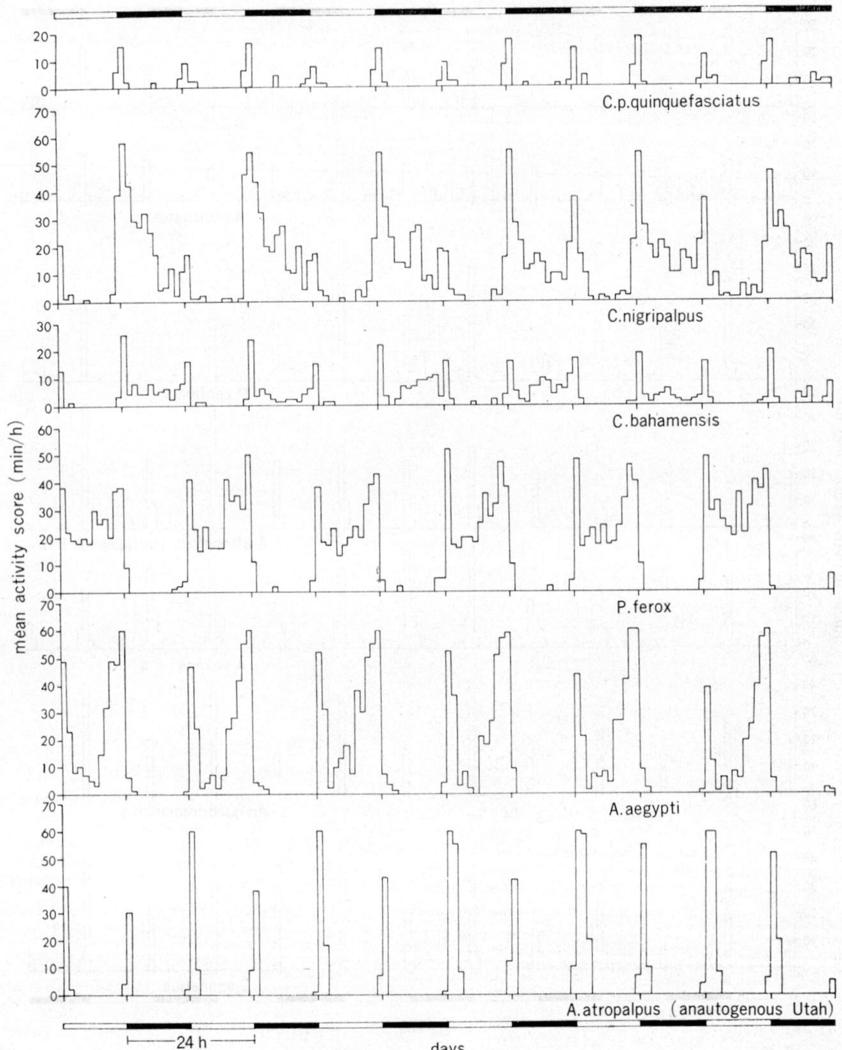

Fig. 2 Mean flight activity patterns in different species of Florida mosquitoes under LD 12:12 at 27°C.

the dark period and almost none during the light period. *C. nigripalpus* and *C. bahamensis* usually exhibited an alternans pattern, with a major peak at light-off times followed by continuous activity throughout the dark period at a slightly lower level, and then a sizable peak at light-on lasting for about an hour. There was almost no activity during the light period except for about an hour at light-on. *P. ferox* and *A. aegypti* showed a bimodal alternans pattern which started with light-on with a peak lasting for about 2 hours, followed by continuous activity for several hours at a lower level and increasing to a maximum during the last 4–5 hours of the light period. This activity stopped abruptly with light-off and there was no activity during the dark period of the LD 12:12 cycle. Only one species, *A. atropalpus* (an anautogenous strain, Utah) exhibited a 'bigeminus' flight pattern, with a flight activity peak of about 2 hours at light-on and a much smaller activity peak at light-off, with almost no flight during the remainder of either light or dark periods.

In all the cases the entrained flight activity patterns continued in continuous darkness. However, in some cases there was no delay in appearance of the endogenous flight activity peaks, e.g., in *C. nigripalpus, A. sollicitans, C. salinarius, C. p. quinquefasciatus* and *P. confinnis*; whereas in other cases appearance of endogenous flight activity peaks under continuous DD was delayed for 2–4 days.

In conclusion, it is suggested that the basic bimodal pattern of flight activity is a persistent property of the circadian oscillating system and it provides a framework for other activities which involve flight, like mating, host-seeking, sugar feeding and oviposition, and which seems to occur in some of the species so far studied [10–12]. Also, it is evident that understanding of these circadian rhythms has given us a better concept of the behavior of different species of mosquitoes under different environmental conditions in nature [14–16].

REFERENCES

1. Provost M. W. and Lum P. T. M. (1967): Ann. Ent. Soc. Amer., *60*: 138–49.
2. Nayar J. K. (1967): Ann. Ent. Soc. Amer., *60*: 946–71.
3. Nayar J. K. (1968): J. Med. Ent., *5*: 39–46.
4. Nayar J. K. (1968): Ann. Ent. Soc. Amer., *61*: 1408–17.
5. Nayar J. K. and Sauerman D. M. (1970): J. Med. Ent., *7*: 163–174.
6. Chiba Y. (1966): Sci. Rep. Tohoku Univ. Ser. IV (Biol.), *32*: 97–104.
7. Jones M. D. R., Hill M. and Hope A. M. (1967): J. exp. Biol., *47*: 503–511.
8. Taylor B. and Jones M. D. R. (1969): J. exp. Biol., *51*: 59–71.
9. Nayar J. K. and Sauerman D. M. (1971): J. exp. Biol., *54*: 745–756.
10. Haddow A. J., Gillett J. D. and Corbet P. S. (1961): Ann. Trop. Med. Parasit., *55*: 343–56.
11. Haddow A. J. and Ssenkubuge Y. (1962): Ann. Trop. Med. Parasit., *56*: 332–35.
12. Gillett J. D., Haddow A. J. and Corbet P. S. (1962): Entomologia exp. app., *5*: 223–32.
13. Aschoff J. (1965): *In*: Circadian Clocks, (J. Aschoff, ed.), pp. 95–111, North-Holland Publishing Co., Amsterdam.
14. Provost M. W. (1971): Proc. 13th Int. Congr. Entomol. Moscow, 1968 (1972), *3*: 228.
15. Mattingly P. F. (1971): Proc. 13th Int. Congr. Entomol. Moscow, 1968, *1*: 416–17.
16. Nayar J. K. and Sauerman D. M. (1969): Ent. exp. app., *12*: 365–375.

CIRCADIAN PATTERN IN SUSCEPTIBILITY OF *AEDES AEGYPTI* (L.) LARVAE TO DURSBAN®

Donald R. ROBERTS*, Michael H. SMOLENSKY, Bartholomew P. HSI and John E. SCANLON

The University of Texas School of Public Health
Houston, Texas, U.S.A.

Circadian rhythms have been defined as oscillations in function having periods in the range from 20–28 hours. Occurrence of an exact 24-hour period length usually results from synchronization of endogenous functions with external events. In a sense, a given organism can be viewed as a multitude of circadian systems organized with temporal interplay. Since circadian patterns are known at all levels of organismic function, it is not inconsistent to expect responses by rhythmic organisms, when challenged, to differ as a function of circadian phase.

The role of circadian function as a determinant of response has become important in the evaluation of medications [6, 10], toxicants [3, 5, 7] and pesticides [11]. With respect to the latter, circadian susceptibility rhythms have been demonstrated in adult boll weevils [1], two-spotted spider mites [8, 9] house flies [11] and madeira cockroaches [11]. These investigations indicate that the same concentration of a given pesticide may result in drastically different effects dependent on circadian phase of testing.

Since a considerable amount of routine susceptibility testing is done on mosquitoes it is desirable that information on temporal variations in susceptibility to insecticides be available. The purpose of this investigation was to determine the existence of a temporal pattern of response. Since mosquito larval susceptibility testing procedures are unusual in that the immature stage of the insects are tested and the 24-hour end-point of mortality is observed following continuous exposure to the toxicant, this investigation was carried out to determine whether there was a detectable daily pattern in susceptibility of *Aedes aegypti* (L.) larvae to Dursban®**,0,0-diethyl 0-(3,5,6-trichloro-2-pyridyl) phosphorothioate, an organic phosphorus insecticide, when standard testing procedures were followed.

The *Aedes aegypti* larvae used in this investigation were obtained from a laboratory colony that originated from wild stock collected in Corpus Christi, Texas in 1969. Adults were fed a 5% sucrose solution; at intervals blood meals were provided by guinea pigs, with occasional feedings on humans and chickens. The insectary was kept at 76±1°F and approximately 80% R.H. The adults, like the larvae, were entrained by L (0600–2000): D (2000–0600). Eggs required for each study were collected two weeks in advance and were properly conditioned before they were dried and stored [2].

---

* Medical Entomologist, Captain, Medical Service Corps, U.S. Army. Presently assigned to the University of Texas School of Public Health at Houston, Texas 77025.

** Mention of a proprietary product does not necessarily imply endorsement of this product by the United States Army.

Temporal patterns of susceptibility were evaluated during two studies. In these studies a susceptibility test was performed every 4 hour during continuous 24-hour spans. Each test consisted of four treatment levels and an acetone control with two replicates at all treatment levels. Each replicate consisted of 20 larvae per 250 ml of treated distilled water in a 400 ml glass beaker. Larvae that could not submerge from the surface or rise from the container bottom following a sitmulus were considered dead. All mortality counts were done by the same investigator.

During the photoperiod entrainment and test periods the lights were shut off at 2000 and turned on at 0600 hours. This 14 L:10D photoperiod approximated day length in the Houston, Texas area during mid-summer. Susceptibility tests were carried out at 4-hour intervals starting at 1700. The 0500 test was omitted in the first study, consequently for convenience in presenting graphical data the results are not given in order of testing. Tests were carried out in an isolated room of the laboratory where the temperature was stabilized at 76±1°F and 78±1°F for first and second studies, respectively.

Standardized methods of hatching, rearing, and picking larvae for testing were developed to eliminate as many random variables as possible. A single paper containing eggs provided the required larvae for each study. To reduce variation in age of larvae exposed to the insecticide, eggs were hatched at 4-hour intervals during a single 24-hour span. Larvae obtained from each of the eclosion times were used for testing at the same clock hour 5 days later, i.e. all larvae were exactly 120 hours old when exposed to the insecticide.

Eggs were hatched by placing strips of paper containing the required number of eggs in distilled water and exerting a 20 pound vacuum. This stimulus served to synchronize hatching so that an adequate hatch was consistently obtained within 30 minutes. After eclosion, 300 larvae were removed from the hatching container and placed in the insectary in a large white enamel pan with 3 liters of distilled water. The larvae were fed on pulverized Wayne Rabbit Ration. The feeding schedule for the two studies differed. Although the total amount of food for both studies was the same, larvae for the first study received on days 0, 1, 2, 3, 4, and 5 a total of 0.2, 0.0, 0.6, 0.8, 0.8 and 0.8 mg food per larvae, respectively and 0.3, 0.4, 0.4, 0.7, 0.7, and 0.7 mg food per larva in the second study.

A standardized method was devised to transfer larvae from the rearing pans to test containers. First, groups of 20 larvae were placed in small disposable medicine cups in 3 ml of water. Next, larvae were poured into the Dursban® treated waters. This procedure was carried out at each time of testing. The total number of larvae needed during the dark phase of the first study were picked at 1900 (one hour prior to the onset of the dark span); however, this procedure was changed in the second study in that larvae were picked and placed in the transfer cups 4 hours prior to each test time.

The test concentrations were prepared by serial dilutions (W/V) in distilled water commencing with a stock solution of 1 mg of 97 percent technical Dursban® to 1 ml of acetone. A 1 ppm solution in water was prepared first and test concentrations were subsequently obtained by simple volumetric dilutions. The entire formulation procedure, except for the stock solution in acetone, was done for each test by the same person, Preliminary investigation of *A. aegypti* susceptibility of larvae to Dursban® resulted in four concentrations: 0.004, 0.0025,

0.00156, and 0.00098 ppm for selection as test dose levels for expected mortalities of approximately 95, 60, 40 and 20 percent, respectively. No control mortality occurred during this investigation. Probit analyses served to determine the maximum likely estimate for the LC_{50} and LC_{95} (lethal concentrations—LC) values based upon data collected at each time of testing [4].

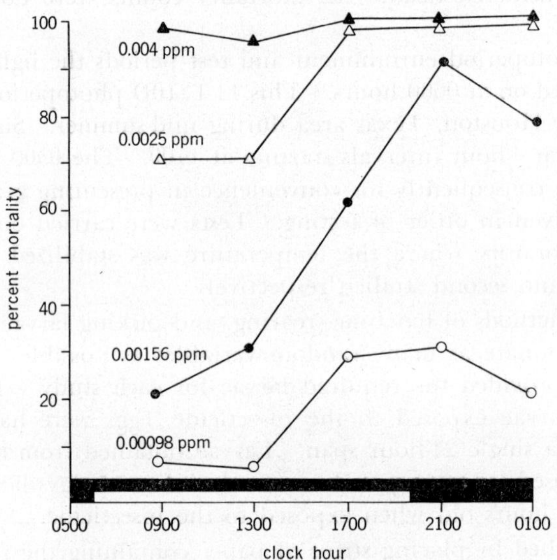

Fig. 1 Temporal variations in percent mortality of *Aedes aegypti* (L.) Larvae to four concentrations of Dursban® (Study 2). Results from tests done every 4 hours for a 24-hour period.

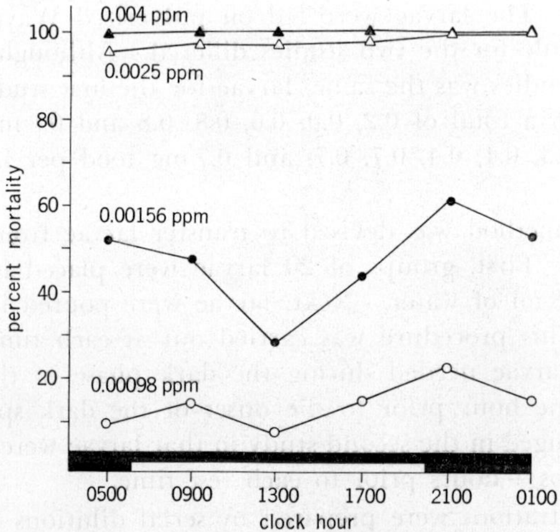

Fig. 2 Temporal variations in percent mortality of *Aedes aegypti* (L.) Larvae to four concentrations of Dursban® (Study 1). Results from tests done every 4 hours for a 24-hour period.

Figs. 1 and 2 illustrate the 24-hour mortalities at each of the four treatment levels for each test time. In the two studies, mortality from Dursban® was greatest between 2100–0100 (one to 5 hours following initiation of the dark period). Lowest mortality generally occurred around 0900 to 1300 (3 and 7 hours following initiation of the light period). For example in Study 1 at 2100 a dose of 0.00156 ppm, produced in the fourth instar mosquito larvae, a mortality of 90 percent. Identical tests carried out at 0900 resulted in mortality of only 20 percent. A similar circadian response rhythm, although of lesser amplitude, was found for a concentration of 0.00098 ppm such that, at 2100 35 percent mortality resulted; whereas, at 0900 mortality was only 2.5 percent. In the second study (Fig. 2), the crest and trough times of susceptibility were quite similar; although, the peak to trough differences were less. This may reflect the influence of nutrition and/or temperature.

The calculated LC_{50} and LC_{95} values and standard errors are presented in Table 1. The LC_{95} values varied with time of exposure by about 2.3-fold in study 1 and roughly 1.1-fold for study 3. Similar findings were evident for the LC_{50} concentrations, i.e. the peak to trough differences in the LC_{50} amounted to a 1.9-fold difference in study 1 and a 1.3-fold difference in study 2. There were indications from the LC_{50} values obtained in the second study that the susceptibility pattern may be at least bimodal.

Table 1 Temporal variations in susceptibility of *Aedes aegypti* (L.) larvae to Dursban® as demonstrated by maximum likely estimates of LC_{50} and LC_{95} values from tests done every 4 hours during a 24-hour span for each of 2 studies (standard errors included).

Time of exposure		Study one		Study two	
		LC_{50}[1] ± s.e.[1]	LC_{95}[1] ± s.e.[1]	LC_{50}[1] ± s.e.[1]	LC_{95}[1] ± s.e.[1]
0500*	(4)	—	—	152±1.09	250±1.09
0900	(5)	207±1.05	370±0.09	149±1.10	258±1.09
1300	(6)	201±1.15	392±0.10	171±1.04	259±1.10
1700	(1)	127±1.15	230±0.40	153±1.09	258±1.09
2100	(2)	109±1.24	171±0.08	134±1.11	228±1.09
0100	(3)	123±1.17	191±0.12	139±1.12	227±1.08

* Test at 0500 was omitted in Study one.

[1] LC value and s.e. $\times 10^{-5}$

( ) Order of testing

Since these rhythmic changes are prominent it seems worthwhile to continue and expand this preliminary study of circadian susceptibilities to include both adult and larvae in order to 1) determine the value of standardizing laboratory testing to take into account circadian rhythmicity of response and 2) to discover the timing of greatest susceptibility (greatest kill-rate) for various insecticides, especially those which differ in chemical composition and thus in physiologic effect, for ultimate application to field problems. The objective for the immediate future will be to determine whether there is agreement in the susceptibility pattern obtained with one-hour exposure to insecticide with the response rhythm obtained with the 24-hour mortalities resulting from continuous exposure.

REFERENCES

1. COLE C. L. and ADKISSON P. L. (1964): Daily rhythm in the susceptibility of an insect to a toxic agent. Science, *144*: 1148–1149.
2. CHRISTOPHERS Sir S. RICKARD (1960): *Aedes aegypti* (L.) the yellow fever mosquito. Cambridge University Press, 739.
3. ERTEL R. J., UNGAR F. and HALBERG F. (1963): Circadian rhythm in susceptibility of mice to toxic doses of SU-4885. Fed. Proc., 1963.
4. FINNEY D. J. (1964): Probit Analysis. 2nd ed., p. 318, Cambridge at the University Press, London.
5. JONES F., HAUS E. and HALBERG F. (1963): Murine circadian susceptibility-resistance cycle to acetylcholine. Proc. Minn. Acad. of Sci., *31*: 61–62.
6. KIRCHOFF H. W. (1970): The impact of diurnal rhythm on drug dosing and drug evaluation. Int. J. Clin. Pharmacol., *4 (1)*: 68–71.
7. MATTHEW J. H., MARTE E. and HALBERG F. (1964): A circadian susceptibility resistance cycle to fluothane in male B_1 mice. Can. Anes. Soc. J., *11*: 280–290.
8. NOWOSIELSKI J. W., PATTON R. L. and NAEGELE J. A. (1964): Daily rhythm of narcotic sensitivity in the house cricket, Gryllus domesticus L., and the two-spotted spider mite, *Tetranychus urticae* Kuch. J. Cell. Comp. Physiol., *63*: 393–398.
9. POLLICK B., NOWOSIELSKI J. W. and NAEGELE J. A. (1964): Daily sensitivity rhythm of the two-spotted spider mite, *Tetranychus urticae* to DDVP. Science, *145*: 405.
10. SMOLENSKY M., HALBERG F. and SARGENT II F. (1972): Chronobiology of the life sequence. *In*: Advances in Climatic Physiology, (S. ITOH, K. OGATA and H. YOSHIMURA, eds.), pp. 281–318, Igaku Shoin Ltd., Tokyo.
11. SULLIVAN, W. N., CAWLEY B., HAYES D. K., ROSENTHAL J. and HALBERG F. (1970): Circadian rhythms in susceptibility of house flies and madeira cockroaches to pyrethrum. J. Econ. Entomol., *63* (v): 159–163.

THE USE OF ARTIFICIAL LIGHT OUTDOORS TO CONTROL AGRICULTURAL INSECTS

MILTON S. SCHECHTER, WILLIAM N. SULLIVAN and DORA K. HAYES

Entomology Research Division, Agricultural Research Service, USDA
Beltsville, Maryland, U.S.A.

Since the publication of the classical work of GARNER and ALLARD [1] and BÜNNING [2], intensive research in laboratories all over the world has yielded valuable information concerning the nature of biological rhythms in animals and plants. The literature is documented in the proceedings of symposia on the subject of biological clocks [3–5] and in several books [6–8]. This basic knowledge about periodicity has been used practically in the control of seed germination and flowering of plants [9] and has been proposed as a guide in the choice of the optimum time of day to administer drugs to humans [10] and to determine the best time to apply pesticides to control insects [11]. In addition, researchers are now investigating the control of pest insects by manipulating diapause in the field [12, 13]. The present paper reports the results of 2 years of field studies at Beltsville, Md. in which artificial light was used in the fall to simulate "long day" light conditions to prevent diapause and thus reduce insect populations the following spring.

Three isolated areas (A, B, and C) were selected for the field trials, divided into plots (3.7×3.7 m), and planted (two 3.7 m rows) to 'Seneca Chief' corn during the early part of June 1969 and 1970. Initially, 4 plots were located in Area A, 4 in Area B, and 8 in Area C. In early July, each of the 16 plots was enclosed in a Saran* screen cage (3.7×3.7×1.9 m), and 2 light fixtures containing a total of eight 4-ft (1.22 m) fluorescent tubes (combinations of 40-watt blue F40B and 38.5-watt daylight F 48T12/D types) were hung from the aluminum frame of each cage in Areas A and B 20.3 cm from the top. The fixtures in Area A were all connected to one time clock switch; those in Area B were similarly connected with another switch. The 8 plots in Area C served as the controls since no artificial light was provided and in these plots, the exposure to natural daylight (sunrise to sunset) decreased from 13 hr:55 min. of light on August 9 to 9 hr: 55 min. of light on December 5.

In addition, in 1970, another open plot (17.36 m×17.36 m) in Area B was planted with 10 56-ft (17.36 m) rows of Seneca Chief corn. Two telephone poles were located on either side of the plot 56 ft (17.36 m) equidistant from the center, and one 1000-watt mercury lamp was mounted on each pole. These lamps also were connected to a time clock switch. The intensity of the artificial light at various locations in the open plot was measured and is shown in Fig. 1.

All 3 timer switches were reset to compensate for each 15 minute of change in the time of sunrise at Washington, D.C. [14] i.e., about every 2 weeks. The LD (light-dark) regimens used are stated in Tables 1 and 2.

* Mention of a proprietary product in this paper does not constitute an endorsement of this product by the USDA.

On August 9, 1969 and on August 18, 1970, the corn in all plots (it was then in the early silk stage) was infested with European corn borers, *Ostrinia nubilalis* (Hübner), by pinning a wax paper disk 10 mm in diameter containing one egg mass (10 to 30 eggs in the blackhead stage furnished by the European Corn Borer Laboratory, Ankeny, Iowa) on the underside of a leaf near the stalk 3 nodes up from the lowest ear of each corn plant. Also, on August 20, 1969 and on September 10, 1970, a total of 1000 young apples artificially infested with 1st-instar larvae of the codling moth, *Laspeyresia pomonella* (L.), (obtained from the Entomology Research Division Laboratory at Yakima, Washington) in separate, open 2-oz (59.2 ml) plastic cups were placed in clear plastic sweater boxes (28×41× 5 cm), and cotton was packed between the cover and the rim of each box so the larvae could not escape (larvae either pupated inside the apples or migrated to the cotton packing). Then the box lids were put in place, and the boxes were set on the shelves of metal racks inside cages in Areas A, B and C.

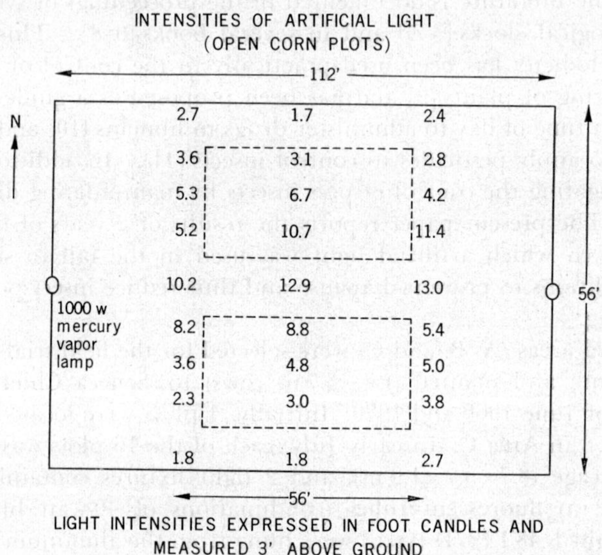

Fig. 1

Daily observations of the number of corn borer and codling moth adults in all 3 areas were made from mid September, when the adult corn borers started to emerge, until the killing frosts of early December (artificial lighting was discontinued at that time). After the killing frosts, every fourth hill of corn in all 3 areas was harvested, and the ears, foliage, and stalks were examined by dissection for diapausing 5th-instar corn borer larvae; the number of larvae was then multiplied by 4 and added to the number of adults collected to obtain the total number of insects in each plot. Also, final counts of codling moths were obtained in December by dissecting the apples and inspecting the cotton packing to determine the number of larvae and pupae present; this number was added to the number of adults collected earlier to obtain the total number of insects. (Those in the pupal and adult stages obviously did not diapause.)

The effects of the manipulated photoperiods on the incidence of diapause for European corn borers and the codling moths in 1969 and 1970 are shown in Table 1.

Table 1 Effect of manipulated photoperiods on the incidence of diapause in the European corn borer and in the codling moth in 1969 and 1970.

Light regimen	Year	European corn borer		Codling moth	
		Total no.	Percent not diapausing	Total no.	Percent not diapausing
		Caged test plots			
LD 17:7	1969	570	76	23	70
LD 17:7	1970	458	95	35	90
Natural with 1 22-min light break 3 hr after sunset and 1 22-min break 2 hr before dawn	1969	819	1	35	6
Natural light + light pulses 1 min light 3 min dark for 2 hr 15 to 17 hr after dawn	1970	447	3	30	0
Natural light	1969	304	0	51	8
Natural light	1970	533	1	42	0
		Open test plot			
17 hr light	1970	175	70		
Natural light	1970	103	9		

Table 2 The number and percentage of European corn borers surviving the winter in photoperiod manipulation tests.

Light regimen	Year	Initial fall population	No. found by cutting in April	% of initial population in fall	No. emerging in the spring	% of initial fall population
LD 17:7	1969	570	150	26	78	14
	1970	458			43	9
Natural LD with 1 22-min light break 3 hr after sunset and 1 22-min break 2 hr before dawn	1969	819	520	64	399	47
Light pulses 1 min light 3 min dark for 2 hr 15 to 17 hr after dawn	1970	447	not done		217	49
Natural light	1969	304	310	102	311	102
Natural light	1970	533	not done		644	121

The results were similar both years though the percentage of European corn borers that did not diapause increased from 76 in the fall of 1969 to 95 in the fall of 1970, and the percentage of codling moths that did not diapause increased from 70 to 90. This increase may have occurred because of the comparatively mild weather in the fall of 1970. In the open plot illuminated with the mercury lamps, 70% of the European corn borers did not diapause. No dramatic differences were observed in the response of insects from different areas in this open plot. Almost all European corn borers and codling moths held in the control cages (Area C) diapaused.

The number and percentage of corn borers that survived the winters of 1969 and 1970 are given in Table 2. In 1969, only 14% and in 1970, only 9% of the original population of the European corn borers (or 51% in 1969 and 100% in 1970 based on the diapausing population) survived the winter when the fall population was subjected to the regimen of LD 17:7; about 100% of the control populations survived.

Interesting but less conclusive results were obtained in some plots (Area B) when the natural outdoor light was supplemented with light pulses introduced during the dark phase. These pulses did not appreciably affect the diapause of the European corn borer, but they did seem to have an adverse effect on their viability. The result is further evidence that some changes in photoperiod can cause damage to or mortality of organisms [11, 15].

The economics involved in the installation of the equipment necessary to extend the light period and the use of power was not considered because the minimum duration and intensity of the light required has not yet been determined. However, it is not likely that such manipulation of the photoperiod will alone provide sufficient control of insects. If the method becomes economically feasible, it will most likely have to be used in integrated control with other methods such as the use of insecticides, sterile insects, attractants and repellents, parasites and predators, insect pathogens, or hormones.

ACKNOWLEDGMENT

Appreciation is expressed for the excellent cooperation given by Lowell CAMPBELL and other members of the staff of the Agricultural Engineering Research Division, ARS, and Karl Norris, Market Quality Research Division, ARS, USDA.

REFERENCES

1. GARNER W. W. and ALLARD H. A. (1920): Effect of relative length of day and night and other factors on growth and reproduction in plants. J. Agr. Res., *18*: 553–606.
2. BÜNNING E. (1935): Zur Kenntnis der endogenen Tagesrhythmik bei Insecten und Pflanzen. Ber. Deut. Bot. Ges., *53*: 594–623.
3. BÜNNING E. (1960): Biological Clocks, Cold Spring Harbor Symposia on Quantitative Biology, pp. 1–524, Vol. XXV, Waverly Press Inc., Baltimore.
4. ASCHOFF J. (1965): Circadian Clocks. Proceedings of the Feldafing Summer School, pp. 1–479, North Holland Publishing Co., Amsterdam.
5. MENAKER M. (1971): Biochronometry. pp. 1–662. Natl. Acad. Sci.-Natl. Res. Council.
6. DANILEVSKII A. S. (1965): Photoperiodism and Seasonal Development of Insects. pp. 1–283, Robert Cunningham & Sons Ltd.
7. BÜNNING E. (1968): The Physiological Clock, pp. 1–145, Academic Press, Inc., New York.
8. BECK S. D. (1968): Insect Photoperiodism. pp. 1–288, Academic Press, New York.
9. BORTHWICK H. A. and HENDRICKS S. B. (1960): Photoperiodism in plants. Science, *132*: 1223–1228.
10. HALBERG F., HAUS E. and STEPHENS A. (1959): Susceptibility to ouabain and physiologic 24-hour periodicity. Fed. Proc., *18*: 63.
11. SULLIVAN W. N., CAWLEY B., HAYES D. K., ROSENTHAL J. and HALBERG F. (1970): Circadian rhythm in susceptibility of house flies and Madeira cockroaches to pyrethrum. J. Econ. Entomol., *63*: 159–163.

12. Hayes D. K., Sullivan W. N., Oliver M. Z. and Schechter M. S. (1970): Photoperiod manipulation of insect diapause in the field. Science, *169*: 382–383.
13. Ankersmit G. W. (1968): The photoperiod as a control agent against *Adoxophyes reticulana* (Lepidoptera: Tortricidae) Entomol. Exp. Appl., *11*: 231–240.
14. Ankersmit G. W. (1962): Tables of Sunrise and Sunset. United States Naval Observatory, Washington, D.C., U.S. Govt. Printing Office, Washington.
15. Pittendrigh C. S. (1961): Harvey Lectures 56, 93–125. Academic Press, New York.

BIOCHRONOMETRY AND BIRD MIGRATION

Chairman: Donald S. FARNER

Co-Chairman: James R. KING

BIOCHRONOMETRY AND BIRD MIGRATION: GENERAL PRESPECTIVE

JAMES R. KING and DONALD S. FARNER

Department of Zoology, Washington State University
Pullman, Washington, U.S.A.
Department of Zoology, University of Washington
Seattle, Washington, U.S.A.

Bird migration in its most typical form is ". . . a regular, seasonal, large-scale, long-distance movement of a population twice a year between a fixed breeding area and a fixed non-breeding area" [1]. In accomplishing these movements, chronometric information is required (1) for direction-finding when celestial cues are used (for review, see GWINNER [2]), at least in some species; (2) for corrcet seasonal timing of the initiation of migration movements; and (3) possibly for timing the duration of migratory movement [3]. Unfortunately the panel contains no expert on the biochrometric aspects of direction finding. The following résumé, therefore, concerns the second and third categories of chronometric aspects.

Migration is only one subcycle of an annual cycle that includes also subcycles of molt, fat storage, reproduction, territorialism, flock-formation, and so on. For a given species, the phase relationships of these subcycles are precisely arranged as an adaptive unit, and the analysis of temporal regulation involves not only chronometric functions for seasonal phasing of the entire cycle, but also for the various subcycles. An isolated consideration of one subcycle is, therefore, basically arbitrary even though it may be justified by convenience. It is highly probable that the several subcycles have common regulatory components and that interpretations about the control of any one of them may be extrapolated, with due caution, to the others.

Chronometric aspects of migratory behavior can be investigated in both the field and the laboratory. The temporal features of migration as studied by either of these methods include not only the seasonal release and duration of migratory movement, but also the prior development of migratory urge (*Zugdisposition*) which is manifested in many but not all species by the accumulation of large fat reserves before the onset of migratory movement [4–6]. Statitsical information about the temporal precision and seasonal timing of migration in free-living populations of birds yields clues about the accuracy of the regulatory processes, and correlations with environmental variables may suggest sources of chronometric information. Several aspects of migratory behavior can be further investigated in controlled laboratory conditions in caged birds by the recording of nocturnal locomotor activity (migratory restlessness, or *Zugunruhe*) that reflects the urge to migrate (for review, see BERTHOLD [7]).

The following temporal aspects of migration and associated events are among those that must be rationalized by any general or special theory of induction and regulation: (1) The development of the migratory urge and the beginning of migration commonly anticipate by weeks or even months the ultimate purposes of migration (in spring, movement to an area favorable for reproduction; in autumn, escape from inadequate food supplies and inclement weather). This has been described by FARNER [8] as the "predictive function" of the regulatory system. (2) The migration schedule of many species is annually precise, astonishingly so in some species [4, 9], relatively so in others. (3) The external manifestation of the urge to migrate, as exemplified by the accumulation of fat reserves and the onset of migratory restlessness in captives, develops suddenly rather than gradually in many species [10, 11], suggesting a "threshold" or strongly non-linear function. (4) Stimulation of migration occurs twice each year, commonly in environments that change conspicuously in weather and photoperiod. (5) Migration occurs in young-of-the-year, often partly or completely separated from adults. The immature birds must therefore inherit or develop independently the necessary chronometric capacity. (6) Individuals of many species are known to migrate between traditional breeding places and winter quarters (philopatry). This becomes a temporal aspect of migration if the alternation between fixed sites is determined or aided by a migration program with components of direction and duration, as indicated by the experiments of GWINNER [3] and of BERTHOLD et al. [12], rather than direction and distance as suggested by EVANS [13].

The foregoing temporal aspects suggest some of the major constraints on the range of plausible theories about the temporal regulation of migration and associated events. Such theories must have in common a chronometric function that phases migration adaptively with the seasons. This "clock" could conceivably be: (1) wholly or principally exogenous, controlling internal events through induction of behavioral and physiological processes and (or) regulation of their rates; (2) wholly or principally endogenous, so that the annual cycle, including migration, is the outcome of a programmed internal rhythm with a periodicity of exactly one year; or (3) a combination of the foregoing hypotheses, in which the annual sequence of events is the outcome of an oscillating endogenous function, with a period of about one year (circannual), that is synchronized with the seasons by environmental agents. These hypotheses from a conceptual continuum that reflects the range of options conceivably available in the evolution of regulatory systems for migration. In the extreme interpretation of hypothesis (2) the birds would be uncoupled from their environment, which is virtually unimaginable for biological systems. The realistic alternatives thus appear to be hypotheses (1) and (3). Hypothesis (1) is a viable interpretation particularly for species that are irruptive migrants, partial migrants, or irregular transients whose movements are influenced by unpredictable environmental extremes or food sources, but is less likely for typical migrants. Hypothesis (3), invoking an endogenous quasi-annual clock or program synchronized with the seasons by environmental cues, is the simplest rationale for the widest spectrum of the attributes of bird migration. The existence of an "innate rhythm" of migration and associated events was suggested on the basis of circumstantial evidence for trans-equatorial migrants [9] or migratory species wintering in tropical and subtropical regions [14, 15] (for review, see LOFTS and MURTON [16]; BERTHOLD [7] and on the basis of experimental evidence from several songbird species [17–19]. Recently, more convincing evidence of a persistent internal annual rhythm has been supplied by GWINNER [3] and BERTHOLD et al. [20] for several species of Old World warblers kept in constant photoperiods for two years or more. The subcycles (e.g., migratory restlessness) continued and tended to be less than one year in duration, indicating by analogy with circadian systems that the cycle was free-running, or circannual. It is a reasonable hypothesis that these periodicities are endogenous, even though their persistence in LL or DD conditions has not yet been demonstrated. From an operational viewpoint (as contrasted with a conceptual viewpoint) this omission is of no consequence, since it can be shown that the rhythms persist in the range of photoperiods (LD 10:14 to LD 16:8) to which the species in question are exposed under natural conditions [12]. The 7- to 9-month periodicity of the reproductive season and movements of several seabird species on equatorial or subequatorial islands can be interpreted as the product of an internal rhythm that has become uncoupled from environmental synchronizers (or never has been coupled with them) in surroundings in which there is no selective advantage associated with a fixed seasonal breeding period (for review, see Lofts and Murton [16]).

In the assessment of hypotheses concerning the chronometric bases and aspects of migration it must be borne in mind that it is, in all probability, a phenomenon multiple evolutionary origin (e.g., Berthold [7]; Farner [8, 21]; Merkel [22]).

It may be meritorious at this point to examine the possibility that hypotheses (1) and (3) above, when applied to real cases, represent merely the extremes of a family of basically similar control schemes. To us this becomes possible on the basis of a further hypothesis for which we do not claim novelty since it exists, at least implicitly, in earlier discussions by others [15, 23, 24]. It is stated here in terms that are doubtless overly simplistic with the hope that it may have some heuristic value. It has not been subjected to rigid analysis, either theoretically or experimentally. This further hypothesis assumes (a) that the control systems of species with annual cycles have natural periods that are circannual and are entrainable by environmental factors to a natural period of one year; (b) that these systems behave as oscillators with specific or group differences in damping constants so that the number of free-running cycles may vary from a fraction of one cycle to several cycles; and (c) that annually recurring environmental events may function in *entrainment* (*Zeitgeber* function in the strict sense) of endogenous circannual cycles (especially important in species in which the damping constant is small, negligible, or negative) or in *induction* or *amplitude boosting* of cycles (especially important in species in which the damping constant is such that the oscillation function "terminates" in less than one to a very few periods under free-running conditions); not to be precluded is the possibility that a single annually periodic environmental variable may have both of these functions in the same species.

It is widely agreed, on the basis of experimental data, that the annual photocycle is a major source of temporal information in the control of migration and associated events [6–8, 25]. There is no other form of environmental variation to which birds could conceivably be sensitive that recurs with sufficient regularity to fulfill reliably a predictive role [8, 26] and to account for the temporal precision exhibited by the annual cycle of various species. Secondary synchronizers and modifiers, such as temperature, wind, and nutritional status, obviously have effects on the release and pace of migration [7], and may replace or supersede photoperiodic controls in non-typical migrants that exhibit erratic movements.

It is not very clear at present how photoperiodic information is coupled with the functional state of typical migrants. The following speculations can supply a framework for debate. It is possible that birds in effect, measure daylength or some other aspect of the annual photocycle through reference to their circadian clock. The migratory restlessness of caged birds of at least one species shows a circadian periodicity in continuous darkness [27]. The circadian time-reference system may take the form of a response curve whose phase-angle with respect to the photophase is variable and modulated annually by daylength or some other aspect (e.g., rate of change) of the annual photocycle. Induction of the migratory state or other events of the annual cycle may occur when some critical fraction of the relevent response curve falls within the range of the photophase. This is a statement of the Bünning hypothesis, the validity of which for gonadotropic stimulation in birds is suggested by data from several species [26, 28, 29]. It is at this point that, as suggested above, there may be, a convergence of circadian theory, traditional photoperiod theory (induction period, threshold, rate control, photorefractory phase, summation of day-lengths; see WOLESON [30, 31], and circannual output.

The response curve to photostimulation may itself be a product of the phase angles between cycles of hormone release. As will be discussed by ALBERT MEIER and his associates elsewhere in this volume, the phase angle between peak release of prolactin and of corticosterone appears to be of critical importance in the induction and of corticosterone appears to be of critical importance in the induction of migratory restlessness and fat deposition in White-throated Sparrows. These observations can be accommodated by the foregoing version of the Bünning hypothesis if it can be shown that the annual photocycle somehow modulates separately the phases of prolactin release and corticosterone release to provide variable phase angles consistent with the biannual periods of migration.

In conclusion, it should be noted that birds have evolved a variety of adaptive strategies that assure appropriate physiological responses to temporal variation in the environment (for an extensive treatment see LOFTS and MURTON [16]), and they appear to have exploited many combinations within the continuum between principally exogenous and principally endogenous controls. It is pertinent to reiterate our earlier admonition [32] that extrapolations of data and conclusions from one avian group to another (and perhaps even among closely related species, in some cases) should be made cautiously. The experimental data currently available have been obtained almost entirely from a few species of passerine birds, and the extent to which conclusions based on these data apply to other groups is not yet apparent.

REFERENCES

1. LACK D. (1968): Bird migration and natural selection. Oikos, *19*: 1–9.
2. GWINNER E. (1971): Orientierung. *In*: Grundriss des Vogelzugskunde, 2 Aufl., (E. SCHÜZ, ed.), pp. 299–348, Verlag P. Parey, Berlin.
3. GWINNER E. (1968): Circannuale Periodik als Grundlage des jahreszeitlichen Funktionswandels bei Zugvögeln. J. Ornithol., *109*: 70–95.
4. KING J. R. and FARNER D. S. (1965): Studies of fat deposition in migratory birds. Ann. New York Acad. Sci., *131*: 422–440.
5. KING J. R. (1972): Adaptive periodic fat storage by birds. Proc. 15th Intern. Ornithol. Congr., pp. 200–217, E. J. Brill, Leiden.
6. EVANS P. R. (1970): Timing mechanisms and the physiology of bird migration. Sci. Progr. (Oxford), *58*: 263–275.
7. BERTHOLD P. (1971): Physiologie des Vogelzugs. *In*: Grundriss der Vogelzugskunde, 2 Aufl., (E. SCHÜZ, ed.), pp. 257–299, Verlag P. Parey, Berlin.

8. Farner D. S. (1970): Predictive functions in the control of annual cycles. Env. Res., *3*: 119–131.
9. Marshall A. J. and Serventy D. L. (1956): Breeding periodicity in the Short-tailed Shearwater (*Puffinus tenuirostris* Temminck) in relation to trans-equatorial migration and its environment. Proc. Zool. Soc. London, *127*: 489–510.
10. King J. R. and Farner D. S. (1963): The relationship of fat deposition to Zugunruhe and migration. Condor, *65*: 200–223.
11. Weise C. M. (1963): Annual physiological cycles in captive birds of differing migratory habits. Proc. 13th Intern. Ornithol. Congr., Ithaca, pp. 983–993.
12. Berthold P., Gwinner E., Kein H. and Westrich P. (1972): Beziehungen zwischen Zugunruhe und Zugablauf bei Garten- und Mönchsgrasmücke (*Sylvia borin* und *S. atricapilla*). Z. Tierpsychol., *30*: 26–35.
13. Evans P. R. (1968): Reorientation of passerine night migrants after displacement by the wind. British Birds, *61*: 281–303.
14. Curry-Lindahl K. (1958): Internal timer and spring migration in an equatorial migrant, the Yellow Wagtail (*Motacilla flava*). Ark. Zool. (Ser. 2), *11*: 541–557.
15. Marshall A. J. and Williams M. C. (1959): The pre-nuptial migration of the Yellow Wagtail (*Motacilla flava*) from latitude 0.04′N. Proc. Zool. Soc. London, *132*: 313–320.
16. Lofts B. and Murton R. K. (1968): Photoperiodic and physiological adaptations regulating avian breeding cycles and their ecological significance. J. Zool., *155*: 327–394.
17. Merkel F. W. (1963): Long-term effects of constant photoperiods on European Robins and Whitethroats. Proc. 13th Intern. Ornithol. Congr., Ithaca, pp. 950–959.
18. Zimmermann J. L. (1966): Effects of extended tropical photoperiod and temperature on the Dickcissel. Condor, *68*: 377–387.
19. King J. R. (1968): Cycles of fat deposition and molt in White-crowned Sparrows in constant environmental conditions. Comp. Biochem. Physiol., *24*: 827–837.
20. Berthold P., Gwinner E. and Klein H. (1971): Circannuale Periodik bei Grasmücken (*Sylvia*). Experientia, *27*: 399.
21. Farner D. S. (1966): Über die photoperiodische Steuerung der Jahreszyklen bei Zugvögeln. Biologische Rundschau, *4*: 228–241.
22. Merkel F. W. (1960): Stoffwechselvorgänge regeln den Wandertrieb der Zugvögel. Die Umschau in Wissenschaft und Technik, *8*: 243–246.
23. Immelmann K. (1963): Tierische Jahresperiodik in ökologischer Sicht. Zool. Jb. Syst., *91*: 91–200.
24. Immelmann K. (1967): Periodische Vorgänge in der Fortpflanzung tierischer Organismen. Studium Generale, *20*: 15–33.
25. Aschoff J. (1955): Jarhesperiodik der Fortpflanzung bei Warmblütern. Studium Generale, *8*: 742–776.
26. Farner D. S. (1964): Photoperiodic control of reproductive cycles in birds. Am. Sci., *52*: 137–156.
27. McMillan J. P., Gauthreaux S. A., Jr. and Helms C. W. (1970): Spring migratory restlessness in caged birds: A circadian rhythm. BioScience, *20*: 1259–1260.
28. Hamner W. M. (1963): Diurnal rhythm and photoperiodism in testicular recrudescence of the House Finch. Science, *142*: 1294–1295.
29. Menaker M. and Eskin A. (1967): Circadian clock in photoperiodic time measurement: A test of the Bünning hypothesis. Science, *157*: 1182–1185.
30. Wolfson A. (1966): Environmental and neuroendocrine regulation of annual gonadal cycles and migratory behavior in birds. Recent Prog. Hormone Res., *22*: 177–244.
31. Wolfson A. (1970): Light and darkness and circadian rhythms in the regulation of annual reproductive cycles in birds. *In*: La Photorégulation de la Reproduction chez les Oiseaux et les Mammifères (J. Benoit and I. Assenmacher, eds.), pp. 93–119, Colloques Intern. du C.N.R.S., Paris.
32. Farner D. S. (1950): The annual stimulus for migration. Condor, *52*: 104–122.

SOME NEUROENDOCRINE AND ENDOCRINE CORRELATES IN THE TIMING OF BIRD MIGRATION

MILTON H. STETSON and JOSEPH P. McMILLAN

Department of Zoology, University of Texas
Austin, Texas, U.S.A.

Seasonal migration is part of an annual cycle of interrelated ethological and physiological events which in many species of birds have long been recognized to be under photoperiodic control. While the descriptive aspects of migration have received considerable attention, the internal mechanisms involved remain under investigation [1–3]. We present here recent findings on the physiology and timing of spring migration in caged nocturnal migrants of the genus *Zonotrichia*.

PREMIGRATORY EVENTS AND *ZUGDISPOSITION*

In this section of our presentation we do not survey all existing data on hormonal interactions in the development of physiological and ethological changes (*Zugdisposition*) resulting in migration, but rather restrict ourselves to an examination of the testicular control system with respect to timing the onset of *Zugdisposition* and migration. ROWAN [4] was among the first to consider the relationship between day length, gonadal growth, and migration. His gonadal hypothesis states that photoperiodic stimulation results in the production and release of gonadal steroid hormones that are at least in part responsible for the induction of hyperphagia, subsequent fat deposition and increase in body weight, and migration (expressed in caged individuals as *Zugunruhe*).

Although it may be presumptuous to assume that the gonadal hypothesis may be valid for all species of migratory birds, and indeed for both vernal and autumnal migration in a single species, recent experiments with three migratory species of the genus *Zonotrichia* (White-throated Sparrows, *Z. albicollis;* White-crowned Sparrows, *Z. leucophrys;* Golden-crowned Sparrows, *Z. atricapilla*) demonstrate testicular regulation of vernal hyperphagia, fat deposition and *Zugunruhe.* Castration (WEISE [5]; STETSON and ERICKSON [6]; GWINNER et al. [7]; STETSON and McMILLAN, unpublished), and alteration of normal photoperiodically induced testicular function by ablation of the gonadotropic region of the hypothalamus (STETSON [8]; YOKOYAMA, unpublished) or blocking the photoreceptor(s) from light [7] prevent premigratory fattening and *Zugunruhe* (Fig. 1). ROWAN's hypothesis suggests that testosterone is the factor responsible for the initiation of ethophysiological changes overtly expressed as hyperphagia and nocturnal restlessness. Preliminary experiments indicate that testosterone (propionate) therapy promotes fattening and *Zugunruhe* in White-crowned Sparrows in which testicular function was abolished by castration (STETSON and McMILLAN, unpublished) or hypothalamic lesions (YOKOYAMA, unpublished).

Of interest is the yet unresolved temporal relationship between photoperiodic stimulation, castration (or other means of suppressing testicular activity) and the

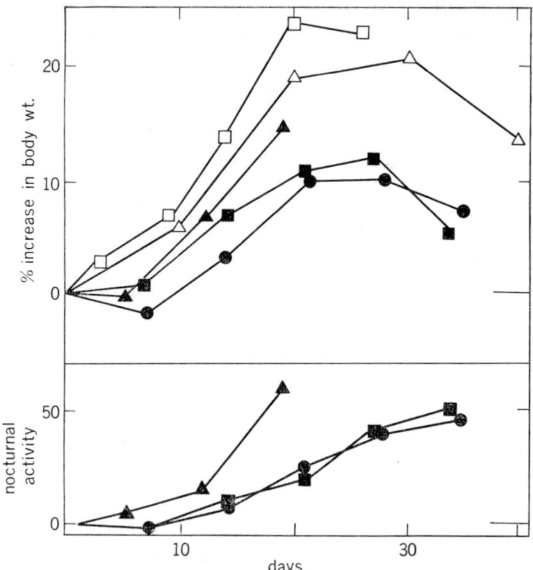

Fig. 1 Effect of suppressed testicular activity on photoperiodically induced fat deposition (above) and *Zugunruhe* (below).
Upper panel: Percent initial body weight of controls minus that of operated birds in 3 species of *Zonotrichia*.
Lower panel: Nocturnal activity units of controls minus those of experimentals in 3 species of *Zonotrichia*. One activity unit=1 hop/3 minutes (WEISE [5]) or 1 hop/2 minutes (GWINNER et al. [7]) of the scotophase.
● =*Z. albicollis*, intact and castrated birds (from WEISE, [5])
■ =*Z. atricapilla*, intact and birds with India ink injected beneath the scalp (from GWINNER et al. [7])
▲ =*Z. leucophrys*, intact and castrated birds (from GWINNER et al. [7])
△ =*Z. leucophrys*, intact and castrated birds (from STETSON and ERICKSON, [6])
□ =*Z. leucophrys*, intact and lesioned birds (from STETSON [8]).

appearance of fattening and *Zugunruhe*. WHEISE [5] has demonstrated that in *Z. albicollis*, fattening and *Zugunruhe* are prevented if the testes are removed *before* but *not after* the onset of photostimulation. In every report of suppression of fattening and *Zugunruhe*, surgery was performed prior to photostimulation. Reconciliation of negative findings from investigations of castration in Golden-crowned [9] and White-crowned Sparrows [10] with the results described above probably rests on the fact that in each case castration was performed *after* the winter solstice. In White-crowned Sparrows subjected to natural day length, castration in September prevented fattening the following spring (WILSON, unpublished; MATTOCKS, unpublished). Continued investigation of testicular physiology, especially from the winter solstice to the onset of migration, may indicate the precise means by which the testes affect *Zugdisposition*. In this regard, integration of histological and gravimetric data [11] indicates that testicular recrudescence begins shortly after the winter solstice in White-crowned Sparrows wintering in Washington.

We feel that there is little doubt that the testicular control system is an important element in the internal regulation of vernal *Zugdisposition* and migration in *Zonotrichia* (Fig. 1). Females should be studied to determine the relative importance of the ovary in the timing of migration. Since seasonal migration is obviously an adaptation to facilitate reproduction, the gonads likely constitute an integral component of the system controlling migration, at least in the spring, though reduced organ size in the fall is not necessarily indicative of endocrine inactivity. Further elucidation of the role, if any, of the testes of adult and immature birds in autumnal migration is desired. The decreasing effect of castration after photostimulation on migratory phenomena suggests an "initiating" or "synchronizing" role for the gonads in the preliminary stages of the complex internal changes that culminate in migration and reproduction. The data suggest that ROWAN's hypothesis is valid for migratory male *Zonotrichia*; the gonads appear to be intimately involved in the timing and control of migration.

MIGRATION AND *ZUGUNRUHE*

The proximate factors that influence nocturnal migration are not well understood. Efforts to analyze mathematically radar and weather data in an attempt to isolate the environmental variables responsible for the night-to-night fluctuations in the magnitude of migration have been only moderately successful [12]. Radar and direct visual observation of spring migration in North America reveal small birds migrating in a wide range of weather conditions [13]. Studies of caged migrants exhibiting *Zugunruhe* may provide a better understanding of the physiological and behavioral parameters that the environment might affect. *Zugunruhe* occurs seasonally (spring and fall) in birds housed under natural photoperiods and shows orientation similar to the migration of free-living populations [14]. Since migration is thought to have evolved independently several times in birds, the relative importance of environmental factors or the factors themselves may differ among migratory species. In the White-throated Sparrow, nocturnal *Zugunruhe* persists in constant conditions of dim light (Fig. 2A) and displays a circadian rhythm [15]. This suggests that no direct environmental stimuli are required for the nightly expression of migratory activity in the species. Internal mechanisms are probably the exclusive cause of the circadian rhythm of *Zugunruhe*, but in migrating birds the internal machinery is also a substrate on which the environment may act, perhaps to inhibit migration temporarily, though leaving the endogenous rhythmicity unaffected. Thus the resultant behavior of birds in their natural environment is doubtless best viewed as a complex, nonadditive mixture of responses to exogenous and endogenous phenomena. ENRIGHT [16] has cautioned, however, that to the extent that internal timing can affect behavior, elementary stimulus-response interpretations of the field situation could prove frustratingly unsuccessful, and if apparently successful, may be positively misleading.

Removal of the pineal complex eliminates the endogenous locomotor activity rhythm of the nonmigratory House Sparrow, *Passer domesticus* [17]. Pinealectomy has recently been shown to abolish the circadian rhythms of vernal *Zugunruhe* (Fig. 2B) and summer "daytime" activity of White-throated Sparrows in constant dim light [18]. Although the effect of removal of the pineal body is clear, it is at present difficult to interpret physiologically.

The biosynthetic events in the avian pineal organ are poorly known and assumed by some, perhaps mistakenly, to be similar to those in mammals. Melatonin, a suspected pineal hormone, serotonin, and the enzyme hydroxyindole-O-methyl transferase (HIOMT) have received the greatest attention [19]. The con-

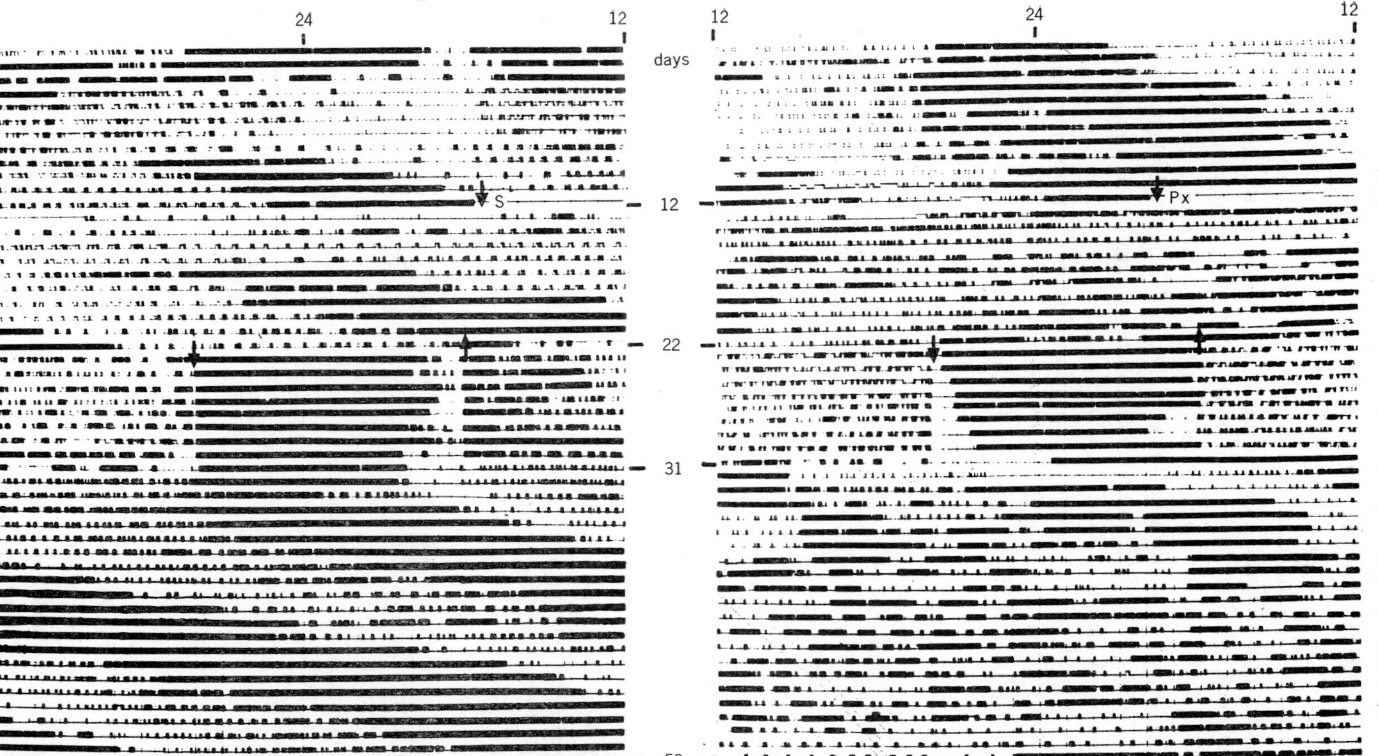

Fig. 2 A and **B** The perch-hopping records of two White-throated Sparrows showing spring *Zugunruhe*. The birds received constant dim light (LL. 1.0 lux) except days 21–31 during an LD14:10 (120:1.0 lux) light cycle (arrows). (A) A circadian rhythm of migratory restlessness persists in LL before and after sham pinealectomy (S) on day 12. The nocturnal restlessness is entrained by the light cycle and free-runs again in LL after day 31. (B) The *Zugunruhe* rhythm decays to apparent arrhythmicity in LL following pinealectomy (Px) on day 12. The rhythm reappears entrained to the light cycle in proper phase with "daytime" activity but arrhythmicity resumes in the subsequent LL free-run,

centrations of indoleamines vary with the photoperiod to which the birds are subjected. In the pineal organs of pigeons [20] and Japanese quail [21] serotonin concentrations are maximal near the onset of light and minimal in the middle of the dark period, whereas in Japanese quail, several species of African weaver birds [22] and chickens [23] melatonin concentrations are greatest in the dark and least during the light period. Alterations of the phase of the imposed light cycle cause corresponding alterations in pineal indoleamine concentrations. A melatonin rhythm persists in the pineal organ of chickens under conditions of continuous light (LL) or continuous darkness (DD), and is phase-locked with the rhythm of locomotor activity [24]; pineal melatonin is maximal at the midpoint of the inactive phase of chickens held in DD or LL.

In Japanese quail undergoing sexual maturation, but not in sexually mature birds, pineal HIOMT activity varies diurnally with greater activity in the scotophase [25]. In continuous light, significant fluctuations in enzyme activity were observed for a period of three days. This rhythm was cited as support for the sustained oscillator hypothesis of GASTON and MENAKER [17] based on data obtained from *Passer domesticus*. Such an assertion seems premature, if not unwarranted, for several reasons; the taxonomic difference of the two species, the restricted age class (older quail showed no diurnal HIOMT cycle), the shortness of the observation period (72 hrs), and the ambiguity surrounding the role of the enzyme. HIOMT activity has been shown to be unrelated to melatonin levels in quail and chickens [26] and to exhibit quite diverse substrate specificity in the pineal organs quail, turkey, duck, chicken and pigeon [27].

Pharmacological and neurophysiological studies suggest a relationship between brain biogenic amines and sleep [28]. Intravenous injection of serotonin [29] and intraperitoneal injection of melatonin [30, 31] lead to slow-wave sleep in chickens, as indicated by behavior and by electrophysiological recording. The birds are easily aroused by auditory and tactile stimuli.

Considering the foregoing discussion in aggregate, it is tempting to speculate that (i) avian pineal organ influences locomotor activity and possibly participates in the regulation of sleep-wake cycles, and (ii) the machinery of the physiological clock in timing locomotor rhythms (both *Zugunruhe* and "daytime" activity) of birds in constant conditions (LL or DD) resides in the pineal and may involve melatonin or some related biogenic amine. If this simple interpretation is valid, then the development of seasonal nocturnal restlessness in migrants should be accompanied by a coincidental change in pineal physiology. Unfortunately, data on pineal rhythms in passerines and studies of locomotor activity of pinealectomized gallinaceous species in constant conditions are lacking.

CONCLUDING REMARKS

Solutions to many fundamental problems in the physiology of bird migration remain elusive. ROWAN's gonadal hypothesis, proposed nearly half a century ago, appears valid for the species and sex in which the hypothesis was tested. A thorough examination of ROWAN's hypothesis, including males and females, immatures and adults, could provide very valuable information on the internal mechanisms initiating vernal and autumnal *Zugdisposition* and migration. Many recent investigations have shown that a biological clock(s) is a major feature in many physiological processes of birds. The relationship between the pineal complex and the circadian rhythm of *Zugunruhe* presents many intriguing possibilities amenable to testing with controlled experiments. Perhaps from seemingly diverse experimental approaches, such as those presented herein, a unifying concept will emerge to assist our understanding of the migratory physiology of birds.

REFERENCES

1. FARNER D. S. (1955): *In*: Recent Studies in Avian Biology, (A. WOLFSON, ed.), pp. 198–273, U. Ill. Press, Urbana.
2. DOLNICK V. R. and BLYMENTAL T. I. (1967): Condor, *69*: 435–468.
3. EVANS P. R. (1970): Sci. Progr. (Oxford), *58*: 263–275.
4. ROWAN W. (1925): Nature, *115*: 494–495.
5. WEISE C. M. (1967): Condor, *69*: 49–68.
6. STETSON M. H. and ERICKSON J. E. (1972): Gen. Comp. Endocrinol., *19*: 355–362.
7. GWINNER E., TUREK F. W. and SMITH Z. D. (1971): Z. vergl. Physiol., *75*: 323–331.
8. STETSON M. H. (1971): J. Exp. Zool., *176*: 409–414.
9. MORTON M. L. and MEWALDT L. R. (1962): Physiol. Zool., *35*: 237–247.
10. KING J. R. and FARNER D. S. (1963): Condor, *65*: 200–223.
11. BLANCHARD B. D. and ERICKSON M. M. (1949): Univ. Calif. Publ. Zool., *47*: 255–318.
12. NISBET I. C. T. and DRURY, W. H. (1968): Anim. Behav., *16*: 496–530.
13. GAUTHREAUX S. A. Jr. (1971): Auk, *88*: 343–365.
14. MEWALDT L. R., MORTON M. L. and BROWN I. L. (1964): Condor, *66*: 377–417.
15. MCMILLAN J. P., GAUTHREAUX S. A. Jr. and HELMS C. W. (1970): BioScience, *20*: 1259–1260.
16. ENRIGHT J. T. (1970): Ann. Rev. Ecol. Systematics, *1*: 221–238.
17. GASTON S. and MENAKER M. (1968): Science, *160*: 1125–1127.
18. MCMILLAN J. P. (1971): Ph.D. Dissertation. University of Georgia, Athens.
19. RALPH C. L. (1970): Amer. Zool., *10*: 217–235.
20. QUAY W. B. (1966): Gen. Comp. Endocrinol., *6*: 371–377.
21. HEDLUND L., RALPH C. L., CHEPKO J. and LYNCH H. J. (1971): Gen. Comp. Endocrinol., *16*: 52–58.
22. RALPH C. L., HEDLUND L. and MURPHY W. A. (1967): Comp. Biochem. Physiol., *22*: 591–599.
23. LYNCH H. J. (1971): Life Sciences I, *10*: 791–795.
24. LYNCH H. J. and RALPH C. L. (1970): Amer. Zool., *10*: 491.
25. SAYLER A. and WOLFSON A. (1969): Neuroendocrinol., *5*: 322–332.
26. LYNCH H. J. and RALPH C. L. (1970): Amer. Zool., *10*: 300.
27. AXELROD J. and LAUBER J. K. (1968): Biochem. Pharmacol., *17*: 828–830.
28. JOUVET M. (1969): Science, *163*: 32–41.
29. SPOONER C. E. and WINTERS W. D. (1965): Experientia, *21*: 256–258.
30. BARCHAS J., DA COSTA F. and SPECTOR S. (1967): Nature, *214*: 919–920.
31. HISHIKAWA Y., CRAMER H. and KUHLO W. (1969): Exp. Brain Res., *7*: 84–94.

AVIAN REPRODUCTIVE SYSTEM: DAILY VARIATIONS IN RESPONSES TO HORMONES

ROBERT MACGREGOR III

*Department of Zoology and Physiology, Louisiana State University
Baton Rouge, Louisiana, U.S.A.*

In many species of birds, long day lengths, initiate the development of the reproductive system (see reviews: FARNER [1]; LOFTS and MURTON [2]). The photoperiodic stimulation of gonadal growth depends on the entrainment of a circadian rhythm of responsiveness to light [3–5]). When the photoperiod is sufficiently long so that light is received during the responsive period, the gonads normally increase in size. It is assumed that light during the daily responsive period causes the release of gonadotropins [6].

In addition to the time of year (photosensitive period) when the reproductive system of many birds may be stimulated by long photoperiods, there is also a photorefractory period that occurs after the breeding season in the summer when the reproductive system is no longer responsive to long photoperiods. The maintenance of the photorefractory period also involves circadian systems [7, 8].

Prolactin has an inhibitory effect on the reproductive system of many birds (review and data: MEIER and DUSSEAU) [9]. Because of this antigonadal effect, LOFTS and MARSHALL [10] proposed that the initiation of the photorefractory period resulted from the release of high levels of prolactin.

Inasmuch as the inhibitory effect of prolactin can be reversed by injections of follicle stimulating hormone (FSH) [11, 12], it is generally believed that prolactin inhibits the release of FSH. However, it has now been demonstrated that prolactin can inhibit the effectiveness of the gonadotropic hormones when these hormones are given in certain temporal relations [13]. Prolactin inhibits the ovarian response to the gonadotropins when these hormones are given early during a 16-hour photoperiod but not when they are given late during the light. The oviducal response to the gonadotropins, however, is completely inhibited by prolactin at either time of day.

Recent studies in our laboratory have demonstrated that prolactin may have a stimulatory as well as inhibitory influence on the reproductive system of the White-throated Sparrow, *Zonotrichia albicollis,* depending on its temporal relations with corticosterone [14]. In photosensitive birds that were maintained in continuous light, corticosterone entrains a daily rhythm of gonadal responses to prolactin. Prolactin has a strong inhibitory effect on gonadal growth when injected 8 hours after the injection of corticostreone but it augments gonadal growth when the injections of prolactin follow the injections of corticosterone by 12 hours.

The temporal synergism of corticosterone and prolactin also regulates gonadal growth in photorefractory House Sparrows, *Passer domesticus* [14]. Gonadal growth is stimulated in birds maintained in continuous light by daily injections of prolactin that follow injections of corticosterone by 4 to 8 hours. Other temporal patterns of the two hormones are ineffective.

The temporal variations in the responses of the reproductive system to prolactin may account for some of the observations that prolactin has little or no inhibitory effect in some birds. For example, ALEXANDER and WOLFSON [15] reported that prolactin was largely ineffective in the Japanese quail, *Coturnix coturnix*, although it did delay the accumulation of egg yolk. In order to determine the relationship of prolactin to corticosterone, two experiments were performed with the Japanese quail.

In one experiment, female quail were maintained in continuous light and treated daily for 2 weeks with prolactin (1 μg/g body wt.) injected at 0, 4, 8 or 12 hours after injections of corticosterone (1 μg/g body wt.). The birds were 51 days old when they were killed. The weight of the ovaries and oviducts in the birds that received prolactin at 8 or 12 hours after corticosterone were much lower (verified statistically by an analysis of variance) than those in the birds that were given corticosterone and prolactin together (Fig. 1). Although there were small amounts of yolk in some of the follicles, there was no egg-laying in

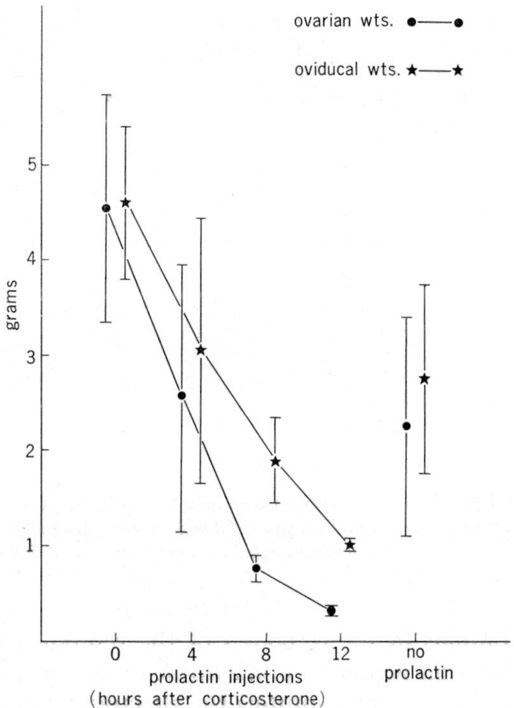

Fig. 1 Temporal relations of corticosterone and prolactin affecting gonadal development in the female Japanese quail.

the group that received injections of prolactin 12 hours after injections of corticosterone, whereas eggs were layed regularly in the group that received corticosterone and prolactin injections together. When compared with a group that received corticosterone alone, injections of prolactin administered at the same time as the injections of corticosterone stimulated ovarian and oviducal growth ($P<0.01$), as well as egg-laying.

In the second experiment, cloacal gland development (an index of testosterone production; SACHS [16]) was observed over a 7-day period in maturing male quail.

These birds also were maintained in constant light and treated with prolactin (1 μg/g body wt.) and corticosterone (1 μg/g body wt.) in one of 6 relationships (Fig. 2). Cloacal gland development equal to or greater than that in untreated controls was found in those groups that received prolactin injections at 0, 4, 12, 16 or 20 hours after the time of corticosterone injections. However, injections of prolactin 8 hours after the time of corticosterone injections had a marked inhibitory effect on the cloacal protuberance ($p<0.001$: Kruskal-Wallis analysis of variance test).

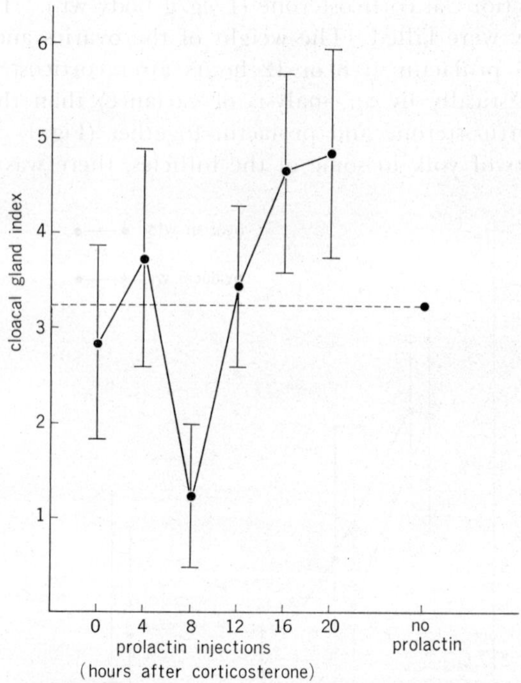

Fig. 2 Temporal relations of corticosterone and prolactin affecting cloacal protuberance in the male Japanese quail (Cloacal Gland Index is determined by the visual comparison of the diameter, vascularization, and foam quantity. Increasing development: 0 to 6.)

The temporal variations in the responses of the reproductive system to prolactin could result from an action of prolactin at several levels. In the White-throated Sparrow, prolactin clearly inhibits the effectiveness of the gonadotropic hormones in stimulating oviducal growth, by inhibiting the production of estrogen, the respones to it, or both [13]. We performed an experiment with photosensitive House Sparrows, *Passer domesticus,* to explore the possibility of a daily variation in the oviducal response to estrogen when estradiol injections (0.5 μg/g body wt.) are accompanied by injections of prolactin (1 μg/g body wt.) (Fig. 3). This experiment has been described in a preliminary report [17]. After 12 daily injections, the weights of the oviducts in the group that received the hormone injections at the onset of a 16-hour photoperiod were greater than those in the group that was injected at the end of the light period.

These studies support the idea that prolactin is involved in the inhibition and stimulation of the reproductive system in several avian species. However,

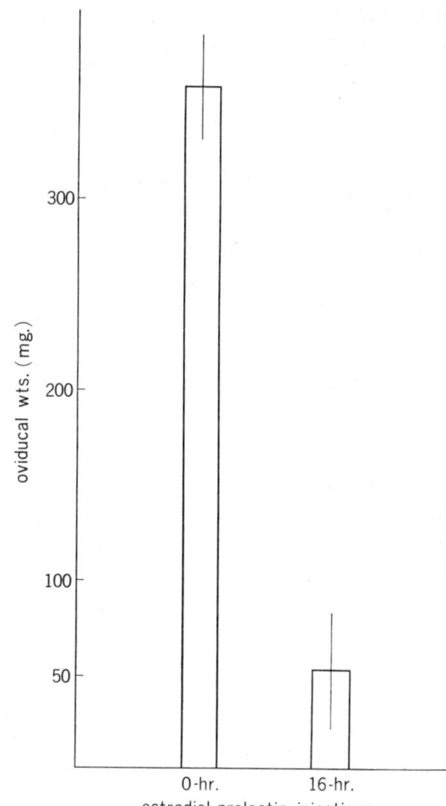

Fig. 3 Daily variation (mean ±SE) in oviducal weight response to estradiol and prolactin in the House Sparrow,

the timing of the daily release of pituitary prolactin with respect to the periodicity in plasma corticosterone is more important, apparently, than the amounts that are released [14]. The importance of the synergism of these two hormones in one species, *Zonotrichia albicollis,* has been described in another paper presented at this symposium (Meier: Temporal synergism of corticosterone and prolactin controlling seasonal conditions in the White-throated Sparrow, *Zonotrichia albicollis*).

The site of action of the antigonadal and progonadal activities of prolactin cannot be designated specifically. The finding that gonadal growth can be induced in photorefractory House Sparrows by a particular pattern of corticosterone and prolactin [14] suggests that prolactin may influence the release of gonadotropins. However, an antigonadal activity has also been demonstrated at the level of the gonad [13], and it may even influence the tissue (oviducal) responses to the steroids (estrogen) (Fig. 3). Our findings emphasize the importance of other hormones besides the gonadotropic hormones in the regulation of the avian reproductive system, and they demonstrate that circadian systems are involved in the organization.

ACKNOWLEDGMENTS

The work described in this report was in part supported by a research grant (GB-20913) from NSF to Dr. Albert H. MEIR. Also, I am grateful to the Endocrinology Study Section of the Public Health Service for its kindness in supplying prolactin and gonadotropins.

REFERENCES

1. FARNER D. S. (1967): Day lengths as environmental information in the control of reproduction of birds. (J. BENOIT and I. ASSENMACHER, eds.), pp. 71–88, Coll. Intern. C.N.R.S. No. 172, Paris.
2. LOFTS B. and MURTON R. K. (1968): Photoperiodic and physiological adaptations regulating avian breeding cycles and their ecological significance. J. Zool. (London), *155*: 327–394.
3. HAMNER W. H. (1963): Diurnal rhythm and photoperiodism in testicular recrudescence of the House Finch. Science, *142*: 1294–1295.
4. FARNER D. S. (1965): Circadian systems in the photoperiodic responses of vertebrates. *In*: Circadian Clocks, (J. ASCHOFF, ed.), pp. 357–369, North-Holland Publ., Amsterdam.
5. MENAKER M. (1965): Circadian rhythms and photoperiodism in *Passer domesticus*. *In*: Circadian Clocks, (J. ASCHOFF, ed.), pp. 385–395, North-Holland Publ., Amsterdam.
6. FOLLETT B. K. and SHARP R. J. (1969): Circadian rhythmicity in photoperiodically induced gonadotrophin release and gonadal growth in the quail. Nature, *223*: 968–971.
7. HAMNER W. H. (1968): The photorefractory period of the House Finch. Ecology, *49*: 211–227.
8. MURTON R. K., LOFTS B. and WESTWOOD N. J. (1970): Manipulation of photorefractoriness in the House Sparrow, (*Passer domesticus*) by circadian light regimens. Gen. Comp. Endocrinol., *14*: 107–113.
9. MEIER A. H. and DUSSEAU J. W. (1968): Prolactin and the photoperiodic gonadal response in several avian species. Physiol. Zool., *41*: 95–103.
10. LOFTS B. and MARSHALL J. J. (1956): The effects of prolactin administration on the internal rhythm of reproduction in male birds. J. Endocrinol., *13*: 101–106.
11. BATES R. W., RIDDLE O. and LAHR E. L. (1937): Mechanism of the antigonadal action of prolactin in adult pigeons. Amer. J. Physiol., *119*: 610–614.
12. NALBANDOV A. V. (1945): A study of the effects of prolactin on broodiness and cock testes. Endocrinology, *36*: 251–258.
13. MEIER A. H. (1969): Antigonadal effects of prolactin in the White-throated Sparrow, *Zonotrichia albicollis*. Gen. Comp. Endocrinol., *13*: 222–225.
14. MEIER A. H., MARTIN D. D. and MACGREGOR R. III (1971): Temporal synergism of corticosterone and prolactin controlling gonadal growth in sparrows. Science, *173*: 1240–1242.
15. ALEXANDER B. and WOLFSON A. (1970): Prolactin and sexual maturation in the Japanese quail, *Coturnix coturnix japonica*. Poultry Sci., *49*: 632–640.
16. SACHS B. D. (1967): Photoperiodic control of the cloacal gland of the Japanese quail. Science, *157*: 201–203.
17. MACGREGOR R. III and MEIER A. H. (1971): Daily variations of the oviducal responses in the House Sparrow, *Passer domesticus*. (Abstract) Assoc. Southeastern Biol. Bull., *18*: 44.
18. SIEGEL S. (1956): Kraskal-Wallis one-way analysis of variance by rank. *Non-parametric Statistic for Behavioral Sciences,* McGraw-Hill publ.

HORMONAL CONTROL OF ORIENTATION IN THE WHITE-THROATED SPARROW, *ZONOTRICHIA ALBICOLLIS*

DONN D. MARTIN

Department of Zoology and Physiology
Louisiana State University
Baton Rouge, Louisiana, U.S.A.

Prior to both spring and fall migration, the White-throated Sparrow deposits large amounts of fat and becomes increasingly active at night. Gonadal recrudescence accompanies spring migration but does not normally occur in the fall. The timing of these premigratory events in the White-throated Sparrow is generally thought to be regulated by hormonal secretions which are stimulated by certain environmental cues such as the photoperiod. Gonadal [1] as well as thyroidal [2, 3] hormones held great favor as potential regulators of migration but recent results [4, 5] argue against these specific hormones having more than secondary roles in the initiation of migratory events. Two other hormones, prolactin and corticosterone, have been implicated in the regulation of the physiological and behavioral events of migration.

Prolactin induces fattering [6] and nocturnal restlessness [7] in the White-crowned Sparrow, *Zonotrichia leucophrys gambelii*. Studies with the White-throated Sparrow, *Zonotrichia albicollis*, have demonstrated that the time of day when prolactin is given is critical. When given in the middle of a sixteen hour photoperiod, prolactin stimulates fattening whereas injections given early in the photoperiod induce losses in fat reserves [8].

Corticosterone stimulates nocturnal restlessness in the White-crowned Sparrow when given simultaneously with prolactin [7]. When given alone at the end of a 6 hour photoperiod, corticosterone induces both fattening and nocturnal restlessness in the White-throated Sparrow [9].

These studies demonstrate that prolactin and corticosterone play a role in migratory events. To what extent and in what fashion these hormones regulate migration may be ascertained from the results of recent experiments in our laboratory. It has been demonstrated that the levels of plasma corticosterone [10] and pituitary prolactin [11] exhibit daily rhythms that differ seasonally with respect to the photoperiod. In the spring, the daily peaks of the two hormones differ by 12 hours whereas in the summer, during the photorefractory period, pituitary prolactin content peaks about 6 hours after plasma corticosterone. When the daily rhythms of the two hormones are simulated by injection, the administration of prolactin at 12, 8 and 4 hours after injections of corticosterone induce physiological and behavioral events that typify conditions found during the spring migratory period, the summer photorefractory period and the fall migratory period, respectively [9, 12]. These results prompted the hypothesis that the seasonal changes in the physiological and behavioral events associated with migration are the result of seasonal changes in the temporal relationship between corticosterone and prolactin [12].

Seasonal changes in the preferred direction of orientation is another behavioral event of migration and is at present not clearly understood. Do birds migrate north at one time of the year and south at another in response to seasonally distinct celestial cues or does the physiological state of the migrant determine the direction of orientation? It is known that some species of migratory birds use the position of the sun for guidance [13, 14] and other species, that migrate at night, use the stars in orientation [15–17]. Inasmuch as celestial information derived from the spring sky may differ from that derived from the fall sky, some investigators believe that seasonal changes in orientation are direct results of seasonally distinct celestial cues [18]. Although this may be a reasonable explanation for some migratory species, other species apparently rely upon an entirely different mechanism. Evidence of this comes from the experiments of ROWAN [1], AAGAARD and WOLFSON [16], and EMLEN [17, 19]. By artificially manipulating the photoperiod, these investigators induced the spring state of migratory readiness in the fall and the fall state of migratory readiness in the spring. When tested in the fall, birds in the spring state flew [1] or oriented [16], north and when tested in the spring [19], birds in the fall state oriented south. These directional choices are exactly opposite those exhibited during spring and fall migration and suggest that changes in the internal physiological state of the migrant are responsible for the seasonal changes in the direction of migratory orientation.

If, as our studies suggest, seasonal changes in the physiological state of the White-throated Sparrow are regulated by a seasonally specific temporal synergism of corticosterone and prolactin then it should be possible to regulate the preferred direction of orientation of this migrant by injections of corticosterone and prolactin that simulate the seasonal patterns of these two hormones. This paper summarizes the results of two experiments that were designed to test this possibility (parts of these experiments have been submitted for publication by MARTIN and MEIER).

The birds were collected from wintering flocks near Baton Rouge, Louisiana, brought indoors in January, 1971, and housed (two per cage) in metal cages (23×36×28 cm) in continuous dim light (0.25 lumens/m^2 at perch level) supplied by a seven watt incandescent bulb. Temperature was maintained at 25±3°C. EMLEN's "footprint technique" was used to record orientational tendencies [20]. From March 1–21, the experimental birds were injected (s.c.) with corticosterone and prolactin in three different relations. Prolactin (25 μg per injection) was injected daily at four, eight or 12 hours after the time (0800) of corticosterone injections (25 μg per injection). Several control groups were included. One group consisted of birds maintained in large outdoor holding aviaries with no treatment. A second group was maintained indoors with the experimentals and given no treatment. A third group of birds was maintained indoors and given prolactin 12 hours after injections of saline given in place of corticosterone.

Only those birds that received prolactin either at four or 12 hours after corticosterone became fat and exhibited nocturnal activity (Fig. 1). The birds that received prolactin 12 hours after injections of corticosterone exhibited nocturnal activity that was oriented in a north-northeast direction. By contrast, the birds that received prolactin four hours after corticosterone exhibited nocturnal activity that was oriented to the south-southwest. Apparently the two groups were orienting in different directions even though exposed to the same celestial cues.

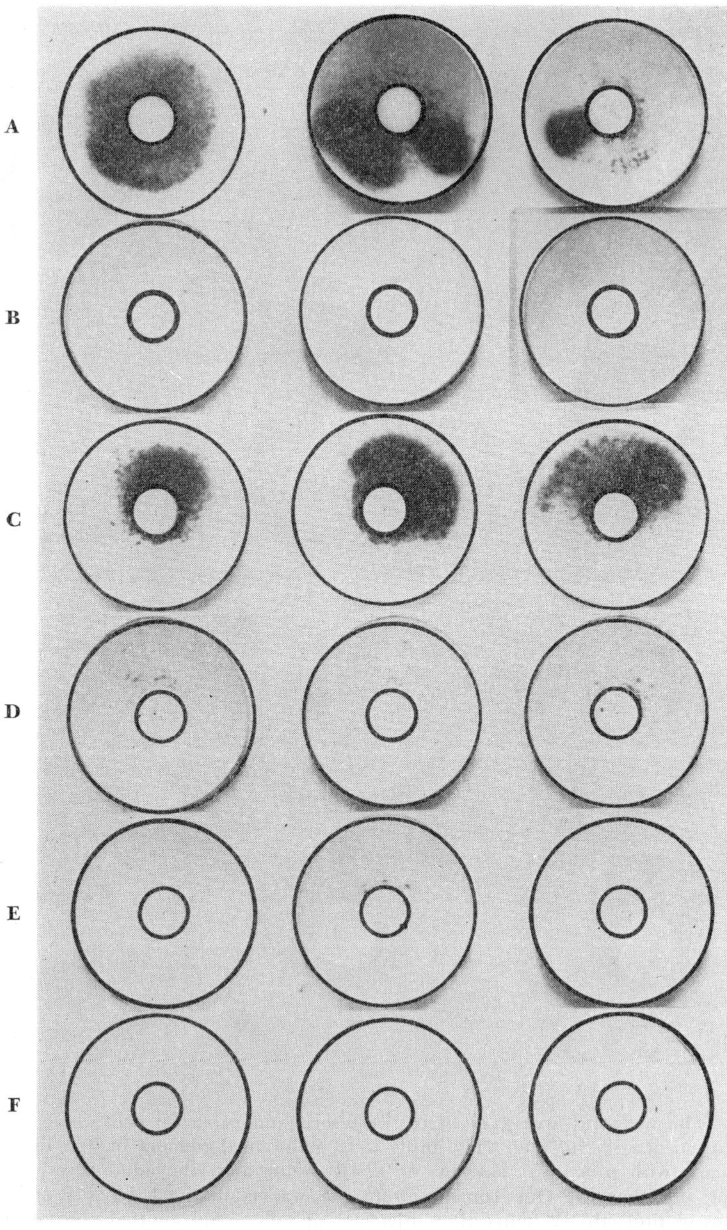

Fig. 1 The temporal synergism of corticosterone and prolactin controlling orientation of nocturnal activity in March. The birds were maintained indoors in continuous dim light and injected daily with prolactin at four (A), eight (B), or 12 (C) hours after the administration of corticosterone injections. Two control groups were maintained indoors, one received prolactin injections 12 hours after injections of saline (D) and the other remained untreated (E). In addition, another control group (F) was kept outdoors and untreated. The birds were brought outdoors and tested for orientation between 2100 and 2400.

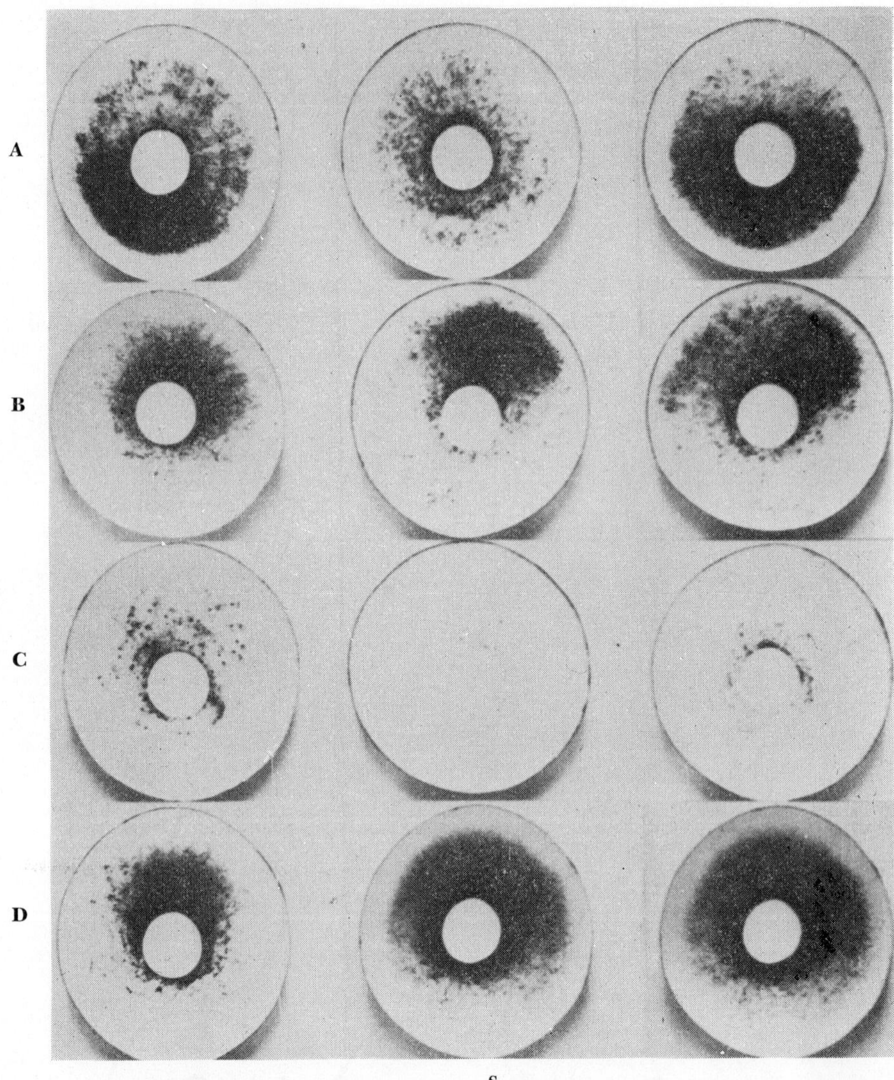

Fig. 2 The temporal synergism of corticosterone and prolactin controlling orientation of nocturnal activity in May. The birds were maintained indoors in dim light and injected daily with proctin at four (A) or 12 (B) hours after the administration of corticosterone injections. One control group (C) was maintained indoors and received injections of prolactin 12 hours after injections of saline, and another control group (D) was kept outdoors and untreated. The birds were brought outdoors and tested for orientation between 2100 and 2400.

Another study was performed during the vernal migratory season in May (Fig. 2). The controls that were kept outdoors had become fat and exhibited some gonadal recrudescence. They exhibited nocturnal activity that was oriented to the north-northeast (toward the breeding grounds). The controls that were kept indoors and given prolactin 12 hours after saline injections were lean and exhibited low levels of random nocturnal activity. To test further the effectiveness of the hormonal pattern in controlling orientation, the experimental treatment was reversed among the groups so that the group which received prolactin four hours after corticosterone in the first test (Fig. 1) was given prolactin 12 hours after corticosterone in the second test (Fig. 2). Similarly, those birds that received prolactin 12 hours after corticosterone in the first test were injected with prolactin four hours after corticosterone in the second test. After two weeks of treatment, birds in the 12 hour group exhibited increased fattening and nocturnal activity that was oriented to the north-northeast. Though the birds in the four hour group exhibited fattening and nocturnal activity, the attainment of directed orientation was noticeably delayed and never 100%. After three weeks, the dominant orientation of two of the three birds tested was to the south-southwest. The orientation of the third bird remained random.

Although a rather small number of birds was used in each of the groups, repetitive testing of each group gave results that were consistent enough to suggest that seasonal changes in the preferred direction of orientation of the White-throated Sparrow may be the result of seasonal changes in the temporal pattern of corticosterone and prolactin and give additional evidence for a basic role for the temporal synergism of corticosterone and prolactin in regulating seasonal conditions in the White-throated Sparrow.

ACKNOWLEDGMENTS

The author is a National Science Foundation Predoctoral Trainee. This study was supported by a grant (Gb-20913) to Dr. Albert H. Meier from the National Science Foundation.

REFERENCES

1. Rowan W. (1932): Experiments in bird migration. III. The effects of artificial light, castration and certain extracts on the autumn movements of the American crow (*Corvus brachyrhychos*). Proc. nat. Acad. Sci., *18*: 659–664.
2. Merkel F. W. (1958): Untersuchungen zur künstlichen Beeinflussung der Aktivität gekäfigter Zugvogel. Vogelwarte, *19*: 173–185.
3. Merkel F. W. (1960): Zur Physiologie der Zugunruhe nachtlich ziehender Kleinvögel: eine Arbeitshypothese. Proc. Intern. Ornith. Congr., *XII*: 507–512.
4. Morton M. L. and Mewaldt L. R. (1962): Some effects of castration on a migratory sparrow, (*Zonotrichia atricapilla*). Physiol. Zool., *35*: 237–247.
5. Wilson A. C. and Farner D. S. (1960): The annual cycle of thyroid activity in White-crowned Sparrows of eastern Washington. Condor, *62*: 414–425.
6. Meier A. H. and Farner D. S. (1964): A possible endocrine basis for premigratory fattening in the White-crowned Sparrow, *Zonotrichia leucophrys gambelii* (Nuttall). Gen. Comp. Endocrinol., *4*: 584–595.
7. Meier A. H., Farner D. S. and King J. R. (1965): A possible endocrine basis for migratory behavior in the White-crowned Sparrow, *Zonotrichia leucophrys gambelii* (Nuttall). Anim. Behav., *13*: 453–465.
8. Meier A. H. and Davis K. B. (1967): Diurnal variation of the fattening response to prolactin in the White-throated Sparrow, *Zonotrichia albicolis*. Gen. Comp. Endocrinol., *8*: 110–114.
9. Meier A. H. and Martin D. D. (1971): Temporal synergism of corticosterone and prolactin controlling fat storage in the White-throated Sparrow, *Zonotrichia albicolis*. Gen. Comp. Endocrinol., *17*: 311–318.
10. Dusseau J. and Meier A. H. (1971): Diurnal and seasonal variations of plasma adrenal steroid hormone in the White-throated Sparrow, *Zonotrichia albicollis*. Gen. Comp. Endocrinol., *16*: 399–408.

11. Meier A. H., Burns J. T. and Dusseau J. W. (1969): Seasonal variations in the diurnal rhythm of pituitary prolactin content in the White-throated Sparrow, *Zonotrichia albicollis*. Gen. Comp. Endocrinol., *12*: 282–289.
12. Meier A. H., Martin D. D. and MacGregor R. III. (1971): Temporal synergism of corticosterone and prolactin controlling gonadal growth in sparrows. Science, *173*: 1240–1242.
13. Kramer G. (1953): Die Sonnenorientierung der Vögel. Verh. Deut. Zool. Ges. Freiburg, *1952*: 72–84.
14. Kramer G. (1957): Experiments on bird orientation and their interpretation. Ibis, *99*: 196–227.
15. Mewaldt L. R. and Rose R. G. (1960): Orientation of migratory restlessness in the White-crowned Sparrow. Science, *131*: 105–106.
16. Aagaard J. S. and Wolfson A. (1962): Transistor equipment for continuous recording of oriented migratory behavior in birds. IRE *Transactions on Bio-Medical Electronics*, BME-9, 204–208.
17. Emlen S. T. (1967): Migratory orientation in the Indigo Bunting, *Passerina cyanea*. I. Evidence for use of celestial cues. Auk., *84*: 309–342.
18. Sauer E. G. and Sauer E. M. (1960): Star navigation of nocturnal migratory birds. Cold Spring Harbor Symp. Quant. Biol., *25*: 463–473.
19. Emlen S. T. (1969): Bird migration: Influence of physiological state upon celestial orientation. Science, *165*: 716–718.
20. Emlen S. T. and Emlen J. T. (1966): A technique for recording migratory orientation of captive birds. Auk., *83*: 361–367.
21. Meier A. H., Trobec T. N., Joseph M. M. and John T. M. (1971): Temporal synergism of prolactin and adrenal steroids in the regulation of fat stores. Proc. Soc. exp. Biol. and Med., *137*: 408–415.

TEMPORAL SYNERGISM OF CORTICOSTERONE AND PROLACTIN CONTROLLING SEASONAL CONDITIONS IN THE WHITE-THROATED SPARROW, *ZONOTRICHIA ALBICOLLIS**

ALBERT H. MEIER**

Department of Zoology and Physiology
Louisiana State University
Baton Rouge, Louisiana, U.S.A.

An essential role for organization is clearly necessary during the annual cycle of migratory birds, such as the White-throated Sparrow, *Zonotrichia albicollis*. This nocturnal migrant follows a very precise series of physiological and behavioral changes during the year. It winters in the southern United States and breeds in northern North America [1, 2]. During the spring and fall migratory periods, large amounts of body fat are stored as an energy source during flight. Reproductive development occurs in the spring and reaches full maturity after the birds arrive at the breeding grounds. After the end of the breeding season in late July, the gonads regress and remain small until the following spring.

The timing of the vernal events (fattening, migratory restlessness, and gonadal recrudescense) is set in many temperate-zone migrants by an increasing day length (for reviews, see LOFTS and MURTON [3], FARNER [4], KING [5]). However, the photoperiodic stimulation of these events does not persist. Within several months, many birds become refractory to long daily photoperiods when the body fat stores are low and as the gonads regress. Investigations using ahemeral light cycles have demonstrated that circadian systems are involved in the photoperiodic stimulation of the avian reproductive system [6–11] as well as in the maintenance of reproductive refractoriness to light [12, 13].

Our laboratory has been trying to determine the principal physiological mechanisms involved in coordinating the seasonal events in the White-throated Sparrow, especially as they relate to migration and reproduction. We have found that prolactin has marked effects on fat stores, nocturnal restlessness (an index of migratory activity in caged birds) and gonadal weights [14–17]. The time of day when the injections are administered is critically important. Injections at certain times stimulate fattening, nocturnal restlessness, and gonadal growth, whereas injections at other times may cause losses in fat stores and gonadal weights, and be ineffective in stimulating nocturnal activity.

The discoveries of daily variations in responses to prolactin suggested that another system mediates the photoperiodic effect and entrains the daily rhythms of responses. Deriving inspiration from our studies of other vertebrate species [18], we explored the possibility that corticosterone, a principal adrenal steroid in birds [19, 20], might entrain rhythms of responses to prolactin in the White-throated Sparrow. In birds maintained in continuous light, we found that in-

* Supported by National Science Foundation grant GB 20913.
** Recipient of Public Health Service Research Career Development Award GM-17,898.

jections of corticosterone entrain rhythms of fattening [17], gonadal weight [21], and locomotor activity (MARTIN and MEIER, unpublished) responses to prolactin (Fig. 1). Peaks of fattening, gonadal growth, and locomotor activity result from daily injections of prolactin administered 12 hours after the time of corticosterone injections. These effects are comparable to those conditions found in the White-throated Sparrow during the vernal migratory period. Secondary peaks of fattening and locomotor activity result from injections of prolactin given 0 to 4 hours after the time of injections of corticosterone. These effects are comparable to those conditions found in birds during fall migration. Contrariwise, prolactin has strong inhibitory effects on fat storage, locomotor activity, and gonadal weights when it is injected 8 hours after the time of injections of corticosterone. These effects resemble the conditions found during the summer photorefractory period. Injections of prolactin in other relations (16 or 20 hours after corticosterone) have less influence on the variables measured.

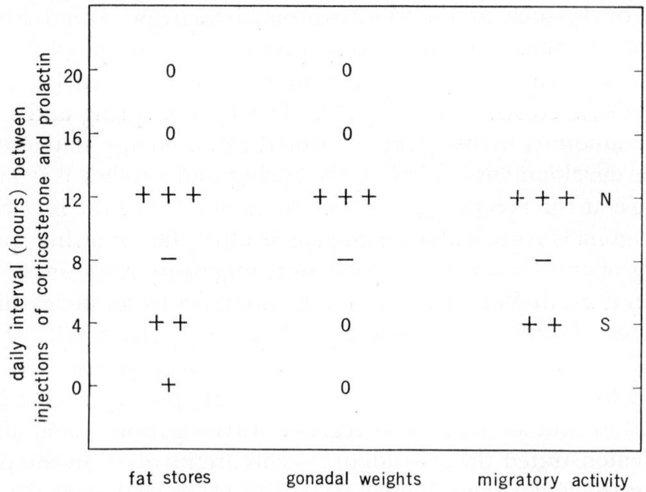

Fig. 1 Temporal relations of corticosterone and prolactin controlling levels of body fat stores, gonadal weights, and migratory activity. Symbols: (+) represents stimulation, (—) represents inhibition, and (O) represents little or no effect; N represents northward orientation of activity, and S represents southward orientation. See text for details and references.

The levels of corticosterone in the plasma [22] and prolactin in the pituitary [23] have been investigated in the White-throated Sparrow. There are marked daily variations in the levels of both hormones. In addition, the phases of the rhythms with respect to the photoperiod change from one season to another so that the relationship between the 2 hormone rhythms also changes (Fig. 2). In the winter (11 February), the interval between the daily rise of plasma corticosterone and the daily release of pituitary prolactin is about 0 to 1 hour. In the spring (5 May) during the migratory period, the interval is about 12 hours; and, in the summer (7 August) during the photorefractory period, the interval is about 6 hours.

These studies have convinced us that several seasonal conditions in the White-throated Sparrow are regulated by a temporary synergism of corticosterone and prolactin (Fig. 3). Winter conditions, including moderate levels of body fat

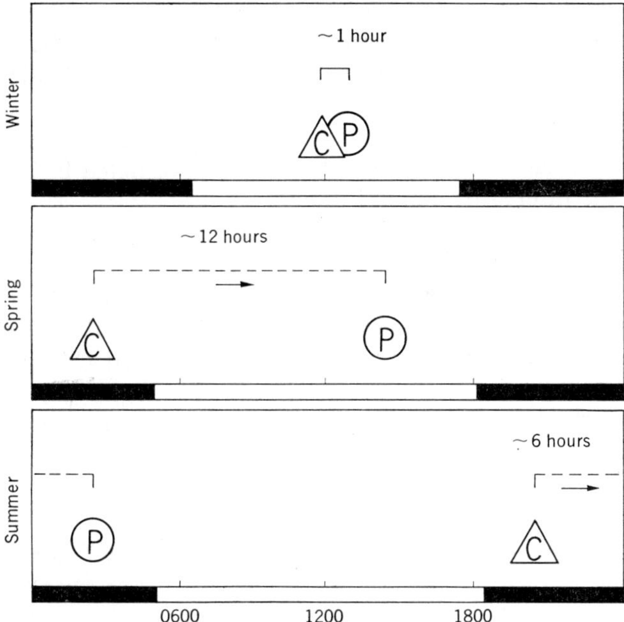

Fig. 2 Temporal relations of the daily rhythms of endogenous corticosterone and prolactin at three seasons. Symbols: The dark bar represents night and the light bar represents day (sunrise—30 minutes to sunset +30 minutes). C represents the time of daily increase in concentration of plasma corticosterone, and P is the time of daily release of pituitary prolactin. See text for details and references.

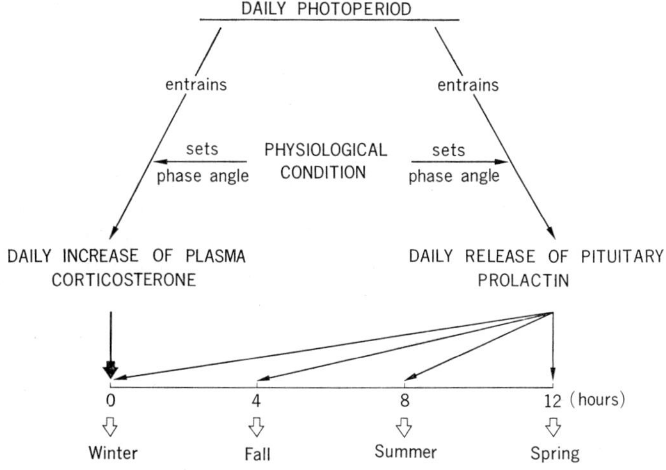

Fig. 3 Temporal patterns of corticosterone and prolactin release controlling seasonal conditions in the White-throated Sparrow.

stores, are apparently controlled by a temporal pattern in which pituitary prolactin is released at about the same time of day as the daily increase in plasma corticosterone. Summer conditions, including low levels of body fat, reproductive photorefractoriness to light, and absence of nocturnal restlessness, are regulated by a temporal pattern in which prolactin is released 6 to 8 hours after the daily rise in plasma concentrations of corticosterone. Spring conditions, including high levels of body fat, reproductive sensitivity to light, and intense locomotor restlessness are regulated by a hormonal pattern in which the interval between the rise in plasma levels of corticosterone and the release of pituitary prolactin is about 12 hours. Fall conditions, including high levels of body fat and moderate to intense locomotor restlessness, are regulated by a hormonal pattern in which the interval is about 4 hours.

The differentiation of the hormonal patterns that control the spring and fall migratory periods is substantiated further by the findings that the nocturnal restlessness induced in birds by the 12-hour pattern of hormone injections is oriented northward under the open night sky, whereas the nocturnal restlessness induced by the 4-hour pattern of injections is oriented southward (Martin and Meier, unpublished, a résumé of these results is given by Martin at this Symposium).

Although the rhythms of both corticosterone and prolactin appear to be entrained by the daily photoperiod, the phase angles or time intervals between a point in the photoperiod and points in the hormone rhythms vary seasonally and may depend on previous environmental, developmental, and physiological experiences of the bird. For example, although the duration of daylight is similar during May and August in Baton Rouge, Louisiana, where the experiments were performed, the phases of the hormonal rhythms with respect to the photoperiod and with respect to one another are altered. Seasonal differences in these phase angles would influence the manner in which similar photoperiods would be interpreted physiologically, and they could account for the conditions of reproductive photosensitivity and photorefractoriness to light [21, 24]. The role of the temporal synergism in controlling gonadal growth is treated in another paper presented by Robert MacGregor at this Symposium. A role for the temporal synergism of corticosterone and prolactin is not limited to the White-throated Sparrow, but appears to be important in many vertebrates (review, Meier [24]).

REFERENCES

1. Wolfson A. (1959): The role of light and darkness in the regulation of spring migration and reproductive cycles in birds. *In*: Photoperiodism and Related Phenomena in Plants and Animals, (R. B. Withrow, ed.), pp. 679–716, AAAS Publ. 55.
2. Helms C. W. (1968): Food, fat and feathers. Amer. Zoologist, *8*: 151–167.
3. Lofts B. and Murton R. K. (1968): Photoperiodic and physiological adaptations regulating avian breeding cycles and their ecological significance. J. Zool., Lond., *155*: 327–394.
4. Farner D. S. (1970): Day length as environmental information in the control of reproduction of birds. Coll. Intern. C.N.R.S., *172*: 365–385.
5. King J. R. (1970): Photoregulation of food intake and fat metabolism in relation to avian sexual cycles. Coll. Intern. C.N.R.S., *172*: 365–385.
6. Hamner W. H. (1963): Diurnal rhythm and photoperiodism in testicular recrudescense of the House Finch. Science, *142*: 1294–1295.
7. Hamner W. H. (1964): Circadian control of photoperiodism in the House Finch demonstrated by interrupted night experiments. Nature, Lond., *203*: 1400–1401.
8. Farner D. S. (1965): Circadian systems in the photoperiodic responses of vertebrates. *In*: Circadian Clocks, (J. Aschoff, ed.), pp. 357–369, North-Holland Publ., Amsterdam.
9. Wolfson A. (1965): Circadian rhythm and the photoperiodic regulation of the annual reproductive cycle in birds. *In*: Circadian Clocks, (J. Aschoff, ed.), North Holland Publ., Amsterdam.
10. Menaker M. (1965): Circadian rhythms and photoperiodism in *Passer domesticus*.

In: Circadian Clocks, (J. Aschoff, ed.), pp. 385–395, North Holland Publ., Amsterdam.

11. Follett B. K. and Sharp P. J. (1969): Circadian rhythmicity in photoperiodically induced gonadotrophin release and gonadal growth in the quail. Nature, Lond., *223*: 968–971.
12. Hamner W. H. (1968): The photorefractory period of the House Finch. Ecology, *49*: 211–227.
13. Murton R. K., Lofts B. and Westwood N. J. (1970): Manipulation of photorefractoriness in the House Sparrow, (*Passer domesticus*) by circadian light regimens. Gen. Comp. Endocrinol., *14*: 107–113.
14. Meier A. H. and Davis K. B. (1967): Diurnal variations of the fattening response to prolactin in the White-throated Sparrow, *Zonotrichia albicollis*. Gen. Comp. Endocrinol., *12*: 282–289.
15. Meier A. H. (1969): Diurnal variations of metabolic responses to prolactin in lower vertebrates. Gen. Comp. Endocrinol. Suppl., *2*: 55–62.
16. Meier A. H. (1969): Antigonadal effects of prolactin in the White-throated Sparrow, *Zonotrichia albicollis*. Gen. Comp. Endocrinol., *13*: 222–225.
17. Meier A. H. and Martin D. D. (1971): Temporal synergism of corticosterone and prolactin controlling fat storage in the White-throated Sparrow, *Zonotrichia albicollis*. Gen. Comp. Endocrinol., *17*: 311–318.
18. Meier A. H., Trobec T. N., Joseph M. M. and John T. M. (1971): Temporal synergism of prolactin and adrenal steroids in the regulation of fat stores. Proc. Soc. Exp. Biol. Med., *137*: 408–415.
19. Nagra C. L., Baum G. J. and Meyer R. K. (1960): Corticosterone levels in adrenal effluent blood of some gallinaceous birds. Proc. Soc. Exp. Biol. Med., *105*: 68–70.
20. DeRoos R. (1962): The physiology of the avian interrenal glands: a review. Proc. Intern. Ornithol. Congr., XIII, 1041–1058.
21. Meier A. H., Martin D. D. and MacGregor R. (1971): Temporal synergism of corticosterone and prolactin controlling gonadal growth in sparrows. Science, *173*: 1240–1242.
22. Dusseau J. W. and Meier A. H. (1971): Diurnal and seasonal variations of plasma adrenal steroid hormone in the White-throated Sparrow, *Zonotrichia albicollis*. Gen. Comp. Endocrinol., *16*: 399–408.
23. Meier A. H., Burns J. T. and Dusseau J. W. (1969): Seasonal variations in the diurnal rhythm of pituitary prolactin content in the White-throated Sparrow, *Zonotrichia albicollis*. Gen. Comp. Endocrinol., *12*: 282–289.
24. Meier A. H. (1971): Temporal synergism of adrenal steroids and prolactin. Presented at Sixth Int. Symp. of Comp. Endocrinol., Banff, Canada. Gen. Comp. Endocrinol. Suppl., *3*: 499–508.

SEASONAL CHANGES IN THE DAILY RHYTHMS OF BRAIN ELECTROLYTES IN THE WHITE-THROATED SPARROW, *ZONOTRICHIA ALBICOLLIS**

Kenneth B. DAVIS

Department of Biology, Memphis State University
Memphis, Tennessee, U.S.A.

Daily variations in electrolyte levels have been reported in a considerable number of vertebrates (for references see Sollberger [1]; Luce [2]). Many of these studies demonstrate that the electrolyte rhythms are entrained by the photoperiod. Although evidence is largely derived from nonseasonal animals (laboratory mice and rats, and humans), it is often assumed that the daily rhythms of electrolytes bear a fixed temporal relation (phase angle) with respect to the photoperiod.

We have been studying daily rhythms of electrolytes in the White-throated Sparrow, *Zonotrichia albicollis* [3]. This nocturnal migrant has marked annual cycles in migration and reproduction. While migration and reproduction are timed in the spring by an increasing day length, many other aspects of the annual cycle, such as the reproductive photorefractory period that occurs during the summer after the breeding season, are not direct reflections of the day length.

This report deals with daily rhythms of brain electrolytes (sodium, potassium, and chloride) in the White-throated Sparrow at two times of the year, in May and in August. Because the day lengths are about the same in Baton Rouge, Louisiana, in May and August, any significant difference in the rhythms could not be directly attributed to the changes in the daily photoperiod.

METHODS

White-throated Sparrows were collected by trapping and netting from wintering flocks near Baton Rouge, Louisiana. They were maintained in large outdoor aviaries and fed chick starter crumbles for at least 2 months before being killed for study.

In May the birds are physiologically prepared for migration (heavy fat stores and nocturnally restless) and the reproductive system is developing. In August, the birds are lean and inactive at night, and the reproductive system is totally regressed [4].

The daily patterns of locomotor activity were determined for several days during 5 days prior to the kill dates of 15 May, and 7 August. These dates were used because of the comparative lengths of the photoperiod. The May photoperiod is only 14 minutes longer than that in August.

At each of the kill dates, groups of five or six birds were killed every six hours beginning at sunrise and continuing for 24 hours. Two additional groups were

* The investigations were done at Louisiana State University, Baton Rouge, and supported by NSF Grants, GB-7277 and GB-20913, to Albert H. Meier.

killed in May, 9 and 15 hours after sunrise. The carcasses, gonads, and whole brains were weighed. The fat content of the carcasses was determined by extraction with petroleum ether using a Soxhlet apparatus.

Whole brains were dried in a vacuum oven and the fat was extracted with petroluem ether. Extracts of each fat-free dried brain were prepared with 10 ml of 0.1 N nitric acid for 48 hours. Chloride concentrations of the extract were determined with a Buchler-Cotlove automatic titrator. Sodium and potassium concentrations were analyzed by flame photometry. Brain electrolytes were converted to meq/kg fat-free fresh brain weight.

Statistical evaluations of the daily variations of the electrolytes were made by a one-way analysis of variance. Differences at the 95% confidence level were considered verification of the presence of daily rhythms. Differences between mean seasonal levels were evaluated using Student's t test.

RESULTS

Seasonal behavioral and physiological differences were apparent. Body weights, fat stores, gonad and oviduct weights were all elevated in May and nocturnal locomotor activity was present. In August the birds were lean, inactive at night, and the reproductive system was regressed.

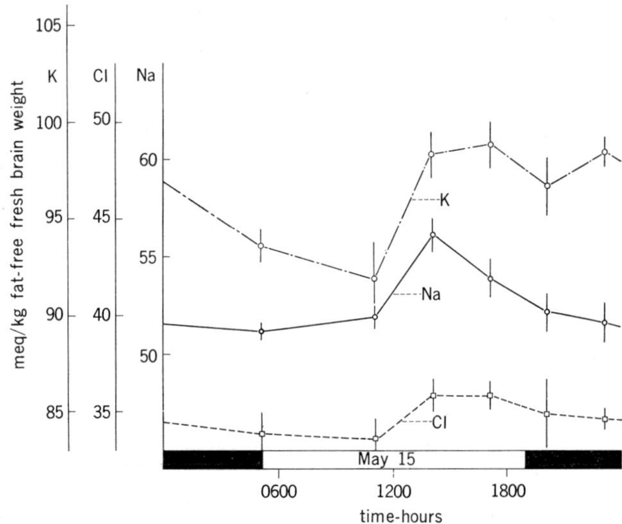

Fig. 1 Daily levels of brain sodium, potassium, and chloride content of White-throated Sparrows on 15 May. The values are expressed as meq/kg fat-free fresh brain weight ± SE. The photoperiod is indicated on the abscissa. Analysis of variance indicates significant temporal differences at the 95% confidence interval for sodium (F—4.7444), and potassium (F=3.9170. No daily group differences were found for chloride (F=1.0190).

Daily rhythms of brain sodium and potassium content were present in May, but the apparent rhythm of chloride content did not vary significantly (Fig. 1). A single peak of brain sodium occurred at 9 hours after sunrise. Brain potassium content also rose at 9 hours after sunrise and remained elevated until about midnight. In August no rhythms of brain sodium or chloride were apparent, and a bimodal peak of brain potassium was evident (Fig. 2). The two peaks of brain potassium occurred at 6 and 18 hours after sunrise.

Comparisons of the seasonal electrolyte levels were made using the seasonal daily mean (Table 1). Brain potassium and chloride content were higher in August than those in May. There was no difference in brain sodium content between the two months.

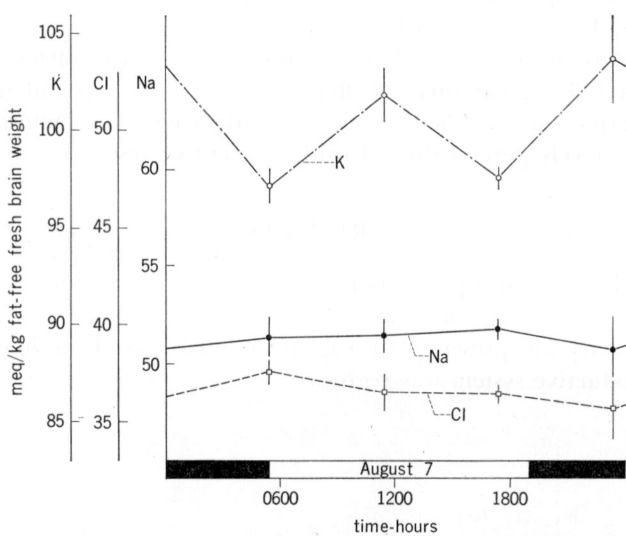

Fig. 2 Daily levels of brain sodium, potassium, and chloride content of White-throated Sparrows on 7 August. The values are expressed as meq/kg fat-free fresh brain weight±SE. The photoperiod is indicated on the abscissa. Analysis of variance indicates significant daily group differences at the 95% confidence interval for potassium (F=6.1571). No daily group differences were apparent for sodium (F=1.0285) or chloride (F=0.8137).

Table 1 Daily means[1)] of electrolyte content[2)] in the brain of White-throated Sparrows in May and August.

	Na^+	K^+	Cl^-
15 May (n=33)	52.67±0.44	96.19±0.71	34.68±0.37
7 August (n=23)	51.33±0.55	99.62±0.84**[3)]	36.64±0.46*

1) Daily means are calculated as the mean of the 24 hour samples in each month.
2) meq/kg fat-free fresh brain weight.
3) Significantly different from levels in May at least at the 95% (*) or 99% (**) interval by Student's t test.

DISCUSSION

Birds in August and May were exposed to a similar photoperiod, but the birds differed greatly in physiological condition. The daily rhythms of brain electrolyte levels were also markedly different between the two seasons. The most striking seasonal differences were not quantitative when daily means were used, but rather they were temporal, indicating seasonal shifts in phases of these rhythms. No direct relation was found between brain electrolyte rhythms and the time of locomotor activity in the two months; further, there was no direct temporal correlation between daily levels of brain and plasma electrolytes [3].

The daily rhythms of electrolytes are probably entrained by the photopreiod, but the relation between the endogenous rhythm and the photoperiod differs from one season to another. The differences of electrolyte rhythms between spring and late summer suggest that these rhythms reflect the physiological state of the animal rather than a simple, direct relation to the photoperiod. These results complement other studies of daily rhythms in the White-throated Sparrow. Prolactin injections during the middle of a 16-hour photoperiod cause fattening, but similar injections at the beginning of the photoperiod cause a loss in fat stores [5]. The rhythms of pituitary prolactin [6] and plasma corticosterone [7] change with respect to the photoperiod from one season to another.

The most notable and best documented studies that correlate physiological and behavioral rhythms with the photoperiod are those of Halberg [8]. The temporal relationships of the peak of 17-hydroxycorticosteroids in the plasma and urine remain rather fixed with respect to locomotor activity and the lighting regime on photoperiods of different lengths and on photoperiods shifted by as much as 180°. The constancy of phase angles of daily rhythms of an organism with the photoperiod may be misleading when studying annual behavior and physiological adjustments to the photoperiod. It should be noted that Halberg's work concerns laboratory rats and mice, and man. All of these species are highly protected from seasonal changes in the photoperiod and temperature. Their ability to survive does not depend on adaptation to the annual cycle.

The results of this study suggest that in a seasonally regulated animal, daily rhythms found at one time of the year may be different from those occurring at other times of the year. Further, seasonal studies should consider seasonal shifts in the phases of daily rhythms. Apparent seasonal changes may only reflect changes in the phase of the daily rhythms and not the mean daily levels of the variables measured.

ACKNOWLEDGMENTS

I am grateful to Dr. Albert H. Meier for his help in the preparation of this manuscript.

REFERENCES

1. Sollberger A. (1965): Biological Rhythm Research. Elsevier, Amsterdam.
2. Luce G. G. (1970): Biological Rhythms in Psychiatry and Medicine. Public Health Service Pub. No. 2088, U. S. Government Printing Office, Washington.
3. Davis K. B. and Meier A. H. (1974): Seasonal and daily variations of sodium, potassium, and chloride levels in the plasma and brain of the migratory White-throated Sparrow, *Zonotrichia albicollis*. Physiol. Zool. (Accepted for publication).
4. Wolfson A. (1954): Weight and fat deposition in relation to spring migration in transient white-throated sparrows. Auk, *71*: 413–434.
5. Meier A. H. and Davis K. B. (1967): Diurnal variations of the fattening response to prolactin in the White-throated Sparrow, *Zonotrichia albicollis*. Gen. Comp. Endocrinol., *8*: 110–114.
6. Meier A. H., Burns J. T. and Dusseau J. W. (1969): Seasonal variations in the diurnal rhythms of pituitary prolactin content in the White-throated Sparrow, *Zonotrichia albicollis*. Gen. Comp. Endocrinol., *12*: 282–289.

7. Dusseau J. W. and Meier A. H. (1971): Diurnal and seasonal variations of plasma adrenal steroid hormone in the White-throated Sparrow, *Zonotrichia albicollis*. Gen. Comp. Endocrinol., *16*: 399–408.
8. Halberg F. (1969): Chronobiology. Ann. Rev. Physiol., *31*: 673–725.

CHRONOBIOLOGY AND PLANTS

Chairman: Solon A. Gordon

PHOTOPERIODISM AND TIMING MECHANISMS IN THE CONTROL OF FLOWERING

RODNEY W. KING**

Department Plant Sciences, University of Western Ontario
London, Ontario, Canada

Many organisms have evolved the ability to use daylength as a source of environmental information in the control of behavioural and developmental responses. In 1936, BÜNNING proposed that circadian rhythms provided the basic timekeeping mechanism in these photoperiodic responses of organisms [1]. A strong argument, although not proof of a timekeeping role of circadian rhythmicity, is the fact that rhythms are displayed in many responses of plants, including flowering.

In the past, a lack of flowering in unfavourable photoperiods has generally frustrated attempts to critically study how the phasing, *i.e.* timing, of an endogenous rhythm could determine time measurement in the photoperiodic control of flowering. However, the short-day plant *Chenopodium rubrum* (origin 60°47'N 137°32'W) offers several advantages in studying this question. It shows a quantitative flowering response to short days. An experiment can be completed in less than two weeks. The capacity of seedlings to flower varies rhythmically in response to a single dark period of varied duration. And, the seedlings can be grown easily and maintained in large numbers in Petri dishes.

PHOTOPERIOD DURATION AND REPHASING OF THE RHYTHM OF FLOWERING

A rhythm in the capacity of *C. rubrum* to flower was obtained when seedlings were given one dark period of varied duration that interrupted continuous light (Fig. 1). As established by CUMMING et al. [2] and CUMMING [3, 4], cycle length is about 30 h. Peaks of the rhythm occur at about 13, 43 and 73 h of darkness. The transition from a long period of continuous light to darkness initiates the oscillation.

The phase of the rhythm can be reset by exposure of the seedlings to 6-h red (115 μw/cm^2) or fluorescent light (1037 μw/cm^2) at different times over the first 36 h of darkness and, hence, at different phases of the oscillation (Fig. 1). After the light pulse, the phasing of the rhythm was assayed by allowing the rhythm to run free in a subsequent dark period of varied duration before the plants were returned to continuous light until assayed for flowering. Rhythm rephasing was assessed on the basis of the timing of the peaks of the oscillation.

* This work was supported by a grant to Dr. B. G. Cumming from the National Research Council of Canada.

** Present address: MSU/AEC Plant Research Laboratory, Michigan State University, East Lansing, Michigan.

The amount (hours) and direction (advance or delay) of rephasing relative to the corresponding peak of the control has been estimated for each 30-h cycle of the oscillation. From this data, it has been possible to derive a phase response curve that relates to the timing of the 6-h light exposure (Fig. 2).

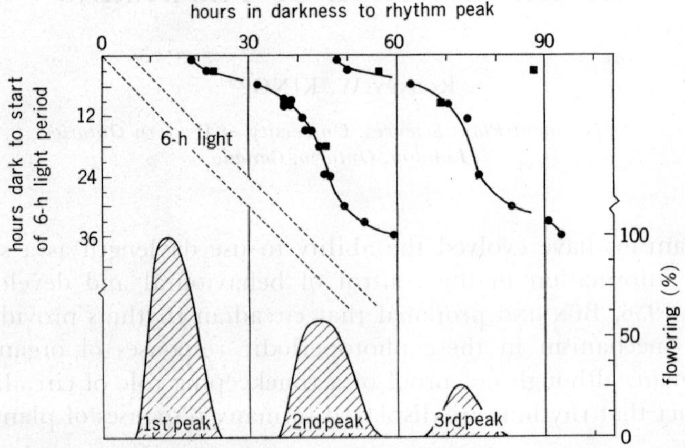

Fig. 1 Rephasing of the rhythmic flowering response of *C. rubrum* (upper curves) when a 6-h light exposure was imposed at different phase of the free-running oscillation (lower curve). Rephasing estimated from the timing of the first, second, third or fourth peaks of the oscilation. Vertical lines distinguish each 30-h cycle of the free-running oscillation. Red (●) or fluorescent (■) light.

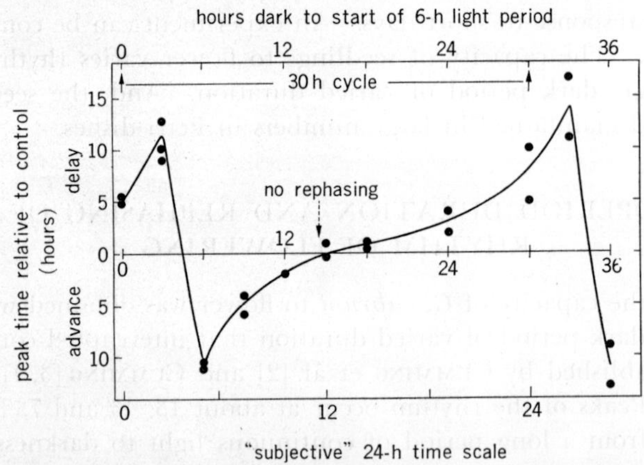

Fig. 2 Phase response curve for rephasing the rhythm of flowering in *C. rubrum* by a 6-h photoperiod. Values derived from Figure 1. The "subjective" 24-h time scale assesses rephasing in daily photoperiods in which instance the circadian period of the rhythm (30 h) will be entrained to a 24-h cycle length.

Relative to the phase of the free-running oscillation, rhythm phase resetting by 6-h light was greatest when light impinged on the positive slope of either the first or second peaks, the third, sixth and ninth hours, or the thirtieth, thirty-third and thirty-sixth hours in darkness, respectively (Figs. 1, 2). The same rephasing resulted when red or fluorescent light was applied for 6 h (Fig. 1). Rephasing altered the timing of the peak immediately following the light interruption and also all subsequent peaks of the oscillation. The period of the rhythm after rephasing remained close to 30 h. Comparable data has been reported previously for a number of plants and animals (see summary in WINFREE, 1970). However, in contrast to earlier observations [5–7], in C. *rubrum*, the next peak of the rhythm, following a light exposure, was always displaced by at least 12 h in darkness (Fig. 1). Such a displacement in darkness suggests that there are reactions which occur over the early hours of darkness that are obligatory for flowering, but not for some of the other rhythmic responses. As discussed later, over the early hours of darkness the presence of the P_{fr} form of phytochrome darkness may prevent expression of the rhythm of flowering.

In a further series of experiments, single 12- or 18-h light periods were administered during darkness and at different phases of the oscillation. The results obtained were quite unexpected. In contrast to the response to a 6-h light exposure, after 12- or 18-h light periods the rhythm was always reinitiated at a fixed phase in the subsequent dark period. Irrespective of the phase of the oscillation when the light period began, following periods of 12- or 18-h light, the next peak of the rhythm always occurred 13 h in darkness after the light-off signal.

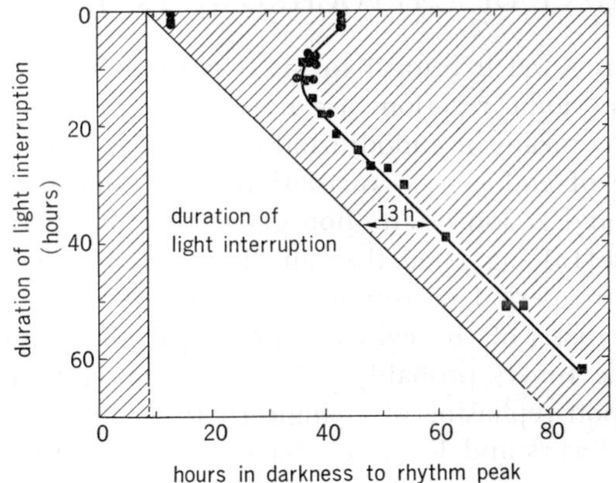

Fig. 3 Rephasing of the rhythm of flowering (time in darkness to the first peak) by a red (●) or fluorescent (■) light period of 5 min to 62 h in duration.

In another experiment (Fig. 3) the rephasing light period began at a fixed time in darkness—the ninth hour—but the duration of the light interruption was varied from 5 min to 62 h. Again, provided the length of the light period was 12-h or longer, the next peak of the rhythm occurred after 13 h in darkness. In *Drosophila* a similar response has been reported by PITTENDRIGH [8, 9]. He found that, when brief light periods were employed, rephasing of the rhythm interacted with the phase of the oscillation at which the light exposure was administered. However, after longer light periods (12 to 34 hours) the rhythm was reinitiated at a fixed phase in a subsequent dark period.

The foregoing data with C. *rubrum* can be related directly to its photoperiodic response and to time measurement.

As a timekeeper, the pattern of changing sensitivity of the rhythm to light (*e.g.* Fig. 2) provides a continuously consulted clock that can be synchronized with solar and, hence, seasonal time. Once synchronized, the capacity of seedlings to flower then depends on whether or not light impinges on the light sensitive (skotophil) or light insensitive (photophil) phase of the oscillation (see Fig. 1).

In terms of the photoperiodic response, in daily photoperiods of 12 h or longer, the rhythm peak will occur 13 h in darkness after the dusk signal (Fig. 3). Thus, flowering should be maximal in a 12-h photoperiod with its associated 12-h dark period, but with longer photoperiods flowering should gradually decrease until the critical dark period length (about 6 h) is reached. An 18-h daily photoperiod, *i.e.* 6-h darkness per day, should be non-inductive. With photoperiods shorter than 12-h, such as in a 6-h daily photoperiod, flowering will result since, at an equilibrium value of rephasing, *i.e.* no net advance or delay (Fig. 2), the rhythm peak will occur at the beginning of the daily 6-h photoperiod *i.e.* 18 h in darkness on a "subjective" 24-h time scale. This has been confirmed experimentally and the justification for a "subjective" 24-h time scale has been presented by KING [10].

The above predictions based on phasing of the rhythm account completely for the observed photoperiodic induction of flowering in *C. rubrum* [10, 11]. Clearly, as first postulated by BÜNNING [1], the phasing of an endogenous rhythm controls photoperiodic time measurement and determines capacity to flower.

THE INVOLVEMENT OF PHYTOCHROME IN TIME MEASUREMENT

BÜNNING [12] considered it central to his hypothesis that "the time-measuring processes in photoperiodic reactions are not carried out by the hourglass principle, but, rather, by means of endodiurnal oscillations." However, evidence for a timekeeping role of endogenous oscillations in regulating flowering does not exclude the additional action of an "hourglass" type of reaction. Indeed, TAKIMOTO and HAMNER [13] suggested that an "hourglass" timer and two rhythmic timers regulated dark period measurement and flowering in the short-day plant *Pharbitis.* On the basis of the evidence presented above for *C. rubrum,* their postulated rhythmic timers probably reflect a changing pattern of response to the photoperiod upon rephasing of a single rhythm. As for the influnece of an "hourglass" timer, EVANS and KING [14] have established that the timing of disappearance of phytochrome P_{fr} during darkness, *i.e.* an "hourglass" reaction, can influence dark period time measurement in *Pharbitis.*

In *C. rubrum* the involvement of dual photoperiodic timers could explain why, on rephasing, the rhythm was always displaced by at least 12 h in darkness from the light period (Fig. 1). Possibly, phytochrome P_{fr} disappearance had not proceeded sufficiently over the early hours of darkness with the result that flowering, *i.e.* expression of the rhythm, was suppressed.

Measurements of the timing of disappearance of phytochrome P_{fr} in seedlings of C. rubrum during a 9.5 h dark period are illustrated in Fig. 4. The use of a null response method for estimating changes in the proportion of phytochrome as P_{fr} avoids any complication that could result from time dependent changes in sensitivity to phytochrome [2, 10, 14]. Thus, the slow disappearance of phytochrome P_{fr} during the early hours of darkness (Fig. 4) suggests a potential for a timekeeping role in flowering. (Although not a typical decay phenomenon this response is referred to subsequently as an "hourglass" reaction.)

If the timing of disappearance of P_{fr} is important to time measurement in *C. rubrum,* it should be possible to delay or suppress the expression of the rhythm by reintroducing P_{fr} after, but not before, its normal time of disappearance. This possibility was readily tested by interrupting darkness at different times with 5 min of red light to recover the P_{fr} form of phytochrome to the P_r

form. Such treatment does not cause any rephasing of the rhythm [2] (Fig. 3). However, delaying P_{fr} disappearance—the "hourglass"—delayed dark period time measurement and partially prevented expression of the next peak of the rhythm of flowering (Fig. 4). This data leaves little doubt that an "hourglass" operates along with a rhythmic timer to regulate photoperiodic time measurement in *C. rubrum*.

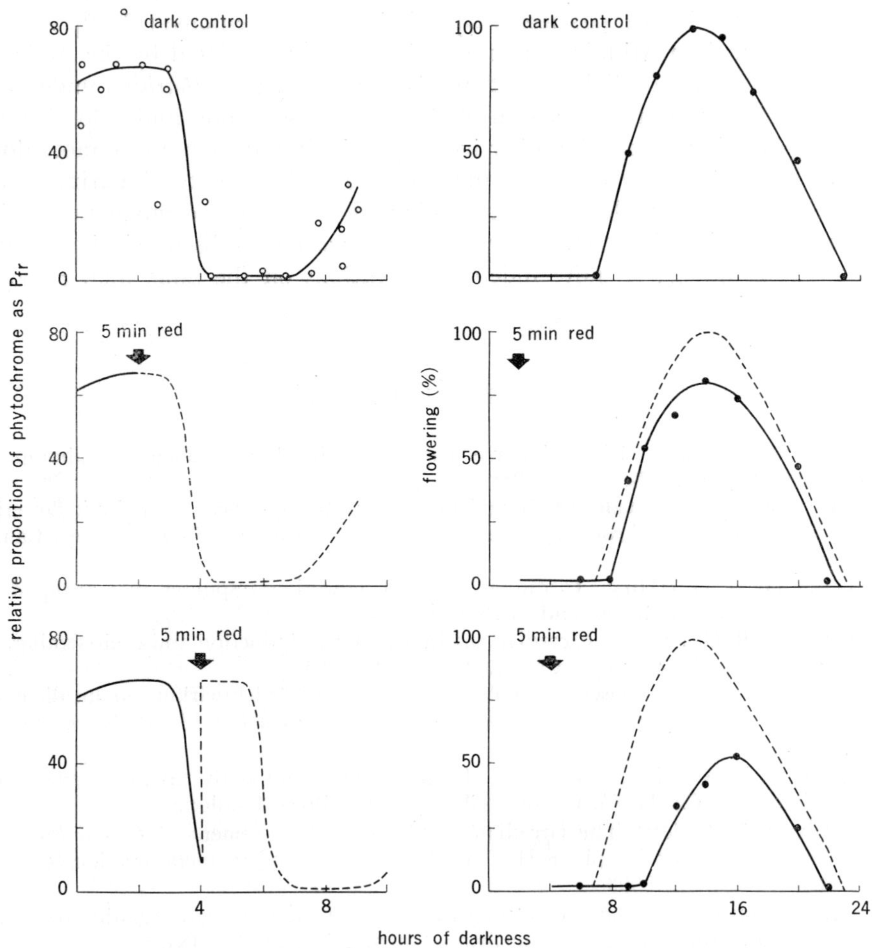

Fig. 4 Disappearance of phytochrome P_{fr} during a dark period of 9.5 h and the effect that a 5 min red irradiation during darkness (second or fourth hour) would have on P_{fr} disappearance. The critical dark period for flowering was lengthened 3 to 4 h by delaying P_{fr} disappearance (5 min red at fourth hour) but phasing of the second and third peaks of the rhythm remained unaltered (results not shown).

Neither a single 5 min light interruption of darkness nor a double light break *i.e.* a skelton photoperiod of 6 or 12 h (5 min. red-dark-5 min. red) altered the phasing of the rhythm in *C. rubrum*. Thus these results provide additional, but indirect, support for the arguments developed by Evans and King [14] for the importance of an "hourglass" component of time measurement in *Pharbitis*. A number of other reports might also be best interpreted in terms of red light interruptions influencing the timing of phytochrome P_{fr} disappearance during darkness. For instance, the results of PAPENFUSS and SALISBURY [15] with *Xanthium* can be understood on this basis. This interpretation also explains findings by DENNEY and SALISBURY [16] with *Xanthium.* They showed that, in this species, the status of a rhythm of leaf movement is not a valid indicator of the photoperiodic clock controlling flowering.

Variation between species in their dependence on the action of "hourglass" and rhythmic timers could explain why flowering in *Xanthium* and *Pharbitis* can go to completion in darkness, a typical "hourglass" response [17–19], whereas, in *C. rubrum,* the rhythmic response of flowering is regulated by the timing of returning plants to the light. Probably in *Xanthium* and *Pharbitis* under most experimental conditions the action of the "hourglass" alone limits floral induction. In *C. rubrum,* on othe other hand, the rhythm appears to be more dominant and, therefore, together with the "hourglass", it regulates flowering in this species. It remains unclear how rhythmic and "hourglass" components of time measurement might interact. However, the experiments discussed above for *C. rubrum* (Fig. 4) suggest quite independent actions on the same final response, flower formation.

REFERENCES

1. BÜNNING E. (1936): Die endonome Tagesrhythmik als Grundlage der photoperiodischen Reaktion. Ber. dtsch. Bot. Ges., *54*: 590–607.
2. CUMMING B. G., HENDRICKS S. B. and BORTHWICK H. A. (1965): Rhythmic flowering responses and phytochrome changes in a selection of *Chenopodium rubrum.* Can. J. Bot., *43*: 825–853.
3. CUMMING B. G. (1967): Circadian rhythmic flowering responses in *Chenopodium rubrum*: effects of glucose and sucrose. Can. J. Bot., *45*: 2173–2193.
4. CUMMING B. G. (1969): Circadian rhythms of flower induction and their significance in photoperiodic response. Can. J. Bot., *47*: 309–324.
5. ZIMMER R. (1962): Phasenverschiebung und andere Störlichtwirkungen auf die endogen tagesperiodischen Blütenblattbewegungen von *Kalanchöe blossfeldiana.* Planta, *58*: 283–300.
6. HASTINGS J. W. (1964): The role of light in persistent daily rhythms. *In*: Photophysiology (I). Ch. 11, (A. C. GIESE ed.), Academic Press, London.
7. HALABAN R. (1968): The circadian rhythm of leaf movement of *Coleus blumei* x *C. frederici,* a short-day plant. II. The effects of light and temperature signals. Plant Physiol., *43*: 1887–18993.
8. PITTENDRIGH C. S. (1960): Circadian rhythms and the circadian organization of living systems. Cold Spring Harbor Symp. Quant. Biol., *25*: 159–184.
9. PITTENDRIGH C. S. (1966): The circadian oscillation in *Drosophila pseudoobscura* pupae: a model for the photoperiodic clock. Z. Pflanzenphysiol., *54*: 275–307.
10. KING R. W. (1971): Time measurement in the photoperiodic control of flowering. Ph. D. Thesis, Univ. of Western Ontario, London, Ontario, Canada.
11. CUMMING B. G. (1963): Evidence of the requirement for phytochrome-Pfr in the floral initiation of *Chenopodium rubrum.* Can. J. Bot., *41*: 901–936.
12. BÜNNING E. (1960): Circadian rhythms and the time measurement in photoperiodism. Cold Spring Harbor Symp. Quant. Biol. *25*: 1–9.
13. TAKIMOTO A. and HAMNER K. C. (1964): Effect of temperature and preconditioning on photoperiodic responses of *Pharbitis nil.* Plant Physiol., *39*: 1024–1030.
14. EVANS L. T. and KING R. W. (1969): Role of phytochrome in photoperiodic induction of *Pharbitis nil.* Z. Pflanzenphysiol., *60*: 277–288.
15. PAPENFUSS H. D. and SALISBURY F. B. (1967): Aspects of clock resetting in flowering of *Xanthium.* Plant Physiol., *42*: 1562–1568.

16. Denney A. and Salisbury F. B. (1970): Separate clocks for leaf movements and photoperiodic flowering in *Xanthium strumarium* L. (Cocklebur). Plant Physiol., *46* (suppl.): 26.
17. Searle N. E. (1961): Persistence and transport of flowering stimulus in *Xanthium*. Plant Physiol., *36*: 656–662.
18. Zeevaart J. A. D. (1963): Climatic control of reproductive development. *In*: Environmental control of plant growth. (L. T. Evans, ed.), pp. 289–310, Academic Press, New York.
19. King R. W., Evans L. T. and Wardlaw I. F. (1968): Translocation of the floral stimulus in *Pharbitis nil* in relation to that of assimilates. Z. Pflanzenphysiol., *59*: 377–388.

PHOTOPERIODIC TIME MEASUREMENT OF FLOWERING IN A SHORT-DAY-PLANT: EFFECTS OF THE EXTERNAL MEDIUM*

RUTH HALABAN** and WILLIAM S. HILLMAN

Biology Department, Brookhaven National Laboratory
Upton, N.Y., U.S.A.

The sensitive phase of flowering of the short day plant *Lemna perpusilla* is affected by at least 3 known factors: a) The form of phytochrome, which is a function of the light quality schedule [7–8], b) pulses of high temperature [6], and c) the nature of the growth medium [3, 4, 9, 12, 13]. By studying the role of these 3 factors separately and together, one can hope to obtain a better understanding of periodic time measurement. We have previously reported [3] that the flowering of *L. perpusilla* is inhibited by daily transfers to water for short period of time during a sensitive phase as demonstrated in Fig. 1. The inhibition of flowering by water transfers can be overcome by transferring the plants to low concentrations of Ca^{++} solutions (Table 1) by lowering the temperature to 20°C while the plants are in water (Table 2), or by supplementing the water with flower-promoting material that is found in water after incubation with plants under dark conditions (Fig. 2). Since the flower promoting activity can be detected in the water after incubation with plants at any time during the dark period, its rate of production cannot be regarded as crucial for the time measurement process. Nor can one explain the inhibitory effect of light signals on flowering by assuming that light destroys the flower-promoting-activity. Very low light intensities are needed to saturate the light signal effect, while on the other hand, high light intensities are needed to prevent the appearance of flower promoting activity in the water (Fig. 3). Results of experiments under different light qualities showed no difference in flower-promoting activity between water incubated with plants in either darkness, blue, red and far-red light (Fig. 4). In fact, it seems evident that photosynthesis is involved in preventing the appearance of flower promoting activity since plants containing little or no chlorophyll, as a result of having been grown in continuous darkness, except for 15 min. light breaks 8 hr apart for 3 weeks, yielded flower promoting activity irrespective of the presence or absence of light.

The flower promoting activity is destroyed by autoclaving and is thus probably not due to the presence of ions. On the other hand, activity is retained after 5 min. of boiling, which suggests that a macromolecule is not involved or, at least, that the natural form of the macromolecule is not crucial for the flower promoting activity. Ultraviolet absorbance curves of water samples obtained after

* Research carried out at Brookhaven National Laboratory under the auspices of the U.S. Atomic Energy Commission.

** Present address: Department of Dermatology, Yale University School of Medicine, New Haven, Conn. 06510.

Table 1 Effect of repeated 4-hr treatment with solutions of macronutrient salts on Flowering of *L. perpusilla*.

Macronutrient	Concn. mM	pH	Percentage flowering (Fl%)			Time
			Experi-mental	Control		
				Water	Medium	
a. Macronutrient salts in Hutner's medium						
$Ca(NO_3)_2 4H_2O$	0.75	6.4	25.9**	4.7	68.3	15–19
	0.75	9.8	52.8**	15.2	66.3	16–20
K_2HPO_4	1.15	6.4	7.4	4.7	68.3	15–19
	1.15	7.3	28.7**	4.7	68.3	15–19
	1.15	7.3	40.5**	15.2	66.3	16–20
NH_4NO_3	1.25	6.4	0.0	4.7	68.2	15–19
	1.25	6.4	1.3**	15.2	66.7	16–20
	1.25	6.4	3.9**	49.6	82.4	15–19
$MgSO_4$	1.00	6.4	0.0**	15.2	66.7	16–20
	1.00	6.4	11.2**	49.6	82.4	15–19
b. Test for the active ions in a—all at pH 6.4						
NH_4NO_3	1.25		3.9**	49.6	82.4	15–19
KNO_3	1.25		60.2	57.9	84.0	15–18
NH_4Cl	1.25		36.1**	49.6	82.4	15–19
$MgSO_4$	1.00		11.2**	49.6	82.4	15–19
MgCl	1.00		4.0**	49.6	82.4	15–19
K_2SO_4	1.00		6.6**	49.6	82.4	15–19
K_2HPO_4	1.15		62.7	57.9	84.0	15–18
KCl	0.92		37.7	37.2	70.4	15–19
$Ca(NO_3)_2$	0.75		72.5**	49.6	82.4	15–19
$CaCl_2$	0.75		73.3**	49.6	82.4	15–19

Water control: The percentage of flowering in plants transferred to twice distilled water at the same time for the same duration. Medium control: The plants were transferred instead to half-strength Hutner's+sucrose. Time: Hours after the lights of the 8-hr photoperiod were turned on. Values marked with two asterisks are significantly different from the water control at the 1% level. Values are means of five replicates.

Table 2 Flowering of *Lemna perpusilla* as affected by various treatments given for 5-hr at the sensitive phase. Letters indicate groups of significant difference at the 1% probability level.

Treatment	21°	25°	29°
H_2O	74.8 A	29.4 C	23.4 C
$MgSO_4$ 0.3mM	78.3 A	30.7 C	5.2 D
$Ca(NO_3)_2$ 1mM	81.5 A	57.0 B	52.1 B
Medium control	85.2 A	77.8 A	80.3 A

incubation with plants under dark or light conditions showed a peak between about 265 and 270 nanometers, without any distinctive qualitative differences. The samples did not absorb visible light; the pH of such samples showed no significant differences dependent upon light or dark incubation conditions.

In order to explain the effect of water on flowering we might assume that the flower promoting activity is lost when the plants were transferred to water, and that the absence of this activity is crucial for the photoperiodic effect when it coincides with the sensitive phase. Calcium can be viewed as overcoming the inhibition by water by preventing this loss. The effect of Ca^{++} on selective ion transport has been documented in several root systems [11, 14] and it is generally

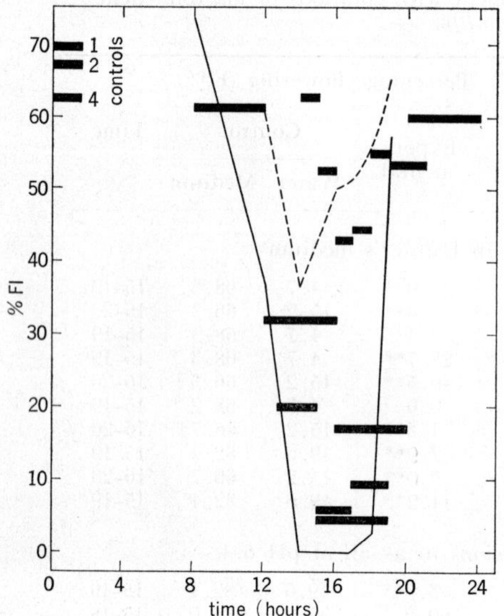

Fig. 1 Flowering response of *L. perpusilla* periodically transferred to water for 1, 2, or 4 hr during different phases of the 16 hr dark period. Each treatment was repeated for the initial four consecutive cycles. The length of each bar indicates the duration of the water treatment, and its height indicates the level of flowering by the end of 7 days of growth. The bars at upper left indicate the percentage of flowering in control cultures for the 1-, 2-, and 4-hr treatments. Solid lines represent the flowering response of cultures exposed to 10 min of white light (500 ft-c) at different times during the initial four consecutive dark periods; the broken line represents the effect of such treatment for only two consecutive cycles. Standard deviations, calculated for each experiment by the analysis of variance, ranged between 2.6 and 7.1.

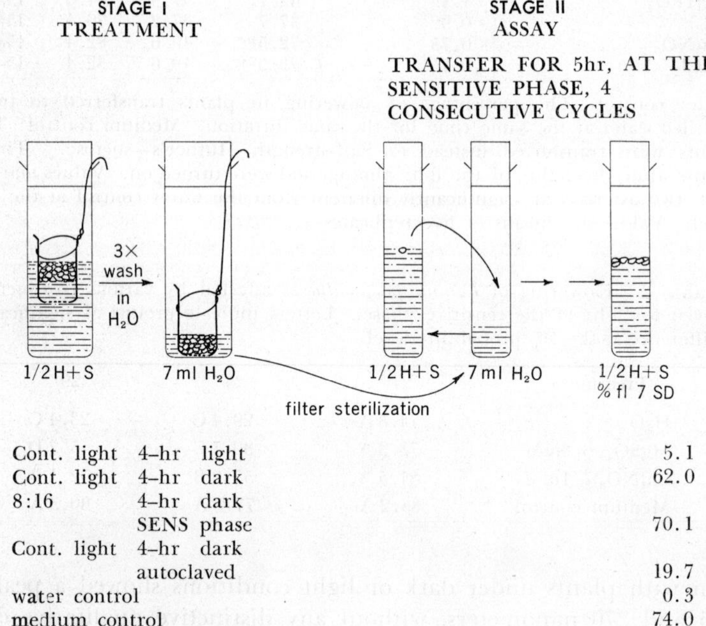

Cont. light	4–hr light	5.1
Cont. light	4–hr dark	62.0
8:16	4–hr dark SENS phase	70.1
Cont. light	4–hr dark autoclaved	19.7
water control		0.3
medium control		74.0

Fig. 2 Diagrammatic summary of an experiment on the assay of flower promoting substance obtained from water incubated plants under different light conditions. The stock plants were grown in perforated vials under either continuous (Cont.) light or short days (8:16—8 hrs of light followed by 16 hrs of darkness) and in half-strength Hutner's medium supplemented with sucrose (1/2 H+S). After 3 washes in water they were incubated in 7 ml water under different light conditions. One set of samples were autoclaved after the plants were taken out. The "incubated" water was then tested for flower promoting activity in stage II of the experiment, in the same way as described in Fig. 1.

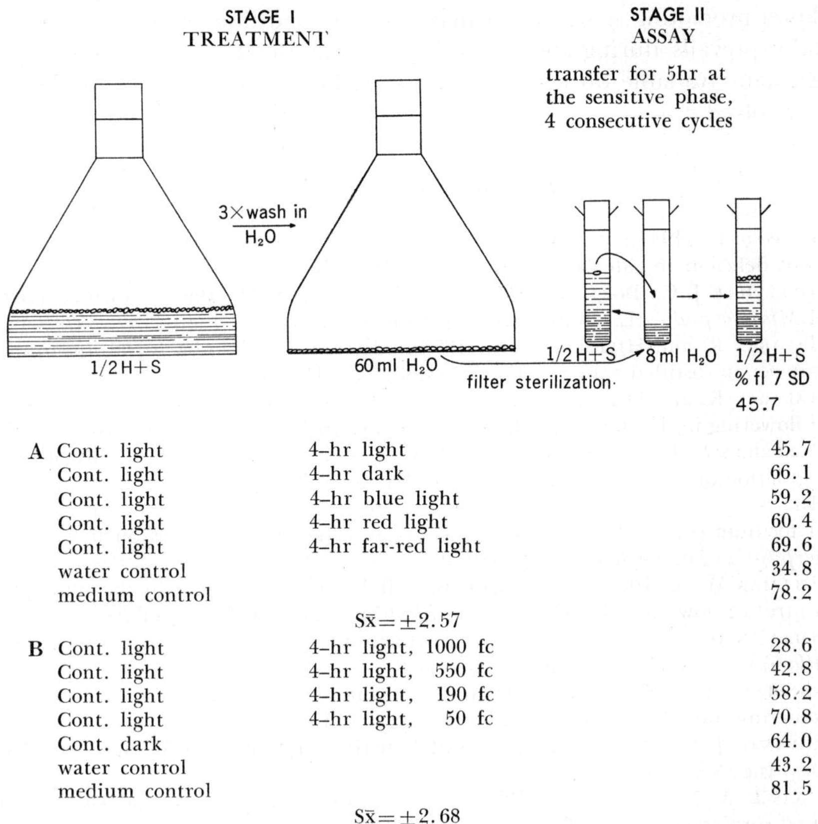

	Stage I	Treatment	% fl 7 SD
A	Cont. light	4-hr light	45.7
	Cont. light	4-hr dark	66.1
	Cont. light	4-hr blue light	59.2
	Cont. light	4-hr red light	60.4
	Cont. light	4-hr far-red light	69.6
	water control		34.8
	medium control		78.2
		$S\bar{x}=\pm 2.57$	
B	Cont. light	4-hr light, 1000 fc	28.6
	Cont. light	4-hr light, 550 fc	42.8
	Cont. light	4-hr light, 190 fc	58.2
	Cont. light	4-hr light, 50 fc	70.8
	Cont. dark		64.0
	water control		43.2
	medium control		81.5
		$S\bar{x}=\pm 2.68$	

Fig. 3 Diagrammatic summary of an experiment on the effect of light quality (A), or intensity (B) on the presence of flower-promoting activity in water. The plants were grown in 1L Erlenmeyer flasks containing 400 ml medium until they formed a continuous layer. The medium was then discarded, the plants were washed 3x with 200 ml water each, and then incubated with 60 ml H_2O. The "incubated" water was tested for flowering promoting activity as described before.

considered to have a positive effect on the integrity of membrane structure [1, 10]. However, the temperature experiment indicates that the loss of flower promoting activity still occurs at 20°C, whereas transferring the plants to water at this temperature did not result in inhibition of flowering. If one assumes that the flower-promoting substance is an essential metabolite for photoperiodic induction, this apparent contradiction could be explained by postulating that at low temperature the rate of metabolism is slower and therefore less of the substance is needed to satisfy the requirements for photoperiodic induction. However, one can speculate that the effect of water, as well as of temperature, Ca^{++}, $MgSO_4$ and the "flower-promoting material" itself are on membrane structure, thereby mimicking the effect of phytochrome. Recent studies on phytochrome action in roots and on nyctinastic movements [2, 5, 15, 16] support the theory that one of the primary actions of phytochrome is probably to control the directional transport of ions through the cell. It is possible, therefore, that by transferring the plants to water or to high temperatures one induces changes in the membranes equivalent to the changes induced by high P_{fr}, which in turn inhibit photoperiodic induction of flowering. Correspondingly, low temperature, Ca^{++} and the flower-promoting material all may favor a condition of the membrane that normally prevails during the sensitive phase in the presence of low P_{fr}. The limited data available do not permit, as yet, firm conclusions about the mechanisms involved.

REFERENCES

1. Bamford R. (1931): Changes in root tips of wheat and corn grown in nutrient solutions deficient in calcium. Bull. Torrey Bot. Club, *58*: 149–178.
2. Fondeville J. C., Borthwick H. A. and Hendricks S. D. (1966): Leaflet movement of *Mimosa pudica* L. Indicative of phytochrome action. Planta, *69*: 357–364.
3. Halaban R. and Hillman W. S. (1970): Response of *Lemna perpusilla* to periodic transfer to distilled water. Plant Physiol., *46*: 641–644.
4. Halaban R. and Hillman W. S. (1971): Factors affecting the water sensitive phase of flowering in the short day plant *Lemna perpusilla*. Plant Physiol., *48*: 760–764.
5. Hendricks S. B. and Borthwick H. A. (1967): The function of phytochrome in regulation of plant growth. Proc. Nat. Acad. Sci., U.S.A., *58*: 2125–2130.
6. Hillman W. S. (1959): Experimental control of flowering in *Lemna*. II. Some effects of medium composition, chelating agents and high temperatures on flowering in *L. perpusilla* 6746. Amer. J. Bot., 489–495.
7. Hillman W. S. (1965): Red light, blue light and copper ion in the photoperiodic control of flowering in *Lemna perpusilla* 6746. Plant Cell Physiol., *3*: 415–417.
8. Hillman W. S. (1967): Blue light, phytochrome and the flowering of *Lemna perpusilla* 6746. Plant Cell Physiol., *8*: 467–473.
9. Kandeler R. (1970): Die Wirkung von Lithium and ADP auf die Phytochromsteuerung der Blütendidung. Planta, *90*: 203–207.
10. Kavanau J. L. (1965): Structure and Function in Biological Membranes. Holden-Day, Inc., San Francisco.
11. Maas E. V. and Ogata G. (1971): Absorption of magnesium and chloride by excised corn roots. Plant Physiol., *47*: 357–360.
12. Posner H. B. (1969): Inhibitory effect of carbohydrate on flowering in *Lemna perpusilla*. I. Interaction of sucrose with calcium and phosphate ions. Plant Physiol., *44*: 562–566.
13. Posner H. B. (1970): Inhibitory effect of carbohydrate on flowering in *Lemna perpusilla*. II. Reversal by glycine and L-aspartate. Correlation with reduced levels of β-carotene and chlorophyll. Plant Physiol., *45*: 687–690.
14. Rains D. W. and Epstein E. (1967): Sodium absorption by barley roots: Role of the dual mechanism of alkali cation transport. Plant Physiol., *42*: 314–318.
15. Satter R. L., Marinoff P. and Galston A. (1970): Phytochrome controlled nyctinasty in *Albizzia julibrissin*. II. Potassium flux as a basis for leaflet movement. Amer. J. Bot., *57*: 916–926.
16. Tanada T. (1968): A rapid photoreversible response of barley root tips in the presence of 3-indolacetic acid. Proc. Nat. Acad. Sci., U.S.A., *59*: 376–380.

MULTIPLICITY OF BIOLOGICAL CLOCKS

Karl C. HAMNER and Takashi HOSHIZAKI

Department of Botanical Sciences, University of California
Los Angeles, California, U.S.A.

There has been considerable speculation as to whether or not there is a single "master" biological clock determining the various overt circadian rhythms. This problem has been clearly presented and discussed by Hastings [2]. In the following discussion we will consider only those data obtained in our own laboratory. It seems clear to us that there are at least four mechanisms within the organism which may be used to meter the passage of time, at least three of which could be associated with circadian rhythms. These are as follows: 1) an hourglass type of clock, 2) "light-on" rhythm, 3) a "light-off" rhythm and 4) a rhythm representing the summation of 2) and 3). All four of these mechanisms may be illustrated in the flowering responses of the Japanese Morning Glory, *Pharbitis nil* [4, 7–9]. When this short day plant is grown under long day conditions it will be induced to flower upon exposure to a single short day. A critical ingredient of the short day is the long dark period. This dark period has a critical length below which the plant will not flower, regardless of other conditions. At dark periods longer than the critical length, the number of flowers is proportional to the length of that segment of the dark period beyond the critical length. The quantity of flowers produced, therefore, is controlled by a mechanism on the hourglass principle. This mechanism is very senistive to temperature, as is shown in Fig. 1.

The critical length of the dark period is determined, however, not by a mechanism on the hourglass principle but by a circadian rhythm of sensitivity to light. This is illustrated in Fig. 2. When the plant is exposed to a single long dark period and the dark period is interrupted at various times with brief exposures to light, it is found [7] that such exposures inhibit flowering to different degrees, depending upon the time during the dark period at which the exposure is made. This sensitivity to brief light interruptions exhibits a circadian rhythm whose periodicity is not affected by temperature (*i.e.*, it is temperature compensated). Since the plants were previously exposed to continuous light, it is apparent from Fig. 2 that this circadian rhythm of sensitivity to light is initiated by the onset of darkness, and so we call it the "light-off" rhythm.

The "light-on" rhythm may be demonstrated by introducing a photoperiodic treatment prior to exposure to the long experimental dark period. If the plant is exposed to an 8-hour non-inductive dark period prior to the longer experimental dark period (which causes floral induction), and this is followed by either a 12-hour light period or an 8-hour light period, the results shown in Fig. 3 are obtained. These results show that the increase in flowering with the increasing length of the dark period is stepwise, indicating not only the effects of the hourglass mechanism but also the working presence of a circadian rhythm. The two curves shown in Fig. 3 become superimposed if they are plotted in relation to the

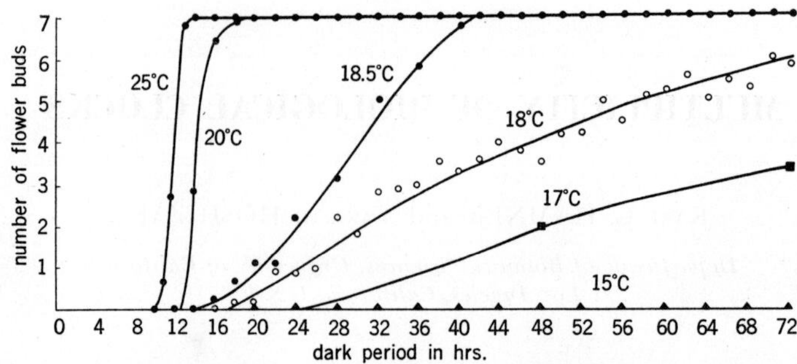

Fig. 1 Flowering response of *Pharbitis nil* exposed to a single dark period of various durations at different temperatures. The plants were exposed to continuous illumination before the dark treatments. After Takimoto and Hamner (1964).

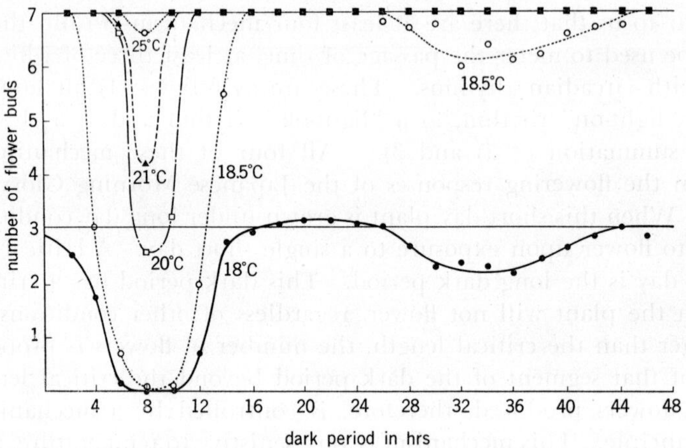

Fig. 2 Flowering responses of *Pharbitis nil* at different temperatures when exposed to five minutes of red light at different times in a 48-hour dark period. The plants were exposed to continuous illumination before the dark treatment. After Takimoto and Hamner (1964).

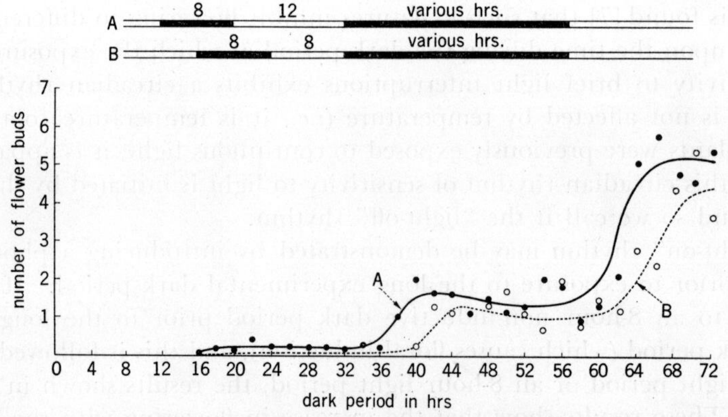

Fig. 3 Flowering responses of *Pharbitis nil* at 18°C, exposed to a single dark period of various durations preceded by different light condition. Group A and B were exposed to an 8-hour dark period followed by 12- and 8-hour light periods respectively preceding the main dark period. After Takimoto and Hamner (1964).

beginning of the preconditioning light period. Thus, the rhythmic response is relative to the time the light was turned on.

Through a series of experiments too complex to be outlined here but which involve a pretreatment with a "light-on" signal and scanning a long dark period with brief light exposures, it has been shown [7] that the overt rhythm of sensitivity to light which is involved in the reaction of this plant to light interruptions during a long dark period is a summation of the above two circadian rhythms, the "light-on" and the "light-off" rhythms. Disregarding the hourglass mechanism, one may say therefore that manifestations of the biological clock may be related to three distinct mechanisms: 1) An internal circadian rhythm initiated by turning the light on, 2) an internal circadian rhythm initiated by turning the light off, and 3) an internal circadian rhythm resulting from the summation of the first two.

It might be argued that we are dealing here with a single rhythm which is phase shifted by both the "light-on" and the "light-off" signals. However such does not seem to be the case, at least in *Pharbitis nil.* A comparison of Figs. 1, 2 and 3 indicates that the "light-on" rhythm is qualitatively different from the "light-off" rhythm. When the plants are exposed only to the "light-off" signal by giving them continuous light from germination until the experimental period and then simply exposing them to dark periods of various lengths there is no evidence of a rhythm in the flowering response (Fig. 1). However such a treatment does initiate an internal rhythm as is shown in Fig. 2 when a long dark period is scanned with a brief exposure to light. Thus, this "light-off" rhythm does not express itself in a length of dark experiment. On the other hand, if a "light-on" signal is introduced prior to the experimental dark period, exposing the plants to various lengths of dark periods does result in a step-wise increase in flowering (Fig. 3), showing the presence of a circadian rhythm qualitatively different from that caused by the "light-off" signal.

We have evidence [1, 5, 6] that the above three mechanisms are also involved in the rhythmic leaf movements of plants. In a study of *Xanthium* leaf movements in light and dark we found that plants transferred from various photoperiodic treatments to continuous light exhibited rhythmic leaf movements with a period of about 24 hours (*i.e.*, circadian). The movement of the leaf was characteristically from the horizontal position downward to an epinastic position and then back upward to the horizontal. On the other hand, plants transferred from continuous light to continuous darkness exhibited a less pronounced rhythmic leaf movement which also had a period of about 24 hours. In this case the movement of the leaf was upward from the horizontal and the return movement backward did not reach the horizontal, so that the leaves gradually assumed a more upright position. It appears therefore that there is a characteristic leaf movement rhythm in continuous light initiated by the transfer of the plant from darkness to light and another characteristic leaf movement in continuous darkness initiated by the transfer of the plant from light to darkness. Additional work has indicated that the leaf movements observed under various photoperiodic treatments on 24-hour cycles represent the summation of these two rhythms. In leaf movements there is therefore also a "light-on" rhythm, a "light-off" rhythm, and a rhythm resulting from the summation of these two.

It seems possible that the functional biological clock in any particular organism may be represented by any one of the three circadian rhythms. In other words, some clock phenomena may be related primarily to "dawn", others primarily to "dusk" and still others related to both "dawn" and "dusk". It seems clear that the timing mechanism used, for example, by birds during their migrations must be related to the third mechanism since, if the biological clock of birds were related primarily to either "dawn" or "dusk", the ability to tell the precise time of day would be impaired by changing day lengths. Since the third mechanism is the summation of the other two, one may consider that the former two are "master" clocks. These "master" clocks will be referred to as the "dawn" and the "dusk" clocks.

It seems that these two clocks are affected differently by temperature. The "dusk" clock is clearly temperature compensated, as is indicated in Fig. 2. On the other hand, there is some evidence [3] that the "dawn" clock may have a relatively high temperature coefficient, as is seen in Fig. 4. The so-called transients which are observed when organisms are transferred from one temperature to another may result from the differing effects of temperature on the two master clocks. In any event, it seems highly desirable to obtain additional data concerning the specific effects of temperature changes on each of these two clocks.

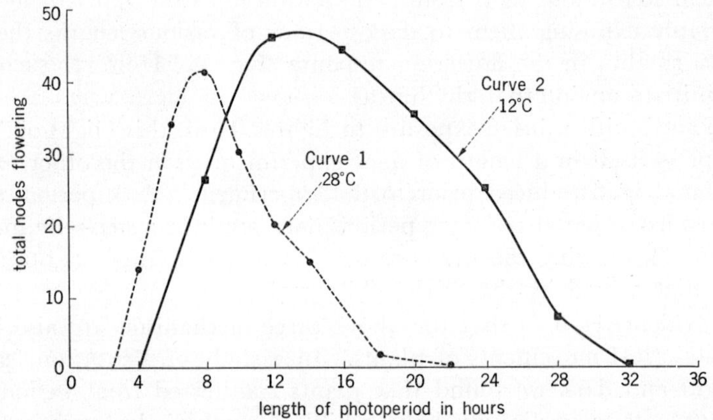

Fig. 4 Floral response of Biloxi Soybean to different lengths of photoperiod given in 7 repeated 72-hour cycles. Curve 1 represents the response to photoperiods given at a normal temperature of 28°C. Curve 2 shows the response with the temperature reduced to 12°C during the photoperiod. In both curves, each photoperiod was followed by continuous darkness at 22°C for the remainder of the cycle.

There is some evidence that the intensity and quantity of light necessary to bring about a phase shift in each of the two clocks differs. For example, during a long dark period for *Pharbitis nil* [8, 9], a small amount of light given about 16 to 18 hours before the end of the dark period produces an inhibitory effect, regardless of the length of the dark period. This would indicate that the small amount of light has produced a "light-on" rhythm which reaches its maximum inhibitory phase about 18 hours after it had been induced. In other words, a small amount of light produces a measurable "light-on" rhythm. On the other hand, the "light-off" rhythm reaches its maximum effectiveness in inhibiting flowering when it is interrupted by light about 8 hours after the beginning of the dark period. In order to reinduce this "light-off" rhythm, the plant must be exposed to 6 to 8 hours of high intensity light [7], indicating that a phase shift of the "light-off" rhythm requires much more light than a phase shift of the "light-on" rhythm.

It is clear, therefore, that if an organism responds primarily to the "dawn" clock its behavior under various conditions of light and temperature will differ from that of one responding primarily to the "dusk" clock. It also seems possible that one or the other clock may predominate in its effectiveness, even in organisms where the overt rhythm is controlled by both the "dawn" and "dusk" clocks, and therefore the responses to various light and temperature conditions will differ from one organism to another. It also seems possible that the desynchronization of various overt rhythms observed occasionally in organisms may result from the fact that some overt rhythms are under the control of one of the "master" clocks and other overt rhythms under the control of the other "master" clock, while still other rhythms are under the control of both.

While there will doubtless be serious questions regarding the validity of our arguments for the two "master" clocks described above, it seems clear that there is sufficient data to justify further experimentation in test of the hypothesis, and to determine its applicability to other living organisms.

REFERENCES

1. Brest D. E., Hoshizaki T. and Hamner K. C. (1970): Circadian leaf movements in Biloxi soybeans. Plant Physiol., *45*: 647–648.
2. Brown F. A., Hastings J. W. and Palmer J. D. (1970): The Biological Clock: Two Views. Academic Press, Inc., New York.
3. Coulter M. W. and Hamner K. C. (1964): Photoperiodic flowering response of Biloxi soybean on 72-hour cycles. Plant Physiol., *39*: 848–856.
4. Hamner K. C. and Takimoto A. (1964): Circadian rhythms and plant photoperiodism. The American Naturalist, *98*: 295–322.
5. Hoshizaki T. and Hamner K. C. (1970): Interaction of 'light-on' and 'light-off' circadian rhythms in *Xanthium strumarium* leaf movements. Plant Physiol. (Suppl.), 46 p. 26, 1970.
6. Hoshizaki T. and Hamner K. C.: Unpublished data.
7. Takimoto A. and Hamner K. C. (1964): Effect of temperature and preconditioning on photoperiodic response of *Pharbitis nil*. Plant Physiol., *39*: 1024–1030.
8. Takimoto A. and Hamner K. C. (1965): Studies on red light interruption in relation to timing mechanisms involved in the photoperiodic response of *Pharbitis nil*. Plant Physiol., *43*: 852–854.
9. Takimoto A. and Hamner K. C. (1965): Effect of double red light interruptions on the photoperiodic response of *Pharbitis nil*. Plant Physiol., *40*: 855–858.

RHYTHMIC MOVEMENTS OF *ALBIZZIA JULIBRISSIN* PINNULES

WILLARD L. KOUKKARI

Department of Botany, University of Minnesota
St. Paul, Minnesota, U.S.A.

Rhythmic leaf movements are very common, especially in legumes, and have been extensively studied [1, 2]. Commencing with studies that depended almost entirely on visual examination of data recorded on time plots, progress has been made towards methods of analyzing rhythms that include a more quantitative and/or statistical basis for the interpretation of data [1, 3–5]. Serial cosinors [6, 7] provide a method for analyzing the time-course of circadian pinnule movements rhythms in *Albizzia julibrissin*. Demonstrated here is the application of the cosinor method to studies of phase-shifting of rhythms following advances or delays in lighting schedule.

Three values (mesor M, amplitude A, and acrophase ϕ) obtained from a 24-hour approximation of data, were used to describe and compare pinnule movement rhythms on the same pinna, different pinnae, separate plants, or even between intact and excised pinnule pairs. The locations of leaves, pinnae, pinnules and angle formed by a pair of pinnules are illustrated in Fig. 1.

The ϕ, which located the peak of the rhythm as approximated by fit of the cosine function, showed very little variation between pinnules on the same plant.

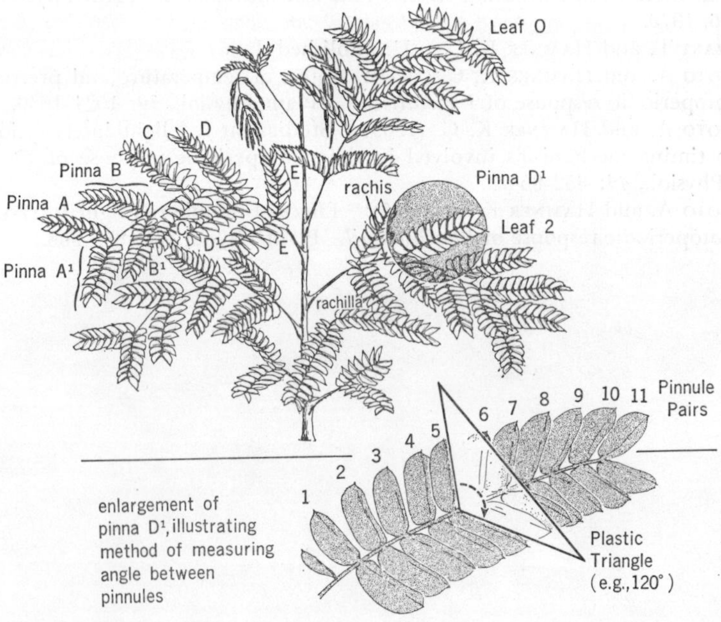

Fig. 1 Diagram of *Albizzia julibrissin* (Silk-Tree).

In fact, there was very little difference in pinnule movement ϕs between plants within the same group (Table 1).

While rhythm parameters, regardless of pinnule position, were very similar, changes in lighting regimen produced subsequent changes in rhythm characteristics. For example, the A decreased significantly when plants on LD 16:8 were placed in DD (Table 2).

Experiments conducted in both DD and LL verified the persistence of the rhythm for at least four days. Although the rhythm continued, the period length was reducde to approximately 23 hours in DD, a result very similar to that which has been observed in *Coleus* [4].

Table 1 Circadian parameters (M, A and ϕ) for movement rhythms of selected pinnule-pairs.

Plant	Pinna	Pinnule-pair	M+SE	A	ϕ	(0.95 confidence arc)
1	A′	4	69±6	80	−185°	(−175° to −196°)
	A′	10	63±5	77	−188°	(−178° to −198°)
	B	4	49±11	85	−197°	(−177° to −218°)
	B	10	57±6	62	−184°	(−169° to −199°)
2	A	4	78±6	73	−204°	(−189° to −218°)
	A′	4	83±6	74	−204°	(−188° to −219°)

Table 2 Circadian parameters for movement rhythm of groups of pinnules on LD16:8 and DD. Period length determined by the best fitting cosine.

	M±SE	A±SE	ϕ	(95% C.I.)	Period hours
LD16:8	81±4	83±6	−200°	(−191° to −208°)	24.3
	80±4	81±5	−201°	(−193° to −209°)	24.2
DD	74±4	43±6	−170°	(−156° to −184°)	23.1
	66±4	51±6	−175°	(−161° to −188°)	23.3

The results from a number of our studies indicate that phase-shifting of *A. julibrissin* pinnule movement rhythms occurs rapidly. When a lighting regimen was advanced 4 hours, the movement rhythm shifted more rapidly than when the lighting regimen was delayed by 4 hours.

Pinule movements of *A. julibrissin* are also controlled by phytochrome, in that pinnules which are briefly illuminated with red light prior to D close faster than those exposed to far-red light [8, 9]. However, the response is synchronized with the light period and depends upon the phase of the rhythm. In addition to phytochrome, auxins, gibberellic acid, and the chelating agent EDTA, also influenced pinnule movements [10].

Pinnule movements are the result of turgor or other changes which occur in ventral and dorsal pulvinule cells (near the base of pinnules). Of particular interest is the fact that these changes have been associated with rhythmic variations of potassium in pulvinules [11].

SUMMARY

Three parameters, mesor, amplitude and acrophase of a cosine function, when fitted to pinnule angle measurements, provide a quantative method for studying the rhythmic movements of leaf pinnules.

Phase-shifting in *Albizza* pinnules, compared to results from mammalian studies, occurs very rapidly. However, differences in type of rhythm as well as regulatory systems must be considered when comparing diverse processes or organisms.

The effects of phytochrome, auxins, etc. plus the movements of potassium, indicate the rhythmic nature of membrane permeability in leaf pinnules.

REFERENCES

1. BÜNNING E. (1964): The Physiological Clock. Springer-Verlag, Berlin.
2. CUMMING B. G. and WAGNER E. (1968): Rhythmic processes in plants. Ann. Rev. Plant Physiol., *19*: 381–416.
3. BÜNNING E. and MOSER I. (1966): Response-Kurven bei der circadianen Rhythmik von *Phaseolus*. Planta, *69*: 101–110.
4. HALABAN R. (1968): The circadian rhythm of leaf movement of *Coleus blumei* × *C. frederici,* a short day plant. II. The effects of light and temperature signals. Plant Physiology, *43*: 1887–1893.
5. HOSHIZAKI R. and HAMNER K. C. (1969): Computer analysis of the leaf movements of pinto beans. Plant Physiol., *44*: 1045–1050.
6. HALBERG F., TONG Y. L. and JOHNSON E. A. (1967): Circadian system-phase—an aspect of temporal morphology; procedures and illustrative examples. Proc. Internat. Congress of Anatomists. *In*: The Cellular Aspects of Biorhythms, pp. 20–48, Springer-Verlag, Berlin.
7. HALBERG F., NELSON W., RUNGE W. J., SCHMITT O. H., PITTS G. C., TREMOR J. and REYNOLDS O. E. (1971): Plans for orbital study of rat biorhythms. Results of interest beyond the Biosatellite program. Space Life Sciences, *2*: 437–471.
8. HILLMAN W. S. and KOUKKARI W. L. (1967): Phytochrome effects in the nyctinastic leaf movements of *Albizzia julibrissin* and some other legumes. Plant Physiol., *42*: 1413–1418.
9. JAFFE M. J. and GALSTON A. W. (1967): Phytochrome control of rapid nyctinastic movements and membrane permeability in *Albizzia julibrissin*. Planta, *77*: 135–141.
10. MCEVOY R. C. and KOUKKARI W. L. (1972): Effects of ethylenediaminetetraacetic acid, auxin and gibberellic acid on phytochrome controlled nyctinasty in *Albizzia julibrissin*. Physiol. Plant., *26*: 143–147.
11. SATTER R. L. and GALSTON A. W. (1971): Potassium flux: a common feature of *Albizzia* leaflet movement controlled by phytochrome or endogenous rhythm. Sci., *174*: 518–520.
12. KOUKKARI W. L. and HILLMAN W. S. (1968): Pulvini as the photoreceptors in the phytochrome effect on nyctinasty in *Albizzia julibrissin*. Plant Physiol., *43*: 698–704.

NONCORRELATION OF LEAF MOVEMENTS AND PHOTOPERIODIC CLOCKS IN *XANTHIUM STRUMARIUM* L.

FRANK B. SALISBURY and ALICE DENNEY

Plant Science Department, Utah State University
Logan, Utah, U.S.A.

Xanthium strumarium L. (cocklebur) is an extremely sensitive short-day plant that remains vegetative as long as it is not exposed to a dark period longer than 8.3 hours but that flowers after exposure to a single dark period exceeding this critical limit. Level of flowering is measured using assigned numerical values to progressive developmental stages of the bud. (See summary of extensive research with this plant in SALISBURY [1] and 1971.)

THE PHOTOPERIODIC FLOWERING CLOCK

Clearly one manifestation of time measurement is the critical dark period for flower induction [2]. Its length is highly resistant to temperature change (Q_{10}= 1.02 from 15° to 30°C). The effects of an inductive dark period are nullified by exposure of the plant to red light, with maximum inhibition about 8 hours after the beginning of the dark period. The inhibitory effects of red light are at least partially overcome by subsequent exposure to far-red light, implicating the phytochrome system [3]. Flowering is noticeably inhibited when plants are illuminated during the entire dark period with about 0.4 microwatts/cm^2 of red light, but this illumination does not influence the length of the critical night nor the time of maximum sensitivity to an interruption with high-intensity red light. That is, the low-intensity light, though it inhibits flowering, does not inhibit time measurement in the flowering process.

A "critical day" for cocklebur can also be observed by exposing plants to a phasing dark period less than 8.3 hours, an intervening light period of varying length, and finally a test dark period long enough to induce flowering (*e.g.*, 12 hours—SALISBURY [4]). In such an experiment, flowering after 9 days (an arbitrary time interval) can be plotted as a function of the length of the intervening light period. With a 12-hr test dark period, the critical day (intervening light period) is about 5 hours, with flowering increasing when light periods are longer than this. In such an experiment, the critical day also proves to be temperature insensitive, low-intensity red light is most promotive, and far-red light is most inhibitory. Hence red light promotes flowering when given at one time during the cycle and inhibits when given at another time, far-red light acting in an opposite way.

If a single light interruption is given about 2 hours after the beginning of the inductive dark period, the critical dark period is extended—but only about an hour [5]. In this sense, a light flash seems to delay the clock. Time of maximum sensitivity to a second light interruption, however, remains about the same following an interruption given 2 or 4 hours after the beginning of the inductive dark period, but following a first interruption at 6 hours, time of maximum sensitivity to a second interruption is shifted 10 hours (Fig. 1). This may be interpreted as an example of phase shifting of the photoperiodic flowering clock in cocklebur.

If time of maximum sensitivity to a light interruption given during a test dark period is studied, it can be seen that this time depends upon the beginning of the test dark period rather than the end of the phasing dark period, provided the intervening light period is longer than 5 hours; that is, the photoperiodism clock seems to go into a suspended state that is maintained as long as the plants are in the light, the clock being restarted when plants are moved to darkness.

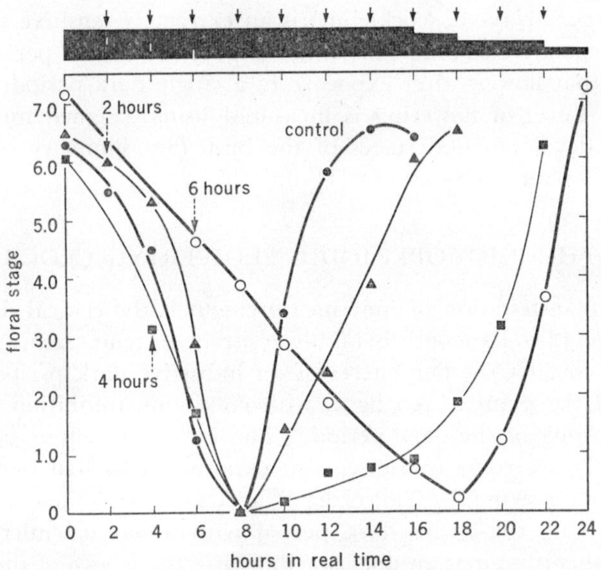

Fig. 1 Effects of a short light interruption (1 min) given at various times during long test dark periods, following phasing dark periods of 0, 2, 4, or 6 hrs. Arrows above the bars at the top indicate times when light interruptions were given and correspond to the data point. Light breaks were a mixture of incandescent and fluorescent light at ca. 20,000 lux; temperatures 25°C light, 21°C dark. (From PAPENFUSS and SALISBURY, 1967.)

These results are similar to those obtained with *Pharbitis nil* by TAKIMOTO and HAMNER (for summary, see paper of HAMNER and HOSHIZAKI in this symposium), and they have been confirmed with *Chenopodium rubrum* (see KING, this symposium).

To summarize: The photoperiodism clock in cocklebur is relatively temperature insensitive, not influenced by low light intensities that can inhibit flowering, apparently controls a changing sensitivity to light such that red promotes at one time and inhibits at another, can be slightly delayed by light interruptions given less than 6 hours after the beginning of the inductive dark period, is rephased by light interruptions after 6 hours, is suspended during long light periods, and is restarted by the change from light to darkness. These characteristics are not unlike those observable in the circadian rhythms of numerous plant and animal species. In any case, the flowering clock has clear characteristics of an oscillator. There are hourglass properties in the flowering process, such as the synthesis of flowering hormone, but it seems appropriate at this time to think of the primary timer as an oscillator. A good working hypothesis is that flowering hormone synthesis is "permitted" when the status of the oscillator clock coincides with that of the phytochrome system.

THE LEAF MOVEMENT CLOCK

Young leaves in cocklebur exhibit a circadian rhythm in movement [6, 7] that can be observed for 6 to 8 days, both in the light and in the dark under constant temperature conditions (by traditional attachment to a kymograph, or by time-lapse photography with infrared film or silhouetted against a dim green screen during darkness—see methods in SALISBURY, 1971).

The pattern of leaf movements is strongly influenced by environment (unpublished). It damps out rapidly under incandescent light but continues strongly under fluorescent light (which has very little energy in the far-red wavelengths). At high temperatures, the leaves are approximately horizontal during the day, *dropping* toward the vertical at night. At lower temperatures (e.g., 20°C), leaves are also about horizontal during the day but *raise* to a nearly vertical position at night. Cross-over temperature depends upon time of year, condition of the plants, etc., but at that temperature, virtually no leaf movement is exhibited. If a diurnal cycle of high and low temperatures is given under constant fluorescent light, the rhythms become entrained to the low-temperature pattern with the leaves usually in a raised, vertical position during the low-temperature part of the cycle. As in other circadian leaf-movement rhythms, the phases may be shifted by changes from darkness to light or light to darkness and by a relatively brief light interruption during continuous darkness.

THE TWO CLOCKS COMPARED

Experiments were initiated to search for the characteristics of the photoperiodism clock in the clock controlling circadian leaf movements. In a number of preliminary experiments, correlations between the two clocks were not apparent, although it appeared that high levels of flowering might occur when leaf movements were particularly extensive during the inductive dark period (the clock functioning optimally?). To settle the question, two large experiments were carried out based upon the pattern of those described above (phasing dark period, intervening light period, and test dark period).

In the first experiment [8], all plants were given two consecutive light-dark cycles including phasing dark periods of 7.5 hours each. The second of these cycles was followed by an intervening light period that varied in duration from 5 minutes to 18 hours, and this was followed by a standard 12-hr test dark period (Fig. 2). In general, short intervening light periods repress the leaf movement rhythms, while longer ones do not, and this is in broad agreement with the flowering results (poor flowering with short intervening light periods, better with longer ones). Careful analysis of the data, however, shows more noncorrelations than correlations (Fig. 3). The extent of the leaf movement is not an accurate indication of flowering response, nor does the position of the peaks and the troughs in relation to the inductive dark period give an indication of flowering response. HALABAN [9] found a constant phase relationship between sensitivity to a light

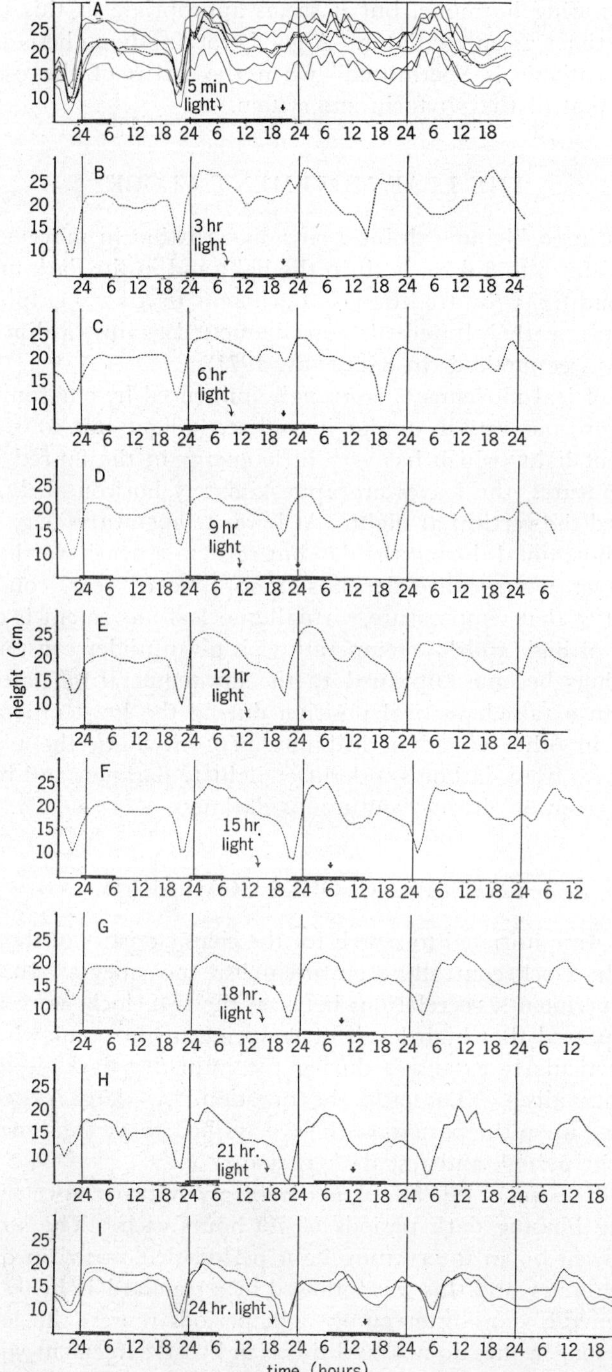

Fig. 2 Effects of various intervening light periods on leaf movements in *Xanthium*. Phasing dark periods were 7.5 hr long, and test dark periods were 16 hr long. Arrows indicate the time of maximum inhibition of flowering produced by a light interruption of the test dark period. Height figures (ordinates) are those on the grid behind the plants. (From SALISBURY and DENNEY, 1791)

interruption during the dark period and leaf position (*Coleus* plants), but this fails in our data. The leaf-movement clock is not suspended during long light periods, although there is some indication that it might be influenced by the change from light to darkness, and from darkness to light. By the second cycle after initiation of the inductive dark period, there may be a constant phase relationship between leaf position and the flowering clock, but this is largely determined by the change from darkness to light *after* induction is complete. Hence leaf position during induction is not an indication of the status of the flowering clock.

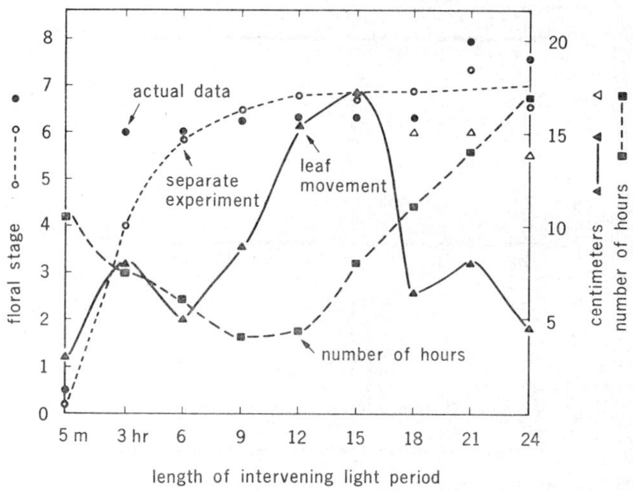

Fig. 3 A summary of some of the features of Fig. 2, including data on flowering from the experiment of Fig. 2 (solid circles)—labeled "actual data"—plus data from a comparable experiment in which all plants were treated at once instead of during several weeks (open circles and broken lines). Solid curve labeled "leaf movement" shows distance moved from lowest to highest points preceeding the test period (triangles). Curve connecting solid squares shows the number of hours between the lowest leaf position and the following time of maximum sensitivity to a light interruption (in flowering). (From SALISBURY and DENNEY, 1971.)

The second experiment (unpublished) followed essentially the same pattern as the first, except that the duration of the phasing dark period also varied, and plants were left in continuous darkness from the beginning of the inductive dark period until the end of the experiment. To accommodate this regimen, fewer intervening light period durations were tested (Fig. 4) (not all treatments are shown). In this experiment it becomes apparent that a dip in leaf position occurs between 18 and 22 hours (real time) on the day following the phasing dark period. This dip only occurs when the test dark period begins after about the 16th or 17th hour. This might suggest that only those treatments permitting the dip permit good functioning of the clock, so only those treatments should show high levels of flowering. Fig. 5 shows that this is not the case. Some treatments exhibiting virtually no leaf movement have high levels of flowering.

The other noncorrelations observable in the first experiment are also apparent here, and in this experiment, it is possible to look for the rephasing (or something analogous to it) following a 6-hr interruption. The 5-minute or 3-hour intervening light period treatments would show this, but no correlation appears.

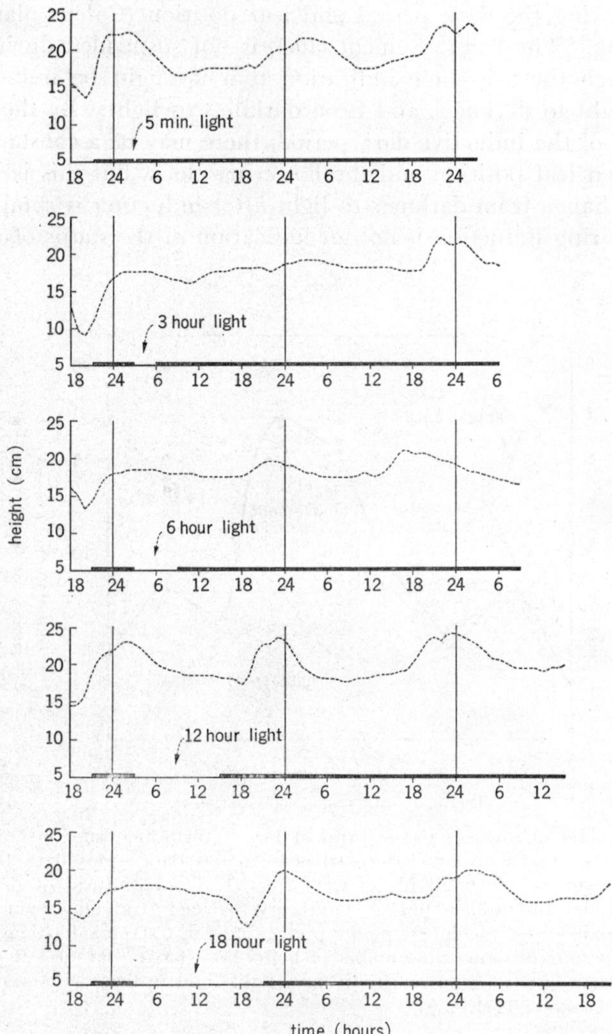

Fig. 4 An experiment such as that of Fig. 2, except that plants remained in darkness for 24 hr following initiation of the test dark period (the inductive dark period). Data shown are for 6-hr phasing dark periods. Data for the 2-hr and 4-hr phasing dark periods not shown, are similar.

In a preliminary experiment (unpublished) designed to test the safety of our green background light used to silhouette plants for photography during the night, it was found that light intensities quite capable of inhibiting flowering have virtually no observable effect upon the circadian leaf movements. It was found that only light of relatively high intensity could disturb the leaf movements, whereas low light intensity inhibits flowering.

CONCLUSIONS

The photoperiodism clock in cocklebur has some characteristics of circadian clocks: it is influenced by the change from light to darkness, it is phase shifted

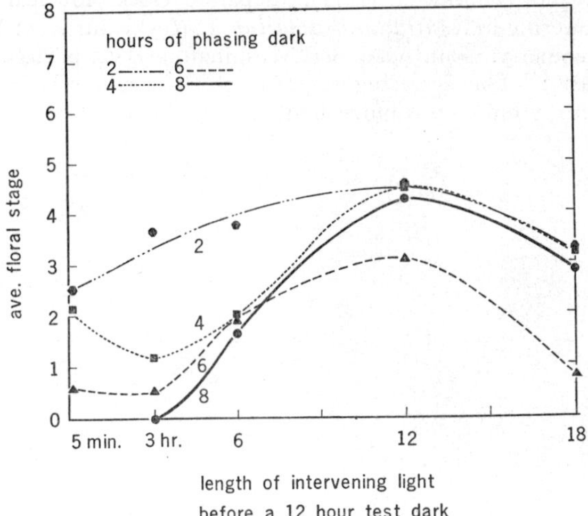

Fig. 5 Flowering results for an experiment such as that of Figure 4. When the phasing dark period is long enough (e.g. 8 hr) flowering is completely inhibited by short intervening light periods. Note noncorrelation of flowering response with rhythms shown in Fig. 4.

by light breaks, it exhibits a 24-hr periodicity in sensitivity to light, and it is relatively temperature independent. The phytochrome system is implicated, and this may prove to be of importance in some circadian systems. On the other hand, several characteristics of the photoperiodism clock fail to appear in leaf movement studies: There is no qualitative difference between a light interruption given at 4 and at 6 hours after the beginning of a long dark period, there is no suspension of the rhythms during long light periods, there is no correlation between extent of the leaf movements during the inductive dark period and subsequent flowering, there is no relationship between peaks or troughs in the leaf-movement rhythms and sensitivity of flowering to a light break, and light intensities quite capable of inhibiting flowering have no obvious effect upon the leaf-movement rhythms.

REFERENCES

1. Salisbury F. B. (1969): *Xanthium strumarium* L. *In*: The Induction of Flowering: Some Case Histories. Chapter 2, (L. T. Evans, ed.), The Macmillan Company of Australia. South Melbourne, Victoria, Australia.
2. Salisbury F. B. (1963): Biological timing and hormone synthesis in flowering of *Xanthium*. Planta, *59*: 518.
3. Borthwick H. A., Hendricks S. B. and Parker M. W. (1952): The reaction controlling floral initiation. Proc. Nat. Acad. Sci., *38*: 929.
4. Salisbury F. B. (1965): Time measurement and the light period in flowering. Planta, *66*: 1.
5. Papenfuss H. D. and Salisbury F. B. (1967): Properties of clock resetting in flowering of *Xanthium*. Plant Physiol., *42*: 1562.
6. Christensen O. V. (1967): Leaf-movement rhythms and flowering in *Xanthium*. M. S. Thesis. Colorado State University, Ft. Collins, Colorado.
7. Hoshizaki T., Brest D. E. and Hamner K. C. (1969): *Xanthium* leaf movements in light and dark. Plant Physiol., *44*: 151.

8. Salisbury F. B. and Denney A. (1971): Separate clocks for leaf movements and photoperiodic flowering in *Xanthium strumarium* L. (cocklebur). *In*: Michael Menaker (editor): Biochronometry. Nat. Acad. Sci., Washington, D.C., p. 292–311.
9. Halaban R. (1968): The flowering response of *Coleus* in relation to photoperiod and the circadian rhythm of leaf movement. Plant Physiol., *43*: 1894.

ELECTRIC AND MAGNETIC FIELDS AND COMPARATIVE PHYSIOLOGIC TOPICS

Chairman: FRANK A. BROWN, Jr.

WHY IS SO LITTLE KNOWN ABOUT THE BIOLOGICAL CLOCK?*

FRANK A. BROWN, Jr.

Morrison Professor of Biology
Department of Biological Science, Northwestern University
Evanston, Illinois, U.S.A.

Living things behave as if they employed 'clocks' to harmonize their adaptive rhythmic variations in bodily processes with the passages of days, ocean tides, phases of moon, and seasons.

In a lead editorial in *Nature New Biology* in May 1971 (231:97–98) the article started, "Why is so little known about the biological clock?" Indeed, little or no solid progress has been made over the past half century or so on the nature of the biological clock despite the best efforts of many very able scientists. This is perhaps very surprising in view of the spectacular advances in our knowledge in molecular biology carrying even to its foundations our understanding of the gene and the generation of the individual organism. The genetic code has been broken and the role of DNA in the synthesis of RNA and, in turn, protein has been resolved in principle. Our advancing knowledge of the physiology and even detailed biochemistry of numerous phenomena that comprise observable biological-clock-timed rhythms provides a substantial pool of information concerning the chain of causally related events constituting the rhythmically recurring cycles. Yet despite this revolution in our understanding of life, our understanding of the 'clocks' that underlie and provide for the extraordinary relative stabilities of biological rhythms in the face of such ordinary metabolic modifiers as temperature and chemical agents seems to have advanced hardly a single step. Why? The answer probably resides in the recent spectacular discoveries and advances in our knowledge of the extraordinary influences of extremely weak electromagnetic-field parameters like those of the earth's natural ones, reflecting perceptive capacities that scientists have long relegated to the realm of fantasy and subjects only for science-fiction writers. The parameters which are concerned appear to span the gamut of the electromagnetic spectrum.

Could it be that our efforts have been misdirected? Could it be that we are seeking an imaginary mechanism, a 'fountain of youth,' in our classical approach of expecting to find an inherited biophysico-chemical oscillator system, present in virtually all living things and even in parts of them, a system whose clock properties are independent of the ambient rhythmicity of the geophysical environment but able to simulate close to the same frequencies? Should we perhaps have been seeking instead the biological clock as a phenomenon which is a consequence of the rhythmic organism being steadily *dependent* upon rhythmicities of the physical environment? Perhaps we have failed to discover the biological clock because of a single-minded drive to discover a particular kind of clock mechanism which is non-existent.

* Preparation of this paper was aided by a grant, #GB-31040, from the National Science Foundation.

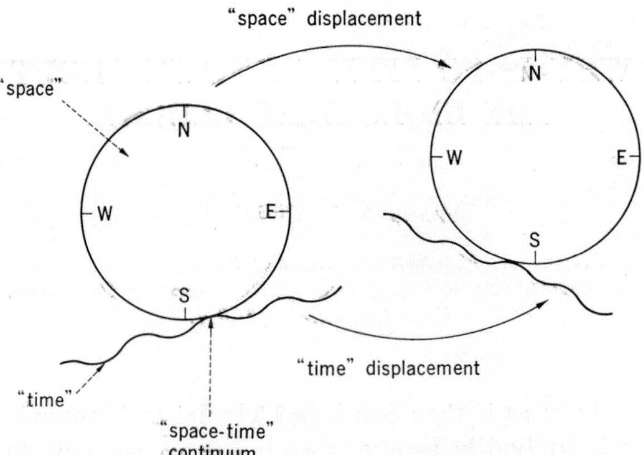

Fig. 1 Space-time variations in the ambient, subtle geophysical fields.

Birds, Salmon, eels, and many other creatures appear to inherit behavioral patterns which include migrations in particular directions at specific times in their lives. No one questions that a bird inheriting a tendency to migrate, say, southwest from a specific site in the fall, has inherited not only the specific behavioral pattern but also the capacity to employ spatial cues, obvious or subtle, to determine the southwest direction. Clearly it can not inherit southwest as a purely endogenous feature. Both heredity and environment are essential, continuously operating components. Just as there is a *geographically*-differentiated environmental field, so is there simultaneously a fully comparable *temporally-differentiated* one. Indeed, the determinations of points both in terrestrial space and time depend ultimately upon celestial relationships. These comprise a space-time continuum.

What is the evidence for independent internal timers?

1. The rhythm persists in an environment in which all obvious parameters are held constant.

Comment: The organism is still in a rhythmic geophysical environment including parameters to which it is sensitive.

2. The recurring patterns may be experimentally phase-shifted and afterward regain their former frequency.

Comment: The adaptive rhythmic patterns timed by the biological clock are known to be labile and plastic and often of complex form, but we are not forced to assume that the clock upon which they depend is similarly labile, plastic, and alterable.

3. A single brief environmental perturbation can sometimes initiate a rhythm.

Comment: The biological clock has probably been continuing to run despite absence of the specific, observed rhythmic variation; here we have probably simply initiated by the environmental perturbation an activity which has become coupled to the clock.

4. Some chemical or physical agents that greatly retard metabolism can sometimes delay the rhythm phase.

Comment: These effects could readily result from phase-shifting of the rhythmic pattern relative to the clock phase and not be influencing in any manner the clock itself.

5. In constant illumination and temperature the ongoing rhythm usually assumes a slightly different frequency and is then termed 'free-running'.

Comment: It is only an arbitrary assumption that the biological clock also free-runs under these conditions with a new period corresponding to that of the observed rhythm. It can not be denied that under these special circumstances the observed rhythmic pattern may be steadily and systematically phase-shifting (autophasing) relative to an accurate biological clock which retains the same period it possessed under natural conditions.

Examined critically, therefore, none of these nor any other known property of biological rhythms, or even all collectively, establish that the observed rhythmic variations with all the well-known 'clock' properties can proceed independently of all ambient geophysical rhythms, nor compel the conclusion that an environmentally-independent clock-system exists.

Let us review very briefly the history of our classical, long-held but unproductive paradigm of an autonomous, endogenous timing system. Perhaps most responsible for its firm establishment was its enunciation by the renowned plant physiologist, Wm. PFEFFER, about the turn of the century. Viewing the results of his numerous experiments with plant circadian rhythms PFEFFER concluded that his organisms must each have its own independent internal clock. The conclusion was *not compelled* by the evidence but simply chosen as the preferred hypothesis. Shortly afterward, Rose STOPPEL, on the basis of her findings that organisms were probably sensitive to a subtle atmospheric 24-hour rhythm related to atmospheric conductivity, soundly and earnestly, but futilely, questioned PFEFFER's postulate over a number of years. Had the doubt been similarly expressed with the same evidence by a *man* instead of a woman, and one of comparable prestige to PFEFFER himself I believe that the outcome would have been different and subsequent progress more rapid.

Indeed when the question of the rhythm timer in the late 1920's and early 1930's became a concern of Professor Erwin BÜNNING the nagging doubts raised by STOPPEL and others led him to do experiments that demonstrated that rhythm periods when free-running were not completely independent of temperature. He reported a Q_{10} of about 1.2 which later investigations in his laboratory reduced to a much lower but still real temperature dependence. Furthermore, BÜNNING discovered that observed free-running periods could reflect genetic differences. The results were arbitrarily interpreted as demonstrating that the period of the clocks was inherited, and that the inherited clock was not quite perfectly compensated for temperature.

The apparent role of the circadian rhythmic system as the timer upon which the measurement of relative lengths of light and darkness depended for the seasonal adjustment of the activities of numerous plants and animals, and especially experiments disclosing an apparent essential role of the circadian system in enabling geographic navigation using the sun as a celestial reference, were considered to clinch the case for an endogenous timing system.

However, none of these later findings, as well, demanded the specific postulation of autonomous individual timers with the extraordinary 'clock' properties. All could be still explained by an environmentally-dependent biological clock employing subtle geophysical cues. The question still remained unresolved. Were the clock characteristics, therefore, really autonomous, intrinsic, inherited properties of organisms and their component parts, or were they a consequence of the inherited behavior of living systems taking advantage of their sensitivity to their ambient rhythmic physical environments of both obvious and subtle factors? Pointing increasingly strongly to the latter being the case has been the very productive studies over the past decade or so on orientational, homing, or navigational phenomena of a variety of animals.

The classical concept of a geographic orientation through the employment of the postulated endogenous timers in collaboration with celestial references has been supplemented by the demonstration of capacity of many kinds of animals to distinguish geographic directions independently of every previously contemplated directional cue. Pigeons have been shown able to home without contribution of any celestial or other obvious or familiar spatial cues and *uninfluenced* by biological clocks. Some sharks have been found to locate with oriented precision hidden prey without benefit of any ordinary senses. Appropriate studies have disclosed that the information which is used is derived from specific ambient electromagnetic-field parameters. In short, the organism deprived of all obvious cues as to its position and orientation in space is capable of deriving information concerning these by means of some subtle electromagnetic-field parameters alone.

As we have stated earlier, the same electromagnetic parameters which are operative in providing information about the three dimensions of space are also varying with time, and contain periodic components of the natural geophysical frequencies. The biological rhythms that reflect patterns adjusting the organisms to the environmental cycles and which are termed geophysically correlated ones are, it will be recalled, the only ones that exhibit the characteristic clock properties.

Could it be that the biological system possesses the capacity to 'read' time from its four-dimensional environment and employ this information in the synthesis of its genetically determined and environmentally modified rhythmic physiological and behavioral patterns? More and more students of biological rhythms, in the face of the mounting evidence of biologically significant roles of the very weak ambient geoelectromagnetic fields, are beginning to refer to *subtle* Zeitgeber, or synchronizers, of the postulated independent internal 'clocks' in much the same manner as the former references to Zeitgeber action of such well-known ones as light and temperature changes. But *is* this step which still continues to assume the existence of an autonomous rhythmic system with all the extraordinary 'clock' properties a sufficiently large one? Remaining for perhaps futile search would still be a clock system of a type that has not only completely eluded discovery for many, many years but for which still today there has not been even a plausible suggestion as to the nature of its mysterious mechanism.

Any periodic geophysical variation which can play the role of a Zeitgeber is obviously a potential contributor to the biological-clock-timed rhythmic system. Less obvious, but more important, is that for a periodic subtle geophysical variation to serve as a period-giver or timer, it should *not* be a Zeitgeber at all, or at best, an extremely weak one else the organism unfortunately becomes a slave of its clock, a situation which natural selection would undoubtedly have disallowed. One rather striking kind of suggestion that the 'clock' involves a continuous organismic 'reading' of time from the environment has been derived from investigations of lunar-related light responsivenesses. Patterns have been reported which appear to be remarkably similar for organisms as diverse as planarian

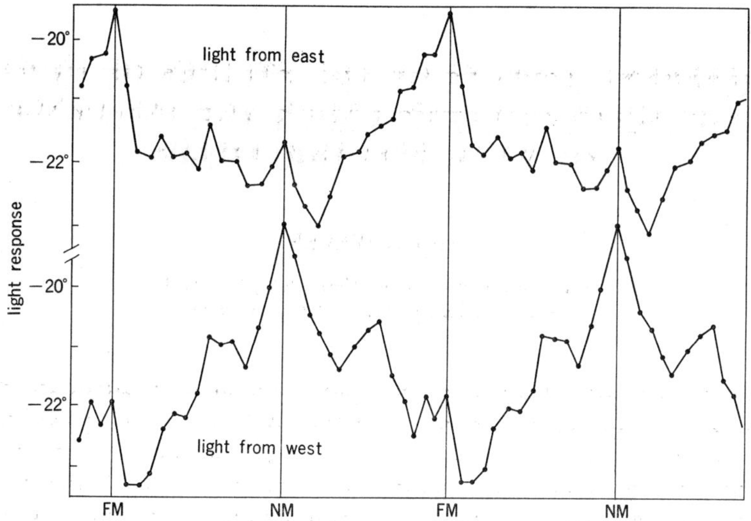

Fig. 2 Synodic monthly variations in strength of negative phototaxis in planarians.

worms, clams, guppies, and rats. For the worms, to light coming from 'magnetic east' a characteristic monthly pattern of variation is disclosed, a pattern which becomes *instantaneously* phase-shifted by 180° by the simple expedient of shifting the light source by 180° spatially to 'magnetic west,' either relative to the earth's natural field, or to an artifically imposed magnetic field. It is important to note that the 180° phase-shift has not been simply the cycle inversion of a simple sine or cosine variation but rather appears to be the equivalent of a temporal phase displacement of a characteristic asymmetrical cycle. The organism appears to be employing the horizontal magnetic vector as a temporal reference for 'reading' the monthly period from still unidentified subtle, temporally-varying environmental fields, phase-relating the 360° monthly cycle in each instance to the corresponding phase angle of the relationship of the light source to the horizontal magnetic vector.

In summary, may I submit that we should have been seeking a mechanism by which living creatures are able to derive subtle timing information from the rhythmic geophysical environment and transduce it into useful biological rhythms, rather than seeking an autonomous, inherited physical oscillator system inside each organism possessing the known extraordinary biological-clock properties. Perhaps like primitive man viewing a rainbow and convinced of the physical reality of the 'structure', with possibly even the legendary 'pot of gold' at its base, we have similarly been seeking the timer of the biological clock in the wrong place, and as a quite different kind of mechanism than it actually is. Perhaps it, too, like the rainbow, depends upon 'relative positions of the sun and observer, and *non-rainbow* characteristics of the atmosphere.' Perhaps, comparably, there are no "circadian clocks," but only *autophasing* behavioral patterns plastically associated with the natural environmental rhythms.

DIFFERENT ASPECTS OF THE STUDIES OF HUMAN CIRCADIAN RHYTHMS UNDER THE INFLUENCE OF WEAK ELECTRIC FIELDS

RÜTGER WEVER

Max-Planck-Institut für Verhaltensphysiologie
Seewiesen und Erling-Andechs, Germany

One of the primary questions in the study of human circadian rhythms concerns the general properties of the oscillating system. To evaluate these, one possible way is to study the dependence of the system on external stimuli [32]. Light which has been proven to be the most effective external stimulus for circadian rhythms of many organisms [1–3], is nearly ineffective in man. This has been demonstrated in experiments under constant conditions with varying light intensities [35] as well as with light-dark cycles [38]. On the other hand, a weak electric 10-cps field influences human circadian rhythms in a predictable manner [29, 31, 36]. Therefore, using electric fields, a systematic study of the properties of human circadian rhythms is possible.

A subject kept without time cues in an isolation chamber which is shielded against the natural electric fields and in which an artificial 10-cps field can be switched on and off by the experimenter, shows a shorter free-running circadian period during those days when the field is in operation as compared to days without a field. This effect is independent of the temporal sequence of the two conditions. In the statistical analysis of many experiments, the shortening effect of the weak electric 10-cps field is significant with $p<0.0005$ [36]. Moreover, the shortening effect has been proven to be the stronger the longer the period; the correlation between the shortening effect of the 10-cps field and the period without the influence of the field is likewise highly significant (r—0.928) [34].

In a few preliminary experiments with green finches, the same shortening effect of a weak electric 10-cps field has been seen. Although no exception from this rule has been observed up to the present, the small number of experiments does not yet allow a statistical statement as significant in the bird experiments as in the human experiments.

Besides the period, several other parameters of human circadian rhythms have been proven, in the experiments mentioned, to be regularly correlated with the state of the 10-cps field [33]. This means, that the values of these parameters are correlated with the value of the free-running period in a regular manner. In the rhythm of activity, for instance, the ratio of activity time to rest time, the precision of the free-running period, and the ratio between standard deviations of end of activity and of onset of activity increase with decreasing period. Similarily, in the rhythm of rectal temperature, mean value, amplitude, and shape of the rhythm as defined by the form-factor (steepness of increase in temperature in relation to steepness of decrease) [33] increase with decreasing period. All these changes depending on the period are statistically significant [39]. The results mentioned are in agreement with results of corresponding animal experiments

[2, 3, 9], and they confirm predictions of a special mathematical model of circadian rhythms [24–26]. In addition, these experiments give insight into the interaction between different rhythms under constant conditions; for instance, with decreasing period, the rectal temperature rhythm is rather advanced relative to the activity rhythm [33].

The artificial electric 10-cps field has been applied not only continuously but also in a periodically changing manner like a Zeitgeber [34]. In all of 10 experiments with Zeitgeber periods between 23.5 and 26.0 hrs, human circadian rhythms tended to be entrained for at least a few days. To be precise, in most experiments, the Zeitgeber was not strong enough for entrainment lasting continuously, but 'relative coordination' [13] was apparent indicating a continuous interaction with the Zeitgeber. The results of these experiments show that the external phase relationship between the biological rhythms and the Zeitgeber as well as the internal phase relationship between different biological rhythms change regularly, depending on the period. For instance, with decreasing period, the rhythm of rectal temperature is delayed relative to the activity rhythm [33, 34]. This result is in agreement with corresponding results obtained with other Zeitgebers [5, 8, 16, 37]. It is, however, in contrast to that which has been derived under constant conditions as discussed above. From this discrepancy, conclusions can be drawn concerning the different inertia of the two rhythms against external stimuli [33].

A field which is switched on and off periodically by the experimenter, represents a Zeitgeber. A completely different situation is given when the field is switched on and off by the subject's own activity. Under these conditions of 'self-control', the periodical state of the field does not represent a Zeitgeber, and the circadian system is still in an autonomous state, although interacting with the field. If the field is in operation during the activity time of the subject, and switched off during its rest time, the free-running circadian period is significantly longer than under constant conditions without the field at any time [33]. Moreover, under the condition of self-control, the internal phase relationship between different biological rhythms changes regularly depending on the period [33], with the same correlation as under the influence of a field Zeitgeber. This result is in agreement with model predictions [27] as well as with results of 'self-control' experiments with light as the controlling stimulus in birds [7] and in man [35]. These results show that the effect of self-control which may be of special importance in the interpretation of results obtained under conditions being only apparently constant [41], is not limited to light as the stimulus.

Not in all human experiments, the different measurable rhythms are internally synchronized to each other. Up to the present, in 20 subjects out of 112 with autonomous rhythms, internal desynchronization [6] has been observed, with different period values of different biological rhythms in the steady state. Because this state is of special interest with regard to the interaction between different rhythms, it should be known under which conditions internal desynchronization occurs. From the model predictions, it can be derived that the tendency towards internal desynchronization increases with increasing period value [39]. Indeed, the intensity of an electric 10-cps field which influences the free-running period, has likewise been proven to influence the internal coupling between different biological rhythms within one organism, and thus, the tendency towards internal desynchronization [29, 31, 36]. Only when this artificial field as well as the natural field is absent, internal desynchronization occurs. This has been demonstrated significantly, with the comparison between all experiments without any field and all experiments with the 10-cps field in operation [34]. Moreover, it can be demonstrated in single subjects that the switching on and off of the 10-cps field influences the state of the internal interaction immediately [36]. In some experiments under constant conditions, after the field has been switched off, internal desynchronization occurs immediately; and in all experiments in which internal desynchronization is present spontaneously in the absence of the field, a re-coupling of different rhythms occurs immediately when the field is switched on, resulting either in full normal internal synchronization or in an internal 1:2-synchronization (circa-bi-dian activity periods) [29, 36]. Thus, the application of electric 10-cps fields opens the possibility to control the tendency towards internal desynchronization experimentally.

The state of internal desynchronization gives first evidence that within one organism different oscillators are present which are separately self-sustained. All other examples of multi-oscillator systems known so far, can be interpreted by more simple hypotheses which need only one self-sustained oscillator. This is the case in the interpretation of special animal experiments [12, 18] with the assumption of one self-sustained master clock and different auxiliary clocks which do not need to be self-sustained. And this is the case in the interpretation of special human experiments [17] with the assumption of one self-sustained clock which controls some physiological variables while other variables are synchronized by a weak persisting Zeitgeber. Beyond this, the analysis of experiments with internal desynchronization leads to a new concept in multi-oscillator systems which is also able to describe all other experimental results [42]. In case of internal desynchronization, period analyses show different main frequencies in different variables. However, in addition, the same (minor) frequencies are to be seen in all variables, although with different amplitudes. This result leads to the following concept: There are some different self-sustained oscillators, and there are many different measurable variables, but there is not a simple 1:1-relationship between oscillators and variables. Each oscillator contributes to the control of all physiological and psychological variables, and each variable is controlled by all oscillators; only the interrelationships by which the different oscilators contribute to the different variables, vary from one variable to another.

So far, the way by which the field acts, is obscure. It cannot be excluded that not the fundamental 10-cps frequency is effective but the higher harmonics of the used square waves [29, 34]; and it cannot even be excluded that the field influences the rhythm not directly but indirectly, e.g. via changes in the air ionization [40]. This uncertainty about the specific way of action of the stimulus does not restrict the usefulness of the 10-cps field as an experimental tool. A further advantage of this stimulus is that it cannot be perceived consciously; in addition, the equipment for generating the stimulus is invisible, and its existence is unknown to the subjects [34]. Yet another aspect of the results mentioned is the study of the stimulus itself. In this respect, human circadian rhythms are taken only for an indicator, and a most sensitive one to study the effects of weak electro-magnetic fields on human beings in general [40]. With this indicator, the influence of weak electric fields with extremely low frequencies on man has been demonstrated for the first time significantly. The question arises whether similar influences can be detected with other kinds of fields.

Some preliminary experiments have been performed under constant conditions, with static electric as well as magnetic fields [36]. In none of these experiments, a change in period has been observed when a static fields has been switched on or off. In other experiments, internal desynchronization had occured spontaneously, in spite of the fact that a static field was in operation [40]. Therefore, it can be stated even from the few experiments that static fields neither influence the free-running period nor prevent internal desynchronization as electric 10-cps fields do [34]. However, the statement that static fields are ineffective, is valid only with weak fields, and especially, it is valid only with homogeneous fields. If static but inhomogeneous fields effect a mechanically vibrating subject, the field strength alternates relative to the subject temporally. It has been proven that the total surface of man as well as that of other homeothermic organisms vibrates mechanically, with a frequency of about 10 cps [19, 20]. Thus, in a static field which is inhomogeneous, a human subject is exposed to a 10-cps field, and that field has been proven to be effective. This means that static fields which are ineffective when homogeneous, may influence human being when inhomogeneous.

Moreover, the question arises to what amount the influence on human circadian rhythms depends on the frequency of an alternating field. Appropriate experiments are in preparation. However, to answer this question, the indicator used has the great disadvantage that years of experiments would be necessary. Therefore, only very preliminary results can be expected.

The final question may be actual whether the natural electro-magnetic field influences human being. This question can be answered by comparing the results obtained in the two experimental rooms available; only one of these rooms is shielded against electric and magnetic fields, and the two rooms differ from each other in no further recognizable respect [29, 34]. In fact, there are statistically significant differences in the results. In the shielded room, the mean free-running period is longer, and the inter-individual standard deviation around this mean is greater than in the non-shielded room; moreover, all experiments with internal desynchronization have occured exclusively in the shielded room [35, 40]. From this result, it must be concluded that the natural field which penetrates the non-shielded room but not the shielded room, shortens the period, reduces the inter-individual differences, and prevents internal desynchronization. With this, a significant proof has been offered that the natural electro-magnetic field influences human beings at all [15]. The result with the natural field is exactly the same as that obtained with the artificial 10-cps field. Therefore, the hypothesis offers itself that the natural 10-cps radiation which is present in the earth's atmosphere [14] as a resonance phenomenon [21], but with an intensity which is about 1000 times smaller than that of the artificial field used, may be an important component in the natural electro-magnetic field influencing human circadian rhythms.

The natural 10 cps field is not constant in field strength but changes periodically, with a higher intensity during day time than during night [14]. Therefore, the question arises whether this change in intensity can act as a Zeitgeber for circadian rhythms. However, considering all results discussed so far, it can clearly be excluded, at least in human circadian rhythms, that this periodic change in the intensity of the natural field can have any synchronizing effect. A rough estimation shows that the range of entrainment of such a Zeitgeber cannot extend more than about one minute [34]; this small range is due to the very low intensity of the natural field and especially its small change, and it is due to its noisy course which allows a determination of the diurnal change only in the long time average. To prove the effectiveness of a Zeitgeber with such a small range, a proper experiment had to last for many years, and that with a standard deviation in the free-running period of less than one minute. Indeed, it cannot be excluded a priori that other organisms are more sensitive against the natural field Zeitgeber than man. However, up to the present, not one case of entrainment of any organism is known in which the entraining Zeitgeber could not be analyzed. In all cases in which entrainment to 24 hrs has been observed under apparently constant conditions, less subtle Zeitgeber could be demonstrated, based for instance, on insufficient shielding against periodical laboratory noise, or on insufficient stabilization of the voltage used for illumination [28].

On the other hand, there is a hypothesis under discussion which is based on periodical changes of subtle stimuli like the natural electro-magnetic field, influencing organisms in another manner than Zeitgebers [10, 11], and the question arises whether the results discussed have any meaning with regard to this hypothesis. The hypothesis postulates a sensitivity of organisms against subtle periodic stimuli unknown so far, with a period of 24 hrs, which are able to induce, via frequency transformation, even in apparently constant conditions persisting rhythms with varying periods deviating from 24 hrs. The discovery of the sensitivity of human circadian rhythms against stimuli as subtle as the natural electromagnetic field, and the fact that this field changes periodically in intensity, seems to support this hypothesis. However, a closer inspection of the results discussed in this paper shows unambiguously that, on the contrary, even these results disprove the hypothesis finally. The reason is that the experiments discussed above allow a determination of the effects of such subtle stimuli, and these effects are opposite to those demanded by the hypothesis. Especially, it has been proven that periodically changing electro-magnetic fields influence circadian rhythms via phase control [22] and not via frequency transformation, just as all other known Zeitgebers do; this statement follows from the dependency of the phase relationship to the Zeitgeber on the period value [23], and it follows from the existence of a phase response curve [4] against the stimulus used. Even if the periodical change in field intensity is too weak for absolute entrainment, this statement has been proven to be valid, by the occurrence of relative coordination [40]. This finding excludes absolutely, in addition to many other experimental results [30], the presuppositions of the hypothesis. In other words: when influenced by a periodic change in the intensity of an electric field, circadian rhythms are either entrained with a period which is, in the steady state, exactly like that of the Zeitgeber, or they are not entrained. When not entrained, or correspondingly, when the period deviates from that of the external periodicity, the persistence of the rhythm is unambiguous proof for its endogenous origin, or its self-sustainment. In summary, the hypothesis mentioned may open, indeed, another way of thinking which is logically consistent, but it is in contradiction to all known experimental results.

REFERENCES

1. Aschoff J. (1960): Cold Spr. Harb. Symp. Quant. Biol., *25*: 11–28.
2. Aschoff J. (1963): Annual Rev. Physiol., *25*: 581–600.

3. Aschoff J. (1964): Rev. Suisse Zool., *71*: 528–558.
4. Aschoff J. (1965): *In*: Circadian Clocks, (J. Aschoff, ed.), pp. 95–111, Amsterdam.
5. Aschoff J., Gerecke U. and Wever R. (1967): Pflügers Arch., *295*: 173–183.
6. Aschoff J., Gerecke U. and Wever R. (1967): Jap. J. Physiol., *17*: 450–457.
7. Aschoff J., Saint Paul U. v., Wever R. (1968): Z. vergl. Physiol., *58*: 304–321.
8. Aschoff J., Pöppel E. and Wever R. (1969): Pflügers Arch., *306*: 58–70.
9. Aschoff J., and coworkers (1971): *In*: Biochronometry, (M. Menaker, ed.), Washington.
10. Brown F. A. Jr. (1965): *In*: Circadian Clocks, (J. Aschoff, ed.), pp. 231–261, Amsterdam.
11. Brown F. A. Jr. (1974): *In*: Chronobiology, (L. E. Scheving, F. Halberg and J. E. Pauly, eds.), pp. 689–693, Igaku Shoin Ltd., Tokyo.
12. Hoffmann K. (1969): Zool. Anz. Suppl.-Bd., *33*: 171–177.
13. Holst E. v. (1939): Ergebn. Physiol., *42*: 228–306.
14. König H. (1959): Z. angew. Physik, *11*: 264–274.
15. König H. and Ankermüller F. (1960): Naturwissensch., *47*: 486–490.
16. Kriebel J. (1970): Pflügers Arch., *319*: R 123.
17. Lobban M. C. (1965): *In*: Circadian Clocks, (J. Aschoff, ed.), pp. 219–227, Amsterdam.
18. Pittendrigh C. S. (1960): Cold Spr. Harb. Symp. Quant. Biol., *25*: 159–182.
19. Rohracher H. (1949): Mechanische Mikroschwingungen des menschlichen Körpers. Wien.
20. Rohracher H. and Inanaga K. (1969): Die Mikrovibration. Bern.
21. Schumann W. O. (1952): Z. Naturforschg., *7a*: 149–154.
22. Wever R. (1960): Cold Spr. Harb. Symp. Quant. Biol., *25*: 197–201.
23. Wever R. (1962): Kybernetik, *1*: 139–154.
24. Wever R. (1964): Kybernetik, *2*: 127–144.
25. Wever R. (1965): *In*: Circadian Clocks, (J. Aschoff, ed.), pp. 47–63, Amsterdam.
26. Wever R. (1966): Z. angew. Math. Mech., *46*: T 148–157.
27. Wever R. (1967): *In*: La Distribution Temporelle des Activités Animales et Humaines, (J. Médioni, ed.), pp. 3–17, Paris.
28. Wever R. (1967): Z. vergl. Physiol., *55*: 255–277.
29. Wever R. (1967): Z. vergl. Physiol., *56*: 111–128.
30. Wever R. (1967): Nachr. Akad. Wiss. Göttingen, *10*: 129–131.
31. Wever R. (1968): Naturwissensch., *55*: 29–32.
32. Wever R. (1968): *In*: Cycles Biologiques et Psychiatrie, (J. de Ajuriaguerra, ed.), pp. 61–72, Paris.
33. Wever R. (1968): Pflügers Arch., *302*: 97–122.
34. Wever R. (1969): Bundesminst. wiss. Forschg., Forschungsber., W 69-31.
35. Wever R. (1969): Pflügers Arch., *306*: 71–91.
36. Wever R. (1970): Life Sciences and Space Research, *8*: 177–187.
37. Wever R. (1970): Pflügers Arch., *319*: R 122.
38. Wever R. (1970): Pflügers Arch., *321*: 133–142.
39. Wever R. (1971): *In*: Biochronometry, (M. Menaker, ed.), pp. 117–132, Washington.
40. Wever R. (1971): Z. physik. Med., *2*: 439–471.
41. Wever R. (1973): Z. physik. Med., (in press).
42. Wever R. (1972): J. Interdisc. Cycle Res., *3*: 253–265.

GEOMAGNETISM AND CIRCADIAN ACTIVITY IN EARTHWORMS

MIRIAM F. BENNETT

Department of Biology, Colby College
Waterville, Maine, U.S.A.

Information about persistent circadian cycles of locomotion in common earthworms, *Lumbricus terrestris*, L., has been available since late in the nineteenth century when Charles Darwin published his now classic monograph on the activities of those animals [1]. That earthworms are generally more active at night than during the daytime when kept under laboratory conditions was also reported by BALDWIN [2] and SZYMANSKI [3]. The later studies of ARBIT [4] and RALPH [5] confirmed and extended the observations of those earlier investigators of persistent cycles of activities in *Lumbricus*. However, not until recently have we had information regarding the mediation of rhythms in that earthworm or evidence that environmental factors affect the circadian organization of the species.

In my laboratory during the last seven years, it has been established that the *rates* of locomotion, light-withdrawal and oxygen-consumption of earthworms, maintained under constant conditions of light, temperature and humidity, vary during different periods of the solar day. In addition, there is an annual variation in those daily differences. We have also shown that a completely intact anterior central nervous system is demanded for the mediation of the circadian variations in the activities of the worms. A review of those investigations is in press [6]. Finally, we have ascertained that alterations in the geomagnetic force to which earthworms are exposed affect the expression of one of the indicators of solar-day rhythmicity in *Lumbricus* [7, 8].

For our studies of the effects of geomagnetism on earthworms, their light-withdrawal reaction was selected as the indicator of the worms' circadian functioning. Tests of that reaction are conducted near midday (1200 to 1300 hours) and again in the evening (1900 to 2000 hours). To time the reflex itself, a circle of white light (25 by 25 mm and 5 lux) is shined on the anterior end of a worm as it crawls across a horizontal wooden platform diffusely illuminated by red light of less than 0.5 lux. The amount of time necessary for the complete withdrawal of the earthworm from the white light is determined with a stopwatch. On the basis of approximately 5,000 single trials, it is known that normal or control earthworms which live in constant darkness and at constant temperature and humidity, withdraw from light 20 to 25 per cent faster during evening test-periods than during trials conducted early in the afternoon.

In October and November (Autumn-winter series), the difference in the rates of the light-withdrawal reaction at midday and in the evening was compared for control worms, maintained in darkness, at a temperature of 20°C and in the earth's magnetic field, and for experimental animals living and being tested in the same laboratory, but within a Helmholtz coil where the magnetic force was essentially zero. A similar series was conducted during March and April (spring series). Control animals were again exposed to normal geomagnetic forces. However, the experimentals lived and were timed in the Helmholtz coil where the magnetic force was approximately twice that of the earth, as measured in our laboratory.

Table 1 Results of the two studies of light-withdrawal reactions of earthworms exposed to normal or altered geomagnetic forces.

Series:	Midday tests:		Evening tests:		Per cent difference:
Autumn-winter:	Number:	Average:	Number:	Average:	
Controls:	555	8.9 sec.	555	7.2 sec.	19.1
Experimentals:	548	8.7 sec.	548	8.8 sec.	1.1
Spring:					
Controls:	220	7.7 sec.	220	4.9 sec.	36.3
Experimentals:	232	7.4 sec.	232	6.8 sec.	8.1

The results of both series (Table 1) convince us that geomagnetism plays some role in effecting the circadian variations in the rates of the light-withdrawal reactions of earthworms. While in the autumn and early winter, control animals were reacting 19 per cent faster ($p<0.005$) in the evening than around midday, the worms living in the greatly reduced magnetic field were an average 1 per cent slower between 1900 and 2000 than between 1200 and 1300. The data for the controls correlate closely with those of earthworms studied in our earlier investigations. During the fall and winter, normal earthworms were generally 20 per cent speedier in the evening. During the spring and early summer, earthworms moved, relatively speaking, even faster—30 per cent—between 1900 and 2000 hours than at midday [9]. The controls of the springtime study of the effects of magnetism behaved in a comparable manner. They were, on the average, 36 per cent faster ($p<0.005$) in the evening than between 1200 and 1300. During that series, the experimental animals, exposed to a magnetic field twice the force of the earth's, were also faster at night. However, the difference was only 8 per cent, and was not statistically significant. Hence, the circadian variation in the rates of light-withdrawal at noon and in the evening, which is so obvious in earthworms living in the earth's typical magnetic fields, is lost or is greatly reduced in *Lumbricus* maintained in experimental fields. Such a loss is also found for worms whose anterior central nervous systems have been manipulated surgically [6].

Major questions remain. Where and how does geomagnetism affect the circadian organization of this earthworm? Does it modulate the annelids' cellular clocks? Is it involved in the generation of the clocks' oscillations? Does it regulate the mediating pathways which are primarily neural and which lie between the clocks and the indicator process we timed? Possibly geomagnetic forces affect receptor or effector reactions of the light-withdrawal reflex itself. Is geomagnetic flux merely a Zeitgeber for *Lumbricus*? On the basis of work of the past summer, the average rates of the light-withdrawal of normal earthworms are available for all hours of the solar day. It may now be possible to find whether the loss of the circadian difference in the reaction times seen in earthworms exposed to reduced or augmented magnetic forces was based on changes in the phasing or the frequency of the cycle of light-withdrawal or whether the alterations in the animals' environments actually disrupted the worms' cellular clocks.

The investigations have been supported by grants from the Committee on Faculty Research, Sweet Briar College and from the National Science Foundation (GY-7661) to Sweet Briar College.

REFERENCES

1. Darwin C. (1900): The Formation of Vegetable Mould Through The Action of Worms. pp. 1–316, D. Appleton and Co., New York.
2. Baldwin F. M. (1917): Diurnal activity of the earthworm. Jour. Anim. Behavior, *7*: 187–190.
3. Szymanski J. S. (1918): Die Verteilung von Ruhe-und Aktivitätsperioden bei einigen Tierarten. Pflüger's Arch., *172*: 430–438.
4. Arbit J. (1957): Diurnal cycles and learning in earthworms. Science, *126*: 654–655.
5. Ralph C. L. (1957): Persistent rhythms of activity and O_2-consumption in the earthworm. Physiol. Zoöl., *30*: 41–55.
6. Bennett M. F. (1973): The central nervous system and circadian differences in the earthworm. *In*: Invertebrate Neurobiology. Mechanisms of Rhythm Regulation. Hungarian Academy of Sciences, Budapest (in press).
7. Bennett M. F. and Huguenin J. (1969): Geomagnetic effects on a circadian differenre in reaction times in earthworms. Z. vergl. Physiol., *63*: 440–445.
8. Bennett M. F. and Huguenin J. (1971): Geomagnetism and circadian organization in earthworms. Proc. Inter. Union Physiol. Sci., *9*: 53.
9. Bennett M. F. (1968): Persistent seasonal variations in the diurnal cycle of earthworms. Z. vergl. Physiol., *60*: 34–40.

CIRCANNUAL BIOCHEMICAL AND GONADAL INDEX RHYTHMS OF MARINE INVERTEBRATES*

FRANCINE HALBERG, FRANZ HALBERG and ARTHUR GIESE

Chronobiology Laboratories, Department of Laboratory Medicine and Pathology
University of Minnesota
Minneapolis, Minnesota, U.S.A.
Department of Life Science, Stanford University
Stanford, California, U.S.A.

ABSTRACT AND SUMMARY

Water, protein, carbohydrate, lipid, and other determinations were carried out on heart, ovary, testis, digestive gland, gill, siphon and adductor muscle as well as foot and shell of *Tivela stultorum* (*Pismo* Clam) and on ovary or testis, digestive gland, gill, foot and shell of *Katharina tunicata* (Black Chiton) collected monthly on California ocean beaches over spans of 6 to 12 months. A 365.25–day cosine curve was fitted by least squares to these limited biochemical data and a rhythm thus tentatively described for 21 of the 71 time series analyzed. Some of these circannual rhythms, e.g., those in gonadal carbohydrate and protein exhibit statistically significant differences in acrophase, others show similar internal timing. Circannual acrophase maps for more complete examination of the timing and the extent of rhythms gauge a facet of chronoecology in the littoral marine habitat.

BACKGROUND

The Pismo Clam, *Tivela stultorum,* found from Halfmoon Bay, Central California to Sacorro Island, off Mexico—perhaps the best known West Coast clam—lives in the surf zone of sandy beaches (moving into deeper water during rough weather). A filter feeder, it siphons water and filters off minute organisms and organic detritus. A five inch clam is about five years old. A large one, 15 years of age, may weigh 6 pounds. Its cream-colored shell is quite thick and may weigh four pounds. As a "broadcast spawner" passing its gametes into sea water, the Pismo Clam produces very many gametes, its reproductive organs being correspondingly large. It has relatively few natural enemies apart from man who has greatly reduced the population by overfishing. The other organisms here studied or discussed have been described earlier [1].

METHOD

The values from biochemical and other determinations on various components from about 8–10 animals of each life form studied, were used for computing

* Supported by grants from the USPHS (5-K6-GM-13,981) and NASA.

monthly averages. Employing the least squares method a 365.25-day cosine curve was fitted to such series covering data spans of 6 to 12 months. Detection and quantification of circannual rhythms in gonadal or hepatic indices was carried out by fitting, one at a time, curves with periods ranging from 395.25 to 335.25 days, consecutive trial periods differing by one-day intervals. Upon completion of such a "window" the parameters, mesor, M, amplitude, A, and acrophase, ϕ, as well as their corresponding dispersion indices, were examined not only at the 365.25-day period, but also at the period with the best fitting curve, *i.e.*, with the least percent error. A statistically significant difference from a 1-year synchronized circannual rhythm is not here reported.

The extremely limited biochemical data prompt us to designate the initial analytical results only as imputations. In chronobiology, the term imputation has been assigned to a procedure for deriving amplitude and acrophase values by the least squares fit of cosine curves to series covering less than twice the period of a rhythm investigated. For instance, when no prior information is available on certain rhythms in a given species, and curve fitting is done on single short series, the (A, ϕ) pairs thus obtained are regarded as imputations until such time as evidence supporting the occurrence of a rhythm with the frequency fitted becomes available. Retrospectively, on the basis of aggregate experience across several time series, the original (A, ϕ) values can become actual estimations rather than imputations of amplitude and acrophase.

Even when a cosine curve well describes the data, the brevity of the series constitutes a real shortcoming of these analyses; yet it must be kept in mind that each of these particular sinusoidal time series here analyzed represents the behavior of synchronized rhythms in relatively large numbers of animals (rather than the behavior of single individuals). For instance, when each average represents 10 individuals and the series consists of eleven monthly averages, 110 invertebrates will have contributed to a single series! Most series cover 11 or 12 months and it is pertinent that many of the series are based upon data from 8–10 organisms per month.

Under these conditions the degree of generality of inference from these short time series is much greater than that from a series describing observations on a single animal. Moreover, by the monthly averaging process some biological noise has been filtered out. It should be recognized, however, that the averaging process has disadvantages as well. Rhythms that differ in phase—although they may be synchronized in (circannual) frequency—will not be apparent in a time series consisting of means such as those here examined. This circumstance may contribute to the failure of describing any rhythm, for 50 of the 71 series analyzed. However, the foregoing finding may be stated in reverse: the occurrence of a circannual rhythm can be suggested for 21 of the 71 time series analyzed; a proportion considerably smaller than 21 "significant" cosine fits out of 71 would be anticipated from chance alone.

Once the occurrence of a rhythm is suggested by results from curve fitting, several rhythm characteristics can be described along with the percentage of the overall variability accounted for by the cosine curve fit, the circannual percent rhythm (PR). Fifty-five to 99 percent of the overall variability was thus accounted for in the 21 time series.

The amplitude, A, another characteristic of the rhythm, is a measure of half the extent of change predictable by the approximation (cosine curve) used. Herein, the A obtained by the fit of a 365.25-day cosine curve approximates one-half the extent of rhythmic change over the 6 to 12 months data spans. The computative acrophase, ϕ, in turn, is an index locating in time (here in terms of time of year) the peak of the cosine function best approximating the rhythm.

In certain cosinor analyses as well as in tables and figures (see Figs. 1 and 2) showing amplitudes computed from single series, the A can be expressed as percentage of its mesor, a rhythm-adjusted mean. Some such transformation becomes necessary when separate sets of data are originally collected in dissimilar units, (for instance in certain chemical determinations involving different methods).

Transformation of the A's into relative values offers statistical advantages; it enables direct comparison of amplitudes representing, say, a change in the number of cases exhibiting a certain property—in time series of differing sample size. Whenever the amplitude is somehow dependent upon the mesor, e.g., if it is comparable from one series to the next when expressed as percentage of the mesor though not in absolute values, a more appropriate weighting of acrophases becomes possible by the relative amplitudes of individual series summarized by cosinor.

RESULTS

Parameter estimations for the 21 series imputed to reveal circannual biochemical rhythms at several different body sites, indicate changes of considerable extent, Figs. 1 and 2. For some of the biochemical variables revealing a circannual rhythm the amplitudes were 20 percent of the mesor or larger. It should be noted that this measure of half the extent of predictable change, the amplitude, differs from the double amplitude, 2 A. The double amplitude—or the total extent of change predictable by the cosine curve used —exceeds 40 percent in several cases. It also should be kept in mind that the total range of variation of the original values may be much greater than 2 A.

The acrophases, all listed for the trial period of 365.25 days, are expressed as negative values or delays from a chosen time reference. The period, τ, of a given cosine curve, serving to approximate circannual rhythms, equals 360°. Since $\tau = 365.25$ days $\equiv 360°$, each 0.986° is equivalent to one calendar day for rhythms completing one cycle per year ($1° \approx 1$ day).

In order to compare acrophases for data collected from various geographic sites and on different calendar dates, a suitable zero reference must at least compensate for differences in the geophysical (and, when pertinent, the social) schedules of the various populations under study. In choosing a time reference for circannual rhythms, the inter-hemispheric disparity of calendar date and season must be compensated for by using as reference points comparable dates on some regular yearly environmental cycles. Within-year changes in the number of hours of sunlight per day are more consistent from one year to the next than are corresponding changes in other environmental factors such as temperature. Hence, it seems reasonable to select for each hemisphere a date approximating, as an average over the years, the day associated with the longest span between sunset and sunrise. Thus ϕ is referred to midnight (0000) of December 22 for data from the northern hemisphere and to 0000 of June 22 for data from the southern hemisphere.

A minus sign in front of a ϕ indicates the delay from the time reference.

Thus, an acrophase, ϕ, of $-213°$ from January 1 of the year preceding the start of sampling locates the peak of the cosine best approximating all data close to August 1. However, our use of December 22 of the year preceding start of data collection for data from the northern hemisphere as zero reference does not imply that the change in daily photofraction along the 1-year scale is the synchronizer. The nature of possible synchronizing agents remains to be elucidated. Hence one can refer to this measure of timing referred to the day with the shortest photofraction only as a computative (rather than external) acrophase. Interesting differences in computative acrophases are revealed by such simple analyses, in Figs. 1 and 2.

In the Pismo Clam's ovaries, the circannual carbohydrate acrophase differs by about 146° (~146 days) from those of protein and lipid, possibly representing a store of carbohydrate utilized to synthesize lipid and protein for gametogenesis. The confidence interval for the carbohydrate ϕ does not overlap that of the other two variables, a finding attesting to the statistical significance of this difference. It does so in a conservative fashion, since overlapping confidence arcs do not necessarily rule out a statistically significant difference in the timing of two rhythms. In the testes of the Pismo Clam, protein and carbohydrate acrophases also differ with statistical significance as revealed again by non-overlapping acrophases.

In the Black Chiton, as in the Pismo Clam, testicular protein and carbohydrates are differently timed. All in all, the circannual acrophase charts in Figs. 1 and 2 for biochemical variables from these two species complement information on differences reported earlier for the circannual timing of rhythms in gonadal and hepatic indices and summarized in the acrophase chart of Fig. 3. For the Short-spined Starfish a grossly visible peak can be seen in April—in the original gonad index data published earlier and hence not shown herein. By contrast, the .95 confidence arc of the acrophase extends from late December to early February. A curve connecting the original data, covering only one year, ascends slowly and descends rapidly. This nonsinusoidality accounts for the difference between on the one hand, the temporal location of the peak in the raw data and, on the other hand, the acrophase representing the highest value of the cosine curve best fitting *all* data. In this small sample of non-sinusoidal data, the curve best fitting *all* values has an acrophase in January.

Indices of waveform, complementing the approximating fundamental cosine curve, can and eventually should be obtained from the fit of harmonics to large samples of non-sinusoidal discrete data. However, it should also be realized that curve fitting has limitations. For instance, a biologic phenomenon may consist of single prominent short-lived yet periodic events bracketed by relatively long spans of random "base level" variability. Such spikes of very short duration may be missed at the time points used for sampling discrete data or if not missed in the samples they may not be detected by fitting only a fundamental cosine curve and a limited number of its harmonics to the spikes overlayered with high noise levels.

Another display, the multiple cosinor, presented elsewhere, portrays, for comparative physiologic goals, overall differences in the circannual acrophases of gonadal indices in three marine invertebrates. These findings are given in Fig. 3 and deserve ecologic follow-up; they represent a "gonadal-index-calendar" of marine invertebrates, reminiscent of Linnaeus' clock of Flora. Changes in these

CIRCANNUAL SYSTEM OF BLACK CHITON (Katharina tunicata)

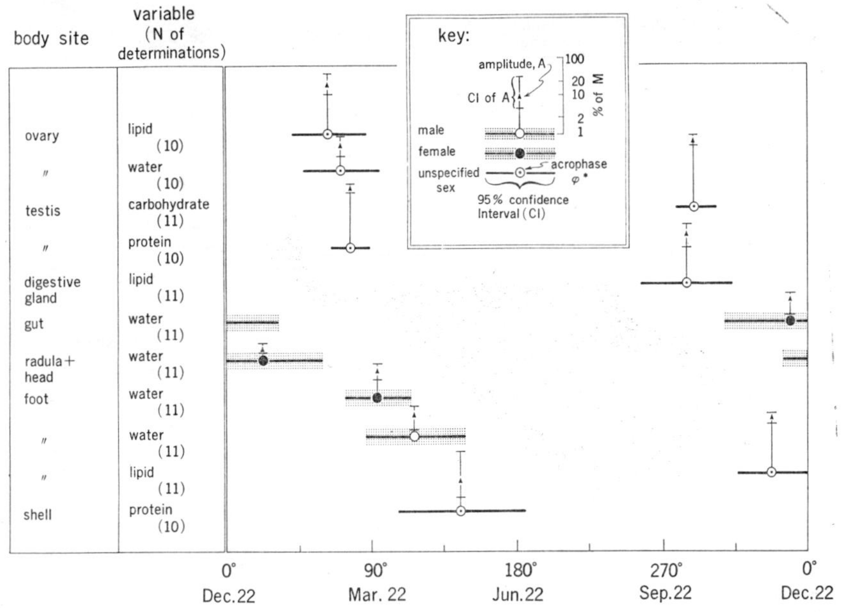

*Acrophase reference=0000, Dec.22 of year preceding data collection

Fig. 1 Circannual acrophase chart displaying for the Black Chiton the timing of circannual rhythms with the corresponding amplitudes (expressed in % of mesor). The columns on the left list the determinations and the body sites (components) studied. The lipid, carbohydrate, and protein indices are defined as percentage, of the particular chemical constituent in respective body components, of the organisms' dry weight. Water is given as percentage of wet weight (see key and text).

CIRCANNUAL SYSTEM OF PISMO CLAM (Tivela stultorum)*

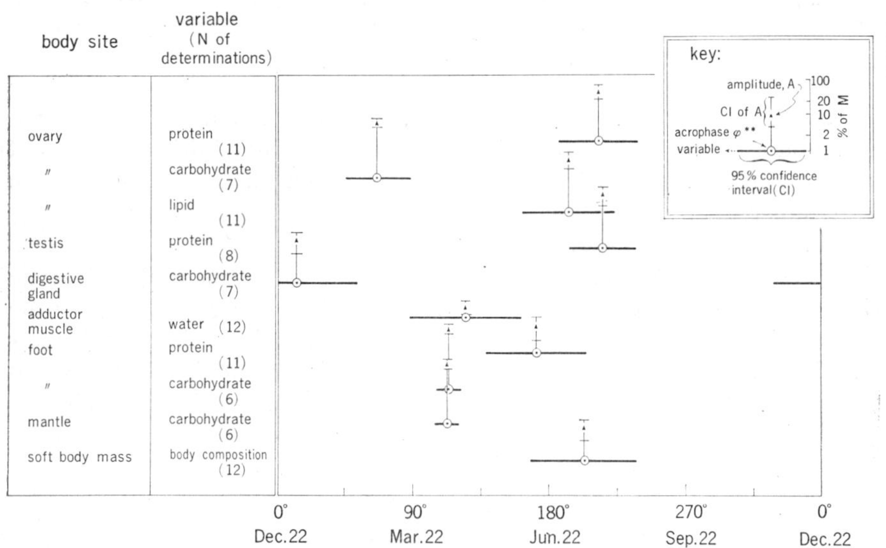

*sex unspecified

**acrophase reference=0000, Dec.22 of year preceding data collection

Fig. 2 Circannual acrophase chart of Pismo Clam. The body composition index (last entry) refers to percent of soft body mass (or other component) of the animals' wet weight (see key and text).

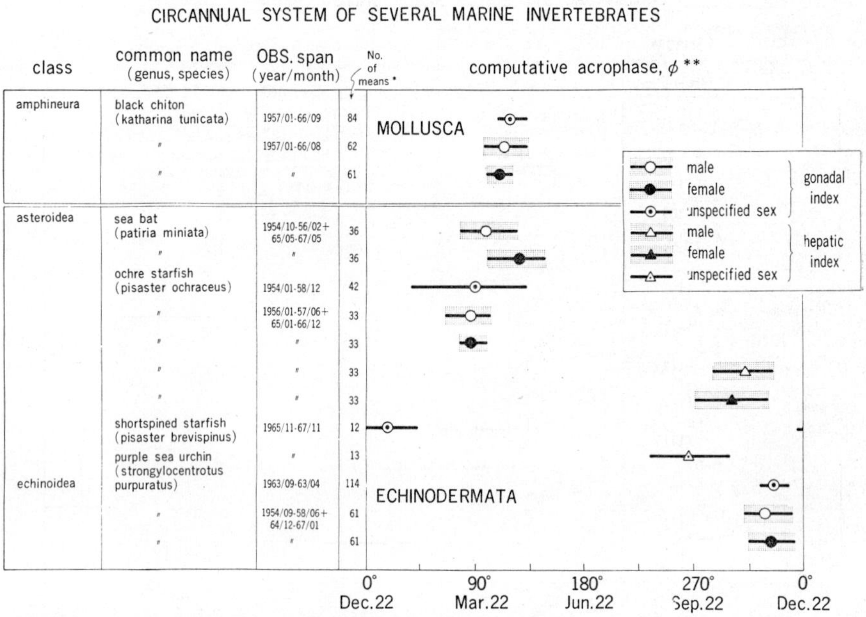

CIRCANNUAL SYSTEM OF SEVERAL MARINE INVERTEBRATES

class	common name (genus, species)	OBS. span (year/month)	No. of means*
amphineura	black chiton (katharina tunicata)	1957/01-66/09	84
	"	1957/01-66/08	62
	"	"	61
asteroidea	sea bat (patiria miniata)	1954/10-56/02+ 65/05-67/05	36
	"	"	36
	ochre starfish (pisaster ochraceus)	1954/01-58/12	42
	"	1956/01-57/06+ 65/01-66/12	33
	"	"	33
	"	"	33
	"	"	33
	shortspined starfish (pisaster brevispinus)	1965/11-67/11	12
	purple sea urchin (strongylocentrotus	"	13
echinoidea	purpuratus)	1963/09-63/04	114
	"	1954/09-58/06+ 64/12-67/01	61
	"	"	61

*Each mean based upon ~5 or usually more organ weights/mo.; average interval between consecutive observations was ~1 month

**Acrophase reference=0000, Dec.22 of year preceding start of data collection in northern hemisphere; p≤.01 in all instances

Fig. 3 Circannual acrophase chart portraying the circannual timing of rhythms in gonadal and hepatic indices for the Black Chiton, Ochre and Shortspined Starfish, as well as the Purple Sea Urchin. The gonadal index acrophase of the Shortspined Starfish (with a .95 confidence interval extending from Dec. 30 to Feb. 12) is similar to that of the Ochre Starfish (.95 confidence interval extending from early February to mid-April); it differs, with statistical significance, from the gonadal index acrophase of the Black Chiton (.95 confidence from late March to late May). In the Ochre and Shortspined Starfish hepatic rhythm acrophases differ by ~180° from those of gonadal rhythms, clearly demonstrating a biologic division of labor along the yearly scale (see text).

acrophases, apart from their basic interest, should be tested as potentially sensitive monitors of marine pollution, a task apparently not initiated as yet. In a one-year synchronized environment, the circannual rhythm-adjusted mean or mesor rather than the acrophase or period may perhaps represent the gauge of immediate practical interest.

The acrophase differences along the 1-year scale complement abundant earlier evidence of acrophase differences in various functions along the 24-hour scale for many species at different levels of organization. A major point of this report is to indicate the possibility of quantifying a physiologic division of labor along the yearly scale. The findings themselves await further scrutiny on their possible degree of generality and their underlying mechanisms, apart from any utility in monitoring the chronoecology of the sea.

REFERENCE

1. Halberg F., Halberg F. and Giese A. C. (1969): Estimation of objective parameters for circannual rhythms in marine invertebrates. Rass. Neur. Veg., *23*: 173–186.

CIRCADIAN AND LOW FREQUENCY RHYTHMS IN THE TEMPERATURE PREFERENCE OF A LIZARD

PHILIP J. REGAL

Museum of Natural History and Department of Zoology
University of Minnesota
Minneapolis, Minnesota, U.S.A.

Previous work has shown that a wide variety of day-active lizards select higher temperatures in laboratory thermal gradients during the "lights-on" phase of a controlled photoperiod, but surprisingly select low and incapacitating temperatures during the "lights-off" phase [1]. In such behavior we may see the antecedent of physiologically maintained temperature depression associated with sleep and inactivity in the higher vertebrate groups that evolved from the reptiles—i.e. birds and mammals. An essential question to be asked is, "Does the change in temperature preference have a circadian rhythm that persists in the laboratory under conditions of constant light and temperature as does the change in mammalian body temperature, or is this behavioral orientation altered simply as a function of light conditions?" Circadian rhythms in lizard *locomotor activity* have been studied from the time of BARDEN [2] through the recent work of HOFFMANN [3].

This question may be studied by placing the lizard, *Klauberina riversiana,* in a 20°C box with food and shelter available as well as a microswitch that activates a heat lamp and recording equipment [4, 5]. The lamp is placed close enough to the microswitch so that the lizard cannot remain long under it and must move away in a few minutes. The regulation of body temperature, then, requires repeated return trips to the microswitch and lamp, which shuts off as soon as the lizard leaves the area. In the original experiments the thermoregulatory depressions of the microswitch were recorded on an Esterline Angus event recorder (see Fig. 1). With this procedure it was possible to say that temperature regulation was not continuous under constant light conditions. The lizards would raise their temperatures with the use of the heat lamp for a number of hours each day at a time of day that seemed to be related to the previous episode of thermal activity. Thus, a circadian rhythm was strongly suggested, although the records were too uneven to allow a more definite statement without some statistical treatment [4].

The next step was to determine the relative thermoregulatory activity of each lizard at particular times of day and from day to day, by taking the number of microswitch depressions in each ½ hour of recording for the 54 days of recording in constant light (some 48×54 data points/lizard), transfer the data to computer punch cards and with least squares spectral analysis search the data for periodicities [6].

Statistically significant circadian rhythms with periods of 24.2 hours (lizard D) and 24.6 hours (lizard CC) were found for both lizards examined. The analysis also suggested the occurrence of certain rhythms with lower frequencies (e.g. a 7 and/or 10 day rhythm). These lower frequency rhythms each account for only

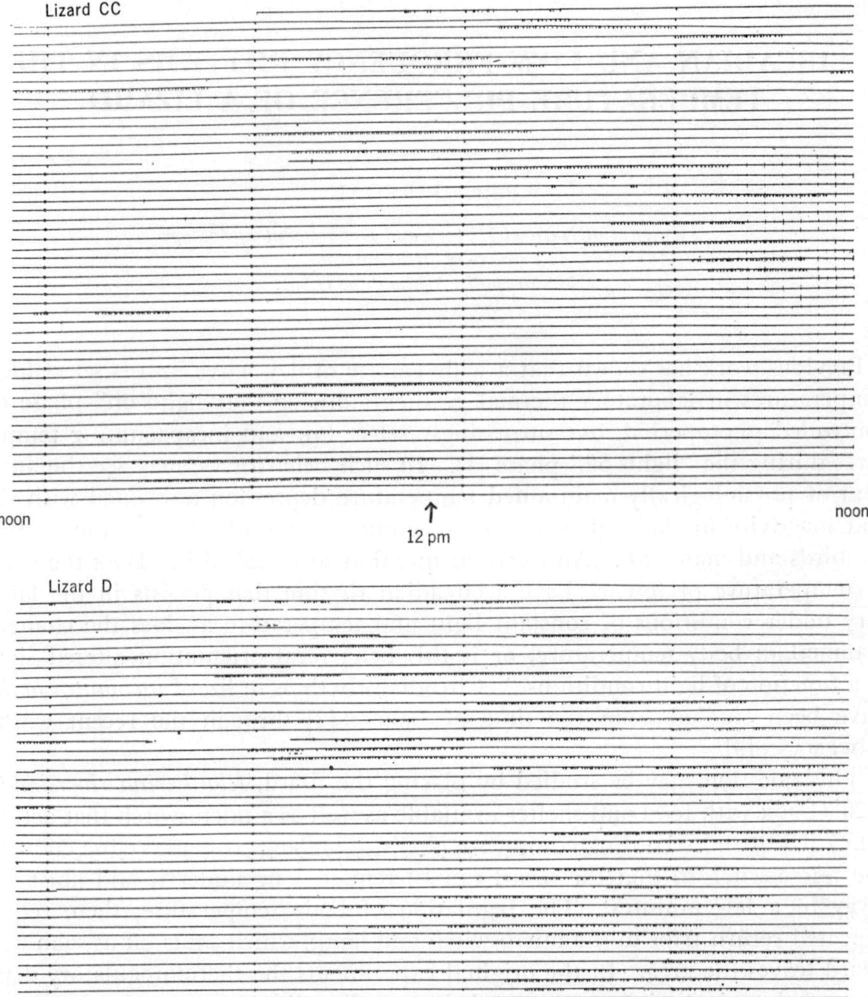

Fig. 1 Actual thermal activity records of two lizards that depress a microswitch for heat. Each deflection of the event recorder pen represents a time when the lizard is heating. The sequence of events is from left to right and from top to bottom.

1%–2% of the variability in the data, however, they are statistically significant and will require further study.

The low frequency rhythms may be consistent with the suspicion of many field biologists that an entire population of lizards is never above ground and active on any particular day. If this widely held belief is valid, then the individual lizard could be active in a random pattern or in a periodic pattern or in some combination. The present detection of periodic elements in the day to day activity could certainly reflect the metabolic economy, but possibly it is involved in social organization as well. If the entire population is never active on any particular day and if we assume day to day constancy of resources, including space as well as food, then it may be to the advantage of each individual that the population distributes itself evenly in time so as to minimize competition on any particular day. An arsenal of low frequency rhythms could give an individual a variety of options for adjusting its activity to the activity patterns of other lizards in its immediate vicinity. A totally random pattern of day to day variation in activity would not allow any adjustment.

Lower mammals such as echidnas may retire into hypothermia and torpor for days at a time. Gross, macroscopic examination of a 92 day record for one echidna [7], shows 5–10 day epochs of hypothermia alternating with 1.4–11 day epochs of high body temperature with only about 4°C diurnal variation at these times. This pattern of activity is in many ways similar to the activity of the reptiles reported on here.

It remains to be seen that epochs of high temperature preference of *Klauberina* are related to epochs of locomotor activity in any direct manner since the lizards were seen to be active at hours and on days when they did not use the microswitch (they were observed through a periscope arrangement and some observations were made on other individual lizards in a separate apparatus without a shelter box, using time-lapse photography). At an ambient of 10°C trained lizards were only active when using the heat lamp. Locomotor activity was abolished in untrained lizards at 10°C. So there does seem to be some independence of temperature preference and locomotor activity. Certainly, data on the precise relationship between these factors would be very highly desirable.

The circadian component of variation in the temperature preference of these lizards makes good ecological sense. It would be dangerous if in an abortive attempt to regulate body temperature after sundown, the lizards with high thermal requirements would move about looking for the warmest areas and expose themselves to predators. Certainly there are also metabolic savings in this alternative to maintaining late into the evening a high, constant, body temperature, *i.e.* in lowering body temperature. It is interesting that variation in the control of body temperature is closely tied to the circadian system at least as far back in vertebrate evolution as the reptilian level of organization.

REFERENCES

1. Regal P. (1967): Voluntary hypothermia in reptiles. Science, *155*: 1551–1553.
2. Barden A. (1942): Activity of the lizard *Cnemidophorus sexlineatus*. Ecology, *23*: 336–344.
3. Hoffmann K. (1969): Zum Einfluss der Zeitgeberstärke auf die Phasenlage der synchronisierten circadianen Periodik. Z. vergl. Physiologie, *62*: 93–110.
4. Regal P. (1968): An analysis of heat-seeking in a lizard. University Microfilms, 69–7258, (1969).
5. Regal P. (1971): Long term studies with operant conditioning techniques of temperature regulation studies in reptiles. Journal de Physiologie, *63*: 403–406.
6. Halberg F. (1967): Physiologic considerations underlying rhythmometry with special reference to emotional illness. *In*: Cycles Biologiques et Psychiatrie. Masson, Paris.
7. Augee M. L., Ealey E. H. M. and Spencer H. (1970): Biotelemetric studies of temperature regulation and torpor in the echidna, *Tachyglossus aculeatus*. J. Mammalogy, *51*: 561–570.

CHRONOBIOLOGICAL TECHNIQUES

Chairman: William GRUEN

Co-Chairmen: Heinz von MAYERSBACH
Heinz P. WOLFF

SUMMARY OF SESSION ON CHRONOBIOLOGICAL TECHNIQUES

William GRUEN

Calibrated Instruments
Ardsley, New York, U.S.A.

In chronobiology, the purpose of instrumentation is the accumulation of useful data. Because change over time is the key factor, instrumentation should be designed so as to make possible the elimination of the effects of other factors on the accumulated data, such as, environmental change or change of position of the subject.

The devices must reduce the measured phenomena to reliable numbers or graphs. Since the physiological changes being measured are often small, the instrumentation must not introduce or record any apparent but non-physiological variations. Calibration of the instruments is essential. With the advent of electronic miniaturization, it is becoming possible to meet other key requirements, that is, that the instrumentation be small, unobtrusive, non-impeding, and with a time scale. Cost of manufacture should become reasonable.

In the session on chronobiological instrumentation, we showed the current state of the art, which has advanced greatly in recent years due to the work of Robert M. GOODMAN of The Franklin Institute in Philadelphia, under contract with the Space Program. He reported that implants in primates are not yet suitable for large population studies, but the data so far acquired on animals does provide a base line for discovering patterns of chronobiological behavior.

Dr. Robert El. SMITH of University of California at Davis told of his remarkable discovery that apparent circadian rhythms of the investigator can cause a variation of as much as fifteen (15%) percent in the calibration of the instrumentation, which obviously, can have a significant effect on the final measurements.

Dr. Lothar WERTHEIMER of the New York Medical College, New York City, described the intensive care ward as an ideal place to gather data, if those in charge are cognizant of the rather stringent requirements of chronobiological measurement. He showed the repeat measurements correctly spaced for rhythmic studies, properly annotated, can yield data that is of immediate value to the physician for the timing of drug administration. Better instrumentation for data gathering, analyzing and recording is needed to further this most practical aspect of chronobiology.

It was pointed out by Heinz von MAYERSBACH of the Medical College of Hannover, Germany, that in taking histological specimens, their "time position" as well as their location in the organ must be noted. Because the sections are fixed, he said, the mistaken notion exists that the conditions of the specimens are not related to the time at which the biopsy is taken. Cells are slightly different at different times of the day, and the time of the taking of the biopsy and the time at fixing should be noted in order to bring about a better understanding of chronobiology at the cellular level.

William A. REDDAN of the University of Wisconsin at Madison explained his extensive measurements in factories that may produce key information for the study of the effects of circadian rhythms when workers have to change shifts periodically.

Finally, Heinz P. WOLFF of the Medical Research Council, London, England, presented his concept of small "socially acceptable" devices for monitoring heart rate, deep body temperature and other physiological phenomena. Such devices, worn unobtrusively by the subject, will be ideal for measuring circadian rhythms in real life situations over extensive periods.

From this session we conclude that there are great instrumentation needs in chronobiology for self-measurement, for data acquisition and for data reduction. There is even greater need to make the data compatible with good statistical capabilities to allow conclusions to be drawn from properly conceived experiments that have been correctly recorded, annotated and reported in a uniform manner. Only in this way can we acquire further understanding in this area of investigation.

INSTRUMENTATION FOR CHRONOBIOLOGIC STUDIES

ROBERT M. GOODMAN*

Laboratory Manager, Biodynamics Laboratory
Franklin Institute Research Laboratories
Philadelphia, Pennsylvania, U.S.A.

Instrumentation for the accumulation of quantified biological data is becoming available at a rate which is both exciting and bewildering. The researcher faces the formidable task of assessing each new device and/or system for its appropriateness to his specific study. This brief paper deals with implantable telemetric devices and specialized, subject-carried recorders for biologic data.

Before I speak of specific devices I urge all who will consider the application of the sort of "intimate" equipment discussed here, to weigh *carefully* in each case, the physiological acceptability of a foreign device whether implanted in, or carried by the subject. All of us are certainly aware of the potential to modify either or both the biological performance or behavioral characteristics of our subjects because of the ways in which we apply instruments and sensors to them.

Implantable, biological telemetric devices provide a means for transmitting internally sensed data from unrestrained subjects. Such equipment is usually designed to be physically small and of *modest* weight in comparison with the subject. These characteristics improve physiologic acceptability and permit considerable versatility in the choice of implantation sites.

These implants are generally classified as "passive" devices, "active" devices and "bio-powered" devices; the differences simply relate to the method by which each is powered. The first, obtains operating power via an externally applied, pervasive, electromagnetic radiation field. The second, utilizes an internal power cell or battery. The latter device, employs special electrodes implanted in the host body-fluids for the generation of operational power. Each of the foregoing has special advantages and certain disadvantages. The use of passive devices which must be irradiated necessarily raises questions concerning the possible stimulus effect(s) of the irradiation field. Further, the subject must be fairly closely confined so as to avoid the necessity to use extremely high level irradiation fields. On the other hand, passive implants have, theoretically at least, infinite life. The active device avoids the need for an irradiating field, but has a limited operational lifetime (months to years). The bio-powered implant is a relatively new class of device with much yet to be learned about power-electrode life, local effects of the electrodes on body fluid and tissue, etc.

In addition to classification based on power sources, telemeters are further categorized as devices for field use and devices for use in the laboratory. Generally speaking, field devices are those (implanted or surface mounted) operating at relatively high power and with the capability for transmitting over ranges measured in hundreds of meters to kilometers. Laboratory devices are usually designed for micropower operation and with transmission ranges on the order of centimeters to several meters.

The telemeter accepts information from a sensor or transducer, conditions that informaton for transmission and transmits it. Irrespective of the ingenuity which many of us in the field have applied to telemetry design, the *sensing element* is a component of cardinal importance. Today it is possible to sense and telemeter: pressure, EKG, EMG, EEG, pH, blood flow and temperature—more or less. Each type of sensor has special and practical problems associated with its application. Generally the thermistor, a semiconductor with a resistance characteristic responsive to temperature, is used to sense temperature. Highly localized measurements can be made and rapid response times obtained. There is however, the problem that with time, the sensor may be isolated from the biologic area of interest by the buildup of tissue caused by the foreign body reaction of the host. Potential sensing such as EKG, etc., can be accomplished quite nicely, but again we remain concerned about tissue reaction. We have done poorly, to date in attempts to obtain chronic, accurate measurements of pH on an implanted basis. Limited effort on this problem is being pursued at The Franklin Institute [1] in the design of long-life, stable, miniature, implantable pH electrodes for use with telemeters. The problem is difficult and in spite of its importance, funding is sparse. Blood flow has been measured using doppler systems [2] and now a new approach is being studied wherein a *permanent magnet* (crossed field) flow sensor, coupled to a low duty-cycle doppler system is expected to make possible micropower flow measurement of unusual long-term accuracy [1]. Pressure is usually sensed via the application of sub-miniature piezo-resistive bridge devices. Unfortunately these devices require substantial operating power and make difficult the design of implants for chronic studies where *continuous* data may be desired. A brilliantly designed system to partially circumvent this problem has been worked out by RADER [3] at the University of Southern California. In my own Laboratory we are studying the application of piezo-junction transistors as pressure sensors. They appear to hold great promise for the evolution of true micropower pressure transducers [1].

Many additional and special forms of sensors are evolving. A number of stimulating descriptions will be found in Mackay's latest book [4]. In any case, the researcher *himself* must always assume the responsibility to become thoroughly aware of the characteristics and performance of the sensors applied in his study. They are critically important to the quality of the data which will be obtained.

Typical of developments here are the telemetric implants shown in Figs. 1, 2, 3 and 4. The first two devices are temperature telemeters with lifetimes (continuous operation) from ca. 7 months to 1.5 years. Their 63% response time to a temperature step lies between 20 and 40 seconds. These designs were selected for the Biosatellite Program. Figs. 3 and 4 illustrate more sophisticated units. The first being a two-channel telemetr for potential and temperature and the latter a four-channel fm/fm implant with an ultimate capability to handle almost any sort of transducer.

Some years ago, my staff and I decided that many situations exist wherein biological data should be recorded periodically from *totally* unrestrained subjects. We were thinking in terms of humans and were fully aware of the boundary conditions which precluded the use of telemetry with the unrestrained human: what would happen if he went swimming? rode the subway? left town by airplane?, etc.? These difficulties simply meant that the subject should *carry* his own recorder which should be unobstrusive, shock resistant, waterproof, etc.

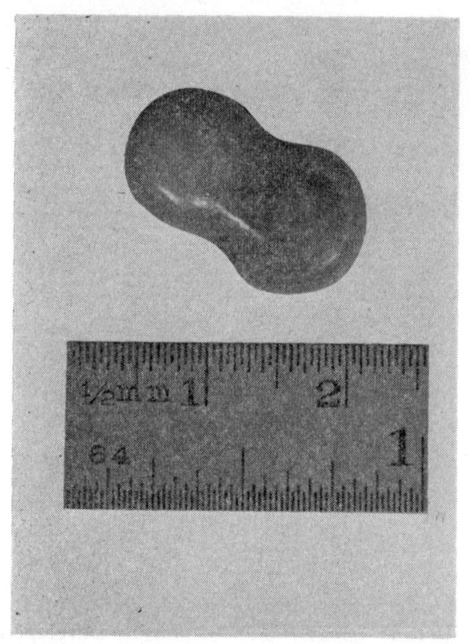

1

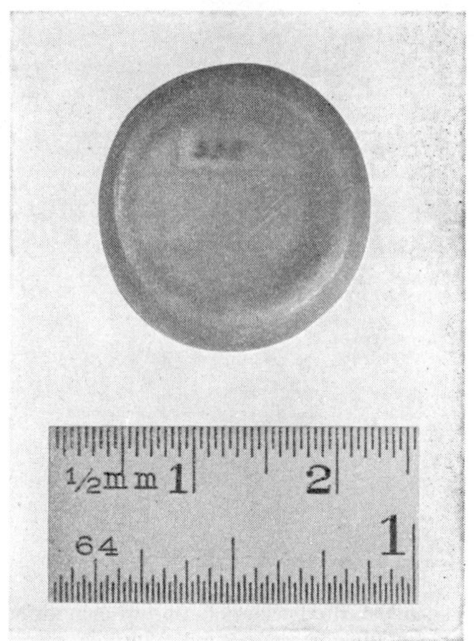

2

Fig. 1 A temperature-sensitive implantable telemeter. Weight: ca. 1.2 gm, Volume: ca. 1 cc, lifetime (continuous) 7 months. Developed with support of NASA Contract NSR-39-005-018.

Fig. 2 A temperature-sensitive implantable telemeter. Weight: ca. 3 gms, Volume: ca. 1.5 cc, lifetime ca. 1.5 years. Developed with support of NASA Contract NSR-39-005-018.

We also decided to settle for biologic data with minimum bandwidth requirements and which could be sensed by exteriorized sensors (*i.e.* axillary temperature, heart rate, etc.). It should be noted here that Heinz Wolff of The British Medical Research Council triggered our thinking in this regard with the design of his SAMI devices.

A recorder system was designed using the Minox film cassette as a non-volatile recording medium. We believe that such a recorder, including self-contained batteries, signal conditioning circuits, timers, etc. will occupy no more than ca. 100–200 cc. The Minox film in a standard cassette has a width of 0.9 cm and a useful length of 52 cm. Its data storage capacity, using relatively simple recording techniques is ca. 7×10^3 binary bits. Assuming our simple approach, Table 1 indicates the relationships between records of say, axillary temperature, sample interval time and experiment duration. Note that we assume the subject to wear the recorder at *all times* for the course of the experiment. Use of fiber optics techniques will permit storage capacities in excess of 10^5 binary bits. Presently we are actively developing such a film recorder [1] on a larger scale (35 mm film). In this latter case we are interested in its application to juvenile humpback whales in a study of whale biology, behavioral characteristics, migration tracks and the like. The planned storage capacity for the whale recorder is in excess of 10^6 binary bits.

All of the foregoing descriptive material is of academic interest *except* as it relates to present and potential application of these sensors, telemeters and recorders to animals and to humans. In this regard, we must raise cogent questions for the biological community to consider–and certainly such questions are reasonable in the light of the literature which already exists [5–18]:

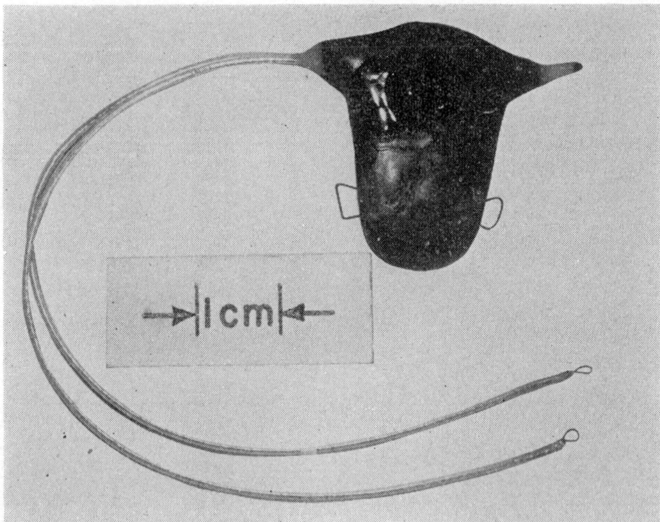

Fig. 3 A dual channel, fm-pulse modulated implantable telemeter (EKG and temperature). Weight: ca. 5.5 gm, Volume: ca. 5 cc, lifetime: greater than 6 months. Developed with support of NASA Contract NSR-39-005-018.

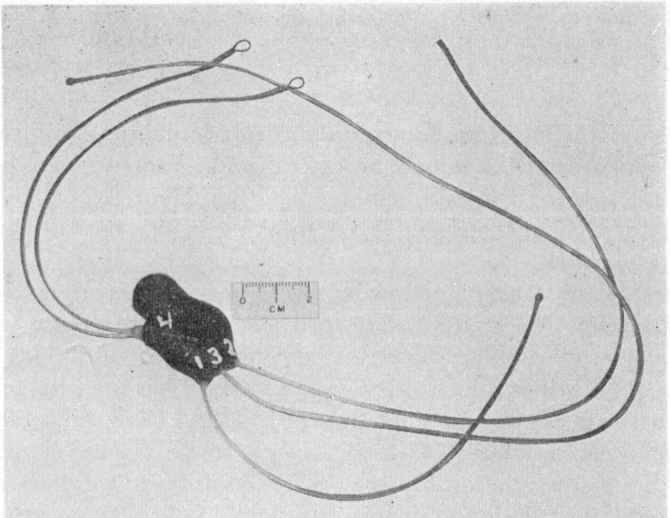

Fig. 4 A four channel, true fm/fm implantable telemeter (EKG and three isolated temperatures). Weight: ca. 8 gm, Volume: ca. 6.5 cc, lifetime: ca. 6 months for continuous operation. Developed with supp. of NASA Contract NSR-39-005-018.

1. Can acceptable rhythmic data for normal human populations be established using exteriorized sensors and long-life subminiature biological recorders?
2. Can variations be detected in the foregoing data which can be shown to be related to pathology?—and are such variations precursors to the overt syndrome we will eventually observe?
3. Should not toxicologic studies be extended to specifically include drug administration time within the circadian period?—and are we not ready to begin to explore the characteristics of drug side effects from the same viewpoint?

Table 1

Minutes between readings	Total Nbr. of readings	Duration of Exp't. in hours	Duration of Exp't. in days	Duration of Exp't. in 30-day months
1	700	11.67	0.49	.02
2	"	23.33	0.97	.03
5	"	58.33	2.43	.08
10	"	116.67	4.86	.16
20	"	233.33	9.72	.32
30	"	350.00	14.58	.49
40	"	466.67	19.44	.65
60	"	700.00	29.17	.97

4. Should not drug efficacy and dosage levels be evaluated as a function of administration time within the circadian period?

The instrument research to permit us to attack the foregoing questions has, for the most part, been accomplished. A number of devices actually exist and others will become available through modest development effort. Biological rhythms appear to be too fundamental to our physical health and mental well-being to continue to be overlooked as additional data from which to generate new diagnostic techniques, new insights to chemotherapy and a better understanding of man as a rhythmic organism.

REFERENCES

1. Gibson R. J. and Goodman R. M. (1970): Space related biological and instrumentation studies. Annual Report to NASA, No. A-B2299-5, March 1970–March 1971.
2. Franklin D., Watson N. W., Pierson K. S. and Van Citters R. L. (1966): Technique for radio telemetry of blood-flow from unrestrained animals. Am. J. Med. Electro., *5*: 24–28.
3. Radar R. D. (1971): Cariovascular Telemetry Implants. Telemetry J., 15–20, April/May.
4. Mackay R. S. (1970): Bio-Medical Telemetry, 2nd Edition. pp. 111–115, 135–147, 148–160, 161–164, 173–182, 198–224, John Wiley & Sons, Inc., New York.
5. Pittendrigh C. S. (1950): The ecoclimatic divergence of *Anopheles* bellator and *A. homunculus*. Evolution, *4*: 43–63.
6. Pittendrigh C. S. and Bruce V. G. (1957): An oscillator model for biological clocks. *In*: Rhythmic and Synthetic Processes in Growth, (Rudnick, ed.), pp. 75–109, Princeton University Press, Princeton.
7. Pittendrigh C. S., Bruch V. G. and Kaus P. (1958): On the significance of transients in daily rhythms. Proc. Nat. Acad. Sci., *44*: 965–975.
8. Bruce V. G. (1960): Environmental entrainment of circadian rhythms. Cold Spring Harbor Symposium on Quant. Biol., *25*: 29–48.
9. Pittendrigh C. S. (1960): Circadian rhythms and the circadian organization of living systems. Cold Spring Harbor Symposium on Quant. Biol., *25*: 159–184.
10. Pitendrigh C. S. (1961): On the Temporal Organization of Living Systems. Harvey Lectures, Series *56*: 93–125, Academic Press, New York.
11. Jores E. A. (1938): First Conf. Ronneby, Sweden, August 13–14, 1937. D. Med. Wschr. 64(21, 28)-1938-737-989.
12. Lyon Conference (1931): Group Lyonnais d'Etudes Medicales, Philosophiques et Biologiques Les Rythmes et La Vie. Lavandier, Lyon.
13. Halberg F. and Stephens A. N. (1958): Twenty-four hour periodicity in mortality of C mice from E. coli lepopolysaccharide. Fed. Proc., 17/1.

14. Cole C. L. and Adkisson P. L. (1964): Daily rhythm in the susceptibility of an insect to a toxic agent. Science, 29 May 1964, 1148.
15. Halberg F. (1960): Circadian temporal organization and experimental pathology. 7th International Conf. for the Soc. for Biol. Rhythms, 52.
16. Halberg F. and E. and Bittner J. J. (1961): Daily periodicity of convulsions in man and in mice. 5th International Conf. Soc. Biol. Rhythms, ACO–Print Stockholm, 97.
17. Halberg F., Haus E. and Stephens A. N. (1959): Susceptibility to oubain and physiologic 24-hour periodicity. Fed. Proc. 18/1.
18. Halberg F. and Howard R. B. (1958): 24-hour periodicity and experimental medicine; examples and interpretations. Postgrad. Med., *24*: 349.

CORRELATION COEFFICIENTS FOR RANKED ANGULAR VARIATES

DEWAYNE C. HILLMAN

Chronobiology Laboratories, Department of Laboratory Medicine and Pathology
University of Minnesota
Minneapolis, Minnesota, U.S.A.

ABSTRACT

A method is shown for extension of Spearman's ρ and Kendall's τ rank correlation coefficients to two angular variates, and for further modification of a similar extension suggested by Batschelet for the correlation of an angular variate with a linear variate. Sample calculations are shown for an example having five data points, and tests of dependency are shown using the null hypothesis that each possible permutation is equally likely.

Introduction

Using data over several replicate pair of time functions, one asks whether or not the timing (phase, ϕ_1) of the first rhythm is related to the timing (ϕ_2) of a second rhythm, or whether or not the timing of a rhythm is related to some non-circular concomitant parameter. The use of order statistics (ranking) makes the significance tests distribution-free.

Designation

The several correlation coefficients are denoted in the matrix:

	linear-linear (11)		jump-ang-lin (jal)		continuous ang-lin (cal)		angular-angular (aa)
r: ordinary	(r_{11}):	?	(r_{jal}):	?	(r_{cal}):	?	(r_{aa}):
ρ: Spearman	(ρ_{11}):	Batschelet	(ρ_{jal}):	suggesting	(ρ_{cal}):	suggesting	(ρ_{aa}):
τ: Kendall	(τ_{11}):	mentioning	(τ_{jal}):	mentioning	(τ_{cal}):	mentioning	(τ_{aa}):

The coefficients are not directly comparable in that each coefficient is a separate entity having its own special distribution depending on the sample size, n, under the null hypothesis so that each coefficient is significant at a different value of correlation.

Coefficients (τ_{11}, ρ_{11}, r_{11}) for linear variates for intermediate calculations.

KENDALL, chapter 2, summarizes the definitions of these coefficients (his τ, ρ, r) for sample of size n using the formula:

$$\Gamma=(\sum_{i=1}^{n}\sum_{j=1}^{n}a_{ij}b_{ij})/\text{sqrt}\{(\sum_{i=1}^{n}\sum_{j=1}^{n}a_{ij}^2)(\sum_{i=1}^{n}\sum_{j=1}^{n}b_{ij}^2)\}$$

where $a_{ij}=-a_{ji}$, $b_{ij}=-b_{ji}$. (Note that Γ is unchanged if the sum on each j is taken from 1 to $i-1$). Numbers a_{ij} and b_{ij} are scores allotted to each pair (i, j) of individuals within the first and second variate, respectively, of the variate pair.

For $\Gamma=\tau$, a_{ij} and b_{ij} become the sign (magnitude=1) of the difference in rank. For $\Gamma=\rho$, they are the actual differences on the rank scale, and for $\Gamma=r$ they are the differences of scores on the actual variate values. For τ, there does not seem to be any alternative to checking $n(n-1)/2$ pairs; but for ρ, (in the special case of no ties when discriminating powers are so great that members of every pair within each variate can actually be ordered) the above formula can be reduced to n comparisons. Several short cuts are given in Kendall.

Sample calculations for ρ

Fig. 1 represents a visualization, for a specific ranked sample of five pair, showing the various rotations to be considered in the intermediate calculations. It also shows some calculations specifically for extensions of ρ, the Spearman rank correlation coefficient. Each of the 25 small graphs represents a ranking for calculation of an original Spearman coefficient. It is convenient to consider the upper right graph first. The first column on the right side of each small graph contains the difference, d, of the rank on the vertical axis minus the rank on the horizontal axis. The second column is the square, d^2, of these differences and is summed, S, in each case at the bottom of the column. The ρ for any graph is obtained by linear transformation of the interval from 40 (the largest possible sum of d^2 over all possible samples) to zero (the smallest possible) transformed to the interval from negative one to one denoted $[-1, 1]$. The value of ρ for the upper right graph is $1-6S/(n^3\text{-}n)=-0.1$.

The Batschelet extension

For this particular coefficient, ρ_{jal}, the second variate (vertical axis) is the rank of the linear variate while the first variate (horizontal axis) is the rank of the angular variate. All rotations of the angular variate appear in the five (=n) graphs in the left box at the center of the figure. The maximum ρ is obtained for the minimum sum of d^2 which is 2, chosen from the left hand box. The right hand box shows the sum of d^2 for corresponding graphs, were the ranks on the horizontal axes reversed, i.e., if "correlated in the other direction". The maximum "correlation in the other direction" corresponds to the minimum sum of d^2 equal to 8 from the right hand box. Since there is a special kind of symmetry in which the sum of the two members of "reversed" pairs equals $(n^3\text{-}n)/3=40$ (for n=5), one may alternatively ignore the right hand box and consider the minimum correlation corresponding to the maximum sum of d^2 which is $32=40-8$ chosen from the left hand box. The two Batschelet coefficients are then $+0.9$ (transformation of min sum of d^2 of 2) and -0.6 (transformation of max sum of d^2 of 32). Transforming to the coefficient is interchangeable in sequence with taking the max and min values if one is aware of the inversion of scale direction.

Extension to two angular variates

The suggested extension, ρ_{aa}, to two angular variates requires trial rotation in *each* axis while max and min sums of d^2 are taken over all position combinations. Thus for the example (n=5) the left hand box containing the rightmost 5 graphs becomes extended to the left hand 25 graphs. The min and max sum of d^2 are again (coincidently for this sample) 2 and 32 giving coefficients of $+0.9$ and -0.6. (Note that such coincidence would not have occurred if the leftmost column of graphs happened to be the sample for the angular-linear variates.)

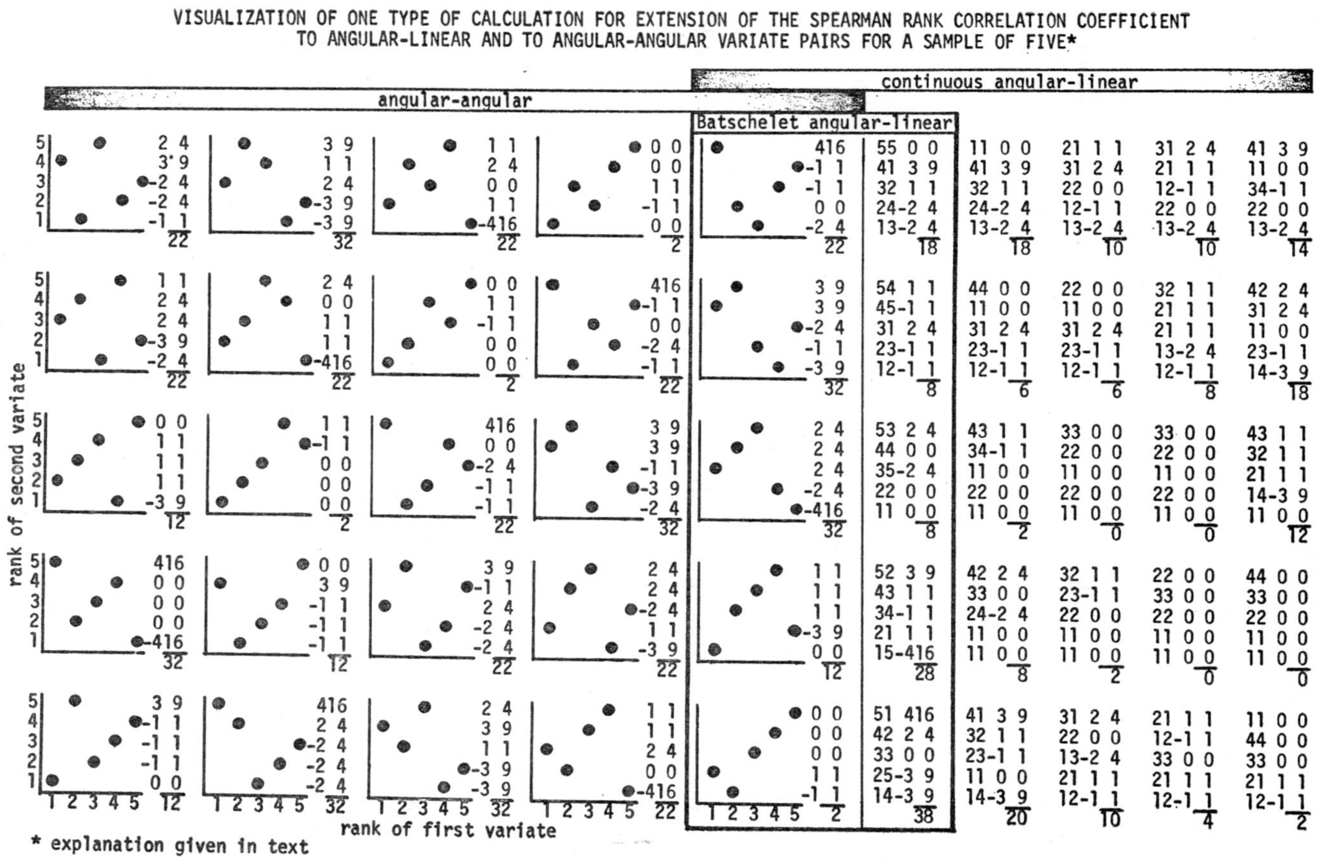

VISUALIZATION OF ONE TYPE OF CALCULATION FOR EXTENSION OF THE SPEARMAN RANK CORRELATION COEFFICIENT TO ANGULAR-LINEAR AND TO ANGULAR-ANGULAR VARIATE PAIRS FOR A SAMPLE OF FIVE*

Batschelet angular-linear	continuous angular-linear			
55 0 0 41 3 9 32 1 1 24-2 4 13-2 4 18	11 0 0 41 3 9 32 1 1 24-2 4 13-2 4 18	21 1 1 31 2 4 22 0 0 12-1 1 13-2 4 10	31 2 4 21 1 1 12-1 1 22 0 0 13-2 4 10	41 3 9 11 0 0 34-1 1 22 0 0 13-2 4 14
54 1 1 45-1 1 31 2 4 23-1 1 12-1 1 8	44 0 0 11 0 0 31 2 4 23-1 1 12-1 1 6	22 0 0 11 0 0 31 2 4 23-1 1 12-1 1 6	32 1 1 21 1 1 21 1 1 13-2 4 12-1 1 8	42 2 4 31 2 4 11 0 0 23-1 1 14-3 9 18
53 2 4 44 0 0 35-2 4 22 0 0 11 0 0 8	43 1 1 34-1 1 11 0 0 22 0 0 11 0 0 2	33 0 0 22 0 0 11 0 0 22 0 0 11 0 0 0	33 0 0 22 0 0 11 0 0 22 0 0 11 0 0 0	43 1 1 32 1 1 21 1 1 14-3 9 11 0 0 12
52 3 9 43 1 1 34-1 1 21 1 1 15-416 28	42 2 4 33 0 0 24-2 4 11 0 0 11 0 0 8	32 1 1 23-1 1 22 0 0 11 0 0 11 0 0 2	22 0 0 33 0 0 22 0 0 11 0 0 11 0 0 0	44 0 0 33 0 0 22 0 0 11 0 0 11 0 0 0
51 416 42 2 4 33 0 0 25-3 9 14-3 9 38	41 3 9 32 1 1 23-1 1 11 0 0 14-3 9 20	31 2 4 22 0 0 13-2 4 21 1 1 12-1 1 10	21 1 1 12-1 1 33 0 0 21 1 1 12-1 1 4	11 0 0 44 0 0 33 0 0 21 1 1 12-1 1 2

* explanation given in text

Fig. 1 Illustrative calculation for extension of the Spearman rank correlation coefficient to angular-linear and to angular-angular variate pairs.

What goes up must come down for angular-linear

The highest value for ρ_{jal} is obtained from samples whose ranks "climb up" around the "topologically cylindrical" sample space and then "drop back down" to the bottom again. Applicability to some kinds of data may be questionable, especially when the data is distributed rather uniformly around the circle for the angular variate. If one objects to the "dropping down", a next attempt might be to relax the restrictions and to minimize the sum of d^2 not only over all rotations in the angular variate, which determines a, say, "bottom cut", but also over all positions of a "top cut", for each "bottom cut", thus dividing the circumference into two arcs or subsets. The subsets can be individually reranked, the angular variates being ranked forward in one subset and in reverse order in the other subset, and the sum of d^2 summed over both arc subsets. The correlation is $+1$ if there is a division wherein the ranks "climb up" in one subset and "slide down" in the other. This might be called the continuous-angular-linear rank correlation coefficient. Referring again to Fig. 1, the minimization is extended from the two boxes in the center to the entire right side of the figure. That is, for any horizontal row, the sum of d^2 in the left box represents a "top cut" at the extreme right of the graph, while that in the right box represents a "top cut" at the left side (recall this was a reverse correlation for ρ_{jal}). Then successively to the right, the top cut is moved one position to the right. There is one more "top cut" than the sample size for each row. The minimum sum of d^2 for the given sample is zero giving a correlation of one. (However for $n=5$, $\rho_{cal}=1$ is not significant at any reasonable level). A larger sample size is needed because the hypothesis alternate to randomness includes a large set of possibilities so that much information must come from the sample.

The three τ coefficients

The graphs and "cuts" are identical to the corresponding ones for ρ. Instead of a sum of d^2 one uses the definitional formula above or some minor short cuts described in KENDALL's book. However, choosing of the extremes over the various "linear-linear" coefficients for the graphs is identical for ρ and τ. KENDALL gives suggestions for choosing between ρ and τ.

Tied ranks

Even though comparisons for rank be on a continuum with theoretical probability of tied ranks equal to zero, indistinguishability of objects to the observer, e.g., experimental error and rounding of digits, creeps abundantly into the experimenter's real life. KENDALL shows two methods of dealing with ties, discarding some of the shortcuts for computing the sample correlation coefficient, but using the test distributions based upon there being no equivalences. Utility depends upon being able to handle ties.

Computational feasability

None of these coefficients lend themselves to easy hand calculation, but the advent of the electronic computer with memory to store the data makes them feasible.

Significance tests

The underlying basis for the significance test is that if there is no relationship existing between the variates from which the sample was drawn, then each ordering of one variate is just as likely as any other ordering (KENDALL). Thus under a null hypothesis of equal probability for each permutation, the probability distribution of the test statistic is given for any particular sample size. The test static may be, e.g., the correlation coefficient or the difference in the absolute value of the max and min coefficients (i.e., the sum of the positive and negative coefficients). If, for the sample drawn, the test statistic lies among cases with few occurrences under the null hypothesis and has a high correlation, then one tends to reject the hypothesis of complete randomness in favor of an alternative hypothesis. Exact distributions for small samples for some of the coefficients are given in the tables attached. For large samples an asymptotic formula hopefully can be developed similarly to those shown in KENDALL to be asymptotically normal with variance depending on sample size. This could be a challenging problem, the algebra of circular distributions being more difficult than that of ordinary permutations. With the advent of still faster computers and possibly some donated computer time, exact distributions can be calculated for larger samples. A plot on probability paper of the distribution of the difference of absolute values of positive and negative r_{aa} did not discourage the hope for asymptotic normality.

Number of sample points

Even ordinary correlation coefficients require seemingly large sample sizes. It is important to note that tests for nonrandomness using rank coefficients are not dependent on the distribution of the population. In this connection we have not put so much information into the problem so that it takes more data to show nonrandomness. It also takes information from the data to determine the

Table 1

Biologic data sample

	Hematocrit		Hemoglobin		Cortisol	
	AMP	ϕ	AMP	ϕ	AMP	ϕ
	.38	162	.09	60	3.60	−148
	1.45	−192	.49	−204	4.35	− 96
	1.91	−216	.61	−217	6.84	− 97
Females	1.03	−222	.31	−213	5.23	−110
	.68	−117	.34	−144	7.75	− 80
	.49	− 99	.05	− 30	7.51	− 92
	1.38	−189	.39	−192	4.84	− 56
	1.04	−153	.42	−168	3.90	− 86
	.75	−186	.43	−216	6.74	− 99
	.72	−203	.37	−202	7.36	−130
Males	.85	−102	.08	−220	5.40	−110
	1.65	−202	.63	−213	5.26	−115
	.79	−160	.48	−176	8.76	− 98
	.48	−288	.29	−261	6.87	−169

Table 2*

Rho	Angular	Angular	Difference of absolute values (i.e. sum of coefficients)			
4	(.400, .1667)	(0 , .8333)				
5	(.500, .0417)	(.300, .2500)	(.100, .4583)	(0 , .5417)		
6	(.457, .0083)	(.343, .0583)	(.286, .0750)	(.229, .1500)	(.114, .2500)	(.057, .3500)
	(0 , .6500)					
7	(.500, .0014)	(.429, .0111)	(.357, .0306)	(.286, .0597)	(.250, .0792)	(.214, .1472)
	(.179, .1861)	(.143, .2542)	(.107, .3125)	(.071, .4097)	(.036, .4583)	(0, .5417)
8	(.476, .0002)	(.429, .0018)	(.381, .0058)	(.333, .0153)	(.286, .0315)	(.238, .0601)
	(.190, .1204)	(.143, .1982)	(,095, .3056)	(.048, .4167)	(0 , .5833)	
9	(.500, .0000)	(.467, .0002)	(.433, .0009)	(.400, .0021)	(.367, .0048)	(.350, .0058)
	(.333, .0101)	(.317, .0132)	(.300, .0190)	(.283, .0239)	(.267, .0331)	(.250, .0412)
	(.233, .0543)	(.217, .0686)	(.200, .0865)	(.183, .1061)	(.167, .1309)	(.150, .1557)
	(.133, .1907)	(.117, .2260)	(.100, .2638)	(.083, .3019)	(.067, .3461)	(.050, .3884)
	(.033, .4326)	(.017, .4763)	(0 , .5237)			
10	(.485, .0000)	(.461, .0000)	(.436, .0001)	(.424, .0001)	(.412, .0003)	(.388, .0008)
	(.376, .0009)	(.364, .0017)	(.352, .0019)	(.339, .0031)	(.327, .0038)	(.315, .0057)
	(.303, .0069)	(.291, .0101)	(.279, .0123)	(.267, .0167)	(.255, .0200)	(.242, .0269)
	(.230, .0333)	(.218, .0429)	(.206, .0520)	(.194, .0660)	(.182, .0789)	(.170, .0954)
	(.158, .1121)	(.145, .1354)	(.133, .1564)	(.121, .1813)	(.109, .2056)	(.097, .2374)
	(.085, .2653)	(.073, .2980)	(.061, .3298)	(.048, .3702)	(.036, .4044)	(.024, .4424)
	(.012, .4774)	(0 , .5226)				
11	(.500, .0000)	(.482, .0000)	(.464, .0000)	(.445, .0000)	(.427, .0001)	(.409, .0002)
	(.400, .0003)	(.391, .0004)	(.382, .0005)	(.373, .0008)	(.364, .0011)	(.355, .0015)
	(.345, .0019)	(.336, .0026)	(.327, .0033)	(.318, .0042)	(.309, .0053)	(.300, .0068)
	(.291, .0085)	(.282, .0107)	(.273, .0131)	(.264, .0163)	(.255, .0198)	(.245, .0241)
	(.236, .0288)	(.227, .0347)	(.218, .0410)	(.209, .0486)	(.200, .0566)	(.191, .0664)
	(.182, .0765)	(.173, .0883)	(.164, .1007)	(.155, .1151)	(.145, .1301)	(.136, .1471)
	(.127, .1642)	(.118, .1842)	(.109, .2038)	(.100, .2258)	(.091, .2475)	(.082, .2716)
	(.073, .2956)	(.064, .3217)	(.055, .3468)	(.045, .3746)	(.036, .4012)	(.027, .4297)
	(.018, .4569)	(.009, .4859)	(0 , .5141)			

* key for table entries: n (correlation coefficient, cumulative probability)

amount of rotation for the extremum. Furthermore it takes more data if we don't know which "relative direction the angular variates are going" before hand.

A sample of biologic data

Use of data in Table 1 illustrates: 1) how real biologic data might be analysed, 2) the need to be able to handle ties, 3) the need for large samples when using correlation coefficients, 4) the need for a) further expansion of the significance tables and/or b) development of asymptotic formulas for large sample sizes to replace the tables and/or c) asymptotic inequalities showing how one might be "safe" using tables for smaller sample sizes. The positive and negative correlation coefficients for phase, ϕ, between hematocrit and hemoglobin for the 14 subjects are $+0.866$ and -0.642. If the sample size were 11 one could enter Table 2 with a difference of 0.224 where one finds 0.0410. The area under both tails together would be twice this or 0.0820 and these data would not support rejection of the hypothesis of randomness. However, it might be determined beforehand on other grounds that the phases could not "go in opposite directions".

Table 3

Rho Angular Angular			One sided			
4	(1.000, .1667)	(.800, .8333)	(.600,1.0000)			
5	(1.000, .0417)	(.900, .2500)	(.800, .4583)	(.700, .6667)	(.600, .8750)	(.500,1.0000)
6	(1.000, .0083)	(.943, .0583)	(.886, .1333)	(.829, .2500)	(.771, .5000)	(.714, .7000)
	(.657, .8750)	(.600, .9750)	(.543,1.0000)			
7	(1.000, .0014)	(.964, .0111)	(.929, .0306)	(.893, .0597)	(.857, .1278)	(.821, .1958)
	(.786, .2931)	(.750, .3903)	(.714, .5264)	(.679, .6528)	(.643, .7694)	(.607, .8472)
	(.571, .9250)	(.536, .9736)	(.500,1.0000)			
8	(1.000, .0002)	(.976, .0018)	(.952, .0058)	(.929, .0121)	(.905, .0268)	(.881, .0458)
	(.857, .0728)	(.833, .1109)	(.810, .1688)	(.786, .2276)	(.762, .3149)	(.738, .4006)
	(.714, .5125)	(.690, .6093)	(.667, .7125)	(.643, .7950)	(.619, .8891)	(.595, .9415)
	(.571, .9804)	(.548, .9962)	(.524, .9996)	(.429, .9998)	(.333,1.0000)	
9	(1.000, .0000)	(.983, .0002)	(.967, .0009)	(.950, .0021)	(.933, .0048)	(.917, .0090)
	(.900, .0148)	(.883, .0240)	(.867, .0372)	(.850, .0529)	(.833, ·0759)	(.817, .1029)
	(.800, .1363)	(.783, .1749)	(.767, .2224)	(.750, .2706)	(.733, .3320)	(.717, .3945)
	(.700, .4619)	(.683, .5305)	(.667, .6048)	(.650, .6711)	(.633, .7423)	(.617, .8005)
	(.600, .8543)	(.583, .8956)	(.567, .9325)	(.550, .9575)	(.533, .9782)	(.517, .9887)
	(.500, .9962)	(.467, .9986)	(.450, .9993)	(.433,1.0000)	(.400,1.0000)	
10	(1.000, .0000)	(.988, .0000)	(.976, .0001)	(.964, .0003)	(.952, .0007)	(.939, .0015)
	(.927, .0026)	(.915, .0044)	(.903, .0071)	(.891, .0107)	(.879, .0156)	(.867, .0222)
	(.855, .0307)	(.842, .0416)	(.830, .0553)	(.818, .0715)	(.806, .0921)	(.794, .1159)
	(.782, .1435)	(.770, .1757)	(.758, .2133)	(.745, .2533)	(.733, .2987)	(.721, .3468)
	(.709, .3984)	(.697, .4527)	(.685, .5096)	(.673, .5635)	(.661, .6207)	(.648, .6758)
	(.636, .7294)	(.624, .7772)	(.612, .8256)	).600, .8663)	(.588, .9015)	(.576, .9321)
	(.564, .9551)	(.552, .9710)	(.539, .9820)	(.527, .9890)	(.515, .9924)	(.503, .9954)
	(.491, .9970)	(.479, .9979)	(.467, .9989)	(.455, .9992)	(.442, .9996)	(.430, .9997)
	(.418, .9999)	(.394,1.0000)				
11	(1.000, .0000)	(.991, .0000)	(.982, .0000)	(.973, .0000)	(.964, .0001)	(.955, .0002)
	(.945, .0004)	(.936, .0007)	(.927, .0012)	(.918, .0019)	(.909, .0028)	(.900, .0041)
	(.891, .0058)	(.882, .0082)	(.873, .0111)	(.864, .0148)	(.855, .0194)	(.845, .0252)
	(.836, .0319)	(.827, .0403)	(.818, .0501)	(.809, .0617)	(.800, .0747)	(.791, .0904)
	(.782, .1076)	(.773, .1275)	(.764, .1491)	(.755, .1736)	(.745, .1996)	(.736, .2291)
	(.727, .2595)	(.718, .2928)	(.709, .3277)	(.700, .3652)	(.691, .4023)	(.682, .4425)
	(.673, .4822)	(.664, .5237)	(.655, .5640)	(.645, .6058)	(.636, .6455)	(.627, .6863)
	(.618, .7241)	(.609, .7615)	(.600, .7958)	(.591, .8291)	(.582, .8578)	(.573, .8848)
	(.564, .9075)	(.555, .9284)	(.545, .9449)	(.536, .9593)	(.527, .9701)	(.518, .9791)
	(.509, .9852)	(.500, .9900)	(.491, .9931)	(.482, .9956)	(.473, .9969)	(.464, .9981)
	(.455, .9987)	).445, .9992)	(.436, .9995)	(.427, .9997)	(.418, .9998)	(.409, .9999)
	(.400,1.0000)	(.391,1.0000)	(.373,1.0000)	(.300,1.0000)		

Then the hypothesis of randomness could be rejected at the 0.05 level, *i.e.* using a 5% chance of rejecting randomness when it actually exists, by using the 0.0410. (However, for this sample, n=14, and one needs extension of the tables, though it is expected by extrapolation that the use of the table for n=11 is "safe"). Alternatively, in the latter case, one may take only the positive correlation coefficient of 0.866 and enter Table 3 to find 0.0148 and again reject randomness. There is still another possible distribution, not included here, in case relative direction is not predetermined. One would choose the coefficient larger in absolute value. The tails of this distribution appear to be approximately twice

Table 4

Rho continuous-angular-linear						
4	(1.000, .6667)	(.800, 1.0000)				
5	(1.000, .3333)	(.900, 1.0000)				
6	(1.000, .1333)	(.943, .5333)	(.886, .9333)	(.829, 1.0000)		
7	(1.000, .0444)	(.964, .2222)	(.929, .4556)	(.893, .6556)	(.857, .9333)	(.821, 1.000)
8	(1.000, .0127)	(.976, .0762)	(.952, .1810)	(.929, .2984)	(.905, .5139)	(.881, .7004)
	(.857, .8516)	(.833, .9468)	(.810, .9948)	(.786, .9968)	(.762, 1.0000)	
9	(1.000, .0032)	(.983, .0222)	(.967, .0611)	(.950, .1127)	(.933, .2071)	(.917, .3147)
	(.900, .4327)	(.883, .5674)	(.867, .7016)	(.850, .8029)	(.833, .9186)	(.817, .9769)
	(.800, .9988)	(.783, 1.0000)				

that of Table 3 through n=11. The amplitude for hemoglobin is a linear variate which when correlated with its phase gives −0.774 and +0.550. Using Batschelet's table for n=12 and doubling because the relative direction is not predetermined, one finds a value of 0.0582. Thus with proper tables this might also be significant at 5%. The continuous-angular-linear coefficient is 0.925. Table 4 computed thus far goes only to n=9, uncomfortably far from n=14.

REFERENCES

1. Kendall M. G. (1962): Rank correlation methods, third edition. Griffin, London.
2. Batschelet E., Hillman D., Smolensky M. and Halberg F. (1973): Angular-linear correlation coefficient for rhythmometry and circannually changing human birth rates at different geographic latitudes. International Journal of Chronobiology *1*.

RECORDING OF SELFMEASUREMENTS OR AUTOMATICALLY COLLECTED DATA FOR HEALTH ASSESSMENTS

DERRICK A. JONES, JOHN A. MUNKBERG and FRANKLIN C. LARIMORE

Electromechanical Research, Central Research Laboratory, 3 M Company
St. Paul, Minnesota, U.S.A.

The application of computer technology to collection and transfer of data is essential to the future progress of medical diagnosis based on analysis of rhythmic functions. The objective of the project reported here is to interface chronobiological measurements directly with the computer, specifically output from the Godart-Statham haematonograph via an interface into a magnetic tape cassette. A second sub-system would transfer data from the magnetic tape cassette into the computer in the University of Minnesota Chronobiology Laboratories. These two sub-systems have been assembled. Number one is an analog to audio tone recording sub-system. Number two is an audio tone to digital playback sub-system.

Audio tone recording was chosen for several reasons. First because we wished to use an inexpensive audio cassette recorder. Second, audio tone recording gives the redundancy necessary to prevent bit dropouts that sometimes happen even in high quality digital tape systems. Another reason is the low cost and convenience of the Philips type cassette compared to reel tape transports. A fourth reason which may be quite important in the future, is compatibility of Bell system audio tones for transmission over an unconditioned phone line to remote recording equipment.

The audio tone recording sub-system shown in Fig. 1 comprises 1) an analog to digital convertor with 100 discrete steps, 2) a pre-scaler to set the range of the A to D converter, 3) a tone generator, 4) a real time clock and 5) an audio cassette recorder modified for incremental operation. The logic is of the COS/MOS type for high noise immunity and low power requirement for future battery operation, and compactness for portability.

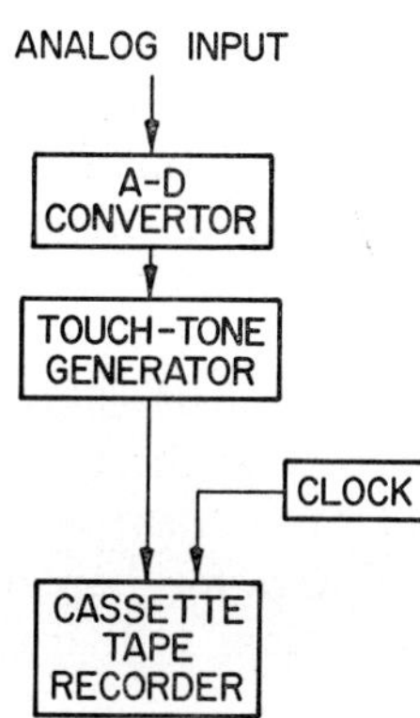

Fig. 1 Data acquisition subsystem.

In operation, the analog voltage from the Godart/Statham unit is scaled for input to the A to D converter which outputs a two digit B.C.D. number from 0 through 99. These two digits are sequentially entered into the tone generator which outputs two tones simultaneously for each B.C.D. digit entered, one high frequency tone and one low frequency tone. These pairs of tones are subsequently recorded on one channel of the cassette tape recorder. First the 1's digit tone pair and then the 10's digit tone pair. At the same time as the tone pair representing blood pressure is being recorded on one channel of the tape the real time clock register is sampled and the tones representing the clock time are recorded on the other channel of the cassette. The whole process requires about 1.2 sec. including the tape transport start and stop times. Therefore, the total time for recording systolic and diastolic pressure would be about 2.4 seconds so that on a 2-hour tape over 2800 entries can be made. In this case, if readings were taken every 10 minutes, a tape would last for more than 19 days.

The audio tone to digital playback sub-system shown in Fig. 2 consists of an audio cassette player, a tone decoder, a B.C.D. to binary converter and a digital interface.

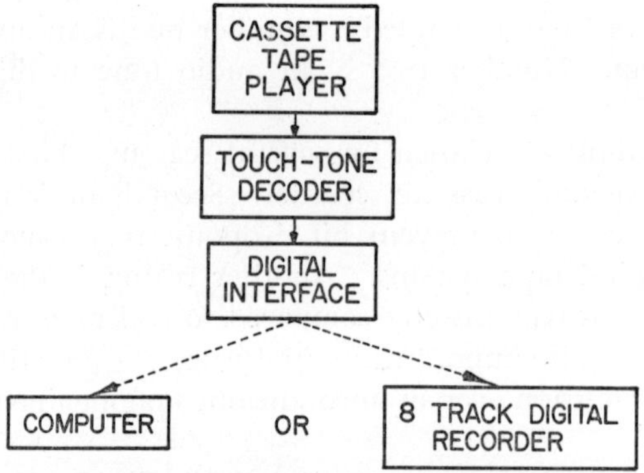

Fig. 2 Playback-subsystem.

The cassette player reproduces both channels of the pre-recorded tape and sends the tone pair bursts to the tone decorder for conversion back to B.C.D. signal information. The binary coded decimal signals are then converted to I.B.M. compatible binary code along with the binary number representing the real time clock signal. This binary information is then reformatted by the logic circuits of the interface unit to drive a Digi-data type of incremental digital tape recorder. The resulting 7 track tape may then he analyzed on any computer using an I.B.M. tape format.

The system is flexible and will accept analog or digital signals ranging from a thermistor temperature probe to a manual or mental dexterity timing device taking advantage of the real time clock. These functions will be researched during 1972. Our initial effort has not included miniaturization of the system. However, we recognize the future need for a pocket sized recorder as part of an automatic self-measurement system. Also, for the future, we envision micro telemetering transducers for implantation to allow continuous recording of many physiological functions so that chronobiology can be then made available to the masses.

STUDIES OF PERIPHERAL CIRCULATORY RHYTHMS IN RESTING AND EXERCISING HUMANS

ROBERT EL. SMITH and WASYL MALYJ, Jr.

Department of Human Physiology, University of California
Davis, California, U.S.A.

INTRODUCTION

Circadian rhythms have been observed in key variables of the cardiovascular system, including heart rate, blood pressure and blood flow. The data of KANEKO, et al. [1] appeared further to imply a circadian variation in peripheral resistance, a result consistent with the circadian shifts in blood distribution inferred from heat flow and conductance data by SMITH [2] and ASCHOFF and POHL [3]. The present study investigates these inferences, obtaining simultaneous measurements of pressure, flow and related cardiovascular and thermoregulatory variables. To evaluate peripheral circulatory control under loading, data have been obtained at rest and at a standardized exercise level. Of primary concern in the present paper will be an analysis of potential errors and discussion of the methodology employed to insure reliability of the variety of data obtained in this study.

METHODS AND ERROR ANALYSIS

Basic to the methods used in this study was the elimination of extraneous circadian rhythms which might otherwise contaminate the data obtained. Sources of such error include rhythms in the environment, the conduct of the experiment by the investigator, and the instruments themselves. The first category includes variations in ambient temperature, ambient air flow (and so convective heat losses) and subject emotional stresses. In the second category would be investigator variations in alertness and visual and auditory acuity. In the last category would be instrumental drift and line voltage changes.

Variables measured in this study included systemic arterial (systolic) pressure, forearm arterial flow and venous compliance, heart rate, arm heat flows, arm and forehead skin temperatures and rectal temperature.

Systemic pressure measurement by conventional sphygmomanometry lends itself to two sources of error: difficulty in obtaining comparable readings under resting and exercise conditions, and variations in the auditory acuity and alertness of the investigator. Elimination of all these problems was accomplished by substituting an ultrasonic flow meter for the stethoscope and recording simultaneously the occluding cuff pressure and the signal detected by the flow meter. This procedure proved highly reproducible and precise within 1 to 2 mmHg in the measurement of systolic pressure under either resting or exercise conditions. Fig. 1 shows the information flow in this system and also shows another major step in the elimination of operator error or variations: the use of a programmer

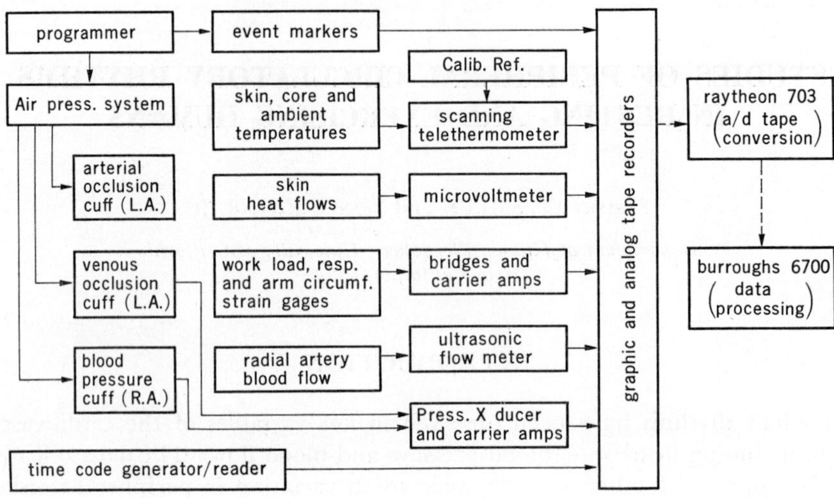

Fig. 1

to control the automatic pressure cuff inflation system and to signal the recorders with event markers indicating the operation in progress. These same event markers are also valuable in computer record reduction.

While measuring systemic pressure in our subject's right arm, we measure forearm arterial flow in his left arm. The procedure used is venous occlusion plethysmography, the hand isolated from the arm by a wrist arterial occlusion cuff. The transducer is a mercury-in-rubber strain gauge, and the key procedure insuring freedom from instrument or transducer drift is rigorous attention to careful calibration of the strain gauge through the system (*i.e.*, physically stretch the gauge at tension and lengths comparable to those of the experimental measurement) both before and after each experimental run (six runs per day). Again, cuffs are inflated automatically, and cuff pressures are recorded, to insure repeatability of the measurement procedure. Strain gauge location is marked on the subject's arm for the same reason. Venous compliance is measured by the same system, except that a pressure ramp, instead of step, is imposed.

Measurement of heat flows (Hatfield-Turner discs) and temperatures is straightforward; however, careful control of ambient temperature and air flow is mandatory. Both have been accomplished by isolating the subject from both the instruments and the rest of the laboratory in a chamber just large enough to house subject, transducers and bicycle ergometer.

The last stress, unfortunately often outside experimental control, is that of unanticipated emotional stress on the subject. Laboratory conditions (e.g., no visitors) and the subject's confidence in the investigator are important here, but the investigator remains obliged to observe carefully the subject's condition and to note any potential problems which might be evident.

Quantitative analysis of some of the error sources and measurement system variations encountered was conducted. Two examples will be considered: the heat flow and the venous occlusion plethysmographic forearm blood flow measurements. It was found that the heat flow noise signals were reduced by better than an order of magnitude by the use of the isolation chamber when compared with those obtained by the same equipment and investigator in an open room. Shifts in the blood flow system's measurement performance were analyzed by observing variations in the slopes of the mercury-in-rubber strain gauge calibration curves. Most probable causes of variations in this measure would be laboratory ambient temperature, instrumental drift and line voltage changes, and possibly operator alertness. These calibration coefficients were examined in five different experiments, the results being summarized in Table 1. Not only was there an instrumental variation up to 15% of the daily mean, but in four of the five experiments these variations exhibited "typical circadian rhythms", peaking in the afternoon or early evening and having minima in the late evening. Inasmuch as each of these "circadian rhythms" is as great or greater in magnitude than several important physiological rhythms, the critical nature of careful calibration is apparent.

Table 1 Circadian rhythm in strain gauge calibration coefficients.

Time of day	0700	1000	1300	1600	1900	2200	% Variation
Subject C							
Calib. Coeff.	.585	.599	.608	.614*	.599	.526**	
% of mean	99	102	103	104	102	89	15
Subject E							
Calib. Coeff.	.606	.623	.616	.630	.640*	.605**	
% of mean	98	100	99	101	103	97	6
Subject R							
Calib. Coeff.	—	.630	.634	.641*	.626	.625**	
% of mean	—	100	100	102	99	99	3
Subject M							
Calib. Coeff.	.597	.597	.603	.610*	.594	.569**	
% of mean	100	100	101	103	100	96	7
Subject B							
Calib. Coeff.	.616**	.638*	.628	.622	—	.631	
% of mean	98	102	100	99	—	101	4

* Maximum ** Minimum

RESULTS

The methods described above have proven the most reliable of any used in our laboratory to date. Both inter-subject and intra-subject variations in the observed rhythms are found, but the existence of the rhythms themselves has been well documented. As one example, Fig. 2 illustrates forearm blood flows in one of our subjects, her circadian blood flow rhythms in both the resting and exercised states being quite apparent.

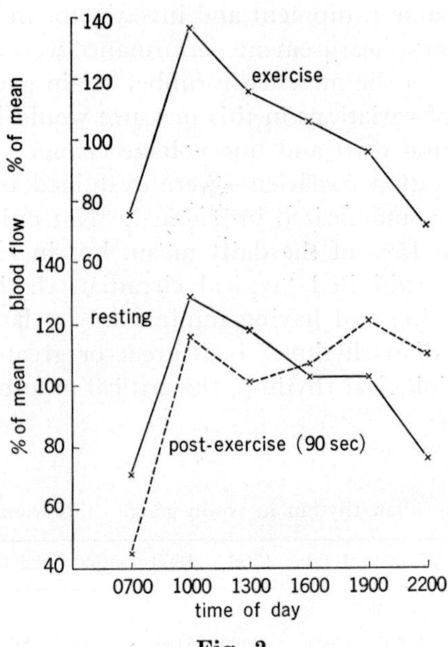

Fig. 2

DISCUSSION

It is the primary purpose of this paper to stress the critical importance of maximizing the reliability of the data obtained. Maximum environmental control, automatic control of the experimental procedures, ultrasonic blood pressure detection, pre- and post-experiment calibration through transducers to recorder, simultaneous graphic and magnetic recording, and computer data reduction have all made essential contributions to sifting valid physiological variations from experimental artifact, with repeated through-the-system calibration undoubtedly being the most important of all.

The study of circadian variations in circulatory state and responses requires the utmost attention to experimental accuracy and reproducibility, for just as in the past ignorance of physiological circadian rhythms has led investigators astray, so now can ignorance of instrumental variations lead us astray.

REFERENCES

1. Kaneko M., Zechman F. W. and Smith R. E. (1968): Circadian variation in human peripheral blood flow levels and exercise responses. J. Appl. Physiol., *25*: 109–114.
2. Smith R. El. (1969): Circadian variations in human thermoregulatory responses. J. Appl. Physiol., *26*: 554–560.
3. Aschoff J. and Pohl H. (1970): Rhythmic variations in energy metabolism. Fed. Proc., *29*: 1541–1552.

PULMONARY PATHO-PHYSIOLOGY OF INDUSTRIAL DISABILITY

WILLIAM G. REDDAN, J. A. DEMPSEY, G. A. doPICO, L. CHOSY and J. RANKIN

Department of Preventive Medicine, Department of Medicine
University of Wisconsin
Madison, Wisconsin, U.S.A.

Industrial workers who had previously experienced varying degrees of respiratory impairment, as a result of abnormal environmental conditions, provided an opportunity and a model for the study of disability status. A multiplicity of factors ultimately determine the degree of industrial disability, whereas *impairment,* an important contributing factor to disability, is essentially a medical condition and can be evaluated as such [6]. In order to determine the degree of disability, our laboratory has combined an initial assessment of impairment with an evaluation of the additional stresses associated with the work environment.

Following a suitable period of recovery out of the environment, the most prevalent symptom remaining was one of exertional dyspnea, followed by individuals with some degree of restricted lung function who were studied to determine their ability to return to work. The current status of the individual was reviewed in terms of the medical history and measurements of resting ventilatory capacity and alveolar-capillary gas exchange over an eight-year period following the onset of impairment. It was observed that those individuals who had never returned to work followed a similar time course in all measurements of lung function, as demonstrated in Fig. 1 with pulmonary diffusing capacity, but were consistently below a group of workers of comparable age who were not initially affected. The

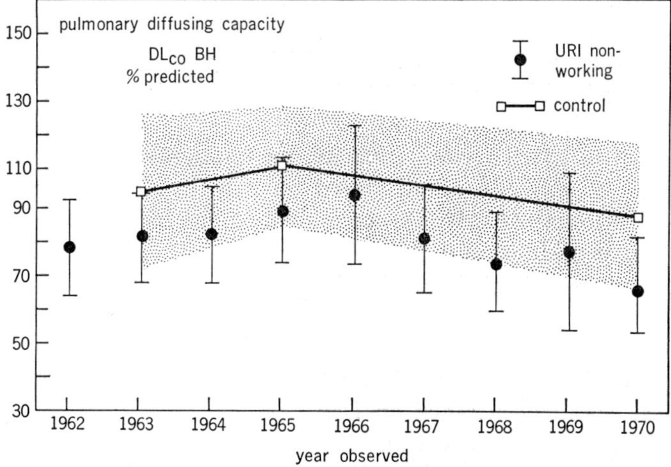

Fig. 1 Pulmonary diffusing capacity, predicted from age and height, in employees not returning to work following an industrial accident (URI) and otherwise healthy employees (controls).

overall annual decline in diffusing capacity noted in both groups is comparable to a reported .6 percent decline in healthy non-smokers [9].

However, the consequences of impaired lung function can most often, and perhaps only, be evaluated in response to some type of work stress. To determine the degree of impairment, standard laboratory procedures were employed in a two-day test regime. On Day 1, an evaluation of the exercise "tolerance" to progressively increasing work on a motor-driven treadmill as well as specific adaptation of pulmonary gas transport system (the onset of hyper- or hypoventilation, dyspnea, tachycardia, marked changes in systemic pressures and the electrocardiogram) was made for purposes of defining zones of work intensity for each patient which represented mild, moderate, and severe physiologic stress. On Day 2, specific studies were carried out, with the addition of arterial blood-gas and metabolic measurements, to determine the possible site of resistance to gas transport. The patient walked at steady-state work intensities chosen to approximate those energy demands which were comparable to and considerably above (if tolerable) the occupational requirements.

Several factors associated with the work environment affect an individual's successful adaptation to a task, such as, intermittent or continuous work, nature of the task, physical environment (dusts, etc.), as well as the amount and type of muscle mass involved. Superimposed upon these factors in many industrial settings are the effects of a disturbance in normal rhythms with periodic shift changes. In order to determine the "return to work" status of these individuals, specific "in-plant" studies were conducted in an attempt to define the effect of these variables on otherwise healthy workers in twelve job categories and then followed with studies on individuals who had returned to work following varying lengths of recovery from illness.

The basic protocol outlined in Figs. 2 and 3 attempted to combine day-long on- and off-the-job assessments of cardio-pulmonary response to work. A physical examination and baseline measurements of resting pulmonary function (lung volume, ventilatory flow rates and alveolar-capillary gas exchange) were administered on Day 1. The resting studies were repeated at three times during the eight-hour shift to examine the effects of possible exposure to industrial contaminants and were shown to vary slightly throughout the day (slight elevation at midday, decline at late afternoon) which coincided with reported diurnal rhythms in both healthy workers [1, 3, 5, 7, 8, 10] and in workers with increased airway resistance [1, 5, 8, 10].

A measure of functional work capacity, or "tolerance" via response to a progressive stress test was also administered on Day 1. On subsequent days, the individual's functional capacity was compared to one of three submaximal workloads prior to, at midday, and following a work shift. As seen in Fig. 3, there was no apparent effect in the stress level throughout the work shift as determined by cardio-pulmonary and venous lactate changes, although diurnal increases in both respiratory [2] and cardiovascular [4] response to moderate exercise over the corresponding time period have been reported. These values may be compared to the highest workload attained on Day 1.

The specific on-the-job work cycle shown in Fig. 3 was intermittent in nature which was typical of most categories studied; although little variation was seen between work cycles, a slight variation in peak heart rate and ventilation may be seen within an operational cycle or "phase". The level of energy expenditure

Protocol

DAY I
1. Resting pulmonary function
   a. Standard spirometry
      VC, $FEV_{1.0}$, MVV, $DL_{CO}BH$
2. Work capacity
   a. Progressive T-mill Exercise

DAY II
1. Pre-shift prolonged T-mill exercise
   a. 3′ at 2 submax. loods
   b. 25′ at ≈ 40% max.
2. On-job
   a. Open circuit
   b. Q_C (N_2O)
   c. ECG-standard

DAY III
1. Post-shift prolonged T-mill exercise

Measurements:

V_{O_2} – open circuit
P_{ETCO_2} – continuous sampling
D_{LCOSS} – pulmonary diffusing capacity-steady state
Q_C – pulmonary blood flow-N_2O method
ECG – standard 3 lead
BP – auscultatory

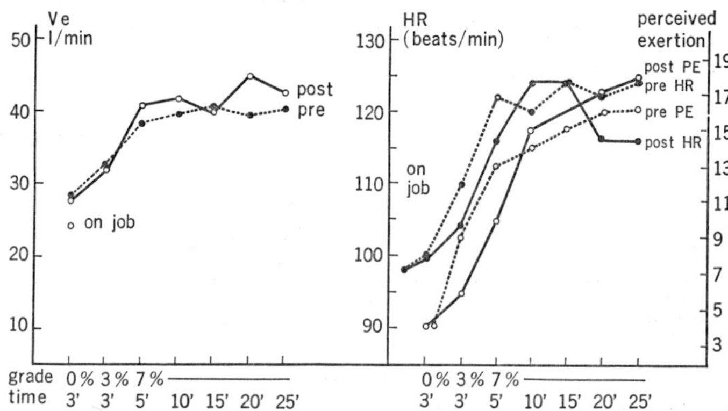

Fig. 2 Top: Protocol for in-plant study. Bottom: The response of ventilation, heart rate and level of perceived exertion during a prolonged treadmill test in one patient prior to and following an eight-hour work shift. The identical tests were performed on separate days.

($\dot{V}_{O_2}$ l/min) "on-the-job" in mid-morning and afternoon was comparable only to a light or moderate work load on a bicycle ergometer. The metabolic requirements for 12 different job categories were equivalent to approximately 3–5 times the resting metabolic rate which corresponded to 20–30% of the functional capacity in these individuals.

It is known that the respiratory distress experienced on the job in many patients is associated with activity of prolonged duration at a low level of intensity. Therefore, individuals who had claimed disability and returned to work were given an extended treadmill test (over a 30–35 min period) prior to and following a normal eight-hour work shift (on separate days) (Fig. 2). Specific cardio-respiratory responses (as previously described) were observed at a level of energy expenditure slightly in excess of that encountered on the job. Although the responses were highly variable between individuals, the work shift had no apparent effect on the post-test ventilatory (slightly elevated) or cardiac response in the example shown. However, it is interesting to note that, although the workload was identical, the individual perceived the degree of stress as greater at the conclusion of the work shift.

Thus, we have attempted to evaluate the degree of respiratory impairment in a defined industrial population by means of a patient profile relative to his pulmonary and metabolic stress response and then practically to assess the importance of this impairment upon the capacity to perform intermittent prolonged work in an industrial environment. The modifying effects of the work

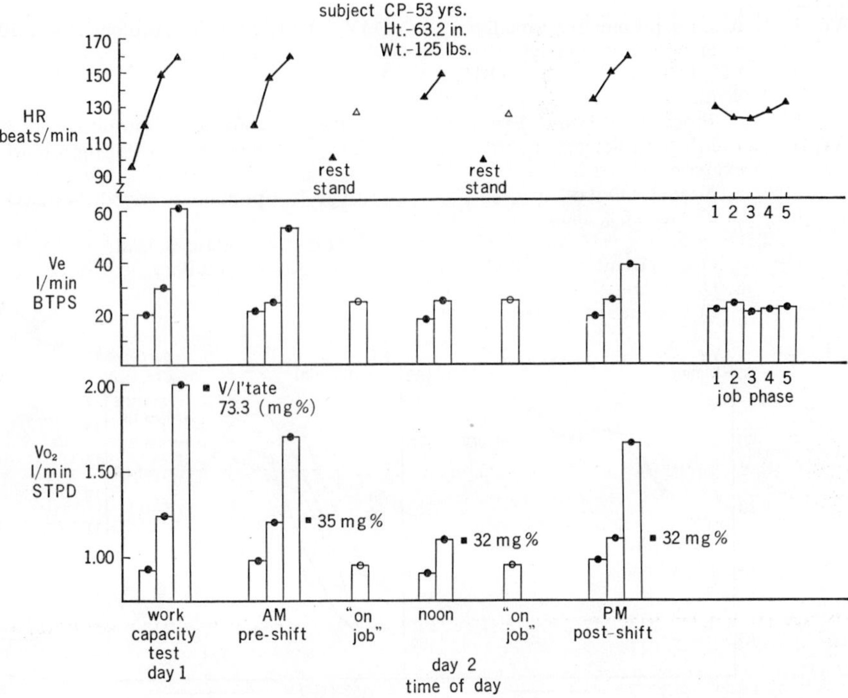

Fig. 3 Metabolic and cardio-pulmonary responses to progressive (Day 1) and on-and-off the job work stress (Day 2) in one subject.

environment were assessed through a series of day long on- and off-the-job measurements to determine: (1) the pulmonary and metabolic requirements of several defined types of work, (2) the independent effects of 8 hour rhythms, (3) changes over the work shift in tolerance to stress at exercise levels bracketing the occupational requirement, and (4) the relationship of physiologic efficiency to productivity. It is recognized however, that although physiological tests may provide precise estimates of respiratory impairment, demands of the job, and physical factors modifying the job stress, the total occupational disability depends upon additional factors, such as, motivation, economic and social background and management- labor attitudes, all of which were not included in this survey.

REFERENCES

1. Batawi M. A. et al. (1964): Byssinosis in the Egyptian cotton industry: changes in ventilatory capacity during the day. Brit. J. Indust. Med., *21*: 13–19.
2. Burger G. C. E. et al. (1957): Human problems in shift work. Proc. 12th Int. Congr. Occup. Health. Helsinki, *3*: 126–128.
3. Bouhuys A., Hartogensis F. and Korfage H. (1963): Byssinosis prevalence and flax processing. Brit. J. Industr. Med., *23*: 320–324.
4. Kaneko M., Zechman F. and Smith R. E. (1968): Circadian variation in human peripheral blood flow levels and exercise responses. J. Appl. Physiol., *25*: 109-114.
5. Lewinsohn H. C., Capel and Smart J. (1960): Changes in forced expiratory volumes throughout the day. Brit. Med. J., *1*: 462–464.
6. Lingren I., Müller B. and Gaensler E. A. (1965): Pulmonary impairment and disability claims. J.A.M.A., *194*: 499–506.

7. McKerrow C. B. (1964): Respiratory disease in Great Britain. Arch. of Environ. Health., *8*: 174–179.
8. McKerrow C. B. et al. (1958): Respiratory function during the day in cotton workers: A study of byssinosis. Brit. J. Industr. Med., *15*: 75–83.
9. Rankin J., Gee J. B. L. and Chosy L. W. (1965): The influence of age and smoking on pulmonary diffusing capacity in healthy subjects. Med. Thorac., *22*: 366–374.
10. Walford J. et al. (1966): Diurnal variation in ventilatory capacity: An epidemiological study of cotton and other factory workers employed in shift work. Brit. J. Industr. Med., *23*: 142–148.

CARDIOVASCULAR CIRCADIAN RHYTHM IN MAN

LOTHAR WERTHEIMER, ANSARI HASSEN, ABNER DELMAN and AKHTAR YASEEN

Departments of Medicine and Pathology, New York Medical College Center for Chronic Disease, Welfare Island New York, U.S.A.

Previous studies have shown a circadian rhythm of arterial pressure and pulse rate in man [1, 2, 3, 4]. Systolic time intervals (STI), including the left ventricular ejection time (LVET) and the pre-ejection period (PEP), are an effective indirect method of measuring ventricular function [5, 6] and cardiac hemodynamics [7] in man. Alterations in ventricular function and cardiac dynamics can be accurately assessed by determining serial changes in the STI.

The level of catecholamine stimulation of the heart is an important determinant of myocardial contractility and ventricular performance. Increased catecholamine stimulation results in increased contractility and improved ventricular function, with concomittant shortening of the LVET and PEP. In addition, an increase in heart rate and blood pressure may occur. With decreased catecholamine stimulation, there is a diminution of contractility and ventricular function, lengthening of the LVET and PEP, and a fall in heart rate and blood pressure.

The present study investigates the circadian rhythm of catecholamine excretion in man, with concomitant alterations in myocardinal contractility (as measured by the STI of the LVET and PEP), heart rate, and blood pressure.

Ten patients, age 18–48 years (mean 28.5), with no evidence of cardiovascular disease, were studied for a 24-hour period. Heart rate, systolic and diastolic pressures (SAP and DAP), and STI were recorded at 2 to 3 hour intervals with patients in a supine resting state. In five of the patients, urines were collected simultaneously at 2 to 3 hour intervals, and volume, creatinine, and catecholamine excretion were measured. All patients were at bedrest for the full 24-hour period and received no medications. During the study period, hospital routine was as follows: 6:00 a.m.–lights on, patients awakened and morning care; 7:30 a.m.–breakfast; 12:00 Noon–lunch; 5:30 p.m.–dinner; 9:30 p.m.–lights off.

Vanilla, chocolate, banana, coffee, tea, and alcohol were excluded during the 24-hour period prior to and during the study. No smoking was allowed.

The duration of the phases of ventricular systole were measured from simultaneous recordings of the electrocardiogram, the phonocardiogram, and the carotid arterial pulse tracing on a multichannel Cambridge photographic recorder at a paper speed of 100 mm/sec (Fig. 1). Heart sounds (phonocardiogram) were recorded with a type 3–364 crystal microphone placed over the anterior chest wall, along the left sternal border, on a Cambridge phono channel. The carotid arterial pulsation was recorded with a funnel shaped pickup attached to polyethylene tubing to a 53642 crystal transducer, through a Cambridge pulse channel. The total electromechanical systole (QS_2), LVET and PEP were measured as per WEISSLER et al. [5, 6] and the indices of the QS_2, LVET and PEP (corrected

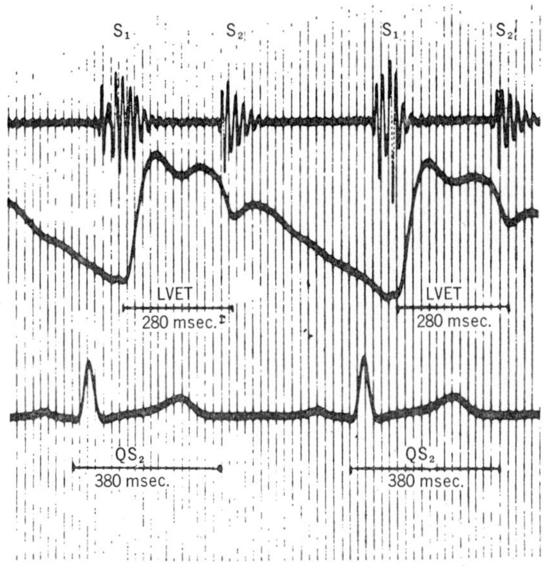

QS_2 interval measures total duration of left ventricular systole from the onset of ventricular depolarization to the closure of aortic valve.

LVET reflects events in the proximal aorta; measures the duration of left ventricular ejection.

PEP (pre-ejection period)=QS_2–LVET, represents the time interval from the beginning of ventricular depolarization until the opening of the aortic valve.

PEP/LVET represents ratio of pre-ejection period to left ventricular ejection time.

Fig. 1 Simultaneous recording of the heart sounds, the carotid artery pulse tracing, and the electrocardiogram (paper speed 100 mm/sec, time markers 0.02 sec). QS2=total electromechanical systole; LVET=left ventricular ejection time; PEP=pre-ejection period.

for heart rate) were determined from previously established regression formulae [5, 6]. SAP and DAP were recorded indirectly by Korofkoff sounds.

Total urine catecholamines (free and conjugated) were assayed by using an ion exchange resin procedure.* Creatinine was assayed on an Autoanalyzer.**

RESULTS

The means of the change of the parameters of cardiovascular function (HR, SAP, DAP and STI) and urinary findings are presented in Table 1 and 2, and Fig. 2. Fig. 2 shows the chronograms of the cardiac and urinary parameters of the ten patients. For each parameter, the 24-hour mean value of all patients was determined. The chronogram value at each time interval in the 24-hour period was obtained by dividing the absolute value of the mean of each parameter by the 24-hour mean value and multiplying by 100. The standard error of the mean of each time interval was measured during the 24-hour cycle.

Table 3 shows the mean and standard error of the mean of the maximal circadian changes of the cardiac parameters. The most significant alterations in the

* Bio-Rad Laboratories, Richmond, California.

** Methodology N-116, Technicon Corporation, Tarrytown, N.Y.

Table 1 Circadian changes in cardiac parameters.

Time (hour)	Baseline (absolute values) *	Mean values ± standard error of changes from the baseline							
	9–12	12–15	15–18	18–21	21–24	24–2	2–4	4–6	6–9
HR (b/min)	73.9 (4.1)	−0.9 (4.2)	+ 3.9 (4.4)	+ 3.5 (3.7)	+2.7 (3.7)	+3.1 (4.4)	−2.7 (3.0)	− 2.1 (3.9)	− 2.8 (3.2)
SAP (mm Hg)	121.0 (3.0)	+2.5 (2.4)	− 0.5 (2.5)	+ 1.5 (2.9)	+0.5 (3.5)	−6.0 (2.6)	−6.0 (3.5)	− 5.0 (3.0)	−10.0 (2.2)
DAP (mm Hg)	74.0 (1.8)	+3.0 (1.5)	− 0.5 (2.7)	+ 1.0 (1.9)	+2.0 (2.5)	+0.5 (2.4)	−2.0 (2.6)	− 0.5 (2.1)	− 3.0 (1.9)
QS_2I (msec)	536.5 (5.9)	+3.6 (6.0)	−11.5 (5.8)	−15.9 (6.0)	−0.5 (6.9)	+7.9 (9.3)	+6.7 (8.9)	+13.7 (7.6)	+10.1 (8.3)
LVETI (msec)	413.4 (5.6)	+1.2 (4.5)	− 7.8 (3.7)	−11.2 (3.9)	−3.8 (4.0)	−1.0 (6.4)	+0.7 (7.8)	+ 8.5 (5.0)	+ 0.8 (8.0)
PEPI (msec)	123.1 (5.7)	+2.4 (6.5)	− 3.7 (5.8)	− 4.7 (5.9)	+3.3 (6.2)	+8.9 (6.0)	+6.0 (6.8)	+ 5.2 (6.9)	+ 9.3 (7.0)

* Mean values and ± standard error of the means ( )

Table 2 Circadian changes in urinary catecholamines and creatinine excretion.

Time (hour)	Mean values and ± standard error of the means ( )								
	9–12	12–15	15–18	18–21	21–24	24–2	2–4	4–6	6–9
Catecholamines (μg)	77 (24)	112 (37)	96 (26)	119 (38)	35 (13)	45 (25)	28 (16)	35 (11)	108 (45)
Percent of total excretion	11.8%	17.0%	14.6%	18.1%	5.4%	6.9%	4.3%	5.5%	16.5%
Creatinine (mg)	192 (61)	224 (75)	129 (15)	183 (60)	139 (50)	98 (32)	37 (15)	90 (33)	172 (52)
Percent of total excretion	15.2%	17.8%	10.2%	14.5%	11.0%	7.7%	2.9%	7.1%	13.6%
Volume (ml)	238 (119)	523 (208)	273 (105)	216 (64)	108 (49)	244 (127)	43 (15)	71 (33)	253 (82)
Percent of total excretion	12.1%	26.6%	13.9%	11.0%	5.5%	12.4%	2.2%	3.5%	12.8%

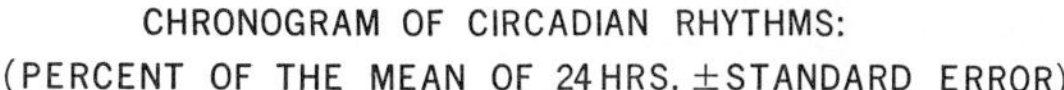

CHRONOGRAM OF CIRCADIAN RHYTHMS:
(PERCENT OF THE MEAN OF 24 HRS. ±STANDARD ERROR)

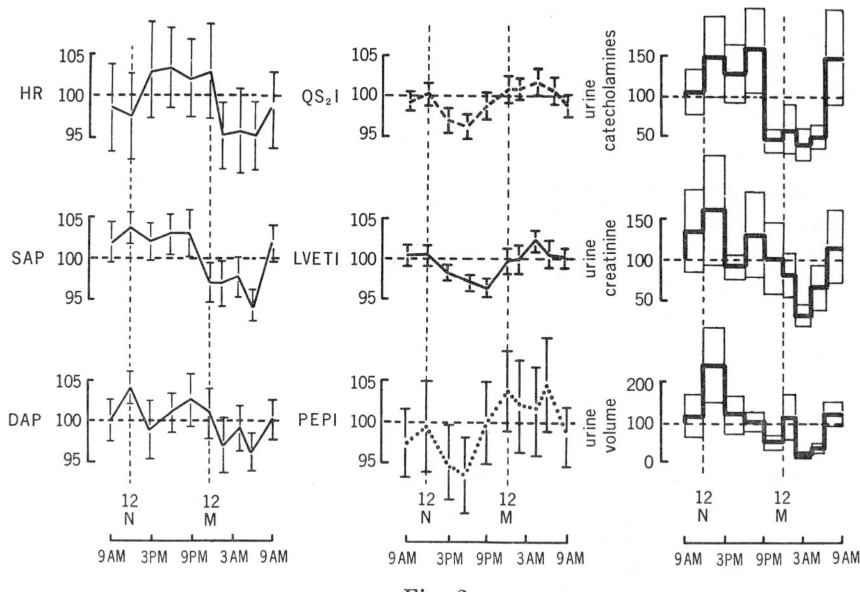

Fig. 2

Table 3 Means and standard error of the means of maximal circadian changes of cardiovascular parameters.

	Maximal		Minimal		p Value*
	Time		Time		
HR (b/min)	6 PM	77.4±3.7	6 AM	71.1±3.2	0.02
SAP (mmHg)	12 N	123.5±2.4	6 AM	111.0±2.2	<0.01
DAP (mmHg)	12 N	77.0±1.5	6 AM	71.0±1.9	<0.01
QS_2I (msec)	4 AM	550.2±7.6	6 PM	520.6±6.1	<0.001
LVETI (msec)	4 AM	421.9±5.6	6 PM	402.2±3.9	<0.01
PEPI (msec)	6 AM	132.4±7.0	6 PM	118.4±5.9	0.02

* P values were obtained using Student's t-test for small sample sizes.

24-hour period were the OS_2I and LVETI systolic intervals and in the systolic and diastolic arterial pressure.

The data show important serial alterations in the parameters for the 24-hour period. During the 6 a.m. to 9 p.m. period there was a high level of catecholamine excretion accompanied by a significant shortening of the LVET and PEP, a slight increase in heart rate, and no significant blood pressure change. In contrast, during the 9 p.m. to 6 a.m. period, there was a distinct reversal in the trend of all parameters, with a marked decrease in catecholamine excretion, a lengthening of LVET and PEP, a decreased heart rate, and a fall in systolic and diastolic pressure. The slowest heart rate, the lowest systolic and diastolic pressures, the most prolonged LVET and PEP, and the lowest urinary excretion of creatinine and catecholamines occurred between 12 midnight and 6 a.m.

DISCUSSION

The present study confirms Weissler's observations on the diurnal variation in ventricular function (as measured by STI) and extends these observations through a 24-hour period. The circadian rhythm of arterial pressure and pulse rate in man is associated with a corresponding circadian alteration in ventricular function and catecholamine secretion. During maximum daytime catecholamine secretion, there is a peak in ventricular performance. Associated with the reduced nocturnal catecholamine secretion, there is a significant fall in myocardial functional capacity, with an increased vulnerability of the heart to stress, particularly from 12 midnight to 6 a.m.

The potential importance of circadian rhythm alterations is usually not considered in the present medical treatment of cardiac patients. Cardiovascular drugs such as digitalis and diuretics are given by a standardized dose schedule (usually in the a.m.), that ignores the patient's circadian variations. However, nocturnal cardiac decompensation, including acute congestive heart failure, arrhythmias and sudden death, is an observed frequent occurrence in patients with significant heart disease. Circadian changes in ventricular function may play an important role in the occurrence of these events, and should be considered in the pharmacologic management of cardiac patients.

Our results strongly suggest that knowledge of these circadian alterations must be incorporated into the effective total treatment of cardiovascular disease.

SUMMARY

A significant circadian rhythm of catecholamine excretion in man, with associated alterations in myocardial contractility, blood pressure and heart rate, has been demonstrated. During the 6 a.m. to 9 p.m. period there was an increase in catecholamine excretion, contractility, and heart rate. During the 9 p.m. to 6 a.m. period, there was a marked decrease in catecholamine excretion, with concomitant decrease in cardiac contractility, blood pressure, and heart rate. These observations suggest that the response to stress of the cardiovascular system in man may be significantly altered by the phase of the circadian rythm. Knowledge of these rhythmic changes may be of great importance in the understanding and proper treatment of the patient with cardiovascular disease.

REFERENCES

1. Menzel W. (1941): Der 24-Stunden-Rhythmus des menschlichen Blutkreislaufes. Ergebn. inn. Med. Kinderheilk., *61*: 1.
2. Reinberg A., Ghata J., Halberg F., Gervais P., Abulker Ch., DuPont J. and Gaudeau Cl. (1970): Rhythmes circadiens du pouls, de la pression arterielle, des excretions urinaires en 17-hydroxycorticosteroides, catecholamines et potassium chez l'homme adulte sain, actif et au repos. Ann. Endocrinol., *31*: 277–287.
3. Richardson D. W., Honour A. J., Fenton G. W., Stott F. H., and Pickering G. W. (1964): Variation in arterial pressure throughout the day and night. Clin. Sci., *26*: 445–460.
4. Schroder R., Dennert J., Heumann H., Prokein E., Ramdohr B., Schachinger H. and Schuren K. P. (1969): 24-Stunden-Rhythmus im Inotropiezustand des Herzmuskels bei gesunden jungen Mannern. Verh. Deutsch. Ges. Kreislaufforsch, *35*: 370–375.

5. WEISSLER A. M., KAMES A. R., BORNSTEIN R. S., SCHOENFELD C. D. and COHEN S. (1965): The effect of deslanoside on the duration of the phases of ventricular systole in man. Amer. J. Cardiol., 15: 153–161.
6. WEISSLER A. M., HARRIS W. S. and SCHOENFELD C. D. (1969): Bedside technics for the evaluation of ventricular function in man. Amer. J. Cardiol., *23*: 577–583.
7. MARTIN C. E., SHAVER J. A., THOMPSON M. E., REDDY P. S. and LEONARD J. J. (1971): Direct correlation of external systolic time intervals with internal studies of left ventricular function in man. Circulation, *44*: 419–431.

RHYTHMOMETRY ON SMALL SAMPLES ILLUSTRATED BY ASSESSMENT OF CIRCADIAN CHANGE IN RAT LIVER GLYCOGEN*

ERNA HALBERG, FRANZ HALBERG and WALTER NELSON

Chronobiology Laboratories, Department of Pathology
University of Minnesota
Minneapolis, Minnesota, U.S.A.

INTRODUCTION AND BACKGROUND

Pioneering work of investigators in Sweden [1–6] and elsewhere [7–10] revealed the occurrence of very marked changes in the glycogen content of livers from several species under usual and special conditions. A long series of follow-up studies elsewhere involving small or large groups of rats and other animals has again demonstrated, under various circumstances, a grossly apparent circadian variation in liver glycogen.

The reproducibility of the bioperiodicity in time plots accompanying reports on such work varied greatly even under presumably comparable conditions of controlled temperature and humidity [11]. By contrast, a rather high macroscopic [10] reproducibility was found in studies carried out in Minnesota under standardized conditions of lighting and single housing, in averages of values obtained four hours apart during a 24-hour span on separate groups of mice [12, 13] or rats [14].

The fit of a 24-hour cosine curve to such data demonstrates and quantifies the parameters of a nearly sinusoidal large-amplitude rhythm in the glycogen content of rodent liver. The results of such least squares fits to the glycogen rhythm mapped by Minnesota investigators on different occasions on different rodents show good agreement among such "microscopic" results; they have been incorporated into acrophase charts of rodents [14, 15], summarizing the timing of circadian rhythms in liver glycogen in relation to the timing of many other rhythms for the rat and mouse.

The present methodologic paper is prompted by the question whether a circadian rhythm can be demonstrated reproducibly and its parameters estimated reliably with smaller numbers of animals than were conventionally used in prior work in Minnesota as well as elsewhere. This question will be examined in systematically collected data from as yet unpublished work in the Chronobiology Laboratories in Minnesota and from studies carried out earlier in the same laboratory [14].

* Supported by the U.S. Public Health Service (5-K6-GM-13,981 and NCI 1RO1-CH-14445-01)

EXTENT OF REPRODUCIBILITY WITH REDUCTION IN NUMBER OF TIME POINTS AND ANIMALS SAMPLED

Can a pool of only two animals at each of six time points, 4 hours apart, suffice for the detection of the circadian rhythm in rat liver glycogen; if so, can it serve for a reliable estimation of rhythm parameters? If one summarizes analyses on 10 times series, each representing data on a pair of livers from separate pairs of rats killed at six consecutive 4-hourly time points along the 24-hour scale, one finds that the variability of the original pairs in any series is large. If one considers all values from all 6 pairs, the smallest within-series difference is 2.7 gm%, e.g., between 2.2 and 4.9 gm%. The coefficient of variation, i.e., the standard deviation divided by the mean, ranges from 35 to 46%. Because of this great dispersion in values of liver glycogen, it is of interest to see how much of the overall variability, if anything, can be accounted for by an analysis involving the fit of a single 24-hour cosine function [16]. For this purpose, one computes the percentage rhythm as the PR=percent of total variability contributed by fitted cosine curve=(SS of values derived from fitted curve at sampling times/total SS) ×100. In the particular series examined, 92 to 99% of the variability on hand is accounted for as circadian rhythmic by the fit of a single cosine curve. The sinusoidal rhythm description is significant below the 2% level for every single series of pairs as well as for each of the five composite pools (of two series of pairs).

It should be emphasized that a circadian rhythm in liver glycogen studied under the standardized conditions practiced herein [14] constitutes one of the most conspicuous rhythms. By synchronization with the lighting regimen, the use of inbred animals, the selection of a relatively long interval of four hours between consecutive time points and the pooling of values from two animals, some variability already has been removed. The results gain in importance if one realizes that this rhythm has previously been mapped under similar conditions and analyzed by the same "microscopic" procedures. The mesors, amplitudes and acrophases here obtained can be compared with similar information already incorporated in a standard handbook of biologic data [17]. However, the intra-experimental and inter-experimental comparison of new values constitutes a major point of this report.

Against this background, one can proceed to estimate the rhythm parameters, focusing first upon the mesor of the rhythm. The mesors (overall rhythm-adjusted means) vary by no more than .45 gm%, i.e., from 3.36 to 3.81, when 6 time points at 4-hour intervals are documented by no more than two livers/time. When the sampling is doubled so that four livers contribute information at each of six time points 4 hours apart to cover a 24-hour span, little is gained for the mesor. It now varies between 3.45 and 3.80 gm%, i.e., by .35 gm%. There is thus a difference of .10 gm% only between the extremes in mesor when sample size is increased from two to four animals at each of 6 time points.

In turning to estimates of amplitudes, these varied from 1.69 to 2.27—a 0.58 gm% variation for two animals/time point. With four animals/time point, there is a .43 gm% inter-series difference in amplitude. The acrophases, in their turn, are rather close, between 140° and 165°—the 25° difference corresponding to one hour and 40 minutes. With four liver samples at each of the six 4-hourly time points, the spread of acrophases is 21° or one hour and 24 minutes.

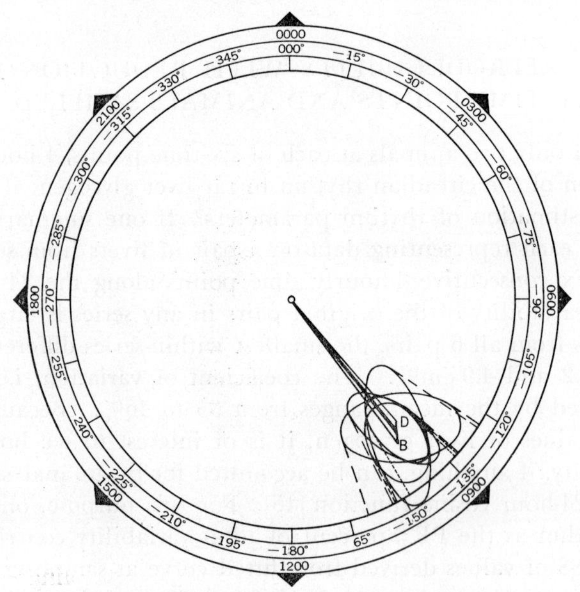

	N	ϕ	(.95 CA)		Amplitude	(.95 CI)		Mesor	(.95 CI)		P
A	12	−140	−132	−147	1.64	1.29	1.98	3.40	2.63	4.16	.001
B	12	−142	−133	−152	1.74	1.39	2.08	3.33	2.51	4.15	.001
C	12	−141	−123	−151	1.67	1.23	2.11	3.31	2.48	4.14	.001
D	12	−138	−122	−159	1.65	1.32	1.97	3.48	2.64	4.32	.001
E	12	−140	−131	−152	1.65	1.23	1.97	3.39	2.26	4.51	.001

Fig. 1 Excellent agreement among cosinor summaries of parameters obtained by fitting 24-hour cosine curves to liver glycogen series of different length.

A = 0400, 0800, 1600, 2000, 0000;
B = 0400, 0800, 1200, 1600, 2000;
C = 0400, 0800, 1600, 2000;
D = 0400, 1200, 1600, 0000;
E = 0800, 1200, 2000, 0000.

Against this background, it is of interest to turn to the results obtained on series consisting of sets of livers removed at each of 6 consecutive time points four hours apart—on the one hand from random-bred CFE rats (purchased from Carworth Farms, 216 Congers Road, New City, Rockland County, New York 10956) and on the other hand from an inbred strain of Minnesota Sprague Dawley rats, maintained in our laboratories. Subgroups of singly housed animals from each of the two stocks—CFE and MSD—were kept under specific conditions. Some animals were kept in "hives", a specially fabricated unit in the so-called periodicity room, with 36 single compartments, each controlled for light, temperature and humidity, eliminating visual contact among the rats and reducing their interactions by noise or odor. One hive was used with closed doors, another hive had an opened compartment door. Some animals were kept with maximal temperature control possible in the hive, whereas for others minor temperature changes in the periodicity room were reproduced to occur in the hive as well (less than ± .5°C). Some animals were not in a hive but on the shelf of the periodicity room, controlled in terms of lighting and temperature (±1.0°C). The same lighting regimen, L0600–1800; D1800–0600, was imposed on all animals.

Thus, one can compare results from animals kept in special sight-, hearing- and

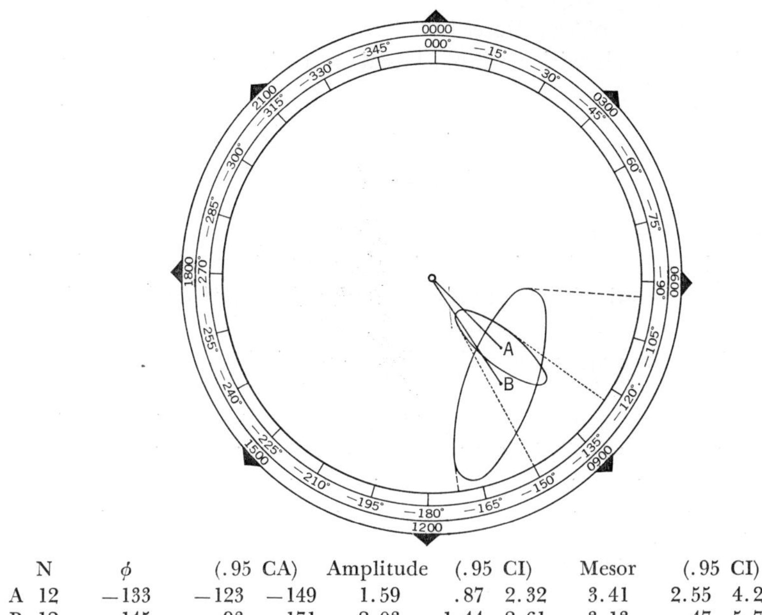

	N	ϕ	(.95 CA)		Amplitude	(.95 CI)		Mesor	(.95 CI)		P
A	12	−133	−123	−149	1.59	.87	2.32	3.41	2.55	4.28	.001
B	12	−145	− 93	−171	2.03	1.44	2.61	3.13	.47	5.79	.001

Fig. 2 Agreement between two cosinor summaries of parameters computed for 12 series each based on three time points.
A=0800, 1600, 0000, B=0800, 1200, 1600.

smell-controlled units with differing extent of control of noise, odors, visual interaction and environmental temperature variation. The data stem from 126 rats; 3 animals were killed at each of six time points from each of the above mentioned subgroups, kept on an LD 12:12 lighting regimen; the liver glycogen rhythm stands out satisfactorily under all of these conditions both in CFE and MSD rats. More drastic variations in specific environmental factors and, perhaps, a noisier rhythm as test criterion will be needed to detect any differences among the conditions tested.

There were limitations to using only a single animal per each of six time points as compared to a pool of two or more livers per time point. The fit of a cosine curve to the single livers may account for as little as 16 or as much as 96% of the variability.

Another way to attempt economizing in sampling on the liver glycogen rhythm is to restrict the number of time points chosen for killing. While it is desirable at the outset of work on a given variable to cover a 24-hour span at 4-hour intervals, it appeared of interest to explore the results from curve fitting to the liver glycogen rhythm with samples available at only 5, 4 or 3 timepoints, Figs. 1 and 2. Thereby one reduces most or all work at inconvenient odd times. One then may inquire into the extent of information loss stemming from such economy in both time and expense. First, data from the noon or midnight sample were omitted (sets A and B in Fig. 1). Thereafter, analyses were done on values from only four time points—set C, 0400, 0800, 1600 and 2000; set D, 0400, 1200, 1600, and 0000, and set E, 0800, 1200, 2000 and 0000. Finally, the analysis was repeated on just 3 values–set A, 0800, 1600 and 0000, and also with a second set B of three values, 0800, 1200 and 1600 (Fig. 2).

No estimates of reliability are obtained by fitting a cosine curve to single series consisting of only three observations but a cosinor summary of the (A, ϕ) pairs thus obtained agrees well with cosinors based upon larger samples, as shown in Figs. 1 and 2 respectively.

SUMMARY

In the case of a most prominent rhythm, previously known to be amenable to the effect of a known synchronizer—as in the case of the circadian rhythm in liver glycogen, synchronized in certain rodents by light and darkness alternating at 12-hour intervals—quite restricted sampling can lead to a rather reliable rhythm detection and parameter estimation. In this instance and some others available prior information facilitates the sampling design. In each case of work on a rhythm one may well search for the minimal sampling requirements. However, it is beyond the scope of this paper to extend the results here presented to other variables and species; our purpose herein is only to show the extent of reproducibility of certain rhythm parameters when they are determined by relatively limited sampling.

REFERENCES

1. Forsgren E. (1928): On the relationship between the formation of bile and glycogen in the liver of rabbit. Scand. Arch. Physiol., *53*: 137.
2. Holmgren Hj. (1931): Beitrag zur Kenntnis der Leberfunktion. Z. f. mikr.-anat. Forschung, *24*: 632.
3. Agren G., Wilander O. and Jorpes E. (1931): Cyclic changes in the glycogen content of the liver and the muscles of rats and mice: their bearing upon the sensitivity of the animals to insulin and their influence on the urinary output of nitrogen. Biochem. J., *25*: 777.
4. Petren T. (1939): Die 24-Stundenrhythmik des Leberglykogens bei cavia cobaya nebst Studien über die Einwirkung der "chronischen" Muskelarbeit auf diese Rhythmik, Morphologisches Jahrbuch, *83*: 256.
5. Sollberger A., Elfvin L.-G. and Petren T. (1955): Influence of some endogenous and exogenous factors in diurnal glycogen rhythm in chicken. Acta Anatomica, *25*: 286.
6. Ekman C. A. and Holmgren Hj. (1949): The effect of alimentary factors on liver glycogen in the liver lobule. Anat. Rec., *104*: 189.
7. Hirsch G. C. and van Pelt R. F. J. (1937): Der Rhythmus des Glykogengehaltes der Leber der weissen Maus, dargestellt durch die Stufenzahlmethode. Proc. Kon. Ac. van Wetensch., *40*: 11.
8. Higgins G. M., Berkson J. and Flock E. (1932): The diurnal cycle in the liver: I. Periodicity of the cycle, with analysis of chemical constituents involved. Am. J. Physiol., *102*: 673.
9. Beringer A. (1950): Ueber das Glykogen und seinen Einfluss auf den Stoffwechsel der Leber beim Gesunden und Diabetiker. Deutsche med. Wchnsch., *75*: 1715.
10. Halberg F. (1969): Chronobiology. Annual Review of Physiology, *31*: 675.
11. von Mayersbach H. and Leske R. (1966): Physiologische Schwankungen des Glycogentagesrhythmus. Acta Morphologica *12*: 33, 1963; cf. also Biorhythmic changes of liver glycogen. Acta Morph. Neer.-Scand., *6*: 343.
12. Halberg F., Halberg E., Barnum C. P. and Bittner J. J. (1959): Physiologic 24-hour periodicity in human beings and mice, the lighting regimen and daily routine. *In*: Photoperiodism and Related Phenomena in Plants and Animals, (Robert B. Withrow, ed.), pp. 803–878, Publ. 55 of the AAAS, Washington, D. C.
13. Halberg F., Albrecht P. G. and Barnum C. P. Jr. (1960): Phase shifting of liver glycogen rhythm in intact mice. Am. J. Physiol., *199*: 400.

14. Halberg F., Nelson W., Runge W. J., Schmitt O. H., Pitts G. C., Tremor J. and Reynolds O. E. (1971): Plans for orbital study of rat biorhythms. Results of interest beyond the Biosatellite program. Space Life Sciences, *2*: 437.
15. Stupfel M., Nelson W., Halberg J. and Halberg F. (1970): Multiple-purpose monitoring of carbon dioxide in closed organism-environment systems, including biosatellites. Space Life Sciences, *2*: 33.
16. Halberg F., Johnson E. A., Nelson W., Runge W. and Sothern R. (1972): Autorhythmometry—procedures for physiologic self-measurements and their analysis. Physiology Teacher, *1*: 1–11.
17. Nelson W. and Halberg F. (1966): Phase relations of circadian rhythms: animals. *In*: Handbook of Environmental Biology, (P. L. Altman and D. S. Dittmer, eds.), pp. 586–596, Fed. Am. Soc. Exptl. Biol., Bethesda, Maryland.

STATISTICAL CONSIDERATIONS

Chairman: Eugene K. HARRIS

COMMENTS ON STATISTICAL METHODS FOR ANALYZING BIOLOGICAL RHYTHMS

EUGENE K. HARRIS

Laboratory of Applied Studies
Division of Computer Research and Technology
National Institutes of Health, Public Health Service
Department of Health, Education and Welfare
Bethesda, Maryland, U.S.A.

Over the past 25 and more years, various statistical tools have been developed to help the analyst recover a periodic signal imbedded in random background. Without doubt, the most powerful of these have been auto- and cross-correlation, particularly the latter [1]. Since the auto-correlation function of random noise dampens to almost zero after some time lag t_0, the autocorrelation of signal plus noise after a time lag $t > t_0$ will be the same as that of the pure signal against which it may be matched for identification. In cross-correlating the observed time series with a known periodic function, the noise component (assumed independent of signal) does not enter at all, so that the entire cross-correlation can be used to extract the pure periodic signal.

These techniques have been applied to such biological phenomena as EEG's, arterial pulsatile bow, and ECG's (in atrial fibrillation) [2–4]. The physiological significance of results, I leave to experts in these areas. This, of course, is the acid test for the value of these methods to biology.

Two characteristics of these correlation techniques of signal recovery should be noted. First, they work for all periodic signals, sinusoidal and non-sinusoidal. They are essentially averaging procedures. For example, cross-correlating a periodic signal of known period with a unit impulse function simply means averaging observations separated by intervals equal to the period. The averages give a representation of the signal in which the variance of the noise has been reduced by a factor of k, the number of cycles over which observations have been collected. Second, the form of the underlying periodic signal will be identifiable regardless of whether we know its mathematical formula. It is possible, incidentally, to separate two or more periodic signals from the same random background. These are not curve-fitting, but curve-finding procedures; hence the question of estimating parameters is not particularly relevant. One drawback is that a large number of equally-spaced observations over many cycles is necessary. Moreover, the period of the signal should be nearly constant; phase shifts degrade the recovered signal.

Now, any periodic function of known period may be approximated by a finite Fourier series expansion, that is, a sum of independent pairs of sine and cosine terms whose frequencies are integral multiples (harmonics) of the frequency of the original function. Indeed, one may often account for a large proportion of the variance of the original function by the first few sinusoidal terms in the

Fourier series. This does not necessarily mean, however, that the original function is itself sinusoidal or that the secondary harmonic terms help us to understand the driving forces generating the original function. A square wave or a parabola within a finite interval can also be approximated fairly well by a Fourier series of a few harmonics. To cite a biological example, serum iron in many individuals [e.g., 5, 6] seems to rise rapidly in the early morning hours before awakening and taper off slowly during the rest of the day, rising slowly again in late evening. This is a periodic but not a sinusoidal pattern of variation. Analyzing a time series of serum iron determinations by Fourier series approximation may impede rather than assist us in understanding the essential characteristics of this variable.

Suppose, however, that we have good reason to believe that the biological variable follows a sinusoidal time course, governed by one frequency or perhaps several widely separated frequencies. Then the method of choice is the least squares fitting of a fixed-frequency cosine curve(s) such as Professor HALBERG and his co-workers have done under the name "cosinor analysis". A great advantage of least squares fitting of a cosine model (or any mathematical model) is that it may be performed on unequally spaced points. As with any mathematical model, parameters are involved and methods of estimation become pertinent. I have one objection in principle to the cosinor estimating procedure, as usually done. When cosines are fitted to separate segments of data, either from different individuals or different time spans from the same individual, the respective estimates of amplitude and phase are not simply averaged to obtain final estimates. Instead the mathematics of the method (vector operations) weights each amplitude estimate by the associated phase to produce a phase-weighted average amplitude. (Similarly, the average phase value is an amplitude-weighted average). As long as any variation exists in the separate estimates of phase angle, the phase-weighted average amplitude must be less than an unweighted average of amplitudes. An analogous situation occurs in averaging the amplitudes of evoked responses in neurophysiological experiments. When the latency (*i.e.*, time between stimulus and response) varies from one experiment to another, time-locked averaging produces smaller average amplitudes than when responses are suitably shifted to a common latency [7]. The same problem was alluded to earlier as one of the drawbacks of auto- and cross-correlation techniques.

The use of unweighted average amplitudes and phases would not affect the size of joint confidence ellipses since these are computed from observed variances in the individual amplitude and phase estimates.

It is worth repeating that a large proportion of the variance of a periodic, skewed function may be accounted for by a symmetrical (and, therefore, misleading) cosine model. One way of testing this model is to examine the time series of residual deviations. If they follow a non-random course, we might be better advised simply to refine the true signal by cross-correlating the observed series with a known periodic function.

When the signal, as a whole, is non-periodic but contains sinusoidal rhythms together with a large amount of random background variation, power spectrum analysis can reveal these periodicities and assess their significance. However, to secure both high resolution and acceptable precision (*i.e.*, to obtain stable estimates of the variance contributions of narrow bandwidths), measurements may have to be taken more often than is feasible. In practice, then, one may have to accept undesirably wide bandwidths and, moreover, face the possibility that high frequencies beyond the ability of the observations to resolve will inflate the variance attributed to lower frequency bands.

More serious, I believe, than these inevitable technical limitations, is a conceptual one affecting all procedures which recover or fit sine curves. The variables we measure are the outputs of complicated biological systems. Usually, we cannot identify all the inputs to these systems, let alone know their frequency characteristics. Nor can we be sure that these systems operate in a linear fashion, transmitting the input frequencies undisturbed to the output variables. Yet unless this assumption is valid, a sinusoidal analysis of the output time series can teach us nothing about the physiological processes involved. Moreover, even if the linearity assumption is valid, or at least a good enough approximation, we cannot tell whether the system has introduced amplitude or phase shifts unless we know the characteristics of the input rhythms. Without adequate knowledge of either input variables or system properties, we are left after laborious "number-crunching" of the output with some empirical statistics which may be useful because they correlate with other findings, but cannot of themselves provide insights to the inner working of the body.

In closing, let me mention two statistical methods of time series analysis which make no pretense of providing clues to physiology but may be useful in practical decision-making. One of these is auto-regression to investigate the possibility that a present or future variate-value may be predicted from a linear function of past values [e.g., 8]. The second is analysis of the variance in repeated measurements of multiple variables in each of a number of individuals, for the purpose of judging whether such individuals, or groups of individuals, have essentially similar "profiles" with respect to these variables [9]. As far as I know, neither of these statistical techniques has been much applied to physiology (however, see [10]); autoregression has been used mostly in economics and profile analysis in psychological testing. In my opinion, both methods, particularly the latter, offer attractions to the analyst of biological time series.

REFERENCES

1. Lee Y. W. (1960): Statistical Theory of Communication. John Wiley & Sons, New York, (Especially Ch. 12).
2. Barlow J. S. (1961): Autocorrelation and crosscorrelation techniques in EEG analysis. *In*: Computer Techniques in EEG Analysis, pp. 31–36, Elsevier Publishing Company, Amsterdam.
3. Randall J. E. (1958): Statistical properties of pulsatile pressure and flow in the femoral artery of the dog. Circulation Research, *6*: 689–698.
4. Braunstein J. R. and Franke E. K. (1961): Autocorrelation of ventricular response in atrial fibrillation. Circulation Research, *9*: 300–304.
5. Hoyer K. (1944): Physiologic variations in the iron content of human blood serum. Acta Med. Scand., *99*: 577–585.
6. Stengle J. M. and Schade A. L. (1957): Diurnal-nocturnal variations of certain blood constituents in normal human subjects: plasma iron, siderophilin, . . .; Brit. J. Haemat., *3*: 117–124.
7. Woody C. D. (1967): Characterization of an adaptive filter for the analysis of variable latency neuroelectric signals. Med. & Biol. Engng., *5*: 539–553.

8. Kendall M. G. and Stuart A. (1966): The Advanced Theory of Statistics. V. 3, Ch. 47, Hafner Publishing Company, New York.
9. Greenhouse S. W. and Geisser S. (1959): On methods in the analysis of profile data. Psychometrika, *24*: 95–112, 1959.
10. Buchsbaum M. and Harris E. K. (1971): Diurnal variation in serum and urine electrolytes. J. Appl. Physiol., *33*: 27–35.

THEORY AND APPLICATIONS OF PERIODIC MARKOV CHAINS*

V. R. R. UPPULURI

Mathematics Division, Oak Ridge National Laboratory
Oak Ridge, Tennessee, U.S.A.

THEORY

Suppose we are observing the weather at a particular place and confine our attention to the fact that is a dry day or a wet day. For a rough analysis, if one wishes to predict the weather pattern, one may assume that the observation made today is statistically independent of the observation made yesterday and analyze the data collected. The assumption of *statistical independence* reduces the technical problem to that of inference of the probability of success (or failure) of a coin in independent trials. All the intuitive answers turn out to be the best. But we all know that this is not fully adequate to predict whether it is going to be dry or not the next day.

From a technically tractable point of view the next best situation is to assume that the observation made today statistically depends only on the observation made yesterday. This is the so called *Markov dependence*. More specifically, if we are observing a random phenomenon at equally spaced intervals of time Δt, let X_n denote the observation at time $n\Delta t$. Such a sequence of random variables $\{X_n, n=1, 2\cdots\}$ is called a *discrete time Stochastic Process*.

If we further assume that at each time unit we can only realize one of a finite number of events $E_1, E_2, \cdots, E_k$, we call $\{X_n\}$ a discrete time, discrete state stochastic process. The Markov condition is characterized by

$$P[X_n \in E_j | X_{n-1} \in E_i, X_{n-2} \in E_{l_{n-2}}, \cdots, X_2 \in E_{l_2}, X_1 \in E_{l_1}]$$
$$= P[X_n \in E_j | X_{n-1} \in E_i] = \pi_{ij}(n),$$

and such a sequence $\{X_n\}$ is said to be a *Finite Markov Chain*. If $\pi_{ij}(n)=\pi_{ij}$ is independent of n, for all i and j, then the Markov chain is said to be a *Finite Stationary Markov Chain*.

In fact, a finite stationary Markov chain is also characterized by the transition probability matrix $\pi=(\pi_{ij})$ given by

Present State \ Next State	E_1	$E_2 \cdots E_k$	Total
E_1	π_{11}	$\pi_{12} \cdots \pi_{1k}$	1.0
E_2	π_{21}	$\pi_{22} \cdots \pi_{2k}$	1.0
$\vdots$	$\vdots$	$\vdots$	$\vdots$
E_k	π_{k1}	$\pi_{k2} \cdots \pi_{kk}$	1.0

There are several things that can be said about finite stationary Markov Chains.

* Research sponsored by the U. S. Atomic Energy Commission under contract with the Union Carbide Corporation.

If one prescribes the initial probability vector $P[X_0=E_i]=\alpha_i$, then one can find answers to several questions, such as: (i) the probability of finding an observation in state l at time $m\Delta t$, or (ii) the expected number of steps needed to reach state l, and so on.

Definition: The state E_j is said to have *period* $d>1$ if $\pi_{jj}^{(m)}=0$ unless $m=\nu d$ is a multiple of d, and d is the largest integer with this property.

One defines a Markov Chain to be *irreducible* if, and only if, every state can be reached from every other state. In an irreducible Markov Chain if one state has period d, then all the states have the same period d, and we shall refer to the chain as a *Markov Chain with period d.* The simplest example of a chain with period 3 is a chain with three states in which only the transitions $E_1 \to E_2 \to E_3 \to E_1$ are possible.

Then

$$\pi=\begin{bmatrix}0&1&0\\0&0&1\\1&0&0\end{bmatrix},\ \pi^2=\begin{bmatrix}0&0&1\\1&0&0\\0&1&0\end{bmatrix},\ \pi^3=\begin{bmatrix}1&0&0\\0&1&0\\0&0&1\end{bmatrix},$$

with $\pi^4=\pi$, $\pi^5=\pi^2$, $\pi^6=I$, $\pi^7=\pi$ etc.

One of the important results to note is the following

Theorem: In an irreducible stationary Markov Chain with period d, we have

$$\lim_{n\to\infty} \pi_{il}^{(nd+1)}=du_l$$

where

$$u_l=\sum_{v} u_v \pi_{vl}\,.$$

For more information on Finite Markov Chains, one may refer to FELLER [2] or KEMENY and SNELL [3].

AN APPLICATION

I am sure that among others the members of the Society for Biorhythm Research are interested in recognizing cycles in observed data and disentangling the components when several cycles are present simultaneously. One common approach seems to be the spectral analysis of time series based on observed correlations of different lags. An example where this technique yielded good results is discussed by SIDDIQUI [4] in a rather lucid paper. However, the skepticism shown by La Mont C. COLE [1] in his article: Biological clock in the Unicorn, should be borne in mind. In the rest of this paper I do not pretend to resolve these problems but present an elementary approach to seek for periodicities in the data.

If one is interested in analyzing the noise in reactor data, it is fair to use sophisticated mathematical tools like Fast Fourier Transforms. To my mind, in such a context, the problem is a second order problem where the first order errors are reasonably under control. But in several areas, though one may be collecting data continuously in time, it is not clear whether the first order errors are under control. So, it seems to be not too inappropriate to use coarser methods to get a rough idea of the underlying phenomenon. Further, if we find tools like the period of Markov Chains are within the reach of more investigators, one can use them to look for periodicities in the data. An added advantage is if the investigator is not satisfied with the conclusions drawn, he can easily question the validity of the assumptions made since the methods are quite elementary by nature.

Let us consider the radio propagation data for which spectral analysis of time series was applied by SIDDIQUI [4]. In Table 1, we have 1240 monthly medians of the noon values of the critical radio frequency of F_2 layer (f_0F_2, measured in megacycles per second) at Washington, D.C., during May 1934 to April 1954.

Table 1 Monthly medians of the noon values of the critical radio frequency of F_2 layer (f_0 F_2, measured in Megacycles per second) at Washington D. C.

	Jan.	Feb.	Mar.	Apr.	May	June	July	Aug.	Sept.	Oct.	Nov.	Dec.
1934					5.4	5.4	5.1	4.8	5.5	6.1	6.4	6.8
1935	6.4	6.0	6.1	6.1	5.3	5.5	5.7	5.0	5.8	9.0	10.4	9.9
1936	10.7	9.7	8.8	8.1	8.4	6.4	6.7	7.4	9.1	11.9	12.5	11.8
1937	11.1	13.1	12.3	9.7	7.7	7.6	8.1	8.9	10.4	13.0	13.2	12.3
1938	11.5	12.8	12.2	10.0	8.1	7.3	7.4	7.9	9.6	12.1	13.1	12.4
1939	11.1	11.7	10.3	9.3	8.0	7.4	6.9	7.6	8.9	12.2	11.4	10.3
1940	9.5	9.6	9.8	8.4	7.4	6.7	6.4	7.1	9.2	10.7	10.6	9.3
1941	9.0	8.9	8.1	7.2	5.7	5.8	5.8	6.2	6.9	9.1	9.0	8.7
1942	8.0	7.8	8.7	7.4	6.1	5.3	4.9	5.1	6.0	7.4	7.5	7.0
1943	6.4	6.6	6.2	5.6	5.1	5.2	5.2	4.8	5.2	6.0	6.6	7.1
1944	6.4	6.0	5.9	5.2	4.9	4.7	4.7	5.0	5.6	7.0	6.6	7.1
1945	6.5	6.6	6.2	6.3	5.7	5.8	5.5	5.7	6.4	9.3	9.3	8.6
1946	7.7	9.4	9.7	9.2	6.2	6.3	6.1	7.2	8.6	11.5	12.2	12.0
1947	12.0	13.0	12.5	11.2	10.0	6.9	7.1	7.7	10.6	13.0	13.5	12.6
1948	12.3	11.3	11.2	10.8	8.2	7.6	7.1	6.8	9.2	11.5	12.4	11.7
1949	12.1	12.5	12.5	10.6	8.3	7.0	7.1	7.8	10.1	12.6	13.1	12.0
1950	11.3	11.0	10.5	8.8	7.2	6.8	6.0	6.1	7.0	9.2	9.0	8.6
1951	8.4	8.7	7.8	6.3	6.1	6.0	5.6	5.8	7.0	8.6	9.2	8.8
1952	8.0	7.5	6.5	5.5	4.8	5.2	5.0	5.3	6.0	7.2	7.6	7.3
1953	6.4	6.2	5.2	5.2	4.8	4.8	4.3	4.7	6.0	6.2	6.5	6.3
1954	5.9	5.7	5.5	5.0								

We may define the integer part of the observation to be the observed state and the time unit to be a month. From the data, for a given lag (in months), we can estimate the transition matrix by counting the number of transitions from state i to state j. For this data such matrices are given in Tables 2, 3, 4 and 5 for lags 4, 6, 12 and 120, respectively. By looking at Table 5 we could say that our data indicates that the observations are almost on the same wave after a lag of 120 months (=10 years), since almost all the transitions are on the diagonal of the matrix. The same conclusion is drawn by SIDDIQUI [4] based on spectral analysis. If one wishes to decide whether a given transition matrix is almost diagonal or not, one can develop powerful statistical tests. As far as this author knows, statistical inference of the period of a Markov Chain is a lacuna, and the members of this conference may give an impetus to this area for further study.

Table 2 Observed transitions for lag 4.

State	4	5	6	7	8	9	10	11	12	13
4	0	1	7	3	0	0	0	0	0	0
5	4	5	9	6	2	7	2	0	0	0
6	4	18	8	2	5	2	2	4	2	1
7	0	8	7	1	2	2	2	2	8	4
8	2	3	6	4	1	5	1	0	2	1
9	0	1	6	6	4	4	0	2	1	0
10	0	0	3	2	5	0	1	2	1	0
11	0	0	1	6	2	2	1	0	1	2
12	0	0	0	5	4	2	3	3	3	0
13	0	0	1	1	0	0	2	2	2	0

Table 3 Observed transitions for lag 6.

State	4	5	6	7	8	9	10	11	12	13
4	0	2	7	2	0	0	0	0	0	0
5	0	2	14	7	5	5	2	0	0	0
6	5	15	4	2	6	7	0	2	5	0
7	3	7	3	3	1	3	2	6	5	3
8	1	6	5	0	0	3	1	3	4	2
9	1	2	6	4	3	2	1	1	3	1
10	0	1	2	2	2	0	1	2	3	1
11	0	0	4	5	1	2	1	1	0	1
12	0	0	2	9	3	2	4	0	0	0
13	0	0	0	2	4	0	2	0	0	0

Table 4 Observed transitions for lag 12.

State	4	5	6	7	8	9	10	11	12	13
4	1	6	0	0	0	0	0	0	0	0
5	8	14	9	2	1	1	0	0	0	0
6	1	13	10	6	4	7	3	0	0	0
7	0	2	17	12	3	1	0	0	1	0
8	0	0	4	11	5	2	1	0	2	0
9	0	0	1	3	7	2	3	4	2	2
10	0	0	0	2	2	7	1	1	1	0
11	0	0	0	0	2	2	2	2	5	2
12	0	0	0	0	1	1	4	5	6	3
13	0	0	0	0	0	1	0	3	3	1

Table 5 Observed transitions for lag lag 120.

State	4	5	6	7	8	9	10	11	12	13
4	1	2	0	0	0	0	0	0	0	0
5	6	11	4	0	0	0	0	0	0	0
6	1	4	14	4	0	0	0	0	0	0
7	0	1	5	10	0	0	1	0	0	0
8	0	0	2	3	6	2	1	0	0	0
9	0	0	0	1	5	4	2	3	0	0
10	0	0	0	1	0	3	2	0	2	0
11	0	0	0	0	0	0	0	1	5	1
12	0	0	0	0	0	0	0	4	4	0
13	0	0	0	0	0	0	0	0	1	3

ACKNOWLEDGMENT

Thanks are due to Professor M. M. SIDDIQUI for making available the data included in this paper.

REFERENCES

1. COLE LA MONT C. (1957): Biological clock in the Unicorn. Science, *125*: 874–876.
2. FELLER W. (1968): An Introduction to Probability Theory and its Applications. 3rd edition, John Wiley, New York.
3. KEMENY J. G. and SNELL J. L. (1960): Finite Markov Chains. Van Nostrand Co., Princeton, N. J.
4. SIDDIQUI M. M. (1962): Some statistical theory for the analysis of radio propagation data. J. Research National Bureau of Standards, D. Radio propagation, *66D*: 571–580.

AUTHOR INDEX

(*Continued*)

SUBJECT INDEX